Handbook of
EMERGENCY CHEMICAL
MANAGEMENT

Handbook of
EMERGENCY CHEMICAL MANAGEMENT

David R. Quigley

CRC Press
Boca Raton   Ann Arbor   London   Tokyo

Library of Congress Cataloging-in-Publication Data

Quigley, D. R. (David R.)
    Handbook of emergency chemical management / by David R. Quigley.
       p.   cm.
    Includes index.
    ISBN 0-8493-8908-9
    1. Hazardous substances--Accidents--Management--Handbooks,
  manuals, etc.   I. Title.
T55.3.H3Q54 1994
660—dc20

94-10960
CIP

No claim to original U.S. Government works
International Standard Book Number 0-8493-8908-9
Library of Congress Card Number 94-10960
Printed in the United States of America 1 2 3 4 5 6 7 8 9 0
Printed on acid-free paper

I would like to dedicate this book to Jenna. Without her patience,understanding, love and computer skills, this book would not been possible.

I would also like to dedicate this book to some of the true heros in the world - the hazardous materials responder. They put their lives on the line to clean events that no else will touch. God bless you.

David R. Quigley, Ph.D., is the founder of D.R. Quigley and Associates which specializes in chemical safety and emergency response. He is also a Senior Engineering Specialist at the Idaho National Engineering Laboratory, an instructor at the Idaho Hazardous Materials Training Center and an Adjunct Professor at both the University of Idaho and Idaho State University.

Dr. Quigley received his ACS certified B.S. in chemistry from Florida Atlantic University in Boca Raton, Florida. In 1979 and 1982, he received his M.S. and Ph.D. in chemistry from the University Of Missouri - Rolla. Dr. Quigley spent 5 years on the research faculty at the University of Colorado Health Sciences Center in Denver, Colorado before moving to the Idaho National Engineering Laboratory.

Dr. Quigley is a member of the American Chemical Society and the American Association for the Advancement of Science. He has over 50 research publications in biochemistry, enzymology, organic chemistry and toxicology and has presented numerous invited lectures in the United States and other countries. His research interests include enzymology, toxicology and the chemistry of recalcitrant chemicals.

Dr. Quigley has been active in chemical safety for numerous years. He has been an emergency responder for several years, is a Hazardous Materials Specialist A and respondes with several fire departments in Eastern Idaho. He has also performed support work for numerous law enforcement organizations.

Handbook of Emergency Chemical Management

While there are many good books written about concerning various aspects of responding to a hazardous materials event, there are few written addressing the entire event and at a level that all find easy to understand. This book encompasses the entire hazardous materials event. Each of approximately 1000 entries is divided into four sections. The first section contains identification information (e.g., DOT #, CAS #, pseudonyms) and physical data. The second section contains fire information (e.g., LEL, UEL, flash point). The third section contains toxicological data. Data on toxicity is given in both the original form and on a relative scale to provide the responder with a better idea on what the data mean. Symptoms and first aid are also in this section. The last section contains spill information (e.g., clean up methodologies, protective clothing compatibilities). Indicies for CAS #, RTECS #, DOT #, smell, and unusual symptoms are provided to aid in the identification of chemical spills of unknown materials.

HOW TO USE THIS BOOK

This book is divided into 4 major sections. The first section consists of complete hazardous materials entries, the second is a series of shorter entries for those chemicals that are either much less hazardous or have one primary hazard, the third is a series of indices to help find or identify chemicals that are involved in a hazardous materials event, and the fourth is a glossary of terms used in this text.

Complete Hazardous Materials Entries

Each complete entry occupies one full page and is divided into four major sections: Physical Data, Flammability Data, Toxicological Data and Recommended Clean-Up Procedures.

Physical Data

Chemical Name and Formula. This will indicate the name of the product being described. When possible, chemical structures will be given.

Synonyms. Only the most common synonyms will be given. Typically, trade names will not be included. An index of chemical names and synonyms listed will be included in the third section.

CAS Number. Each chemical has a discrete Chemical Abstracts Service (CAS) registry number. This number will accurately indicate the chemical being described. An index of CAS numbers is present in the third section.

Description. Each chemical will be described as to their color and state (gas, liquid or solid) at room temperature. Ocassionally, when a chemical has a melting or boiling point near room temperature, the description of the material and temperature constraints will be given.

DOT Number. Some hazardous materials that are commonly transported are given identification numbers by the Department of Transportation. These numbers are many times present on placards that are prominantly displayed on truck and railcars and are present on shipping manifests. In all cases, the UN (United Nations) and NA (North America) prefixes are not included in the listing. An index of DOT numbers is included in section 3.

DOT Classification. Many chemical that are transported are classified as to the primary hazard associated with it. This classification may be present on the placard and will be present on shipping manifests.

Molecular Weight. The molecular weight of the product.

Melting Point. The temperature at which a chemical will change from a solid to a liquid at room pressure is given in bothFahrenheit and Celcius. Abbreviations used are s (sublimes) and d (decomposes).

Boiling Point. The temperature at which a chemical will change at room pressure from a liquid to a gaseous state is given in both Fahrenheit and Celcius. Abbreviations used are the same as those used for Melting Point.

Vapor Density. The density of the chemical's vapor (air = 1.0) will be given. Densities greater than 1.0 means that the vapor is more dense than air and will tend to collect in low lying areas. Densities less than 1.0 indicate the vapor is lighter than air and will rise.

Specific Gravity. Density is given as grams per cubic centimeter at standard room temperature and pressure. A value greater than 1.0 indicates that the chemical is more dense than water and will tend to sink when placed in water. A values less than 1.0 indicates that the chemical is less dense than water and will tend to float.

Vapor Pressure. This is the pressure of the chemical's vapor at room pressure and at a temperature between 20 and 30 C. Values are given as mm of mercury (parenthetically) and on a scale of 1 - 10 where 1 is essentially no vapor pressure and 10 is virtually a gas. The greater the number, the greater the volatility of the chemical.

Water Solubility. The solubility of the chemical is given as a weight/weight percentage. Values of less than 0.1% are simply listed as insoluble and values greater than 50% will be listed as soluble. The abbreviation "dec." will indicate that the chemical decomposes upon contact with water. Unless water is also listed as an incompatible, the reaction between the chemical and water will not release signifiant amounts of energy or hazardous material.

Chemical Incompatibilities. Chemicals that may cause a hazardous reaction with the subject product will be listed. Hazardous reactions include explosions, reactions that release large amounts of energy over a short duration, or reactions that generate toxic explosive or otherwise hazardous materials.

Flammability Data

NFPA Classification. Some liquids are classified by the National Fire Protection Association (NFPA) according to their degree of flammability. These classifications are defined by NFPA 45 and 29 CFR 1910.106.

NFPA Hazard Code. This is a three digit code that quantifies the relative degree of hazards associated with a chemical and may be folloewed by a special symbol. The first digit corresponds to the health hazard, the second is fire hazard and the third is reactivity. Each hazard is rated on a 0 (benign) to 4 (extremely hazardous) scale. Special symbols used are ox for oxidizer and W for water sensitive comounds.

Flash Point This is the temperature at which enough vapor is released to support combustion when an ignition source is present.

Autoignition Temperature. This is the temperature at which combustion will occur in the absence of an ignition source.

LEL (Lower Explosion Limit). The minimum concentration of vapor in air that will support combustion. Concentrations lower than this value will be too lean to burn.

UEL (Upper Explosion Limit) The maximum concentration of vapor in air that will support combustion. Concentration greater than this will be too rich to burn. The greater the range between the LEL and UEL, the greater the danger for ignition when the flash point is exceeded.

Fire Extinguishing Methods. This is a listing of materials that may be used to extinguish a fire involving the chemical in question.

Special Fire Fighting Considerations. This is a listing of issues that might become important during a fire that involves the chemical in question. Since every fire has its own unique set of circumstances, this is not a comprehensive listing of every fire related concern, rather it should be used as a guide to increase the safety of firefighters involved. Certain issues are assumed to be standard operating practices so they are not listed. These include removing the container from the fire only if it can safely be done, using a flow rate of 500 gallons per minute for each flame inpingement to cool exposed tanks and stopping leaks only if it can safely be done.

DOT Recommended Isolation Zones. Values are taken from DOT P 5800.5. Isolation zones are circular areas that should be evacuated and are applicable only a) for the first 30 minutes after a chemical spill and b) when the chemical spilled is not involved in a fire. Small spills are defined as those spills of 55 gallons or less while large spills are greater than 55 gallons.

DOT Recommendedn Down Wind Take Cover Distance. Values were taken from the same source as the DOT Recommended Isolation Zones and have the same limitations. Values given are distances downwind that the public should take cover. If these two sections are not present, then no values were available.

Toxicology Data

Carcinogenicity. If any data is present that provides evidence of carcinogenicity, a statement will be made directly under the section heading. A confirmed carcinogen is a chemical that has been proven to cause cancer in humans. A suspected carcinogen is a chemical for which there is some evidence that it may cause cancer in humans. A questionable carcinogen is a chemical for which there limited evidence of causing cancer in humans.

Odor. This is the typical description given for the odor of the chemical. Variations in chemical odor descriptions should be expected from person to person.

Odor Threshold. Values given are the minimum reported concentrations at which people were able to detect the presence of the chemical. Values should be compared with exposure limits to determine if one's exposure is above permissable exposure limits if one can smell the chemical. It should be noted that women are typically able to detect chemicals at lower concentrations than men.

Physical Contact. These are expected effects that will be observed when the chemical comes into contact with human tissues.

TLV (Threshold Limit Values). These are the time-weighted average concentrations of the chemical that a worker can be exposed to for either 8 hours or a 40 hour work week. Values are those defined by OSHA.

STEL (Short Term Exposure Limit). Values are concentrations that should not be allowed for longer than 15 minutes. Exceeding these values may result in a) irritation, b) tissue damage or c) narcosis to a sufficient degree to result in a significant safety hazard.

IDLH(Immediately Dangerous to Life or Health). Values given are the maximum concentrations that an individual can indure without either impairing their ability to escape or cuasing irreversible health effects.

RTECS #. The number present is that assigned by NIOSH for entry into the Registry of Toxic Effects of Chemical Substances (RTECS). RTECS is a comprehensive listing of toxicological data that is compiled by NIOSH.

Routes of Entry and Relative LD_{50} (LC_{50}). Values given are the concentrations or quantities of a chemical that will cause the death of 50% of a population. Parenthetical values are relative toxicities on a 1 to 10 scale. A 10 on this scale correspondes to poisons defined by the DOT to be in Packing Group I. A 9 correspondes to poisons defined by the DOT to be in Packing Group II and defined by OSHA to be highly toxic as defined in 29 CFR 1910.1200. A 5 correspondes to poisons that the DOT defines as Packing Group III and OSHA defines as toxic as defined in 29 CFR 1910.1200. Models used for these determinations are the rat model for inhalation and ingestion and the rabit model for skin absorption as described in 29 CFR 1910.1200.

Symptoms of Exposure. Listed are some of the more common symptoms that may be observed if one is exposed to the chemical. Variations of symptoms from person to person should be expected.

Emergency/First Aid Treatment. Recommended first aid treatment is given. First actions should be to remove any victims to a noncontaminated area and to remove any contaminated clothing. It is especially important to remove any belts, shoes or other contaminated leather articles. Since a given chemical may effect different individuals differently, treatment recommendations are kept fairly general. Most can be deduced from the symptoms. A consistant recommendation for any chemical exposure is to transport the victim to a medical care facility as quickly as possible. Appropriate MSDSs should eithr transported with the victim or transmitted (faxed) to the medical care facility prior to the victim's arrival.

Recommended Clean-Up Procedures

Personnal Protection. The recommended level of personal protection is given for a moderate size outdoor spill of 30 - 50 gallons. Levels of personal protection may either be upgraded or downgraded if conditions warrent it. Levels of protection used are those defined in 29 CFR 1910.120.

Recommended Material. Materials offering the best protection to chemical permeation or degradation is provided. Recommendations followed by a (+) indicate that available data indicate that this material offers excellent resistance to the chemical. A (0) indicates that the material offers good protection from the chemical. A (-) indicates that no data could be found on the material's ability to provide protection, but that reasonably good protection would be expected based on the material's resistance to other chemicals with similar properties. A N.D. indicates that there were no data upon which a recommendation could be based.

RCRA Waste #. Certain harardous chemicals have been identified and given an identification number for reporting spills. Numbers were obtained from 40 CFR 302.4.

Reportable Quantities (RQs). When certain chemical are spilled into the environment, various government agencies must be notified if the amount spilled is above the reportable quantity. Two types of reportable quantities exist: statutory and final. Final RQs are those quantities that are reportable according to RCRA in 40 CFR 302.4 and are the first values listed. Statutory RQs are given parenthetically and are those RQs established by Section 102 of CERCLA.

Spills. Recommended methods and materials that should be used to clean up a spill are given. These may be modified according to various circumstances involved with the spill.

Special Emergency Information. This section provides an overview of of various hazards that may present themselves during a hazardous materials event. Also provided in this section is a listing of chemicals that are extremely similar to the subject of the listing. The intent is that information present may be used for those chamicals listed. These chemicals are present in all indices.

CAUTION

Information provided in this book is based upon relatively pure materials. Some variations in chemical properties, appearence, etc. should be expected when when an impure or contaminated material is encountered. Technical experts should be used in conjunction with this book to ensure that proper guidance is provided for proper and safe HazMat responses.

ACETAL - CH3CH(OCH2CH3)2

Synonyms: acetaladehydediethylacetal; 1,1-diethoxyethane; diethylacetal; ethylidene diethyl ether

CAS Number: 105-57-7   **Description:** Colorless Liquid
DOT Number: 1088   **DOT Classification:** Flammable Liquid

Molecular Weight: 118.2
Melting Point: 19°C   (66°F)   **Vapor Density:** 4.1   **Vapor Pressure:** 5 (30 mmHg)
Boiling Point: 102°C (215°F)   **Specific Gravity:** 0.8   **Water Solubility:** 5%

Chemical Incompatibilities or Instabilities: Oxidizers

FLAMMABILITY

NFPA Hazard Code: 2-3-0   **NFPA Classification:** Class 1B Flammable
Flash Point: -21°C   (-5°F)   **LEL:**   1.7%
Autoignition Temp: 230°C (446°F)   **UEL:**   10.4%

Fire Extinguishing Methods: Alcohol Resistant Foam, Carbon Dioxide, Dry Chemical, Water Spray

Special Fire Fighting Considerations: Water spray should be used to keep closed containers cool. Fight fire from a distance or protected location, if possible. Immediately withdraw if rising sound from venting device is heard or if fire is causing discoloration to the tank. For large fires, if possible, withdraw and allow to burn. Isolate for 1/2 mile if rail or tank truck is involved in a fire.

TOXICOLOGY

Odor: **Odor Threshold:** N.D.
Physical Contact: Irritant
TLV = N.D.   **STEL** = N.D.   **IDLH** = N.D.

Routes of Entry and Relative LD$_{50}$ (or LC$_{50}$)   **RTECS # AB2800000**
    Inhalation   1   80,000 mg/m^3/H
    Ingestion   4   4570 mg/kg
    Skin Absorption   N.D.

Symptoms of Exposure: Possible irritation to skin, eyes and upper respiratory tract; headache; nausea.

Emergency/First Aid Treatment: Remove to ventilated area; immediately remove any contaminated clothing and wash contaminated areas for 15 minutes using water. Treat supportively and observe for possible shock. If ingested, seek immediate medical aid.

Recommended Clean-Up Procedures

Personal Protection: Level B Ensemble   **Recommended Material** Butyl Rubber (-)

RCRA Waste # None   **Reportable Quantity:** None

Spills: Remove all potential ignition sources. Absorb with non-combustible absorbent and take up using non-sparking tools. Decontaminate spill area using soapy water. Keep area isolated until vapors have dissipated.

Special Emergency Information: May be harmful is inhaled, swallowed, or absorbed through the skin. Vapors are heavier than air and may travel some distance to an ignition source. Vapors are more dense than air and may settle in low lying areas. May form unstable peroxides upon prolonged exposure to air.

ACETALDEHYDE - CH3CHO

Synonyms: acetic aldehyde; ethylaldehyde; ethanal

CAS Number: 75-07-0	**Description:** Colorless liquid that rapidly volatilizes at room temperature.
DOT Number: 1089	**DOT Classification:** Flammable Liquid

Molecular Weight: 44.1

Melting Point:	-123°C (-190°F)	**Vapor Density:** 1.5	**Vapor Pressure:**	10 (750 mm Hg)
Boiling Point:	21°C (69°F)	**Specific Gravity:** 0.78	**Water Solubility:**	Soluble

Chemical Incompatibilities or Instabilities: Acid anhydrides, alcohols, ketones, ammonia, cyanides, hydrogen sulfide, halogens, phosphorous, isocyanates, oxidizers, cobalt chloride.

FLAMMABILITY

NFPA Hazard Code:	2-4-2	**NFPA Classification:**	Class 1A Flammable
Flash Point:	-38°C (-36°F)	**LEL:**	4%
Autoignition Temp:	175°C (°F)	**UEL:**	60%

Fire Extinguishing Methods: Alcohol Resistant Foam, Carbon Dioxide, Dry Chemical

Special Fire Fighting Considerations: Water spray should be used to keep closed containers cool. Continue to cool container after fire is extinguished. For large fires, if possible, withdraw and allow to burn. Fight fire from a distance or protected location, if possible. Immediately withdraw if rising sound from venting device is heard or if fire is causing discoloration to the tank. Isolate for 1/2 mile if rail or tank truck is involved in a fire.

TOXICOLOGY
SUSPECTED CARCINOGEN

Odor: Fruity	**Odor Threshold:** 0.2 - 0.4 mg/m^3
Physical Contact: Irritant	
TLV = 180 mg/m^3 **STEL** = 270 mg/m^3	**IDLH** = 20,000 mg/m^3

Routes of Entry and Relative LD$_{50}$ (or LC$_{50}$)   **RTECS # AB1925000**

Inhalation	5	(16,000 mg/m^3)
Ingestion	3	(1930 mg/m^3)
Skin Absorption	N.D.	

Symptoms of Exposure: Possible irritation to skin, eyes and upper respiratory tract; headache; nausea. May cause an allergic response. Exposure may result in chemical pneumonitis or pulmonary edema.

Emergency/First Aid Treatment: Remove to ventilated area; immediately remove any contaminated clothing and wash contaminated areas for 15 minutes using water. Treat supportively and observe for possible shock. If ingested, seek immediate medical aid.

Recommended Clean-Up Procedures

Personal Protection: Level B Ensemble	**Recommended Material**	Butyl Rubber (+)
RCRA Waste # U001	**Reportable Quantity:**	1000 lb(1000 lb)

Spills: Remove all potential ignition sources. Dike to contain spill, absorb with non-combustible absorbent and take up using non-sparking tools. Decontaminate spill area using water. Keep area isolated until vapors have dissipated. Treat all materials used or generated and equipment involved as contaminated by hazardous waste.

Special Emergency Information: May be harmful if inhaled, swallowed or absorbed through the skin. Vapors are heavier than air and may travel some distance to an ignition source. Vapors are more dense than air and may settle in low lying areas. May undergo hazardous polymerization. May form unstable peroxides upon prolonged exposure to air.

ACETALDEHYDE OXIME

Synonyms: aldoxime; ethylidenehydroxylamine

CAS Number: 107-29-9		**Description:** Colorless Solid		
DOT Number: 2332		**DOT Classification:** Flammable Liquid		

Molecular Weight: 59.1

Melting Point:	46°C (115°F)	**Vapor Density:** N.D.	**Vapor Pressure:**	1 (<1 mm Hg)
Boiling Point:	115°C (239°F)	**Specific Gravity:** 0.97	**Water Solubility:**	100%

Chemical Incompatibilities or Instabilities: Acid, oxidizers.

FLAMMABILITY

NFPA Hazard Code:		**NFPA Classification:** Class II Combustible
Flash Point:	40°C (104°F)	**LEL:** N.D.
Autoignition Temp:	N.D.	**UEL:** N.D.

Fire Extinguishing Methods: Alcohol Resistant Foam, Carbon Dioxide, Dry Chemical, Water Spray or Fog.

Special Fire Fighting Considerations: Isolate for 1/2 mile if rail or tank truck is involved in a fire. Water spray should be used to keep closed containers cool. Continue to cool container after fire is extinguished. Fight fire from a distance or protected location, if possible. Immediately withdraw if rising sound from venting device is heard or if fire is causing discoloration to the tank. For large fires, if possible, withdraw and allow to burn.

TOXICOLOGY

Odor: **Odor Threshold:** N.D.

Physical Contact: Irritant

TLV = N.D.   **STEL** = N.D.   **IDLH** = N.D.

Routes of Entry and Relative LD$_{50}$ (or LC$_{50}$)   **RTECS # AB2975000**

Inhalation	N.D.
Ingestion	N.D.
Skin Absorption	N.D.

Symptoms of Exposure: Possible irritation to skin, eyes and upper respiratory tract; headache; nausea.

Emergency/First Aid Treatment: Remove to ventilated area; immediately remove any contaminated clothing and wash contaminated areas for 15 minutes using water. Treat supportively and observe for possible shock. If ingested, seek immediate medical aid.

Recommended Clean-Up Procedures

Personal Protection:	Level B Ensemble	**Recommended Material**	N.D.
RCRA Waste #	None	**Reportable Quantity:**	None

Spills: Remove all potential ignition sources. Absorb with non-combustible absorbent and take up using non-sparking tools. Decontaminate spill area using acidic solution. Keep area isolated until vapors have dissipated. Treat all materials used or generated and equipment involved as contaminated by hazardous waste.

Special Emergency Information: May be harmful if inhaled, swallowed or absorbed through the skin. Mixing with acid may result in decomposition to form acetaldeyde and toxic hydoxylamine.

ACETIC ACID, GLACIAL - CH3COOH

Synonyms: ethanoic acid; vinegar acid; methane carboxylic acid; glacial acetic acid

CAS Number:	64-19-7	**Description:**	Colorless Liquid
DOT Number:	2789, 2790	**DOT Classification:**	Corrosive Material

Molecular Weight: 60.1

Melting Point:	17°C	(62°F)	**Vapor Density:** 2.1	**Vapor Pressure:**	4 (11 mm Hg)
Boiling Point:	118°C	224(°F)	**Specific Gravity:** 1.05	**Water Solubility:**	Soluble

Chemical Incompatibilities or Instabilities: Strong Oxidizers.

FLAMMABILITY

NFPA Hazard Code: 2-2-0   **NFPA Classification:** Class II Combustible

Flash Point:	39°C	(103°F)	**LEL:**	5.4%
Autoignition Temp:	516°C	(961°F)	**UEL:**	16%

Fire Extinguishing Methods: Carbon Dioxide, Dry Chemical, Water Spray, or Regular Foam.

Special Fire Fighting Considerations: Isolate for 1/2 mile if rail or tank truck is involved in a fire. Water spray should be used to keep closed containers cool. Continue to cool container after fire is extinguished. Immediately withdraw if rising sound from venting device is heard or if fire is causing discoloration to the tank. Do not get water inside container.

TOXICOLOGY

Odor: Vinegar   **Odor Threshold:** 0.03 mg/m^3

Physical Contact: Material is extremely destructive to human tissues.

TLV = 25 mg/m^3 **STEL** = 37 mg/m^3 **IDLH** = 2500 mg/m^3

Routes of Entry and Relative LD$_{50}$ (or LC$_{50}$)   **RTECS # AF1225000**

Inhalation	N.D.	
Ingestion	3	(3530 mg/m^3)
Skin Absorption	4	(1060 mg/kg)

Symptoms of Exposure: Possible burns or irritation to skin, eyes and upper respiratory tract; headache; nausea. Inhalation of vapors may be fatal by causing glottis to spasm and suffocation. Exposure may result in chemical pneumonitis or pulmonary edema.

Emergency/First Aid Treatment: Remove to ventilated area; immediately remove any contaminated clothing and wash contaminated areas for 15 minutes using water. Treat supportively and observe for possible shock. If ingested, seek immediate medical aid.

Recommended Clean-Up Procedures

Personal Protection:	Level B Ensemble	**Recommended Material**	Butyl Rubber (+), Teflon (+), Saranex (+)
RCRA Waste #	None	**Reportable Quantity:**	5000 lb (1000 lb)

Spills: Remove all potential ignition sources. Absorb with non-combustible absorbent and take up using non-sparking tools. Decontaminate spill area using alkaline solution. Keep area isolated until vapors have dissipated.

Special Emergency Information: May be harmful if inhaled, swallowed or absorbed through the skin. Vapors are more dense than air and may settle in low lying areas.

ACETIC ANHYDRIDE - CH3C(O)O(O)CCH3

Synonyms: acetic oxide; acetyl oxide; ethanoic anhydride; acetic acid anhydride

CAS Number: 108-24-7	**Description:** Colorless Liquid	
DOT Number: 1715	**DOT Classification:** Corrosive Material	

Molecular Weight: 102.1

Melting Point: -74°C (-101°F)	**Vapor Density:** 3.5	**Vapor Pressure:** 2 (4 mm Hg)	
Boiling Point: 139°C (282°F)	**Specific Gravity:** 1.08	**Water Solubility:** Decomposes	

Chemical Incompatibilities or Instabilities: Water, alcohols, strong oxidizers, amines, ethylene diamine, oleum.

FLAMMABILITY

NFPA Hazard Code: 2-2-1- ~~W~~	**NFPA Classification:** Class II Combustible
Flash Point: 52°C (126°F)	**LEL:** 2.9%
Autoignition Temp: 316°C (600°F)	**UEL:** 10.3%

Fire Extinguishing Methods: Carbon Dioxide, Dry Chemical, or flooding quantities of Water.

Special Fire Fighting Considerations: Structural fire fighter protective clothing will not provide adequate protection. Isolate for 150' if rail or tank truck is involved in a fire. Do not allow water to enter container.

TOXICOLOGY

Odor: Sour

Odor Threshold: mg/m^3

Physical Contact: Material is extremely destructive to human tissues.

$TLV = 5\ mg/m^3$ STEL $= 21\ mg/m^3$ IDLH $= 4250\ mg/m^3$

Routes of Entry and Relative LD$_{50}$ (or LC$_{50}$)   RTECS # AK1925000

Inhalation	N.D.	
Ingestion	3	($1790\ mg/m^3$)
Skin Absorption	2	($4000\ mg/kg$)

Symptoms of Exposure: Possible burns or irritation to skin, eyes and upper respiratory tract; headache; nausea; lachramator. Inhalation of vapors may be fatal by causing glottis to spasm and suffocation. Exposure may result in chemical pneumonitis or pulmonary edema.

Emergency/First Aid Treatment: Remove to ventilated area; immediately remove any contaminated clothing and wash contaminated areas for 15 minutes using water. Treat supportively and observe for possible shock. If ingested, seek immediate medical aid.

Recommended Clean-Up Procedures

Personal Protection: Level A Ensemble	**Recommended Material** Butyl Rubber (0), Teflon (0)
RCRA Waste # None	**Reportable Quantity:** 5000 lb (1000 lb)

Spills: Remove all potential ignition sources. Absorb with non-combustible absorbent and take up using non-sparking tools. Decontaminate spill area using alkaline solution.

Special Emergency Information: May be harmful if inhaled, swallowed or absorbed through the skin. Will exothermically decompose upon exposure to water.

ACETOIN- CH3CH(OH)COCH3

Synonyms: 3-hydroxy-2-butanone; 2,3-butolone; acetyl methyl carbinol

CAS Number: 513-86-0   **Description:** Colorless liquid
DOT Number: 2621   **DOT Classification:** Flammable Liquid

Molecular Weight: 88.1
Melting Point: 15°C   (59°F)   **Vapor Density:** N.D.   **Vapor Pressure:** N.D.
Boiling Point: 148°C (298°F)   **Specific Gravity:** 0.99 **Water Solubility:** 100%

Chemical Incompatibilities or Instabilities: Strong oxidizers.

FLAMMABILITY

NFPA Hazard Code: **NFPA Classification:**
Flash Point: N.D.   **LEL:** N.D.
Autoignition Temp: N.D.   **UEL:** N.D.

Fire Extinguishing Methods: Carbon Dioxide, Dry Chemical, Water Spray or Fog.

Special Fire Fighting Considerations: Isolate for 1/2 mile if rail or tank truck is involved in a fire. Water spray should be used to keep closed containers cool. Continue to cool container after fire is extinguished. For large fires, if possible, withdraw and allow to burn. Immediately withdraw if rising sound from venting device is heard or if fire is causing discoloration to the tank.

TOXICOLOGY

Odor: Butter   **Odor Threshold:** N.D.
Physical Contact: Irritant
TLV = N.D.   **STEL** = N.D.   **IDLH** = N.D.

Routes of Entry and Relative LD50 (or LC50)   **RTECS # EL8790000**
    Inhalation   N.D.
    Ingestion   N.D.
    Skin Absorption   N.D.

Symptoms of Exposure: Possible irritation to skin, eyes and upper respiratory tract; headache; nausea.

Emergency/First Aid Treatment: Remove to fresh air, remove contaminated clothing, and wash contacted areas for 15 minutes with water.

Recommended Clean-Up Procedures

Personal Protection: Level B Ensemble   **Recommended Material** N.D.

RCRA Waste #   None   **Reportable Quantity:**   None

Spills: Remove all potential ignition sources. Absorb with non-combustible absorbent and take up using non-sparking tools. Decontaminate spill area using water. Keep area isolated until vapors have dissipated.

Special Emergency Information: May be harmful if inhaled, swallowed or absorbed through the skin. Vapors are heavier than air and may travel some distance to an ignition source. Vapors are more dense than air and may settle in low lying areas.

ACETONE - CH₃COCH₃

Synonyms: 2-Propanone; dimethyl ketone; β-ketopropane; pyroacetic ether

CAS Number: 67-64-1	**Description:** Colorless Liquid		
DOT Number: 1091, 1090	**DOT Classification:** Flammable Liquid		

Molecular Weight: 58.1

Melting Point: -94°C (-137°F)	**Vapor Density:** 2.0	**Vapor Pressure:** 7 (180 mm Hg)
Boiling Point: 56°C (133°F)	**Specific Gravity:** 0.8	**Water Solubility:** Soluble

Chemical Incompatibilities or Instabilities: Oxidizers, nitric acid, sulfuric acid, haloforms with alkali.

FLAMMABILITY

NFPA Hazard Code: 1-3-0	**NFPA Classification:** Class 1B Flammable
Flash Point: -20°C (-4°F)	**LEL:** 2.5%
Autoignition Temp: 465°C (869°F)	**UEL:** 12.8%

Fire Extinguishing Methods: Alcohol Resistant Foam, Carbon Dioxide, Dry Chemical, Water Spray or Fog.

Special Fire Fighting Considerations: Isolate for 1/2 mile if rail or tank truck is involved in a fire. Water spray should be used to keep closed containers cool. Continue to cool container after fire is extinguished. For large fires, if possible, withdraw and allow to burn. Immediately withdraw if rising sound from venting device is heard or if fire is causing discoloration to the tank.

TOXICOLOGY

Odor: Nail Polish Remover **Odor Threshold:** 1 mg/m³
Physical Contact: Irritant
TLV = 1800 mg/m³   **STEL** = 2400 mg/m³   **IDLH** = 50,000 mg/m³

Routes of Entry and Relative LD₅₀ (or LC₅₀)   **RTECS # AL3150000**
  Inhalation   1   (40,000 mg/m³/H)
  Ingestion   2   (5800 mg/kg)
  Skin Absorption 1   (20,000 mg/kg)

Symptoms of Exposure: Erythema, dryness, headache, fatigue, excitement, bronchial irritation, narcosis, paralysis.

Emergency/First Aid Treatment: Remove to ventilated area; immediately remove any contaminated clothing and wash contaminated areas for 15 minutes using water. Treat supportively and observe for possible shock. If ingested, seek immediate medical aid.

Recommended Clean-Up Procedures

Personal Protection: Level B Ensemble	**Recommended Material** Butyl Rubber (+), Teflon (+)
RCRA Waste # U002	**Reportable Quantity:** 5000 lb (1 lb)

Spills: Remove all potential ignition sources. Absorb with non-combustible absorbent and take up using non-sparking tools. Decontaminate spill area using water. Keep area isolated until vapors have dissipated.

Special Emergency Information: May be harmful if inhaled, swallowed or absorbed through the skin. Vapors are heavier than air and may travel some distance to an ignition source. Vapors are more dense than air and may settle in low lying areas.

ACETONE CYANOHYDRIN- (CH3)2C(OH)CN

Synonyms: 2-hydroxy-2-methylpropanenitrile; 2-methyllacetonitrile

CAS Number: 75-86-5   **Description:** Colorless liquid
DOT Number: 1541   **DOT Classification:** Poison B

Molecular Weight: 85.1
Melting Point: -19°C (165°F)   **Vapor Density:** 2.9   **Vapor Pressure:** 2 (1 mm Hg)
Boiling Point: 95°C (203°F)   **Specific Gravity:** 0.93 **Water Solubility:** 100%

Chemical Incompatibilities or Instabilities: Acids, strong alkalis, oxidizers.

FLAMMABILITY

NFPA Hazard Code: 4-1-2   **NFPA Classification:** Class III A Combustible Liquid
Flash Point: 74°C (165°F)   **LEL:** 2%
Autoignition Temp: 688°C (1270°F) **UEL:** 12%

Fire Extinguishing Methods: Alcohol Resistant Foam, Carbon Dioxide, Dry Chemical, Water Spray or Fog.

Special Fire Fighting Considerations: Structural fire fighter protective clothing will not provide adequate protection. Fight fire from a distance or protected location, if possible. Water spray should be used to keep closed containers cool. Continue to cool container after fire is extinguished.

DOT Recommended Isolation Zones: Small Spill: 150 ft   Large Spill: 150 ft
DOT Recommended Down Wind Take Cover Distance: Small Spill: 0.2 miles   Large Spill: 0.2 miles

TOXICOLOGY

Odor: **Odor Threshold:** N.D.
Physical Contact: Irritant
TLV = N.D.   **STEL** = N.D.   **IDLH** = N.D.

Routes of Entry and Relative LD$_{50}$ (or LC$_{50}$)   **RTECS # OD9275000**
  Inhalation   N.D.
  Ingestion   9   17 mg/kg
  Skin Absorption   10   17 mg/kg

Symptoms of Exposure: Headache, dizziness, nausea, cyanosis, convulsions, unconsciousness.

Emergency/First Aid Treatment: Remove to ventilated area; immediately remove any contaminated clothing and wash contaminated areas for 15 minutes using water. Treat supportively and observe for possible shock. If ingested, seek immediate medical aid. Treat for cyanide poisoning.

Recommended Clean-Up Procedures

Personal Protection: Level A Ensemble   **Recommended Material** Butyl Rubber (-). Teflon (-)

RCRA Waste # P069   **Reportable Quantity:** 10 lb (10 lb)

Spills: Dike to contain spill and collect liquid for later disposal. Decontaminate spill area using alkaline hypochlorite. Treat all materials used or generated and equipment involved as contaminated by hazardous waste.

Special Emergency Information: May be fatal in inhaled, ingested or absorbed through the skin. The material will readily decompose to form hydrogen cyanide. The rate of decomposition increases with temperature. Vapors are more dense than air and may settle in low lying areas.

ACETONITRILE- CH$_3$CN

Synonyms: methyl cyanide; cyanomethane; ethyl nitrile; ethane nitrile

CAS Number: 75-05-8   **Description:** Colorless liquid with ethereal odor
DOT Number: 1648   **DOT Classification:** Flammable Liquid

Molecular Weight: 41.1
Melting Point: -46°C (-50°F)   **Vapor Density:** 1.4   **Vapor Pressure:** 6 (73 mm Hg)
Boiling Point: 82°C (179°F)   **Specific Gravity:** 0.79   **Water Solubility:** Soluble

Chemical Incompatibilities or Instabilities: Fluorine, n-fluoro compounds, oxidants, nitric acid, sulfuric acid.

FLAMMABILITY

NFPA Hazard Code: 3-3-0   **NFPA Classification:** Class 1B Flammable
Flash Point: 6°C (42°F)   **LEL:** 4.4%
Autoignition Temp: 524°C (975°F)   **UEL:** 16.0%

Fire Extinguishing Methods: Alcohol Resistant Foam, Carbon Dioxide, Dry Chemical, Water Spray

Special Fire Fighting Considerations: Isolate for 1/2 mile if rail or tank truck is involved in a fire. Water spray should be used to keep closed containers cool. Continue to cool container after fire is extinguished. Immediately withdraw if rising sound from venting device is heard or if fire is causing discoloration to the tank. Structural fire fighter protective clothing will not provide adequate protection. Fight fire from a distance or protected location, if possible. Treat all materials used or generated and equipment involved as contaminated by hazardous waste.

TOXICOLOGY

Odor: Ethereal   **Odor Threshold:** 70 mg/m^3
Physical Contact: Irritant
TLV = 70 mg/m^3 **STEL** = 105 mg/m^3 **IDLH** = 7000 mg/m^3

Routes of Entry and Relative LD$_{50}$ (or LC$_{50}$)   **RTECS # AL7700000**
   Inhalation   N.D.
   Ingestion   3   (3800 mg/kg)
   Skin Absorption   4   (1250 mg/kg)

Symptoms of Exposure: Headache, rapid pulse, nausea, unconsciousness, convulsions, cyanosis.

Emergency/First Aid Treatment: Remove to ventilated area; immediately remove any contaminated clothing and wash contaminated areas for 15 minutes using water. Treat supportively and observe for possible shock. If ingested, seek immediate medical aid. Treat for cyanide poisoning.

Recommended Clean-Up Procedures

Personal Protection: Level A Ensemble   **Recommended Material:** Butyl Rubber (+), Teflon (+), Polyvinyl Alcohol (0)

RCRA Waste # U003   **Reportable Quantity:** 5000 lb (1 lb)

Spills: Remove all potential ignition sources. Absorb with non-combustible absorbent and take up using non-sparking tools. Decontaminate spill area using water. Decontaminate spill area using alkaline hypochlorite. Treat all materials used or generated and equipment involved as contaminated by hazardous waste. Keep area isolated until vapors have dissipated.

Special Emergency Information: May be harmful if inhaled, swallowed or absorbed through the skin. Vapors are heavier than air and may travel some distance to an ignition source. Vapors are more dense than air and may settle in low lying areas.

ACETOPHENONE - $C_6H_5COCH_3$

Synonyms: phenyl methyl ketone; acetylbenzene; phenyl methyl ketone; methyl phenyl ketone

CAS Number: 98-86-2	**Description:** Colorless liquid	
DOT Number: None	**DOT Classification:** None	

Molecular Weight: 120.2

Melting Point:	21°C (68°F)	**Vapor Density:** 4.1	**Vapor Pressure:**	3 (10 mm Hg)
Boiling Point:	202°C (396°F)	**Specific Gravity:** 1.0+	**Water Solubility:**	Insoluble

Chemical Incompatibilities or Instabilities: Oxidizers.

FLAMMABILITY

NFPA Hazard Code:	1-2-0	**NFPA Classification:**
Flash Point:	77°C (170°F)	**LEL:** N.D.
Autoignition Temp:	570°C (1058°F)	**UEL:** N.D.

Fire Extinguishing Methods: Alcohol Resistant Foam, Carbon Dioxide, Dry Chemical

Special Fire Fighting Considerations: Water spray should be used to keep closed containers cool. Continue to cool container after fire is extinguished. For large fires, if possible, withdraw and allow to burn. Immediately withdraw if rising sound from venting device is heard or if fire is causing discoloration to the tank.

TOXICOLOGY

Odor:
Physical Contact: Irritant

TLV = N.D.	**STEL** = N.D.	**IDLH** = N.D.

Odor Threshold: N.D.

Routes of Entry and Relative LD$_{50}$ (or LC$_{50}$)

Inhalation		N.D.
Ingestion	4	(900 mg/Kg)
Skin Absorption		N.D.

RTECS # AM5250000

Symptoms of Exposure: Dermatitis, narcosis, depression of CNS.

Emergency/First Aid Treatment: Remove to ventilated area; immediately remove any contaminated clothing and wash contaminated areas for 15 minutes using water. Treat supportively and observe for possible shock. If ingested, seek immediate medical aid.

Recommended Clean-Up Procedures

Personal Protection:	Level C Ensemble	**Recommended Material** Teflon (+)
RCRA Waste #	U004	**Reportable Quantity:** 5000 lb (1 lb)

Spills: Take up in non-combustible absorbent and dispose. Decontaminate spill area using soapy water

Special Emergency Information: May be harmful if inhaled, swallowed or absorbed through the skin. Vapors are more dense than air and may settle in low lying areas.

ACETOZONE - $C_6H_5C(O)OO(O)CCH_3$

Synonyms: acetyl benzoyl peroxide

CAS Number: 644-31-5
DOT Number: 2081

Description: Colorless Solid
DOT Classification: Solid - Forbidden, Solution - Organic Peroxide

Molecular Weight: 180.2

Melting Point:	36°C (97°F)	**Vapor Density:**	N.D.	**Vapor Pressure:**	1 (< 1 mm Hg)
Boiling Point:	Decomposes	**Specific Gravity:**	N.D.	**Water Solubility:**	Decomposes

Chemical Incompatibilities or Instabilities: Water, reducing agents, heat, shock.

FLAMMABILITY

NFPA Hazard Code: N.D.
Flash Point: N.D.
Autoignition Temp: N.D.

NFPA Classification:
LEL: N.D.
UEL: N.D.

Fire Extinguishing Methods: Carbon Dioxide, Dry Chemical, Water Spray or Fog, and Regular Foam.

Special Fire Fighting Considerations:

TOXICOLOGY

Odor:

Odor Threshold: N.D.

Physical Contact: Material is extremely destructive to human tissues.

TLV = N.D.　　　　　　**STEL** = N.D.　　　　　　**IDLH** = N.D.

Routes of Entry and Relative LD$_{50}$ (or LC$_{50}$)

RTECS # SD7860000

Inhalation	N.D.
Ingestion	N.D.
Skin Absorption	N.D.

Symptoms of Exposure: Possible burns or irritation to skin, eyes and upper respiratory tract; headache; nausea. Inhalation of vapors may be fatal by causing glottis to spasm and suffocation. Exposure may result in chemical pneumonitis or pulmonary edema.

Emergency/First Aid Treatment: Remove to ventilated area; immediately remove any contaminated clothing and wash contaminated areas for 15 minutes using water. Treat supportively and observe for possible shock. If ingested, seek immediate medical aid.

Recommended Clean-Up Procedures

Personal Protection: Level B Ensemble
Recommended Material Butyl Rubber (-)

RCRA Waste # None
Reportable Quantity: None

Spills: Take up spill for later disposal. Decontaminate spill area using water.

Special Emergency Information: May be harmful if inhaled, swallowed or absorbed through the skin. subjecting solid material to heat, shock, or friction may result in an explosion.

ACETYL BROMIDE - CH₃COBr

Synonyms: acotic bromide; ethanoyl bromide

CAS Number: 506-96-7		**Description:** Fuming Liquid	
DOT Number: 1716		**DOT Classification:** Corrosive Material	

Molecular Weight: 123.0

Melting Point:	-96°C (-141°F)	**Vapor Density:** 2.7	**Vapor Pressure:** 7 (135 mm Hg)
Boiling Point:	76°C (169°F)	**Specific Gravity:** 1.52	**Water Solubility:** Reacts violently

Chemical Incompatibilities or Instabilities: Water, alcohols, amines, alkalis, strong oxidizers

FLAMMABILITY

NFPA Hazard Code:		**NFPA Classification:**	
Flash Point:	N.D.	**LEL:**	N.D.
Autoignition Temp:	N.D.	**UEL:**	N.D.

Fire Extinguishing Methods: Carbon Dioxide or Dry Chemical, ONLY.

Special Fire Fighting Considerations: Violently reacts with water and alcohols to form hydrogen bromide gas. Water spray should be used to keep closed containers cool. Continue to cool container after fire is extinguished. Do not allow water to enter container.

TOXICOLOGY

Odor: Acrid		**Odor Threshold:** N.D.

Physical Contact: Material is extremely destructive to human tissues.

TLV = N.D.	**STEL** = N.D.	**IDLH** = N.D.

Routes of Entry and Relative LD₅₀ (or LC₅₀)   **RTECS # AO5955000**

Inhalation	N.D.
Ingestion	N.D.
Skin Absorption	N.D.

Symptoms of Exposure: Possible burns or irritation to skin, eyes and upper respiratory tract; headache; nausea. Inhalation of vapors may be fatal by causing glottis to spasm and suffocation. Exposure may result in chemical pneumonitis or pulmonary edema.

Emergency/First Aid Treatment: Remove to fresh air, remove contaminated clothing, and wash contacted areas for 15 minutes with water.

Recommended Clean-Up Procedures

Personal Protection:	Level A Ensemble	**Recommended Material**	Teflon (0)
RCRA Waste #	None	**Reportable Quantity:**	5000 lb (5000 lb)

Spills: Remove all potential ignition sources. Absorb with non-combustible absorbent and take up using non-sparking tools. Decontaminate spill area using water. Decontaminate spill area using alkaline solution. Keep area isolated until vapors have dissipated.

Special Emergency Information: May be harmful if inhaled, swallowed or absorbed through the skin. Vapors are heavier than air and may travel some distance to an ignition source. Vapors are more dense than air and may settle in low lying areas. Sodium carbonate/acetyl bromide wastes may generate carbon dioxide gas.

ACETYL CHLORIDE - CH3C(O)Cl

Synonyms: ethanoyl chloride; acetic chloride

CAS Number: 75-36-5
DOT Number: 1717

Description: Colorless to pale yellow fulling liquid
DOT Classification: Flammable Liquid

Molecular Weight: 78.5
Melting Point: -112°C (-170°F)
Boiling Point: 51°C (124°F)

Vapor Density: 2.7
Specific Gravity: 1.1

Vapor Pressure: 7 (135 mm Hg)
Water Solubility: Decomposes

Chemical Incompatibilities or Instabilities: Water, alcohols, alkalis, amines, strong oxidizers

FLAMMABILITY

NFPA Hazard Code: 3-3-2-W
Flash Point: 4°C (40°F)
Autoignition Temp: 390°C (734°F)

NFPA Classification: Class 1B Flammable
LEL: 5.0%
UEL: N.D.

Fire Extinguishing Methods: Carbon Dioxide or Dry Chemical.

Special Fire Fighting Considerations: Do not use water. Normally decomposes to from hydrogen chloride and phosgene gas. Isolate for 1/2 mile if rail or tank truck is involved in a fire. Do not allow water to enter container. May release toxic fumes upon heating (HCl and Phosgene).

TOXICOLOGY

Odor: Strong pungent
Physical Contact: Material is extremely destructive to human tissues.
TLV = N.D.   **STEL** = N.D.

Odor Threshold: N.D.

IDLH = N.D.

Routes of Entry and Relative LD$_{50}$ (or LC$_{50}$)
    Inhalation   N.D.
    Ingestion   N.D.
    Skin Absorption   N.D.

RTECS # AO6390000

Symptoms of Exposure: Possible irritation and burns to skin, eyes and upper respiratory tract; headache; nausea. Inhalation of vapors may be fatal by causing glottis to spasm and suffocation. Exposure may result in chemical pneumonitis or pulmonary edema.

Emergency/First Aid Treatment: Remove to ventilated area; immediately remove any contaminated clothing and wash contaminated areas for 15 minutes using water. Treat supportively and observe for possible shock. If ingested, seek immediate medical aid.

Recommended Clean-Up Procedures

Personal Protection: Level A Ensemble   **Recommended Material** Teflon (0)

RCRA Waste # U006   **Reportable Quantity:** 5000 lb (5000 lb)

Spills: Remove all potential ignition sources. Absorb with non-combustible absorbent and take up using non-sparking tools. Decontaminate spill area using alkaline. Keep area isolated until vapors have dissipated.

Special Emergency Information: May be harmful if inhaled, swallowed or absorbed through the skin. Vapors are heavier than air and may travel some distance to an ignition source. Vapors are more dense than air and may settle in low lying areas.

ACETYLACETONE - CH3C(O)CH2C(O)CH3

Synonyms: 2,4-pentanedione; diacetyl methane

CAS Number: 123-54-6	**Description:** Colorless liquid
DOT Number: 2310	**DOT Classification:** Flammable Liquid

Molecular Weight: 100.1

Melting Point:	-23°C (°F)	**Vapor Density:** 3.5	**Vapor Pressure:**
Boiling Point:	141°C (°F)	**Specific Gravity:** 0.98	**Water Solubility:** 12%

Chemical Incompatibilities or Instabilities: Oxidizers.

FLAMMABILITY

NFPA Hazard Code: 2-2-0	**NFPA Classification:** Class 1C Flammable Liquid
Flash Point: 34°C (93°F)	**LEL:** N.D.
Autoignition Temp: 340°C (644°F)	**UEL:** N.D.

Fire Extinguishing Methods: Carbon Dioxide, Dry Chemical, Foam, Water Spray or Fog.

Special Fire Fighting Considerations: Isolate for 1/2 mile if rail or tank truck is involved in a fire. Water spray should be used to keep closed containers cool. Continue to cool container after fire is extinguished. For large fires, if possible, withdraw and allow to burn. Immediately withdraw if rising sound from venting device is heard or if fire is causing discoloration to the tank.

TOXICOLOGY

Odor:	**Odor Threshold:**
Physical Contact: Irritant	
TLV = N.D. **STEL** = N.D.	**IDLH** = N.D.

Routes of Entry and Relative LD$_{50}$ (or LC$_{50}$)   **RTECS # SA1925000**

Inhalation	N.D.	
Ingestion	4	(1000 mg/kg)
Skin Absorption	2	(5000 mg/kg)

Symptoms of Exposure: Possible irritation to skin, eyes and upper respiratory tract; headache; nausea.

Emergency/First Aid Treatment: Remove to ventilated area; immediately remove any contaminated clothing and wash contaminated areas for 15 minutes using water. Treat supportively and observe for possible shock. If ingested, seek immediate medical aid.

Recommended Clean-Up Procedures

Personal Protection: Level B Ensemble	**Recommended Material** N.D.
RCRA Waste # None	**Reportable Quantity:** None

Spills: Remove all potential ignition sources. Absorb with non-combustible absorbent and take up using non-sparking tools. Decontaminate spill area using water. Keep area isolated until vapors have dissipated.

Special Emergency Information: May be harmful if inhaled, swallowed or absorbed through the skin. Vapors are heavier than air and may travel some distance to an ignition source. Vapors are more dense than air and may settle in low lying areas. May form unstable peroxides upon prolonged exposure to air.

ACETYLENE - HC≡CH

Synonyms: ethyne; ethine

CAS Number: 74-86-2   **Description:** Colorless gas with slight garlic-like odor
DOT Number: 1001   **DOT Classification:** Flammable Gas

Molecular Weight: 26.0
Melting Point: -82°C (116°F)   **Vapor Density:** 0.9   **Vapor Pressure:** Gas
Boiling Point: -83°C (-118°F)   **Specific Gravity:** ---   **Water Solubility:** 0%

Chemical Incompatibilities or Instabilities: Pressure, oxidizers, copper, mercury, silver, and their salts, halogens, sodium hydride, alkali metals.

FLAMMABILITY

NFPA Hazard Code: 1-4-3   **NFPA Classification:** Flammable Gas
Flash Point: Gas   **LEL:** 2.5%
Autoignition Temp: 305°C (581°F)   **UEL:** 100%

Fire Extinguishing Methods: Carbon Dioxide, Dry Chemical, Water Spray, Regular Foam. Stop gas flow prior to fighting fire.

Special Fire Fighting Considerations: Acetylene has a very low ignition energy and may travel some distance to an ignition source.

TOXICOLOGY

Odor: Slight, garlic-like   **Odor Threshold:** 230 mg/m^3
Physical Contact: A simple asphyxiant
TLV = 2500 mg/m^3 (ceiling)   **STEL** = N.D.   **IDLH** = N.D.

Routes of Entry and Relative LD$_{50}$ (or LC$_{50}$)   **RTECS # AO9600000**
    Inhalation   N.D.
    Ingestion   N.D.
    Skin Absorption   N.D.

Symptoms of Exposure: Possible irritation to skin, eyes and upper respiratory tract; headache; nausea.

Emergency/First Aid Treatment: Remove to fresh air, give artificial respiration and oxygen if necessary.

Recommended Clean-Up Procedures

Personal Protection: Level B Ensemble   **Recommended Material** N.D.

RCRA Waste #   None   **Reportable Quantity:**   None

Spills: Remove all potential ignition sources. Stop leak if it can safely be done. Keep area isolated until vapors have dissipated.

Special Emergency Information: May be harmful if inhaled, swallowed or absorbed through the skin. Vapors are heavier than air and may travel some distance to an ignition source. Vapors are more dense than air and may settle in low lying areas.

ACETYLENE TETRABROMIDE - $Br_2CHCHBr_2$

Synonyms: 1,1,2,2 tetrabromoethane

CAS Number: 79-27-6   **Description:** Pale yellow liquid
DOT Number: 2504   **DOT Classification:** ORM-A

Molecular Weight: 345.7
Melting Point: 0°C (32°F)   **Vapor Density:** 11.9   **Vapor Pressure:**   1 (< 1 mm Hg)
Boiling Point: D 245°C (474°F)   **Specific Gravity:** 3.00   **Water Solubility:**   < 1%

Chemical Incompatibilities or Instabilities: Strong alkali, dinitrogen tetroxide, alkali metals, oxidizers.

FLAMMABILITY

NFPA Hazard Code: 3-0-1   **NFPA Classification:** Non-Combustible Liquid
Flash Point: N.A.   **LEL:**   N.A.
Autoignition Temp: 335°C (635°F)   **UEL:**   N.A.

Fire Extinguishing Methods: Use agent suitable for surrounding fire.

Special Fire Fighting Considerations: Structural fire fighter protective clothing will not provide adequate protection.

TOXICOLOGY
QUESTIONABLE CARCINOGEN

Odor: **Odor Threshold:**
Physical Contact: Irritant
TLV = 14 mg/m^3 **STEL** = N.D.   **IDLH** = 140 mg/m^3

Routes of Entry and Relative LD$_{50}$ (or LC$_{50}$)   **RTECS # KI8225000**
　　Inhalation   N.D.
　　Ingestion   5   (400 mg/kg)
　　Skin Absorption   N.D.

Symptoms of Exposure: Possible irritation to skin, eyes and upper respiratory tract; headache; nausea.

Emergency/First Aid Treatment: Remove to ventilated area; immediately remove any contaminated clothing and wash contaminated areas for 15 minutes using water. Treat supportively and observe for possible shock. If ingested, seek immediate medical aid.

Recommended Clean-Up Procedures

Personal Protection: Level B Ensemble   **Recommended Material**   Viton (-)

RCRA Waste #   None   **Reportable Quantity:**   None

Spills: Dike to contain spill and collect liquid for later disposal. Decontaminate spill area using soapy water.

Special Emergency Information: May be harmful if inhaled, swallowed or absorbed through the skin. Vapors are more dense than air and may settle in low lying areas.

ACRIDINE

Synonyms: benzo(b)quinoline; 2,3 benzoquinoline; diazo(b,e)pyridine

CAS Number: 260-94-6
DOT Number: 2713

Description: White powder
DOT Classification: IMO

Molecular Weight: 179.2
Melting Point: 110°C (230°F)
Boiling Point: 340°C (644°F)

Vapor Density: N.D.
Specific Gravity: 1.01

Vapor Pressure: 1 (<1 mm Hg)
Water Solubility: None

Chemical Incompatibilities or Instabilities: Oxidizers.

FLAMMABILITY

NFPA Classification:
Flash Point: N.D.
Autoignition Temp: N.D.

NFPA Hazard Code:
LEL: N.A.
UEL: N.A.

Fire Extinguishing Methods: Use agent suitable for surrounding fire.

Special Fire Fighting Considerations: Water spray should be used to keep closed containers cool. Continue to cool container after fire is extinguished. For large fires, if possible, withdraw and allow to burn. Treat all materials used or generated and equipment involved as contaminated by hazardous waste.

TOXICOLOGY
Suspected CARCINOGEN

Odor: None
Physical Contact: Irritant
TLV = 0.2 mg/m^3

STEL = N.D.

Odor Threshold: N.A.

IDLH = N.D.

Routes of Entry and Relative LD$_{50}$ (or LC$_{50}$)
Inhalation   N.D.
Ingestion   3   (2000 mg/kg)
Skin Absorption   N.D.

RTECS # AR7175000

Symptoms of Exposure: Possible irritation to skin, eyes and upper respiratory tract; headache; nausea.

Emergency/First Aid Treatment: Remove to fresh air, remove contaminated clothing, and wash contacted areas for 15 minutes with water.

Recommended Clean-Up Procedures

Personal Protection: Level B Ensemble
RCRA Waste # None

Recommended Material Butyl Rubber (-), Viton (-), Teflon (-)
Reportable Quantity: None

Spills: Take up spill and dispose. Decontaminate spill area using soapy water. Treat all materials used or generated and equipment involved as contaminated by hazardous waste.

Special Emergency Information: May be harmful if inhaled, swallowed or absorbed through the skin.

ACROLEIN - CH₂=CHCHO

Synonyms: 2-propenal; acrylaldehyde; acrylic aldehyde; acraldehyde

CAS Number:	107-02-8	**Description:** Colorless to light yellow liquid	
DOT Number:	1092	**DOT Classification:** Flammable Liquid	

Molecular Weight: 56.1

Melting Point:	-88°C (126°F)	**Vapor Density:**	1.9	**Vapor Pressure:**	**8** (210 mm Hg)
Boiling Point:	53°C (127°F)	**Specific Gravity:**	0.84	**Water Solubility:**	Soluble

Chemical Incompatibilities or Instabilities: Oxidizers, acids, alkalis, ammonia, amines

FLAMMABILITY

NFPA Hazard Code:	3-3-3	**NFPA Classification:** Class 1B Flammable	
Flash Point:	-15°C (-26°F)	**LEL:**	2.8%
Autoignition Temp:	235°C (445°F)	**UEL:**	31%

Fire Extinguishing Methods: Alcohol Resistant Foam, Carbon Dioxide, Dry Chemical, Water Spray or Fog.

Special Fire Fighting Considerations: Isolate for 1/2 mile if rail or tank truck is involved in a fire. Water spray should be used to keep closed containers cool. Continue to cool container after fire is extinguished. For large fires, if possible, withdraw and allow to burn. Immediately withdraw if rising sound from venting device is heard or if fire is causing discoloration to the tank.

DOT Recommended Isolation Zones:	Small Spill: 900 ft	Large Spill: 1200 ft
DOT Recommended Down Wind Take Cover Distance:	Small Spill: 3 miles	Large Spill: 4 miles

TOXICOLOGY

Odor: Pungent, disagreeable odor   **Odor Threshold:** 0.05 mg/m³

Physical Contact: Material is extremely destructive to human tissues.

TLV = 0.23 mg/m³   **STEL** = 0.7 mg/m³   **IDLH** = 11.6 mg/m³

Routes of Entry and Relative LD₅₀ (or LC₅₀)   **RTECS #** AS1050000

Inhalation	N.D.	
Ingestion	7	(46 mg/kg)
Skin Absorption	7	(562 mg/kg)

Symptoms of Exposure: Possible irritation and burns to skin, eyes and upper respiratory tract; headache; nausea. Inhalation of vapors may be fatal by causing glottis to spasm and suffocation. Exposure may result in chemical pneumonitis or pulmonary edema. Lachrymation. Sensitizer.

Emergency/First Aid Treatment: Remove to ventilated area; immediately remove any contaminated clothing and wash contaminated areas for 15 minutes using water. Treat supportively and observe for possible shock. If ingested, seek immediate medical aid.

Recommended Clean-Up Procedures

Personal Protection:	Level A Ensemble	**Recommended Material**	Butyl Rubber (+)
RCRA Waste #	P003	**Reportable Quantity:**	1 lb (1 lb)

Spills: Remove all potential ignition sources. Absorb with non-combustible absorbent and take up using non-sparking tools. Decontaminate spill area using water. Keep area isolated until vapors have dissipated. Treat all materials used or generated and equipment involved as contaminated by hazardous waste.

Special Emergency Information: May be harmful if inhaled, swallowed or absorbed through the skin. Low ignition energy, vapors are heavier than air and may travel some distance to an ignition source. May undergo hazardous polymerization. May form unstable peroxides upon prolonged exposure to air. Vapors are more dense than air and may settle in low lying areas.

ACRYLAMIDE - CH$_2$CHCONH$_2$

Synonyms: 2-propenamide; acrylic amide; ethelene carboxamide

CAS Number: 79-06-1   **Description:** White powder
DOT Number: 2074   **DOT Classification:** IMO

Molecular Weight: 71.1
Melting Point: 85°C (184°F)   **Vapor Density:** 2.5   **Vapor Pressure:**   1 (< 1 mm Hg)
Boiling Point: N.D.   **Specific Gravity:** 1.22   **Water Solubility:**   Soluble

Chemical Incompatibilities or Instabilities: Strong oxidizers

FLAMMABILITY

NFPA Classification: **NFPA Hazard Code:**
Flash Point: °C   (°F)   **LEL:** N.A.
Autoignition Temp: 240°C (464°F)   **UEL:** N.A.

Fire Extinguishing Methods: Alcohol Resistant Foam, Carbon Dioxide, Dry Chemical, Water Spray or Fog.

Special Fire Fighting Considerations: May decompose. Polymerizes releasing ammonia and hydrogen upon melting.
Closed containers may rupture violently.

TOXICOLOGY
CARCINOGEN

Odor: None   **Odor Threshold:** N.A.
Physical Contact: Irritant
TLV = 0.03 mg/m^3 (Skin)   **STEL** = N.D.   **IDLH** = N.D.

Routes of Entry and Relative LD$_{50}$ (or LC$_{50}$)   **RTECS #** AS3325000
    Inhalation   N.D.
    Ingestion   5   (170 mg/m^3)
    Skin Absorption   5   (1000 mg/m^3)

Symptoms of Exposure: Muscular atrophy, tremors, hallucination. Possible irritation to skin, eyes and upper respiratory
tract; headache; nausea.

Emergency/First Aid Treatment: Remove to ventilated area; immediately remove any contaminated clothing and wash
contaminated areas for 15 minutes using water. Treat supportively and observe for possible shock. If ingested, seek
immediate medical aid.

Recommended Clean-Up Procedures

Personal Protection: Level A Ensemble   **Recommended Material**   Butyl Rubber (-), Polyethylene (0),
                                Ethylene Vinyl Alcohol (0)

RCRA Waste #   U007   **Reportable Quantity:**   5000 lb (1 lb)

Spills: Take up spill and dispose. Decontaminate spill area using water. Treat all materials used or generated and
equipment involved as contaminated by hazardous waste.

Special Emergency Information: May be harmful if inhaled, swallowed or absorbed through the skin.

ACRYLIC ACID - $CH_2=CHCO_2H$

Synonyms: acroleic acid; propenoic acid; vinyl formic acid

CAS Number: 79-10-7   **Description:** Colorless liquid
DOT Number: 2218   **DOT Classification:** Corrosive Material

Molecular Weight: 72.1

Melting Point:	14°C (56°F)	**Vapor Density:**	2.5	**Vapor Pressure:**	2 (3 mm Hg)
Boiling Point:	141°C (285°F)	**Specific Gravity:**	1.06	**Water Solubility:**	100%

Chemical Incompatibilities or Instabilities: Acids, Alkalis, peroxides, oxidizers

FLAMMABILITY

NFPA Hazard Code: 3-2-2   **NFPA Classification:** Class IIIA Combustible Liquid
Flash Point: 68°C (155°F)   **LEL:**   2%
Autoignition Temp: 438°C (820°F)   **UEL:**   8%

Fire Extinguishing Methods: Alcohol Resistant Foam, Carbon Dioxide, Dry Chemical, Water Spray or Fog.

Special Fire Fighting Considerations: Isolate for 1/2 mile if rail or tank truck is involved in a fire. Water spray should be used to keep closed containers cool. Continue to cool container after fire is extinguished. Immediately withdraw if rising sound from venting device is heard or if fire is causing discoloration to the tank. Closed containers may explode violently when heated.

TOXICOLOGY
Questionable CARCINOGEN

Odor: Plastic-like, sweet   **Odor Threshold:** 0.3 mg/m^3
Physical Contact: Material is extremely destructive to human tissues.
TLV = 30 mg/m^3 (skin)   **STEL** = N.D.   **IDLH** = N.D.

Routes of Entry and Relative LD$_{50}$ (or LC$_{50}$)   **RTECS # AS4375000**

Inhalation	N.D.	
Ingestion	5	(340 mg/kg)
Skin Absorption	8	(280 mg/kg)

Symptoms of Exposure: Possible irritation and burns to skin, eyes and upper respiratory tract; headache; nausea. Exposure may result in chemical pneumonitis or pulmonary edema. Inhalation of vapors may be fatal by causing glottis to spasm and suffocation.

Emergency/First Aid Treatment: Remove to ventilated area; immediately remove any contaminated clothing and wash contaminated areas for 15 minutes using water. Treat supportively and observe for possible shock. If ingested, seek immediate medical aid.

Recommended Clean-Up Procedures

Personal Protection: Level A Ensemble   **Recommended Material** Butyl Rubber (+), Saranex (+), Teflon (0)

RCRA Waste #   U008   **Reportable Quantity:**   5000 lb (1 lb)

Spills: Remove all potential ignition sources. Absorb with non-combustible absorbent and take up using non-sparking tools. Decontaminate spill area using water. Decontaminate spill area using alkaline. Treat all materials used or generated and equipment involved as contaminated by hazardous waste.

Special Emergency Information: May be harmful if inhaled, swallowed or absorbed through the skin. May undergo hazardous polymerization.

ACRYLONITRILE - CH₂=CHCN

Synonyms: Vinyl cyanide; propenenitrile; cyanoethylene

CAS Number: 107-13-1

DOT Number: 1093

Description: Colorless to pale yellow liquid

DOT Classification: Flammable Liquid

Molecular Weight: 53.1

Melting Point: -84°C (-118°F)

Boiling Point: 77°C (171°F)

Vapor Density: 1.8

Specific Gravity: 0.81

Vapor Pressure: 6 (83 mm Hg)

Water Solubility: 7%

Chemical Incompatibilities or Instabilities: Halogens, acids, bases, amines, strong oxidizers, peroxides

FLAMMABILITY

NFPA Hazard Code: 4-3-2

Flash Point: 0°C (32°F)

Autoignition Temp: 481°C (898°F)

NFPA Classification: Class 1B Flammable

LEL: 3%

UEL: 17%

Fire Extinguishing Methods: Alcohol Resistant Foam, Carbon Dioxide, Dry Chemical, Water Spray or Fog.

Special Fire Fighting Considerations: Structural fire fighter protective clothing will not provide adequate protection. Isolate for 1/2 mile if rail or tank truck is involved in a fire. Water spray from an unmanned device should be used to keep closed containers cool. Continue to cool container after fire is extinguished. For large fires, if possible, withdraw and allow to burn. Immediately withdraw if rising sound from venting device is heard or if fire is causing discoloration to the tank. Treat all materials used or generated and equipment involved as contaminated by hazardous waste. Fight fire from a distance or protected location, if possible.

TOXICOLOGY
Suspected CARCINOGEN

Odor: garlic, onion

Physical Contact: Material is extremely destructive to human tissues.

TLV = 4.3 mg/m³   **STEL** = 22 mg/m³ (skin)

Odor Threshold: N.D.

IDLH = 1200 mg/m³

RTECS # AT5250000

Routes of Entry and Relative LD₅₀ (or LC₅₀)

Inhalation	N.D.	
Ingestion	6	(82 mg/kg)
Skin Absorption	8	(250 mg/kg)

Symptoms of Exposure: Possible irritation to skin, eyes and upper respiratory tract; headache; nausea. Inhalation of vapors may be fatal by causing glottis to spasm and suffocation. Exposure may result in chemical pneumonitis or pulmonary edema. Cyanosis. Lachrymation.

Emergency/First Aid Treatment: Remove to ventilated area; immediately remove any contaminated clothing and wash contaminated areas for 15 minutes using water. Treat supportively and observe for possible shock. If ingested, seek immediate medical aid. Treat for cyanide poisoning as necessary.

Recommended Clean-Up Procedures

Personal Protection: Level A Ensemble

RCRA Waste # U009

Recommended Material Butyl Rubber (0)

Reportable Quantity: 100 lb (100 lb)

Spills: Remove all potential ignition sources. Absorb with non-combustible absorbent and take up using non-sparking tools. Decontaminate spill area using water. Decontaminate spill area using alkaline hypochlorite. Vent to dissipate vapors. Treat all materials used or generated and equipment involved as contaminated by hazardous waste.

Special Emergency Information: May be fatal in inhaled, ingested or absorbed through the skin. Acrylonitrile has a very low ignition energy. Vapors are heavier than air and may travel some distance to an ignition source. Vapors are more dense than air and may settle in low lying areas.

ADIPONITRILE - CN(CH$_2$)$_4$CN

Synonyms: adipic acid dinitrile; 1,4,-dicyanobutane; hexanedinitrile

CAS Number: 111-69-3	**Description:** White liquid
DOT Number: 2205	**DOT Classification:** IMO; Poison B

Molecular Weight: 108.1

Melting Point:	1°C (34°F)	**Vapor Density:** 3.7	**Vapor Pressure:**	1 (< 1 mm Hg)
Boiling Point:	295°C (563°F)	**Specific Gravity:** 0.97	**Water Solubility:**	100%

Chemical Incompatibilities or Instabilities: Oxidizers, acids.

FLAMMABILITY

NFPA Hazard Code:	4-2-1	**NFPA Classification:**	Class III B Combustible Liquid
Flash Point:	93°C (200°F)	**LEL:**	1%
Autoignition Temp:	550°C (1022°F)	**UEL:**	N.D.

Fire Extinguishing Methods: Carbon Dioxide, Dry Chemical, Water Spray or Fog.

Special Fire Fighting Considerations: Structural fire fighter protective clothing will not provide adequate protection. Fight fire from a distance or protected location, if possible. Treat all materials used or generated and equipment involved as contaminated by hazardous waste.

TOXICOLOGY

Odor: None	**Odor Threshold:** N.A.
Physical Contact: Irritant	
TLV = N.D. **STEL** = N.D.	**IDLH** = N.D.

Routes of Entry and Relative LD$_{50}$ (or LC$_{50}$)   **RTECS # AV2625000**

Inhalation	N.D.	
Ingestion	5	(300 mg/kg)
Skin Absorption	N.D.	

Symptoms of Exposure: Possible irritation to skin, eyes and upper respiratory tract; headache; nausea. Cyanosis.

Emergency/First Aid Treatment: Remove to ventilated area; immediately remove any contaminated clothing and wash contaminated areas for 15 minutes using water. Treat supportively and observe for possible shock. If ingested, seek immediate medical aid. Treat for cyanide poisoning as necessary.

Recommended Clean-Up Procedures

Personal Protection:	Level A Ensemble	**Recommended Material** Butyl Rubber (-), Teflon (0)
RCRA Waste #	None	**Reportable Quantity:** None

Spills: Remove all potential ignition sources. Absorb with non-combustible absorbent and take up using non-sparking tools. Decontaminate spill area using water. Decontaminate spill area using alkaline hypochlorite.

Special Emergency Information: May be fatal in inhaled, ingested or absorbed through the skin.

ALDOL - CH₃CH(OH)CH₂CHO

Synonyms: 3-hydroxybutanal; 3- hydroxybuteraldehyde

CAS Number: 107-89-1	**Description:** Colorless, viscous liquid
DOT Number: 2839	**DOT Classification:** Poison B

Molecular Weight: 88.1

Melting Point:	°C	(°F)	**Vapor Density:** 3.0	**Vapor Pressure:** N.D.
Boiling Point: D	85°C	(185°F)	**Specific Gravity:** 1.11	**Water Solubility:** 100%

Chemical Incompatibilities or Instabilities: Oxidizers.

FLAMMABILITY

NFPA Hazard Code: 3-2-2	**NFPA Classification:** Class III A Combustible Liquid
Flash Point: 66°C (150°F)	**LEL:** N.D.
Autoignition Temp: 250°C (482°F)	**UEL:** N.D.

Fire Extinguishing Methods: Alcohol Resistant Foam, Carbon Dioxide, Dry Chemical, Water Spray or Fog.

Special Fire Fighting Considerations: Structural fire fighter protective clothing will not provide adequate protection. Fight fire from a distance or protected location, if possible. Treat all materials used or generated and equipment involved as contaminated by hazardous waste.

TOXICOLOGY

Odor:	**Odor Threshold:** N.D.
Physical Contact: Mild Irritant	
TLV = N.D. **STEL** = N.D.	**IDLH** = N.D.

Routes of Entry and Relative LD_{50} (or LC_{50})   **RTECS # ES3150000**

Inhalation	N.D.	
Ingestion	3	(2180 mg/kg)
Skin Absorption	8	(140 mg/kg)

Symptoms of Exposure: Disorientation, fatigue. Possible irritation to skin, eyes and upper respiratory tract; headache; nausea.

Emergency/First Aid Treatment: Remove to ventilated area; immediately remove any contaminated clothing and wash contaminated areas for 15 minutes using water. Treat supportively and observe for possible shock. If ingested, seek immediate medical aid.

Recommended Clean-Up Procedures

Personal Protection: Level A Ensemble	**Recommended Material** Butyl Rubber (0), Teflon (-)
RCRA Waste # None	**Reportable Quantity:** None

Spills: Take up spill and dispose. Decontaminate spill area using water. Treat all materials used or generated and equipment involved as contaminated by hazardous waste.

Special Emergency Information: May be harmful if inhaled, swallowed or absorbed through the skin. Vapors are heavier than air and may travel some distance to an ignition source. Vapors are more dense than air and may settle in low lying areas.

ALDRIN

Synonyms: 1,2,3,4,10,10-hexachloro-1,4,4a,5,8,8a-hexahydro-endo,exo-1,4:5,8-dimethanonaphthlene

CAS Number: 309-00-2	**Description:** Colorless to yellow brown solid	
DOT Number: 2761, 2762	**DOT Classification:** Poison B	

Molecular Weight: 364.9

Melting Point: 104°C (219°F)	**Vapor Density:** N.D.	**Vapor Pressure:**	1 (< 0 mm Hg)
Boiling Point: D °C (°F)	**Specific Gravity:** 1.6	**Water Solubility:**	Insoluble

Chemical Incompatibilities or Instabilities: Acids, oxidizers, phenol.

FLAMMABILITY

NFPA Hazard Code: -----	**NFPA Classification:**	
Flash Point: N.A.	**LEL:** N.A.	
Autoignition Temp: N.A.	**UEL:** N.A.	

Fire Extinguishing Methods: Use agent suitable for surrounding fire.

Special Fire Fighting Considerations: Structural fire fighter protective clothing will not provide adequate protection. Fight fire from a distance or protected location, if possible. Treat all materials used or generated and equipment involved as contaminated by hazardous waste. Some aldrin formations are made using flammable (F.P. < 100°F) solvents. Solvents used will determine how fire should be extinguished.

TOXICOLOGY
SUSPECTED **CARCINOGEN**

Odor: Mild, chemical	**Odor Threshold:** N.D.
Physical Contact: Irritant	

$TLV = 0.25$ mg/m^3 (skin) $STEL = $ N.D. $IDLH = 100$ mg/m^3

Routes of Entry and Relative LD$_{50}$ (or LC$_{50}$)   **RTECS # IO2100000**

Inhalation	N.D.	
Ingestion	8	(39 mg/kg)
Skin Absorption	N.D.	

Symptoms of Exposure: Headache, nausea, vomiting, blurred vision, convulsions, unconscious.

Emergency/First Aid Treatment: Remove to ventilated area; immediately remove any contaminated clothing and wash contaminated areas for 15 minutes using water. Treat supportively and observe for possible shock. If ingested, seek immediate medical aid.

Recommended Clean-Up Procedures

Personal Protection: Level A Ensemble	**Recommended Material**	N.D.
RCRA Waste # P004	**Reportable Quantity:**	1 lb (1 lb)

Spills: If solid, then sweep up spill and decontaminate area by washing area with soapy water. If liquid, remove all potential ignition sources. Absorb with non-combustible absorbent and take up using non-sparking tools. Decontaminate spill area using water. Decontaminate spill area using soapy water. Treat all materials used or generated and equipment involved as contaminated by hazardous waste. Vent to dissipate vapors.

Special Emergency Information: May be fatal in inhaled, ingested or absorbed through the skin. If aldrin/flammable solvent formulation is spilled, then compatibility of the solvent with personal protective equipment must be determined.

ALLYL ACETATE - $CH_3CO_2CH_2CH=CH$

Synonyms: 3-acetoxypropene; 2-propenyl acetate

CAS Number: 591-87-7   **Description:** Colorless liquid
DOT Number: 2833   **DOT Classification:** Flammable Liquid

Molecular Weight: 100.1
Melting Point: °C   (°F)   **Vapor Density:** 3.45   **Vapor Pressure:** N.D.
Boiling Point: 104°C (219°F)   **Specific Gravity:** 0.9   **Water Solubility:** Insoluble

Chemical Incompatibilities or Instabilities: Oxidizers, peroxides.

FLAMMABILITY

NFPA Hazard Code: 1-3-0   **NFPA Classification:** Class 1B Flammable
Flash Point: 22°C   (72°F)   **LEL:** N.D.
Autoignition Temp: 374°C (705°F)   **UEL:** N.D.

Fire Extinguishing Methods: Alcohol Resistant Foam, Carbon Dioxide, Dry Chemical, Water Spray or Fog.

Special Fire Fighting Considerations: Water spray should be used to keep closed containers cool. Continue to cool container after fire is extinguished. Isolate for 1/2 mile if rail or tank truck is involved in a fire. Water spray should be used to keep closed containers cool. Continue to cool container after fire is extinguished. Immediately withdraw if rising sound from venting device is heard or if fire is causing discoloration to the tank. Treat all materials used or generated and equipment involved as contaminated by hazardous waste.

TOXICOLOGY

Odor: **Odor Threshold:** N.D.
Physical Contact: Irritant
TLV = N.D.   **STEL** = N.D.   **IDLH** = N.D.

Routes of Entry and Relative LD$_{50}$ (or LC$_{50}$)   **RTECS #** AF1750000
   Inhalation   8   (4100 mg/m^3/H)
   Ingestion   7   (130 mg/kg)
   Skin Absorption   4   (1100 mg/kg)

Symptoms of Exposure: Possible irritation to skin, eyes and upper respiratory tract; headache; nausea.

Emergency/First Aid Treatment: Remove to ventilated area; immediately remove any contaminated clothing and wash contaminated areas for 15 minutes using water. Treat supportively and observe for possible shock. If ingested, seek immediate medical aid.

Recommended Clean-Up Procedures

Personal Protection: Level B Ensemble   **Recommended Material** N.D.

RCRA Waste #   None   **Reportable Quantity:** None

Spills: Remove all potential ignition sources. Absorb with non-combustible absorbent and take up using non-sparking tools. Decontaminate spill area using water. Decontaminate spill area using soapy water. Keep area isolated until vapors have dissipated. Treat all materials used or generated and equipment involved as contaminated by hazardous waste.

Special Emergency Information: May be fatal in inhaled, ingested or absorbed through the skin. May undergo hazardous polymerization.

Allyl formate (CAS # 1838-59-1, DOT # 2336, RTECS # LQ9800000) has similar chemical, physical and toxicological properties.

ALLYL ALCOHOL - $CH_2=CHCH_2OH$

Synonyms: 1-propen-3-ol, propenol; 2-proenyl alcohol; 2-propen-1-ol

CAS Number: 107-18-6
DOT Number: 1098

Description: Colorless liquid
DOT Classification: Flammable Liquid

Molecular Weight: 58.1
Melting Point: -129°C (-200°F)
Boiling Point: 97°C (207°F)

Vapor Density: 2.0
Specific Gravity: 0.85

Vapor Pressure: 4 (17 mm Hg)
Water Solubility: 100%

Chemical Incompatibilities or Instabilities: Acids (sulfuric), carbon tetrachloride, oxidizers, peroxides.

FLAMMABILITY

NFPA Hazard Code: 3-3-1
Flash Point: 22°C (72°F)
Autoignition Temp: 378°C (713°F)

NFPA Classification: Class 1B Flammable
LEL: 2.5%
UEL: 18%

Fire Extinguishing Methods: Alcohol Resistant Foam, Carbon Dioxide, Dry Chemical, Water Spray or Fog.

Special Fire Fighting Considerations: Structural fire fighter protective clothing will not provide adequate protection. Water spray from an unmanned device should be used to keep closed containers cool. Continue to cool container after fire is extinguished. Fight fire from a distance or protected location, if possible. Treat all materials used or generated and equipment involved as contaminated by hazardous waste.

DOT Recommended Isolation Zones: Small Spill: 150 ft   Large Spill: 150 ft
DOT Recommended Down Wind Take Cover Distance: Small Spill: 0.8 mile   Large Spill: 0.8 mile

TOXICOLOGY

Odor: Pungent, mustard-like odor
Physical Contact: Material is extremely destructive to human tissues
TLV = 5 mg/m^3 **STEL** = 10 mg/m^3

Odor Threshold: 1.2 mg/m^3

IDLH = 330 mg/m^3

RTECS # BA5075000

Routes of Entry and Relative LD$_{50}$ (or LC$_{50}$)
Inhalation	6	(1565 mg/m^3/H)
Ingestion	8	(64 mg/kg)
Skin Absorption	9	(45 mg/kg)

Symptoms of Exposure: Possible burns or irritation to skin, eyes and upper respiratory tract; headache; nausea. Inhalation of vapors may be fatal by causing glottis to spasm and suffocation. Exposure may result in chemical pneumonitis or pulmonary edema.

Emergency/First Aid Treatment: Remove to ventilated area; immediately remove any contaminated clothing and wash contaminated areas for 15 minutes using water. Treat supportively and observe for possible shock. If ingested, seek immediate medical aid.

Recommended Clean-Up Procedures

Personal Protection: Level A Ensemble
RCRA Waste # P005

Recommended Material Butyl Rubber (+), Teflon (+), Viton (0)
Reportable Quantity: 100 lb (100 lb)

Spills: Remove all potential ignition sources. Absorb with non-combustible absorbent and take up using non-sparking tools. Decontaminate spill area using water. Decontaminate spill area using water. Keep area isolated until vapors have dissipated. Treat all materials used or generated and equipment involved as contaminated by hazardous waste.

Special Emergency Information: May be fatal in inhaled, ingested or absorbed through the skin. Vapors are heavier than air and may travel some distance to an ignition source. May undergo hazardous polymerization at elevated temperatures. Vapors are more dense than air and may settle in low lying areas.

ALLYLAMINE - $CH_2CHCH_2NH_2$

Synonyms: 2-propenamine; 3-aminopropene; 3-aminopropylamine

CAS Number: 107-11-9	**Description:** Colorless Liquid	
DOT Number: 2334	**DOT Classification:** Flammable Liquid	

Molecular Weight:	57.1			
Melting Point:	-88°C (-127°F)	**Vapor Density:** 1.97	**Vapor Pressure:**	N.D.
Boiling Point:	53°C (128°F)	**Specific Gravity:** 0.76	**Water Solubility:**	Soluble

Chemical Incompatibilities or Instabilities: Acids, oxidizers, halogens, hologenated compounds.

FLAMMABILITY

NFPA Hazard Code:	3-3-1	**NFPA Classification:**	Class 1B Flammable
Flash Point:	-20°C (-9°F)	**LEL:**	2%
Auto ignition Temp:	374°C (705°F)	**UEL:**	22%

Fire Extinguishing Methods: Alcohol Resistant Foam, Carbon Dioxide, Dry Chemical, Water Spray or Fog.

Special Fire Fighting Considerations: Structural fire fighter protective clothing will not provide adequate protection. Water spray from an unmanned device should be used to keep closed containers cool. Continue to cool container after fire is extinguished. Immediately withdraw if rising sound from venting device is heard or if fire is causing discoloration to the tank. Treat all materials used or generated and equipment involved as contaminated by hazardous waste.

DOT Recommended Isolation Zones:	Small Spill: 150 ft	Large Spill: 600 ft	
DOT Recommended Down Wind Take Cover Distance:	Small Spill: 0.8 mile	Large Spill: 2 miles	

TOXICOLOGY

Odor: Sharp   **Odor Threshold:** N.D.

Physical Contact: Severe skin and eye burns.

TLV = N.D.   **STEL** = N.D.   **IDLH** = N.D.

Routes of Entry and Relative LD_{50} (or LC_{50})   **RTECS #** BA5425000

Inhalation	7	(3600 mg/m^3/ 4H)
Ingestion	5	(106 mg/kg)
Skin Absorption	10	(35 mg/kg)

Symptoms of Exposure: Possible burns or irritation to skin, eyes and upper respiratory tract; headache; nausea. Lachrymation. Inhalation of vapors may be fatal by causing glottis to spasm and suffocation. Exposure may result in chemical pneumonitis or pulmonary edema.

Emergency/First Aid Treatment: Remove to ventilated area; immediately remove any contaminated clothing and wash contaminated areas for 15 minutes using water. Treat supportively and observe for possible shock. If ingested, seek immediate medical aid.

Recommended Clean-Up Procedures

Personal Protection:	Level A Ensemble	**Recommended Material** Allylamine (Poor)
RCRA Waste #	None	**Reportable Quantity:** None

Spills: Remove all potential ignition sources. Absorb with non-combustible absorbent and take up using non-sparking tools. Decontaminate spill area using water. Decontaminate spill area using Water spray should be used to keep closed containers cool. Continue to cool container after fire is extinguished. Vent to dissipate vapors. Treat all materials used or generated and equipment involved as contaminated by hazardous waste.

Special Emergency Information: May be fatal in inhaled, ingested or absorbed through the skin. Vapors are heavier than air and may travel some distance to an ignition source. Vapors are more dense than air and may settle in low lying areas.

ALLYL CHLORIDE - $CH_2=CHCH_2Cl$

Synonyms: 3-chloropropene, 1-chloro-2-propene

CAS Number: 107-05-1	**Description:** Colorless Liquid
DOT Number: 1100	**DOT Classification:** Flammable Liquid

Molecular Weight: 76.5

Melting Point: -134°C (-209°F)	**Vapor Density:** 2.6	**Vapor Pressure:** 8 (295 mm/Hg)
Boiling Point: 45°C (113°F)	**Specific Gravity:** 0.94	**Water Solubility:** Insoluble

Chemical Incompatibilities or Instabilities: Oxidizers, alkyl aluminum chlorides, strong acids, aluminum chloride, boron trifluoride, aluminum, magnesium, zinc, ethylene imine, ethylene diamine.

FLAMMABILITY

NFPA Hazard Code: 3-3-1	**NFPA Classification:** Class 1B Flammable
Flash Point: -29°C (-20°F)	**LEL:** 3%
Auto ignition Temp: 392°C (737°F)	**UEL:** 11%

Fire Extinguishing Methods: Carbon Dioxide, Dry Chemical, Foam, or Water Fog.

Special Fire Fighting Considerations: Structural fire fighter protective clothing will not provide adequate protection. Water spray from an unmanned device should be used to keep closed containers cool. Continue to cool container after fire is extinguished. Fight fire from a distance or protected location, if possible. Treat all materials used or generated and equipment involved as contaminated by hazardous waste. May release toxic fumes (phosgene) upon heating

TOXICOLOGY
Suspected CARCINOGEN

Odor: **Odor Threshold:** N.D.

Physical Contact: Material is extremely destructive to human tissues.

TLV = 3 mg/m^3 **STEL** = 6 mg/m^3 **IDLH** = 900 mg/m^3

Routes of Entry and Relative LD$_{50}$ (or LC$_{50}$)   **RTECS #** UC735000

Inhalation	3	(22,000 mg/m^3/H)
Ingestion	4	(700 mg/kg)
Skin Absorption	1	(2066 mg/kg)

Symptoms of Exposure: Possible burns or irritation to skin, eyes and upper respiratory tract; headache; nausea. Lachrymation. Inhalation of vapors may be fatal by causing glottis to spasm and suffocation. Exposure may result in chemical pneumonitis or pulmonary edema.

Emergency/First Aid Treatment: Remove to ventilated area; immediately remove any contaminated clothing and wash contaminated areas for 15 minutes using water. Treat supportively and observe for possible shock. If ingested, seek immediate medical aid.

Recommended Clean-Up Procedures

Personal Protection: Level A Ensemble	**Recommended Material** Polyvinyl Alcohol (poor) Teflon (+)
RCRA Waste # None	**Reportable Quantity:** 1000 lb (1000 lb)

Spills: Remove all potential ignition sources. Absorb with non-combustible absorbent and take up using non-sparking tools. Decontaminate spill area using water. Decontaminate spill area using soapy water. Keep area isolated until vapors have dissipated. Water spray should be used to keep closed containers cool. Treat all materials used or generated and equipment involved as contaminated by hazardous waste.

Special Emergency Information: May be harmful if inhaled, swallowed or absorbed through the skin. Vapors are heavier than air and may travel some distance to an ignition source. Vapors are more dense than air and may settle in low lying areas. May undergo hazardous polymerization. Allyl Bromide (CAS 106-95-6, DOT 1099, RTECS UC7090000) has very similar chemical, physical, and toxicological properties.

ALLYL CHLOROCARBONATE - CH=CHCH$_2$OCOCl

Synonyms: allyl chloroformate

CAS Number: 2937-50-2   **Description:** Colorless Liquid
DOT Number: 1722   **DOT Classification:** Flammable Liquid

Molecular Weight: 120.5
Melting Point: °C   (°F)   **Vapor Density:** 4.2   **Vapor Pressure:** **4** (20 mg Hg)
Boiling Point: 113°C (235°F)   **Specific Gravity:** 1.1   **Water Solubility:** Decomposes

Chemical Incompatibilities or Instabilities: Oxidizers, rust, water, strong acids, amines, alcohols.

FLAMMABILITY

NFPA Hazard Code: 3-3-1   **NFPA Classification:** Class 1C Flammable
Flash Point: 31°C (88°F)   **LEL:** N.D.
Auto ignition Temp: N.D.   **UEL:** N.D.

Fire Extinguishing Methods: Carbon Dioxide, Dry Chemical, Foam, Water Spray or Fog.

Special Fire Fighting Considerations: Structural fire fighter protective clothing will not provide adequate protection. Water spray from an unmanned device should be used to keep closed containers cool. Continue to cool container after fire is extinguished. Fight fire from a distance or protected location, if possible. Treat all materials used or generated and equipment involved as contaminated by hazardous waste.

DOT Recommended Isolation Zones: Small Spill: 105 ft   Large Spill: 150 ft
DOT Recommended Down Wind Take Cover Distance   Small Spill: 0.2 miles   Large Spill: 0.2 miles

TOXICOLOGY

Odor: **Odor Threshold:** N.D.
Physical Contact: Material is extremely destructive to human tissues.
TLV = N.D.   **STEL** = N.D.   **IDLH** = N.D.

Routes of Entry and Relative LD$_{50}$ (or LC$_{50}$)   **RTECS #** LQ5775000
  Inhalation   10   (32 mg/m^3/H)
  Ingestion   6   (244 mg/kg)
  Skin Absorption   N.D.

Symptoms of Exposure: Possible burns or irritation to skin, eyes and upper respiratory tract; headache; nausea. Lachrymation. Inhalation of vapors may be fatal by causing glottis to spasm and suffocation. Exposure may result in chemical pneumonitis or pulmonary edema.

Emergency/First Aid Treatment: Remove to ventilated area; immediately remove any contaminated clothing and wash contaminated areas for 15 minutes using water. Treat supportively and observe for possible shock. If ingested, seek immediate medical aid.

Recommended Clean-Up Procedures

Personal Protection: Level A Ensemble   **Recommended Material** N.D.

RCRA Waste #   None   **Reportable Quantity:**   None

Spills: Remove all potential ignition sources. Absorb with non-combustible absorbent and take up using non-sparking tools. Decontaminate spill area using water. Decontaminate spill area using alkaline. Keep area isolated until vapors have dissipated. Treat all materials used or generated and equipment involved as contaminated by hazardous waste.

Special Emergency Information: May be fatal in inhaled, ingested or absorbed through the skin. Vapors are heavier than air and may travel some distance to an ignition source. Vapors are more dense than air and may settle in low lying areas. reacts with water to yield allyl alcohol and chloroformic acid.

ALLYL GLYCIDYL ETHER

Synonyms: 1-(allyloxy)-2,3-epoxypropane

CAS Number: 106-92-3   **Description:** Colorless Liquid
DOT Number: 2219   **DOT Classification:** Flammable Liquid

Molecular Weight:
Melting Point: -100°C (-148°F)   **Vapor Density:**   3.9   **Vapor Pressure:**   2 (2 mm Hg)
Boiling Point: 154°C (309°F)   **Specific Gravity:** 0.9   **Water Solubility:**   Insoluble

Chemical Incompatibilities or Instabilities: Strong oxidizers.

FLAMMABILITY

NFPA Hazard Code: **NFPA Classification:**   Class II Combustible
Flash Point: 57°C (135°F)   **LEL:** N.D.
Auto ignition Temp: N.D.   **UEL:** N.D.

Fire Extinguishing Methods: Carbon Dioxide, Dry Chemical, Foam, Water Spray or Fog.

Special Fire Fighting Considerations: Isolate for 1/2 mile if rail or tank truck is involved in a fire. Water spray should be used to keep closed containers cool. Continue to cool container after fire is extinguished. Immediately withdraw if rising sound from venting device is heard or if fire is causing discoloration to the tank.

TOXICOLOGY

Odor: Pleasant   **Odor Threshold:** N.D.
Physical Contact: Material is extremely destructive to human tissues.
TLV = 22 mg/m^3 **STEL** = 44 mg/m^3 **IDLH** = 1300 mg/m^3

Routes of Entry and Relative LD$_{50}$ (or LC$_{50}$)   **RTECS # RR0875000**
   Inhalation   N.D.
   Ingestion   4   (922 mg/kg)
   Skin Absorption   2   (2550 mg/kg)

Symptoms of Exposure: Possible burns or irritation to skin, eyes and upper respiratory tract; headache; nausea. Inhalation of vapors may be fatal by causing glottis to spasm and suffocation. Exposure may result in chemical pneumonitis or pulmonary edema.

Emergency/First Aid Treatment: Remove to ventilated area; immediately remove any contaminated clothing and wash contaminated areas for 15 minutes using water. Treat supportively and observe for possible shock. If ingested, seek immediate medical aid.

Recommended Clean-Up Procedures

Personal Protection: Level B Ensemble   **Recommended Material**   N.D.

RCRA Waste #   None   **Reportable Quantity:**   None

Spills: Dike to contain spill and collect liquid for later disposal. Decontaminate spill area using soapy water.

Special Emergency Information: May be harmful if inhaled, swallowed or absorbed through the skin.

ALLYL ISOTHIOCYANATE - CH$_2$=CHCH$_2$N=C=S

Synonyms: mustard oil; 3-isothiocyanato-1-propene

CAS Number: 57-06-7

DOT Number: 1545

Description: Colorless to Pale Yellow Liquid

DOT Classification: Poison B

Molecular Weight: 99.2

Melting Point: -80°C (-112°F)

Boiling Point: 151°C (305°F)

Vapor Density: 3.4

Specific Gravity: 1.01

Vapor Pressure: 2 (5 mm Hg)

Water Solubility: 1%

Chemical Incompatibilities or Instabilities: Water, alcohols, amines, oxidizers, heat.

FLAMMABILITY

NFPA Hazard Code: 3-2-0

Flash Point: 46°C (115°F)

Autoignition Temp: N.D.

NFPA Classification: Class II Combustible

LEL: N.D.

UEL: N.D.

Fire Extinguishing Methods: Carbon Dioxide, Dry Chemical, Foam, Water Spray or Fog.

Special Fire Fighting Considerations: Structural fire fighter protective clothing will not provide adequate protection. Water spray from an unmanned device should be used to keep closed containers cool. Continue to cool container after fire is extinguished. Fight fire from a distance or protected location, if possible. Treat all materials used or generated and equipment involved as contaminated by hazardous waste.

TOXICOLOGY
CARCINOGEN (Suspect)

Odor: Irritating, mustard-like

Physical Contact: Material is extremely destructive to human tissues.

TLV = N.D.   **STEL** = N.D.

Odor Threshold: N.D.

IDLH = N.D.

Routes of Entry and Relative LD$_{50}$ (or LC$_{50}$)

Inhalation	N.D.	
Ingestion	5	(148 mg/kg)
Skin Absorption	N.D.	

RTECS: NX8225000

Symptoms of Exposure: Possible burns or irritation to skin, eyes and upper respiratory tract; headache; nausea. Lachrymation. Inhalation of vapors may be fatal by causing glottis to spasm and suffocation. Exposure may result in chemical pneumonitis or pulmonary edema.

Emergency/First Aid Treatment: Remove to ventilated area; immediately remove any contaminated clothing and wash contaminated areas for 15 minutes using water. Treat supportively and observe for possible shock. If ingested, seek immediate medical aid.

Recommended Clean-Up Procedures

Personal Protection: Class A Ensemble

RCRA Waste # None

Recommended Material N.D.

Reportable Quantity: None

Spills: Remove all potential ignition sources. Dike to contain spill, absorb with non-combustible absorbent and take up using non-sparking tools. Decontaminate spill area using soapy water. Keep area isolated until vapors have dissipated. Treat all materials used or generated and equipment involved as contaminated by hazardous waste.

Special Emergency Information: May be fatal in inhaled, ingested or absorbed through the skin. Vapors are heavier than air and may travel some distance to an ignition source. Vapors are more dense than air and may settle in low lying areas. May undergo hazardous polymerization.

ALLYL THRICHLOROSILANE - $CH_2=CHCH_2SiCl_3$

Synonyms: trichloroallylsilane

CAS Number: 107-37-9   **Description:** Colorless Liquid
DOT Number: 1724   **DOT Classification:** Corrosive Material

Molecular Weight: 175.5
Melting Point: N.D.   **Vapor Density:** 6.1   **Vapor Pressure:** 4 (15 mm Hg)
Boiling Point: 118°C (243°F)   **Specific Gravity:** 1.2   **Water Solubility:** Decomposes

Chemical Incompatibilities or Instabilities: Water, acids, alkali.

FLAMMABILITY

NFPA Hazard Code: 3-3-2-W   **NFPA Classification:** Class 1C Flammable
Flash Point: 35°C   (95°F)   **LEL:** N.D.
Autoignition Temp: N.D.   **UEL:** N.D.

Fire Extinguishing Methods: Carbon Dioxide, Dry Chemical, Foam.

Special Fire Fighting Considerations: DO NOT USE WATER. Isolate for 1/2 mile if rail or tank truck is involved in a fire. Do not allow water to enter container. Water spray should be used to keep closed containers cool. Continue to cool container after fire is extinguished. Immediately withdraw if rising sound from venting device is heard or if fire is causing discoloration to the tank.

TOXICOLOGY

Odor: Sharp, irritating   **Odor Threshold:** N.D.
Physical Contact: Material is extremely destructive to human tissues
TLV = N.D.   **STEL** = N.D.   **IDLH** = N.D.

Routes of Entry and Relative LD$_{50}$ (or LC$_{50}$)   **RTECS:** VV1530000
  Inhalation   N.D.
  Ingestion   N.D.
  Skin Absorption   N.D.

Symptoms of Exposure: Possible burns or irritation to skin, eyes and upper respiratory tract; headache; nausea. Inhalation of vapors may be fatal by causing glottis to spasm and suffocation. Exposure may result in chemical pneumonitis or pulmonary edema.

Emergency/First Aid Treatment: Remove to ventilated area; immediately remove any contaminated clothing and wash contaminated areas for 15 minutes using water. Treat supportively and observe for possible shock. If ingested, seek immediate medical aid.

Recommended Clean-Up Procedures

Personal Protection: Level B Ensemble   **Recommended Material** N.D.

RCRA Waste #   None   **Reportable Quantity:**   None

Spills: Remove all potential ignition sources. Take up using non-sparking tools. Decontaminate spill area using water. Keep area isolated until vapors have dissipated.

Special Emergency Information: May be harmful if inhaled, or swallowed or absorbed through the skin. May react violently with water to yield hydrogen chloride gas.

ALUMINUM - Al
(Powdered)

Synonyms: ---

CAS Number: 7429-90-5   **Description:** Gray, White Powder

DOT Number: 1383, 1309, 1396   **DOT Classification:** Flammable Solid

Molecular Weight: 27.0

Melting Point: 660°C (1220°F)   **Vapor Density:** N.A.   **Vapor Pressure:** 1 (< 0 mm Hg)

Boiling Point: 2330°C (4226°F)   **Specific Gravity:** 2.7   **Water Solubility:** Insoluble

Chemical Incompatibilities or Instabilities: Oxidizers, halocarbons, alcohols, alkali, sodium sulfate, phosphorous, selenium, sulfur, sulfur dibromide, acids, halogens, interhalogens.

FLAMMABILITY

NFPA Hazard Code: 0-1-1   **NFPA Classification:** --

Flash Point: N.A.   **LEL:** N.A.

Auto ignition Temp: 760°C (1400°F)   **UEL:** N.A.

Fire Extinguishing Methods: Smother with dry sand or clay, Class D fire extinguisher.

Special Fire Fighting Considerations: Do not use CO_2, water, or halogenated agents. Water spray should be used to keep closed containers cool. Continue to cool container after fire is extinguished. For large fires, if possible, withdraw and allow to burn. Do not allow water to enter container.

TOXICOLOGY

Odor: None   **Odor Threshold:** N.A.

Physical Contact: Irritant

$TLV = 5$ mg/m^3 **STEL** = N.D.   **IDLH** = N.D.

Routes of Entry and Relative LD$_{50}$ (or LC$_{50}$)   **RTECS #** BD0330000

Inhalation	N.D.
Ingestion	N.D.
Skin Absorption	N.D.

Symptoms of Exposure: Irritation.

Emergency/First Aid Treatment: Remove to ventilated area; immediately remove any contaminated clothing and wash contaminated areas for 15 minutes using water. Treat supportively and observe for possible shock. If ingested, seek immediate medical aid.

Recommended Clean-Up Procedures

Personal Protection: Class C Ensemble   **Recommended Material** Any

RCRA Waste # None   **Reportable Quantity:** None

Spills: Take up. Place in suitable dry container and dispose.

Special Emergency Information: May be harmful if inhaled, swallowed or absorbed through the skin. Dust is easily ignited. May explode if fine dust clouds are allowed to form. Moist air or water can cause a spontaneous, exothermic reaction.

ALUMINUM CHLORIDE (ANHYDROUS) - AlCl$_3$

Synonyms: aluminum trichloride (anhydrous)

CAS Number: 7446-70-0   **Description:** White crystals that fume in moist air
DOT Number: 1725, 1726   **DOT Classification:** Corrosive Material

Molecular Weight: 133.3
Melting Point: (S) 181°C (389°F)   **Vapor Density:** N.D.   **Vapor Pressure:** None
Boiling Point: °C   (°F)   **Specific Gravity:** 2.44   **Water Solubility:** Soluble

Chemical Incompatibilities or Instabilities: Water

FLAMMABILITY

NFPA Classification: **NFPA Hazard Code:**
Flash Point: N.A.   **LEL:** N.A.
Auto ignition Temp: N.A.   **UEL:** N.A.

Fire Extinguishing Methods: Use Dry Chemical or Foam for surrounding fire.

Special Fire Fighting Considerations: If involved in fire, **DO NOT USE WATER** or anhydrous AlCL$_3$ will react explosively.

TOXICOLOGY

Odor: None   **Odor Threshold:** N.A.
Physical Contact: Skin and eye burns.
TLV = N.D.   **STEL** = N.D.   **IDLH** = N.D.

Routes of Entry and Relative LD$_{50}$ (or LC$_{50}$)   **RTECS #**
   Inhalation   N.D.
   Ingestion   N.D.
   Skin Absorption   N.D.

Symptoms of Exposure: Contact burns.

Emergency/First Aid Treatment: Remove to fresh air, remove contaminated clothing, and wash contacted areas for 15 minutes with water.

Recommended Clean-Up Procedures

Personal Protection: Level B Ensemble   **Recommended Material**   N.D.

RCRA Waste #   **Reportable Quantity:**

Small Spills: Cover with sodium bicarbonate, take up and dispose, wash spill area with dilute alkaline solution.

Special Emergency Information: Physical, chemical and toxicological properties of anhydrous AlBr$_3$ are very similar.

ALUMINUM NITRATE (Anhydrous) - Al(NO3)3
ALUMINUM NITRATE (9-hydrate) - Al(NO3)3•9H2O

Synonyms:

CAS Number: 13473-90-0, 7784-27-2   **Description:** White Powder
DOT Number: 1438   **DOT Classification:** Oxidizer

Molecular Weight: 213.0
Melting Point: 73 C (163 F)   **Vapor Density:** N.D.   **Vapor Pressure:** 1 (< 1 mm Hg)
Boiling Point: (D) 135 C (F)   **Specific Gravity:** N.D.   **Water Solubility:** 64%

Chemical Incompatibilities or Instabilities: Finely divided metals, reducing agents.

FLAMMABILITY

NFPA Hazard Code: **NFPA Classification:**
Flash Point: N.A.   **LEL:** N.A.
Autoignition Temp: N.A.   **UEL:** N.A.

Fire Extinguishing Methods: WATER ONLY.

Special Fire Fighting Considerations: For large fires, use flooding quantities of water. Water spray should be used to keep closed containers cool. Continue to cool container after fire is extinguished. For large fires, if possible, withdraw and allow to burn.

TOXICOLOGY

Odor: None   **Odor Threshold:** N.A.
Physical Contact: Irritant
TLV = N.D.   **STEL** = N.D.   **IDLH** = N.D.

Routes of Entry and Relative LD$_{50}$ (or LC$_{50}$)   **RTECS # BD1050000**
   Inhalation   N.D.
   Ingestion   **5**   (264 mg/kg)
   Skin Absorption   N.D.

Symptoms of Exposure: Possible irritation to skin, eyes and upper respiratory tract; headache; nausea.

Emergency/First Aid Treatment: Remove to ventilated area; immediately remove any contaminated clothing and wash contaminated areas for 15 minutes using water. Treat supportively and observe for possible shock. If ingested, seek immediate medical aid.

Recommended Clean-Up Procedures

Personal Protection: Level B Ensemble   **Recommended Material** N.D.

RCRA Waste # None   **Reportable Quantity:** None

Spills: Take up in non-combustible absorbent and dispose. Decontaminate spill area using water.

Special Emergency Information: May be harmful if inhaled, swallowed or absorbed through the skin. May oxidize combustibles at elevated temperatures.

ALUMINUM PHOSPHIDE - AlP

Synonyms: celphos; detia; phostoxin

CAS Number: 20859-73-8   **Description:** Dark gray to yellow powder
DOT Number: 1397   **DOT Classification:** Flammable Solid

Molecular Weight: 58.0
Melting Point: >1000 C   **Vapor Density:** N.D.   **Vapor Pressure:** 1 (< 1 mm Hg)
Boiling Point: N.D.   **Specific Gravity:** 2.4   **Water Solubility:** Decomposes

Chemical Incompatibilities or Instabilities: Water, acid.

FLAMMABILITY

NFPA Hazard Code: 3-4-2-W̶   **NFPA Classification:**
Flash Point: Pyrophoric   **LEL:** N.A.
Autoignition Temp: N.A.   **UEL:** N.A.

Fire Extinguishing Methods: Dry Sand or Clay, Class D extinguisher.

Special Fire Fighting Considerations: DO NOT USE WATER. If controlled, allow fire to burn.

TOXICOLOGY

Odor: None   **Odor Threshold:** N.A.
Physical Contact: Material is extremely destructive to human tissues.
TLV = 2 mg(Al)/m^3 **STEL** = N.D.
                                                       IDLH = N.D.

Routes of Entry and Relative LD$_{50}$ (or LC$_{50}$)   **RTECS # BD1400000**
   Inhalation   N.D.
   Ingestion   N.D.
   Skin Absorption   N.D.

Symptoms of Exposure: Possible burns or irritation to skin, eyes and upper respiratory tract; headache; nausea.

Emergency/First Aid Treatment: Remove to ventilated area; immediately remove any contaminated clothing and wash contaminated areas for 15 minutes using water. Treat supportively and observe for possible shock. If ingested, seek immediate medical aid.

Recommended Clean-Up Procedures

Personal Protection: Level A Ensemble   **Recommended Material** N.D.

RCRA Waste # P006   **Reportable Quantity:** 100 lb (1 lb)

Spills: Clean up only under expert supervision.

Special Emergency Information: May be fatal in inhaled, ingested or absorbed through the skin. Reacts with water, especially if acidic, to form highly toxic phospine gas and phosphoric acid.

AMINOETHOXYETHANOL - $H_2NCH_2CH_2OCH_2CH_2OH$

Synonyms: 2-aminoethoxyethanol; 2-(2-aminoethoxy)ethanol, diglycolamine

CAS Number: 929-06-6	**Description:** Colorless Liquid	
DOT Number: 1760	**DOT Classification:** Corrosive Material	

Molecular Weight: 105.1

Melting Point:	N.D.	**Vapor Density:** N.D.	**Vapor Pressure:**	N.D.
Boiling Point:	224 C (435 F)	**Specific Gravity:** 1.5	**Water Solubility:**	Soluble

Chemical Incompatibilities or Instabilities: Oxidizers, acids.

FLAMMABILITY

NFPA Hazard Code:		**NFPA Classification:**
Flash Point:	N.D.	**LEL:** N.D.
Autoignition Temp:	N.D.	**UEL:** N.D.

Fire Extinguishing Methods: Carbon Dioxide, Dry Chemical, Foam, Water Spray or Fog.

Special Fire Fighting Considerations: Water spray should be used to keep closed containers cool. Continue to cool container after fire is extinguished.

TOXICOLOGY

Odor: **Odor Threshold:** N.D.

Physical Contact: Material is extremely destructive to human tissues.

TLV = N.D.   **STEL** = N.D.   **IDLH** = N.D.

Routes of Entry and Relative LD$_{50}$ (or LC$_{50}$)   **RTECS # KJ6125000**

Inhalation	N.D.	
Ingestion	2	(5560 mg/kg)
Skin Absorption	4	(1190 mg/kg)

Symptoms of Exposure: Possible burns or irritation to skin, eyes and upper respiratory tract; headache; nausea. Lachrymation. Inhalation of vapors may be fatal by causing glottis to spasm and suffocation. Exposure may result in chemical pneumonitis or pulmonary edema.

Emergency/First Aid Treatment: Remove to ventilated area; immediately remove any contaminated clothing and wash contaminated areas for 15 minutes using water. Treat supportively and observe for possible shock. If ingested, seek immediate medical aid.

Recommended Clean-Up Procedures

Personal Protection:	Level B Ensemble	**Recommended Material** Butyl Rubber (-), Viton (-), Nitrile Rubber (-)
RCRA Waste #	None	**Reportable Quantity:** None

Spills: Remove all potential ignition sources. Dike to contain spill, absorb with non-combustible absorbent and take up using non-sparking tools. Decontaminate spill area using water.

Special Emergency Information: May be harmful if inhaled, swallowed or absorbed through the skin. Vapors are heavier than air and may travel some distance to an ignition source. Vapors are more dense than air and may settle in low lying areas.

AMINOETHYLPIPERAZINE

Synonyms: n-aminoethylpiperazine; 1-(2-aminoethyl)piperazine

CAS Number: 140-31-8   **Description:** Colorless Liquid
DOT Number: 2815   **DOT Classification:** Corrosive Material

Molecular Weight: 129.2
Melting Point: -19 C   (-2 F)   **Vapor Density:** 4.4   **Vapor Pressure:** N.D.
Boiling Point: 220 C   (428 F)   **Specific Gravity:** 1.0   **Water Solubility:** Soluble

Chemical Incompatibilities or Instabilities: Oxidizers.

FLAMMABILITY

NFPA Hazard Code: 2-2-0   **NFPA Classification:** Class III A Combustible
Flash Point: 95 C   (200 F)   **LEL:** N.D.
Autoignition Temp: N.D.   **UEL:** N.D.

Fire Extinguishing Methods: Carbon Dioxide, Dry Chemical, Foam, Water Spray or Fog.

Special Fire Fighting Considerations: Water spray should be used to keep closed containers cool. Continue to cool container after fire is extinguished.

TOXICOLOGY

Odor: **Odor Threshold:** N.D.
Physical Contact: Material is extremely destructive to human tissues.
TLV = N.D.   **STEL** = N.D.   **IDLH** = N.D.

Routes of Entry and Relative LD$_{50}$ (or LC$_{50}$)   **RTECS # TK8050000**
    Inhalation   N.D.
    Ingestion   3   (2140 mg/kg)
    Skin Absorption   5   (880 mg/kg)

Symptoms of Exposure: Possible burns or irritation to skin, eyes and upper respiratory tract; headache; nausea. Lachrymation. Inhalation of vapors may be fatal by causing glottis to spasm and suffocation. Exposure may result in chemical pneumonitis or pulmonary edema.

Emergency/First Aid Treatment: Remove to ventilated area; immediately remove any contaminated clothing and wash contaminated areas for 15 minutes using water. Treat supportively and observe for possible shock. If ingested, seek immediate medical aid.

Recommended Clean-Up Procedures

Personal Protection: Level B Ensemble   **Recommended Material**   Butyl Rubber (0)

RCRA Waste #   None   **Reportable Quantity:**   None

Spills: Dike to contain spill and collect liquid for later disposal. Decontaminate spill area using water.

Special Emergency Information: May be harmful if inhaled, swallowed or absorbed through the skin.

p-AMINOPHENOL

Synonyms: 4-aminophenol, *p*-hydroxyaniline, 4-amino-1-hydroxybenzene

CAS Number: 123-30-8   **Description:** White powder

DOT Number: 2512   **DOT Classification:** Poison B

Molecular Weight: 109.1

Melting Point:	190 C (374 F)	**Vapor Density:** N.D.	**Vapor Pressure:** N.D.
Boiling Point: (D)	284 C (544 F)	**Specific Gravity:** N.D.	**Water Solubility:** < 1%

Chemical Incompatibilities or Instabilities: Strong oxidizers.

FLAMMABILITY

NFPA Hazard Code: **NFPA Classification:**

Flash Point: N.D.   **LEL:** N.D.

Autoignition Temp: N.D.   **UEL:** N.D.

Fire Extinguishing Methods: Carbon Dioxide, Dry Chemical, Water Spray.

Special Fire Fighting Considerations: Structural fire fighter protective clothing will not provide adequate protection. Fight fire from a distance or protected location, if possible. Treat all materials used or generated and equipment involved as contaminated by hazardous waste.

TOXICOLOGY

Odor: **Odor Threshold:** N.D.

Physical Contact: Irritant.

TLV = N.D.   **STEL** = N.D.   **IDLH** = N.D.

Routes of Entry and Relative LD$_{50}$ (or LC$_{50}$)   **RTECS #** SJ5075000

     Inhalation   N.D.

     Ingestion   5   (375 mg/kg)

     Skin Absorption   N.D.

Symptoms of Exposure: Possible irritation to skin, eyes and upper respiratory tract; headache; nausea. Allergic reaction. Absorption may lead to the formation of methemoglobin resulting in cyanosis several hours after exposure.

Emergency/First Aid Treatment: Remove to ventilated area; immediately remove any contaminated clothing and wash contaminated areas for 15 minutes using water. Treat supportively and observe for possible shock. If ingested, seek immediate medical aid.

Recommended Clean-Up Procedures

Personal Protection: Level B Ensemble   **Recommended Material** Butyl Rubber (-), Polyvinyl Alcohol (-)

RCRA Waste # None   **Reportable Quantity:** None

Spills: Take up spill. Decontaminate spill area using water. Water spray should be used to keep closed containers cool. Continue to cool container after fire is extinguished. Treat all materials used or generated and equipment involved as contaminated by hazardous waste.

Special Emergency Information: May be harmful if inhaled, swallowed or absorbed through the skin.

m-aminophenol (CAS # 591-27-5, DOT # 2521, RTECS # SJ4900000) and o-aminophenol (CAS # 95-55-6, RTECS # SJ4950000), have similar chemical, physical and toxicological properties.

AMINOPROPYLDIETHANOLAMINE

Synonyms: 4-aminopropylmorpholine

CAS Number: 123-00-2   **Description:** Colorless Liquid
DOT Number: 1760   **DOT Classification:** Corrosive Material

Molecular Weight: 144.3
Melting Point: -15 C   (5 F)   **Vapor Density:** 5.0   **Vapor Pressure:**   1 (< 1 mm Hg)
Boiling Point: 226 C   (438 F)   **Specific Gravity:** 1.0-   **Water Solubility:**   Soluble

Chemical Incompatibilities or Instabilities: Oxidizers.

FLAMMABILITY

NFPA Hazard Code: 2-1-0   **NFPA Classification:** Class III B Combustible
Flash Point: 104 C (220 F) **LEL:** N.D.
Autoignition Temp: N.D.   **UEL:** N.D.

Fire Extinguishing Methods: Alcohol Resistant Foam, Carbon Dioxide, Dry Chemical, Water Spray or Fog.

Special Fire Fighting Considerations: Water spray should be used to keep closed containers cool. Continue to cool container after fire is extinguished.

TOXICOLOGY

Odor: **Odor Threshold:** N.D.
Physical Contact: Material is extremely destructive to human tissues.
TLV = N.D.   **STEL** = N.D.   **IDLH** = N.D.

Routes of Entry and Relative LD$_{50}$ (or LC$_{50}$)   **RTECS # QD7700000**
  Inhalation   N.D.
  Ingestion   3   (3560 mg/kg)
  Skin Absorption   4   (1230 mg/kg)

Symptoms of Exposure: Possible burns or irritation to skin, eyes and upper respiratory tract; headache; nausea. Laryngitis. Inhalation of vapors may be fatal by causing glottis to spasm and suffocation. Exposure may result in chemical pneumonitis or pulmonary edema.

Emergency/First Aid Treatment: Remove to ventilated area; immediately remove any contaminated clothing and wash contaminated areas for 15 minutes using water. Treat supportively and observe for possible shock. If ingested, seek immediate medical aid.

Recommended Clean-Up Procedures

Personal Protection: Level B Ensemble   **Recommended Material**   N.D.

RCRA Waste #   None   **Reportable Quantity:**   None

Spills: Dike to contain spill and collect liquid for later disposal. Decontaminate spill area using water.

Special Emergency Information: May be harmful if inhaled, swallowed or absorbed through the skin.

2-AMINOPYRIDINE - $C_5H_5NNH_2$

Synonyms: α-aminopyridine

CAS Number: 504-29-0
DOT Number: 2671

Description: White to Yellow Powder
DOT Classification: IMO: Poison B

Molecular Weight: 94.1
Melting Point: 58 C (F)
Boiling Point: 211 C (F)

Vapor Density: N.D.
Specific Gravity: N.D.

Vapor Pressure: N.D.
Water Solubility: Soluble

Chemical Incompatibilities or Instabilities: Oxidizers.

FLAMMABILITY

NFPA Hazard Code:
Flash Point: 92 C (198 F)
Autoignition Temp: N.D.

NFPA Classification: Class III A Combustible
LEL: N.D.
UEL: N.D.

Fire Extinguishing Methods: Dry Chemical, Foam, or Water Spray.

Special Fire Fighting Considerations: Structural fire fighter protective clothing will not provide adequate protection. Fight fire from a distance or protected location, if possible. Treat all materials used or generated and equipment involved as contaminated by hazardous waste.

TOXICOLOGY

Odor:
Physical Contact: Irritant.
TLV = 2 mg/m^3

STEL = N.D.

Odor Threshold: N.D.

IDLH = 20 mg/m^3

Routes of Entry and Relative LD$_{50}$ (or LC$_{50}$)

Inhalation	N.D.
Ingestion	N.D.
Skin Absorption	N.D.

RTECS # US1575000

Symptoms of Exposure: Possible irritation to skin, eyes and upper respiratory tract; headache; nausea. Convulsions.

Emergency/First Aid Treatment: Remove to ventilated area; immediately remove any contaminated clothing and wash contaminated areas for 15 minutes using water. Treat supportively and observe for possible shock. If ingested, seek immediate medical aid

Recommended Clean-Up Procedures

Personal Protection: Level B Ensemble
RCRA Waste # None

Recommended Material N.D.
Reportable Quantity: None

Spills: Take up in non-combustible absorbent and dispose. Decontaminate spill area using water.

Special Emergency Information: May be fatal in inhaled, ingested or absorbed through the skin.

3-aminopyridine (CAS # 462-08-8, RTECS # US1650000) and 4-aminopyridine (CAS # 504-24-5, RTECS # US1750000) have similar chemical, physical and toxicological properties.

AMMONIA (anhydrous) - NH₃

Synonyms: liquid ammonia

CAS Number: 7664-91-7

DOT Number: 1005

Molecular Weight: 17.0

Melting Point: -78 C (-108 F)

Boiling Point: N.A.

Description: Colorless Gas (when escaping from a pressurized system, may appear as a white plume)

DOT Classification: Nonflammable Gas

Vapor Density: 0.6

Specific Gravity: N.D.

Vapor Pressure: 760 mm Hg

Water Solubility: Soluble

Chemical Incompatibilities or Instabilities: Halogens, oxidizers, acids, interhalogenes.

FLAMMABILITY

NFPA Hazard Code: 3-1-0

Flash Point: N.D.

Autoignition Temp: 1203 C (2197 F)

NFPA Classification: ---

LEL: 16%

UEL: 25%

Fire Extinguishing Methods: Use agent suitable for surrounding fire.

Special Fire Fighting Considerations: Structural fire fighter protective clothing will not provide adequate protection. Isolate for 150 ft if rail or tank truck is involved in a fire. Water spray should be used to keep closed containers cool. Continue to cool container after fire is extinguished. Keep area isolated until vapors have dissipated.

DOT Recommended Isolation Zones: Small Spill: 150 ft   Large Spill: 300 ft
DOT Recommended Down Wind Take Cover Distance: Small Spill: 0.2 mile   Large Spill: 1 mile

TOXICOLOGY

Odor: Pungent

Odor Threshold: 0.05 mg/m^3

Physical Contact: (Liquid) frostbite; material is extremely destructive to human tissues.

TLV = 17 mg/m^3 **STEL** = 24 mg/m^3 **IDLH** = 350 mg/m^3

Routes of Entry and Relative LD$_{50}$ (or LC$_{50}$)   **RTECS # BO0875000**

Inhalation	5	(5600 mg/m^3/H)
Ingestion	5	(350 mg/kg)
Skin Absorption	N.D.	

Symptoms of Exposure: GAS: Possible irritation to skin, eyes and upper respiratory tract; headache; nausea. Inhalation of vapors may be fatal by causing glottis to spasm and suffocation. Exposure may result in chemical pneumonitis or pulmonary edema. **LIQUID:** Frostbite, severe burns.

Emergency/First Aid Treatment: Remove to fresh air. Give artificial respiration if necessary and obtain prompt medical aid.

Recommended Clean-Up Procedures

Personal Protection: Level B Ensemble

RCRA Waste #   None

Recommended Material   Butyl Rubber (+), Teflon (+)

Reportable Quantity: 100 lb (100 lb)

Spills: Stop leak if it can safely be done. Keep area isolated until vapors have dissipated. Water spray can be used to keep vapors down.

Special Emergency Information: May be harmful if inhaled, swallowed or absorbed through the skin.

AMMONIUM ARSENATE - NH$_4$AsO$_2$

Synonyms:

CAS Number: 7784-44-3   **Description:** White Powder
DOT Number: 1546   **DOT Classification:** Poison B

Molecular Weight: 125.0
Melting Point: (D)   C   (F)   **Vapor Density:** N.D.   **Vapor Pressure:** N.D.
Boiling Point: C   (F)   **Specific Gravity:** 1.99   **Water Solubility:** Soluble

Chemical Incompatibilities or Instabilities:

FLAMMABILITY

NFPA Hazard Code: **NFPA Classification:**
Flash Point: N.A.   **LEL:**   N.A.
Autoignition Temp: N.A.   **UEL:**   N.A.

Fire Extinguishing Methods: Use agent suitable for surrounding fire.

Special Fire Fighting Considerations: Treat all materials used or generated and equipment involved as contaminated by hazardous waste.

TOXICOLOGY
CARCINOGEN

Odor: None   **Odor Threshold:** N.A.
Physical Contact: Irritant.
TLV = 0.2 mg (As)/m^3 **STEL** = N.D.   **IDLH** = N.D.

Routes of Entry and Relative LD$_{50}$ (or LC$_{50}$)   **RTECS #** CG0850000
    Inhalation   N.D.
    Ingestion   N.D.
    Skin Absorption   N.D.

Symptoms of Exposure: Headache, nausea, numbness in the extremities, vertigo, garlic odor on breath.

Emergency/First Aid Treatment: Remove to ventilated area; immediately remove any contaminated clothing and wash contaminated areas for 15 minutes using water. Treat supportively and observe for possible shock. If ingested, seek immediate medical aid.

Recommended Clean-Up Procedures

Personal Protection: Level B Ensemble   **Recommended Material**   Natural Rubber (-), Neoprene (-),
                                                     Nitrile Rubber (-)

RCRA Waste #   None   **Reportable Quantity:**   None

Spills: Take up spill. Decontaminate spill area using water. Treat all materials used or generated and equipment involved as contaminated by hazardous waste.

Special Emergency Information: May be harmful if inhaled, swallowed or absorbed through the skin. Symptoms of arsenic poisoning may be delayed for several days. Anyone potentially exposed should be examined by a physician and kept under observation if necessary.

AMMONIUM BIFLUORIDE - NH4HF2

Synonyms: ammonium hydrogen fluoride; acid ammonium fluoride

CAS Number:	1341-49-7	**Description:**	Colorless crystals or white powder		
DOT Number:	1727, 2817	**DOT Classification:**	ORM-B		

Molecular Weight: 57.1

Melting Point:	125°C (225°F)	**Vapor Density:**	N.D.	**Vapor Pressure:**	N.D.
Boiling Point:	°C (°F)	**Specific Gravity:**	1.5	**Water Solubility:**	Soluble

Chemical Incompatibilities or Instabilities: Acids, alkalis, calcium salts.

FLAMMABILITY

NFPA Classification:		**NFPA Hazard Code:**	
Flash Point:	N.A.	**LEL:**	N.A.
Autoignition Temp:	N.A.	**UEL:**	N.A.

Fire Extinguishing Methods: Carbon Dioxide, Dry Chemical, Foam or Water.

Special Fire Fighting Considerations: May release hydrogen fluoride and ammonia. Both are very toxic and may be absorbed through the skin. Treat all materials used or generated and equipment involved as contaminated by hazardous waste.

TOXICOLOGY

Odor: None   **Odor Threshold:** N.D.

Physical Contact: Material is extremely destructive to human tissues.

$TLV = 2.5$ mg (F) $/m^3$ **STEL** = N.D. $IDLH = 500$ mg (F) $/m^3$

Routes of Entry and Relative LD$_{50}$ (or LC$_{50}$)   **RTECS # BQ9200000**

Inhalation	N.D.
Ingestion	N.D.
Skin Absorption	N.D.

Symptoms of Exposure: Possible irritation to skin, eyes and upper respiratory tract; headache; nausea. Salivation, convulsions, partial paralysis.

Emergency/First Aid Treatment: Remove to ventilated area; immediately remove any contaminated clothing and wash contaminated areas for 15 minutes using water. Treat supportively and observe for possible shock. If ingested, seek immediate medical aid.

Recommended Clean-Up Procedures

Personal Protection:	Level A Ensemble	**Recommended Material**	Butyl Rubber (-), Saranex (-), Neoprene (-)
RCRA Waste #	None	**Reportable Quantity:**	100 lb (5000 lb)

Spills: If powder, take up and dispose. If solution, absorb into sodium bicarbonate, take up, and dispose. Decontaminate spill area using bicarbonate solution. Treat all materials used or generated and equipment involved as contaminated by hazardous waste.

Special Emergency Information: May be fatal if inhaled, swallowed, or absorbed through skin.

AMMONIUM DICHROMATE - $(NH_4)_2Cr_2O_7$

Synonyms: ammonium bichromate

CAS Number: 7789-09-5	**Description:** Red-Orange Crystals	
DOT Number: 1439	**DOT Classification:** Oxidizer	

Molecular Weight: 252.1

Melting Point: (D) 170°C (338°F)	**Vapor Density:** N.D.	**Vapor Pressure:** N.D.
Boiling Point: °C (°F)	**Specific Gravity:** 2.15	**Water Solubility:** Soluble

Chemical Incompatibilities or Instabilities: Acids, reducing agents, alcohols, hydrazines.

FLAMMABILITY

NFPA Hazard Code: 2-1-1-OX	**NFPA Classification:**	
Flash Point: N.A.	**LEL:** N.A.	
Autoignition Temp: N.A.	**UEL:** N.A.	

Fire Extinguishing Methods: Water Only.

Special Fire Fighting Considerations: Material swells upon heating and may violently rupture container. Water spray from an unmanned device should be used to keep closed containers cool. Continue to cool container after fire is extinguished. For large fires, if possible, withdraw and allow to burn.

TOXICOLOGY
CARCINOGEN

Odor: None

Odor Threshold: N.A.

Physical Contact: Irritant, burns

TLV = 0.05 mg/m^3 **STEL** = N.D.   **IDLH** = N.D.

Routes of Entry and Relative LD$_{50}$ (or LC$_{50}$)

Inhalation	N.D.	**RTECS # HX7650000**
Ingestion	N.D.	**HX3500000**
Skin Absorption	N.D.	

Symptoms of Exposure: Possible burns or irritation to skin, eyes and upper respiratory tract; headache; nausea.

Emergency/First Aid Treatment: Remove to fresh air, remove contaminated clothing, and wash contacted areas for 15 minutes with water.

Recommended Clean-Up Procedures

Personal Protection: Level B Ensemble	**Recommended Material** Polyethylene (-), Saranex (-)
RCRA Waste # None	**Reportable Quantity:** None

Spills: Take up spill. Decontaminate spill area using water.

Special Emergency Information: May be fatal in inhaled, ingested or absorbed through the skin. Violent decomposition occurs at 225°C (437°F). May be absorbed through the skin.

AMMONIUM FLUORIDE - NH4F

Synonyms:

CAS Number: 12125-01-8   **Description:** Colorless to white powder
DOT Number: 2505   **DOT Classification:** ORM-B

Molecular Weight: 37
Melting Point: (D)   °C   (°F)   **Vapor Density:** N.D.   **Vapor Pressure:**   N.D.
Boiling Point: °C   (°F)   **Specific Gravity:** 1.01   **Water Solubility:**   Soluble

Chemical Incompatibilities or Instabilities: Alkalis

FLAMMABILITY

NFPA Hazard Code: 3-0-0   **NFPA Classification:**
Flash Point: N.A.   **LEL:** N.A.
Autoignition Temp: N.A.   **UEL:** N.A.

Fire Extinguishing Methods: Use agent suitable for surrounding fire.

Special Fire Fighting Considerations: Use water spray to keep vapors down. Combustion products include hydrogen fluoride and nitrogen oxides.

TOXICOLOGY

Odor: None   **Odor Threshold:** N.D.
Physical Contact: Irritant
TLV = 2.5 mg(F)/m^3 **STEL** = N.D.   **IDLH** = 500 mg (F) /m^3

Routes of Entry and Relative LD$_{50}$ (or LC$_{50}$)   **RTECS # BQ6300000**
    Inhalation   N.D.
    Ingestion   N.D.
    Skin Absorption   N.D.

Symptoms of Exposure: Possible irritation to skin, eyes and upper respiratory tract; headache; nausea. salivation, vomiting, diarrhea, convulsions.

Emergency/First Aid Treatment: Remove to ventilated area; immediately remove any contaminated clothing and wash contaminated areas for 15 minutes using water. Treat supportively and observe for possible shock. If ingested, seek immediate medical aid.

Recommended Clean-Up Procedures

Personal Protection: Level A Ensemble   **Recommended Material**   Natural Rubber (-), Neoprene (0), Nitrile Rubber (0)

RCRA Waste #   None   **Reportable Quantity:**   100 lb (5000 lb)

Spills: Take up spill. Decontaminate spill area using bicarbonate solution.

Special Emergency Information: Very toxic when ingested, inhaled, or absorbed through the skin. Extremely corrosive when wet.

AMMONIUM HYDROXIDE - NH$_4$OH

Synonyms: ammonia water

CAS Number: 1336-21-6
DOT Number: 2672, 2073

Description: Colorless, fuming liquid
DOT Classification: Corrosive

Molecular Weight: 35.1
Melting Point: °C (°F)
Boiling Point: °C (°F)

Vapor Density: 0.77
Specific Gravity: 0.90

Vapor Pressure: 4 (12 mm Hg)
Water Solubility: Soluble

Chemical Incompatibilities or Instabilities: Halogens and their derivatives, oxidizers, numerous metals.

FLAMMABILITY

NFPA Hazard Code:
Flash Point: N.A.
Autoignition Temp: N.A.

NFPA Classification:
LEL: N.A.
UEL: N.A.

Fire Extinguishing Methods: Use agent suitable for surrounding fire.

Special Fire Fighting Considerations: May form flammable fumes of ammonia. Water spray should be used to keep closed containers cool. Continue to cool container after fire is extinguished.

TOXICOLOGY

Odor: Characteristic, pungent odor
Physical Contact: Material is extremely destructive to human tissues.
TLV = 17 mg/m^3 **STEL** = 24 mg/m^3

Odor Threshold: 0.4 mg/m^3

IDLH = 350 mg/m^3

Routes of Entry and Relative LD$_{50}$ (or LC$_{50}$)
    Inhalation   N.D.
    Ingestion   5   (350 mg/kg)
    Skin Absorption   N.D.

RTECS # BQ9625000

Symptoms of Exposure: Possible burns or irritation to skin, eyes and upper respiratory tract; headache; nausea. Inhalation of vapors may be fatal by causing glottis to spasm and suffocation. Exposure may result in chemical pneumonitis or pulmonary edema.

Emergency/First Aid Treatment: Remove to ventilated area; immediately remove any contaminated clothing and wash contaminated areas for 15 minutes using water. Treat supportively and observe for possible shock. If ingested, seek immediate medical aid.

Recommended Clean-Up Procedures

Personal Protection: Level B Ensemble
RCRA Waste # None

Recommended Material Butyl Rubber (+), Neoprene (0), Nitrile Rubber (0)
Reportable Quantity: 1000 lb (1000 lb)

Spills: Take up spill. Decontaminate spill area using dilute acid solution. Isolate for 1/2 mile if rail or tank truck is involved in a fire.

Special Emergency Information: May be fatal in inhaled, ingested or absorbed through the skin.

AMMONIUM NITRATE - NH4NO3

Synonyms: nitram, nitric acid, ammonium salt

CAS Number: 6484-52-2		**Description:** White to Gray/Brown Solid	
DOT Number: 1942, 0222, 2426		**DOT Classification:** Oxidizer	

Molecular Weight: 80.0

Melting Point:	169°C (336°F)	**Vapor Density:** N.D.	**Vapor Pressure:** N.D.
Boiling Point: (D)	210°C (410°F)	**Specific Gravity:** 1.7	**Water Solubility:** Soluble

Chemical Incompatibilities or Instabilities: Hydrocarbons, acetic acid, sawdust, sulfide ores, urea, sulfur, powdered metals, reducing agents, ordinary combustibles.

FLAMMABILITY

NFPA Hazard Code:	1-0-3-OX	**NFPA Classification:**	
Flash Point:	N.A.	**LEL:**	N.A.
Autoignition Temp:	N.A.	**UEL:**	N.A.

Fire Extinguishing Methods: Water Only.

Special Fire Fighting Considerations: Structural fire fighter protective clothing will not provide adequate protection. Fight fire from a distance or protected location, if possible. Water spray from an unmanned device should be used to keep closed containers cool. Continue to cool container after fire is extinguished. For large fires, if possible, withdraw and allow to burn. May explode upon heating.

TOXICOLOGY

Odor: None		**Odor Threshold:** N.D.

Physical Contact: Irritant.

TLV = N.D.	**STEL** = N.D.	**IDLH** = N.D.

Routes of Entry and Relative LD$_{50}$ (or LC$_{50}$)   **RTECS # BR9050000**

Inhalation	N.D.
Ingestion	N.D.
Skin Absorption	N.D.

Symptoms of Exposure: Possible irritation to skin, eyes and upper respiratory tract; headache; nausea. Allergic reaction.

Emergency/First Aid Treatment: Remove to ventilated area; immediately remove any contaminated clothing and wash contaminated areas for 15 minutes using water. Treat supportively and observe for possible shock. If ingested, seek immediate medical aid.

Recommended Clean-Up Procedures

Personal Protection:	Level B Ensemble	**Recommended Material**	N.D.
RCRA Waste #	None	**Reportable Quantity:**	None

Spills: Sweep up spill. Decontaminate spill area using water.

Special Emergency Information: May be harmful if inhaled, swallowed or absorbed through the skin. Use extreme caution when ammonium nitrate is combined with organics or ordinary combustibles; it becomes a powerful unstable explosive.

AMMONIUM PERCHLORATE - NH4ClO4

Synonyms:

CAS Number: 7790-98-9	**Description:** White Powder
DOT Number: 0402, 1442	**DOT Classification:** Oxidizer

Molecular Weight: 117.5

Melting Point: (D) 240°C (464°F)	**Vapor Density:** N.D.	**Vapor Pressure:** N.D.
Boiling Point: °C (°F)	**Specific Gravity:** 1.95	**Water Solubility:** Soluble

Chemical Incompatibilities or Instabilities: Sulfur, organics, ordinary combustibles, finely divided metals, aluminum, copper, phosphorous, chlorine, chlorine dioxide, metal oxides.

FLAMMABILITY

NFPA Hazard Code: 1-0-4-OX		**NFPA Classification:**
Flash Point: N.A.		**LEL:** N.A.
Autoignition Temp: N.A.		**UEL:** N.A.

Fire Extinguishing Methods: Water only.

Special Fire Fighting Considerations: Fight fire from a distance or protected location, if possible. Isolate for 1 mile if rail or tank truck is involved in a fire. Water spray from an unmanned device should be used to keep closed containers cool. Continue to cool container after fire is extinguished.

TOXICOLOGY

Odor: None	**Odor Threshold:** N.D.
Physical Contact: Irritant.	

TLV = N.D.	**STEL** = N.D.	**IDLH** = N.D.

Routes of Entry and Relative LD$_{50}$ (or LC$_{50}$)   **RTECS #** SC7520000

Inhalation	N.D.
Ingestion	N.D.
Skin Absorption	N.D.

Symptoms of Exposure: Possible irritation to skin, eyes and upper respiratory tract; headache; nausea.

Emergency/First Aid Treatment: Remove to ventilated area; immediately remove any contaminated clothing and wash contaminated areas for 15 minutes using water. Treat supportively and observe for possible shock. If ingested, seek immediate medical aid.

Recommended Clean-Up Procedures

Personal Protection: Level B Ensemble	**Recommended Material** N.D.
RCRA Waste # None	**Reportable Quantity:** None

Spills: Take up spill. Decontaminate spill area using dilute bisulfate solution.

Special Emergency Information: May be harmful if inhaled, swallowed or absorbed through the skin. Contamination or mixing with incompatible materials may form a shock, friction sensitive explosives.

AMMONIUM PERMANGANATE - NH_4MnO_4

Synonyms: permanganic acid, ammonium salt

CAS Number: 13446-10-1	**Description:** Purple Powder	
DOT Number: 9190	**DOT Classification:** Oxidizer	

Molecular Weight: 137

Melting Point: 110°C (230°F)	**Vapor Density:** N.D.	**Vapor Pressure:** N.D..
Boiling Point: --°C (--°F)	**Specific Gravity:** 2.21	**Water Solubility:** Soluble

Chemical Incompatibilities or Instabilities: Ordinary combustibles, organics, reducing agents, finely divided metals.

FLAMMABILITY

NFPA Hazard Code: 1-0-3-OX	**NFPA Classification:**	
Flash Point: N.A.	**LEL:** N.A.	
Autoignition Temp: N.A.	**UEL:** N.A.	

Fire Extinguishing Methods: Water only.

Special Fire Fighting Considerations: Structural fire fighter protective clothing will not provide adequate protection. Fight fire from a distance or protected location, if possible. Isolate for 1/2 mile if rail or tank truck is involved in a fire. Water spray from an unmanned device should be used to keep closed containers cool. Continue to cool container after fire is extinguished. For large fires, if possible, withdraw and allow to burn.

TOXICOLOGY

Odor: None   **Odor Threshold:** N.D.

Physical Contact: Material is extremely destructive to human tissues.

TLV = 5 mg(Mn)/m^3 **STEL** = N.D.   **IDLH** = N.D.

Routes of Entry and Relative LD$_{50}$ (or LC$_{50}$)   **RTECS # SD6400000**

Inhalation	N.D.
Ingestion	N.D.
Skin Absorption	N.D.

Symptoms of Exposure: Possible irritation to skin, eyes and upper respiratory tract; headache; nausea. Inhalation of vapors may be fatal by causing glottis to spasm and suffocation. Exposure may result in chemical pneumonitis or pulmonary edema.

Emergency/First Aid Treatment: Remove to ventilated area; immediately remove any contaminated clothing and wash contaminated areas for 15 minutes using water. Treat supportively and observe for possible shock. If ingested, seek immediate medical aid.

Recommended Clean-Up Procedures

Personal Protection: Level C Ensemble	**Recommended Material**	N.D.
RCRA Waste # None	**Reportable Quantity:**	None

Spills: Take up spill. Decontaminate spill area using water.

Special Emergency Information: May be harmful if inhaled, swallowed or absorbed through the skin. May explosively decompose when heated. Explosion danger increases when incompatible materials are present.

AMMONIUM PERSULFATE - $(NH_4)_2S_2O_8$

Synonyms: ammonium peroxydisulfate

CAS Number: 7727-54-0   **Description:** White Powder
DOT Number: 1444   **DOT Classification:** Oxidizer

Molecular Weight:
Melting Point: (D)   120°C (248°F)   **Vapor Density:**   N.D.   **Vapor Pressure:**   1 (< 1 mm Hg)
Boiling Point: N.A.   **Specific Gravity:** 1.98   **Water Solubility:**   Soluble

Chemical Incompatibilities or Instabilities: Iron, ammonia, silver, aluminum, sulfuric acid, zinc, reducing agents, heat, finely divided metals.

FLAMMABILITY

NFPA Hazard Code: **NFPA Classification:**
Flash Point: N.A.   **LEL:** N.A.
Autoignition Temp: N.A.   **UEL:** N.A.

Fire Extinguishing Methods: Water only.

Special Fire Fighting Considerations: Fight fire from a distance or protected location, if possible. Water spray from an unmanned device should be used to keep closed containers cool. Continue to cool container after fire is extinguished. For large fires, if possible, withdraw and allow to burn.

TOXICOLOGY

Odor: None   **Odor Threshold:** N.A.
Physical Contact: Material is extremely destructive to human tissues.
TLV = 2 mg(S_2O_8)/m^3 **STEL** = N.D.   **IDLH** = N.D.

Routes of Entry and Relative LD$_{50}$ (or LC$_{50}$)   **RTECS #** SE0350000
   Inhalation   N.D.
   Ingestion   4   (820 mg/kg)
   Skin Absorption   N.D.

Symptoms of Exposure: Possible irritation to skin, eyes and upper respiratory tract; headache; nausea. Inhalation of vapors may be fatal by causing glottis to spasm and suffocation. Exposure may result in chemical pneumonitis or pulmonary edema.

Emergency/First Aid Treatment: Remove to ventilated area; immediately remove any contaminated clothing and wash contaminated areas for 15 minutes using water. Treat supportively and observe for possible shock. If ingested, seek immediate medical aid.

Recommended Clean-Up Procedures

Personal Protection: Level B Ensemble   **Recommended Material**   N.D.

Spills: Take up spill. Decontaminate spill area using bisulfite solution.

Special Emergency Information: May be harmful if inhaled, swallowed or absorbed through the skin. May explosively decompose at elevated temperatures. Danger of explosion increases when incompatible materials are present.

AMMONIUM PICRATE

Synonyms: ammonium carbozoate; ammonium picronitrate; 2,4,6-trinitrophenol; ammonium salt

CAS Number: 131-74-8	**Description:** Yellow Crystals or Powder	
DOT Number: 0004 (dry)	**DOT Classification:** dry - Class A Explosive	
1310 (wet)	wet - Flammable Solid	

Molecular Weight: 246.1

Melting Point: °C (°F)	**Vapor Density:** N.D.	**Vapor Pressure:** 1 (< 1 mm Hg)
Boiling Point: °C (°F)	**Specific Gravity:** 1.72	**Water Solubility:** 1%

Chemical Incompatibilities or Instabilities: Reducing materials, heat, shock.

FLAMMABILITY

NFPA Hazard Code:	**NFPA Classification:**
Flash Point: N.A.	**LEL:** N.A.
Autoignition Temp: N.A.	**UEL:** N.A.

Fire Extinguishing Methods: Water only.

Special Fire Fighting Considerations: Readily explodes from shock or friction if dry or if fire reaches cargo. Evacuation of up to a mile may be required when fire reaches cargo.

TOXICOLOGY

Odor: None	**Odor Threshold:** N.D.
Physical Contact: N.D.	

TLV = N.D.	**STEL** = N.D.	**IDLH** = N.D.

Routes of Entry and Relative LD$_{50}$ (or LC$_{50}$)   **RTECS #** BS3856000

Inhalation	N.D.
Ingestion	N.D.
Skin Absorption	N.D.

Symptoms of Exposure: Possible irritation to skin, eyes and upper respiratory tract; headache; nausea. Allergic reaction.

Emergency/First Aid Treatment: Remove to ventilated area; immediately remove any contaminated clothing and wash contaminated areas for 15 minutes using water. Treat supportively and observe for possible shock. If ingested, seek immediate medical aid.

Recommended Clean-Up Procedures

Personal Protection: Level B Ensemble	**Recommended Material** N.D.	
RCRA Waste # P009	**Reportable Quantity:** 10 lb (1 lb)	

Spills: Take up spill. Decontaminate spill area using soapy water.

Special Emergency Information: May be harmful if inhaled, swallowed or absorbed through the skin.

AMMONIUM VANADATE - NH$_4$VO$_3$

Synonyms: ammonium metavanadate; vandanic acid, ammonium salt

CAS Number: 7803-55-6   **Description:** White to Yellow Powder
DOT Number: 2859   **DOT Classification:** IMO: Poison B

Molecular Weight: 117
Melting Point: (D) 200°C (392°F)   **Vapor Density:** N.D.   **Vapor Pressure:** 1 (< 1 mm Hg)
Boiling Point: °C (°F)   **Specific Gravity:** 2.3   **Water Solubility:** 1%

Chemical Incompatibilities or Instabilities:

FLAMMABILITY

NFPA Hazard Code: **NFPA Classification:**
Flash Point: N.A.   **LEL:** N.A.
Autoignition Temp: N.A.   **UEL:** N.A.

Fire Extinguishing Methods: Use agent suitable for surrounding fire.

Special Fire Fighting Considerations: Treat all materials used or generated and equipment involved as contaminated by hazardous waste.

TOXICOLOGY

Odor: None   **Odor Threshold:** N.D.
Physical Contact:
TLV = 0.05 mg(V)/m^3/15 min   **STEL** = N.D.   **IDLH** = 70 mg(V)/m^3

Routes of Entry and Relative LD$_{50}$ (or LC$_{50}$)   **RTECS # YW0875000**
   Inhalation   N.D.
   Ingestion   7   (160 mg/kg)
   Skin Absorption   N.D.

Symptoms of Exposure: Skin inflammation, tremors in extremities. Possible irritation to skin, eyes and upper respiratory tract; headache; nausea.

Emergency/First Aid Treatment: Remove to ventilated area; immediately remove any contaminated clothing and wash contaminated areas for 15 minutes using water. Treat supportively and observe for possible shock. If ingested, seek immediate medical aid

Recommended Clean-Up Procedures

Personal Protection: Level B Ensemble   **Recommended Material** N.D.
RCRA Waste # P119   **Reportable Quantity:** 1000 lb (1 lb)

Small Spills: Take up spill. Decontaminate spill area using water. Treat all materials used or generated and equipment involved as contaminated by hazardous waste.

Special Emergency Information: May be harmful if inhaled, swallowed or absorbed through the skin.

AMYL ACETATE - $CH_3(CH_2)COOCH_3$

Synonyms: n-amyl acetate

CAS Number: 628-63-7   **Description:** Colorless Liquid.
DOT Number: 1104   **DOT Classification:** Flammable Liquid

Molecular Weight: 130.2
Melting Point: -65°C (-95°F)   **Vapor Density:** 4.5   **Vapor Pressure:** 2 (5 mg Hg)
Boiling Point: 149°C (300°F)   **Specific Gravity:** 0.9   **Water Solubility:** 0.2%

Chemical Incompatibilities or Instabilities: Strong oxidizers.

FLAMMABILITY

NFPA Hazard Code: 1-3-0   **NFPA Classification:** Class 1C Flammable
Flash Point: 16°C (60°F)   **LEL:** 1.0%
Autoignition Temp: 360°C (680°F)   **UEL:** 7.5%

Fire Extinguishing Methods: Carbon Dioxide, Dry Chemical, Foam.

Special Fire Fighting Considerations: Water spray should be used to keep closed containers cool. Isolate for 1/2 mile if rail or tank truck is involved in at fire.

TOXICOLOGY

Odor: Banana   **Odor Threshold:** 0.05 mg/m^3
Physical Contact: Irritation
TLV = 525 mg/m^3.   **STEL** = N.D.   **IDLH** = 21,500 mg/m^3

Routes of Entry and Relative LD$_{50}$ (or LC$_{50}$)   **RTECS # AJ1925000**
    Inhalation   N.D.
    Ingestion   2   (6500 mg/kg)
    Skin Absorption   N.D.

Symptoms of Exposure: Possible irritation to skin, eyes and upper respiratory tract; headache; nausea.

Emergency/First Aid Treatment: Remove to ventilated area; immediately remove any contaminated clothing and wash contaminated areas for 15 minutes using water. Treat supportively and observe for possible shock. If ingested, seek immediate medical aid.

Recommended Clean-Up Procedures

Personal Protection: Level B Ensemble   **Recommended Material**   Polyvinyl Alcohol (0)

RCRA Waste #   U012   **Reportable Quantity:**   5000 lb (100 lb)

Spills: Remove all potential ignition sources. Dike to contain spill, absorb with non-combustible absorbent and take up using non-sparking tools. Decontaminate spill area using soapy water.

Special Emergency Information: May be harmful if inhaled, swallowed or absorbed through the skin. Vapors are heavier than air and may travel some distance to an ignition source. Dike to contain spill and collect liquid for later disposal.

Sec-amyl acetate (CAS # 626-38-0, DOT # 1104, RTECS # AJ2100000) has similar chemical, physical and toxicological properties.

AMYL ALCOHOL - $CH_3(CH_2)_3CH_2OH$

Synonyms: n-pentanol, pentyl alcohol

CAS Number: 71-41-0   **Description:** Colorless Liquid
DOT Number: 1105   **DOT Classification:** N.D.

Molecular Weight: 88.2
Melting Point: -79°C (-110°F)   **Vapor Density:**   3.0   **Vapor Pressure:**   2(5 mg Hg)
Boiling Point: 138°C (280°F)   **Specific Gravity:** 0.8   **Water Solubility:**   2%

Chemical Incompatibilities or Instabilities: Strong oxidizers.

FLAMMABILITY

NFPA Hazard Code: 1-3-0   **NFPA Classification:** Class 1C Flammable
Flash Point: 33°C   (91°F)   **LEL:**   1.2%
Autoignition Temp: 138°C (572°F)   **UEL:**   10.0%

Fire Extinguishing Methods: Alcohol Resistant Foam, Carbon Dioxide, Dry Chemical, Water..

Special Fire Fighting Considerations: Water spray should be used to keep closed containers cool. Continue to cool container after fire is extinguished. Water spray from an unmanned device should be used to keep closed containers cool. Continue to cool container after fire is extinguished. For large fires, if possible, withdraw and allow to burn.

TOXICOLOGY

Odor: **Odor Threshold:** N.D.
Physical Contact: Irritation
TLV = N.D.   **STEL** = N.D.   **IDLH** = N.D.

Routes of Entry and Relative LD$_{50}$ (or LC$_{50}$)   **RTECS # SB9800000**
    Inhalation   2   (84,000 mg/m^3/H)
    Ingestion   3   (3,030 mg/kg)
    Skin Absorption   1   (4490 mg/kg)

Symptoms of Exposure: Possible irritation to skin, eyes and upper respiratory tract; headache; nausea. Double vision.

Emergency/First Aid Treatment: Remove to ventilated area; immediately remove any contaminated clothing and wash contaminated areas for 15 minutes using water. Treat supportively and observe for possible shock. If ingested, seek immediate medical aid.

Recommended Clean-Up Procedures

Personal Protection: Level B Ensemble   **Recommended Material**   Butyl Rubber (+), Teflon (+), Viton (0)

RCRA Waste #   None   **Reportable Quantity:**   None

Spills: Remove all potential ignition sources. Dike to contain spill, absorb with non-combustible absorbent and take up using non-sparking tools. Decontaminate spill area using water.

Special Emergency Information: May be harmful if inhaled, swallowed or absorbed through the skin. Vapors are heavier than air and may travel some distance to an ignition source. Vapors are more dense than air and may settle in low lying areas.

2-pentanol (CAS # 6032-29-7, RTECS # SA4900000) and 3-pentanol (CAS # 584-02-1, RTECS # SA5075000) have similar chemical, physical and toxicological properties.

AMYLALDEHYDE - $CH_3(CH_2)_3CHO$

Synonyms: pentanaldehyde, pentanal, valeraldehyde

CAS Number: 110-62-3
DOT Number: 2058

Description: Colorless Liquid
DOT Classification: Flammable Liquid

Molecular Weight: 86.1
Melting Point: -91°C (-132°F)
Boiling Point: 103°C (217°F)

Vapor Density: 3.0
Specific Gravity: 0.8

Vapor Pressure: N.D.
Water Solubility: Insoluble

Chemical Incompatibilities or Instabilities: Strong oxidizers.

FLAMMABILITY

NFPA Hazard Code: 1-3-0
Flash Point: 12°C (54°F)
Autoignition Temp: 222°C (432°F)

NFPA Classification: Class 1B Flammable
LEL: N.D.
UEL: N.D.

Fire Extinguishing Methods: Alcohol Resistant Foam, Carbon Dioxide, Dry Chemical, Water.

Special Fire Fighting Considerations: Isolate for 1/2 mile if rail or tank truck is involved in at fire. Water spray from an unmanned device should be used to keep closed containers cool. Continue to cool container after fire is extinguished. Immediately withdraw if rising sound from venting device is heard or if fire is causing discoloration to the tank. For large fires, if possible, withdraw and allow to burn.

TOXICOLOGY

Odor: Rancid, decayed
Physical Contact: Irritation
TLV = 176 mg/m^3 **STEL** = N.D.

Odor Threshold: 0.02 mg/m^3

IDLH = N.D.

Routes of Entry and Relative LD$_{50}$ (or LC$_{50}$)
    Inhalation   N.D.
    Ingestion   3   (3200 mg/kg)
    Skin Absorption 1   (6000 mg/kg)

RTECS # YV3600000

Symptoms of Exposure: Possible irritation to skin, eyes and upper respiratory tract; headache; nausea.

Emergency/First Aid Treatment: Remove to ventilated area; immediately remove any contaminated clothing and wash contaminated areas for 15 minutes using water. Treat supportively and observe for possible shock. If ingested, seek immediate medical aid.

Recommended Clean-Up Procedures

Personal Protection: Level B Ensemble

Recommended Material Butyl Rubber (-), Teflon (0)

RCRA Waste # None

Reportable Quantity: None

Spills: Remove all potential ignition sources. Dike to contain spill, absorb with non-combustible absorbent and take up using non-sparking tools. Decontaminate spill area using soapy water. Vent to dissipate vapors.

Special Emergency Information: May be harmful if inhaled, swallowed or absorbed through the skin. Vapors are heavier than air and may travel some distance to an ignition source. Vapors are more dense than air and may settle in low lying areas.

AMYL CHLORIDE - CH$_3$(CH$_2$)$_3$CH$_2$Cl

Synonyms: 1-chloropentane; pentyl chloride

CAS Number: 543-59-9   **Description:** Colorless Liquid
DOT Number: 1107   **DOT Classification:** Flammable Liquid

Molecular Weight: 106.6
Melting Point: -99°C (-146°F)   **Vapor Density:** 3.7   **Vapor Pressure:** N.D.
Boiling Point: 106°C (223°F)   **Specific Gravity:** 0.9 **Water Solubility:** Insoluble

Chemical Incompatibilities or Instabilities: Strong oxidizers.

FLAMMABILITY

NFPA Hazard Code: 1-3-0   **NFPA Classification:** Class 1C Flammable
Flash Point: 13°C (53°F)   **LEL:**   1.6%
Autoignition Temp: 260°C (500°F)   **UEL:**   8.6%

Fire Extinguishing Methods: Alcohol Resistant Foam, Carbon Dioxide, Dry Chemical, Water.

Special Fire Fighting Considerations: Isolate for 1/2 mile if rail or tank truck is involved in a fire. Water spray from an unmanned device should be used to keep closed containers cool. Continue to cool container after fire is extinguished. For large fires, if possible, withdraw and allow to burn. Immediately withdraw if rising sound from venting device is heard or if fire is causing discoloration to the tank. May release toxic fumes (phosgene) upon heating

TOXICOLOGY

Odor: Sweet   **Odor Threshold:** N.D.
Physical Contact: Mild Irritant
TLV = N.D.   **STEL** = N.D.   **IDLH** = N.D.

Routes of Entry and Relative LD$_{50}$ (or LC$_{50}$)   **RTECS # RZ9865000**
  Inhalation   N.D.
  Ingestion   N.D.
  Skin Absorption   N.D.

Symptoms of Exposure: Possible irritation to skin, eyes and upper respiratory tract; headache; nausea.

Emergency/First Aid Treatment: Remove to ventilated area; immediately remove any contaminated clothing and wash contaminated areas for 15 minutes using water. Treat supportively and observe for possible shock. If ingested, seek immediate medical aid.

Recommended Clean-Up Procedures

Personal Protection: Level B Ensemble   **Recommended Material** Teflon (-), Viton (-)

RCRA Waste #   None   **Reportable Quantity:**   None

Spills: Remove all potential ignition sources. Dike to contain spill, absorb with non-combustible absorbent and take up using non-sparking tools. Decontaminate spill area using soapy water.

Special Emergency Information: May be harmful if inhaled, swallowed or absorbed through the skin. Vapors are heavier than air and may travel some distance to an ignition source. Vapors are more dense than air and may settle in low lying areas.

AMYL FORMATE - CH3(CH2)3CH2O

Synonyms: pentyl formate

CAS Number: 638-49-3   **Description:** Colorless Liquid
DOT Number: 1109   **DOT Classification:** Flammable Liquid

Molecular Weight: 116.2
Melting Point: 74°C (165°F)   **Vapor Density:** 4.0   **Vapor Pressure:** N.D.
Boiling Point: 131°C (267°F)   **Specific Gravity:** 0.9   **Water Solubility:** Insoluble

Chemical Incompatibilities or Instabilities: Strong oxidizers.

FLAMMABILITY

NFPA Hazard Code: 1-3-0   **NFPA Classification:** Class IC Flammable
Flash Point: 26°C (79°F)   **LEL:** N.D.
Autoignition Temp: N.D.   **UEL:** N.D.

Fire Extinguishing Methods: Alcohol Resistant Foam, Carbon Dioxide, Dry Chemical, Water.

Special Fire Fighting Considerations: Isolate for 1/2 mile if rail or tank truck is involved in at fire. Water spray from an unmanned device should be used to keep closed containers cool. Continue to cool container after fire is extinguished. For large fires, if possible, withdraw and allow to burn. Immediately withdraw if rising sound from venting device is heard or if fire is causing discoloration to the tank.

TOXICOLOGY

Odor: Sweet   **Odor Threshold:** N.D.
Physical Contact: Irritant
TLV = N.D.   **STEL** = N.D.   **IDLH** = N.D.

Routes of Entry and Relative LD$_{50}$ (or LC$_{50}$)   **RTECS #** LQ9370000
   Inhalation   N.D.
   Ingestion   N.D.
   Skin Absorption   N.D.

Symptoms of Exposure: Possible irritation to skin, eyes and upper respiratory tract; headache; nausea.

Emergency/First Aid Treatment: Remove to ventilated area; immediately remove any contaminated clothing and wash contaminated areas for 15 minutes using water. Treat supportively and observe for possible shock. If ingested, seek immediate medical aid.

Recommended Clean-Up Procedures

Personal Protection: Level B Ensemble   **Recommended Material** N.D.

RCRA Waste # None   **Reportable Quantity:** None

Spills: Remove all potential ignition sources. Dike to contain spill, absorb with non-combustible absorbent and take up using non-sparking tools. Decontaminate spill area using soapy water. Vent to dissipate vapors.

Special Emergency Information: May be harmful if inhaled, swallowed or absorbed through the skin. Vapors are heavier than air and may travel some distance to an ignition source. Vapors are more dense than air and may settle in low lying areas.

AMYL MERCAPTAN - CH3(CH2)CH2SH

Synonyms: 1-pentanethiol, amylhydrosulfide

CAS Number: 110-66-7

DOT Number: 1111

Description: Colorless to Light Yellow Liquid

DOT Classification: Flammable Liquid

Molecular Weight: 104.2

Melting Point: -76°C (-105°F)	**Vapor Density:** 3.6	**Vapor Pressure:** 4 (14 mm Hg)
Boiling Point: °C (°F)	**Specific Gravity:** 0.8	**Water Solubility:** Insoluble

Chemical Incompatibilities or Instabilities: Nitric acid, strong oxidizers.

FLAMMABILITY

NFPA Hazard Code: 2-3-0

Flash Point: 18°C (68°F)

Autoignition Temp: N.D

NFPA Classification: Class 1B Flammable

LEL: N.D.

UEL: N.D.

Fire Extinguishing Methods: Alcohol Resistant Foam, Carbon Dioxide, Dry Chemical, Water.

Special Fire Fighting Considerations: Isolate for 1/2 mile if rail or tank truck is involved in at fire. Water spray should be used to keep closed containers cool. Water spray from an unmanned device should be used to keep closed containers cool. Continue to cool container after fire is extinguished. For large fires, if possible, withdraw and allow to burn. Immediately withdraw if rising sound from venting device is heard or if fire is causing discoloration to the tank.

TOXICOLOGY

Odor: Disagreeable

Physical Contact: Irritant

TLV = N.D.

STEL = N.D.

Odor Threshold: N.D.

IDLH = N.D.

Routes of Entry and Relative LD50 (or LC50)

Inhalation	N.D.
Ingestion	N.D.
Skin Absorption	N.D.

RTECS # SA3150000

Symptoms of Exposure: Possible irritation to skin, eyes and upper respiratory tract; headache; nausea. Laryngitis.

Emergency/First Aid Treatment: Remove to ventilated area; immediately remove any contaminated clothing and wash contaminated areas for 15 minutes using water. Treat supportively and observe for possible shock. If ingested, seek immediate medical aid.

Recommended Clean-Up Procedures

Personal Protection: Level B Ensemble

RCRA Waste # None

Recommended Material N.D.

Reportable Quantity: None

Spills: Remove all potential ignition sources. Dike to contain spill, absorb with non-combustible absorbent and take up using non-sparking tools. Decontaminate spill area using soapy water. Vent to dissipate vapors.

Special Emergency Information: May be harmful if inhaled, swallowed or absorbed through the skin. Vapors are heavier than air and may travel some distance to an ignition source. Vapors are more dense than air and may settle in low lying areas.

AMYL NITRATE - $CH_3(CH_2)_3CH_2NO_3$

Synonyms: nitric acid, pentyl ester

CAS Number: 1002-16-0	**Description:** Colorless Liquid	
DOT Number: 1112	**DOT Classification:** Flammable Liquid	

Molecular Weight: 133.2

Melting Point: °C (°F)	**Vapor Density:**	**Vapor Pressure:**
Boiling Point: 145°C (293°F)	**Specific Gravity:** 1.0	**Water Solubility:** Insoluble

Chemical Incompatibilities or Instabilities: Reducing agents, strong oxidizers.

FLAMMABILITY

NFPA Hazard Code: 2-2-0-OX	**NFPA Classification:** Class II Combustible
Flash Point: 48°C (118°F)	**LEL:** N.D.
Autoignition Temp: N.D.	**UEL:** N.D.

Fire Extinguishing Methods: Alcohol Resistant Foam, Carbon Dioxide, Dry Chemical, and Water.

Special Fire Fighting Considerations: Isolate for 1/2 mile if rail or tank truck is involved in at fire. Water spray from an unmanned device should be used to keep closed containers cool. Continue to cool container after fire is extinguished. For large fires, if possible, withdraw and allow to burn. Immediately withdraw if rising sound from venting device is heard or if fire is causing discoloration to the tank.

TOXICOLOGY

Odor: Ethereal	**Odor Threshold:** N.D.

Physical Contact: Irritant.

TLV = N.D.	**STEL** = N.D.	**IDLH** = N.D.

Routes of Entry and Relative LD_{50} (or LC_{50})   **RTECS # QV0600000**

Inhalation	N.D.
Ingestion	N.D.
Skin Absorption	N.D.

Symptoms of Exposure: Possible irritation to skin, eyes and upper respiratory tract; headache; nausea. Absorption may lead to the formation of methemoglobin resulting in cyanosis several hours after exposure.

Emergency/First Aid Treatment: Remove to ventilated area; immediately remove any contaminated clothing and wash contaminated areas for 15 minutes using water. Treat supportively and observe for possible shock. If ingested, seek immediate medical aid.

Recommended Clean-Up Procedures

Personal Protection: Level B Ensemble	**Recommended Material** N.D.	
RCRA Waste # None	**Reportable Quantity:** None	

Spills: Remove all potential ignition sources. Dike to contain spill, absorb with non-combustible absorbent and take up using non-sparking tools. Decontaminate spill area using soapy water. Vent to dissipate vapors.

Special Emergency Information: May be harmful if inhaled, swallowed or absorbed through the skin. Vapors are heavier than air and may travel some distance to an ignition source. Vapors are more dense than air and may settle in low lying areas.

AMYL NITRITE - CH$_3$(CH$_2$)$_3$CH$_2$NO$_2$

Synonyms: 1-nitropentane, nitrous acid, pentyl ester

CAS Number: 463-04-7

DOT Number: 1113

Description: Colorless to Yellow Liquid

DOT Classification: Flammable Liquid

Molecular Weight: 117.2

Melting Point: °C (°F)	**Vapor Density:** 4.0	**Vapor Pressure:** N.D.
Boiling Point: 104°C (220°F)	**Specific Gravity:** 0.9	**Water Solubility:** Insoluble

Chemical Incompatibilities or Instabilities: Strong oxidizers, reducing agents.

FLAMMABILITY

NFPA Hazard Code: 1-2-2

Flash Point: 59°C (139°F)

Autoignition Temp: 210°C (410°F)

NFPA Classification: Class II Combustible

LEL: N.D.

UEL: N.D.

Fire Extinguishing Methods: Alcohol Resistant Foam, Carbon Dioxide, Dry Chemical, and Water.

Special Fire Fighting Considerations: Isolate for 1/2 mile if rail or tank truck is involved in a fire. Water spray from an unmanned device should be used to keep closed containers cool. Continue to cool container after fire is extinguished. For large fires, if possible, withdraw and allow to burn. Immediately withdraw if rising sound from venting device is heard or if fire is causing discoloration to the tank.

TOXICOLOGY

Odor: Fruity, ethereal

Physical Contact: Irritant

TLV = N.D.   **STEL** = N.D.

Odor Threshold: N.D.

IDLH = N.D.

Routes of Entry and Relative LD$_{50}$ (or LC$_{50}$)

RTECS # RA1140000

Inhalation	N.D.
Ingestion	N.D.
Skin Absorption	N.D.

Symptoms of Exposure: Possible irritation to skin, eyes and upper respiratory tract; headache; nausea. Flushing of skin, rapid pulse, low blood pressure.

Emergency/First Aid Treatment: Remove to ventilated area; immediately remove any contaminated clothing and wash contaminated areas for 15 minutes using water. Treat supportively and observe for possible shock. If ingested, seek immediate medical aid.

Recommended Clean-Up Procedures

Personal Protection: Level B Ensemble

RCRA Waste # None

Recommended Material N.D.

Reportable Quantity: None

Spills: Remove all potential ignition sources. Dike to contain spill, absorb with non-combustible absorbent and take up using non-sparking tools. Decontaminate spill area using soapy water. Vent to dissipate vapors.

Special Emergency Information: May be harmful if inhaled, swallowed or absorbed through the skin. Vapors are heavier than air and may travel some distance to an ignition source. Vapors are more dense than air and may settle in low lying areas.

AMYLAMINE - $CH_3(CH_2)_3CH_2NH_2$

Synonyms: 1-amino pentane, pentylamine

CAS Number: 110-58-7	**Description:** Colorless Liquid	
DOT Number: 1106	**DOT Classification:** Flammable Liquid	

Molecular Weight: 87.2

Melting Point:	-55°C (-67°F)	**Vapor Density:** 3.0	**Vapor Pressure:**	N.D.
Boiling Point:	104°C (219°F)	**Specific Gravity:** 0.8	**Water Solubility:**	Soluble

Chemical Incompatibilities or Instabilities: Oxidizers.

FLAMMABILITY

NFPA Hazard Code: 3-3-0	**NFPA Classification:** Class 1B Flammable	
Flash Point: -1°C (30°F)	**LEL:** 2%	
Autoignition Temp: N.D.	**UEL:** 22%	

Fire Extinguishing Methods: Alcohol Resistant Foam, Carbon Dioxide, Dry Chemical, Water.

Special Fire Fighting Considerations: Isolate for 1/2 mile if rail or tank truck is involved in a fire. Water spray should be used to keep closed containers cool. Continue to cool container after fire is extinguished. Immediately withdraw if rising sound from venting device is heard or if fire is causing discoloration to the tank.

TOXICOLOGY

Odor:

Physical Contact: Material is extremely destructive to human tissues.

TLV = N.D.   **STEL** = N.D.

Odor Threshold: N.D.

IDLH = N.D.

RTECS # PBV5000000

Routes of Entry and Relative LD$_{50}$ (or LC$_{50}$)

Inhalation	3	(28,000 mg/m^3/H)
Ingestion	5	(470 mg/kg)
Skin Absorption	4	(1120 mg/kg)

Symptoms of Exposure: Possible irritation to skin, eyes and upper respiratory tract; headache; nausea. Inhalation of vapors may be fatal by causing glottis to spasm and suffocation. Exposure may result in chemical pneumonitis or pulmonary edema. Laryngitis.

Emergency/First Aid Treatment: Remove to ventilated area; immediately remove any contaminated clothing and wash contaminated areas for 15 minutes using water. Treat supportively and observe for possible shock. If ingested, seek immediate medical aid.

Recommended Clean-Up Procedures

Personal Protection: Level B Ensemble	**Recommended Material**	Teflon (-)
RCRA Waste # None	**Reportable Quantity:**	None

Spills: Remove all potential ignition sources. Dike to contain spill, absorb with non-combustible absorbent and take up using non-sparking tools. Decontaminate spill area using water. Vent to dissipate vapors.

Special Emergency Information: May be harmful if inhaled, swallowed or absorbed through the skin. Vapors are heavier than air and may travel some distance to an ignition source. Vapors are more dense than air and may settle in low lying areas.

AMYLBUTYRATE - $CH_2(CH_2)_3CH_2OOCCH_2CH_2CH_3$

Synonyms: pentyl butyrate

CAS Number: 540-18-1	**Description:** Colorless Liquid	
DOT Number: 2620	**DOT Classification:** Flammable Liquid	

Molecular Weight: 158.3

Melting Point: -73°C (-99°F)	**Vapor Density:** 5.5	**Vapor Pressure:** N.D.
Boiling Point: 185°C (365°F)	**Specific Gravity:** 0.9	**Water Solubility:** Insoluble

Chemical Incompatibilities or Instabilities: Strong oxidizers.

FLAMMABILITY

NFPA Hazard Code: 1-2-0	**NFPA Classification:** Class II Combustible
Flash Point: 57°C (135°F)	**LEL:** N.D.
Autoignition Temp: N.D.	**UEL:** N.D.

Fire Extinguishing Methods: Alcohol Resistant Foam, Carbon Dioxide, Dry Chemical, and Water.

Special Fire Fighting Considerations: Isolate for 1/2 mile if rail or tank truck is involved in at fire. Water spray from an unmanned device should be used to keep closed containers cool. Continue to cool container after fire is extinguished. For large fires, if possible, withdraw and allow to burn. Immediately withdraw if rising sound from venting device is heard or if fire is causing discoloration to the tank.

TOXICOLOGY

Odor:	**Odor Threshold:** N.D.
Physical Contact: Irritant.	
TLV = N.D. **STEL** = N.D.	**IDLH** = .N.D.

Routes of Entry and Relative LD$_{50}$ (or LC$_{50}$)   **RTECS # ET5956000**

Inhalation	N.D.	
Ingestion	1	(12,210 mg/kg)
Skin Absorption	N.D.	

Symptoms of Exposure: Possible irritation to skin, eyes and upper respiratory tract; headache; nausea.

Emergency/First Aid Treatment: Remove to ventilated area; immediately remove any contaminated clothing and wash contaminated areas for 15 minutes using water. Treat supportively and observe for possible shock. If ingested, seek immediate medical aid.

Recommended Clean-Up Procedures

Personal Protection: Level B Ensemble	**Recommended Material** N.D.
RCRA Waste # None	**Reportable Quantity:** None

Spills: Remove all potential ignition sources. Dike to contain spill, absorb with non-combustible absorbent and take up using non-sparking tools. Decontaminate spill area using soapy water.

Special Emergency Information: May be harmful if inhaled, swallowed or absorbed through the skin. Vapors are heavier than air and may travel some distance to an ignition source. Vapors are more dense than air and may settle in low lying areas.

AMYLENE - $CH_2=CH_2CH_2CH_2CH_3$

Synonyms: 1-pentene

CAS Number: 109-67-1   **Description:** Colorless Liquid
DOT Number: 1108   **DOT Classification:** Flammable Liquid

Molecular Weight: 70.6
Melting Point: °C (°F)   **Vapor Density:** 2.4   **Vapor Pressure:** 9 (600 mm Hg)
Boiling Point: 30°C (86°F)   **Specific Gravity:** 0.7   **Water Solubility:** Insoluble

Chemical Incompatibilities or Instabilities: Strong oxidizers.

FLAMMABILITY

NFPA Hazard Code: 1-4-0   **NFPA Classification:** Class 1A Flammable
Flash Point: -18°C (0°F)   **LEL:** 1%
Autoignition Temp: 275°C (527°F)   **UEL:** 9%

Fire Extinguishing Methods: Alcohol Resistant Foam, Carbon Dioxide, Dry Chemical, and Water.

Special Fire Fighting Considerations: Isolate for 1/2 mile if rail or tank truck is involved in a fire. Water spray from an unmanned device should be used to keep closed containers cool. Continue to cool container after fire is extinguished. For large fires, if possible, withdraw and allow to burn. Immediately withdraw if rising sound from venting device is heard or if fire is causing discoloration to the tank.

TOXICOLOGY

Odor: **Odor Threshold:** N.D.
Physical Contact: Irritant
TLV = N.D.   **STEL** = N.D.   **IDLH** = N.D.

Routes of Entry and Relative LD$_{50}$ (or LC$_{50}$)   **RTECS # EM7650000**
    Inhalation   N.D.
    Ingestion   N.D.
    Skin Absorption   N.D.

Symptoms of Exposure: Narcotic at high concentrations. Acts as a simple asphyxiant.

Emergency/First Aid Treatment: Remove to ventilated area; immediately remove any contaminated clothing and wash contaminated areas for 15 minutes using water. Treat supportively and observe for possible shock. If ingested, seek immediate medical aid.

Recommended Clean-Up Procedures

Personal Protection: Level B Ensemble   **Recommended Material** Viton (-)

RCRA Waste # None   **Reportable Quantity:** None

Spills: Remove all potential ignition sources. Dike to contain spill, absorb with non-combustible absorbent and take up using non-sparking tools. Decontaminate spill area using soapy water. Keep area isolated until vapors have dissipated.

Special Emergency Information: May be harmful if inhaled, swallowed or absorbed through the skin. Vapors are heavier than air and may travel some distance to an ignition source. Vapors are more dense than air and may settle in low lying areas.

ANILINE - $C_6H_5NH_2$

Synonyms: amino benzene, phenyl amine, benzenamine

CAS Number: 62-53-3   **Description:** Colorless to Brown Viscous Liquid
DOT Number: 1547   **DOT Classification:** Poison B

Molecular Weight: 93.1
Melting Point: -6°C (21°F)   **Vapor Density:** 3.2   **Vapor Pressure:** 1 (1 mm Hg)
Boiling Point: 184°C (364°F)   **Specific Gravity:** 1.02   **Water Solubility:** Insoluble

Chemical Incompatibilities or Instabilities: Oxidizers, acids, strong alkali.

FLAMMABILITY

NFPA Hazard Code: 3-2-0   **NFPA Classification:** Class IIIA Combustible
Flash Point: 70°C (158°F)   **LEL:** 1%
Autoignition Temp: 615°C (1139°F)   **UEL:** 11%

Fire Extinguishing Methods: Carbon Dioxide, Dry Chemical, Foam, Water Spray.

Special Fire Fighting Considerations: Structural fire fighter protective clothing will not provide adequate protection. Water spray should be used to keep closed containers cool. Continue to cool container after fire is extinguished. Fight fire from a distance or protected location, if possible. Treat all materials used or generated and equipment involved as contaminated by hazardous waste.

TOXICOLOGY
CARCINOGEN

Odor: Pungent   **Odor Threshold:** 0.05 mg/m^3
Physical Contact: Material is extremely destructive to human tissues.
TLV = 8 mg/m^3 **STEL** = N.D.   **IDLH** = 387 mg/m^3

Routes of Entry and Relative LD$_{50}$ (or LC$_{50}$)   **RTECS #** BW6650000
    Inhalation   8   (970 mg/m^3)
    Ingestion   5   (250 mg/kg)
    Skin Absorption   5   (820 mg/kg)

Symptoms of Exposure: Possible irritation to skin, eyes and upper respiratory tract; headache; nausea. Absorption may lead to the formation of methemoglobin resulting in cyanosis several hours after exposure. Cyanosis, rapid pulse.

Emergency/First Aid Treatment:

Recommended Clean-Up Procedures

Personal Protection: Level A Ensemble   **Recommended Material**   Butyl Rubber (+), Polyvinyl Alcohol (+),.
                                                                          Teflon (0), Saranex (0)

RCRA Waste # U012   **Reportable Quantity:**   5000 lb (1000 lb)

Spills: Take up in non-combustible absorbent and dispose. Decontaminate spill area using soapy water. Treat all materials used or generated and equipment involved as contaminated by hazardous waste.

Special Emergency Information: May be fatal in inhaled, ingested or absorbed through the skin.

O-ANISIDINE - $H_2NC_6H_4OCH_3$

Synonyms: 1-amino-2-methoxy benzene; 2-amino anisole; o-methoxyphenylamine

CAS Number: 90-04-0	**Description:** Yellow-Red Liquid	
DOT Number: 2431	**DOT Classification:** Poison B	

Molecular Weight: 123.2

Melting Point: 5°C (41°F)	**Vapor Density:** N.D.	**Vapor Pressure:** 1 (< 1.0 mm Hg)
Boiling Point: 246°C (475°F)	**Specific Gravity:** 1.1	**Water Solubility:** Insoluble

Chemical Incompatibilities or Instabilities: Strong oxidizers.

FLAMMABILITY

NFPA Hazard Code: 2-1-0	**NFPA Classification:** Class IIIB Combustible
Flash Point: 118°C (244°F)	**LEL:** N.D.
Autoignition Temp: N.D.	**UEL:** N.D.

Fire Extinguishing Methods: Carbon Dioxide, Dry Chemical, Foam, and Water.

Special Fire Fighting Considerations: Structural fire fighter protective clothing will not provide adequate protection. Fight fire from a distance or protected location, if possible. Treat all materials used or generated and equipment involved as contaminated by hazardous waste.

TOXICOLOGY
CARCINOGEN

Odor: Amine	**Odor Threshold:** N.D.
Physical Contact: Irritant	
TLV = 0.5 mg/m^3 (skin) **STEL** = N.D.	**IDLH** = 50 mg/m^3

Routes of Entry and Relative LD$_{50}$ (or LC$_{50}$)   **RTECS # BZ5410000**

Inhalation	N.D.	
Ingestion	3	(2000 mg/kg)
Skin Absorption	N.D.	

Symptoms of Exposure: Possible irritation to skin, eyes and upper respiratory tract; headache; nausea. Absorption may lead to the formation of methemoglobin resulting in cyanosis several hours after exposure. Allergic Reaction.

Emergency/First Aid Treatment: Remove to ventilated area; immediately remove any contaminated clothing and wash contaminated areas for 15 minutes using water. Treat supportively and observe for possible shock. If ingested, seek immediate medical aid.

Recommended Clean-Up Procedures

Personal Protection: Level A Ensemble	**Recommended Material**	N.D.
RCRA Waste # None	**Reportable Quantity:**	None

Spills: Take up with a non-combustible absorbent and dispose. Decontaminate spill area using soapy water. Treat all materials used as hazardous waste.

Special Emergency Information: May be harmful if inhaled, swallowed or absorbed through the skin.

p-anisidine (CAS # 104-94-9, RTECS # BZ5450000) and
m-anisidine (CAS # 536-90-3, RTECS # BZ5408000) have similar chemical, physical and toxicological properties.

ANISOLE - C$_6$H$_5$OCH$_3$

Synonyms: methoxybenzene; methyl phenyl ether

CAS Number: 100-66-3	**Description:** Colorless Liquid
DOT Number: 2222	**DOT Classification:** Flammable Liquid

Molecular Weight: 108.2

Melting Point: -37°C (-35°F)	**Vapor Density:** 3.7	**Vapor Pressure:** **7** (155 mm Hg)
Boiling Point: 154°C (309°F)	**Specific Gravity:** 1.0	**Water Solubility:** Insoluble

Chemical Incompatibilities or Instabilities: Strong oxidizers.

FLAMMABILITY

NFPA Hazard Code: 1-2-0	**NFPA Classification:** Class 1A Flammable
Flash Point: 52°C (125°F)	**LEL:** N.D.
Autoignition Temp: 475°C (887°F)	**UEL:** N.D.

Fire Extinguishing Methods: Alcohol Resistant Foam, Carbon Dioxide, Dry Chemical, and Water.

Special Fire Fighting Considerations: Isolate for 1/2 mile if rail or tank truck is involved in a fire. Water spray from an unmanned device should be used to keep closed containers cool. Continue to cool container after fire is extinguished. For large fires, if possible, withdraw and allow to burn. Immediately withdraw if rising sound from venting device is heard or if fire is causing discoloration to the tank.

TOXICOLOGY

Odor: Anise, licorice	**Odor Threshold:** N.D.

Physical Contact: Irritant

TLV = N.D.	**STEL** = N.D.	**IDLH** = N.D.

Routes of Entry and Relative LD$_{50}$ (or LC$_{50}$)   **RTECS # BZ8050000**

Inhalation	N.D.	
Ingestion	3	(3700 mg/kg)
Skin Absorption	N.D.	

Symptoms of Exposure: Possible irritation to skin, eyes and upper respiratory tract; headache; nausea.

Emergency/First Aid Treatment: Remove to ventilated area; immediately remove any contaminated clothing and wash contaminated areas for 15 minutes using water. Treat supportively and observe for possible shock. If ingested, seek immediate medical aid.

Recommended Clean-Up Procedures

Personal Protection: Level B Ensemble	**Recommended Material**	N.D.
RCRA Waste # None	**Reportable Quantity:**	None

Spills: Remove all potential ignition sources. Dike to contain spill, absorb with non-combustible absorbent and take up using non-sparking tools. Decontaminate spill area using soapy water. Keep area isolated until vapors have dissipated.

Special Emergency Information: May be harmful if inhaled, swallowed or absorbed through the skin. Vapors are heavier than air and may travel some distance to an ignition source. Vapors are more dense than air and may settle in low lying areas.

P-ANISOYL CHLORIDE

Synonyms: methoxy benzoyl chloride

CAS Number: 100-07-2   **Description:** Colorless to Yellow Liquid
DOT Number: 1729   **DOT Classification:** Corrosive Material

Molecular Weight: 170.1
Melting Point: 22°C   (72°F)   **Vapor Density:** N.D.   **Vapor Pressure:**   N.D.
Boiling Point: 262°C (504°F)   **Specific Gravity:** N.D.   **Water Solubility:**   Insoluble

Chemical Incompatibilities or Instabilities: Water.

FLAMMABILITY

NFPA Hazard Code: **NFPA Classification:**   Class 3A Combustible
Flash Point: 87°C   (190°F)   **LEL:** N.D.
Autoignition Temp: N.D.   **UEL:** N.D.

Fire Extinguishing Methods: Alcohol Resistant Foam, Carbon Dioxide, Dry Chemical, and Water.

Special Fire Fighting Considerations: Water spray should be used to keep closed containers cool. Continue to cool container after fire is extinguished.

TOXICOLOGY

Odor: **Odor Threshold:**
Physical Contact: Material is extremely destructive to human tissues.
TLV = N.D.   **STEL** = N.D.   **IDLH** = N.D.

Routes of Entry and Relative LD$_{50}$ (or LC$_{50}$)   **RTECS # CA0270000**
   Inhalation   N.D.
   Ingestion   N.D.
   Skin Absorption   N.D.

Symptoms of Exposure: Possible burns or irritation to skin, eyes and upper respiratory tract; headache; nausea. Laryngitis. Lachrymation.

Emergency/First Aid Treatment: Remove to ventilated area; immediately remove any contaminated clothing and wash contaminated areas for 15 minutes using water. Treat supportively and observe for possible shock. If ingested, seek immediate medical aid.

Recommended Clean-Up Procedures

Personal Protection: Level A Ensemble   **Recommended Material**   N.D.

RCRA Waste #   None   **Reportable Quantity:**   None

Spills: Take up with a non-combustible absorbent and dispose. Decontaminate spill area using dilute alkaline solution

Special Emergency Information: May be harmful if inhaled, swallowed or absorbed through the skin. May spontaneously explode at room temperature after prolonged storage.

o-anisoly chloride (CAS #21615-24-9) and
m-anisoyl chloride (CAS # 1711-05-3) have similar chemical, physical and toxicological properties.

ANTIMONY - Sb

Synonyms: stibium

CAS Number: 7440-36-0
DOT Number: 1733

Description: Sliver White, Brittle Metal
DOT Classification: Poison B

Molecular Weight: 121.8
Melting Point: 630°C (1166°F)
Boiling Point: 1635°C (2975°F)

Vapor Density: N.A.
Specific Gravity: 6.68

Vapor Pressure: N.A.
Water Solubility: Insoluble

Chemical Incompatibilities or Instabilities: Hydrogen, halogens, oxidants, alkali nitrates, bromine nitride, peroxides..

FLAMMABILITY

NFPA Hazard Code:
Flash Point: °C   (°F)
Autoignition Temp: °C   (°F)

NFPA Classification:
LEL: %
UEL: %

Fire Extinguishing Methods: Use agent suitable for surrounding fire.

Special Fire Fighting Considerations: Dust may become moderate explosion hazard.

TOXICOLOGY

Odor: N.A.
Physical Contact:
TLV = 0.5 mg/m^3 **STEL** = N.D.

Odor Threshold: N.A.

IDLH = 80 mg/m^3

Routes of Entry and Relative LD$_{50}$ (or LC$_{50}$)
  Inhalation   N.D.
  Ingestion   3   (5,000 mg/kg)
  Skin Absorption N.D.

RTECS # CC4025000

Symptoms of Exposure: Possible irritation to skin, eyes and upper respiratory tract; headache; nausea. Diarrhea, muscular pains.

Emergency/First Aid Treatment: Remove to ventilated area; immediately remove any contaminated clothing and wash contaminated areas for 15 minutes using water. Treat supportively and observe for possible shock. If ingested, seek immediate medical aid.

Recommended Clean-Up Procedures

Personal Protection: Level B Ensemble

RCRA Waste #   None

Recommended Material   N.D.

Reportable Quantity: 5000 lb (1 lb)

Spills: Sweep up and dispose.

Special Emergency Information: May be harmful if inhaled, swallowed or absorbed through the skin. Antimony reacts with hydrogen and acids to spontaneously form stibine (SbH$_3$), an extremely toxic gas.

ANTIMONY PENTACHLORIDE - SbCl$_5$

Synonyms: Butter of antimony, antimony perchloride

CAS Number: 7647-18-9	**Description:** Colorless to Red-Yellow Liquid
DOT Number: 1730, 1731	**DOT Classification:** Corrosive Material

Molecular Weight: 299.0

Melting Point: 3°C (37°F)	**Vapor Density:** N.D.	**Vapor Pressure:** 1 (1 mm Hg)
Boiling Point: (D) 60°C (140°F)	**Specific Gravity:** 2.34	**Water Solubility:** Decomposes

Chemical Incompatibilities or Instabilities: Water.

FLAMMABILITY

NFPA Hazard Code: 3-0-1	**NFPA Classification:**
Flash Point: N.A.	**LEL:** N.A.
Autoignition Temp: N.A.	**UEL:** N.A.

Fire Extinguishing Methods: Use agent suitable for surrounding fire.

Special Fire Fighting Considerations: Water spray should be used to keep closed containers cool. Continue to cool container after fire is extinguished. Use water spray to keep vapors down. Produces HCl upon contact with water.

TOXICOLOGY

Odor:	**Odor Threshold:** N.D.

Physical Contact: Material is extremely destructive to human tissues.

TLV = 0.5 mg (Sb)/m^3	**STEL** = N.D.	**IDLH** = 80 mg/m^3

Routes of Entry and Relative LD$_{50}$ (or LC$_{50}$)   **RTECS # CC5075000**

Inhalation	7	(1440 mg/m^3/H)
Ingestion	3	(1115 mg/kg).
Skin Absorption	N.D.	

Symptoms of Exposure: Possible burns or irritation to skin, eyes and upper respiratory tract; headache; nausea. Inhalation of vapors may be fatal by causing glottis to spasm and suffocation. Exposure may result in chemical pneumonitis or pulmonary edema. Laryngitis.

Emergency/First Aid Treatment: Remove to ventilated area; immediately remove any contaminated clothing and wash contaminated areas for 15 minutes using water. Treat supportively and observe for possible shock. If ingested, seek immediate medical aid.

Recommended Clean-Up Procedures

Personal Protection:	Level A Ensemble	**Recommended Material** N.D.
RCRA Waste #	None	**Reportable Quantity:** 1000 lb (1000 lb)

Spills: Dike to contain spill and collect liquid for later disposal. Decontaminate spill area using dilute alkaline solution.

Special Emergency Information: May be harmful if inhaled, swallowed or absorbed through the skin.

Antimony pentafluoride,SbF$_5$, (CAS - 7783-70-2, DOT # 1732) has similar chemical, physical and toxicological properties. SbF$_5$ reacts violently with water.

ANTIMONY TRICHLORIDE - SbCl$_3$

Synonyms: trichlorostibine, antimony(III) chloride

CAS Number: 10025-91-9   **Description:** Transparent Crystals, Fumes in Air
DOT Number: 1733   **DOT Classification:** Corrosive Material

Molecular Weight: 228.1
Melting Point: 73°C (163°F)   **Vapor Density:** N.D.   **Vapor Pressure:** **3** (10 mm Hg)
Boiling Point: 224°C (435°F)   **Specific Gravity:** 3.14   **Water Solubility:** Decomposes

Chemical Incompatibilities or Instabilities: Aluminum, water, alkali metals.

FLAMMABILITY

NFPA Hazard Code: N.D.   **NFPA Classification:**
Flash Point: N.D.   **LEL:** N.D.
Autoignition Temp: N.D.   **UEL:** N.D.

Fire Extinguishing Methods: Use agent suitable for surrounding fire.

Special Fire Fighting Considerations: Water spray should be used to keep closed containers cool. Continue to cool container after fire is extinguished. Reacts with water to form HCl.

TOXICOLOGY

Odor: **Odor Threshold:** N.D.
Physical Contact: Material is extremely destructive to human tissues.
TLV = 0.5 mg/m^3 **STEL** = N.D.   **IDLH** = 80 mg/m^3

Routes of Entry and Relative LD$_{50}$ (or LC$_{50}$)   **RTECS # CC4900000**

Inhalation	N.D.	
Ingestion	4	(525 mg/m^3)
Skin Absorption	N.D.	

Symptoms of Exposure: Possible burns or irritation to skin, eyes and upper respiratory tract; headache; nausea. Inhalation of vapors may be fatal by causing glottis to spasm and suffocation. Exposure may result in chemical pneumonitis or pulmonary edema. Laryngitis.

Emergency/First Aid Treatment: Remove to ventilated area; immediately remove any contaminated clothing and wash contaminated areas for 15 minutes using water. Treat supportively and observe for possible shock. If ingested, seek immediate medical aid.

Recommended Clean-Up Procedures

Personal Protection: Level A Ensemble   **Recommended Material** N.D.

RCRA Waste # None   **Reportable Quantity:** 1000 lb (1000 lb)

Spills: Sweep up spill and dispose. Wash spill area with dilute alkaline solution. Vent to dissipate vapors.

Special Emergency Information: May be harmful if inhaled, swallowed or absorbed through the skin.

Antimony trifluoride (CAS # 7783-56-4, DOT # 1549, RTECS # CC5150000) and antimony tribromide (CAS # 7789-61-9, DOT # 1549) have similar chemical, physical and toxicological properties.

ARSENIC - As

Synonyms: metallic arsenic

CAS Number: 7440-38-2	**Description:** Silver/Gray metal
DOT Number: 1558, 1561	**DOT Classification:** Poison B

Molecular Weight: 74.9

Melting Point: (S) 613°C (°F)	**Vapor Density:** N.D.	**Vapor Pressure:** 1 (< 1 mm Hg)
Boiling Point: °C (°F)	**Specific Gravity:** 5.7	**Water Solubility:** Insoluble

Chemical Incompatibilities or Instabilities: Halogens, oxidizers, lead, zinc, platinum, alkali metals.

FLAMMABILITY

NFPA Hazard Code:		**NFPA Classification:**
Flash Point:	N.A.	**LEL:** N.A.
Autoignition Temp:	N.A.	**UEL:** N.A.

Fire Extinguishing Methods: Use agent suitable for surrounding fire.

Special Fire Fighting Considerations: Structural fire fighter protective clothing will not provide adequate protection. Fight fire from a distance or protected location, if possible. Treat all materials used or generated and equipment involved as contaminated by hazardous waste.

TOXICOLOGY
CARCINOGEN

Odor: None		**Odor Threshold:** N.A.
Physical Contact: Irritant.		
TLV = 0.01 mg/m^3	**STEL** = N.D.	**IDLH** = 100 mg/m^3

Routes of Entry and Relative LD$_{50}$ (or LC$_{50}$)   **RTECS # CG0525000**

Inhalation	N.D.	
Ingestion	4	(763 mg/kg)
Skin Absorption	N.D.	

Symptoms of Exposure: Possible irritation to skin, eyes and upper respiratory tract; headache; nausea. Numbness. Garlic odor on breath.

Emergency/First Aid Treatment: Remove to ventilated area; immediately remove any contaminated clothing and wash contaminated areas for 15 minutes using water. Treat supportively and observe for possible shock. If ingested, seek immediate medical aid.

Recommended Clean-Up Procedures

Personal Protection: Level a Ensemble	**Recommended Material**	N.D.
RCRA Waste # None	**Reportable Quantity:**	1 lb (1 lb)

Spills: Sweep up spill and dispose. Decontaminate spill area using soapy water. Treat all materials used or generated and equipment involved as contaminated by hazardous waste.

Special Emergency Information: May be fatal in inhaled, ingested or absorbed through the skin. May react with hydrogen to produce arsine gas.

ARSENIC ACID - AsH₃O₄

Synonyms: orthoarsenic acid

CAS Number: 7778-39-4   **Description:** Hygroscopic Crystals
DOT Number: 1553, 1554   **DOT Classification:** Poison B

Molecular Weight: 142.0
Melting Point: 36°C (97°F)   **Vapor Density:** N.D.   **Vapor Pressure:** N.D.
Boiling Point: °C (°F)   **Specific Gravity:** 2.0   **Water Solubility:** Soluble

Chemical Incompatibilities or Instabilities: Reducing Agents.

FLAMMABILITY

NFPA Hazard Code: **NFPA Classification:**
Flash Point: N.A.   **LEL:** N.A.
Autoignition Temp: N.A.   **UEL:** N.A.

Fire Extinguishing Methods: Use agent suitable for surrounding fire.

Special Fire Fighting Considerations: Structural fire fighter protective clothing will not provide adequate protection. Fight fire from a distance or protected location, if possible. Treat all materials used or generated and equipment involved as contaminated by hazardous waste.

TOXICOLOGY
CARCINOGEN

Odor: **Odor Threshold:** N.D.
Physical Contact: Irritant
TLV = 0.01 mg/m³   **STEL** = N.D.   **IDLH** = 100 mg/m³

Routes of Entry and Relative LD₅₀ (or LC₅₀)   **RTECS #** CG0700000
    Inhalation   N.D.
    Ingestion   9   (48 mg/kg)
    Skin Absorption   N.D.

Symptoms of Exposure: Possible irritation to skin, eyes and upper respiratory tract; headache; nausea. Numbness. Garlic odor on breath.

Emergency/First Aid Treatment: Remove to ventilated area; immediately remove any contaminated clothing and wash contaminated areas for 15 minutes using water. Treat supportively and observe for possible shock. If ingested, seek immediate medical aid.

Recommended Clean-Up Procedures

Personal Protection: Level A Ensemble   **Recommended Material** N.D.

RCRA Waste # P010   **Reportable Quantity:** 1 lb (1 lb)

Spills: Sweep up spill and dispose. Decontaminate spill area using water. Treat all materials used or generated and equipment involved as contaminated by hazardous waste.

Special Emergency Information: May be fatal in inhaled, ingested or absorbed through the skin.

Arsenic acid, calcium salt (CAS # 7778-44-1, DOT # 1573, RTECS # CG0830000), arsenic acid, lead salt (CAS # 7645-25-2, DOT # 1617, RTECS # CG0980000), and arsenic acid, magnesium salt (CAS # 10103-50-1, DOT # 1622, RTECS # CG1050000) have similar chemical, physical and toxicological properties.

ARSENIC SULFIDE - As$_2$S$_3$

Synonyms: arsenic trisulfide

CAS Number: 1303-33-9	**Description:** Yellow to Red Solid
DOT Number: 1557	**DOT Classification:** Poison B

Molecular Weight:

Melting Point: 312°C (594°F)	**Vapor Density:** ---	**Vapor Pressure:** ---
Boiling Point: 710°C (1310°F)	**Specific Gravity:** 3.4	**Water Solubility:** Insoluble

Chemical Incompatibilities or Instabilities: Hydrogen peroxide, water, strong acids, oxidizers.

FLAMMABILITY

NFPA Hazard Code:	**NFPA Classification:**
Flash Point: N.A.	**LEL:** N.A.
Autoignition Temp: N.A.	**UEL:** N.A.

Fire Extinguishing Methods: Use agent suitable for surrounding fire.

Special Fire Fighting Considerations: Structural fire fighter protective clothing will not provide adequate protection. Fight fire from a distance or protected location, if possible. Treat all materials used or generated and equipment involved as contaminated by hazardous waste. May react with water to yield toxic and/or flammable gases.

TOXICOLOGY
CARCINOGEN

Odor:	**Odor Threshold:** N.D.
Physical Contact: Irritant.	
TLV = 0.2 mg (As)/m^3 **STEL** = N.D.	**IDLH** = N.D.

Routes of Entry and Relative LD$_{50}$ (or LC$_{50}$)		**RTECS # CG2638000**
Inhalation	N.D.	
Ingestion	N.D.	
Skin Absorption	N.D.	

Symptoms of Exposure: Possible irritation to skin, eyes and upper respiratory tract; headache; nausea. Numbness. Garlic odor on breath.

Emergency/First Aid Treatment: Remove to ventilated area; immediately remove any contaminated clothing and wash contaminated areas for 15 minutes using water. Treat supportively and observe for possible shock. If ingested, seek immediate medical aid.

Recommended Clean-Up Procedures

Personal Protection: Level A Ensemble	**Recommended Material** N.D.
RCRA Waste # None	**Reportable Quantity:** 1 lb (1 lb)

Spills: Take up spill. Decontaminate spill area using soapy water.

Special Emergency Information: May be fatal in inhaled, ingested or absorbed through the skin.

ARSENIC TRICHLORIDE - AsCl₃

Synonyms: trichloroarsine; arsenious chloride; fuming liquid arsenic; arsenic chloride

CAS Number: 7784-34-1	**Description:** Colorless to Pale Yellow Liquid, Fumes in Air
DOT Number: 1560	**DOT Classification:** Poison B

Molecular Weight: 181.3

Melting Point:	-8°C (17°F)	**Vapor Density:**	6.3	**Vapor Pressure:**	3 (10 mg Hg)
Boiling Point:	130°C (266°F)	**Specific Gravity:**	2.2	**Water Solubility:**	Decomposes

Chemical Incompatibilities or Instabilities: Aluminum, water.

FLAMMABILITY

NFPA Hazard Code:	3-0-0	**NFPA Classification:**	
Flash Point:	N.A.	**LEL:**	N.A.
Autoignition Temp:	N.A.	**UEL:**	N.A.

Fire Extinguishing Methods: Use agent suitable for surrounding fire.

Special Fire Fighting Considerations: Structural fire fighter protective clothing will not provide adequate protection. Fight fire from a distance or protected location, if possible. Treat all materials used or generated and equipment involved as contaminated by hazardous waste.

DOT Recommended Isolation Zones:	Small Spill: 1200 ft	Large Spill: 1500 ft
DOT Recommended Down Wind Take Cover Distance	Small Spill: 4 miles	Large Spill: 5 miles

TOXICOLOGY
CARCINOGEN

Odor:	**Odor Threshold:** N.D.

Physical Contact: Material is extremely destructive to human tissues.

TLV = 0.01 mg/m³	**STEL** = N.D.	**IDLH** = 100 mg/m³

Routes of Entry and Relative LD₅₀ (or LC₅₀)   **RTECS # CG1750000**

Inhalation	N.D.
Ingestion	N.D.
Skin Absorption	N.D.

Symptoms of Exposure: Possible irritation to skin, eyes and upper respiratory tract; headache; nausea. Numbness. Garlic odor on breath.

Emergency/First Aid Treatment: Remove to ventilated area; immediately remove any contaminated clothing and wash contaminated areas for 15 minutes using water. Treat supportively and observe for possible shock. If ingested, seek immediate medical aid.

Recommended Clean-Up Procedures

Personal Protection:	Level A Ensemble	**Recommended Material**	N.D.
RCRA Waste #	None	**Reportable Quantity:**	1 lb (1 lb)

Spills: Dike to contain spill and collect liquid for later disposal. Decontaminate spill area using dilute alkaline solution. Keep area isolated until vapors have dissipated. Treat all materials used or generated and equipment involved as contaminated by hazardous waste.

Special Emergency Information: May be fatal in inhaled, ingested or absorbed through the skin.

Arsenic trifluoride (CAS # 7784-35-2, RTECS # CG5775000) has similar chemical, physical and toxicological properties.

ARSENIC TRIOXIDE - As$_2$O$_3$

Synonyms: arsenous acid; arsenic oxide; white arsenic; arsenious anhydride

CAS Number: 1327-53-3   **Description:** White Powder
DOT Number: 1561   **DOT Classification:** Poison B

Molecular Weight: 197.8
Melting Point: 146°C (°F)   **Vapor Density:** N.D.   **Vapor Pressure:** 3 (6.6 mm Hg)
Boiling Point: 465°C (°F)   **Specific Gravity:** 4.7   **Water Solubility:** Insoluble

Chemical Incompatibilities or Instabilities: Acids, halogens, aluminum, zinc, mercury, interhalogens, strong oxidizers..

FLAMMABILITY

NFPA Hazard Code: 2-0-0   **NFPA Classification:**
Flash Point: N.A.   **LEL:** N.A.
Autoignition Temp: N.A.   **UEL:** N.A.

Fire Extinguishing Methods: Use agent suitable for surrounding fire.

Special Fire Fighting Considerations: Structural fire fighter protective clothing will not provide adequate protection. Fight fire from a distance or protected location, if possible. Treat all materials used or generated and equipment involved as contaminated by hazardous waste.

TOXICOLOGY
Suspected CARCINOGEN

Odor: **Odor Threshold:** N.D.
Physical Contact: Irritant.
TLV = 0.01 mg/m^3 **STEL** = N.D.   **IDLH** = 100 mg/m^3

Routes of Entry and Relative LD$_{50}$ (or LC$_{50}$)   **RTECS #** CG3325000
    Inhalation   N.D.
    Ingestion   9   (15 mg/kg)
    Skin Absorption N.D.

Symptoms of Exposure: Possible irritation to skin, eyes and upper respiratory tract; headache; nausea. Numbness. Garlic odor on breath.

Emergency/First Aid Treatment: Remove to ventilated area; immediately remove any contaminated clothing and wash contaminated areas for 15 minutes using water. Treat supportively and observe for possible shock. If ingested, seek immediate medical aid.

Recommended Clean-Up Procedures

Personal Protection: Level A Ensemble   **Recommended Material** N.D.

RCRA Waste #   P012   **Reportable Quantity:**   1 lb (5000 lb)

Spills: Take up spill. Decontaminate spill area using water. Treat all materials used or generated and equipment involved as contaminated by hazardous waste.

Special Emergency Information: May be fatal in inhaled, ingested or absorbed through the skin.

ARSINE - AsH₃

Synonyms: arsenic trihydride; hydrogen ansenide; arsenic anhydride

CAS Number: 7784-42-1	**Description:** Colorless Gas		
DOT Number: 2188	**DOT Classification:** Poison A / Flammable Gas		

Molecular Weight: 77.9

Melting Point: -117°C (-179°F)	**Vapor Density:** 2.66	**Vapor Pressure:** Gas	
Boiling Point: -62°C (-81°F)	**Specific Gravity:**	**Water Solubility:** 20%	

Chemical Incompatibilities or Instabilities: Oxidizers, acids, halogens, light.

FLAMMABILITY

NFPA Hazard Code: 4-4-2	**NFPA Classification:**
Flash Point: N.D.	**LEL:** 5%
Autoignition Temp: 300°C Decomposes	**UEL:** 78%

Fire Extinguishing Methods: Carbon Dioxide, Foam, Water Spray, Water.

Special Fire Fighting Considerations: Isolate for 1/2 mile if rail or tank truck is involved in at fire. Water spray from an unmanned device should be used to keep closed containers cool. Continue to cool container after fire is extinguished. For large fires, if possible, withdraw and allow to burn. Immediately withdraw if rising sound from venting device is heard or if fire is causing discoloration to the tank. Extremely toxic gas. It may be safer to let small fires burn.

DOT Recommended Isolation Zones:	Small Spill: 1500 ft	Large Spill: 1500 ft
DOT Recommended Down Wind Take Cover Distance:	Small Spill: 5 miles	Large Spill: 5 miles

TOXICOLOGY
CARCINOGEN

Odor: Mild Garlic   **Odor Threshold:** 3.5 mg/m³

Physical Contact: Irritant.

$TLV = 0.2$ mg/m³ $STEL = $ N.D. $IDLH = 20$ mg/m³

Routes of Entry and Relative LD$_{50}$ (or LC$_{50}$)   **RTECS # CG6475000**

Inhalation	10	(75 mg/m³/H)
Ingestion	N.D.	
Skin Absorption	N.D.	

Symptoms of Exposure: Possible irritation to skin, eyes and upper respiratory tract; headache; nausea. Numbness. Garlic odor on breath.

Emergency/First Aid Treatment: Remove to ventilated area; immediately remove any contaminated clothing and wash contaminated areas for 15 minutes using water. Treat supportively and observe for possible shock. If ingested, seek immediate medical aid.

Recommended Clean-Up Procedures

Personal Protection: Level A Ensemble	**Recommended Material** N.D.		
RCRA Waste # None	**Reportable Quantity:** None		

Spills: If it can be done safely, stop flow of gas. Remove all potential ignition sources.

Special Emergency Information: May be fatal in inhaled, ingested or absorbed through the skin. Vapors are heavier than air and may travel some distance to an ignition source. Vapors are more dense than air and may settle in low lying areas.

BARIUM - Ba

Synonyms:

CAS Number:	7440-39-3	**Description:**	White to Gray Solid
DOT Number:	1400, 1399, 1854	**DOT Classification:**	Flammable Solid

Molecular Weight: 137.3

Melting Point:	725°C (°F)	**Vapor Density:**	N.A.	**Vapor Pressure:**	N.A.
Boiling Point:	1640°C (°F)	**Specific Gravity:**	3.6	**Water Solubility:**	Decomposes

Chemical Incompatibilities or Instabilities: Oxidizing agents, water, acids, halocarbons.

FLAMMABILITY

NFPA Hazard Code:			**NFPA Classification:**	
Flash Point:	N.A.		**LEL:**	N.A.
Autoignition Temp:	N.A.		**UEL:**	N.A.

Fire Extinguishing Methods: Dry Chemical, Sand, and Soda Ash.

Special Fire Fighting Considerations: DO NOT use CO_2, water or foam. For large fires, if possible, withdraw and allow to burn.

TOXICOLOGY

Odor: None   **Odor Threshold:** N.A.

Physical Contact: Irritant.

TLV = 0.5 mg (Ba)/m^3 **STEL** = N.D.   **IDLH** = 1100 mg/m^3

Routes of Entry and Relative LD$_{50}$ (or LC$_{50}$)   **RTECS # CQ8370000**

Inhalation	N.D.
Ingestion	N.D.
Skin Absorption	N.D.

Symptoms of Exposure: Possible irritation to skin, eyes and upper respiratory tract; headache; nausea. Prolonged contact may cause burns. Colic, diarrhea, slow pulse.

Emergency/First Aid Treatment: Remove to ventilated area; immediately remove any contaminated clothing and wash contaminated areas for 15 minutes using water. Treat supportively and observe for possible shock. If ingested, seek immediate medical aid.

Recommended Clean-Up Procedures

Personal Protection:	Level B Ensemble	**Recommended Material**	N.D.
RCRA Waste #	None	**Reportable Quantity:**	None

Spills: Cover spill with dry sand and dispose.

Special Emergency Information: May be harmful if inhaled, swallowed or absorbed through the skin. Reactions with water and halocarbons may be explosive.

BARIUM ACETATE - Ba(OOCCH3)2

Synonyms: barium diacetate; acetic acid; barium salt

CAS Number: 543-80-6	**Description:** White, Odorless Solid
DOT Number: None	**DOT Classification:** None

Molecular Weight: 255.5

Melting Point: (D) 110°C (230°F)	**Vapor Density:** N.A.	**Vapor Pressure:** 1 (0 mm Hg)
Boiling Point: °C (°F)	**Specific Gravity:** 2.5	**Water Solubility:** Soluble

Chemical Incompatibilities or Instabilities: Strong oxidizers.

FLAMMABILITY

NFPA Hazard Code: N.D.		**NFPA Classification:**
Flash Point: N.D.		**LEL:** N.D.
Autoignition Temp: N.D.		**UEL:** N.D.

Fire Extinguishing Methods: Use agent suitable for surrounding fire.

Special Fire Fighting Considerations: None.

TOXICOLOGY

Odor: None	**Odor Threshold:** N.A.

Physical Contact: Irritant.

TLV = 0.5 mg/m^3	**STEL** = N.D.	**IDLH** = 1100 mg/m^3

Routes of Entry and Relative LD$_{50}$ (or LC$_{50}$)   **RTECS # AF4550000**

Inhalation	N.D.	
Ingestion	4	(921 mg/kg)
Skin Absorption	N.D.	

Symptoms of Exposure: Possible irritation to skin, eyes and upper respiratory tract; headache; nausea. Colic, vomiting, diarrhea, slow pulse.

Emergency/First Aid Treatment: Remove to ventilated area; immediately remove any contaminated clothing and wash contaminated areas for 15 minutes using water. Treat supportively and observe for possible shock. If ingested, seek immediate medical aid.

Recommended Clean-Up Procedures

Personal Protection: Level B Ensemble	**Recommended Material**	N.D.
RCRA Waste # None	**Reportable Quantity:**	None

Spills: Take up spill. Decontaminate spill area using water

Special Emergency Information: May be harmful if inhaled, swallowed or absorbed through the skin.

Barium chloride (CAS # 10361-37-2, RTECS # CQ8750000), barium hydroxide (CAS # 12230-71-6), and barium sulfate (CAS # 7727-43-7, RTECS # CR0600000) have similar chemical, physical and toxicological properties.

BARIUM BROMATE - Ba(BrO3)2

Synonyms:

CAS Number: 13967-90-3   **Description:** White Powder
DOT Number: 2719   **DOT Classification:** Oxidizer

Molecular Weight: 393.1
Melting Point: (D) 260°C (°F)   **Vapor Density:** N.A.   **Vapor Pressure:** 1 (< 1 mm Hg)
Boiling Point: °C   (°F)   **Specific Gravity:** 3.99   **Water Solubility:** Soluble

Chemical Incompatibilities or Instabilities: Ammonia, Carbon, Copper, Aluminum, metal sulfides, ordinary combustibles, heat organics.

FLAMMABILITY

NFPA Hazard Code: 2-0-1-OX   **NFPA Classification:**
Flash Point: N.A.   **LEL:** N.A.
Autoignition Temp: N.A.   **UEL:** N.A.

Fire Extinguishing Methods: Water Only.

Special Fire Fighting Considerations: Water spray from an unmanned device should be used to keep closed containers cool. Continue to cool container after fire is extinguished. For large fires, if possible, withdraw and allow to burn.

TOXICOLOGY

Odor: None   **Odor Threshold:** N.A.
Physical Contact: Material is extremely destructive to human tissues.
TLV = 0.5 mg (Ba)/m^3 **STEL** = N.D.   **IDLH** = 1100 mg/m^3

Routes of Entry and Relative LD$_{50}$ (or LC$_{50}$)   **RTECS # EF8715000**
    Inhalation   N.D.
    Ingestion   N.D.
    Skin Absorption   N.D.

Symptoms of Exposure: Possible burns or irritation to skin, eyes and upper respiratory tract; headache; nausea. Colic, diarrhea, slow pulse.

Emergency/First Aid Treatment: Remove to ventilated area; immediately remove any contaminated clothing and wash contaminated areas for 15 minutes using water. Treat supportively and observe for possible shock. If ingested, seek immediate medical aid.

Recommended Clean-Up Procedures

Personal Protection: Level B Ensemble

RCRA Waste #   None   **Reportable Quantity:**   None

Spills: Sweep up spill and dispose. Decontaminate spill area using water.

Special Emergency Information: May be fatal in inhaled, ingested or absorbed through the skin. May explode if heated above 300°C.

Barium Chlorate (CAS # 13477-00-4, DOT #1445, RTECS # FN9770000) has similar chemical, physical and toxicological properties.

BARIUM CYANIDE - Ba(CN)$_2$

Synonyms:

CAS Number: 542-62-1   **Description:** White Powder
DOT Number: 1565   **DOT Classification:** Poison B

Molecular Weight: 189.4

Melting Point:	°C	(°F)	**Vapor Density:** N.A.	**Vapor Pressure:**	1 (< 1 mm Hg)
Boiling Point:	°C	(°F)	**Specific Gravity:** N.D.	**Water Solubility:**	Soluble

Chemical Incompatibilities or Instabilities: Acids, strong oxidizers.

FLAMMABILITY

NFPA Hazard Code: N.D.   **NFPA Classification:**
Flash Point: N.A.   **LEL:** N.A.
Autoignition Temp: N.A.   **UEL:** N.A.

Fire Extinguishing Methods: Carbon Dioxide, Dry Chemical, Foam, Water.

Special Fire Fighting Considerations: Structural fire fighter protective clothing will not provide adequate protection. Fight fire from a distance or protected location, if possible. Treat all materials used or generated and equipment involved as contaminated by hazardous waste.

TOXICOLOGY

Odor: None   **Odor Threshold:** N.A.
Physical Contact: Irritant.
TLV = 0.5 mg (Ba)/m^3 **STEL** = N.D.   **IDLH** = 1100 mg/m^3

Routes of Entry and Relative LD$_{50}$ (or LC$_{50}$)   **RTECS #** CQ8785000
   Inhalation   N.D.
   Ingestion   N.D.
   Skin Absorption   N.D.

Symptoms of Exposure: Possible irritation to skin, eyes and upper respiratory tract; headache; nausea. Cyanosis.

Emergency/First Aid Treatment: Remove to ventilated area; immediately remove any contaminated clothing and wash contaminated areas for 15 minutes using water. Treat supportively and observe for possible shock. If ingested, seek immediate medical aid. Treat for cyanide poisoning as necessary.

Recommended Clean-Up Procedures

Personal Protection: Level A Ensemble   **Recommended Material**   N.D.

RCRA Waste #   P013   **Reportable Quantity:**   10 lb (10 lb)

Spills: Take up with a non-combustible absorbent and dispose. Decontaminate spill area using dilute hypochlorite solution.

Special Emergency Information: May be fatal in inhaled, ingested or absorbed through the skin.

BARIUM NITRATE - Ba(NO3)2

Synonyms: barium dinitrate

CAS Number: 10022-31-8   **Description:** White Powder
DOT Number: 1446   **DOT Classification:** Oxidizer

Molecular Weight: 361.4
Melting Point: (D)   590°C   (°F)   **Vapor Density:** N.A.   **Vapor Pressure:**   1 (< 1 mm Hg)
Boiling Point: °C   (°F)   **Specific Gravity:** 3.2   **Water Solubility:**   Soluble

Chemical Incompatibilities or Instabilities: Combustibles, magnesium, aluminum, zinc, reducing agents, organics, amines, ammonium salts, mercury salts, cyanides, boron, thiosulfates.

FLAMMABILITY

NFPA Hazard Code: **NFPA Classification:**
Flash Point: N.A.   **LEL:** N.A.
Autoignition Temp: N.A.   **UEL:** N.A.

Fire Extinguishing Methods: Water only.

Special Fire Fighting Considerations: Water spray from an unmanned device should be used to keep closed containers cool. Continue to cool container after fire is extinguished. For large fires, if possible, withdraw and allow to burn.

TOXICOLOGY

Odor: None   **Odor Threshold:** N.A.
Physical Contact: Irritant
TLV = 0.5 mg (Ba)/m^3 **STEL** = N.D.   **IDLH** = 1100 mg/m^3

Routes of Entry and Relative LD$_{50}$ (or LC$_{50}$)   **RTECS # CQ9625000**
    Inhalation   N.D.
    Ingestion   5   (355 mg/kg)
    Skin Absorption   N.D.

Symptoms of Exposure: Possible irritation to skin, eyes and upper respiratory tract; headache; nausea. Colic, slow pulse.

Emergency/First Aid Treatment: Remove to ventilated area; immediately remove any contaminated clothing and wash contaminated areas for 15 minutes using water. Treat supportively and observe for possible shock. If ingested, seek immediate medical aid.

Recommended Clean-Up Procedures

Personal Protection: Level B Ensemble   **Recommended Material**   N.D.

RCRA Waste #   None   **Reportable Quantity:**   None

Spills: Take up using polyethylene tools and place in suitable container for later disposal. Decontaminate spill area using water.

Special Emergency Information: May be harmful is inhaled, swallowed, or absorbed through the skin.

BARIUM OXIDE - BaO

Synonyms:

CAS Number: 1304-28-5   **Description:** White Powder
DOT Number: 1884   **DOT Classification:** ORM-B

Molecular Weight: 153.3
Melting Point: 1920°C   (°F)   **Vapor Density:**   N.A.   **Vapor Pressure:**   1 (< 1 mm Hg)
Boiling Point: °C   (°F)   **Specific Gravity:** 5.7   **Water Solubility:**   3% (D)

Chemical Incompatibilities or Instabilities: Acids, reducing agents, H_2S, SO_3, hydroxylamine, ordinary combustibles, organics.

FLAMMABILITY

NFPA Hazard Code: N.D.   **NFPA Classification:**
Flash Point: N.A.   **LEL:** N.A.
Autoignition Temp: N.A.   **UEL:** N.A.

Fire Extinguishing Methods: Use agent suitable for surrounding fire.

Special Fire Fighting Considerations: Reaction with water is exothermic.

TOXICOLOGY

Odor: None   **Odor Threshold:** N.D.
Physical Contact: Material is extremely destructive to human tissues.
TLV = 0.5 mg (Ba)/m^3 **STEL** = N.D.   **IDLH** = 1100 mg/m^3

Routes of Entry and Relative LD$_{50}$ (or LC$_{50}$)   **RTECS # ÇQ9800000**
    Inhalation   N.D.
    Ingestion   N.D.
    Skin Absorption   N.D.

Symptoms of Exposure: Possible burns or irritation to skin, eyes and upper respiratory tract; headache; nausea. Colic, slow pulse. Inhalation of vapors may be fatal by causing glottis to spasm and suffocation. Exposure may result in chemical pneumonitis or pulmonary edema.

Emergency/First Aid Treatment: Remove to ventilated area; immediately remove any contaminated clothing and wash contaminated areas for 15 minutes using water. Treat supportively and observe for possible shock. If ingested, seek immediate medical aid.

Recommended Clean-Up Procedures

Personal Protection: Level B Ensemble   **Recommended Material**   N.D.

RCRA Waste #   None   **Reportable Quantity:**   None

Spills: Take up spill. Decontaminate spill area using water.

Special Emergency Information: May be harmful is inhaled, swallowed, or absorbed through the skin.

BARIUM PEROXIDE - BaO_2

Synonyms: barium dioxide; barium superoxide

CAS Number: 1304-29-6	**Description:** White to Gray Powder	
DOT Number: 1449	**DOT Classification:** Oxidizer	

Molecular Weight: 169.3

Melting Point:	450°C (°F)	**Vapor Density:** N.D.	**Vapor Pressure:**	N.D.
Boiling Point: (D)	800°C (°F)	**Specific Gravity:** 5.0	**Water Solubility:**	Slight

Chemical Incompatibilities or Instabilities: H_2S, water, ordinary combustibles, aluminum, magnesium, acetic anhydride, organics.

FLAMMABILITY

NFPA Hazard Code:		**NFPA Classification:**	
Flash Point:	N.A.	**LEL:**	N.A.
Autoignition Temp:	N.A.	**UEL:**	N.A.

Fire Extinguishing Methods: WATER ONLY.

Special Fire Fighting Considerations: Water spray from an unmanned device should be used to keep closed containers cool. Continue to cool container after fire is extinguished. For large fires, if possible, withdraw and allow to burn.

TOXICOLOGY

Odor: None

Odor Threshold: N.A.

Physical Contact: Irritant

$TLV = 0.5$ mg (Ba)/m^3 STEL = N.D. $IDLH = 1100$ mg/m^3

Routes of Entry and Relative LD_{50} (or LC_{50})   **RTECS #** CR0175000

Inhalation	N.D.
Ingestion	N.D.
Skin Absorption	N.D.

Symptoms of Exposure: Possible burns or irritation to skin, eyes and upper respiratory tract; headache; nausea.

Emergency/First Aid Treatment: Remove to ventilated area; immediately remove any contaminated clothing and wash contaminated areas for 15 minutes using water. Treat supportively and observe for possible shock. If ingested, seek immediate medical aid.

Recommended Clean-Up Procedures

Personal Protection:	Level B Ensemble	**Recommended Material** N.D.
RCRA Waste #	None	**Reportable Quantity:** None

Spills: Cover spill with excess sand or sodium carbonate. Take up using polyethylene tools and dispose. Decontaminate spill area using water.

Special Emergency Information: May be harmful is inhaled, swallowed, or absorbed through the skin. Reactions with incompatible materials may be explosive.

BENZALDEHYDE - C6H5CHO

Synonyms: artificial almond oil

CAS Number: 100-52-7
DOT Number: 1989

Description: Colorless Liquid
DOT Classification: Combustible Liquid

Molecular Weight: 106.1
Melting Point: -26°C (°F)
Boiling Point: 179°C (°F)

Vapor Density: 3.7
Specific Gravity: 1.1

Vapor Pressure: 1 (1 mm Hg)
Water Solubility: Slight

Chemical Incompatibilities or Instabilities: Oxidizers

FLAMMABILITY

NFPA Hazard Code: 2-2-0
Flash Point: 64°C (148°F)
Autoignition Temp: 191°C (377°F)

NFPA Classification: Class IIIA Combustible
LEL: N.D.
UEL: N.D.

Fire Extinguishing Methods: Alcohol Resistant Foam, Carbon Dioxide, Dry Chemical, Foam, Water.

Special Fire Fighting Considerations: Isolate for 1/2 mile if rail or tank truck is involved in a fire. Water spray from an unmanned device should be used to keep closed containers cool. Continue to cool container after fire is extinguished. For large fires, if possible, withdraw and allow to burn. Immediately withdraw if rising sound from venting device is heard or if fire is causing discoloration to the tank.

TOXICOLOGY
QUESTIONABLE CARCINOGEN

Odor: Almond
Physical Contact: Irritant
TLV = N.D.   **STEL** = N.D.

Odor Threshold: N.D.

IDLH = N.D.

Routes of Entry and Relative LD$_{50}$ (or LC$_{50}$)
    Inhalation   N.D.
    Ingestion   3   (1300 mg/kg)
    Skin Absorption N.D.

RTECS # CU4375000

Symptoms of Exposure: Possible irritation to skin, eyes and upper respiratory tract; headache; nausea. May act as a narcotic in high concentrations.

Emergency/First Aid Treatment: Remove to ventilated area; immediately remove any contaminated clothing and wash contaminated areas for 15 minutes using water. Treat supportively and observe for possible shock. If ingested, seek immediate medical aid.

Recommended Clean-Up Procedures

Personal Protection: Level B Ensemble

Recommended Material Butyl Rubber (+), Polyvinyl Alcohol (0), Viton (0)

RCRA Waste # None

Reportable Quantity: None

Spills: Remove all potential ignition sources. Dike to contain spill, absorb with non-combustible absorbent and take up using non-sparking tools. Decontaminate spill area using soapy water.

Special Emergency Information: May be harmful is inhaled, swallowed, or absorbed through the skin. Vapors are heavier than air and may travel some distance to an ignition source. Vapors are more dense than air and may settle in low lying areas.

BENZENE - C_6H_6

Synonyms: cyclohexatriene; benzol; phenyl hydride

CAS Number: 71-43-2   **Description:** Colorless Liquid with Characteristic Odor
DOT Number: 1114   **DOT Classification:** Flammable Liquid

Molecular Weight: 78.1
Melting Point: 6°C (42°F)   **Vapor Density:** 2.8   **Vapor Pressure:** 6 (75 mm Hg)
Boiling Point: 80°C (176°F)   **Specific Gravity:** 0.88 **Water Solubility:** Insoluble

Chemical Incompatibilities or Instabilities: Oxidizing agents, fluorides, nitric acid, perchlorates.

FLAMMABILITY

NFPA Hazard Code: 2-3-0   **NFPA Classification:** Class 1B Flammable Liquid
Flash Point: -11°C (12°F)   **LEL:** 2%
Autoignition Temp: 498°C (928°F) **UEL:** 7%

Fire Extinguishing Methods: Alcohol Resistant Foam, Carbon Dioxide, Dry Chemical, Foam.

Special Fire Fighting Considerations: Water spray from an unmanned device should be used to keep closed containers cool. Continue to cool container after fire is extinguished. For large fires, if possible, withdraw and allow to burn. Immediately withdraw if rising sound from venting device is heard or if fire is causing discoloration to the tank. Treat all materials used or generated and equipment involved as contaminated by hazardous waste.

TOXICOLOGY
CARCINOGEN

Odor: Aromatic   **Odor Threshold:** 3 mg/m^3
Physical Contact: Irritation
TLV = 3 mg/m^3 **STEL** = 18 mg/m^3 **IDLH** = 9750 mg/m^3

Routes of Entry and Relative LD$_{50}$ (or LC$_{50}$)   **RTECS # CY1400000**
    Inhalation   2   (233,000 mg/m^3/H)
    Ingestion   3   (3306 mg/kg)
    Skin Absorption N.D.

Symptoms of Exposure: Possible irritation to skin, eyes and upper respiratory tract; headache; nausea. May act as a narcotic in high concentrations.

Emergency/First Aid Treatment: Remove to ventilated area; immediately remove any contaminated clothing and wash contaminated areas for 15 minutes using water. Treat supportively and observe for possible shock. If ingested, seek immediate medical aid.

Recommended Clean-Up Procedures

Personal Protection: Level A Ensemble   **Recommended Material** Polyvinyl Alcohol (+), Teflon (0), Viton (0)

RCRA Waste #   U109   **Reportable Quantity:**   10 lb (1000 lb)

Spills: Remove all potential ignition sources. Dike to contain spill, absorb with non-combustible absorbent and take up using non-sparking tools. Decontaminate spill area using soapy water. Treat all materials used or generated and equipment involved as contaminated by hazardous waste.

Special Emergency Information: May be harmful is inhaled, swallowed, or absorbed through the skin. Liquid may accumulate a static electric charge. Vapors are heavier than air and may travel some distance to an ignition source. Vapors are more dense than air and may settle in low lying areas.

BENZENE SULFONYL CHLORIDE - $C_6H_5SO_2Cl$

Synonyms: phenyl sulfonyl chloride

CAS Number: 98-09-9

DOT Number: 2225

Description: Colorless Liquid

DOT Classification: Corrosive Material

Molecular Weight: 176.6

Melting Point: 15°C (59°F)

Boiling Point: (D) 253°C (°F)

Vapor Density: N.A.

Specific Gravity: 1.4

Vapor Pressure: 0 (<1 mm Hg)

Water Solubility: (D)

Chemical Incompatibilities or Instabilities: Dimethyl sulfoxide, methyl formamide.

FLAMMABILITY

NFPA Hazard Code:

Flash Point: 110°C (230°F)

Autoignition Temp: N.D.

NFPA Classification:

LEL: N.D.

UEL: N.D.

Fire Extinguishing Methods: Carbon Dioxide, Dry Chemical, Foam, Water.

Special Fire Fighting Considerations: Structural fire fighter protective clothing will not provide adequate protection. Water spray should be used to keep closed containers cool. Continue to cool container after fire is extinguished. Fight fire from a distance or protected location, if possible.

TOXICOLOGY

Odor: Pungent, irritating

Physical Contact: Material is extremely destructive to human tissues.

TLV = N.D.

STEL = N.D.

Odor Threshold: N.D.

IDLH = N.D.

RTECS # DB8750000

Routes of Entry and Relative LD$_{50}$ (or LC$_{50}$)

Inhalation	**10**	(230 mg/m^3/hr)
Ingestion	**3**	(1960 mg/kg)
Skin Absorption	N.D.	

Symptoms of Exposure: Possible burns or irritation to skin, eyes and upper respiratory tract; headache; nausea. Laryngitis. Difficulty breathing.

Emergency/First Aid Treatment: Remove to ventilated area; immediately remove any contaminated clothing and wash contaminated areas for 15 minutes using water. Treat supportively and observe for possible shock. If ingested, seek immediate medical aid.

Recommended Clean-Up Procedures

Personal Protection: Level B Ensemble

RCRA Waste # U020

Recommended Material Polyvinyl Alcohol (-), Viton (-)

Reportable Quantity: 100 lb (1 lb)

Spills: Dike to contain spill and collect liquid for later disposal. Decontaminate spill area using dilute alkaline solution.

Special Emergency Information: May be harmful is inhaled, swallowed, or absorbed through the skin. Reacts exothemically with water to release HCl vapors.

BENZIDINE - $H_2C_6H_4C_6H_8NH_2$

Synonyms: p-diaminodiphenyl,[1,1-biphenyl]-4,4-diamine

CAS Number: 92-87-5   **Description:** White Crystals that darken upon exposure to light
DOT Number: 1885   **DOT Classification:** Poison B

Molecular Weight: 184.2
Melting Point: 127°C (239°F)   **Vapor Density:** N.D.   **Vapor Pressure:** 1 (< 1 mm Hg)
Boiling Point: 400°C (750°F)   **Specific Gravity:** 1.25   **Water Solubility:** 0.05%

Chemical Incompatibilities or Instabilities: Nitric acid.

FLAMMABILITY

NFPA Hazard Code: **NFPA Classification:**
Flash Point: N.D.   **LEL:** N.D.
Autoignition Temp: N.D.   **UEL:** N.D.

Fire Extinguishing Methods: Use agent suitable for surrounding fire.

Special Fire Fighting Considerations: Structural fire fighter protective clothing will not provide adequate protection. Fight fire from a distance or protected location, if possible. Treat all materials used or generated and equipment involved as contaminated by hazardous waste.

TOXICOLOGY
CARCINOGEN

Odor: **Odor Threshold:** N.D.
Physical Contact: Irritant.
TLV = See 29 CFR 1910.1010   **STEL** = N.D.   **IDLH** = N.D.

Routes of Entry and Relative LD$_{50}$ (or LC$_{50}$)   **RTECS # DC9625000**
 Inhalation   N.D.
 Ingestion   **5**   (390 mg/kg)
 Skin Absorption   N.D.

Symptoms of Exposure: Possible irritation to skin, eyes and upper respiratory tract; headache; nausea. Hemolysis, kidney and liver dysfunction.

Emergency/First Aid Treatment: Remove to ventilated area; immediately remove any contaminated clothing and wash contaminated areas for 15 minutes using water. Treat supportively and observe for possible shock. If ingested, seek immediate medical aid.

Recommended Clean-Up Procedures

Personal Protection: Level A Ensemble   **Recommended Material**   N.D.

RCRA Waste #   U021   **Reportable Quantity:**   1 lb (1 lb)

Spills: Take up spill and dispose. Decontaminate spill area using dilute acid solution. Treat all materials used or generated and equipment involved as contaminated by hazardous waste.

Special Emergency Information: May be harmful is inhaled, swallowed, or absorbed through the skin.

BENZOIC ACID - C_6H_5COOH

Synonyms: benzene carboxylic acid; phenylformic acid; dracylic acid

CAS Number: 65-85-0

DOT Number: ----

Description: White, Crystalline Solid

DOT Classification: ORM-E

Molecular Weight: 122.1

Melting Point:	121°C (°F)	**Vapor Density:** 4.2	**Vapor Pressure:**	N.D.
Boiling Point:	250°C (°F)	**Specific Gravity:** 1.3	**Water Solubility:**	3.4%

Chemical Incompatibilities or Instabilities: Strong oxidizers.

FLAMMABILITY

NFPA Hazard Code: 2-1-0

Flash Point: 120°C (250°F)

Autoignition Temp: 570°C (1060°F)

NFPA Classification:

LEL: N.D.

UEL: N.D.

Fire Extinguishing Methods: Use agent suitable for surrounding fire.

Special Fire Fighting Considerations: N.D.

TOXICOLOGY

Odor:

Physical Contact: Irritant

TLV = N.D.   **STEL** = N.D.

Odor Threshold: N.D.

IDLH = N.D.

Routes of Entry and Relative LD_{50} (or LC_{50})

RTECS # DG0875000

Inhalation	N.D.	
Ingestion	3	(2530 mg/kg)
Skin Absorption	N.D.	

Symptoms of Exposure: Possible irritation to skin, eyes and upper respiratory tract; headache; nausea.

Emergency/First Aid Treatment: Remove to ventilated area; immediately remove any contaminated clothing and wash contaminated areas for 15 minutes using water. Treat supportively and observe for possible shock. If ingested, seek immediate medical aid.

Recommended Clean-Up Procedures

Personal Protection: Level B Ensemble   **Recommended Material** N.D.

RCRA Waste # None   **Reportable Quantity:** 5000 lb (5000 lb)

Spills: Take up spill. Decontaminate spill area using dilute alkaline solution.

Special Emergency Information: May be harmful is inhaled, swallowed, or absorbed through the skin.

BENEZOYL PEROXIDE - $C_6H_5COOOOCC_6H_5$

Synonyms: benzoperoxide

CAS Number: 94-36-0
DOT Number: 2085, 2086, 2087
2088, 2089, 2090

Description: White Powder
DOT Classification: Organic Peroxide

Molecular Weight: 242.2
Melting Point: (D) 105°C (°F)
Boiling Point: N.A.

Vapor Density: N.A.
Specific Gravity: 1.33

Vapor Pressure: N.A.
Water Solubility: Insoluble

Chemical Incompatibilities or Instabilities: Reducing agents, amines, aniline, aniline derivatives, organics, ordinary combustibles.

FLAMMABILITY

NFPA Hazard Code:
Flash Point: N.D.
Autoignition Temp: 176°C (349°F)

NFPA Classification:
LEL: N.A.
UEL: N.A.

Fire Extinguishing Methods: Carbon Dioxide, Dry Chemical, Foam, Water.

Special Fire Fighting Considerations: May explosively decompose upon heating. Isolate for 1/2 mile if rail or tank truck is involved in at fire. Fight fire from a distance or protected location, if possible. Water spray from an unmanned device should be used to keep closed containers cool. Continue to cool container after fire is extinguished. For large fires, if possible, withdraw and allow to burn.

TOXICOLOGY
QUESTIONABLE CARCINOGEN

Odor: None
Physical Contact: Irritant
TLV = 5 mg/m^3

STEL = N.D.

Odor Threshold: N.A.

IDLH = 7000 mg/m^3

Routes of Entry and Relative LD$_{50}$ (or LC$_{50}$)

RTECS # DM8575000

Inhalation	N.D.	
Ingestion	2	(7700 mg/kg)
Skin Absorption	N.D.	

Symptoms of Exposure: Possible irritation to skin, eyes and upper respiratory tract; headache; nausea.

Emergency/First Aid Treatment: Remove to ventilated area; immediately remove any contaminated clothing and wash contaminated areas for 15 minutes using water. Treat supportively and observe for possible shock. If ingested, seek immediate medical aid.

Recommended Clean-Up Procedures

Personal Protection: Level B Ensemble

RCRA Waste # None

Recommended Material N.D.

Reportable Quantity: None

Spills: Remove all potential ignition sources. Clean up only under expert supervision. Decontaminate spill area using water.

Special Emergency Information: May be harmful is inhaled, swallowed, or absorbed through the skin. Extreme explosion hazard.

Explosion hazard: heat, shock, or friction may cause an explosion.

BENZONITRILE - C6H5CN

Synonyms: cyanobenzene; phenyl cyanide

CAS Number: 100-47-0
DOT Number: 2224

Description: Colorless Liquid
DOT Classification: Combustible Liquid

Molecular Weight: 103.1
Melting Point: -13°C (9°F)
Boiling Point: 190°C (374°F)

Vapor Density: N.D.
Specific Gravity: 1.2

Vapor Pressure: 2 (1 mm Hg)
Water Solubility: 1%

Chemical Incompatibilities or Instabilities: Strong acids and alkali.

FLAMMABILITY

NFPA Hazard Code:
Flash Point: 75°C (167°F)
Autoignition Temp: N.D.

NFPA Classification: Class IIIA Combustible
LEL: N.D.
UEL: N.D.

Fire Extinguishing Methods: Carbon Dioxide, Dry Chemical, Foam, Water.

Special Fire Fighting Considerations: Structural fire fighter protective clothing will not provide adequate protection. Fight fire from a distance or protected location, if possible. Treat all materials used or generated and equipment involved as contaminated by hazardous waste.

TOXICOLOGY

Odor: Almonds
Physical Contact: Irritant
TLV = N.D.   **STEL** = N.D.

Odor Threshold: N.D.

IDLH = N.D.

Routes of Entry and Relative LD50 (or LC50)
    Inhalation   N.D.
    Ingestion   4   (971 mg/kg)
    Skin Absorption N.D.

RTECS # DI2450000

Symptoms of Exposure: Possible irritation to skin, eyes and upper respiratory tract; headache; nausea. Cyanosis.

Emergency/First Aid Treatment: Remove to ventilated area; immediately remove any contaminated clothing and wash contaminated areas for 15 minutes using water. Treat supportively and observe for possible shock. If ingested, seek immediate medical aid. Treat for cyanide as necessary.

Recommended Clean-Up Procedures

Personal Protection: Level B Ensemble
RCRA Waste # None

Recommended Material Butyl Rubber (+), Polyvinyl Alcohol (+)
Reportable Quantity: 5000 lb (1000 lb)

Spills: Dike to contain spill and collect liquid for later disposal. Decontaminate spill area using soapy water.

Special Emergency Information: May be harmful is inhaled, swallowed, or absorbed through the skin.

BENZOQUINONE - OC_6H_4O

Synonyms: p-benzoquinone; quinone

CAS Number: 106-51-4	**Description:** White Solid	
DOT Number: 2587	**DOT Classification:** Poison B	

Molecular Weight: 108.1

Melting Point: 112°C (234°F)	**Vapor Density:** 3.7	**Vapor Pressure:** 1 (<1 mm Hg)
Boiling Point: Sublimes	**Specific Gravity:** 1.3	**Water Solubility:** Insoluble

Chemical Incompatibilities or Instabilities: Oxidizers.

FLAMMABILITY

NFPA Hazard Code: 1-2-1	**NFPA Classification:**
Flash Point: N.D.	**LEL:** N.D.
Autoignition Temp: 560°C (1040°F)	**UEL:** N.D.

Fire Extinguishing Methods: Carbon Dioxide, Dry Chemical, Water.

Special Fire Fighting Considerations: Structural fire fighter protective clothing will not provide adequate protection. Fight fire from a distance or protected location, if possible. Treat all materials used or generated and equipment involved as contaminated by hazardous waste.

TOXICOLOGY
QUESTIONABLE CARCINOGEN

Odor: Penetrating, chlorine-like   **Odor Threshold:** N.D.
Physical Contact: Material is extremely destructive to human tissues.
TLV = 0.44 mg/m^3 (AICGH)   **STEL** = N.D.   **IDLH** = N.D.

Routes of Entry and Relative LD$_{50}$ (or LC$_{50}$)   **RTECS # DK2625000**
    Inhalation   N.D.
    Ingestion   7   (130 mg/kg)
    Skin Absorption   N.D.

Symptoms of Exposure: Possible burns or irritation to skin, eyes and upper respiratory tract; headache; nausea. Laryngitis.

Emergency/First Aid Treatment: Remove to ventilated area; immediately remove any contaminated clothing and wash contaminated areas for 15 minutes using water. Treat supportively and observe for possible shock. If ingested, seek immediate medical aid.

Recommended Clean-Up Procedures

Personal Protection: Level B Ensemble	**Recommended Material** Saranex (+)	
RCRA Waste # U197	**Reportable Quantity:** 10 lb (1 lb)	

Spills: Take up spill. Decontaminate spill area using soapy water. Treat all materials used or generated and equipment involved as contaminated by hazardous waste.

Special Emergency Information: May be fatal in inhaled, ingested or absorbed through the skin.

BENZOTRICHLORIDE - C_6H_5CCl

Synonyms: (trichloromethyl) benzene; phenyl chloroform; α,α,α, trichlotoluene; benzyl trichloride

CAS Number: 98-07-7	**Description:** Fuming Liquid	
DOT Number: 2226	**DOT Classification:** Corrosive Material	

Molecular Weight: 195.5

Melting Point:	-5°C (23°F)	**Vapor Density:** 6.8	**Vapor Pressure:**	1 (< 1 mm Hg)	
Boiling Point:	221°C (430°F)	**Specific Gravity:** 1.4	**Water Solubility:**	Insoluble	

Chemical Incompatibilities or Instabilities: Strong oxidizers.

FLAMMABILITY

NFPA Hazard Code: 3-1-0	**NFPA Classification:** Class III Combustible	
Flash Point: 97°C (207°F)	**LEL:** N.D.	
Autoignition Temp: 411°C (772°F)	**UEL:** N.D.	

Fire Extinguishing Methods: Alcohol Resistant Foam, Carbon Dioxide, Dry Chemical, Water.

Special Fire Fighting Considerations: Structural fire fighter protective clothing will not provide adequate protection. Fight fire from a distance or protected location, if possible. Water spray should be used to keep closed containers cool. Continue to cool container after fire is extinguished. Treat all materials used or generated and equipment involved as contaminated by hazardous waste.

TOXICOLOGY
SUSPECT CARCINOGEN

Odor: Penetrating

Physical Contact: Material is extremely destructive to human tissues.

TLV = N.D.   **STEL** = N.D.

Odor Threshold: N.D.

IDLH = N.D.

RTECS # XT9275000

Routes of Entry and Relative LD$_{50}$ (or LC$_{50}$)

Inhalation	10	(300 mg/m^3/H)
Ingestion	2	(6000 mg/kg)
Skin Absorption	N.D.	

Symptoms of Exposure: Possible burns or irritation to skin, eyes and upper respiratory tract; headache; nausea. Inhalation of vapors may be fatal by causing glottis to spasm and suffocation. Exposure may result in chemical pneumonitis or pulmonary edema. Lachrymation.

Emergency/First Aid Treatment: Remove to ventilated area; immediately remove any contaminated clothing and wash contaminated areas for 15 minutes using water. Treat supportively and observe for possible shock. If ingested, seek immediate medical aid.

Recommended Clean-Up Procedures

Personal Protection: Level A Ensemble	**Recommended Material** N.D.	
RCRA Waste # U203	**Reportable Quantity:** 10 lb (1 lb)	

Spills: Dike to contain spill and collect liquid for later disposal. Decontaminate spill area using soapy water. Treat all materials used or generated and equipment involved as contaminated by hazardous waste.

Special Emergency Information: May be fatal if inhaled, swallowed, or absorbed through skin.

BENZOTRIFLUORIDE - $C_6H_5CF_3$

Synonyms: $\alpha,\alpha,\alpha,$ -trifluorotuluene; phenyl fluoroform; (trifluoromethyl) benzene

CAS Number: 98-08-8
DOT Number: 2338

Description: Colorless Liquid
DOT Classification: Flammable Liquid

Molecular Weight: 146.1
Melting Point: -29°C (-20°F)
Boiling Point: 102°C (216°F)

Vapor Density: 5.0
Specific Gravity: 1.2

Vapor Pressure: **5** (40 mm Hg)
Water Solubility: Insoluble

Chemical Incompatibilities or Instabilities: Strong oxidizers, water.

FLAMMABILITY

NFPA Hazard Code: 4-3-0
Flash Point: -29°C (-20°F)
Autoignition Temp: 102°C (216°F)

NFPA Classification: Class 1B Flammable
LEL: N.D.
UEL: N.D.

Fire Extinguishing Methods: Carbon Dioxide, Dry Chemical, Foam.

Special Fire Fighting Considerations: Structural fire fighter protective clothing will not provide adequate protection. Isolate for 1/2 mile if rail or tank truck is involved in a fire. Water spray from an unmanned device should be used to keep closed containers cool. Continue to cool container after fire is extinguished. For large fires, if possible, withdraw and allow to burn. Immediately withdraw if rising sound from venting device is heard or if fire is causing discoloration to the tank. Treat all materials used or generated and equipment involved as contaminated by hazardous waste.

TOXICOLOGY

Odor: Aromatic
Physical Contact: Material is extremely destructive to human tissues.
TLV = N.D.   **STEL** = N.D.

Odor Threshold: N.D.

IDLH = N.D.

Routes of Entry and Relative LD$_{50}$ (or LC$_{50}$)
  Inhalation   2   (283,000 mg/m^3/H)
  Ingestion   1   (15,000 mg/kg)
  Skin Absorption   N.D.

RTECS # XT9450000

Symptoms of Exposure: Possible burns or irritation to skin, eyes and upper respiratory tract; headache; nausea. Inhalation of vapors may be fatal by causing glottis to spasm and suffocation. Exposure may result in chemical pneumonitis or pulmonary edema. Laryngitis.

Emergency/First Aid Treatment: Remove to ventilated area; immediately remove any contaminated clothing and wash contaminated areas for 15 minutes using water. Treat supportively and observe for possible shock. If ingested, seek immediate medical aid.

Recommended Clean-Up Procedures

Personal Protection: Level A Ensemble
RCRA Waste # None

Recommended Material N.D.
Reportable Quantity: None

Spills: Remove all potential ignition sources. Dike to contain spill, absorb with non-combustible absorbent and take up using non-sparking tools. Decontaminate spill area using dilute alkaline solution. Treat all materials used or generated and equipment involved as contaminated by hazardous waste.

Special Emergency Information: May be harmful is inhaled, swallowed, or absorbed through the skin. Reacts with water to form HF.

BENZOYL CHLORIDE - C$_6$H$_5$COCl

Synonyms: benzenecarbonyl chloride

CAS Number: 98-88-4
DOT Number: 1736

Description: Colorless Liquid
DOT Classification: Corrosive

Molecular Weight: 140.6
Melting Point: -1°C (30°F)
Boiling Point: 197°C (387°F)

Vapor Density: 1.5
Specific Gravity: 1.21

Vapor Pressure: 2 (1 mm Hg)
Water Solubility: Decomposes

Chemical Incompatibilities or Instabilities: Water, dimethyl sulfoxide, oxidizers, AlCl$_3$.

FLAMMABILITY

NFPA Hazard Code: 3-2-2-~~W~~
Flash Point: 88°C (190°F)
Autoignition Temp: N.D.

NFPA Classification: Class IIIA Combustible Liquid
LEL: 1.2%
UEL: 4.9%

Fire Extinguishing Methods: Dry Chemical, Carbon Dioxide, or flooding quantities of Water.

Special Fire Fighting Considerations: Structural fire fighter protective clothing will not provide adequate protection. Isolate for 1/2 mile if rail or tank truck is involved in a fire. Water spray should be used to keep closed containers cool. Continue to cool container after fire is extinguished. DO NOT put water stream directly on spilled material.

TOXICOLOGY
QUESTIONABLE CARCINOGEN

Odor: Pungent
Physical Contact: Material is extremely destructive to human tissues.
TLV = N.D.   **STEL** = N.D.

Odor Threshold: N.D.

IDLH = N.D.

Routes of Entry and Relative LD$_{50}$ (or LC$_{50}$)
Inhalation   8   (3740 mg/m^3/H)
Ingestion   N.D.
Skin Absorption   N.D.

RTECS #

Symptoms of Exposure: Possible burns or irritation to skin, eyes and upper respiratory tract; headache; nausea. Lachrymation. Inhalation of vapors may be fatal by causing glottis to spasm and suffocation. Exposure may result in chemical pneumonitis or pulmonary edema.

Emergency/First Aid Treatment: Remove to ventilated area; immediately remove any contaminated clothing and wash contaminated areas for 15 minutes using water. Treat supportively and observe for possible shock. If ingested, seek immediate medical aid.

Recommended Clean-Up Procedures

Personal Protection: Level B Ensemble

RCRA Waste #   None

Recommended Material Polyvinyl Alcohol (+), Viton (0)

Reportable Quantity: 1000 lb (1000 lb)

Spills: Dike to contain spill and collect liquid for later disposal. Decontaminate spill area using dilute alkaline solution. Keep area isolated until vapors have dissipated.

Special Emergency Information: May be harmful is inhaled, swallowed, or absorbed through the skin. Reaction with water is exothermic and yields hydrochloric acid vapors.

BENZYL ALCOHOL - $C_6H_5CH_2OH$

Synonyms: benzene methanol; α-hydroxytoluene; phenyl methanol

CAS Number: 100-51-6   **Description:** Colorless Liquid
DOT Number: ----   **DOT Classification:** N.D.

Molecular Weight: 108.1
Melting Point: -15°C   (5°F)   **Vapor Density:** 3.7   **Vapor Pressure:**   **2** (1 mm Hg)
Boiling Point: 205°C (403°F)   **Specific Gravity:** 1.05   **Water Solubility:**   4%

Chemical Incompatibilities or Instabilities: Sulfuric acid, strong oxidizers, hydrogen bromide + iron..

FLAMMABILITY

NFPA Hazard Code: 2-1-0   **NFPA Classification:**   Class IIIA Combustible Liquid
Flash Point: 93°C   (200°F)   **LEL:** N.D.
Autoignition Temp: 436°C   (817°F)   **UEL:** N.D.

Fire Extinguishing Methods: Alcohol Resistant Foam, Carbon Dioxide, Dry Chemical.

Special Fire Fighting Considerations: Water spray should be used to keep closed containers cool.

TOXICOLOGY

Odor: Aromatic   **Odor Threshold:** N.D.
Physical Contact: Irritant.
TLV = N.D.   **STEL** = N.D.   **IDLH** = N.D.

Routes of Entry and Relative LD$_{50}$ (or LC$_{50}$)   **RTECS # DN3150000**
    Inhalation   N.D.
    Ingestion   3   (1230 mg/kg)
    Skin Absorption   2   (2000 mg/kg)

Symptoms of Exposure: Possible irritation to skin, eyes and upper respiratory tract; headache; nausea.

Emergency/First Aid Treatment: Remove to ventilated area; immediately remove any contaminated clothing and wash contaminated areas for 15 minutes using water. Treat supportively and observe for possible shock. If ingested, seek immediate medical aid.

Recommended Clean-Up Procedures

Personal Protection: Level B Ensemble   **Recommended Material**   Viton (+), Butyl Rubber (0)

RCRA Waste #   None   **Reportable Quantity:**   None

Spills: Dike to contain spill and collect liquid for disposal. Decontaminate spill area using water.

Special Emergency Information: May be harmful is inhaled, swallowed, or absorbed through the skin.

BENZYL BROMIDE - $C_6H_5CH_2Br$

Synonyms: α-bromotoluene

CAS Number: 100-39-0	**Description:** Colorless Liquid
DOT Number: 1737	**DOT Classification:** Corrosive Material

Molecular Weight: 171.1

Melting Point: -4°C (°F)	**Vapor Density:** 5.8	**Vapor Pressure:** 1 (1 mm Hg)
Boiling Point: 198°C (°F)	**Specific Gravity:** 1.4	**Water Solubility:** Insoluble

Chemical Incompatibilities or Instabilities: Molecular sieves, oxidizers.

FLAMMABILITY

NFPA Hazard Code:	**NFPA Classification:** Class IIIA Combustible
Flash Point: 86°C (188°F)	**LEL:** N.D.
Autoignition Temp: N.D.	**UEL:** N.D.

Fire Extinguishing Methods: Carbon Dioxide, Dry Chemical, Foam, flooding quantities of Water.

Special Fire Fighting Considerations: Structural fire fighter protective clothing will not provide adequate protection. Water spray should be used to keep closed containers cool. Continue to cool container after fire is extinguished.

TOXICOLOGY

Odor: Pleasant   **Odor Threshold:** N.D.

Physical Contact: Material is extremely destructive to human tissues.

TLV = N.D.   **STEL** = N.D.   **IDLH** = N.D.

Routes of Entry and Relative LD_{50} (or LC_{50})   **RTECS # XS7965000**

Inhalation	N.D.
Ingestion	N.D.
Skin Absorption	N.D.

Symptoms of Exposure: Possible burns or irritation to skin, eyes and upper respiratory tract; headache; nausea. Lachrymation. Laryngitis. Inhalation of vapors may be fatal by causing glottis to spasm and suffocation. Exposure may result in chemical pneumonitis or pulmonary edema.

Emergency/First Aid Treatment: Remove to ventilated area; immediately remove any contaminated clothing and wash contaminated areas for 15 minutes using water. Treat supportively and observe for possible shock. If ingested, seek immediate medical aid.

Recommended Clean-Up Procedures

Personal Protection: Level A Ensemble	**Recommended Material** Teflon (-)
RCRA Waste # None	**Reportable Quantity:** None

Spills: Dike to contain spill and collect liquid for disposal. Decontaminate spill area using soapy water.

Special Emergency Information: May be fatal if inhaled, swallowed, or absorbed through skin.

BENZYL CHLORIDE - $C_6H_5CH_2Cl$

Synonyms: chloromethyl benzene; α-chlorotoluene

CAS Number: 100-44-7		**Description:** Refractive Liquid		
DOT Number: 1738		**DOT Classification:** Corrosive		

Molecular Weight: 126.6

Melting Point:	-45°C (-49°F)	**Vapor Density:** 4.4	**Vapor Pressure:**	2 (2 mm Hg)
Boiling Point:	179°C 354°F)	**Specific Gravity:** 1.1	**Water Solubility:**	Insoluble

Chemical Incompatibilities or Instabilities: Water, dimethyl sulfoxide, metals, heat.

FLAMMABILITY

NFPA Hazard Code:	2-2-1	**NFPA Classification:**	Class IIIA Combustible Liquid
Flash Point:	63°C (153°F)	**LEL:**	1.1%
Autoignition Temp:	585°C (1085°F)	**UEL:**	N.D.

Fire Extinguishing Methods: Carbon Dioxide, Dry Chemical, Foam, and Water.

Special Fire Fighting Considerations: Structural fire fighter protective clothing will not provide adequate protection. Water spray should be used to keep closed containers cool. Continue to cool container after fire is extinguished. Treat all materials used or generated and equipment involved as contaminated by hazardous waste. May undergo explosive decomposition at elevated temperatures.

TOXICOLOGY
SUSPECTED CARCINOGEN

Odor: Pungent

Odor Threshold: 0.2 mg/m^3

Physical Contact: Material is extremely destructive to human tissues.

TLV = 5.2 mg/m^3 **STEL** = N.D.   **IDLH** = 53 mg/m^3

Routes of Entry and Relative LD$_{50}$ (or LC$_{50}$)   **RTECS #** XS8925000

Inhalation	9	(1500 mg/m^3/H)
Ingestion	3	(1230 mg/kg)
Skin Absorption	N.D.	

Symptoms of Exposure: Possible burns or irritation to skin, eyes and upper respiratory tract; headache; nausea. Lachrymation. Laryngitis. Inhalation of vapors may be fatal by causing glottis to spasm and suffocation. Exposure may result in chemical pneumonitis or pulmonary edema.

Emergency/First Aid Treatment: Remove to ventilated area; immediately remove any contaminated clothing and wash contaminated areas for 15 minutes using water. Treat supportively and observe for possible shock. If ingested, seek immediate medical aid.

Recommended Clean-Up Procedures

Personal Protection:	Level A Ensemble	**Recommended Material** Teflon (0)
RCRA Waste #	P028	**Reportable Quantity:** None

Spills: Dike to contain spill and collect liquid for disposal. Decontaminate spill area using soapy water. Treat all materials used or generated and equipment involved as contaminated by hazardous waste.

Special Emergency Information: May be fatal if inhaled, swallowed, or absorbed through skin. May react with water to yield hydrochloric acid vapors.

BERYLLIUM - Be

Synonyms: glucinium

CAS Number: 7440-41-7

DOT Number: 1567

Description: Gray Metal or Powder

DOT Classification: Poison B

Molecular Weight: 9.0

Melting Point:	1287°C (°F)	**Vapor Density:** N.A.	**Vapor Pressure:** 1 (0 mm Hg)
Boiling Point:	2500°C (°F)	**Specific Gravity:** 1.87	**Water Solubility:** Insoluble

Chemical Incompatibilities or Instabilities: Acids, bases, halocarbons, halogens, lithium, phosphorous.

FLAMMABILITY

NFPA Hazard Code: 3-1-0

Flash Point: N.A.

Autoignition Temp: N.A.

NFPA Classification:

LEL: N.A.

UEL: N.A.

Fire Extinguishing Methods: Dry Sand, Dry Chemical, Foam, and Water.

Special Fire Fighting Considerations: Structural fire fighter protective clothing will not provide adequate protection. Water spray should be used to keep closed containers cool. Continue to cool container after fire is extinguished. For large fires, if possible, withdraw and allow to burn. Isolate for 1/2 mile if rail or tank truck is involved in a fire. Treat all materials used or generated and equipment involved as contaminated by hazardous waste.

TOXICOLOGY
CARCINOGEN

Odor: None

Physical Contact: Irritant.

$TLV = 0.002$ mg (Be)/m^3

Odor Threshold: N.A.

$STEL = 0.005$ mg (Be)/m^3

$IDLH = 10$ mg/m^3

Routes of Entry and Relative LD$_{50}$ (or LC$_{50}$)

Inhalation	N.D.
Ingestion	N.D.
Skin Absorption	N.D.

RTECS # DS1750000

Symptoms of Exposure: Possible irritation to skin, eyes and upper respiratory tract; headache; nausea. Inhalation of vapors may be fatal by causing glottis to spasm and suffocation. Exposure may result in chemical pneumonitis or pulmonary edema.

Emergency/First Aid Treatment: Remove to ventilated area; immediately remove any contaminated clothing and wash contaminated areas for 15 minutes using water. Treat supportively and observe for possible shock. If ingested, seek immediate medical aid.

Recommended Clean-Up Procedures

Personal Protection: Level A Ensemble

RCRA Waste # P015

Recommended Material N.D.

Reportable Quantity: 10 lb (1 lb)

Spills: Take up spill. Decontaminate spill area using soapy water. Treat all materials used or generated and equipment involved as contaminated by hazardous waste.

Special Emergency Information: May be fatal if inhaled, swallowed, or absorbed through skin. Inhalation of beryllium compounds may lead to berylliosis, a fatal lung disease, several months or years after exposure.

BERYLLIUM FLUORIDE - BeF$_2$

Synonyms: beryllium difluoride

CAS Number: 7787-49-7   **Description:** White to Yellow Deliquescent Crystals
DOT Number: 1566   **DOT Classification:** Poison B

Molecular Weight: 47.0
Melting Point: 555°C (1350°F)   **Vapor Density:** N.A.   **Vapor Pressure:** 1 (< 1 mm Hg)
Boiling Point: °C   (°F)   **Specific Gravity:** 1.99 **Water Solubility:** Soluble

Chemical Incompatibilities or Instabilities: Acids, magnesium.

FLAMMABILITY

NFPA Hazard Code: **NFPA Classification:**
Flash Point: N.A.   **LEL:** N.A.
Autoignition Temp: N.A.   **UEL:** N.A.

Fire Extinguishing Methods: Use agent suitable for surrounding fire.

Special Fire Fighting Considerations: Structural fire fighter protective clothing will not provide adequate protection. Isolate for 1/2 mile if rail or tank truck is involved in a fire. Fight fire from a distance or protected location, if possible. Treat all materials used or generated and equipment involved as contaminated by hazardous waste.

TOXICOLOGY
CARCINOGEN

Odor: None   **Odor Threshold:** N.A.
Physical Contact: Irritation.
TLV = 0.002 mg (Be)/m^3 **STEL** = 0.005 mg (Be)/m^3 **IDLH** = N.D.

Routes of Entry and Relative LD$_{50}$ (or LC$_{50}$)   **RTECS # DS2800000**
  Inhalation   N.D.
  Ingestion   8   (98 mg/kg)
  Skin Absorption   N.D.

Symptoms of Exposure: Possible irritation to skin, eyes and upper respiratory tract; headache; nausea. Laryngitis.

Emergency/First Aid Treatment: Remove to ventilated area; immediately remove any contaminated clothing and wash contaminated areas for 15 minutes using water. Treat supportively and observe for possible shock. If ingested, seek immediate medical aid.

Recommended Clean-Up Procedures

Personal Protection: Level A Ensemble   **Recommended Material** N.D.

RCRA Waste #   None   **Reportable Quantity:**   1 lb (5000 lb)

Spills: Take up spill. Decontaminate spill area using water. Treat all materials used or generated and equipment involved as contaminated by hazardous waste.

Special Emergency Information: May be fatal if inhaled, swallowed, or absorbed through skin. Inhalation of beryllium compounds may lead to berylliosis, a fatal lung disease, several months or years after exposure. Reaction with acid yields hydrofluoric acid vapors.

Beryllium chloride (CAS # 7787-47-5, DOT # 1566, RTECS # DS2675000) has similar chemical, physical and toxicological properties.

BERYLLIUM NITRATE - Be(NO$_3$)$_2$

Synonyms: beryllium dinitrate

CAS Number: 13597-99-4

DOT Number: 2464

Description: White to Yellow Deliquescent Crystals

DOT Classification: Oxidizer and Poison

Molecular Weight: 133.0

Melting Point: 60°C (140°F)

Boiling Point: (D) 100°C (212°F)

Vapor Density: N.A.

Specific Gravity:

Vapor Pressure: N.A.

Water Solubility: Soluble

Chemical Incompatibilities or Instabilities: Combustibles.

FLAMMABILITY

NFPA Hazard Code:

Flash Point: N.A.

Autoignition Temp: N.A.

NFPA Classification: N.A

LEL: N.A.

UEL: N.A.

Fire Extinguishing Methods: WATER ONLY.

Special Fire Fighting Considerations: Structural fire fighter protective clothing will not provide adequate protection. Isolate for 1/2 mile if rail or tank truck is involved in a fire. Water spray from an unmanned device should be used to keep closed containers cool. Continue to cool container after fire is extinguished. For large fires, if possible, withdraw and allow to burn. Treat all materials used or generated and equipment involved as contaminated by hazardous waste.

TOXICOLOGY
CARCINOGEN

Odor: None

Odor Threshold: N.A.

Physical Contact: Irritant.

TLV = 0.002 mg (Be)/m^3 **STEL** = 0.005 mg (Be)/m^3 **IDLH** = 10 mg (Be)/m^3

Routes of Entry and Relative LD$_{50}$ (or LC$_{50}$)   **RTECS # DS3675000**

Inhalation	N.D.
Ingestion	N.D.
Skin Absorption	N.D.

Symptoms of Exposure: Possible irritation to skin, eyes and upper respiratory tract; headache; nausea. Inhalation of vapors may be fatal by causing glottis to spasm and suffocation. Exposure may result in chemical pneumonitis or pulmonary edema.

Emergency/First Aid Treatment: Remove to ventilated area; immediately remove any contaminated clothing and wash contaminated areas for 15 minutes using water. Treat supportively and observe for possible shock. If ingested, seek immediate medical aid.

Recommended Clean-Up Procedures

Personal Protection: Level A Ensemble

RCRA Waste # None

Recommended Material N.D.

Reportable Quantity: 1 lb (5000 lb)

Spills: Take up with a non-combustible tools and dispose. Decontaminate spill area using water. Treat all materials used as contaminated by a hazardous waste.

Special Emergency Information: May be fatal if inhaled, swallowed, or absorbed through skin. Inhalation of beryllium compounds may lead to berylliosis, a fatal lung disease, several months or years after exposure.

BIS(CHLOROMETHYL)ETHER

Synonyms: BIS-CME; sym-dichlorodimethyl ether

CAS Number: 542-88-1   **Description:** Colorless Liquid
DOT Number: 2249   **DOT Classification:** Poison B

Molecular Weight: 115.0
Melting Point: °C (°F)   **Vapor Density:** 4.0   **Vapor Pressure:** N.D.
Boiling Point: 105°C (221°F)   **Specific Gravity:** 1.3   **Water Solubility:** Decomposes

Chemical Incompatibilities or Instabilities: Acids, water.

FLAMMABILITY

NFPA Hazard Code: **NFPA Classification:** Class 1B Flammable
Flash Point: 19°C (66°F)   **LEL:** N.A.
Autoignition Temp: N.D.   **UEL:** N.A.

Fire Extinguishing Methods: Carbon Dioxide, Dry Chemical, Foam, Water.

Special Fire Fighting Considerations: Structural fire fighter protective clothing will not provide adequate protection. Fight fire from a distance or protected location, if possible. Treat all materials used or generated and equipment involved as contaminated by hazardous waste. May decompose in water to form formaldehyde.

TOXICOLOGY
SUSPECTED CARCINOGEN

Odor: Suffocating   **Odor Threshold:** N.D.
Physical Contact: Irritant.
TLV = See 29 CFR 1910.1008   **STEL** = N.D.   **IDLH** = N.D.

Routes of Entry and Relative LD$_{50}$ (or LC$_{50}$)   **RTECS #** KN1575000

Inhalation	10	(30 mg/m^3/7 h)
Ingestion	6	(210 mg/kg)
Skin Absorption	8	(280 mg/kg)

Symptoms of Exposure: Possible irritation to skin, eyes and upper respiratory tract; headache; nausea. Exposure may result in chemical pneumonitis or pulmonary edema. Bloody sputum.

Emergency/First Aid Treatment: Remove to ventilated area; immediately remove any contaminated clothing and wash contaminated areas for 15 minutes using water. Treat supportively and observe for possible shock. If ingested, seek immediate medical aid.

Recommended Clean-Up Procedures

Personal Protection: Level A Ensemble   **Recommended Material** Teflon (-)

RCRA Waste # P016   **Reportable Quantity:** 10 lb (1 lb)

Spills: Remove all potential ignition sources. Dike to contain spill, absorb with non-combustible absorbent and take up using non-sparking tools. Decontaminate spill area using dilute alkaline solution. Keep area isolated until vapors have dissipated. Treat all materials used as contaminated by a hazardous waste.

Special Emergency Information: May be fatal if inhaled, swallowed, or absorbed through skin. May form unstable peroxides upon prolonged exposure to air.

1,1-BIS(4-CHLOROPHENYL)-2,2-DICHLOROETHANE

Synonyms: TDE; 1,1-dichloro-2,2BIS(p-chlorophenyl)ethan; rhothane

CAS Number: 72-54-8 **Description:** Colorless Liquid
DOT Number: 2761 **DOT Classification:** ORM-A

Molecular Weight: 320.0
Melting Point: 110°C (230°F) **Vapor Density:** 1.11 **Vapor Pressure:** 1 (< 1 mm Hg)
Boiling Point: --- **Specific Gravity:** N.D. **Water Solubility:** Insoluble

Chemical Incompatibilities or Instabilities: Strong oxidizers.

FLAMMABILITY

NFPA Hazard Code: **NFPA Classification:**
Flash Point: N.A. **LEL:** N.A.
Autoignition Temp: N.A. **UEL:** N.A.

Fire Extinguishing Methods: Carbon Dioxide, Dry Chemical, Foam, Water.

Special Fire Fighting Considerations: Structural fire fighter protective clothing will not provide adequate protection. Fight fire from a distance or protected location, if possible. Treat all materials used or generated and equipment involved as contaminated by hazardous waste.

TOXICOLOGY
CARCINOGEN

Odor: **Odor Threshold:** N.D.
Physical Contact: Irritation
TLV = N.D. **STEL** = N.D. **IDLH** = N.D.

Routes of Entry and Relative LD$_{50}$ (or LC$_{50}$) **RTECS # KI0700000**
   Inhalation   N.D.
   Ingestion   7   (113 mg/kg)
   Skin Absorption 4   (1200 mg/kg)

Symptoms of Exposure: Possible irritation to skin, eyes and upper respiratory tract; headache; nausea. Lethargy. Muscle spasms.

Emergency/First Aid Treatment: Remove to ventilated area; immediately remove any contaminated clothing and wash contaminated areas for 15 minutes using water. Treat supportively and observe for possible shock. If ingested, seek immediate medical aid.

Recommended Clean-Up Procedures

Personal Protection: Level B Ensemble **Recommended Material** N.D.

RCRA Waste # U060 **Reportable Quantity:** 1 lb (1 lb)

Spills: Dike to contain spill and collect liquid for later disposal. Decontaminate spill area using soapy water. Treat all materials used or generated and equipment involved as contaminated by hazardous waste.

Special Emergency Information: May be harmful is inhaled, swallowed, or absorbed through the skin.

BORNEOL

Synonyms: endo-1,7,7-trimethylbicyclo(2,2,1)heptan-2-ol; 2 camphanol

CAS Number: 507-70-0   **Description:** White Powder
DOT Number: 1312   **DOT Classification:** Flammable Solid

Molecular Weight: 154.2
Melting Point: 208°C (406°F)   **Vapor Density:** 5.3   **Vapor Pressure:** N.D.
Boiling Point: 212°C (434°F)   **Specific Gravity:** 1.01   **Water Solubility:** Insoluble

Chemical Incompatibilities or Instabilities: Strong oxidizers.

FLAMMABILITY

NFPA Hazard Code: 2-2-0   **NFPA Classification:** Class III A Combustible
Flash Point: 66°C (150°F)   **LEL:** N.D.
Autoignition Temp: N.D.   **UEL:** N.D.

Fire Extinguishing Methods: Carbon Dioxide, Dry Chemical, Foam, Water.

Special Fire Fighting Considerations: Water spray from an unmanned device should be used to keep closed containers cool. Continue to cool container after fire is extinguished. For large fires, if possible, withdraw and allow to burn.

TOXICOLOGY

Odor: Pepper   **Odor Threshold:** N.D.
Physical Contact: Irritant
TLV = N.D.   **STEL** = N.D.   **IDLH** = N.D.

Routes of Entry and Relative LD$_{50}$ (or LC$_{50}$)   **RTECS # ED7000000**
   Inhalation   N.D.
   Ingestion   **5**   (500 mg/kg)
   Skin Absorption   N.D.

Symptoms of Exposure: Possible irritation to skin, eyes and upper respiratory tract; headache; nausea.

Emergency/First Aid Treatment: Remove to ventilated area; immediately remove any contaminated clothing and wash contaminated areas for 15 minutes using water. Treat supportively and observe for possible shock. If ingested, seek immediate medical aid.

Recommended Clean-Up Procedures

Personal Protection: Level A Ensemble   **Recommended Material** N.D.

RCRA Waste # None   **Reportable Quantity:** None

Spills: Remove all potential ignition sources. Dike to contain spill, absorb with non-combustible absorbent and take up using non-sparking tools. Decontaminate spill area using soapy water.

Special Emergency Information: May be harmful is inhaled, swallowed, or absorbed through the skin.

BORON TRIBROMIDE - BBr$_3$

Synonyms: boron bromide; trona

CAS Number: 10294-33-4
DOT Number: 2692

Description: Colorless Fuming Liquid
DOT Classification: Corrosive Material

Molecular Weight: 250.5

Melting Point:	-45°C (-49°F)	**Vapor Density:** N.D.	**Vapor Pressure:**	6 (65 mm Hg)
Boiling Point:	92°C (198°F)	**Specific Gravity:** 2.7	**Water Solubility:**	Decomposes

Chemical Incompatibilities or Instabilities: Water, alkali metals, alcohols, tungsten trioxide.

FLAMMABILITY

NFPA Hazard Code: 4-0-2-~~W~~
Flash Point: N.A.
Autoignition Temp: N.A.

NFPA Classification:
LEL: N.A.
UEL: N.A.

Fire Extinguishing Methods: Dry Chemical, Carbon Dioxide, flooding quantities of Water.

Special Fire Fighting Considerations: Structural fire fighter protective clothing will not provide adequate protection. Water spray should be used to keep closed containers cool. Continue to cool container after fire is extinguished. Do not allow water to enter container.

DOT Recommended Isolation Zones: Small Spill: 600 ft   Large Spill: 900 ft
DOT Recommended Down Wind Take Cover Distance: Small Spill: 2 miles   Large Spill: 3 miles

TOXICOLOGY

Odor: Pungent
Physical Contact: Material is extremely destructive to human tissues..
TLV = 10 mg/m^3 **STEL** = 30 mg/m^3

Odor Threshold: N.D.

IDLH = N.D.

Routes of Entry and Relative LD$_{50}$ (or LC$_{50}$)
    Inhalation   N.D.
    Ingestion   N.D.
    Skin Absorption   N.D.

RTECS # ED7400000

Symptoms of Exposure: Possible burns or irritation to skin, eyes and upper respiratory tract; headache; nausea. Laryngitis. Inhalation of vapors may be fatal by causing glottis to spasm and suffocation. Exposure may result in chemical pneumonitis or pulmonary edema.

Emergency/First Aid Treatment: Remove to ventilated area; immediately remove any contaminated clothing and wash contaminated areas for 15 minutes using water. Treat supportively and observe for possible shock. If ingested, seek immediate medical aid.

Recommended Clean-Up Procedures

Personal Protection: Level A Ensemble   **Recommended Material**   N.D.

RCRA Waste #   None   **Reportable Quantity:**   None

Spills: Dike to contain spill and collect liquid for later disposal. Decontaminate spill area using dilute bicarbonate solution. Treat all materials used as contaminated by a hazardous waste.

Special Emergency Information: May be fatal if inhaled, swallowed, or absorbed through skin. Reacts violently to yield hydrobromic acid vapors. Product may be packaged as a solution. Properties of the solvent should be considered before any action is taken.

BORON TRICHLORIDE - BCl₃

Synonyms: boron chloride

CAS Number: 10294-34-5   **Description:** Colorless Fuming Gas or Liquid
DOT Number: 1741   **DOT Classification:** Corrosive Material

Molecular Weight: 117.2
Melting Point: -107 C (-161 F)   **Vapor Density:** 4.0   **Vapor Pressure:** GAS
Boiling Point: 12 C   (54 F)   **Specific Gravity:** 1.4   **Water Solubility:** Decomposes

Chemical Incompatibilities or Instabilities: Water, ordinary combustibles, organics.

FLAMMABILITY

NFPA Hazard Code: **NFPA Classification:**
Flash Point: N.D.   **LEL:** N.D.
Autoignition Temp: N.D.   **UEL:** N.D.

Fire Extinguishing Methods: Carbon Dioxide, Dry Chemical, flooding quantities of Water.

Special Fire Fighting Considerations: Structural fire fighter protective clothing will not provide adequate protection. Do not allow water to enter container. Water spray should be used to keep closed containers cool. Continue to cool container after fire is extinguished. Keep area isolated until vapors have dissipated.

DOT Recommended Isolation Zones: Small Spill: 600 ft   Large Spill: 900 ft
DOT Recommended Down Wind Take Cover Distance: Small Spill: 2 miles   Large Spill: 5 miles

TOXICOLOGY

Odor: Pungent   **Odor Threshold:** N.D.
Physical Contact: Material is extremely destructive to human tissues.
TLV = 3 mg/m³ (ceiling)   **STEL** = N.D.   **IDLH** = N.D.

Routes of Entry and Relative LD₅₀ (or LC₅₀)   **RTECS # ED1925000**
     Inhalation   N.D.
     Ingestion   N.D.
     Skin Absorption   N.D.

Symptoms of Exposure: Possible burns or irritation to skin, eyes and upper respiratory tract; headache; nausea. Laryngitis. Inhalation of vapors may be fatal by causing glottis to spasm and suffocation. Exposure may result in chemical pneumonitis or pulmonary edema.

Emergency/First Aid Treatment: Remove to ventilated area; immediately remove any contaminated clothing and wash contaminated areas for 15 minutes using water. Treat supportively and observe for possible shock. If ingested, seek immediate medical aid.

Recommended Clean-Up Procedures

Personal Protection: Level A Ensemble   **Recommended Material**   N.D.

RCRA Waste #   None   **Reportable Quantity:**   None

Spills: Stop leak if it can safely be done. Decontaminate spill area using dilute alkaline solution.

Special Emergency Information: May be fatal if inhaled, swallowed, or absorbed through skin. Reacts exothermically with water to yield hydrochloric acid vapors. Product may be packaged as a solution. Properties of the solvent should be considered before any action is taken.

BORON TRIFLUORIDE - BF₃

Synonyms: boron fluoride

CAS Number: 7637-07-2
DOT Number: 1009

Description: Fuming Gas
DOT Classification: Non Flammable Gas

Molecular Weight: 67.8
Melting Point: -127°C (-197°F)
Boiling Point: -100°C (-148°F)

Vapor Density: 2.34
Specific Gravity: N.A.

Vapor Pressure: N.D.
Water Solubility: Decomposes

Chemical Incompatibilities or Instabilities: Water, organic matter, amines, alkali metals, calcium oxide, alkaline earth metals..

FLAMMABILITY

NFPA Hazard Code: 4-0-1
Flash Point: N.A.
Autoignition Temp: N.A.

NFPA Classification:
LEL: N.A.
UEL: N.A.

Fire Extinguishing Methods: Dry Chemical, Carbon Dioxide, and Water.

Special Fire Fighting Considerations: Structural fire fighter protective clothing will not provide adequate protection. Do not allow water to enter container. Water spray should be used to keep closed containers cool. Continue to cool container after fire is extinguished. Keep area isolated until vapors have dissipated.

DOT Recommended Isolation Zones:
DOT Recommended Down Wind Take Cover Distance:

Small Spill: 1500 ft
Small Spill: 5 miles

Large Spill: 1500 ft
Large Spill: 5 miles

TOXICOLOGY

Odor: Suffocating
Physical Contact: Material is extremely destructive to human tissues.
TLV = 2.8 mg/m³ (ceiling)   **STEL** = N.D.

Odor Threshold: 5 mg/m³

IDLH = 280 mg/m³

Routes of Entry and Relative LD₅₀ (or LC₅₀)
Inhalation   N.D.
Ingestion   N.D.
Skin Absorption   N.D.

RTECS # ED2275000

Symptoms of Exposure: Possible burns or irritation to skin, eyes and upper respiratory tract; headache; nausea. Laryngitis. Inhalation of vapors may be fatal by causing glottis to spasm and suffocation. Exposure may result in chemical pneumonitis or pulmonary edema.

Emergency/First Aid Treatment: Remove to ventilated area; immediately remove any contaminated clothing and wash contaminated areas for 15 minutes using water. Treat supportively and observe for possible shock. If ingested, seek immediate medical aid.

Recommended Clean-Up Procedures

Personal Protection: Level A Ensemble   **Recommended Material** N.D.

RCRA Waste # None   **Reportable Quantity:** None

Spills: Stop leak if it can safely be done. Decontaminate spill area using dilute alkaline solution.

Special Emergency Information: May be fatal if inhaled, swallowed, or absorbed through skin. Reacts exothermically to yield hydrofluoric acid vapors. Product may be packaged as a solution. Properties of the solvent should be considered before any action is taken.

BROMINE - Br$_2$

Synonyms:

CAS Number: 7726-95-6   **Description:** Dark, Red-Brown Liquid
DOT Number: 1744   **DOT Classification:** Corrosive Material

Molecular Weight: 79.9
Melting Point: -7°C (19°F)   **Vapor Density:** 5.5   **Vapor Pressure:** 7 (175 mm Hg)
Boiling Point: 59°C (138°F)   **Specific Gravity:** 3.10 **Water Solubility:** 16%

Chemical Incompatibilities or Instabilities: Organics, reducing agents, alkali metals, finely divided metals, hydrogen metal azides, organo metalics, ammonia, ordinary combustibles.

FLAMMABILITY

NFPA Hazard Code: 3-0-0-OX   **NFPA Classification:**
Flash Point: N.A.   **LEL:** N.A.
Autoignition Temp: N.A.   **UEL:** N.A.

Fire Extinguishing Methods: Use agent suitable for surrounding fire.

Special Fire Fighting Considerations: Structural fire fighter protective clothing will not provide adequate protection. Water spray should be used to keep closed containers cool. Continue to cool container after fire is extinguished.

DOT Recommended Isolation Zones: Small Spill: 1500 ft   Large Spill: 1500 ft
DOT Recommended Down Wind Take Cover Distance: Small Spill: 5 miles   Large Spill: 5 miles

TOXICOLOGY

Odor: Sharp, irritating   **Odor Threshold:** 0.06 mg/m^3
Physical Contact: Material is extremely destructive to human tissues.
TLV = 0.66 mg/m^3 **STEL** = 2 mg/m^3 **IDLH** = 66 mg/m^3

Routes of Entry and Relative LD$_{50}$ (or LC$_{50}$)   **RTECS #** EF9100000
   Inhalation   N.D.
   Ingestion   N.D.
   Skin Absorption   N.D.

Symptoms of Exposure: Possible burns or irritation to skin, eyes and upper respiratory tract; headache; nausea. Laryngitis. Lachrymation. Inhalation of vapors may be fatal by causing glottis to spasm and suffocation. Exposure may result in chemical pneumonitis or pulmonary edema.

Emergency/First Aid Treatment: Remove to ventilated area; immediately remove any contaminated clothing and wash contaminated areas for 15 minutes using water. Treat supportively and observe for possible shock. If ingested, seek immediate medical aid.

Recommended Clean-Up Procedures

Personal Protection: Level A Ensemble   **Recommended Material**   Teflon (0)

RCRA Waste #   None   **Reportable Quantity:**   None

Spills: Dike to contain spill and collect liquid for disposal. Decontaminate spill area using water.

Special Emergency Information: May be fatal if inhaled, swallowed, or absorbed through skin.

BROMINE PENTAFLUORIDE - BrF₅

Synonyms:

CAS Number: 7789-30-2   **Description:** Colorless, Fuming Liquid
DOT Number: 1745   **DOT Classification:** Oxidizer

Molecular Weight: 174.9
Melting Point: -61°C (-78°F)   **Vapor Density:** 6.05   **Vapor Pressure:** N.D.
Boiling Point: 41°C (108°F)   **Specific Gravity:** 2.46   **Water Solubility:** Explosively Decomposes

Chemical Incompatibilities or Instabilities: Water, organics, acids, halogens, finely divided metals, alkali metals, halogens, phosphorous, ordinary combustibles amines, ammonia salts.

FLAMMABILITY

NFPA Hazard Code: 4-0-3-W-OX   **NFPA Classification:**
Flash Point: N.A.   **LEL:** N.A.
Autoignition Temp: N.A.   **UEL:** N.A.

Fire Extinguishing Methods: Dry Chemical, Sand.

Special Fire Fighting Considerations: Structural fire fighter protective clothing will not provide adequate protection. Water spray should be used to keep closed containers cool. Continue to cool container after fire is extinguished. Do not allow water to enter container.

DOT Recommended Isolation Zones: Small Spill: 1500 ft   Large Spill: 1500 ft
DOT Recommended Down Wind Take Cover Distance: Small Spill: 5 miles   Large Spill: 5 miles

TOXICOLOGY

Odor: **Odor Threshold:** N.D.
Physical Contact: Material is extremely destructive to human tissues.
TLV = 0.7 mg/m³   **STEL** = N.D.   **IDLH** = N.D.

Routes of Entry and Relative LD₅₀ (or LC₅₀)   **RTECS # EF9350000**
  Inhalation   N.D.
  Ingestion   N.D.
  Skin Absorption   N.D.

Symptoms of Exposure: Possible burns or irritation to skin, eyes and upper respiratory tract; headache; nausea. Inhalation of vapors may be fatal by causing glottis to spasm and suffocation. Exposure may result in chemical pneumonitis or pulmonary edema.

Emergency/First Aid Treatment: Remove to ventilated area; immediately remove any contaminated clothing and wash contaminated areas for 15 minutes using water. Treat supportively and observe for possible shock. If ingested, seek immediate medical aid.

Recommended Clean-Up Procedures

Personal Protection: Level A Ensemble   **Recommended Material** N.D.

RCRA Waste # None   **Reportable Quantity:** None

Spills: Dike to contain spill and collect liquid for later disposal. Decontaminate spill area using dilute alkaline solution. Keep area isolated until vapors have dissipated.

Special Emergency Information: May be fatal if inhaled, swallowed, or absorbed through skin. Will react violently with water to liberate toxic and corrosive vapors.

BROMINE TRIFLUORIDE - BrF₃

Synonyms:

CAS Number: 7787-71-5   **Description:** Fuming Liquid
DOT Number: 1746   **DOT Classification:** Oxidizer

Molecular Weight: 136.9
Melting Point: 9°C   (48°F)   **Vapor Density:** N.D.   **Vapor Pressure:** N.D.
Boiling Point: 135°C (275°F)   **Specific Gravity:** 2.8   **Water Solubility:** Decomposes

Chemical Incompatibilities or Instabilities: Water, organics, acids, bases, halogens, metals, ordinary combustibles.

FLAMMABILITY

NFPA Hazard Code: 4-0-3-W-OX   **NFPA Classification:**
Flash Point: N.A.   **LEL:** N.A.
Autoignition Temp: N.A.   **UEL:** N.A.

Fire Extinguishing Methods: Dry Chemical, Sand.

Special Fire Fighting Considerations: Structural fire fighters protective clothing will not provide adequate protection. Water spray should be used to keep closed containers cool. Continue to cool container after fire is extinguished. Do not allow water to enter container.

DOT Recommended Isolation Zones: Small Spill: 150 ft   Large Spill: 150 ft
DOT Recommended Down Wind Take Cover Distance: Small Spill: 0.8 mile   Large Spill: 0.8 mile

TOXICOLOGY

Odor: **Odor Threshold:** N.D.
Physical Contact: Material is extremely destructive to human tissues.
TLV = 2.5 mg (F)/m³   **STEL** = N.D.   **IDLH** = N.D.

Routes of Entry and Relative LD₅₀ (or LC₅₀)   **RTECS # EP9360000**
   Inhalation   N.D.
   Ingestion   N.D.
   Skin Absorption   N.D.

Symptoms of Exposure: Possible burns or irritation to skin, eyes and upper respiratory tract; headache; nausea. Inhalation of vapors may be fatal by causing glottis to spasm and suffocation. Exposure may result in chemical pneumonitis or pulmonary edema.

Emergency/First Aid Treatment: Remove to ventilated area; immediately remove any contaminated clothing and wash contaminated areas for 15 minutes using water. Treat supportively and observe for possible shock. If ingested, seek immediate medical aid.

Recommended Clean-Up Procedures

Personal Protection: Level A Ensemble   **Recommended Material**   N.D.

RCRA Waste #   None   **Reportable Quantity:**   None

Spills: Dike to contain spill and collect liquid for disposal. Decontaminate spill area using dilute alkaline solution. Keep area isolated until vapors have dissipated.

Special Emergency Information: May be fatal if inhaled, swallowed, or absorbed through skin. Will react violently with water to yield toxic and corrosive vapors.

BROMOACETIC ACID - BrCH₂COOH

Synonyms: monobromoacetic acid

CAS Number: 79-08-3   **Description:** Colorless, Hygroscopic Crystals
DOT Number: 1938   **DOT Classification:** Corrosive Materials

Molecular Weight: 139
Melting Point: 50°C (122°F)   **Vapor Density:** N.D.   **Vapor Pressure:** N.D.
Boiling Point: 208°C (406°F)   **Specific Gravity:** 1.9   **Water Solubility:**   Soluble

Chemical Incompatibilities or Instabilities: Strong oxidizers, strong bases.

FLAMMABILITY

NFPA Hazard Code: **NFPA Classification:**
Flash Point: N.D.   **LEL:** N.D.
Autoignition Temp: N.D.   **UEL:** N.D.

Fire Extinguishing Methods: Carbon Dioxide, Dry Chemical, Foam, Water.

Special Fire Fighting Considerations: Water spray should be used to keep closed containers cool. Continue to cool container after fire is extinguished.

TOXICOLOGY

Odor: **Odor Threshold:** N.D.
Physical Contact: Material is extremely destructive to human tissues.
TLV = N.D.   **STEL** = N.D.   **IDLH** = N.D.

Routes of Entry and Relative LD$_{50}$ (or LC$_{50}$)   **RTECS # AF5950000**
    Inhalation   N.D.
    Ingestion   N.D.
    Skin Absorption   N.D.

Symptoms of Exposure: Possible burns or irritation to skin, eyes and upper respiratory tract; headache; nausea. Lachrymation. Inhalation of vapors may be fatal by causing glottis to spasm and suffocation. Exposure may result in chemical pneumonitis or pulmonary edema.

Emergency/First Aid Treatment: Remove to ventilated area; immediately remove any contaminated clothing and wash contaminated areas for 15 minutes using water. Treat supportively and observe for possible shock. If ingested, seek immediate medical aid.

Recommended Clean-Up Procedures

Personal Protection: Level B Ensemble   **Recommended Material**   N.D.

RCRA Waste #   None   **Reportable Quantity:**   None

Spills: Dike to contain spill and collect liquid for later disposal. Decontaminate spill area using water.

Special Emergency Information: May be harmful is inhaled, swallowed, or absorbed through the skin.

BROMOACETONE - BrCH₂COCH₃

Synonyms: bromo-2-propanone

CAS Number: 598-31-2   **Description:** Colorless Liquid
DOT Number: 1569   **DOT Classification:** Poison A

Molecular Weight: 137
Melting Point: -37°C (-35°F)   **Vapor Density:** N.D.   **Vapor Pressure:**   N.D.
Boiling Point: 137°C (279°F)   **Specific Gravity:** 1.6   **Water Solubility:** Slight

Chemical Incompatibilities or Instabilities: Strong oxidizers.

FLAMMABILITY

NFPA Hazard Code: **NFPA Classification:**
Flash Point: N.D.   **LEL:** N.D.
Autoignition Temp: N.D.   **UEL:** N.D.

Fire Extinguishing Methods: Carbon Dioxide, Dry Chemical, Foam, Water.

Special Fire Fighting Considerations: Structural fire fighters protective clothing will not provide adequate protection. Fight fire from a distance or protected location, if possible. Treat all materials used or generated and equipment involved as contaminated by hazardous waste.

DOT Recommended Isolation Zones: Small Spill: 150 ft   Large Spill: 150 ft
DOT Recommended Down Wind Take Cover Distance: Small Spill: 0.2 mile   Large Spill: 0.2 mile

TOXICOLOGY

Odor: **Odor Threshold:** N.D.
Physical Contact: Irritant
TLV = N.D.   **STEL** = N.D.   **IDLH** = N.D.

Routes of Entry and Relative LD$_{50}$ (or LC$_{50}$)   **RTECS # UC0525000**
    Inhalation   N.D.
    Ingestion   N.D.
    Skin Absorption   N.D.

Symptoms of Exposure: Possible irritation to skin, eyes and upper respiratory tract; headache; nausea. Lachrymation.

Emergency/First Aid Treatment: Remove to ventilated area; immediately remove any contaminated clothing and wash contaminated areas for 15 minutes using water. Treat supportively and observe for possible shock. If ingested, seek immediate medical aid.

Recommended Clean-Up Procedures

Personal Protection: Level A Ensemble   **Recommended Material**   N.D.

RCRA Waste #   P017   **Reportable Quantity:**   1000 lb (1 lb)

Spills: Remove all potential ignition sources. Dike to contain spill and collect liquid for disposal. Decontaminate spill area using soapy water. Treat all materials used or generated and equipment involved as contaminated by hazardous waste.

Special Emergency Information: May be fatal if inhaled, swallowed, or absorbed through skin.

BROMOBENZENE - C$_6$H$_5$Br

Synonyms: monobromobenzene, phenyl bromide

CAS Number: 108-86-1   **Description:** Colorless Liquid
DOT Number: 2514   **DOT Classification:** Flammable Liquid

Molecular Weight: 157.0
Melting Point: -31°C (-24°F)   **Vapor Density:** 5.4   **Vapor Pressure:** **3** (10 mm Hg)
Boiling Point: 156°C (313°F)   **Specific Gravity:** 1.5   **Water Solubility:** Insoluble

Chemical Incompatibilities or Instabilities: Sodium, oxidizers.

FLAMMABILITY

NFPA Hazard Code: 2-2-0   **NFPA Classification:** Class II Combustible
Flash Point: 51°C (124°F)   **LEL:** N.D.
Autoignition Temp: 565°C (1050°F)   **UEL:** N.D.

Fire Extinguishing Methods: Carbon Dioxide, Dry Chemical, Foam, Water.

Special Fire Fighting Considerations: Isolate for 1/2 mile if rail or tank truck is involved in a fire. Water spray from an unmanned device should be used to keep closed containers cool. Continue to cool container after fire is extinguished. For large fires, if possible, withdraw and allow to burn. Immediately withdraw if rising sound from venting device is heard or if fire is causing discoloration to the tank.

TOXICOLOGY

Odor: **Odor Threshold:** N.D.
Physical Contact: Irritant.
TLV = N.D.   **STEL** = N.D.   **IDLH** = N.D.

Routes of Entry and Relative LD$_{50}$ (or LC$_{50}$)   **RTECS # CY9000000**
   Inhalation   6   (20,411 mg/m^3)
   Ingestion   3   (2700 mg/kg)
   Skin Absorption   N.D.

Symptoms of Exposure: Possible irritation to skin, eyes and upper respiratory tract; headache; nausea.

Emergency/First Aid Treatment: Remove to ventilated area; immediately remove any contaminated clothing and wash contaminated areas for 15 minutes using water. Treat supportively and observe for possible shock. If ingested, seek immediate medical aid.

Recommended Clean-Up Procedures

Personal Protection: Level B Ensemble   **Recommended Material** Polyvinyl Alcohol (+), Viton (+)

RCRA Waste # None   **Reportable Quantities:** None

Spills: Remove all potential ignition sources. Dike to contain spill and collect liquid for disposal.

Special Emergency Information: May be harmful is inhaled, swallowed, or absorbed through the skin. Vapors are heavier than air and may travel some distance to an ignition source. Vapors are more dense than air and may settle in low lying areas.

Chlorobenzene (DOT #1134, CAS #108-90-7, RTECS #CZ0175000) has similar physical, chemical, and toxicological properties.

1-BROMOBUTANE - BrCH$_2$CH$_2$CH$_2$CH$_3$

Synonyms: butyl bromide

CAS Number: 109-65-9   **Description:** Colorless Liquid
DOT Number: 1126   **DOT Classification:** Flammable Liquid

Molecular Weight: 137.0
Melting Point: -112°C (-170°F)   **Vapor Density:** 4.7   **Vapor Pressure:** **5** (40 mm Hg)
Boiling Point: 101°C (214°F)   **Specific Gravity:** 1.3   **Water Solubility:** Insoluble

Chemical Incompatibilities or Instabilities: Alkali metals, strong oxidizers.

FLAMMABILITY

NFPA Hazard Code: 2-3-0   **NFPA Classification:** Class 1B Flammable
Flash Point: 65°C (149°F)   **LEL:** N.D.
Autoignition Temp: 510°C (950°F)   **UEL:** N.D.

Fire Extinguishing Methods: Carbon Dioxide, Dry Chemical, Foam, and Water.

Special Fire Fighting Considerations: Isolate for 1/2 mile if rail or tank truck is involved in at fire. Water spray should be used to keep closed containers cool. Continue to cool container after fire is extinguished. Immediately withdraw if rising sound from venting device is heard or if fire is causing discoloration to the tank.

TOXICOLOGY

Odor: **Odor Threshold:** N.D.
Physical Contact: Mild Irritant.
TLV = N.D.   **STEL** = N.D.   **IDLH** = N.D.

Routes of Entry and Relative LD$_{50}$ (or LC$_{50}$)   **RTECS # EJ6225000**
    Inhalation   3   (118,500 mg/m^3/H)
    Ingestion   N.D.
    Skin Absorption   N.D.

Symptoms of Exposure: Possible irritation to skin, eyes and upper respiratory tract; headache; nausea. May act as a narcotic in high concentrations.

Emergency/First Aid Treatment: Remove to ventilated area; immediately remove any contaminated clothing and wash contaminated areas for 15 minutes using water. Treat supportively and observe for possible shock. If ingested, seek immediate medical aid.

Recommended Clean-Up Procedures

Personal Protection: Level B Ensemble   **Recommended Material**   Nitrile Rubber (-), Viton (-)

RCRA Waste #   None   **Reportable Quantities:**   None

Spills: Remove all potential ignition sources. Dike to contain spill, absorb with non-combustible absorbent and take up using non-sparking tools. Decontaminate spill area using soapy water.

Special Emergency Information: May be harmful is inhaled, swallowed, or absorbed through the skin.

2-BROMOBUTANE - CH3CHBrCH2CH3

Synonyms: sec-butyl bromide

CAS Number: 78-76-2   **Description:** Colorless Liquid
DOT Number: 2339   **DOT Classification:** Flammable Liquid

Molecular Weight: 137
Melting Point: -112°C (-170°F)   **Vapor Density:** N.D.   **Vapor Pressure:** N.D.
Boiling Point: 91°C (196°F)   **Specific Gravity:** 1.3   **Water Solubility:** Insoluble

Chemical Incompatibilities or Instabilities: Alkali metals, magnesium, strong oxidizers.

FLAMMABILITY

NFPA Hazard Code: **NFPA Classification:** Class 1B Flammable
Flash Point: 70°C (158°F)   **LEL:** N.D.
Autoignition Temp: °C   (°F)   **UEL:** N.D.

Fire Extinguishing Methods: Carbon Dioxide, Dry Chemical, Foam.

Special Fire Fighting Considerations: May release toxic fumes upon heating (HBr). Isolate for 1/2 mile if rail or tank truck is involved in at fire. Water spray should be used to keep closed containers cool.

TOXICOLOGY

Odor: **Odor Threshold:** N.D.
Physical Contact: Mild Irritant
TLV = N.D.   **STEL** = N.D.   **IDLH** = N.D.

Routes of Entry and Relative LD50 (or LC50)   **RTECS # EJ6228000**
  Inhalation   N.D.
  Ingestion   N.D.
  Skin Absorption   N.D.

Symptoms of Exposure: Difficulty breathing, headache, nausea, narcotic in high concentrations.

Emergency/First Aid Treatment: Remove to fresh air, remove contaminated clothing, and wash contacted areas for 15 minutes with water.

Recommended Clean-Up Procedures

Personal Protection: Level B Ensemble   **Recommended Material** Nitrile Rubber (-), Viton (-)

RCRA Waste #   None   **Reportable Quantities:** None

Spills: Remove all potential ignition sources. Dike to contain spill, absorb with non-combustible absorbent and take up using non-sparking tools. Decontaminate spill area using soapy water.

Special Emergency Information: May be harmful is inhaled, swallowed, or absorbed through the skin.

2-Chlorobutane (DOT # 1127, CAS # 78-86-4, RTECS # EJ6475000) and l-Chlorobutane (DOT # 1127, CAS # 109-69-3, RTECS # EJ6300000) have similar chemical, physical, and toxicological properties.

BROMOFORM - CHBr$_3$

Synonyms: tribromomethane

CAS Number: 75-25-2   **Description:** Colorless Liquid
DOT Number: 2515   **DOT Classification:** Poison B

Molecular Weight: 252.8
Melting Point: 8°C (46°F)   **Vapor Density:** N.D.   **Vapor Pressure:** N.D.
Boiling Point: 144°C (291°F)   **Specific Gravity:** 2.9   **Water Solubility:** Insoluble

Chemical Incompatibilities or Instabilities: Alkali metals, acetone, bases, crown ethers, alkali earth metals, zinc..

FLAMMABILITY

NFPA Hazard Code: **NFPA Classification:**
Flash Point: N.A.   **LEL:** N.A.
Autoignition Temp: N.A.   **UEL:** N.A.

Fire Extinguishing Methods: Use agent suitable for surrounding fire.

Special Fire Fighting Considerations: Structural fire fighter protective clothing will not provide adequate protection.

TOXICOLOGY
QUESTIONABLE CARCINOGEN

Odor: Sweet, chloroform-like   **Odor Threshold:** 2 mg/m^3
Physical Contact: Irritant
TLV = 5 mg/m^3 **STEL** = N.D.   **IDLH** = N.D.

Routes of Entry and Relative LD$_{50}$ (or LC$_{50}$)   **RTECS # PB5600000**
    Inhalation   N.D.
    Ingestion   3   (1147 mg/kg)
    Skin Absorption   N.D.

Symptoms of Exposure: Possible irritation to skin, eyes and upper respiratory tract; headache; nausea. Lachrymation.

Emergency/First Aid Treatment: Remove to ventilated area; immediately remove any contaminated clothing and wash contaminated areas for 15 minutes using water. Treat supportively and observe for possible shock. If ingested, seek immediate medical aid.

Recommended Clean-Up Procedures

Personal Protection: Level B Ensemble   **Recommended Material** Viton (-)

RCRA Waste # U225   **Reportable Quantities:** 100 lb (1 lb)

Spills: Dike to contain spill and collect liquid for disposal. Decontaminate spill area using soapy water.

Special Emergency Information: May be harmful is inhaled, swallowed, or absorbed through the skin.

2-BROMOPENTANE - CH3CH(Br)CH2CH2CH3

Synonyms:

CAS Number: 107-81-3

DOT Number: 2343

Description: Colorless Liquid

DOT Classification: Flammable Liquid

Molecular Weight: 151.1

Melting Point: N.D.

Boiling Point: 120°C (248°F)

Vapor Density: N.D.

Specific Gravity: 1.2

Vapor Pressure: N.D.

Water Solubility: Insoluble

Chemical Incompatibilities or Instabilities: Oxidizers.

FLAMMABILITY

NFPA Hazard Code: 1-3-0

Flash Point: 32°C (90°F)

Autoignition Temp: N.D.

NFPA Classification: Class IC Flammable

LEL: N.D.

UEL: N.D.

Fire Extinguishing Methods: Carbon Dioxide, Dry Chemical, Foam, and Water.

Special Fire Fighting Considerations: Water spray from an unmanned device should be used to keep closed containers cool. Continue to cool container after fire is extinguished. For large fires, if possible, withdraw and allow to burn. Immediately withdraw if rising sound from venting device is heard or if fire is causing discoloration to the tank.

TOXICOLOGY

Odor:

Physical Contact: Mild Irritant

TLV = N.D.   **STEL** = N.D.

Odor Threshold: N.D.

IDLH = N.D.

Routes of Entry and Relative LD$_{50}$ (or LC$_{50}$)

   Inhalation   N.D.

   Ingestion   N.D.

   Skin Absorption N.D.

RTECS # RZ9800000

Symptoms of Exposure: Possible irritation to skin, eyes and upper respiratory tract; headache; nausea.

Emergency/First Aid Treatment: Remove to ventilated area; immediately remove any contaminated clothing and wash contaminated areas for 15 minutes using water. Treat supportively and observe for possible shock. If ingested, seek immediate medical aid.

Recommended Clean-Up Procedures

Personal Protection: Level B Ensemble

RCRA Waste # None

Recommended Material N.D.

Reportable Quantities: None

Spills: Remove all potential ignition sources. Dike to contain spill and collect liquid for later disposal. Dike to contain spill, absorb with non-combustible absorbent and take up using non-sparking tools. Decontaminate spill area using soapy water.

Special Emergency Information: May be harmful is inhaled, swallowed, or absorbed through the skin. Vapors are heavier than air and may travel some distance to an ignition source. Vapors are more dense than air and may settle in low lying areas.

1-BROMOPROPANE - $BrCH_2CH_2CH_3$

Synonyms: propyl bromide

CAS Number: 106-94-5
DOT Number: 2344

Description: Colorless Liquid
DOT Classification: Flammable Liquid

Molecular Weight: 123.0
Melting Point: -110°C (-166°F)
Boiling Point: 71°C (160°F)

Vapor Density: 4.3
Specific Gravity: 1.4

Vapor Pressure: 7 (143 mm Hg)
Water Solubility: Insoluble

Chemical Incompatibilities or Instabilities: Oxidizers.

FLAMMABILITY

NFPA Hazard Code: 2-3-0
Flash Point: 19°C   (67°F)
Autoignition Temp: 914°C (1677°F)

NFPA Classification: Class 1A Flammable
LEL: 4.6%
UEL: N.D.

Fire Extinguishing Methods: Carbon Dioxide, Dry Chemical, Foam, and Water.

Special Fire Fighting Considerations: Isolate for 1/2 mile if rail or tank truck is involved in at fire. Water spray from an unmanned device should be used to keep closed containers cool. Continue to cool container after fire is extinguished. For large fires, if possible, withdraw and allow to burn. Immediately withdraw if rising sound from venting device is heard or if fire is causing discoloration to the tank.

TOXICOLOGY

Odor:
Physical Contact: Irritant.
TLV = N.D.   **STEL** = N.D.

Odor Threshold: N.D.

IDLH = N.D.

Routes of Entry and Relative LD$_{50}$ (or LC$_{50}$)
    Inhalation   3   (126,500 mg/m^3/H)
    Ingestion   N.D.
    Skin Absorption   N.D.

RTECS # TX4110000

Symptoms of Exposure: Possible irritation to skin, eyes and upper respiratory tract; headache; nausea.

Emergency/First Aid Treatment: Remove to ventilated area; immediately remove any contaminated clothing and wash contaminated areas for 15 minutes using water. Treat supportively and observe for possible shock. If ingested, seek immediate medical aid.

Recommended Clean-Up Procedures

Personal Protection: Level B Ensemble
RCRA Waste #   None

Recommended Material   N.D.
Reportable Quantities: None

Spills: Remove all potential ignition sources. Dike to contain spill and collect liquid for later disposal. Decontaminate spill area using water.

Special Emergency Information: May be harmful is inhaled, swallowed, or absorbed through the skin. Vapors are heavier than air and may travel some distance to an ignition source. Vapors are more dense than air and may settle in low lying areas.

2-Bromopropane (CAS # 75-26-3, DOT # 2344, RTECS # TX4111000),
1-Chloropropane (CAS # 540-54-5, DOT # 2356, RTECS # TX4400000), and
2-Chloropropane (CAS # 75-29-6, DOT # 2356, RTECS # TX4410000) have similar chemical, physical, and toxicological properties.

3-BROMOPROPYNE - CH≡CCH$_2$Br

Synonyms: propargyl bromide

CAS Number: 106-96-7

DOT Number: 2345

Description: Colorless Liquid

DOT Classification: Flammable Liquid

Molecular Weight:

Melting Point:	-61°C (-76°F)	**Vapor Density:** 6.7	**Vapor Pressure:**	N.D.
Boiling Point:	85°C (185°F)	**Specific Gravity:** 1.6	**Water Solubility:**	Insoluble

Chemical Incompatibilities or Instabilities: Copper, mercury, silver.

FLAMMABILITY

NFPA Hazard Code: 4-3-4

Flash Point: 10°C   (50°F)

Autoignition Temp: 324°C (615°F)

NFPA Classification: Class IB Flammable

LEL: 3%

UEL: N.D.

Fire Extinguishing Methods: Carbon Dioxide, Dry Chemical, Foam.

Special Fire Fighting Considerations: Isolate for 1/2 mile if rail or tank truck is involved in at fire. Fight fire from a distance or protected location, if possible. Water spray from an unmanned device should be used to keep closed containers cool. Continue to cool container after fire is extinguished. Immediately withdraw if rising sound from venting device is heard or if fire is causing discoloration to the tank.

TOXICOLOGY

Odor: Sharp

Physical Contact: Irritant.

TLV = N.D.   **STEL** = N.D.

Odor Threshold: N.D.

IDLH = N.D.

Routes of Entry and Relative LD$_{50}$ (or LC$_{50}$)

    Inhalation   N.D.

    Ingestion   N.D.

    Skin Absorption   N.D.

RTECS # UK4375000

Symptoms of Exposure: Possible irritation to skin, eyes and upper respiratory tract; headache; nausea. Lachrymation.

Emergency/First Aid Treatment: Remove to ventilated area; immediately remove any contaminated clothing and wash contaminated areas for 15 minutes using water. Treat supportively and observe for possible shock. If ingested, seek immediate medical aid.

Recommended Clean-Up Procedures

Personal Protection: Level A Ensemble

RCRA Waste #   None

Recommended Material   N.D.

Reportable Quantities: None

Spills: Remove all potential ignition sources. Dike to contain spill, absorb with non-combustible absorbent and take up using non-sparking tools. Decontaminate spill area using soapy water.

Special Emergency Information: May be fatal if inhaled, swallowed, or absorbed through skin. Vapors are heavier than air and may travel some distance to an ignition source. Vapors are more dense than air and may settle in low lying areas.

BRUCINE

Synonyms: dimethoxystrychnine

CAS Number: 5892-11-5	**Description:** White Powder
DOT Number: 1570	**DOT Classification:** Poison B

Molecular Weight: 394.5

Melting Point: 178°C (352°F)	**Vapor Density:** N.A.	**Vapor Pressure:** 1 (0 mm Hg)
Boiling Point: °C (°F)	**Specific Gravity:** N.D.	**Water Solubility:** Slight

Chemical Incompatibilities or Instabilities: Strong oxidizers.

FLAMMABILITY

NFPA Hazard Code:	**NFPA Classification:**
Flash Point: N.D.	**LEL:** N.D.
Autoignition Temp: N.D.	**UEL:** N.D.

Fire Extinguishing Methods: Carbon Dioxide, Dry Chemical, Foam, Water.

Special Fire Fighting Considerations: Structural fire fighter protective clothing will not provide adequate protection. Fight fire from a distance or protected location, if possible. Treat all materials used or generated and equipment involved as contaminated by hazardous waste.

TOXICOLOGY

Odor:

Physical Contact: Irritant.

Odor Threshold: N. A.

TLV = N.D.   **STEL** = N.D.   **IDLH** = N.D.

Routes of Entry and Relative LD$_{50}$ (or LC$_{50}$)
Inhalation	N.D.
Ingestion	N.D.
Skin Absorption	N.D.

RTECS # EH8925000

Symptoms of Exposure: Possible irritation to skin, eyes and upper respiratory tract; headache; nausea. Irritability, muscle twitching, involuntary screaming, convulsions.

Emergency/First Aid Treatment: Remove to ventilated area; immediately remove any contaminated clothing and wash contaminated areas for 15 minutes using water. Treat supportively and observe for possible shock. If ingested, seek immediate medical aid.

Recommended Clean-Up Procedures

Personal Protection: Level B Ensemble	**Recommended Material** N.D.
RCRA Waste # P018	**Reportable Quantities:** 100 lb (1 lb)

Spills: Take up spill. Decontaminate spill area using soapy water. Treat all materials used or generated and equipment involved as contaminated by hazardous waste.

Special Emergency Information: May be fatal if inhaled, swallowed, or absorbed through skin.

1,3-BUTADIEN - $CH_2=CHCH=CH_2$

Synonyms: bivinyl, pyrrolylene, vinyl ethylene

CAS Number: 106-99-0	**Description:** Colorless Gas		
DOT Number: 1010	**DOT Classification:** Flammable Gas		

Molecular Weight: 54.1

Melting Point:	-109°C (-164°F)	**Vapor Density:** 1.9	**Vapor Pressure:**	Gas
Boiling Point:	-5°C (23°F)	**Specific Gravity:** N.A.	**Water Solubility:**	Insoluble

Chemical Incompatibilities or Instabilities: Halogens, oxygen, copper, and alloys.

FLAMMABILITY

NFPA Hazard Code:	2-4-2	**NFPA Classification:**	
Flash Point:	-76°C (-105°F)	**LEL:**	2%
Autoignition Temp:	420°C (790°F)	**UEL:**	12%

Fire Extinguishing Methods: Carbon Dioxide, Dry Chemical, Foam, Water.

Special Fire Fighting Considerations: Isolate for 1/2 mile if rail or tank truck is involved in at fire. Structural fire fighter protective clothing will not provide adequate protection. Fight fire from a distance or protected location, if possible. Immediately withdraw if rising sound from venting device is heard or if fire is causing discoloration to the tank. For large fires, if possible, withdraw and allow to burn.

TOXICOLOGY
CARCINOGEN

Odor: Rubber-like		**Odor Threshold:** 0.25 mg/m^3
Physical Contact:		
TLV = 2200 mg/m^3	**STEL** = N.D.	**IDLH** = 40,000 mg/m^3

Routes of Entry and Relative LD$_{50}$ (or LC$_{50}$)   **RTECS # EI9275000**

Inhalation	**1**	(1,140,000 mg/m^3/H)
Ingestion	**2**	(5480 mg/kg)
Skin Absorption	N.D.	

Symptoms of Exposure: Possible irritation to skin, eyes and upper respiratory tract; headache; nausea. Lachrymation. May act as a narcotic in high concentrations.

Emergency/First Aid Treatment: Remove to ventilated area; immediately remove any contaminated clothing and wash contaminated areas for 15 minutes using water. Treat supportively and observe for possible shock. If ingested, seek immediate medical aid.

Recommended Clean-Up Procedures

Personal Protection: Level B Ensemble	**Recommended Material** Butyl Rubber (+), Viton (+)
RCRA Waste # None	**Reportable Quantities:** None

Spills: Remove all potential ignition sources. Stop leak if it can safely be done. If tank cannot be sealed using non-sparking tools, allow tank to empty and vapors to dissipate.

Special Emergency Information: May be harmful is inhaled, swallowed, or absorbed through the skin. May undergo hazardous polymerization. May form unstable peroxides upon exposure to air.

BUTANE - CH3CH2CH2CH3

Synonyms:

CAS Number: 106-97-8   **Description:** Colorless Gas
DOT Number: 1011, 1075   **DOT Classification:** Flammable Gas

Molecular Weight: 58.1
Melting Point: -138°C (-216°F)   **Vapor Density:**   2.0   **Vapor Pressure:**   Gas
Boiling Point: -1°C   (30°F)   **Specific Gravity:** N.A.   **Water Solubility:**   Insoluble

Chemical Incompatibilities or Instabilities: Strong oxidizers.

FLAMMABILITY

NFPA Hazard Code: 1-4-0   **NFPA Classification:**
Flash Point: -60°C   (-76°F)   **LEL:**   1.9%
Autoignition Temp: 405°C (761°F)   **UEL:**   8.5%

Fire Extinguishing Methods: Carbon Dioxide, Dry Chemical, Foam, Water.

Special Fire Fighting Considerations: Isolate for 1/2 mile if rail or tank truck is involved in at fire. Water spray should be used to keep closed containers cool. For large fires, if possible, withdraw and allow to burn. Immediately withdraw if rising sound from venting device is heard or if fire is causing discoloration to the tank.

TOXICOLOGY

Odor: Natural gas   **Odor Threshold:** 2822 mg/m^3
Physical Contact: None
TLV = 1900 mg/m^3 **STEL** = N.D.   **IDLH** = N.D.

Routes of Entry and Relative LD$_{50}$ (or LC$_{50}$)   **RTECS # EJ4200000**
    Inhalation   1   (2632,000 mg/m^3/H)
    Ingestion   N.D.
    Skin Absorption   N.D.

Symptoms of Exposure: A simple asphyxiant.

Emergency/First Aid Treatment: Remove to ventilated area; immediately remove any contaminated clothing and wash contaminated areas for 15 minutes using water. Treat supportively and observe for possible shock. If ingested, seek immediate medical aid.

Recommended Clean-Up Procedures

Personal Protection: Level B Ensemble   **Recommended Material**   Viton (-)

RCRA Waste #   None   **Reportable Quantities:**   None

Spills: Remove all potential ignition sources. Stop leak if it can be safely be done. Ventilate to dissipate vapors.

Special Emergency Information: May be harmful is inhaled, swallowed, or absorbed through the skin. Vapors are heavier than air and may travel some distance to an ignition source. Vapors are more dense than air and may settle in low lying areas.

1-Butene (CAS # 25167-67-3, DOT # 1012), and
Cyclobutane (CAS # 287-23-0, DOT # 2601) have similar chemical, physical, and toxicological properties.

1-BUTANOL - $CH_3CH_2CH_2CH_2OH$

Synonyms: n-butyl alcohol

CAS Number: 71-36-3
DOT Number: 1120

Description: Colorless Liquid
DOT Classification: Flammable Liquid

Molecular Weight: 74.1
Melting Point: -90 C (-130 F)   **Vapor Density:** 2.6   **Vapor Pressure:** N.D.
Boiling Point: 117 C (243 F)   **Specific Gravity:** 0.81   **Water Solubility:** 9%

Chemical Incompatibilities or Instabilities: Copper and its alloys, anhydrides.

FLAMMABILITY

NFPA Hazard Code: 1-3-0   **NFPA Classification:** Class 1C Flammable
Flash Point: 37 C (98 F)   **LEL:** 1.4%
Autoignition Temp: 343 C (650 F)   **UEL:** 11.0%

Fire Extinguishing Methods: Alcohol Resistant Foam, Carbon Dioxide, Dry Chemical.

Special Fire Fighting Considerations: Isolate for 1/2 mile if rail or tank truck is involved in at fire. Water spray from an unmanned device should be used to keep closed containers cool. Continue to cool container after fire is extinguished. For large fires, if possible, withdraw and allow to burn. Immediately withdraw if rising sound from venting device is heard or if fire is causing discoloration to the tank.

TOXICOLOGY

Odor: Sweet   **Odor Threshold:** 0.15 mg/m^3
Physical Contact: Irritant
TLV = 300 mg/m^3 **STEL** = N.D.   **IDLH** = N.D.
     150 mg/m^3 (skin)

Routes of Entry and Relative LD$_{50}$ (or LC$_{50}$)   **RTECS # ED1400000**
   Inhalation   2   (96,800 mg/m^3/H)
   Ingestion   4   (7900 mg/kg)
   Skin Absorption   1   (3400 mg/kg)

Symptoms of Exposure: Possible irritation to skin, eyes and upper respiratory tract; headache; nausea.

Emergency/First Aid Treatment: Remove to ventilated area; immediately remove any contaminated clothing and wash contaminated areas for 15 minutes using water. Treat supportively and observe for possible shock. If ingested, seek immediate medical aid.

Recommended Clean-Up Procedures

Personal Protection: Level B Ensemble   **Recommended Material**   Teflon (+), Butyl Rubber (0), Polyethylene (0)

RCRA Waste #   U031   **Reportable Quantities:**   5000 lb (1 lb)

Spills: Remove all potential ignition sources. Dike to contain spill, absorb with non-combustible absorbent and take up using non-sparking tools. Decontaminate spill area using water.

Special Emergency Information: May be harmful is inhaled, swallowed, or absorbed through the skin. Vapors are heavier than air and may travel some distance to an ignition source. Vapors are more dense than air and may settle in low lying areas.

2-Butanol (CAS # 4221-99-2, DOT # 1120, RTECS # ED1750000) and
tert-Butanol (CAS # 75-65-0, DOT # 1120, RTECS # ED1925000) have similar chemical, physical, and toxicological properties.

n-BUTYL ACETATE - CH$_3$COOCH$_2$CH$_2$CH$_2$CH$_3$

Synonyms: acetic acid, butyl ester

CAS Number: 122-86-4

DOT Number: 1123

Description: Colorless Liquid

DOT Classification: Flammable Liquid

Molecular Weight: 116.2

Melting Point:	-77°C (-107°F)	**Vapor Density:** 4.0	**Vapor Pressure:** **3** (8 mm Hg)
Boiling Point:	125°C (257°F)	**Specific Gravity:** 0.88	**Water Solubility:** Insoluble

Chemical Incompatibilities or Instabilities: Oxidizers.

FLAMMABILITY

NFPA Hazard Code: **1-3-0**

Flash Point: 22°C   (72°F)

Autoignition Temp: 425°C (800°F)

NFPA Classification: Class IB Flammable

LEL: 1.7%

UEL: 7.6%

Fire Extinguishing Methods: Alcohol Resistant Foam, Carbon Dioxide, Dry Chemical.

Special Fire Fighting Considerations: Isolate for 1/2 mile if rail or tank truck is involved in at fire. Water spray from an unmanned device should be used to keep closed containers cool. Continue to cool container after fire is extinguished. For large fires, if possible, withdraw and allow to burn. Immediately withdraw if rising sound from venting device is heard or if fire is causing discoloration to the tank.

TOXICOLOGY

Odor: Banana

Physical Contact: Irritant.

TLV = 700 mg/m^3 **STEL** = 950 mg/m^3

Odor Threshold: 3.0 mg/m^3

IDLH = 48,000 mg/m^3

Routes of Entry and Relative LD$_{50}$ (or LC$_{50}$)

Inhalation	1	(360,000 mg/m^3/H)
Ingestion	1	(13,100 mg/kg)
Skin Absorption	N.D.	

RTECS # AF7350000

Symptoms of Exposure: Possible irritation to skin, eyes and upper respiratory tract; headache; nausea. May act as a narcotic in high concentrations.

Emergency/First Aid Treatment: Remove to ventilated area; immediately remove any contaminated clothing and wash contaminated areas for 15 minutes using water. Treat supportively and observe for possible shock. If ingested, seek immediate medical aid.

Recommended Clean-Up Procedures

Personal Protection: Level B Ensemble

RCRA Waste #   None

Recommended Material Teflon (0), Polyvinyl Alcohol (0)

Reportable Quantities: 5000 lb (5000 lb)

Spills: Remove all potential ignition sources. Dike to contain spill, absorb with non-combustible absorbent and take up using non-sparking tools. Decontaminate spill area using soapy water.

Special Emergency Information: May be harmful is inhaled, swallowed, or absorbed through the skin. Vapors are heavier than air and may travel some distance to an ignition source. Vapors are more dense than air and may settle in low lying areas.

n-BUTYL ACRYLATE - $CH_2=CHCOOCH_2CH_2CH_2CH_3$

Synonyms: acrylic acid butyl ester, butyl-2-propenoate

CAS Number: 141-32-3	**Description:** Colorless Liquid
DOT Number: 2348	**DOT Classification:** Flammable Liquid

Molecular Weight: 128.2

Melting Point:	-65°C (-85°F)	**Vapor Density:** 4.4	**Vapor Pressure:** 2 (3 mm Hg)
Boiling Point:	149°C (300°F)	**Specific Gravity:** 0.9	**Water Solubility:** Insoluble

Chemical Incompatibilities or Instabilities: Strong oxidizers.

FLAMMABILITY

NFPA Hazard Code:	2-2-2	**NFPA Classification:** Class II Combustible
Flash Point:	39°C (103°F)	**LEL:** 1.3%
Autoignition Temp:	279°C (534°F)	**UEL:** 9.9%

Fire Extinguishing Methods: Carbon Dioxide, Dry Chemical, Foam, and Water.

Special Fire Fighting Considerations: Isolate for 1/2 mile if rail or tank truck is involved in at fire. Water spray from an unmanned device should be used to keep closed containers cool. Continue to cool container after fire is extinguished. For large fires, if possible, withdraw and allow to burn. Immediately withdraw if rising sound from venting device is heard or if fire is causing discoloration to the tank.

TOXICOLOGY

Odor: Rancid, plastic-like　　　　　　　　　　　**Odor Threshold:** 0.005 mg/m^3
Physical Contact: Irritant
TLV = 52 mg/m^3　　　　　**STEL** = N.D.　　　　　**IDLH** = N.D.

Routes of Entry and Relative LD$_{50}$ (or LC$_{50}$)　　　**RTECS # UD3150000**

Inhalation	2	(56,000 mg/m^3/H)
Ingestion	4	(900 mg/kg)
Skin Absorption	2	(2000 mg/kg)

Symptoms of Exposure: Possible irritation to skin, eyes and upper respiratory tract; headache; nausea.

Emergency/First Aid Treatment: Remove to ventilated area; immediately remove any contaminated clothing and wash contaminated areas for 15 minutes using water. Treat supportively and observe for possible shock. If ingested, seek immediate medical aid.

Recommended Clean-Up Procedures

Personal Protection: Level B Ensemble	**Recommended Material** Teflon (0)
RCRA Waste # None	**Reportable Quantities:** None

Spills: Remove all potential ignition sources. Dike to contain spill, absorb with non-combustible absorbent and take up using non-sparking tools. Decontaminate spill area using soapy water.

Special Emergency Information: May undergo hazardous polymerization. May form unstable peroxides upon exposure to air. May be harmful is inhaled, swallowed, or absorbed through the skin. Vapors are heavier than air and may travel some distance to an ignition source. Vapors are more dense than air and may settle in low lying areas.

n-BUTYLAMINE - $CH_3CH_2CH_2CH_2NH_2$

Synonyms: 1-aminobutane, 1-butanamine

CAS Number: 109-73-9	**Description:** Colorless Liquid	
DOT Number: 1125	**DOT Classification:** Flammable Liquid	

Molecular Weight: 73.3

Melting Point: -50°C (-58°F)	**Vapor Density:** 2.5	**Vapor Pressure:** 6 (88 mm Hg)
Boiling Point: 78°C (172°F)	**Specific Gravity:** 0.73	**Water Solubility:** Soluble

Chemical Incompatibilities or Instabilities: Oxidizers, halogens, copper, aluminum.

FLAMMABILITY

NFPA Hazard Code: 3-3-0	**NFPA Classification:**
Flash Point: -12°C (10°F)	**LEL:** 1.7%
Autoignition Temp: 312°C (594°F)	**UEL:** 9.8%

Fire Extinguishing Methods: Alcohol Resistant Foam, Carbon Dioxide, Dry Chemical, Water.

Special Fire Fighting Considerations: Water spray should be used to keep closed containers cool. Isolate for 1/2 mile if rail or tank truck is involved in at fire. Immediately withdraw if rising sound from venting device is heard or if fire is causing discoloration to the tank.

TOXICOLOGY

Odor: Sour   **Odor Threshold:** 0.3 mg/m^3

Physical Contact: Material is extremely destructive to human tissues.

TLV = 15 mg/m^3 **STEL** = 15 mg/m^3 **IDLH** = 6000 mg/m^3

Routes of Entry and Relative LD$_{50}$ (or LC$_{50}$)   **RTECS # EO2975000**

Inhalation	N.D.	
Ingestion	5	(366 mg/kg)
Skin Absorption	5	(850 mg/kg)

Symptoms of Exposure: Possible burns or irritation to skin, eyes and upper respiratory tract; headache; nausea. Inhalation of vapors may be fatal by causing glottis to spasm and suffocation. Exposure may result in chemical pneumonitis or pulmonary edema.

Emergency/First Aid Treatment: Remove to ventilated area; immediately remove any contaminated clothing and wash contaminated areas for 15 minutes using water. Treat supportively and observe for possible shock. If ingested, seek immediate medical aid.

Recommended Clean-Up Procedures

Personal Protection: Level A Ensemble	**Recommended Material** Teflon (0)	
RCRA Waste # None	**Reportable Quantities:** 1000 lb (1000 lb)	

Spills: Remove all potential ignition sources. Dike to contain spill, absorb with non-combustible absorbent and take up using non-sparking tools. Decontaminate spill area using water.

Special Emergency Information: May be harmful is inhaled, swallowed, or absorbed through the skin. Vapors are heavier than air and may travel some distance to an ignition source. Vapors are more dense than air and may settle in low lying areas.

n-BUTYL ALDEHYDE - CH₃CH₂CH₂CHO

Synonyms: butanal; butyraldehyde

CAS Number: 123-72-8

DOT Number: 1129

Description: Colorless Liquid

DOT Classification: Flammable Liquid

Molecular Weight: 72.1

Melting Point: -100°C (-148°F)   **Vapor Density:** 2.5   **Vapor Pressure:** 6 (89 mm Hg)

Boiling Point: 75°C (167°F)   **Specific Gravity:** 0.9   **Water Solubility:** Soluble

Chemical Incompatibilities or Instabilities: Oxidizers, Strong acids.

FLAMMABILITY

NFPA Hazard Code: 2-3-2   **NFPA Classification:** Class 1B Flammable

Flash Point: -6°C (20°F)   **LEL:** 2.5%

Autoignition Temp: 230°C (446°F)   **UEL:** 12.5%

Fire Extinguishing Methods: Alcohol Resistant Foam, Carbon Dioxide, Dry Chemical, and Water.

Special Fire Fighting Considerations: Isolate for 1/2 mile if rail or tank truck is involved in at fire. Water spray should be used to keep closed containers cool. Fight fire from a distance or protected location, if possible. For large fires, if possible, withdraw and allow to burn. Immediately withdraw if rising sound from venting device is heard or if fire is causing discoloration to the tank.

TOXICOLOGY

Odor: Nutty   **Odor Threshold:** N.D.

Physical Contact: Irritant.

TLV = N.D.   **STEL** = N.D.   **IDLH** = N.D.

Routes of Entry and Relative LD₅₀ (or LC₅₀)   **RTECS #** ES2275000

Inhalation	2	(87,000 mg/m³/H)
Ingestion	3	(2490 mg/kg)
Skin Absorption	1	(3560 mg/kg)

Symptoms of Exposure: Possible irritation to skin, eyes and upper respiratory tract; headache; nausea. Inhalation of vapors may be fatal by causing glottis to spasm and suffocation. Exposure may result in chemical pneumonitis or pulmonary edema.

Emergency/First Aid Treatment: Remove to ventilated area; immediately remove any contaminated clothing and wash contaminated areas for 15 minutes using water. Treat supportively and observe for possible shock. If ingested, seek immediate medical aid.

Recommended Clean-Up Procedures

Personal Protection: Level B Ensemble   **Recommended Material** Butyl Rubber (+), Teflon (+)

RCRA Waste # None   **Reportable Quantities:** None

Spills: Remove all potential ignition sources. Dike to contain spill, absorb with non-combustible absorbent and take up using non-sparking tools. Decontaminate spill area using water.

Special Emergency Information: May be harmful is inhaled, swallowed, or absorbed through the skin. Vapors are heavier than air and may travel some distance to an ignition source. Vapors are more dense than air and may settle in low lying areas.

n-BUTYL BENZENE - $C_6H_5CH_2CH_2CH_2CH_3$

Synonyms: 1-phenyl butane

CAS Number: 104-51-8   **Description:** Colorless Liquid
DOT Number: 2709   **DOT Classification:** Flammable Liquid

Molecular Weight: 134.2
Melting Point: -88°C (-126°F)   **Vapor Density:**   4.6   **Vapor Pressure:**   2 (1 mm Hg)
Boiling Point: 183°C (361°F)   **Specific Gravity:** 0.9   **Water Solubility:**   Insoluble

Chemical Incompatibilities or Instabilities: Strong oxidizers.

FLAMMABILITY

NFPA Hazard Code: 2-2-0   **NFPA Classification:**   Class II Combustible
Flash Point: 59°C (139°F)   **LEL:**   0.8%
Autoignition Temp: 412°C (774°F)   **UEL:**   5.8%

Fire Extinguishing Methods: Carbon Dioxide, Dry Chemical, Foam, and Water.

Special Fire Fighting Considerations: Isolate for 1/2 mile if rail or tank truck is involved in at fire. Water spray should be used to keep closed containers cool. Fight fire from a distance or protected location, if possible. For large fires, if possible, withdraw and allow to burn. Immediately withdraw if rising sound from venting device is heard or if fire is causing discoloration to the tank.

TOXICOLOGY

Odor: **Odor Threshold:** N.D.
Physical Contact: Irritant.
TLV = N.D.   **STEL** = N.D.   **IDLH** = N.D.

Routes of Entry and Relative LD$_{50}$ (or LC$_{50}$)   **RTECS # CY9070000**
    Inhalation   N.D.
    Ingestion   N.D.
    Skin Absorption   N.D.

Symptoms of Exposure: Possible irritation to skin, eyes and upper respiratory tract; headache; nausea.

Emergency/First Aid Treatment: Remove to ventilated area; immediately remove any contaminated clothing and wash contaminated areas for 15 minutes using water. Treat supportively and observe for possible shock. If ingested, seek immediate medical aid.

Recommended Clean-Up Procedures

Personal Protection: Level B Ensemble   **Recommended Material**   N.D.

RCRA Waste #   None   **Reportable Quantities:**   None

Spills: Remove all potential ignition sources. Dike to contain spill, absorb with non-combustible absorbent and take up using non-sparking tools. Decontaminate spill area using soapy water.

Special Emergency Information: May be harmful is inhaled, swallowed, or absorbed through the skin. Vapors are heavier than air and may travel some distance to an ignition source. Vapors are more dense than air and may settle in low lying areas.

sec-butyl benzene (CAS # 135-98-8, DOT # 2709, RTECS # CY9100000) and
tert-butyl benzene (CAS # 98-06-6, DOT # 2709, RTECS # CY9120000) have similar chemical, physical, and toxicological properties.

n-BUTYL CHLORIDE - CH3CH2CH2CH2Cl

Synonyms: 1-chlorobutane

CAS Number: 109-69-3
DOT Number: 1127

Description: Colorless Liquid
DOT Classification: Flammable Liquid

Molecular Weight: 92.6
Melting Point: -123°C (-189°F)
Boiling Point: 78°C (172°F)

Vapor Density: 3.2
Specific Gravity: 0.9

Vapor Pressure: 6 (81 mm Hg)
Water Solubility: Insoluble

Chemical Incompatibilities or Instabilities: Strong oxidizers.

FLAMMABILITY

NFPA Hazard Code: 2-3-0
Flash Point: -6°C (20°F)
Autoignition Temp: 466°C (860°F)

NFPA Classification: Class 1B Flammable
LEL: 1.8%
UEL: 10.1%

Fire Extinguishing Methods: Carbon Dioxide, Dry Chemical, Foam, and Water.

Special Fire Fighting Considerations: Isolate for 1/2 mile if rail or tank truck is involved in at fire. Water spray should be used to keep closed containers cool. Fight fire from a distance or protected location, if possible. For large fires, if possible, withdraw and allow to burn. Immediately withdraw if rising sound from venting device is heard or if fire is causing discoloration to the tank.

TOXICOLOGY

Odor:
Physical Contact: Irritant.
TLV = N.D.

Odor Threshold: N.D.

STEL = N.D.

IDLH = N.D.

RTECS # EJ6300000

Routes of Entry and Relative LD$_{50}$ (or LC$_{50}$)
Inhalation   N.D.
Ingestion   3   (2670 mg/kg)
Skin Absorption   N.D.

Symptoms of Exposure: Possible irritation to skin, eyes and upper respiratory tract; headache; nausea.

Emergency/First Aid Treatment: Remove to ventilated area; immediately remove any contaminated clothing and wash contaminated areas for 15 minutes using water. Treat supportively and observe for possible shock. If ingested, seek immediate medical aid.

Recommended Clean-Up Procedures

Personal Protection: Level B Ensemble

RCRA Waste # None

Recommended Material Polyvinyl Alcohol (+)

Reportable Quantities: None

Spills: Remove all potential ignition sources. Dike to contain spill, absorb with non-combustible absorbent and take up using non-sparking tools. Decontaminate spill area using soapy water.

Special Emergency Information: May be harmful if inhaled, swallowed or absorbed through the skin. Vapors are heavier than air and may travel some distance to an ignition source. Vapors are more dense than air and may settle in low lying areas.

2-Chlorobutane (CAS # 78-86-4, RTECS # EJ6475000) has similar chemical, physical, and toxicological properties.

BUTYL ETHER - $(CH_3CH_2CH_2CH_2)_2O$

Synonyms: dibutyl ether, dibutyl oxide

CAS Number: 142-96-1	**Description:** Colorless Liquid
DOT Number: 1149	**DOT Classification:** Flammable Liquid

Molecular Weight: 130.2

Melting Point: -98°C (-144°F)	**Vapor Density:** 4.5	**Vapor Pressure:** N.D.
Boiling Point: 142°C (288°F)	**Specific Gravity:** 0.8	**Water Solubility:** Insoluble

Chemical Incompatibilities or Instabilities: Strong oxidizers.

FLAMMABILITY

NFPA Hazard Code:	2-3-1	**NFPA Classification:**	Class 1C Flammable
Flash Point:	33°C (91°F)	**LEL:**	1.5%
Autoignition Temp:	194°C (381°F)	**UEL:**	7.6%

Fire Extinguishing Methods: Carbon Dioxide, Dry Chemical, Foam, and Water.

Special Fire Fighting Considerations: Isolate for 1/2 mile if rail or tank truck is involved in at fire. Water spray should be used to keep closed containers cool. Fight fire from a distance or protected location, if possible. For large fires, if possible, withdraw and allow to burn. Immediately withdraw if rising sound from venting device is heard or if fire is causing discoloration to the tank.

TOXICOLOGY

Odor:	**Odor Threshold:** N.D.
Physical Contact: Irritant.	
TLV = N.D. **STEL** = N.D.	**IDLH** = N.D.

Routes of Entry and Relative LD$_{50}$ (or LC$_{50}$)   **RTECS #** EK5425000

Inhalation	N.D.	
Ingestion	2	(7400 mg/kg)
Skin Absorption	1	(10,000 mg/kg)

Symptoms of Exposure: Possible irritation to skin, eyes and upper respiratory tract; headache; nausea.

Emergency/First Aid Treatment: Remove to ventilated area; immediately remove any contaminated clothing and wash contaminated areas for 15 minutes using water. Treat supportively and observe for possible shock. If ingested, seek immediate medical aid.

Recommended Clean-Up Procedures

Personal Protection:	Level B Ensemble	**Recommended Material** Teflon (-)
RCRA Waste #	None	**Reportable Quantities:** None

Spills: Remove all potential ignition sources. Dike to contain spill, absorb with non-combustible absorbent and take up using non-sparking tools. Decontaminate spill area using soapy water.

Special Emergency Information: May be harmful is inhaled, swallowed, or absorbed through the skin. Vapors are heavier than air and may travel some distance to an ignition source. Vapors are more dense than air and may settle in low lying areas. May form unstable peroxides upon prolonged exposure to air.

n-BUTYL ISOCYANATE - $CH_3CH_2CH_2CH_2N=C=O$

Synonyms: BIC; isocyanic acid; butyl ester

CAS Number: 111-36-4		**Description:** Colorless Liquid	
DOT Number: 2485		**DOT Classification:** Flammable Liquid	

Molecular Weight:	99.1				
Melting Point:	°C (°F)	**Vapor Density:** 3.0	**Vapor Pressure:** **4** (12 mm Hg)		
Boiling Point:	115°C (235°F)	**Specific Gravity:** 0.9	**Water Solubility:** Decomposes		

Chemical Incompatibilities or Instabilities: Water, alcohols, acids, amines.

FLAMMABILITY

NFPA Hazard Code:	3-2-2	**NFPA Classification:** Class 1B Flammable	
Flash Point:	9°C (66°F)	**LEL:** N.D.	
Autoignition Temp:	N.D.	**UEL:** N.D.	

Fire Extinguishing Methods: Carbon Dioxide, Dry Chemical.

Special Fire Fighting Considerations: Structural fire fighters protective clothing will not provide adequate protection. May release toxic fumes upon heating (HCN). Material is extremely destructive to human tissues. Water spray from an unmanned device should be used to keep closed containers cool. Continue to cool container after fire is extinguished. Treat all materials used or generated and equipment involved as contaminated by hazardous waste.

DOT Recommended Isolation Zones:		Small Spill: 150 ft	Large Spill: 150 ft
DOT Recommended Down Wind Take Cover Distance:		Small Spill: 0.8 miles	Large Spill: 0.8 miles

TOXICOLOGY

Odor:	**Odor Threshold:** N.D.

Physical Contact: Material is extremely destructive to human tissues.

TLV = N.D.	**STEL** = N.D.	**IDLH** = N.D.

Routes of Entry and Relative LD$_{50}$ (or LC$_{50}$)   **RTECS #** NQ8250000

Inhalation	8	$(3,000 \text{ mg/m}^3/H)$
Ingestion	4	(600 mg/kg)
Skin Absorption	N.D.	

Symptoms of Exposure: Possible burns or irritation to skin, eyes and upper respiratory tract; headache; nausea. Allergic Reaction. Inhalation of vapors may be fatal by causing glottis to spasm and suffocation. Exposure may result in chemical pneumonitis or pulmonary edema.

Emergency/First Aid Treatment: Remove to ventilated area; immediately remove any contaminated clothing and wash contaminated areas for 15 minutes using water. Treat supportively and observe for possible shock. If ingested, seek immediate medical aid.

Recommended Clean-Up Procedures

Personal Protection:	Level A Ensemble	**Recommended Material**	N.D.
RCRA Waste #	None	**Reportable Quantities:**	None

Spills: Remove all potential ignition sources. Dike to contain spill, absorb with non-combustible absorbent and take up using non-sparking tools. Decontaminate spill area using dilute alkaline solution. Treat all materials used or generated and equipment involved as contaminated by hazardous waste.

Special Emergency Information: May be fatal if inhaled, swallowed, or absorbed through skin. Vapors are heavier than air and may travel some distance to an ignition source. Vapors are more dense than air and may settle in low lying areas.

tert-butyl iscyanate (CAS # 1609-86-5, DOT # 2484, RTECS # NQ8300000) has similar chemical, physical, and toxicological properties and DOT isolation zones and take cover distances.

BUTYRIC ACID - CH$_3$CH$_2$CH$_2$COOH

Synonyms: butanoic acid, n-butyric acid

CAS Number: 107-92-6
DOT Number: 2820

Description: Colorless, Viscous Liquid
DOT Classification: Corrosive Material

Molecular Weight: 88.1
Melting Point: -8°C (18°F)
Boiling Point: 164°C (526°F)

Vapor Density: 3.0
Specific Gravity: 0.9

Vapor Pressure: 2 (1 mm Hg)
Water Solubility: Soluble

Chemical Incompatibilities or Instabilities: Oxidizers

FLAMMABILITY

NFPA Hazard Code: 3-2-0
Flash Point: 72°C (161°F)
Autoignition Temp: 452°C (846°F)

NFPA Classification: Class III Combustible
LEL: 2.0%
UEL: 10.0%

Fire Extinguishing Methods: Alcohol Resistant Foam, Carbon Dioxide, Dry Chemical, and Water.

Special Fire Fighting Considerations: Water spray should be used to keep closed containers cool. Continue to cool container after fire is extinguished.

TOXICOLOGY

Odor: Rancid butter
Physical Contact: Material is extremely destructive to human tissues.
TLV = N.D.   **STEL** = N.D.

Odor Threshold: N.D.

IDLH = N.D.

Routes of Entry and Relative LD$_{50}$ (or LC$_{50}$)

Inhalation	N.D.	
Ingestion	3	(2940 mg/kg)
Skin Absorption	6	(530 mg/kg)

RTECS # ES5425000

Symptoms of Exposure: Possible burns or irritation to skin, eyes and upper respiratory tract; headache; nausea. Inhalation of vapors may be fatal by causing glottis to spasm and suffocation. Exposure may result in chemical pneumonitis or pulmonary edema.

Emergency/First Aid Treatment: Remove to ventilated area; immediately remove any contaminated clothing and wash contaminated areas for 15 minutes using water. Treat supportively and observe for possible shock. If ingested, seek immediate medical aid.

Recommended Clean-Up Procedures

Personal Protection: Level B Ensemble

RCRA Waste # None

Recommended Material Butyl Rubber (+), Viton(+)

Reportable Quantities: 5000 lb (5000 lb)

Spills: Dike to contain spill and collect liquid for disposal. Decontaminate spill area using water. Keep area isolated until vapors have dissipated.

Special Emergency Information: May be harmful is inhaled, swallowed, or absorbed through the skin.

BUTYRIC ANHYDRIDE - $CH_3CH_2CH_2COCCH_2CH_2CH_3$

$$\overset{O}{\underset{\|}{}} \quad \overset{O}{\underset{\|}{}}$$

Synonyms: butanoic acid anhydride, butyryl anhydride

CAS Number: 106-31-0   **Description:** Colorless Liquid
DOT Number: 2739   **DOT Classification:** Flammable Liquid

Molecular Weight: 158.2
Melting Point: -75 C (-103 F)   **Vapor Density:** 5.4   **Vapor Pressure:**
Boiling Point: 200 C (392 F)   **Specific Gravity:** 0.9   **Water Solubility:**   Decomposes

Chemical Incompatibilities or Instabilities: Water, strong oxidizers.

FLAMMABILITY

NFPA Hazard Code: 1-2-1-W   **NFPA Classification:** Class IIIA Combustible
Flash Point: 88 C (190 F)   **LEL:**   0.9%
Autoignition Temp: 279 C (535 F) **UEL:**   5.8%

Fire Extinguishing Methods: Alcohol Resistant Foam, Carbon Dioxide, Dry Chemical, Water.

Special Fire Fighting Considerations: Water spray should be used to keep closed containers cool. Continue to cool container after fire is extinguished.

TOXICOLOGY

Odor: **Odor Threshold:** N.D.
Physical Contact: Material is extremely destructive to human tissues.
TLV = N.D.   **STEL** = N.D.   **IDLH** = N.D.

Routes of Entry and Relative LD$_{50}$ (or LC$_{50}$)   **RTECS # ET7090000**
   Inhalation   N.D.
   Ingestion   N.D.
   Skin Absorption   N.D.

Symptoms of Exposure: Possible burns or irritation to skin, eyes and upper respiratory tract; headache; nausea. Inhalation of vapors may be fatal by causing glottis to spasm and suffocation. Exposure may result in chemical pneumonitis or pulmonary edema.

Emergency/First Aid Treatment: Remove to ventilated area; immediately remove any contaminated clothing and wash contaminated areas for 15 minutes using water. Treat supportively and observe for possible shock. If ingested, seek immediate medical aid.

Recommended Clean-Up Procedures

Personal Protection: Level B Ensemble   **Recommended Material**   N.D.

RCRA Waste #   None   **Reportable Quantities:**   None

Spills: Dike to contain spill and collect liquid for disposal. Decontaminate spill area using dilute alkaline solution. Keep area isolated until vapors have dissipated.

Special Emergency Information: May be harmful is inhaled, swallowed, or absorbed through the skin.

$$O$$
$$\|$$
CACODYLIC ACID - CH_3-As-OH
$$|$$
$$CH_3$$

Synonyms: hydroxy dimethyl arsine oxide

CAS Number: 75-60-5   **Description:** White Powder
DOT Number: 1572   **DOT Classification:** Poison B

Molecular Weight: 138.0
Melting Point: 192 C (378 F)   **Vapor Density:** N.A.   **Vapor Pressure:** N.A.
Boiling Point: C   (F)   **Specific Gravity:**   **Water Solubility:** Soluble

Chemical Incompatibilities or Instabilities: Strong oxidizers.

FLAMMABILITY

NFPA Hazard Code: **NFPA Classification:**
Flash Point: N.A.   **LEL:** N.A.
Autoignition Temp: N.A.   **UEL:** N.A.

Fire Extinguishing Methods: Carbon Dioxide, Dry Chemical, Foam, Water.

Special Fire Fighting Considerations: May release toxic fumes upon heating (arsine).

TOXICOLOGY
QUESTIONABLE CARCINOGEN

Odor: **Odor Threshold:** N.D.
Physical Contact: Irritant.
TLV = 0.2 mg (As)/m^3 **STEL** = N.D.   **IDLH** = N.D.

Routes of Entry and Relative LD$_{50}$ (or LC$_{50}$)   **RTECS # CH7525000**
    Inhalation   N.D.
    Ingestion   4   (644 mg/kg)
    Skin Absorption   N.D.

Symptoms of Exposure: Possible irritation to skin, eyes and upper respiratory tract; headache; nausea.

Emergency/First Aid Treatment: Remove to ventilated area; immediately remove any contaminated clothing and wash contaminated areas for 15 minutes using water. Treat supportively and observe for possible shock. If ingested, seek immediate medical aid.

Recommended Clean-Up Procedures

Personal Protection: Level B Ensemble   **Recommended Material**   N.D.

RCRA Waste #   U136   **Reportable Quantities:**   1 lb (1 lb)

Spills: Take up spill. Decontaminate spill area using water. Treat all materials used or generated and equipment involved as contaminated by hazardous waste.

Special Emergency Information: May be fatal if inhaled, swallowed, or absorbed through skin.

CADMIUM - Cd

Synonyms:

CAS Number: 7440-43-9   **Description:** Silver White Metal or Gray-Black Powder
DOT Number: N.D.   **DOT Classification:** N.D.

Molecular Weight: 112.4
Melting Point: 321 C (610 F)   **Vapor Density:** N.A.   **Vapor Pressure:** N.A.
Boiling Point: 765 C (1409 F)   **Specific Gravity:** 8.65   **Water Solubility:** Insoluble

Chemical Incompatibilities or Instabilities: Oxidizers, acids, alkali metals, ammonium nitrate, sulfur, zinc, hydroazoic acid.

FLAMMABILITY

NFPA Hazard Code: **NFPA Classification:**
Flash Point: N.A.   **LEL:** N.A.
Autoignition Temp: N.A.   **UEL:** N.A.

Fire Extinguishing Methods: For dust use CO_2 or dry chemical; for metal use agent suitable for surrounding fire.

Special Fire Fighting Considerations: None.

TOXICOLOGY
SUSPECTED CARCINOGEN

Odor: **Odor Threshold:** N.D.
Physical Contact: Irritant.
TLV = 0.05 mg (Cd)/m^3 **STEL** = N.D.   **IDLH** = 50 mg (Cd)/m^3

Routes of Entry and Relative LD$_{50}$ (or LC$_{50}$)   **RTECS #** EV9800000

Inhalation	10	(50 mg/m^3/H)
Ingestion	5	(225 mg/kg)
Skin Absorption	N.D.	

Symptoms of Exposure: Possible irritation to skin, eyes and upper respiratory tract; headache; nausea. Allergic Reaction.

Emergency/First Aid Treatment: Remove to ventilated area; immediately remove any contaminated clothing and wash contaminated areas for 15 minutes using water. Treat supportively and observe for possible shock. If ingested, seek immediate medical aid.

Recommended Clean-Up Procedures

Personal Protection: Level B Ensemble   **Recommended Material** Neoprene (+), Nitrile Rubber (+)

RCRA Waste # None   **Reportable Quantities:** 10 lb (1 lb)

Spills: Take up spill. Decontaminate spill area using water. Treat all materials used as contaminated by a hazardous waste.

Special Emergency Information: May be harmful is inhaled, swallowed, or absorbed through the skin.

CADMIUM CHLORIDE - CdCl$_2$

Synonyms:

CAS Number: 10108-64-2   **Description:** White Powder
DOT Number: 2570   **DOT Classification:** N.D.

Molecular Weight: 183.3
Melting Point: 568 C (1054 F)   **Vapor Density:**   N.A.   **Vapor Pressure:**   N.A.
Boiling Point: (D)   960 C (1760 F)   **Specific Gravity:** 4.0   **Water Solubility:**   Soluble

Chemical Incompatibilities or Instabilities: Alkali metals.

FLAMMABILITY

NFPA Hazard Code: **NFPA Classification:**
Flash Point: N.A.   **LEL:**   N.A.
Autoignition Temp: N.A.   **UEL:**   N.A.

Fire Extinguishing Methods: Use agent suitable for surrounding fire.

Special Fire Fighting Considerations: May emit toxic fumes.

TOXICOLOGY
CARCINOGEN

Odor: **Odor Threshold:** N.D.
Physical Contact: Irritant.
TLV = 0.05 mg (Cd)/m^3 **STEL** = N.D.   **IDLH** = N.D.

Routes of Entry and Relative LD$_{50}$ (or LC$_{50}$)   **RTECS # EV0175000**
   Inhalation   N.D.
   Ingestion   **8**   (88 mg/kg)
   Skin Absorption   N.D.

Symptoms of Exposure: Possible irritation to skin, eyes and upper respiratory tract; headache; nausea.

Emergency/First Aid Treatment: Remove to ventilated area; immediately remove any contaminated clothing and wash contaminated areas for 15 minutes using water. Treat supportively and observe for possible shock. If ingested, seek immediate medical aid.

Recommended Clean-Up Procedures

Personal Protection: Level B Ensemble   **Recommended Material**   N.D.

RCRA Waste #   None   **Reportable Quantities:**   None (1 lb)

Small Spills: Take up spill. Decontaminate spill area using water. Treat all materials used or generated and equipment involved as contaminated by hazardous waste.

Special Emergency Information: May be fatal if inhaled, swallowed, or absorbed through skin.

CALCIUM CARBIDE - CaC$_2$

Synonyms: calcium acetylide, acetylenogen

CAS Number: 75-20-7	**Description:** Gray Powder
DOT Number: 1402	**DOT Classification:** Flammable Solid

Molecular Weight: 64.1

Melting Point: 2300 C (4170 F)	**Vapor Density:** N.A.	**Vapor Pressure:** N.A.
Boiling Point: C (F)	**Specific Gravity:** 2.2	**Water Solubility:** Decomposes

Chemical Incompatibilities or Instabilities: Water, halogens, halogenated hycrocarbons, oxidizers.

FLAMMABILITY

NFPA Hazard Code: 1-3-2-~~W~~	**NFPA Classification:**
Flash Point: N.A.	**LEL:** N.A.
Autoignition Temp: N.A.	**UEL:** N.A.

Fire Extinguishing Methods: Class D extinguisher, dry sand.

Special Fire Fighting Considerations: DO NOT USE WATER. For large fires, if possible, withdraw and allow to burn.

TOXICOLOGY

Odor:	**Odor Threshold:** N.D.

Physical Contact: Material is extremely destructive to human tissues.

TLV = N.D.	**STEL** = N.D.	**IDLH** = N.D.

Routes of Entry and Relative LD$_{50}$ (or LC$_{50}$)   **RTECS # EV9400000**

Inhalation	N.D.
Ingestion	N.D.
Skin Absorption	N.D.

Symptoms of Exposure: Possible burns or irritation to skin, eyes and upper respiratory tract; headache; nausea. Laryngitis. Inhalation of vapors may be fatal by causing glottis to spasm and suffocation. Exposure may result in chemical pneumonitis or pulmonary edema.

Emergency/First Aid Treatment: Remove to ventilated area; immediately remove any contaminated clothing and wash contaminated areas for 15 minutes using water. Treat supportively and observe for possible shock. If ingested, seek immediate medical aid.

Recommended Clean-Up Procedures

Personal Protection: Level B Ensemble	**Recommended Material** N.D.
RCRA Waste # None	**Reportable Quantities:** 10 lb (5000 lb)

Spills: Remove all potential ignition sources. Take up spill. Decontaminate spill area using water

Special Emergency Information: May be harmful is inhaled, swallowed, or absorbed through the skin. Will react with water to form acetylene gas which is very flammable and unstable.

CALCIUM CHLORATE - Ca(ClO$_3$)$_2$

Synonyms:

CAS Number: 10137-74-3   **Description:** White Powder
DOT Number: 1452, 2429   **DOT Classification:** Oxidizer

Molecular Weight: 207.0
Melting Point: 100 C (212 F)   **Vapor Density:**   N.A.   **Vapor Pressure:**   N.A.
Boiling Point: C   (F)   **Specific Gravity:** 2.7   **Water Solubility:**   Soluble

Chemical Incompatibilities or Instabilities: Organic matter, strong acids, finely divided metals, phosphorous, metal sulfides, aluminum, arsenic, copper, manganese oxide, sulfur.

FLAMMABILITY

NFPA Hazard Code: **1-0-1-OX**   **NFPA Classification:**
Flash Point: N.A.   **LEL:** N.A.
Autoignition Temp: N.A.   **UEL:** N.A.

Fire Extinguishing Methods: Water ONLY.

Special Fire Fighting Considerations: Water spray should be used to keep closed containers cool. Fight fire from a distance or protected location, if possible. For large fires, if possible, withdraw and allow to burn.

TOXICOLOGY

Odor: **Odor Threshold:** N.D.
Physical Contact: Irritant.
TLV = N.D.   **STEL** = N.D.   **IDLH** = N.D.

Routes of Entry and Relative LD$_{50}$ (or LC$_{50}$)   **RTECS # EV9650000**
    Inhalation   N.D.
    Ingestion   **3**   (4500 mg/kg)
    Skin Absorption   N.D.

Symptoms of Exposure: Possible irritation to skin, eyes and upper respiratory tract; headache; nausea.

Emergency/First Aid Treatment: Remove to ventilated area; immediately remove any contaminated clothing and wash contaminated areas for 15 minutes using water. Treat supportively and observe for possible shock. If ingested, seek immediate medical aid.

Recommended Clean-Up Procedures

Personal Protection: Level B Ensemble   **Recommended Material**   N.D.

RCRA Waste #   None   **Reportable Quantities:**   None

Spills: Take up spill. Decontaminate spill area using water.

Special Emergency Information: May be harmful is inhaled, swallowed, or absorbed through the skin.

CALCIUM CHLORITE - $Ca(ClO_2)_2$

Synonyms:

CAS Number: 14674-72-7   **Description:** White Solid
DOT Number: 1453   **DOT Classification:** Oxidizer

Molecular Weight: 175.0

Melting Point:	C	(F)	**Vapor Density:** N.A.	**Vapor Pressure:**	N.A.
Boiling Point:	C	(F)	**Specific Gravity:** 2.1	**Water Solubility:**	Soluble

Chemical Incompatibilities or Instabilities: Potassium thiocyanate, organic matter.

FLAMMABILITY

NFPA Hazard Code: **NFPA Classification:**
Flash Point: N.A.   **LEL:** N.A.
Autoignition Temp: N.A.   **UEL:** N.A.

Fire Extinguishing Methods: Water ONLY.

Special Fire Fighting Considerations: Water spray should be used to keep closed containers cool. Fight fire from a distance or protected location, if possible.

TOXICOLOGY

Odor: **Odor Threshold:** N.D.
Physical Contact: Severe Irritation
TLV = N.D.   **STEL** = N.D.   **IDLH** = N.D.

Routes of Entry and Relative LD_{50} (or LC_{50})   **RTECS # EV9850000**
     Inhalation   N.D.
     Ingestion   N.D.
     Skin Absorption   N.D.

Symptoms of Exposure: Possible irritation to skin, eyes and upper respiratory tract; headache; nausea.

Emergency/First Aid Treatment: Remove to ventilated area; immediately remove any contaminated clothing and wash contaminated areas for 15 minutes using water. Treat supportively and observe for possible shock. If ingested, seek immediate medical aid.

Recommended Clean-Up Procedures

Personal Protection: Level A Ensemble   **Recommended Material** N.D.

RCRA Waste # None   **Reportable Quantities:** None

Spills: Take up spill. Decontaminate spill area using water.

Special Emergency Information: May be harmful is inhaled, swallowed, or absorbed through the skin.

CALCIUM CYANIDE - Ca(CN)$_2$

Synonyms:

CAS Number: 592-01-8   **Description:** White Powder
DOT Number: 1575   **DOT Classification:** Poison B

Molecular Weight: 92.1
Melting Point: (D) 350 C (662 F)   **Vapor Density:**   **Vapor Pressure:** N.A.
Boiling Point: C   (F)   **Specific Gravity:**   **Water Solubility:** Soluble

Chemical Incompatibilities or Instabilities: Acids.

FLAMMABILITY

NFPA Hazard Code: 3-0-1   **NFPA Classification:**
Flash Point: N.A.   **LEL:** N.A.
Autoignition Temp: N.A.   **UEL:** N.A.

Fire Extinguishing Methods: Use agent suitable for surrounding fire. DO NOT use CO$_2$.

Special Fire Fighting Considerations: May release toxic fumes upon heating (HCN). Structural fire fighters protective clothing will not provide adequate protection. Fight fire from a distance or protected location, if possible. Treat all materials used or generated and equipment involved as contaminated by hazardous waste.

TOXICOLOGY

Odor: **Odor Threshold:** N.D.
Physical Contact: Irritant.
TLV = 5 mg (CN)/m^3 **STEL** = N.D.   **IDLH** = 50 mg (CN)/m^3

Routes of Entry and Relative LD$_{50}$ (or LC$_{50}$)   **RTECS # EW0700000**
    Inhalation   N.D.
    Ingestion   **9**   (39 mg/kg)
    Skin Absorption   N.D.

Symptoms of Exposure: Possible irritation to skin, eyes and upper respiratory tract; headache; nausea. Cyanosis.

Emergency/First Aid Treatment: Remove to ventilated area; immediately remove any contaminated clothing and wash contaminated areas for 15 minutes using water. Treat supportively and observe for possible shock. If ingested, seek immediate medical aid.

Recommended Clean-Up Procedures

Personal Protection: Level B Ensemble   **Recommended Material**   Teflon (-)

RCRA Waste #   P021   **Reportable Quantities:**   10 lb (10 lb)

Spills: Take up spill. Decontaminate spill area using dilute alkaline solution. Treat all materials used or generated and equipment involved as contaminated by hazardous waste. Keep area isolated until vapors have dissipated.

Special Emergency Information: May be fatal if inhaled, swallowed, or absorbed through skin. Will react with water or acid to yield hydrogen cyanide gas.

CALCIUM FLUORIDE - CaF$_2$

Synonyms: calcium difluoride

CAS Number: 7789-75-5   **Description:** White Powder
DOT Number: N.D.   **DOT Classification:** N.D.

Molecular Weight: 78.1
Melting Point: 1360 C (2480 F)   **Vapor Density:** N.A.   **Vapor Pressure:** N.A.
Boiling Point: C   (F)   **Specific Gravity:** 3.2   **Water Solubility:** Soluble

Chemical Incompatibilities or Instabilities: Acids (will release HF).

FLAMMABILITY

NFPA Hazard Code: **NFPA Classification:**
Flash Point: N.A.   **LEL:** N.A.
Autoignition Temp: N.A.   **UEL:** N.A.

Fire Extinguishing Methods: Use agent suitable for surrounding fire.

Special Fire Fighting Considerations: May release toxic fumes upon heating (HF). Approach fire from upwind.

TOXICOLOGY

Odor: **Odor Threshold:** N.D.
Physical Contact: Irritant.
TLV = 2.5 mg (F)/m^3 **STEL** = N.D.   **IDLH** = 500 mg (F)/m^3

Routes of Entry and Relative LD$_{50}$ (or LC$_{50}$)   **RTECS # EW1760000**
    Inhalation   N.D.
    Ingestion   3   (4250 mg/kg)
    Skin Absorption N.D.

Symptoms of Exposure: Possible irritation to skin, eyes and upper respiratory tract; headache; nausea.

Emergency/First Aid Treatment: Remove to ventilated area; immediately remove any contaminated clothing and wash contaminated areas for 15 minutes using water. Treat supportively and observe for possible shock. If ingested, seek immediate medical aid.

Recommended Clean-Up Procedures

Personal Protection: Level B Ensemble   **Recommended Material** N.D.

RCRA Waste #   None   **Reportable Quantities:**   None

Spills: Take up spill. Decontaminate spill area using water.

Special Emergency Information: May be harmful is inhaled, swallowed, or absorbed through the skin.

CALCIUM HYDRIDE - CaH$_2$

Synonyms: calcium dihydride

CAS Number: 7789-78-8	**Description:** Gray Powder
DOT Number: 1404	**DOT Classification:** N.D.

Molecular Weight: 42.1

Melting Point: (D) 600 C (1112 F)	**Vapor Density:** N.A.	**Vapor Pressure:** N.A.
Boiling Point: C (F)	**Specific Gravity:** 1.9	**Water Solubility:** Decomposes

Chemical Incompatibilities or Instabilities: Acids, alcohols, water, halogens, strong oxidizers.

FLAMMABILITY

NFPA Hazard Code:	**NFPA Classification:**
Flash Point: N.A.	**LEL:** N.A.
Autoignition Temp: N.A.	**UEL:** N.A.

Fire Extinguishing Methods: Dry Chemical, Sand. **DO NOT USE WATER OR FOAM.**

Special Fire Fighting Considerations: Reacts with water to form H$_2$. Fight fire from a distance or protected location, if possible. For large fires, withdraw and let burn.

TOXICOLOGY

Odor:	**Odor Threshold:** N.D.

Physical Contact: Material is extremely destructive to human tissues.

TLV = N.D.	**STEL** = N.D.	**IDLH** = N.D.

Routes of Entry and Relative LD$_{50}$ (or LC$_{50}$)   **RTECS #** N.D.

Inhalation	N.D.
Ingestion	N.D.
Skin Absorption	N.D.

Symptoms of Exposure: Possible burns or irritation to skin, eyes and upper respiratory tract; headache; nausea. Inhalation of vapors may be fatal by causing glottis to spasm and suffocation. Exposure may result in chemical pneumonitis or pulmonary edema.

Emergency/First Aid Treatment: Remove to ventilated area; immediately remove any contaminated clothing and wash contaminated areas for 15 minutes using water. Treat supportively and observe for possible shock. If ingested, seek immediate medical aid.

Recommended Clean-Up Procedures

Personal Protection: Level B Ensemble	**Recommended Material** None
RCRA Waste # None	**Reportable Quantities:** None

Spills: Remove all potential ignition sources. Take up in non-combustible absorbent and dispose. Decontaminate spill area using water. Keep area isolated until vapors have dissipated.

Special Emergency Information: May be harmful is inhaled, swallowed, or absorbed through the skin.

CALCIUM HYPOCHLORITE - Ca(OCl)$_2$

Synonyms: hypochlorous acid, calcium salt, bleaching powder, HTH

CAS Number: 7778-54-3   **Description:** White Powder
DOT Number: 1748, 2208, 2880   **DOT Classification:** ORM C

Molecular Weight: 143.0
Melting Point: **(D)** 100 C (212 F)   **Vapor Density:** N.A.   **Vapor Pressure:** N.A.
Boiling Point: C   (F)   **Specific Gravity:** 2.4   **Water Solubility:** Decomposes

Chemical Incompatibilities or Instabilities: Acids, alcohols, heat, organics, ordinary combustibles.

FLAMMABILITY

NFPA Hazard Code: **1-0-2-OX**   **NFPA Classification:**
Flash Point: N.A.   **LEL:** N.A.
Autoignition Temp: N.A.   **UEL:** N.A.

Fire Extinguishing Methods: WATER ONLY.

Special Fire Fighting Considerations: Approach fire from upwind. Fight fire from a distance or protected location, if possible. Water spray should be used to keep closed containers cool.

TOXICOLOGY

Odor: **Odor Threshold:** N.D.
Physical Contact: Material is extremely destructive to human tissues.
TLV = N.D.   **STEL** = N.D.   **IDLH** = N.D.

Routes of Entry and Relative LD$_{50}$ (or LC$_{50}$)   **RTECS # NH3485500**
    Inhalation   N.D.
    Ingestion   4   (850 mg/kg)
    Skin Absorption   N.D.

Symptoms of Exposure: Possible burns or irritation to skin, eyes and upper respiratory tract; headache; nausea. Laryngitis.

Emergency/First Aid Treatment: Remove to ventilated area; immediately remove any contaminated clothing and wash contaminated areas for 15 minutes using water. Treat supportively and observe for possible shock. If ingested, seek immediate medical aid.

Recommended Clean-Up Procedures

Personal Protection: Level A Ensemble   **Recommended Material**   Polyvinyl Chloride (-), Natural Rubber (-)
                                                                           Neoprene (-)

RCRA Waste #   None   **Reportable Quantities:**   None

Spills: Remove all potential ignition sources. Take up with a non-combustible absorbent and dispose. Absorb with non-combustible absorbent and take up using non-sparking tools. Wash contaminated areas with large volumes of water.

Special Emergency Information: May be harmful is inhaled, swallowed, or absorbed through the skin.

CALCIUM NITRATE TETRATHYDRATE - Ca(NO₃)₂•4H₂O

Synonyms:

CAS Number: 13477-34-4   **Description:** White Powder
DOT Number: N.D.   **DOT Classification:** N.D.

Molecular Weight: 236.2
Melting Point: 561 C (1042 F)   **Vapor Density:** N.A.   **Vapor Pressure:** N.A.
Boiling Point: C   (F)   **Specific Gravity:** 2.4 **Water Solubility:** Soluble

Chemical Incompatibilities or Instabilities: Aluminum powder, acids, cyanides, thiocyanates, isothiocyanates, alkyl esters, phosphorous, tin (II) chloride, reducing agents.

FLAMMABILITY

NFPA Hazard Code: **NFPA Classification:**
Flash Point: N.A.   **LEL:** N.A.
Autoignition Temp: N.A.   **UEL:** N.A.

Fire Extinguishing Methods: WATER ONLY.

Special Fire Fighting Considerations: Fight fire from a distance or protected location, if possible. Water spray should be used to keep closed containers cool. For large fires, if possible, withdraw and allow to burn.

TOXICOLOGY

Odor: None   **Odor Threshold:** N.D.
Physical Contact: Irritant.
TLV = N.D.   **STEL** = N.D.   **IDLH** = N.D.

Routes of Entry and Relative LD$_{50}$ (or LC$_{50}$)   **RTECS # EW3000000**
   Inhalation   N.D.
   Ingestion   **3**   (3900 mg/kg))
   Skin Absorption N.D.

Symptoms of Exposure: Possible irritation to skin, eyes and upper respiratory tract; headache; nausea.

Emergency/First Aid Treatment: Remove to ventilated area; immediately remove any contaminated clothing and wash contaminated areas for 15 minutes using water. Treat supportively and observe for possible shock. If ingested, seek immediate medical aid.

Recommended Clean-Up Procedures

Personal Protection: Level B Ensemble   **Recommended Material** N.D.

RCRA Waste #   None   **Reportable Quantities:** None

Spills: Take up spill. Decontaminate spill area using water.

Special Emergency Information: May be harmful is inhaled, swallowed, or absorbed through the skin.

Anhydrous calcium nitrate (CAS # 10124-37-5, DOT # 1454, DOT Classification: Oxidizer, RTECS # EW2985000) has similar chemical, physical, and toxicological properties.

CALCIUM OXIDE - CaO

Synonyms: lime, quick lime

CAS Number: 1305-78-8   **Description:** White Powder
DOT Number: 1910   **DOT Classification:** ORM-B

Molecular Weight: 56.1
Melting Point: 2580 C (4737 F)   **Vapor Density:** N.A.   **Vapor Pressure:**   N.A.
Boiling Point: 2850 C (5162 F)   **Specific Gravity:** 3.4   **Water Solubility:**   Reacts

Chemical Incompatibilities or Instabilities: Acids, halogens, alcohols, interholgens, phosphorous pentoxide.

FLAMMABILITY

NFPA Hazard Code: **NFPA Classification:**
Flash Point: N.A.   **LEL:** N.A.
Autoignition Temp: N.A.   **UEL:** N.A.

Fire Extinguishing Methods: Use agent suitable for surrounding fire or flooding quantities of Water.

Special Fire Fighting Considerations: DO NOT use CO_2 or halogenated agents to extinguish fire. Water spray should be used to keep closed containers cool. Continue to cool container after fire is extinguished.

TOXICOLOGY

Odor: None   **Odor Threshold:** N.D.
Physical Contact: Material is extremely destructive to human tissues.
TLV = 5 mg/m^3 **STEL** = N.D.   **IDLH** = N.D.

Routes of Entry and Relative LD$_{50}$ (or LC$_{50}$)   **RTECS # EW3100000**
  Inhalation   N.D.
  Ingestion   N.D.
  Skin Absorption   N.D.

Symptoms of Exposure: Possible burns or irritation to skin, eyes and upper respiratory tract; headache; nausea. Laryngitis.

Emergency/First Aid Treatment: Remove to ventilated area; immediately remove any contaminated clothing and wash contaminated areas for 15 minutes using water. Treat supportively and observe for possible shock. If ingested, seek immediate medical aid

Recommended Clean-Up Procedures

Personal Protection: Level B Ensemble   **Recommended Material**   Natural Rubber (-), Neoprene (-),
                                                                        Nitrile Rubber (-)

RCRA Waste #   None   **Reportable Quantities:**   None

Spills: Take up spill. Decontaminate spill area using dilute acid solution.

Special Emergency Information: May be harmful is inhaled, swallowed, or absorbed through the skin. Will react with water to form calcium hydroxide.

CALCIUM PERMANGANATE - Ca(MnO$_4$)$_2$

Synonyms:

CAS Number: 10118-76-0   **Description:** Purple Powder
DOT Number: 1456   **DOT Classification:** Oxidizer

Molecular Weight: 278
Melting Point: (D)   C   (F)   **Vapor Density:** N.A.   **Vapor Pressure:** N.A.
Boiling Point: C   (F)   **Specific Gravity:** 2.4   **Water Solubility:**   Soluble

Chemical Incompatibilities or Instabilities: Acetic acid, acetic anhydride, ordinary combustibles.

FLAMMABILITY

NFPA Hazard Code: **NFPA Classification:**
Flash Point: N.A.   **LEL:** N.A.
Autoignition Temp: N.A.   **UEL:** N.A.

Fire Extinguishing Methods: WATER ONLY.

Special Fire Fighting Considerations: Fight fire from a distance or protected location, if possible. Water spray should be used to keep closed containers cool. For large fires, if possible, withdraw and allow to burn.

TOXICOLOGY

Odor: None   **Odor Threshold:** N.D.
Physical Contact: Irritant.
TLV = 5 mg (Mn)/m^3 **STEL** = N.D.   **IDLH** = N.D.

Routes of Entry and Relative LD$_{50}$ (or LC$_{50}$)   **RTECS # EW3860000**
    Inhalation   N.D.
    Ingestion   N.D.
    Skin Absorption   N.D.

Symptoms of Exposure: Possible irritation to skin, eyes and upper respiratory tract; headache; nausea.

Emergency/First Aid Treatment: Remove to ventilated area; immediately remove any contaminated clothing and wash contaminated areas for 15 minutes using water. Treat supportively and observe for possible shock. If ingested, seek immediate medical aid.

Recommended Clean-Up Procedures

Personal Protection: Level B Ensemble   **Recommended Material** N.D.

RCRA Waste #   None   **Reportable Quantities:**   None

 Spills: Take up spill. Decontaminate spill area using water.

Special Emergency Information: May be harmful is inhaled, swallowed, or absorbed through the skin.

CALCIUM PEROXIDE - CaO$_2$

Synonyms: calcium dioxide, calcium superoxide

CAS Number: 1305-79-9   **Description:** White to Yellow Powder
DOT Number: 1457   **DOT Classification:** Oxidizer

Molecular Weight: 72.1
Melting Point: **(D)** 275 C (527 F)   **Vapor Density:** N.A.   **Vapor Pressure:** N.A.
Boiling Point: N.A.   **Specific Gravity:** ---   **Water Solubility:** Decomposes

Chemical Incompatibilities or Instabilities: Water, polysulfide polymers, ordinary combustibles.

FLAMMABILITY

NFPA Hazard Code: **NFPA Classification:**
Flash Point: N.A.   **LEL:** N.A.
Autoignition Temp: N.A.   **UEL:** N.A.

Fire Extinguishing Methods: WATER ONLY. DO NOT use CO$_2$ or halogentated agents to extinguish fire.

Special Fire Fighting Considerations: Fight fire from a distance or protected location, if possible. Water spray should be used to keep closed containers cool. For large fires, if possible, withdraw and allow to burn.

TOXICOLOGY

Odor: None   **Odor Threshold:** N.D.
Physical Contact: Irritant.
TLV = N.D.   **STEL** = N.D.   **IDLH** = N.D.

Routes of Entry and Relative LD$_{50}$ (or LC$_{50}$)   **RTECS # EW3865000**
   Inhalation   N.D.
   Ingestion   N.D.
   Skin Absorption   N.D.

Symptoms of Exposure: Possible irritation to skin, eyes and upper respiratory tract, headache, nausea.

Emergency/First Aid Treatment: Remove to ventilated area; immediately remove any contaminated clothing and wash contaminated areas for 15 minutes using water. Treat supportively and observe for possible shock. If ingested, seek immediate medical aid.

Recommended Clean-Up Procedures

Personal Protection: Level B Ensemble   **Recommended Material**   N.D.

RCRA Waste #   None   **Reportable Quantities:**   None

Spills: Take up spill. Decontaminate spill area using water.

Special Emergency Information: May be harmful is inhaled, swallowed, or absorbed through the skin.

CALCIUM PHOSPHIDE - Ca_3P_2

Synonyms: tricalcium diphosphide

CAS Number: 1305-99-3	**Description:** Red Powder	
DOT Number: 1360	**DOT Classification:** Flammable Solid	

Molecular Weight: 182.2

Melting Point: 1600 C (2912 F)   **Vapor Density:** N.A.   **Vapor Pressure:** N.A.

Boiling Point: N.A.   **Specific Gravity:** 2.5   **Water Solubility:** Decomposes

Chemical Incompatibilities or Instabilities: Water, Cl_2O.

FLAMMABILITY

NFPA Hazard Code: **NFPA Classification:**

Flash Point: N.A.   **LEL:** N.A.

Autoignition Temp: N.A.   **UEL:** N.A.

Fire Extinguishing Methods: Dry Chemical, Sand. **DO NOT use Water or Foam.**

Special Fire Fighting Considerations: Structural fire fighter protective clothing will not provide adequate protection. For large fires, if possible, withdraw and allow to burn.

TOXICOLOGY

Odor: None   **Odor Threshold:** N.D.

Physical Contact: Material is extremely destructive to human tissues.

TLV = N.D.   **STEL** = N.D.   **IDLH** = N.D.

Routes of Entry and Relative LD$_{50}$ (or LC$_{50}$)   **RTECS # EW3870000**

Inhalation	N.D.
Ingestion	N.D.
Skin Absorption	N.D.

Symptoms of Exposure: Possible burns or irritation to skin, eyes and upper respiratory tract; headache; nausea.

Emergency/First Aid Treatment: Remove to ventilated area; immediately remove any contaminated clothing and wash contaminated areas for 15 minutes using water. Treat supportively and observe for possible shock. If ingested, seek immediate medical aid.

Recommended Clean-Up Procedures

Personal Protection: Level A Ensemble   **Recommended Material**   N.D.

RCRA Waste #   None   **Reportable Quantities:**   None

Spills: Do not attempt without expert supervision.

Special Emergency Information: May be fatal if inhaled, swallowed, or absorbed through skin. Reacts with water to form phosphine gas which is both highly toxic and pyrophoric.

CAMPHENE

Synonyms:

CAS Number: 79-92-5
DOT Number: 9011

Description: Colorless, Oily Crystals
DOT Classification: ORM-A

Molecular Weight: 136.3
Melting Point: 51 C   (124 F)
Boiling Point: 159 C   (318 F)

Vapor Density: N.D.
Specific Gravity: 0.8

Vapor Pressure: **2** (5 mm Hg)
Water Solubility: Insoluble

Chemical Incompatibilities or Instabilities: Strong oxidizers.

FLAMMABILITY

NFPA Hazard Code:
Flash Point: N.A.
Autoignition Temp: N.A.

NFPA Classification:
LEL: N.A.
UEL: N.A.

Fire Extinguishing Methods: Carbon Dioxide, Dry Chemical, Foam, Water.

Special Fire Fighting Considerations: Structural fire fighters protective clothing will not provide adequate protection.

TOXICOLOGY

Odor: Oily
Physical Contact: Irritant.
TLV = N.D.   **STEL** = N.D.

Odor Threshold: N.D.

IDLH = N.D.

Routes of Entry and Relative LD$_{50}$ (or LC$_{50}$)
  Inhalation   N.D.
  Ingestion   N.D.
  Skin Absorption   N.D.

RTECS # EX1055000

Symptoms of Exposure: Possible irritation to skin, eyes and upper respiratory tract, headache, nausea.

Emergency/First Aid Treatment: Remove to ventilated area; immediately remove any contaminated clothing and wash contaminated areas for 15 minutes using water. Treat supportively and observe for possible shock. If ingested, seek immediate medical aid.

Recommended Clean-Up Procedures

Personal Protection: Level B Ensemble

RCRA Waste #   None

Recommended Material   N.D.

Reportable Quantities: None

Spills: Remove all potential ignition sources. Take up with a non-combustible absorbent and dispose. Absorb with non-combustible absorbent and take up using non-sparking tools. Decontaminate spill area using soapy water.

Special Emergency Information: May be harmful is inhaled, swallowed, or absorbed through the skin.

(+)-camphene (CAS # 5794-03-6, RTECS # EX1035000) and
(-)-camphene (CAS # 5794-04-7, RTECS # EX1055000) have similar chemical, physical and toxicological properties.

CAMPHOR

Synonyms:

CAS Number: 76-22-2
DOT Number: 2717

Description: White Crystals
DOT Classification: Flammable Solid

Molecular Weight: 152.3
Melting Point: 180 C (356 F)
Boiling Point: 204 C (399 F)

Vapor Density: 5.2
Specific Gravity: 0.99

Vapor Pressure:
Water Solubility: Insoluble

Chemical Incompatibilities or Instabilities: Chlorinated solvents, strong oxidizers, p-dichlorobenzene, naphthalene.

FLAMMABILITY

NFPA Hazard Code:
Flash Point: 68 C (155 F)
Autoignition Temp: 466 C (871 F)

NFPA Classification:
LEL: 0.6%
UEL: 3.5%

Fire Extinguishing Methods: Carbon Dioxide, Dry Chemical, Foam, Water.

Special Fire Fighting Considerations: Water spray should be used to keep closed containers cool. For large fires, if possible, withdraw and allow to burn.

TOXICOLOGY

Odor: Pungent, aromatic
Physical Contact: In high concentrations, material is extremely destructive to human tissues.
TLV = 2 mg/m^3 **STEL** = N.D.

Odor Threshold: 0.02 mg/m^3

IDLH = 200 mg/m^3

Routes of Entry and Relative LD$_{50}$ (or LC$_{50}$)
  Inhalation   N.D.
  Ingestion   N.D.
  Skin Absorption   N.D.

RTECS # EX1234000

Symptoms of Exposure: Possible burns or irritation to skin, eyes and upper respiratory tract; headache; nausea. Laryngitis. Vertigo. Confusion.

Emergency/First Aid Treatment: Remove to ventilated area; immediately remove any contaminated clothing and wash contaminated areas for 15 minutes using water. Treat supportively and observe for possible shock. If ingested, seek immediate medical aid.

Recommended Clean-Up Procedures

Personal Protection: Level A Ensemble
RCRA Waste # None

Recommended Material N.D.
Reportable Quantities: None

Spills: Remove all potential ignition sources. Take up with a non-combustible absorbent and dispose. Absorb with non-combustible absorbent and take up using non-sparking tools. Decontaminate spill area using soapy water.

Special Emergency Information: May be harmful is inhaled, swallowed, or absorbed through the skin.

CAMPHOR OIL

Synonyms: oil of camphor

CAS Number: 8008-51-3	**Description:** Pale Yellow, Oily Liquid
DOT Number: 1130	**DOT Classification:** Combustible Liquid

Molecular Weight: N. A.

Melting Point: C (F) **Vapor Density:** **Vapor Pressure:**

Boiling Point: 200 C (392 F) **Specific Gravity:** 0.90 **Water Solubility:** Insoluble

Chemical Incompatibilities or Instabilities: Strong oxidizers.

FLAMMABILITY

NFPA Hazard Code: **NFPA Classification:**

Flash Point: 47 C (117 F) **LEL:** N.D.

Autoignition Temp: N.D. **UEL:** N.D.

Fire Extinguishing Methods: Carbon Dioxide, Dry Chemical, Foam, Water.

Special Fire Fighting Considerations: Isolate for 1/2 mile if rail or tank truck is involved in a fire. Water spray from an unmanned device should be used to keep closed containers cool. Continue to cool container after fire is extinguished. For large fires, if possible, withdraw and allow to burn. Immediately withdraw if rising sound from venting device is heard or if fire is causing discoloration to the tank.

TOXICOLOGY

Odor: Fragrant **Odor Threshold:** N.D.

Physical Contact: Irritant.

TLV = N.D. **STEL** = N.D. **IDLH** = N.D.

Routes of Entry and Relative LD$_{50}$ (or LC$_{50}$) **RTECS # EX1490000**

    Inhalation N.D.

    Ingestion 3 (3730 mg/kg)

    Skin Absorption N.D.

Symptoms of Exposure: Possible irritation to skin, eyes and upper respiratory tract; headache; nausea. Muscle spasms. Convulsions.

Emergency/First Aid Treatment: Remove to ventilated area; immediately remove any contaminated clothing and wash contaminated areas for 15 minutes using water. Treat supportively and observe for possible shock. If ingested, seek immediate medical aid.

Recommended Clean-Up Procedures

Personal Protection: Level B Ensemble **Recommended Material** N.D.

RCRA Waste # None **Reportable Quantities:** None

Spills: Remove all potential ignition sources. Take up with a non-combustible absorbent and dispose. Absorb with non-combustible absorbent and take up using non-sparking tools. Decontaminate spill area using soapy water.

Special Emergency Information: May be harmful is inhaled, swallowed, or absorbed through the skin. Vapors are heavier than air and may travel some distance to an ignition source. Vapors are more dense than air and may settle in low lying areas.

CAPROIC ACID - CH₃CH₂CH₂CH₂CH₂COOH

Synonyms: hexanoic acid

CAS Number: 142-62-1	**Description:** Colorless Liquid
DOT Number: 1760, 2829	**DOT Classification:** Corrosive Material

Molecular Weight: 116.2

Melting Point: -3 C (27 F)	**Vapor Density:** 4.0	**Vapor Pressure:** 1 (<1 mm Hg)
Boiling Point: 203 C (400 F)	**Specific Gravity:** 0.9	**Water Solubility:** 1%

Chemical Incompatibilities or Instabilities: Strong oxidizers.

FLAMMABILITY

NFPA Hazard Code: 2-1-0	**NFPA Classification:** Class IIIB Combustible
Flash Point: 102 C (215 F)	**LEL:** N.D.
Autoignition Temp: 380 C (716 F)	**UEL:** N.D.

Fire Extinguishing Methods: Carbon Dioxide, Dry Chemical, Foam, Water.

Special Fire Fighting Considerations: Water spray should be used to keep closed containers cool.

TOXICOLOGY

Odor: Limburger cheese

Odor Threshold: N.D.

Physical Contact: Material is extremely destructive to human tissues.

TLV = N.D.   **STEL** = N.D.

IDLH = N.D.

Routes of Entry and Relative LD₅₀ (or LC₅₀)

Inhalation	N.D.	
Ingestion	3	(3000 mg/kg)
Skin Absorption	6	(630 mg/kg)

RTECS # MO5250000

Symptoms of Exposure: Possible burns or irritation to skin, eyes and upper respiratory tract; headache; nausea.

Emergency/First Aid Treatment: Remove to ventilated area; immediately remove any contaminated clothing and wash contaminated areas for 15 minutes using water. Treat supportively and observe for possible shock. If ingested, seek immediate medical aid.

Recommended Clean-Up Procedures

Personal Protection: Level B Ensemble	**Recommended Material** Viton (-), Saranex (-)
RCRA Waste # None	**Reportable Quantities:** None

Spills: Dike to contain spill and collect liquid for later disposal.

Special Emergency Information: May be harmful is inhaled, swallowed, or absorbed through the skin.

CAPROLACTAM

Synonyms:

CAS Number: 105-60-2	**Description:** White Powder
DOT Number: N.D.	**DOT Classification:** N.D.

Molecular Weight: 113.2

Melting Point:	68 C (154 F)	**Vapor Density:** N.A.	**Vapor Pressure:**	N.A.
Boiling Point:	180 C (356 F)	**Specific Gravity:** 1.02	**Water Solubility:**	Soluble

Chemical Incompatibilities or Instabilities: Strong oxidizers, acetic acid, nitrogen trioxide.

FLAMMABILITY

NFPA Hazard Code:	**NFPA Classification:**
Flash Point: 125 C (257 F)	**LEL:** N.D.
Autoignition Temp: N.D.	**UEL:** N.D.

Fire Extinguishing Methods: Carbon Dioxide, Dry Chemical, Foam, Water.

Special Fire Fighting Considerations: Water spray should be used to keep closed containers cool. Continue to cool container after fire is extinguished.

TOXICOLOGY

Odor: **Odor Threshold:** N.D.

Physical Contact: Irritant

$TLV = 1$ mg/m^3 (Dust) $STEL = 3$ mg/m^3 (Dust) $IDLH =$ N.D.

Routes of Entry and Relative LD$_{50}$ (or LC$_{50}$)   **RTECS # CM3675000**

Inhalation	7	(7200 mg/m^3/H)
Ingestion	3	(2140 mg/kg)
Skin Absorption	3	(1410 mg/kg)

Symptoms of Exposure: Possible irritation to skin, eyes and upper respiratory tract; headache; nausea.

Emergency/First Aid Treatment: Remove to ventilated area; immediately remove any contaminated clothing and wash contaminated areas for 15 minutes using water. Treat supportively and observe for possible shock. If ingested, seek immediate medical aid.

Recommended Clean-Up Procedures

Personal Protection:	Level B Ensemble	**Recommended Material**	N.D.
RCRA Waste #	None	**Reportable Quantities:**	None

Spills: Take up in non-combustible absorbent and dispose. Decontaminate spill area using water.

Special Emergency Information: May be harmful is inhaled, swallowed, or absorbed through the skin.

CARBARYL

Synonyms: methyl carbamate-1-naphthaleneol; sevin,crag sevin, sevin 50W

CAS Number: 63-25-2	**Description:** White Powder
DOT Number: 2757	**DOT Classification:** ORM-A

Molecular Weight: 201.2

Melting Point:	142 C (288 F)	**Vapor Density:** N.A.	**Vapor Pressure:**	N.A.
Boiling Point:	C (F)	**Specific Gravity:** 1.2	**Water Solubility:**	< 1%

Chemical Incompatibilities or Instabilities: Strong oxidizers.

FLAMMABILITY

NFPA Hazard Code:		**NFPA Classification:**	
Flash Point:	N.A.	**LEL:**	N.A.
Autoignition Temp:	N.A.	**UEL:**	N.A.

Fire Extinguishing Methods: Carbon Dioxide, Dry Chemical, Foam, Water.

Special Fire Fighting Considerations: Structural fire fighters protective clothing will not provide adequate protection. Fight fire from a distance or protected location, if possible. Treat all materials used or generated and equipment involved as contaminated by hazardous waste.

TOXICOLOGY
QUESTIONABLE CARCINOGEN

Odor:	**Odor Threshold:** N.D.
Physical Contact:	
$TLV = 5$ mg/m^3 $STEL = $ N.D.	$IDLH = 600$ mg/m^3

Routes of Entry and Relative LD$_{50}$ (or LC$_{50}$)   **RTECS # FC5950000**

Inhalation	N.D.	
Ingestion	5	(230 mg/kg)
Skin Absorption	1	(2000 mg/kg)

Symptoms of Exposure: Possible irritation to skin, eyes and upper respiratory tract; headache; nausea. Blurred vision.

Emergency/First Aid Treatment: Remove to ventilated area; immediately remove any contaminated clothing and wash contaminated areas for 15 minutes using water. Treat supportively and observe for possible shock. If ingested, seek immediate medical aid.

Recommended Clean-Up Procedures

Personal Protection:	Level A Ensemble	**Recommended Material** Natural Rubber (0), Nitrile Rubber (0), Neoprene (0), Polyvinyl Alcohol (0)
RCRA Waste #	None	**Reportable Quantities:** 100 lb (100 lb)

Spills: Take up spill. Decontaminate spill area using soapy water. Treat all materials used or generated and equipment involved as contaminated by hazardous waste.

Special Emergency Information: May be fatal is inhaled, swallowed, or absorbed through the skin.

CARBAZOLE

Synonyms: dibenzopyrrole; diphenylenimide

CAS Number: 86-74-8 **Description:** Light Brown Powder
DOT Number: N.D. **DOT Classification:** N.D.

Molecular Weight: 167.2
Melting Point: 245 C (473 F) **Vapor Density:** N.A. **Vapor Pressure:** N.A.
Boiling Point: 355 C (671 F) **Specific Gravity:** 1.1 **Water Solubility:** Insoluble

Chemical Incompatibilities or Instabilities: Strong oxidizers.

FLAMMABILITY

NFPA Hazard Code: **NFPA Classification:**
Flash Point: N.A. **LEL:** N.A.
Autoignition Temp: N.A. **UEL:** N.A.

Fire Extinguishing Methods: Carbon Dioxide, Dry Chemical, Foam, Water.

Special Fire Fighting Considerations: Water spray should be used to keep closed containers cool. Continue to cool container after fire is extinguished.

TOXICOLOGY
QUESTIONABLE CARCINOGEN

Odor: **Odor Threshold:** N.D.
Physical Contact: Irritant.
TLV = N.D. **STEL** = N.D. **IDLH** = N.D.

Routes of Entry and Relative LD$_{50}$ (or LC$_{50}$) **RTECS #** FE3150000
    Inhalation   N.D.
    Ingestion   5   (500 mg/kg)
    Skin Absorption   N.D.

Symptoms of Exposure: Possible irritation to skin, eyes and upper respiratory tract; headache; nausea.

Emergency/First Aid Treatment: Remove to ventilated area; immediately remove any contaminated clothing and wash contaminated areas for 15 minutes using water. Treat supportively and observe for possible shock. If ingested, seek immediate medical aid.

Recommended Clean-Up Procedures

Personal Protection: Level B Ensemble **Recommended Material** N.D.

RCRA Waste # None **Reportable Quantities:** None

Spills: Take up spill. Decontaminate spill area using soapy water.

Special Emergency Information: May be harmful is inhaled, swallowed, or absorbed through the skin.

CARBOFURAN

Synonyms: D 1221; FMC 10242; furadan; niagra 10242; yaltox

CAS Number:	1563-66-2	**Description:**	White Powder
DOT Number:	2757	**DOT Classification:**	Poison B

Molecular Weight:

Melting Point:	152 C	(306 F)	**Vapor Density:**	N.A.	**Vapor Pressure:**	1 (0 mm Hg)
Boiling Point:	C	(F)	**Specific Gravity:**	1.2	**Water Solubility:**	< 1%

Chemical Incompatibilities or Instabilities: Alkali, strong oxidizers.

FLAMMABILITY

NFPA Hazard Code:		**NFPA Classification:**	
Flash Point:	N.A.	**LEL:**	N.A.
Autoignition Temp:	N.A.	**UEL:**	N.A.

Fire Extinguishing Methods: Carbon Dioxide, Dry Chemical, Foam, Water.

Special Fire Fighting Considerations: Approach fire from upwind. Structural fire fighters protective clothing will not provide adequate protection. Fight fire from a distance or protected location, if possible. Treat all materials used or generated and equipment involved as contaminated by hazardous waste.

TOXICOLOGY
REPRODUCTIVE TOXIN

Odor:		**Odor Threshold:** N.D.	
Physical Contact:			
TLV = 0.1 mg/m^3	**STEL** = N.D.	**IDLH** = N.D.	

Routes of Entry and Relative LD$_{50}$ (or LC$_{50}$)   **RTECS # FB9450000**

Inhalation	10	(85 mg/m^3/H)
Ingestion	10	(5 mg/kg)
Skin Absorption	5	(885 mg/kg)

Symptoms of Exposure: Possible irritation to skin, eyes and upper respiratory tract; headache; nausea. Blurred Vision. Muscle twitching.

Emergency/First Aid Treatment: Remove to ventilated area; immediately remove any contaminated clothing and wash contaminated areas for 15 minutes using water. Treat supportively and observe for possible shock. If ingested, seek immediate medical aid.

Recommended Clean-Up Procedures

Personal Protection:	Level A Ensemble	**Recommended Material**	N.D.
RCRA Waste #	None	**Reportable Quantities:**	10 lb (10 lb)

Spills: Take up spill. Decontaminate spill area using soapy Water spray should be used to keep closed containers cool. Continue to cool container after fire is extinguished. Treat all materials used or generated and equipment involved as contaminated by hazardous waste.

Special Emergency Information: May be fatal if inhaled, swallowed, or absorbed through skin.

CARBON DIOXIDE - CO$_2$

Synonyms:

CAS Number: 124-38-4   **Description:** Colorless, Odorless Gas

DOT Number: 1013 (gas), 2187 (liquefied)   **DOT Classification:** Non-Flammable Gas (1013, 2187),
1845 (solid)   ORM-A (1845)

Molecular Weight: 44.0

Melting Point: (S) -79 C (-110 F)   **Vapor Density:** 1.5   **Vapor Pressure:** **GAS**

Boiling Point: C   (F)   **Specific Gravity:** ---   **Water Solubility:** Soluble

Chemical Incompatibilities or Instabilities: ---

FLAMMABILITY

NFPA Hazard Code: **NFPA Classification:**

Flash Point: N.A.   **LEL:** N.A.

Autoignition Temp: N.A.   **UEL:** N.A.

Fire Extinguishing Methods: Use agent suitable for surrounding fire.

Special Fire Fighting Considerations: Approach fire from upwind. Water spray should be used to keep closed containers cool and keep vapors down.

TOXICOLOGY

Odor: None   **Odor Threshold:** N.D.

Physical Contact: Frostbite (1845, 2187)

TLV = 9000 mg/m^3 **STEL** = 54,000 mg/m^3 **IDLH** = 90,000 mg/m^3

Routes of Entry and Relative LD$_{50}$ (or LC$_{50}$)   **RTECS # FF6400000**

     Inhalation   N.D.

     Ingestion   N.D.

     Skin Absorption   N.D.

Symptoms of Exposure: Possible irritation to skin, eyes and upper respiratory tract; headache; nausea. Vertigo.

Emergency/First Aid Treatment: Remove to ventilated area; immediately remove any contaminated clothing and wash contaminated areas for 15 minutes using water. Treat supportively and observe for possible shock. If ingested, seek immediate medical aid.

Recommended Clean-Up Procedures

Personal Protection: Level B Ensemble   **Recommended Material**   N.D.

RCRA Waste #   None   **Reportable Quantities:**   None

Spills: Stop leak if it can safely be done. Evacuate area and allow vapors to disperse.

Special Emergency Information: May be harmful is inhaled, swallowed, or absorbed through the skin.

CARBON DISULFIDE - CS$_2$

Synonyms: carbon bisulfide

CAS Number: 75-15-0
DOT Number: 1131

Description: Colorless Liquid
DOT Classification: Flammable Liquid

Molecular Weight: 76.1
Melting Point: -111 C (-168 F)
Boiling Point: 47 C (116 F)

Vapor Density: 2.7
Specific Gravity: 1.3

Vapor Pressure: 8 (400 mm Hg)
Water Solubility: Insoluble

Chemical Incompatibilities or Instabilities: Alkali metals, zinc, oxidizers, amines, metal azides, halogens.

FLAMMABILITY

NFPA Hazard Code: 2-3-0
Flash Point: -30 C (-22 F)
Autoignition Temp: 125 C (257 F)

NFPA Classification: Class 1B Flammable
LEL: 1.3%
UEL: 50.0%

Fire Extinguishing Methods: Alcohol Resistant Foam, Carbon Dioxide, Dry Chemical, Water.

Special Fire Fighting Considerations: Structural fire fighter protective clothing will not provide adequate protection. Isolate for 1/2 mile if rail or tank truck is involved in a fire. Water spray from an unmanned device should be used to keep closed containers cool. Continue to cool container after fire is extinguished. For large fires, if possible, withdraw and allow to burn. Treat all materials used or generated and equipment involved as contaminated by hazardous waste.

TOXICOLOGY
POSSIBLE REPRODUCTIVE TOXIN

Odor: Medicinal, ethereal
Physical Contact: Irritant
TLV = 12 mg/m^3 **STEL** = 36 mg/m^3

Odor Threshold: 0.06 mg/m^3

IDLH = 1500 mg/m^3

Routes of Entry and Relative LD$_{50}$ (or LC$_{50}$)
   Inhalation   2   (50,000 mg/m^3/H)
   Ingestion   3   (3188 mg/kg)
   Skin Absorption N.D.

RTECS # FF6650000

Symptoms of Exposure: Possible irritation to skin, eyes and upper respiratory tract, headache, nausea. May act as a narcotic in high concentrations.

Emergency/First Aid Treatment: Remove to ventilated area; immediately remove any contaminated clothing and wash contaminated areas for 15 minutes using water. Treat supportively and observe for possible shock. If ingested, seek immediate medical aid.

Recommended Clean-Up Procedures

Personal Protection: Level B Ensemble

RCRA Waste # P022

Recommended Material Polyvinyl Alcohol (+), Viton (0)

Reportable Quantities: 100 lb (5000 lb)

Spills: Remove all potential ignition sources. Dike to contain spill, absorb with non-combustible absorbent and take up using non-sparking tools. Decontaminate spill area using soapy water. Treat all materials used or generated and equipment involved as contaminated by hazardous waste.

Special Emergency Information: May be harmful if inhaled, swallowed or absorbed through the skin. Vapors are heavier than air and may travel some distance to an ignition source. Vapors are more dense than air and may settle in low lying areas. Very low ignition energy; static electric sparks or friction may ignite vapors.

CARBON MONOXIDE - CO

Synonyms:

CAS Number: 630-08-0   **Description:** Colorless, Odorless Gas
DOT Number: 1016 (gas), 9202 (liquid)   **DOT Classification:** Flammable Gas

Molecular Weight: 28.0
Melting Point: -207 C (-341 F)   **Vapor Density:** 0.9   **Vapor Pressure:** 10 (> 760 mm Hg)
Boiling Point: -191 C (-312 F)   **Specific Gravity:**   **Water Solubility:** 2%

Chemical Incompatibilities or Instabilities: Halogens, interhalogens (ClF_2), ferric iron, alkali metals.

FLAMMABILITY

NFPA Hazard Code: 2-4-0   **NFPA Classification:**
Flash Point: N.A.   **LEL:** 12.5%
Autoignition Temp: 609 C (1128 F)   **UEL:** 74.8%

Fire Extinguishing Methods: Foam, Water Fog.

Special Fire Fighting Considerations: Isolate for 1/2 mile if rail or tank truck is involved in at fire. Water spray should be used to keep closed containers cool. Fight fire from a distance or protected location, if possible. For large fires, if possible, withdraw and allow to burn. Immediately withdraw if rising sound from venting device is heard or if fire is causing discoloration to the tank. Let small fires burn if the lead cannot safely be stopped.

DOT Recommended Isolation Zones: Small Spill: 150 ft   Large Spill: 150 ft
DOT Recommended Down Wind Take Cover Distance: Small Spill: 0.4 miles   Large Spill: 0.8 miles

TOXICOLOGY

Odor: None   **Odor Threshold:** N.D.
Physical Contact: Frostbite (9202)
TLV = 30 mg/m^3 **STEL** = N.D.   **IDLH** = 1500 mg/m^3

Routes of Entry and Relative LD$_{50}$ (or LC$_{50}$)   **RTECS #** FG3500000
    Inhalation   5   (8250 mg/m^3/H)
    Ingestion   N.D.
    Skin Absorption   N.D.

Symptoms of Exposure: Headache, nausea, vertigo, unconsciousness.

Emergency/First Aid Treatment: Remove to ventilated area; immediately remove any contaminated clothing and wash contaminated areas for 15 minutes using water. Treat supportively and observe for possible shock. If ingested, seek immediate medical aid.

Recommended Clean-Up Procedure

Personal Protection: Level B Ensemble   **Recommended Material** N.D.

RCRA Waste # None   **Reportable Quantities:** None

Spills: Remove all potential ignition sources. Stop leak if it can be safely be done. Evacuate area and allow vapors to disperse.

Special Emergency Information: May be fatal if inhaled, swallowed, or absorbed through skin.

CARBON TETRABROMIDE - CBr$_4$

Synonyms: tetra bromo ethane

CAS Number: 558-13-4
DOT Number: 2516

Description: White Solid
DOT Classification: Poison B

Molecular Weight: 331.7
Melting Point: 90 C (194 F)
Boiling Point: 190 C (374 F)

Vapor Density:
Specific Gravity: 3.4

Vapor Pressure:
Water Solubility: Insoluble

Chemical Incompatibilities or Instabilities: Alkali metals.

FLAMMABILITY

NFPA Hazard Code:
Flash Point: N.A.
Autoignition Temp: N.A.

NFPA Classification:
LEL: N.A.
UEL: N.A.

Fire Extinguishing Methods: Use agent suitable for surrounding fire.

Special Fire Fighting Considerations: May release toxic fumes upon heating (HBr). Approach fire from upwind.

TOXICOLOGY

Odor:
Physical Contact: Irritant.
TLV = 1.4 mg/m^3 **STEL** = 4 mg/m^3

Odor Threshold: N.D.

IDLH = N.D.

Routes of Entry and Relative LD$_{50}$ (or LC$_{50}$)
    Inhalation   N.D.
    Ingestion   **4**   (1000 mg/kg)
    Skin Absorption   N.D.

RTECS # FG4725000

Symptoms of Exposure: Possible irritation to skin, eyes and upper respiratory tract, headache, nausea.

Emergency/First Aid Treatment: Remove to ventilated area; immediately remove any contaminated clothing and wash contaminated areas for 15 minutes using water. Treat supportively and observe for possible shock. If ingested, seek immediate medical aid.

Recommended Clean-Up Procedures

Personal Protection: Level B Ensemble
RCRA Waste # None

Recommended Material Polyvinyl Alcohol (-), Viton (-)
Reportable Quantities: None

Spills: Take up with a non-combustible absorbent and dispose. Decontaminate spill area using soapy water.

Special Emergency Information: May be harmful is inhaled, swallowed, or absorbed through the skin.

CARBON TETRACHLORIDE - CCl₄

Synonyms: tetrachloromethane

CAS Number: 56-23-5   **Description:** Colorless Liquid
DOT Number: 1846   **DOT Classification:** ORM-A

Molecular Weight: 153.8
Melting Point: -23 C (-9 F)   **Vapor Density:** 5.3   **Vapor Pressure:** 6 (90 mm Hg)
Boiling Point: 77 C (171 F)   **Specific Gravity:** 1.6   **Water Solubility:** Insoluble

Chemical Incompatibilities or Instabilities: Alkali metals, powdered metals, fluorine, barium, strong oxidizers, halogens, aluminum, magnesium, zinc, peroxides.

FLAMMABILITY

NFPA Hazard Code: 3-0-0   **NFPA Classification:**
Flash Point: N.A.   **LEL:** N.A.
Autoignition Temp: N.A.   **UEL:** N.A.

Fire Extinguishing Methods: Use agent suitable for surrounding fire.

Special Fire Fighting Considerations: Approach fire from upwind. Structural fire fighters protective clothing will not provide adequate protection. Fight fire from a distance or protected location, if possible. Treat all materials used or generated and equipment involved as contaminated by hazardous waste. May release toxic fumes upon heating (phosgene).

TOXICOLOGY
CONFIRMED CARCINOGEN

Odor: Ethereal, Suffocating   **Odor Threshold:** 10.2 mg/m³
Physical Contact:
TLV = 31 mg/m³   **STEL** = 63 mg/m³   **IDLH** = 2000 mg/m³

Routes of Entry and Relative LD₅₀ (or LC₅₀)   **RTECS # FG4900000**
   Inhalation   3   (230,,000 mg/m³/H)
   Ingestion   3   (2350 mg/kg)
   Skin Absorption   N.D.

Symptoms of Exposure: Possible irritation to skin, eyes and upper respiratory tract, headache, nausea. May act as a narcotic in high concentrations.

Emergency/First Aid Treatment: Remove to ventilated area; immediately remove any contaminated clothing and wash contaminated areas for 15 minutes using water. Treat supportively and observe for possible shock. If ingested, seek immediate medical aid.

Recommended Clean-Up Procedures

Personal Protection: Level B Ensemble   **Recommended Material**   Polyvinyl Alcohol (+), Viton (+), Teflon (0)

RCRA Waste #   U211   **Reportable Quantities:**   10 lb (5000 lb)

Spills: Dike to contain spill and collect liquid for later disposal. Decontaminate spill area using soapy water. Treat all materials used or generated and equipment involved as contaminated by hazardous waste.

Special Emergency Information: May be harmful is inhaled, swallowed, or absorbed through the skin.

CARBONYL FLUORIDE - COF$_2$

Synonyms: carbon oxyfluoride

CAS Number: 353-50-4	**Description:** Colorless Gas		
DOT Number: 2417	**DOT Classification:** Poison A		

Molecular Weight: 66.0

Melting Point: -114 C (-173 F)	**Vapor Density:**	**Vapor Pressure:**	GAS
Boiling Point: -83 C (-117 F)	**Specific Gravity:**	**Water Solubility:**	Decomposes

Chemical Incompatibilities or Instabilities: Hydrolyzes in water to form HF.

FLAMMABILITY

NFPA Hazard Code:	**NFPA Classification:**	
Flash Point: N.D.	**LEL:** N.D.	
Autoignition Temp: N.D.	**UEL:** N.D.	

Fire Extinguishing Methods: Carbon Dioxide, Dry Chemical, Foam, Water.

Special Fire Fighting Considerations: Approach fire from upwind. Structural fire fighters protective clothing will not provide adequate protection. DO NOT allow water to enter container. Water spray should be used to keep closed containers cool. Keep area isolated until vapors have dissipated.

DOT Recommended Isolation Zones:	Small Spill: 1500 ft	Large Spill: 1500 ft
DOT Recommended Down Wind Take Cover Distance:	Small Spill: 5 miles	Large Spill: 5 miles

TOXICOLOGY

Odor: Pungent

Odor Threshold: N.D.

Physical Contact: Material is extremely destructive to human tissues.

$TLV = 5.4$ mg/m^3 $STEL = 13$ mg/m^3 $IDLH = $ N.D.

Routes of Entry and Relative LD$_{50}$ (or LC$_{50}$)

Inhalation	8	(1000 mg/m^3/hr)	**RTECS # FG6125000**
Ingestion	N.D.		
Skin Absorption	N.D.		

Symptoms of Exposure: Possible burns or irritation to skin, eyes and upper respiratory tract; headache; nausea. Inhalation of vapors may be fatal by causing glottis to spasm and suffocation. Exposure may result in chemical pneumonitis or pulmonary edema.

Emergency/First Aid Treatment: Remove to ventilated area; immediately remove any contaminated clothing and wash contaminated areas for 15 minutes using water. Treat supportively and observe for possible shock. If ingested, seek immediate medical aid.

Recommended Clean-Up Procedures

Personal Protection: Level A Ensemble	**Recommended Material**	N.D.
RCRA Waste # U033	**Reportable Quantities:**	1000 lb (1 lb)

Spills: Remove all potential ignition sources. Stop leak if it can be safely be done. Allow vapors to dissipate. Decontaminate spill area using dilute alkaline solution.

Special Emergency Information: May be fatal if inhaled, swallowed, or absorbed through skin.

CARBONYL SULFIDE - COS

Synonyms: carbon oxysulfide

CAS Number: 463-58-1
DOT Number: 2204

Description: Colorless Gas
DOT Classification: Poison A

Molecular Weight: 60.1
Melting Point: -138°C (-216°F)
Boiling Point: -50°C (-58°F)

Vapor Density: 2.1
Specific Gravity: 1.2

Vapor Pressure: GAS
Water Solubility: Soluble

Chemical Incompatibilities or Instabilities: Oxidizers.

FLAMMABILITY

NFPA Hazard Code: 3-4-1
Flash Point: N.A.
Autoignition Temp: N.A.

LEL: 12%
UEL: 28.5%

Fire Extinguishing Methods: Water Spray.

Special Fire Fighting Considerations: May release toxic fumes upon heating (H_2S). Approach fire from upwind. Isolate for 1/2 mile if rail or tank truck is involved in at fire. Structural fire fighters protective clothing will not provide adequate protection. Water spray from an unmanned device should be used to keep closed containers cool. Continue to cool container after fire is extinguished. For large fires, if possible, withdraw and allow to burn. Immediately withdraw if rising sound from venting device is heard or if fire is causing discoloration to the tank.

DOT Recommended Isolation Zones: Small Spill: 150 ft   Large Spill: 150 ft
DOT Recommended Down Wind Take Cover Distance   Small Spill: 0.2 miles   Large Spill: 0.2 miles

TOXICOLOGY

Odor:
Physical Contact: Material is extremely destructive to human tissues.
TLV = N.D.   **STEL** = N.D.

Odor Threshold: N.D.

IDLH = N.D.

Routes of Entry and Relative LD$_{50}$ (or LC$_{50}$)
    Inhalation   7   (2800 mg/m^3/H)
    Ingestion   N.D.
    Skin Absorption   N.D.

RTECS # FG6400000

Symptoms of Exposure: Possible burns or irritation to skin, eyes and upper respiratory tract; headache; nausea. Laryngitis. Inhalation of vapors may be fatal by causing glottis to spasm and suffocation. Exposure may result in chemical pneumonitis or pulmonary edema.

Emergency/First Aid Treatment: Remove to ventilated area; immediately remove any contaminated clothing and wash contaminated areas for 15 minutes using water. Treat supportively and observe for possible shock. If ingested, seek immediate medical aid.

Recommended Clean-Up Procedures

Personal Protection: Level A Ensemble   **Recommended Material** N.D.

RCRA Waste # None   **Reportable Quantities:** None

Spills: Remove all potential ignition sources. Stop leak if it can be safely be done. Allow vapors to dissipate.

Special Emergency Information: May be fatal if inhaled, swallowed, or absorbed through skin. Vapors are heavier than air and may travel some distance to an ignition source. Vapors are more dense than air and may settle in low lying areas.

CERIUM - Ce

Synonyms:

CAS Number: 7440-45-1   **Description:** Dark Gray Solid or Powder
DOT Number: 1333, 3078   **DOT Classification:** Flammable Solid

Molecular Weight: 140.1
Melting Point: 815°C (1499°F)   **Vapor Density:** N.A.   **Vapor Pressure:** N.A.
Boiling Point: 3257°C (5895°F)   **Specific Gravity:** 6.9   **Water Solubility:** Decomposes

Chemical Incompatibilities or Instabilities: Halogens, water, phosphorous, silicon, zinc, heat, antinomy, bismuth.

FLAMMABILITY

NFPA Hazard Code: **NFPA Classification:**
Flash Point: N.A.   **LEL:** N.A.
Autoignition Temp: N.A.   **UEL:** N.A.

Fire Extinguishing Methods: Class D Extinguisher, Dry Sand. **DO NOT USE WATER.**

Special Fire Fighting Considerations: Water spray from an unmanned device should be used to keep closed containers cool. Continue to cool container after fire is extinguished. For large fires, if possible, withdraw and allow to burn.

TOXICOLOGY

Odor: None   **Odor Threshold:** N. A.
Physical Contact: Irritant
TLV = N.D.   **STEL** = N.D.   **IDLH** = N.D.

Routes of Entry and Relative LD$_{50}$ (or LC$_{50}$)   **RTECS #** FK4850000
    Inhalation   N.D.
    Ingestion   N.D.
    Skin Absorption N.D.

Symptoms of Exposure: Possible irritation to skin, eyes and upper respiratory tract, headache, nausea.

Emergency/First Aid Treatment: Remove to ventilated area; immediately remove any contaminated clothing and wash contaminated areas for 15 minutes using water. Treat supportively and observe for possible shock. If ingested, seek immediate medical aid.

Recommended Clean-Up Procedures

Personal Protection: Level B Ensemble   **Recommended Material** N.D.

RCRA Waste #   None   **Reportable Quantities:** None

Spills: Take up and dispose under expert supervision.

Special Emergency Information: May be harmful is inhaled, swallowed, or absorbed through the skin. Cerium will spark from friction and may ignite nearby flammable liquids. Contact with water may result in release of hydrogen.

CESIUM - Cs

Synonyms:

CAS Number: 7440-46-2   **Description:** Silvery Metal, may be a Liquid
DOT Number: 1407, 1383   **DOT Classification:** Flammable Solid

Molecular Weight: 132.9
Melting Point: 29°C   (84°F)   **Vapor Density:** N.A.   **Vapor Pressure:**   N.A.
Boiling Point: 705°C (1301°F)   **Specific Gravity:** 1.9   **Water Solubility:**   Decomposes

Chemical Incompatibilities or Instabilities: Violent reactions with water, acids, halogens, oxidizers.

FLAMMABILITY

NFPA Hazard Code: **NFPA Classification:**
Flash Point: N.A.   **LEL:**   N.A.
Autoignition Temp: N.A.   **UEL:**   N.A.

Fire Extinguishing Methods: Class D Extinguisher, Dry Sand. **DO NOT USE WATER.**

Special Fire Fighting Considerations: Allow large fires to burn if possible.

TOXICOLOGY

Odor: **Odor Threshold:** N.D.
Physical Contact: Material is extremely destructive to human tissues.
TLV = N.D.   **STEL** = N.D.   **IDLH** = N.D.

Routes of Entry and Relative LD$_{50}$ (or LC$_{50}$)   **RTECS # FK9225000**
    Inhalation   N.D.
    Ingestion   N.D.
    Skin Absorption N.D.

Symptoms of Exposure: Possible burns or irritation to skin, eyes and upper respiratory tract; headache; nausea. Inhalation of vapors may be fatal by causing glottis to spasm and suffocation. Exposure may result in chemical pneumonitis or pulmonary edema.

Emergency/First Aid Treatment: Remove to ventilated area; immediately remove any contaminated clothing and wash contaminated areas for 15 minutes using water. Treat supportively and observe for possible shock. If ingested, seek immediate medical aid.

Recommended Clean-Up Procedures

Personal Protection: Level A Ensemble   **Recommended Material**   N.D.

RCRA Waste #   None   **Reportable Quantities:**   None

Spills: Take up and dispose under expert supervision.

Special Emergency Information: May be harmful is inhaled, swallowed, or absorbed through the skin.

CESIUM HYDROXIDE - CsOH

Synonyms: cesium hydrate

CAS Number: 21351-79-1   **Description:** White Powder
DOT Number: 2681, 2682   **DOT Classification:** Corrosive Material

Molecular Weight: 149.9
Melting Point: 272°C (520°F)   **Vapor Density:** N.A.   **Vapor Pressure:** N.A.
Boiling Point: °C   (°F)   **Specific Gravity:** 3.7   **Water Solubility:** Soluble

Chemical Incompatibilities or Instabilities: Acids.

FLAMMABILITY

NFPA Hazard Code: **NFPA Classification:**
Flash Point: N.A.   **LEL:** N.A.
Autoignition Temp: N.A.   **UEL:** N.A.

Fire Extinguishing Methods: Use agent suitable for surrounding fire.

Special Fire Fighting Considerations: Water spray should be used to keep closed containers cool. Continue to cool container after fire is extinguished.

TOXICOLOGY

Odor: None   **Odor Threshold:** N. A.
Physical Contact: Material is extremely destructive to human tissues.
$TLV = 2$ mg/m^3 **STEL** = N.D.   **IDLH** = N.D.

Routes of Entry and Relative LD$_{50}$ (or LC$_{50}$)   **RTECS #** FK9800000
    Inhalation   N.D.
    Ingestion   **4**   (570 mg/kg)
    Skin Absorption   N.D.

Symptoms of Exposure: Possible burns or irritation to skin, eyes and upper respiratory tract; headache; nausea.

Emergency/First Aid Treatment: Remove to ventilated area; immediately remove any contaminated clothing and wash contaminated areas for 15 minutes using water. Treat supportively and observe for possible shock. If ingested, seek immediate medical aid.

Recommended Clean-Up Procedures

Personal Protection: Level B Ensemble   **Recommended Material**   Butyl Rubber (-), Neoprene (-),
                                                                               Polyvinyl Chloride (-)

RCRA Waste #   None   **Reportable Quantities:**   None

Spills: Take up spill. Decontaminate spill area using dilute acid solution.

Special Emergency Information: May be harmful is inhaled, swallowed, or absorbed through the skin.

Cesium hydroxide monohydrate (CAS # 35103-79-8) has similar chemical, physical, and toxicological properties.

CESIUM NITRATE - CsNO3

Synonyms:

CAS Number: 7789-18-6
DOT Number: 1451

Description: White Powder
DOT Classification: Oxidizer

Molecular Weight: 194.9
Melting Point: 414°C (777°F)
Boiling Point: (D) °C (°F)

Vapor Density: N.A.
Specific Gravity: 3.7

Vapor Pressure: N.A.
Water Solubility: Soluble

Chemical Incompatibilities or Instabilities: Aluminum, organics, ordinary combustibles.

FLAMMABILITY

NFPA Hazard Code: N.A.
Flash Point: N.A.
Autoignition Temp: N.A.

LEL: N.A.
UEL: N.A.

Fire Extinguishing Methods: Use Water ONLY.

Special Fire Fighting Considerations: Fight fire from a distance or protected location, if possible. Water spray from an unmanned device should be used to keep closed containers cool. Continue to cool container after fire is extinguished. For large fires, if possible, withdraw and allow to burn.

TOXICOLOGY

Odor:
Odor Threshold: N.D.

Physical Contact: Possible irritation to skin, eyes and upper respiratory tract, headache, nausea.

TLV = N.D.　　　　**STEL** = N.D.　　　　**IDLH** = N.D.

Routes of Entry and Relative LD$_{50}$ (or LC$_{50}$)　　　**RTECS # FL0700000**

Inhalation	N.D.	
Ingestion	3	(2390 mg/kg)
Skin Absorption	N.D.	

Symptoms of Exposure: Possible burns or irritation to skin, eyes and upper respiratory tract; headache; nausea.

Emergency/First Aid Treatment: Remove to ventilated area; immediately remove any contaminated clothing and wash contaminated areas for 15 minutes using water. Treat supportively and observe for possible shock. If ingested, seek immediate medical aid.

Recommended Clean-Up Procedures

Personal Protection: Level B Ensemble　　**Recommended Material** N.D.

RCRA Waste # None　　**Reportable Quantities:** None

Spills: Take up spill. Decontaminate spill area using water.

Special Emergency Information: May be harmful is inhaled, swallowed, or absorbed through the skin.

CETYL PYRIDINIUM CHLORIDE

Synonyms: cepacol; 1-hexadecylpyridiniun chloride

CAS Number: 6004-24-6	**Description:** White Powder
DOT Number: ---	**DOT Classification:** ---

Molecular Weight: 358.1

Melting Point:	83°C (181°F)	**Vapor Density:** N.A.	**Vapor Pressure:** N.A.
Boiling Point:	°C (°F)	**Specific Gravity:**	**Water Solubility:** Soluble

Chemical Incompatibilities or Instabilities: Strong oxidizers.

FLAMMABILITY

NFPA Hazard Code:	**NFPA Classification:**
Flash Point: N.A.	**LEL:** N.A.
Autoignition Temp: N.A.	**UEL:** N.A.

Fire Extinguishing Methods: Carbon Dioxide, Dry Chemical, Foam, Water.

Special Fire Fighting Considerations: May release toxic fumes upon heating (NO_x, HCl).

TOXICOLOGY

Odor:	**Odor Threshold:** N.D.
Physical Contact: Irritant.	
TLV = N.D. **STEL** = N.D.	**IDLH** = N.D.

Routes of Entry and Relative LD_{50} (or LC_{50})   **RTECS # UU5075000**

Inhalation	N.D.
Ingestion	6 (200 mg/kg)
Skin Absorption	N.D.

Symptoms of Exposure: Possible irritation to skin, eyes and upper respiratory tract, headache, nausea.

Emergency/First Aid Treatment: Remove to ventilated area; immediately remove any contaminated clothing and wash contaminated areas for 15 minutes using water. Treat supportively and observe for possible shock. If ingested, seek immediate medical aid.

Recommended Clean-Up Procedures

Personal Protection:	Level B Ensemble	**Recommended Material**	N.D.
RCRA Waste #	None	**Reportable Quantities:**	None

Spills: Take up spill. Decontaminate spill area using water.

Special Emergency Information: May be harmful is inhaled, swallowed, or absorbed through the skin.

CHLORAL HYDRATE - $Cl_3CH(OH)_2$

Synonyms: trichloroacetaldehyde

CAS Number: 302-17-0	**Description:** Colorless Crystals
DOT Number: ---	**DOT Classification:** ---

Molecular Weight: 165.4

Melting Point: 57°C (135°F)	**Vapor Density:** N.A.	**Vapor Pressure:** N.A.
Boiling Point: 98°C (208°F)	**Specific Gravity:** 1.9	**Water Solubility:** Soluble

Chemical Incompatibilities or Instabilities: Carbonates, alkalis, alkali earths, oxidizers.

FLAMMABILITY

NFPA Hazard Code:	**NFPA Classification:**
Flash Point: N.A.	**LEL:** N.A.
Autoignition Temp: N.A.	**UEL:** N.A.

Fire Extinguishing Methods: Dry Chemical, Foam, Water.

Special Fire Fighting Considerations: Structural fire fighters protective clothing will not provide adequate protection. May release toxic fumes upon heating (HCl). Approach fire from upwind.

TOXICOLOGY
QUESTIONABLE CARCINOGEN

Odor:	**Odor Threshold:** N.D.
Physical Contact: Material is extremely destructive to human tissues.	
TLV = N.D. **STEL** = N.D.	**IDLH** = N.D.

Routes of Entry and Relative LD_{50} (or LC_{50})   **RTECS # FM8750000**

Inhalation	N.D.
Ingestion	5 (479 mg/kg)
Skin Absorption	N.D.

Symptoms of Exposure: Possible irritation to skin, eyes and upper respiratory tract; headache; nausea. May act as a narcotic in high concentrations. Symptoms may be delayed.

Emergency/First Aid Treatment: Remove to ventilated area; immediately remove any contaminated clothing and wash contaminated areas for 15 minutes using water. Treat supportively and observe for possible shock. If ingested, seek immediate medical aid.

Recommended Clean-Up Procedures

Personal Protection: Level A Ensemble	**Recommended Material**	Polyvinyl Alcohol (+), Viton (0)
RCRA Waste # U034	**Reportable Quantities:**	5000 lb (1 lb)

Spills: Take up spill. Decontaminate spill area using water. Treat all materials used as contaminated by a hazardous waste.

Special Emergency Information: May be harmful is inhaled, swallowed, or absorbed through the skin.

Chloral (DOT # 2075, CAS # 75-87-6) rapidly reacts with water to form chloral hydrate. Chloral hydrate is a controlled substance.

CHLORDANE

Synonyms:

CAS Number: 57-74-9

DOT Number: 2762

Description: Colorless to Yellow Liquid

DOT Classification: Flammable Liquid

Molecular Weight: 409.8

Melting Point: °C (°F)

Boiling Point: 175°C (347°F)

Vapor Density: N.D.

Specific Gravity: 1.6

Vapor Pressure: N.D.

Water Solubility: Insoluble

Chemical Incompatibilities or Instabilities: Strong oxidizers.

FLAMMABILITY

NFPA Hazard Code:

Flash Point: 55°C (132°F)

Autoignition Temp: N.D.

NFPA Classification: Class II Combustible

LEL: N.D.

UEL: N.D.

Fire Extinguishing Methods: Alcohol Resistant Foam, Carbon Dioxide, Dry Chemical, Water.

Special Fire Fighting Considerations: Isolate for 1/2 mile if rail or tank truck is involved in at fire. Approach fire from upwind. Structural fire fighters protective clothing will not provide adequate protection. Water spray should be used to keep closed containers cool. Immediately withdraw if rising sound from venting device is heard or if fire is causing discoloration to the tank. Treat all materials used or generated and equipment involved as contaminated by hazardous waste.

TOXICOLOGY
SUSPECTED CARCINOGEN

Odor:

Physical Contact:

TLV = 0.5 mg/m^3 (skin)   **STEL** = N.D.

Odor Threshold:

IDLH = 500 mg/m^3

RTECS # PB9800000

Routes of Entry and Relative LD$_{50}$ (or LC$_{50}$)

Inhalation	N.D.	
Ingestion	7	(200 mg/kg)
Skin Absorption	6	(780 mg/kg)

Symptoms of Exposure: Tremors, convulsions, loss of muscle control. Symptoms may be delayed.

Emergency/First Aid Treatment: Remove to ventilated area; immediately remove any contaminated clothing and wash contaminated areas for 15 minutes using water. Treat supportively and observe for possible shock. If ingested, seek immediate medical aid.

Recommended Clean-Up Procedures

Personal Protection: Level A Ensemble

RCRA Waste # U036

Recommended Material Teflon (0)

Reportable Quantities: 1 lb (1 lb)

Spills: Remove all potential ignition sources. Dike to contain spill, absorb with non-combustible absorbent and take up using non-sparking tools. Decontaminate spill area using soapy water. Treat all materials used or generated and equipment involved as contaminated by hazardous waste.

Special Emergency Information: May be harmful is inhaled, swallowed, or absorbed through the skin. Vapors are heavier than air and may travel some distance to an ignition source. Vapors are more dense than air and may settle in low lying areas.

CHLORIC ACID - HClO₃

Synonyms:

CAS Number: 7790-93-4   **Description:** Colorless Liquid
DOT Number: 2626   **DOT Classification:** Oxidizer

Molecular Weight:

Melting Point: -20°C (-4°F)	**Vapor Density:** N.D.	**Vapor Pressure:** N.D.
Boiling Point: (D) 40°C (104°F)	**Specific Gravity:** 1.3	**Water Solubility:** Soluble

Chemical Incompatibilities or Instabilities: Metal sulfides, ammonia, antimony, bismuth, iron, organics, ordinary combustibles, metal chlorides.

FLAMMABILITY

NFPA Hazard Code: **NFPA Classification:**
Flash Point: N.A.   **LEL:** N.A.
Autoignition Temp: N.A.   **UEL:** N.A.

Fire Extinguishing Methods: WATER ONLY.

Special Fire Fighting Considerations: Water spray from an unmanned device should be used to keep closed containers cool. Continue to cool container after fire is extinguished. Fight fire from a distance or protected location, if possible. Strong explosion hazard. For very large fires, withdraw and allow to burn.

TOXICOLOGY

Odor: **Odor Threshold:**
Physical Contact: Material is extremely destructive to human tissues.
TLV = N.D.   **STEL** = N.D.   **IDLH** = N.D.

Routes of Entry and Relative LD$_{50}$ (or LC$_{50}$)   **RTECS # CN9750000**
    Inhalation   N.D.
    Ingestion   N.D.
    Skin Absorption   N.D.

Symptoms of Exposure: Possible burns or irritation to skin, eyes and upper respiratory tract; headache; nausea.

Emergency/First Aid Treatment: Remove to ventilated area; immediately remove any contaminated clothing and wash contaminated areas for 15 minutes using water. Treat supportively and observe for possible shock. If ingested, seek immediate medical aid.

Recommended Clean-Up Procedures

Personal Protection: Level A Ensemble   **Recommended Material** N.D.

RCRA Waste #   None   **Reportable Quantities:**   None

Spills: Dike to contain spill and collect liquid for disposal. Take up with a non-combustible absorbent and dispose. Decontaminate spill area using water.

Special Emergency Information: May be harmful is inhaled, swallowed, or absorbed through the skin.

CHLORHYDRIN - HOCH$_2$CH(OH)CH$_2$Cl

Synonyms: 3-chloro-1,2-propandiol; glycerin-α-monochlorohydrin

CAS Number: 96-24-2
DOT Number: 2689

Description: Colorless Liquid
DOT Classification: Poison B

Molecular Weight: 110.6
Melting Point: °C (°F)
Boiling Point: (D) 213°C (415°F)

Vapor Density: N.A.
Specific Gravity: 1.3

Vapor Pressure: 1 (< 1 mm Hg)
Water Solubility: Soluble

Chemical Incompatibilities or Instabilities: Perchloric acid, strong oxidizers.

FLAMMABILITY

NFPA Hazard Code:
Flash Point: 110°C (230°F)
Autoignition Temp: N.A.

NFPA Classification:
LEL: N.A.
UEL: N.A.

Fire Extinguishing Methods: Dry Chemical, Foam, Water.

Special Fire Fighting Considerations: Approach fire from upwind. Structural fire fighters protective clothing will not provide adequate protection. Fight fire from a distance or protected location, if possible. Treat all materials used or generated and equipment involved as contaminated by hazardous waste.

TOXICOLOGY
QUESTIONABLE CARCINOGEN

Odor:
Physical Contact: Irritant
TLV = N.D.   **STEL** = N.D.

Odor Threshold: N.D.

IDLH = N.D.

RTECS # TY4025000

Routes of Entry and Relative LD$_{50}$ (or LC$_{50}$)
  Inhalation   10   (2250 mg/m^3/H)
  Ingestion   8   (55 mg/kg)
  Skin Absorption   N.D.

Symptoms of Exposure: Possible irritation to skin, eyes and upper respiratory tract, headache, nausea.

Emergency/First Aid Treatment: Remove to ventilated area; immediately remove any contaminated clothing and wash contaminated areas for 15 minutes using water. Treat supportively and observe for possible shock. If ingested, seek immediate medical aid.

Recommended Clean-Up Procedures

Personal Protection: Level A Ensemble
RCRA Waste # None

Recommended Material N.D.
Reportable Quantities: None

Spills: Dike to contain spill and collect liquid for disposal. Take up with a non-combustible absorbent and dispose. Decontaminate spill area using water.

Special Emergency Information: May be fatal if inhaled, swallowed, or absorbed through skin.

CHLORINATED CAMPHENE

Synonyms: toxaphene; camphechlor; chlorocamphene; octachlorochampene; toxaphene

CAS Number: 8001-35-2   **Description:** Yellow, Waxy Solid
DOT Number: 2761   **DOT Classification:** ORM-A

Molecular Weight: 413.8
Melting Point: 90°C (194°F)   **Vapor Density:**   **Vapor Pressure:**   1 (< 1 mm Hg)
Boiling Point: °C   (°F)   **Specific Gravity:**   **Water Solubility:** Insoluble

Chemical Incompatibilities or Instabilities: Strong oxidizers.

FLAMMABILITY

NFPA Hazard Code: **NFPA Classification:** Class IIIB Combustible
Flash Point: 135°C (275°F)   **LEL:** N.D.
Autoignition Temp: °C   (°F)   **UEL:** N.D.

Fire Extinguishing Methods: Use agent suitable for surrounding fire.

Special Fire Fighting Considerations: Structural fire fighters protective clothing will not provide adequate protection. Approach fire from upwind. Fight fire from a distance or protected location, if possible. Treat all materials used or generated and equipment involved as contaminated by hazardous waste.

TOXICOLOGY
CARCINOGEN

Odor: Camphor-like   **Odor Threshold:** N.D.
Physical Contact: Irritant.
TLV = 0.5 mg/m^3 **STEL** = 1 mg/m^3 **IDLH** = 200 mg/m^3.

Routes of Entry and Relative LD$_{50}$ (or LC$_{50}$)   **RTECS # RW5250000**
  Inhalation   N.D.
  Ingestion   9   (50 mg/kg)
  Skin Absorption   4   (1025 mg/kg)

Symptoms of Exposure: Possible irritation to skin, eyes and upper respiratory tract; headache; nausea. Confusion. Convulsions.

Emergency/First Aid Treatment: Remove to ventilated area; immediately remove any contaminated clothing and wash contaminated areas for 15 minutes using water. Treat supportively and observe for possible shock. If ingested, seek immediate medical aid.

Recommended Clean-Up Procedures

Personal Protection: Level A Ensemble   **Recommended Material** N.D.

RCRA Waste #   D015   **Reportable Quantities:**   100 lb (1 lb)

Spills: Take up spill. Decontaminate spill area using soapy water. Treat all materials used or generated and equipment involved as contaminated by hazardous waste.

Special Emergency Information: May be fatal is inhaled, swallowed, or absorbed through the skin.

CHLORINE - Cl$_2$

Synonyms:

CAS Number: 7782-50-5   **Description:** Yellow-Green Gas
DOT Number: 1017   **DOT Classification:** Nonflammable Gas

Molecular Weight: 70.9
Melting Point: -101°C (-150°F)   **Vapor Density:** 2.5   **Vapor Pressure:**   GAS
Boiling Point: -35°C (-31°F)   **Specific Gravity:**   **Water Solubility:**   Soluble

Chemical Incompatibilities or Instabilities: Organics, ammonia, finely divided metals, sulfides, hydrides.

FLAMMABILITY

NFPA Hazard Code: 3-0-0-~~W~~   **NFPA Classification:**
Flash Point: N.A.   **LEL:** N.A.
Autoignition Temp: N.A.   **UEL:** N.A.

Fire Extinguishing Methods: WATER ONLY.

Special Fire Fighting Considerations: Structural fire fighters protective clothing will not provide adequate protection. Water spray should be used to keep closed containers cool. Isolate for 150 feet if rail or tank truck is involved in at fire. For large fires, if possible, withdraw and allow to burn.

DOT Recommended Isolation Zones: Small Spill: 900 ft   Large Spill: 1500 ft
DOT Recommended Down Wind Take Cover Distance: Small Spill: 3 miles   Large Spill: 5 miles

TOXICOLOGY

Odor: Bleach   **Odor Threshold:** 0.06 mg/m^3
Physical Contact: Material is extremely destructive to human tissues.
TLV = 1.5 mg/m^3 **STEL** = 3 mg/m^3 **IDLH** = 90 mg/m^3

Routes of Entry and Relative LD$_{50}$ (or LC$_{50}$)   **RTECS # FO2100000**
    Inhalation   5   (850 mg/m^3/H)
    Ingestion   N.D.
    Skin Absorption N.D.

Symptoms of Exposure: Possible burns or irritation to skin, eyes and upper respiratory tract; headache; nausea. Inhalation of vapors may be fatal by causing glottis to spasm and suffocation. Exposure may result in chemical pneumonitis or pulmonary edema.

Emergency/First Aid Treatment: Remove to ventilated area; immediately remove any contaminated clothing and wash contaminated areas for 15 minutes using water. Treat supportively and observe for possible shock. If ingested, seek immediate medical aid.

Recommended Clean-Up Procedures

Personal Protection: Level A Ensemble   **Recommended Material**   Neoprene (+), Teflon (+), Saranex (+)

RCRA Waste #   None   **Reportable Quantities:**   10 lb (10 lb)

Spills: Stop leak if it can be done safely. Allow vapors to dissipate. Water fog may be used to keep vapors down. Do not allow water to enter container.

Special Emergency Information: May be fatal if inhaled, swallowed, or absorbed through skin.

CHLORINE DIOXIDE - ClO₂

Synonyms:

CAS Number: 10049-04-4   **Description:** Red-Orange Crystals
DOT Number: 9191 (Frozen/hydrated)   **DOT Classification:** Oxidizer (Hydrated and Frozen)

Molecular Weight: 67.5
Melting Point: -59°C (-74°F)   **Vapor Density:**   **Vapor Pressure:** 10 (> 760 mm Hg)
Boiling Point: (EX) 10°C (50°F)   **Specific Gravity:** 3.1   **Water Solubility:** < 1%

Chemical Incompatibilities or Instabilities: Heat, light, combustibles, hydrogen, carbon monoxide, non-metals.

FLAMMABILITY

NFPA Hazard Code: **NFPA Classification:**
Flash Point: N.A.   **LEL:** N.A.
Autoignition Temp: N.A.   **UEL:** N.A.

Fire Extinguishing Methods: WATER ONLY.

Special Fire Fighting Considerations: Isolate for 1/2 mile if rail or tank truck is involved in at fire. Water spray should be used to keep closed containers cool. Keep away from ends of tank. For large fires, if possible, withdraw and allow to burn. May release toxic fumes upon heating (HCl).

DOT Recommended Isolation Zones: Small Spill: 1500 ft   Large Spill: 1500 ft
DOT Recommended Down Wind Take Cover Distance: Small Spill: 5 miles   Large Spill: 5 miles

TOXICOLOGY
POSSIBLE REPRODUCTIVE TOXIN

Odor: Chlorine, bleach   **Odor Threshold:** 40 mg/m³
Physical Contact: Material is extremely destructive to human tissues.
TLV = 0.3 mg/m³   **STEL** = 0.8 mg/m³   **IDLH** = 28 mg/m³

Routes of Entry and Relative LD₅₀ (or LC₅₀)   **RTECS #** FO3000000
    Inhalation   10   (345 mg/m³/H)   **FN9750000** (frozen/hydrated)
    Ingestion   6   (292 mg/kg)
    Skin Absorption   N.D.

Symptoms of Exposure: Possible burns or irritation to skin, eyes and upper respiratory tract; headache; nausea. Inhalation of vapors may be fatal by causing glottis to spasm and suffocation. Exposure may result in chemical pneumonitis or pulmonary edema.

Emergency/First Aid Treatment: Remove to ventilated area; immediately remove any contaminated clothing and wash contaminated areas for 15 minutes using water. Treat supportively and observe for possible shock. If ingested, seek immediate medical aid.

Recommended Clean-Up Procedures

Personal Protection: Level A Ensemble   **Recommended Material** N.D.

RCRA Waste # None   **Reportable Quantities:** None

Spills: Use large amounts of water to dilute spill. Dike to contain spill and collect liquid for disposal. Decontaminate spill area using dilute alkaline solution.

Special Emergency Information: May be fatal if inhaled, swallowed, or absorbed through skin. Reacts with water to form HCl.

CHLORINE TRIFLUORIDE - ClF$_3$

Synonyms:

CAS Number: 7790-91-2   **Description:** Colorless Gas (Yellow-Green Liquid below 53°F)
DOT Number: 1749   **DOT Classification:** Oxidizer

Molecular Weight: 92.5
Melting Point: -83°C (-105°F)   **Vapor Density:**   3.1   **Vapor Pressure:**   10 (Gas)
Boiling Point: 11°C   (53°F)   **Specific Gravity:** 1.8 < 50°F **Water Solubility:**   Decomposes

Chemical Incompatibilities or Instabilities: Ordinary combustibles, organics, water, silicates, halogens, finely divided solids, acids, metal salts, metal oxides, most metals, boron alloys.

FLAMMABILITY

NFPA Hazard Code: 4-0-3-W-OX   **NFPA Classification:**
Flash Point: N.A.   **LEL:**   N.A.
Autoignition Temp: N.A.   **UEL:**   N.A.

Fire Extinguishing Methods: DO NOT USE WATER. Use Dry Chemical or Carbon Dioxide.

Special Fire Fighting Considerations: Structural fire fighters protective clothing will not provide adequate protection. Approach fire from upwind. Fight fire from a distance or protected location, if possible. Water spray from an unmanned device should be used to keep closed containers cool. Continue to cool container after fire is extinguished. For large fires, if possible, withdraw and allow to burn.

TOXICOLOGY

Odor: **Odor Threshold:**
Physical Contact: Material is extremely destructive to human tissues.
TLV = 0.4 mg/m^3 **STEL** = N.D.   **IDLH** = 75 mg/m^3

Routes of Entry and Relative LD$_{50}$ (or LC$_{50}$)   **RTECS # FO2800000**
   Inhalation   10   (750 mg/m^3/H)
   Ingestion   N.D.
   Skin Absorption   N.D.

Symptoms of Exposure: Possible burns or irritation to skin, eyes and upper respiratory tract; headache; nausea. Inhalation of vapors may be fatal by causing glottis to spasm and suffocation. Exposure may result in chemical pneumonitis or pulmonary edema.

Emergency/First Aid Treatment: Remove to ventilated area; immediately remove any contaminated clothing and wash contaminated areas for 15 minutes using water. Treat supportively and observe for possible shock. If ingested, seek immediate medical aid.

Recommended Clean-Up Procedures

Personal Protection: Level A Ensemble   **Recommended Material**   N.D.

RCRA Waste #   None   **Reportable Quantities:**   None

Spills: Use large amounts of water to dilute spill. Dike to contain spill and collect liquid for disposal. DO NOT get water inside container. Decontaminate spill area using dilute alkaline solution.

Special Emergency Information: May be fatal if inhaled, swallowed, or absorbed through skin. May react explosively with fuels and water.

CHLORINE PENTAFLUORIDE - ClF$_5$

Synonyms:

CAS Number: 13637-63-3   **Description:** Gas
DOT Number: 2548   **DOT Classification:** Poison A

Molecular Weight: 130.5
Melting Point: N.D.   **Vapor Density:** N.D.   **Vapor Pressure:** N.D.
Boiling Point: N.D.   **Specific Gravity:** N.D.   **Water Solubility:** Decomposes

Chemical Incompatibilities or Instabilities: Metals, water, nitric acid, ordinary combustibles.

FLAMMABILITY

NFPA Hazard Code: **NFPA Classification:**
Flash Point: N.A.   **LEL:** N.A.
Autoignition Temp: N.A.   **UEL:** N.A.

Fire Extinguishing Methods: Carbon Dioxide or Dry Chemical. **DO NOT USE WATER.**

Special Fire Fighting Considerations: Structural fire fighters protective clothing will not provide adequate protection. Approach fire from upwind. If water must be used, fight fire from a distance or protected location. Do not allow water to enter container. For large fires, if possible, withdraw and allow to burn.

DOT Recommended Isolation Zones: Small Spill: 900 ft   Large Spill: 1500 ft
DOT Recommended Down Wind Take Cover Distance: Small Spill: 3 miles   Large Spill: 5 miles

TOXICOLOGY

Odor: **Odor Threshold:** N.D.
Physical Contact: Material is extremely destructive to human tissues.
TLV = 2.25 mg (F)/m^3 **STEL** = N.D.   **IDLH** = 500 mg/m^3

Routes of Entry and Relative LD$_{50}$ (or LC$_{50}$)   **RTECS # FO2975000**
  Inhalation   10   (650 mg/m^3/H)
  Ingestion   N.D.
  Skin Absorption   N.D.

Symptoms of Exposure: Possible burns or irritation to skin, eyes and upper respiratory tract; headache; nausea. Inhalation of vapors may be fatal by causing glottis to spasm and suffocation. Exposure may result in chemical pneumonitis or pulmonary edema.

Emergency/First Aid Treatment: Remove to ventilated area; immediately remove any contaminated clothing and wash contaminated areas for 15 minutes using water. Treat supportively and observe for possible shock. If ingested, seek immediate medical aid.

Recommended Clean-Up Procedures

Personal Protection: Level A Ensemble   **Recommended Material**   N.D.

RCRA Waste #   None   **Reportable Quantities:**   None

Spills: Stop leak if it can safely be done. Keep area isolated until vapors have dissipated.

Special Emergency Information: May be fatal if inhaled, swallowed, or absorbed through skin.

CHLOROACETALDEHYDE - $ClCH_2CHO$

Synonyms:

CAS Number: 107-20-0   **Description:** Colorless Liquid
DOT Number: 2232   **DOT Classification:** Poison B

Molecular Weight: 78.5
Melting Point: 46°C (115°F)   **Vapor Density:** N.A.   **Vapor Pressure:** 6 (100 mm Hg)
Boiling Point: 86°C (187°F)   **Specific Gravity:** 1.2   **Water Solubility:** Soluble

Chemical Incompatibilities or Instabilities: Strong oxidizers.

FLAMMABILITY

NFPA Hazard Code: **NFPA Classification:** Class IIIA Combustible
Flash Point: 88°C (190°F)   **LEL:** N.D.
Autoignition Temp: N.D.   **UEL:** N.D.

Fire Extinguishing Methods: Dry Chemical, Foam, Water.

Special Fire Fighting Considerations: Structural fire fighters protective clothing will not provide adequate protection. Approach fire from upwind. Water spray from an unmanned device should be used to keep closed containers cool. Continue to cool container after fire is extinguished. Fight fire from a distance or protected location, if possible. Immediately withdraw if rising sound from venting device is heard or if fire is causing discoloration to the tank.

TOXICOLOGY

Odor: Pungent   **Odor Threshold:** N.D.
Physical Contact: Irritant
TLV = 3 mg/m^3 **STEL** = N.D.   **IDLH** = 300 mg/m^3

Routes of Entry and Relative LD$_{50}$ (or LC$_{50}$)   **RTECS # AB2450000**
   Inhalation   N.D.
   Ingestion   8   (75 mg/kg)
   Skin Absorption   8   (224 mg/kg)

Symptoms of Exposure: Irritation to skin, eyes and upper respiratory tract, headache, nausea.

Emergency/First Aid Treatment: Remove to ventilated area; immediately remove any contaminated clothing and wash contaminated areas for 15 minutes using water. Treat supportively and observe for possible shock. If ingested, seek immediate medical aid.

Recommended Clean-Up Procedures

Personal Protection: Level A Ensemble   **Recommended Material**   N.D.

RCRA Waste #   P023   **Reportable Quantities:**   1000 lb (1 lb)

Spills: Dike to contain spill and collect liquid for disposal. Decontaminate spill area using water.

Special Emergency Information: May be fatal if inhaled, swallowed, or absorbed through skin.

CHLOROACETIC ACID - $ClCH_2COOH$

Synonyms: monochloroacetic acid

CAS Number: 79-11-8
DOT Number: 1750 (liquid)
1751 (solid)

Description: White Powder
DOT Classification: Corrosive Material

Molecular Weight: 94.5
Melting Point: 63°C (145°F)
Boiling Point: 189°C (372°F)

Vapor Density: 3.3
Specific Gravity: 1.6

Vapor Pressure: 1 (< 1 mm Hg)
Water Solubility: Soluble

Chemical Incompatibilities or Instabilities: Strong oxidizers.

FLAMMABILITY

NFPA Hazard Code: 3-1-0
Flash Point: 126°C (259°F)
Autoignition Temp: N.D.

NFPA Classification: Class IIIB Combustible
LEL: 8%
UEL: N.D.

Fire Extinguishing Methods: Alcohol Resistant Foam, Carbon Dioxide, Dry Chemical, Water.

Special Fire Fighting Considerations: Structural fire fighters protective clothing will not provide adequate protection. Approach fire from upwind. Water spray should be used to keep closed containers cool. May release toxic fumes upon heating (HCl, phosgene).

DOT Recommended Isolation Zones: Small Spill: 150 ft   Large Spill: 150 ft
DOT Recommended Down Wind Take Cover Distance   Small Spill: 0.2 miles   Large Spill: 0.2 miles

TOXICOLOGY
QUESTIONABLE CARCINOGEN

Odor:
Physical Contact: Material is extremely destructive to human tissues.
TLV = N.D.   **STEL** = N.D.

Odor Threshold: N.D.

IDLH = N.D.

Routes of Entry and Relative LD_{50} (or LC_{50})
    Inhalation   10   (180 mg/m^3)
    Ingestion   N.D.
    Skin Absorption N.D.

RTECS # AF8575000

Symptoms of Exposure: Possible irritation to skin, eyes and upper respiratory tract; headache; nausea. Inhalation of vapors may be fatal by causing glottis to spasm and suffocation. Exposure may result in chemical pneumonitis or pulmonary edema.

Emergency/First Aid Treatment: Remove to ventilated area; immediately remove any contaminated clothing and wash contaminated areas for 15 minutes using water. Treat supportively and observe for possible shock. If ingested, seek immediate medical aid.

Recommended Clean-Up Procedures

Personal Protection: Level B Ensemble

RCRA Waste #   None

Recommended Material   Butyl Rubber (+), Neoprene (+), Viton (+), Polyethylene (+)

Reportable Quantities: None

Spills: Dike to contain spill and collect liquid for disposal. Decontaminate spill area using dilute alkaline solution.

Special Emergency Information: May be fatal if inhaled, swallowed, or absorbed through skin.

CHLOROACETONE - ClCH₂COCH₃

Synonyms: chloracetone; monochloroacetone

CAS Number: 78-95-5
DOT Number: 1695

Description: Colorless Liquid that darkens upon exposure to light.
DOT Classification: Irritating Material

Molecular Weight: 92.5
Melting Point: -45°C (-49°F)
Boiling Point: 120°C (248°F)

Vapor Density: N.D.
Specific Gravity: 1.2

Vapor Pressure: N.D.
Water Solubility: Soluble

Chemical Incompatibilities or Instabilities: Light, strong oxidizers.

FLAMMABILITY

NFPA Hazard Code:
Flash Point: 27°C (81°F)
Autoignition Temp: N.A.

NFPA Classification: Class IC Flammable
LEL: N.A.
UEL: N.A.

Fire Extinguishing Methods: Carbon Dioxide, Dry Chemical, Foam, Water.

Special Fire Fighting Considerations: Structural fire fighters protective clothing will not provide adequate protection. Approach fire from upwind. Water spray should be used to keep closed containers cool.

DOT Recommended Isolation Zones: Small Spill: 150 ft   Large Spill: 150 ft
DOT Recommended Down Wind Take Cover Distance: Small Spill: 0.2 miles   Large Spill: 0.2 miles

TOXICOLOGY

Odor: Pungent
Physical Contact: Material is extremely destructive to human tissues.
TLV = 4 mg/m³   **STEL** = N.D.

Odor Threshold: N.D.

IDLH = N.D.

Routes of Entry and Relative LD₅₀ (or LC₅₀)
Inhalation	8	(985 mg/m³/HR)
Ingestion	8	(100 mg/kg)
Skin Absorption	9	(141 mg/kg)

RTECS # UC0700000

Symptoms of Exposure: Possible burns or irritation to skin, eyes and upper respiratory tract; headache; nausea. Lachrymation. Inhalation of vapors may be fatal by causing glottis to spasm and suffocation. Exposure may result in chemical pneumonitis or pulmonary edema.

Emergency/First Aid Treatment: Remove to ventilated area; immediately remove any contaminated clothing and wash contaminated areas for 15 minutes using water. Treat supportively and observe for possible shock. If ingested, seek immediate medical aid.

Recommended Clean-Up Procedures

Personal Protection: Level A Ensemble   **Recommended Material** N.D.

RCRA Waste #   None   **Reportable Quantities:** None

Spills: Remove all potential ignition sources. Dike to contain spill and collect liquid for disposal. Absorb with non-combustible absorbent and take up using non-sparking tools. Decontaminate spill area using water.

Special Emergency Information: May be fatal if inhaled, swallowed, or absorbed through skin. Vapors are heavier than air and may travel some distance to an ignition source. Vapors are more dense than air and may settle in low lying areas.

CHLOROACETOPHENONE - $C_6H_5C(O)CH_2Cl$

Synonyms: α-chloroacetophenone, mace

CAS Number: 532-27-4
DOT Number: 1697

Description: White to Tan Powder
DOT Classification: Irritating Material

Molecular Weight: 154.6
Melting Point: 56°C (133°F)
Boiling Point: 245°C (473°F)

Vapor Density: 5.3
Specific Gravity: 1.3

Vapor Pressure: 1 (1 < mm Hg)
Water Solubility: Insoluble

Chemical Incompatibilities or Instabilities: Strong oxidizers.

FLAMMABILITY

NFPA Hazard Code: 2-1-0
Flash Point: 118°C (244°F)
Autoignition Temp: N.D.

NFPA Classification:
LEL: N.D.
UEL: N.D.

Fire Extinguishing Methods: Dry Chemical, Foam, Water.

Special Fire Fighting Considerations: Structural fire fighters protective clothing will not provide adequate protection. Approach fire from upwind. Water spray should be used to keep closed containers cool.

DOT Recommended Isolation Zones:
DOT Recommended Down Wind Take Cover Distance:

Small Spill: 900 ft
Small Spill: 3 miles

Large Spill: 1200 ft
Large Spill: 4 miles

TOXICOLOGY
QUESTIONABLE CARCINOGEN

Odor:
Physical Contact: Material is extremely destructive to human tissues.
TLV = 3 mg/m^3 **STEL** = N.D.

Odor Threshold: N.D.

IDLH = 100 mg/m^3

RTECS # AM6300000

Routes of Entry and Relative LD$_{50}$ (or LC$_{50}$)
  Inhalation   10   (417 mg/m^3/15 min)
  Ingestion   7   (127 mg/kg)
  Skin Absorption N.D.

Symptoms of Exposure: Possible burns or irritation to skin, eyes and upper respiratory tract; headache; nausea. Lachrymation. Inhalation of vapors may be fatal by causing glottis to spasm and suffocation. Exposure may result in chemical pneumonitis or pulmonary edema.

Emergency/First Aid Treatment: Remove to ventilated area; immediately remove any contaminated clothing and wash contaminated areas for 15 minutes using water. Treat supportively and observe for possible shock. If ingested, seek immediate medical aid.

Recommended Clean-Up Procedures

Personal Protection: Level B Ensemble

RCRA Waste # None

Recommended Material None

Reportable Quantities: None

Spills: Dike to contain spill and collect liquid for disposal. Decontaminate spill area using soapy water.

Special Emergency Information: May be harmful is inhaled, swallowed, or absorbed through the skin.

CHLOROACETYLCHLORIDE - ClCH₂CClO

Synonyms: chloroacetic acid chloride, chloracetyl chloride

CAS Number: 74-04-9	**Description:** Colorless Liquid
DOT Number: 1752	**DOT Classification:** Corrosive Material

Molecular Weight: 112.9

Melting Point: -22°C (-8°F)	**Vapor Density:** 3.9	**Vapor Pressure:** 4 (19 mm Hg)
Boiling Point: 107°C (225°F)	**Specific Gravity:** 1.4	**Water Solubility:** Decomposes

Chemical Incompatibilities or Instabilities: Water, alcohols.

FLAMMABILITY

NFPA Hazard Code: 3-0-1	**NFPA Classification:**
Flash Point: N. A.	**LEL:** N.D.
Autoignition Temp: N.D.	**UEL:** N.D.

Fire Extinguishing Methods: Use agent suitable for surrounding fire.

Special Fire Fighting Considerations: Structural fire fighters protective clothing will not provide adequate protection. Approach fire from upwind. Water spray should be used to keep closed containers cool. Continue to cool container after fire is extinguished.

TOXICOLOGY

Odor:

Physical Contact: Material is extremely destructive to human tissues.

TLV = 0.2 mg/m³ (skin)   **STEL** = 0.7 mg/m³ (skin)

Odor Threshold: N.D.

IDLH = N.D.

Routes of Entry and Relative LD₅₀ (or LC₅₀)

Inhalation	6	(18,320 mg/m³/H)
Ingestion	7	(120 mg/kg)
Skin Absorption	N.D.	

RTECS # AO6475000

Symptoms of Exposure: Possible burns or irritation to skin, eyes and upper respiratory tract; headache; nausea. Lachrymation. Inhalation of vapors may be fatal by causing glottis to spasm and suffocation. Exposure may result in chemical pneumonitis or pulmonary edema.

Emergency/First Aid Treatment: Remove to ventilated area; immediately remove any contaminated clothing and wash contaminated areas for 15 minutes using water. Treat supportively and observe for possible shock. If ingested, seek immediate medical aid.

Recommended Clean-Up Procedures

Personal Protection: Level B Ensemble	**Recommended Material** None
RCRA Waste # None	**Reportable Quantities:** None

Spills: Dike to contain spill and collect liquid for disposal. Decontaminate spill area using dilute alkaline solution.

Special Emergency Information: May be harmful is inhaled, swallowed, or absorbed through the skin. Reacts with water to yield hydrochloric acid vapors

4-CHLOROANILINE - $H_2NC_6H_4Cl$

Synonyms: parachloroaniline

CAS Number: 106-47-8
DOT Number: 2018 (s)/2019 (l)

Description: White to Tan Powder
DOT Classification: Poison B

Molecular Weight: 127.6
Melting Point: 73°C (163°F)
Boiling Point: °C (°F)

Vapor Density: N.A.
Specific Gravity: 1.2

Vapor Pressure: 1 (< 1 mm Hg)
Water Solubility: Slight

Chemical Incompatibilities or Instabilities: Nitrous acid, strong oxidizers.

FLAMMABILITY

NFPA Hazard Code:
Flash Point: N.D.
Autoignition Temp: N.D.

NFPA Classification:
LEL: N.D.
UEL: N.D.

Fire Extinguishing Methods: Carbon Dioxide, Dry Chemical, Foam, Water.

Special Fire Fighting Considerations: Structural fire fighters protective clothing will not provide adequate protection. Approach fire from upwind. Fight fire from a distance or protected location, if possible. Treat all materials used or generated and equipment involved as contaminated by hazardous waste.

TOXICOLOGY
QUESTIONABLE CARCINOGEN

Odor:
Physical Contact: Irritant.
TLV = N.D.

STEL = N.D.

Odor Threshold: N. A.

IDLH = N.D.

Routes of Entry and Relative LD_{50} (or LC_{50})

Inhalation	N.D.	
Ingestion	6	(310 mg/kg)
Skin Absorption	8	(360 mg/kg)

RTECS # BX0700000

Symptoms of Exposure: Possible irritation to skin, eyes and upper respiratory tract; headache; nausea. Absorption may lead to the formation of methemoglobin resulting in cyanosis several hours after exposure.

Emergency/First Aid Treatment: Remove to ventilated area; immediately remove any contaminated clothing and wash contaminated areas for 15 minutes using water. Treat supportively and observe for possible shock. If ingested, seek immediate medical aid.

Recommended Clean-Up Procedures

Personal Protection: Level B Ensemble
RCRA Waste # P024

Recommended Material Butyl Rubber (-), Polyvinyl Alcohol (-)
Reportable Quantities: 1000 lb (1 lb)

Spills: Dike to contain spill and collect liquid for disposal. Decontaminate spill area using soapy water. Treat all materials used as contaminated by a hazardous waste.

Special Emergency Information: May be fatal if inhaled, swallowed, or absorbed through skin.

2-chloroaniline (CAS # 95-51-2, RTECS # BX0525000) and
3-chloroaniline (CAS # 108-42-9, RTECS # BX0350000) have similar chemical, physical, and toxicological properties.

1-CHLORO-2,4-DINITROBENZENE

Synonyms: dinitrochlorobenzene

CAS Number: 97-00-7   **Description:** Yellow Powder or Crystals
DOT Number: ----   **DOT Classification:** Poison B

Molecular Weight: 202.6
Melting Point: 52°C (126°F)   **Vapor Density:** 6.9   **Vapor Pressure:** N.A.
Boiling Point: 315°C (599°F)   **Specific Gravity:** 1.7   **Water Solubility:** Insoluble

Chemical Incompatibilities or Instabilities: Hydrazine, ammonia, heat.

FLAMMABILITY

NFPA Hazard Code: 3-1-4   **NFPA Classification:**
Flash Point: 194°C (382°F)   **LEL:**   2.0%
Autoignition Temp: 432°C (810°F)   **UEL:**   22%

Fire Extinguishing Methods: Carbon Dioxide, Dry Chemical, Foam, Water.

Special Fire Fighting Considerations: Structural fire fighters protective clothing will not provide adequate protection. Approach fire from upwind. Material is thermally unstable and may detonate upon heating. Fight fire from a distance or protected location, if possible. For large fires, if possible, withdraw and allow to burn.

TOXICOLOGY
SENSITIZER, ALLERGEN

Odor: **Odor Threshold:** N.D.
Physical Contact: Irritant
TLV = N.D.   **STEL** = N.D.   **IDLH** = N.D.

Routes of Entry and Relative LD$_{50}$ (or LC$_{50}$)   **RTECS # CZ0525000**
   Inhalation   N.D.
   Ingestion   3   (1070 mg/kg)
   Skin Absorption   9   (130 mg/kg)

Symptoms of Exposure: Possible irritation to skin, eyes and upper respiratory tract; headache; nausea. Absorption may lead to the formation of methemoglobin resulting in cyanosis several hours after exposure. May cause and allergic reaction.

Emergency/First Aid Treatment: Remove to ventilated area; immediately remove any contaminated clothing and wash contaminated areas for 15 minutes using water. Treat supportively and observe for possible shock. If ingested, seek immediate medical aid.

Recommended Clean-Up Procedures

Personal Protection: Level B Ensemble   **Recommended Material**   Viton (-)

RCRA Waste #   None   **Reportable Quantities:**   None

Spills: Absorb with non-combustible absorbent and take up using non-sparking tools. Decontaminate spill area using soapy water. Treat all materials used or generated and equipment involved as contaminated by hazardous waste.

Special Emergency Information: May be fatal if inhaled, swallowed, or absorbed through skin.

Chlorodinitrobenzene (mixed isomers) (DOT # 1577) has similar chemical, physical, and toxicological properties.

CHLOROFORM - CHCl₃

Synonyms: trichloromethane

CAS Number: 67-66-3
DOT Number: 1888

Description: Colorless Liquid
DOT Classification: ORM-A

Molecular Weight: 119.4
Melting Point: -63°C (-81°F)
Boiling Point: 61°C (142°F)

Vapor Density: 4.1
Specific Gravity: 1.5

Vapor Pressure: 7 (160 mm Hg)
Water Solubility: Insoluble

Chemical Incompatibilities or Instabilities: Alkali metals.

FLAMMABILITY

NFPA Hazard Code: 2-0-0
Flash Point: N.A.
Autoignition Temp: N.A.

NFPA Classification:
LEL: N.A.
UEL: N.A.

Fire Extinguishing Methods: Use agent suitable for surrounding fire.

Special Fire Fighting Considerations: Structural fire fighters protective clothing will not provide adequate protection. Approach fire from upwind. Fight fire from a distance or protected location, if possible. Treat all materials used or generated and equipment involved as contaminated by hazardous waste. May release toxic fumes upon heating (HCl, phosgene).

TOXICOLOGY
CONFIRMED CARCINOGEN

Odor: Sweet, suffocating
Physical Contact: Irritant
TLV = 49 mg/m³   **STEL** = N.D.

Odor Threshold: 3 mg/m³

IDLH = 5000 mg/m³

Routes of Entry and Relative LD₅₀ (or LC₅₀)
  Inhalation   3   (190,800 mg/m³/H)
  Ingestion   4   (908 mg/kg)
  Skin Absorption N.D.

RTECS # FS9100000

Symptoms of Exposure: Possible irritation to skin, eyes and upper respiratory tract; headache; nausea. Nervous disturbances.

Emergency/First Aid Treatment: Remove to ventilated area; immediately remove any contaminated clothing and wash contaminated areas for 15 minutes using water. Treat supportively and observe for possible shock. If ingested, seek immediate medical aid.

Recommended Clean-Up Procedures

Personal Protection: Level A Ensemble
RCRA Waste # U044

Recommended Material Polyvinyl Alcohol (+), Viton (-), Teflon (0)
Reportable Quantities: 10 lb (5000 lb)

Spills: Dike to contain spill and collect liquid for later disposal. Decontaminate spill area using soapy water. Treat all materials used or generated and equipment involved as contaminated by hazardous waste.

Special Emergency Information: May be harmful is inhaled, swallowed, or absorbed through the skin.

1-CHLORO-2-NITRO BENZENE - $ClC_6H_4NO_2$

Synonyms: o-chloronitrobenzene; o-nitrochlorobenzene; 1-nitro-2-chloro benzene

CAS Number: 88-73-3   **Description:** Yellow Powder
DOT Number: 1578   **DOT Classification:** Poison B

Molecular Weight: 157.6

| **Melting Point:** | 35°C (95°F) | **Vapor Density:** | 5.4 | **Vapor Pressure:** | 1 (< 1 mm hg) |
| **Boiling Point:** | 246°C (475°F) | **Specific Gravity:** | 1.4 | **Water Solubility:** | Insoluble |

Chemical Incompatibilities or Instabilities: Strong oxidizer.

FLAMMABILITY

NFPA Hazard Code: 3-1-0   **NFPA Classification:**
Flash Point: 50°C (123°F)   **LEL:** 1.4%
Autoignition Temp: 260°C (500°F)   **UEL:** 8.7%

Fire Extinguishing Methods: Carbon Dioxide, Dry Chemical, Foam, Water.

Special Fire Fighting Considerations: Structural fire fighters protective clothing will not provide adequate protection. Approach fire from upwind. Fight fire from a distance or protected location, if possible. May release toxic fumes upon heating (HCl, NO_x, phosgene). Treat all materials used or generated and equipment involved as contaminated by hazardous waste.

TOXICOLOGY
QUESTIONABLE CARCINOGEN

Odor: **Odor Threshold:** N.D.
Physical Contact: Irritant.
TLV = 0.6 mg/m^3 (skin)   **STEL** = N.D.   **IDLH** = 1000 mg/m^3

Routes of Entry and Relative LD$_{50}$ (or LC$_{50}$)   **RTECS #** CZ0875000

Inhalation	N.D.	
Ingestion	6	(288 mg/kg)
Skin Absorption	N.D.	

Symptoms of Exposure: Possible irritation to skin, eyes and upper respiratory tract; headache; nausea. Absorption may lead to the formation of methemoglobin resulting in cyanosis several hours after exposure.

Emergency/First Aid Treatment: Remove to ventilated area; immediately remove any contaminated clothing and wash contaminated areas for 15 minutes using water. Treat supportively and observe for possible shock. If ingested, seek immediate medical aid.

Recommended Clean-Up Procedures

Personal Protection: Level B Ensemble   **Recommended Material** Butyl Rubber (-), Viton (-)

RCRA Waste # None   **Reportable Quantities:** None

Large Spills: Absorb with non-combustible absorbent and take up using non-sparking tools. Decontaminate spill area using soapy Water spray should be used to keep closed containers cool. Continue to cool container after fire is extinguished. Treat all materials used as contaminated by a hazardous waste.

Special Emergency Information: May be fatal if inhaled, swallowed, or absorbed through skin.

1-chloro-3-nitrobenzene (CAS #121-73-3, RTECS # CZ0940000) and
1-chloro-4-nitro benzene (CAS # 100-00-5, RTECS # CZ1050000) have similar chemical, physical, and toxicological properties.

Synonyms: p-chloronitrobenzene, p-nitrochlorobenzene, m-chloronitrobenzene, m-nitrochlorobenzene.

4-CHLOROPHENOL - ClC6H4OH

Synonyms: p-chlorophenol

CAS Number: 106-48-9

Description: White to Tan Powder

DOT Number: 2020/2021

DOT Classification: Poison B

Molecular Weight: 128.6

Melting Point: 43°C (109°F) **Vapor Density:** **Vapor Pressure:** 1 (<1 mm Hg)

Boiling Point: 220°C (428°F) **Specific Gravity:** 1.2 **Water Solubility:** < 1%

Chemical Incompatibilities or Instabilities: Strong oxidizers.

FLAMMABILITY

NFPA Hazard Code: 3-2-0 **NFPA Classification:**

Flash Point: 121°C (250°F) **LEL:** N.D.

Autoignition Temp: N.D. **UEL:** N.D.

Fire Extinguishing Methods: Carbon Dioxide, Dry Chemical, Foam, Water.

Special Fire Fighting Considerations: Structural fire fighters protective clothing will not provide adequate protection. Approach fire from upwind. Fight fire from a distance or protected location, if possible. Treat all materials used or generated and equipment involved as contaminated by hazardous waste.

TOXICOLOGY
QUESTIONABLE CARCINOGEN

Odor: Unpleasant

Odor Threshold: N.D.

Physical Contact: Material is extremely destructive to human tissues.

TLV = N.D. **STEL** = N.D. **IDLH** = N.D.

Routes of Entry and Relative LD50 (or LC50) **RTECS # SK2800000**

Inhalation	N.D.	
Ingestion	6	(261 mg/kg)
Skin Absorption	10	(11 mg/kg)

Symptoms of Exposure: Possible burns or irritation to skin, eyes and upper respiratory tract; headache; nausea. Laryngitis. Inhalation of vapors may be fatal by causing glottis to spasm and suffocation. Exposure may result in chemical pneumonitis or pulmonary edema.

Emergency/First Aid Treatment: Remove to ventilated area; immediately remove any contaminated clothing and wash contaminated areas for 15 minutes using water. Treat supportively and observe for possible shock. If ingested, seek immediate medical aid.

Recommended Clean-Up Procedures

Personal Protection: Level B Ensemble **Recommended Material** N.D.

RCRA Waste # None **Reportable Quantities:** None

Spills: Dike to contain spill and collect liquid for disposal. Absorb with non-combustible absorbent and take up using non-sparking tools. Decontaminate spill area using soapy water. Treat all materials used as contaminated by a hazardous waste.

Special Emergency Information: May be harmful is inhaled, swallowed, or absorbed through the skin.

2-chlorophenol (CAS # 95-57-8, RTECS # SK2625000, RCRA Waste # U048, Reportable Quantities: 100 lb [1 lb]) and 3-chlorophenol (CAS # 108-43-0, RTECS # SK2450000) have similar chemical, physical, and toxicological properties.

Synonyms: o-chlorophenol; m-chlorophenol.

CHLOROPICRIN - Cl_3CNO_2

Synonyms: trichloronitromethane; nitrochloroform

CAS Number: 76-06-2
DOT Number: 1580, 1583, 1955, 2929

Description: Colorless Liquid
DOT Classification: Poison B

Molecular Weight: 164.4
Melting Point: -64°C (-83°F)
Boiling Point: 112°C (234°F)

Vapor Density: 6.7
Specific Gravity: 1.7

Vapor Pressure: 4 (20 mm Hg)
Water Solubility: Insoluble

Chemical Incompatibilities or Instabilities: Aniline, 3-bromopropyne, strong alkali, heat, propargyl bromide.

FLAMMABILITY

NFPA Hazard Code: 4-0-3
Flash Point: N.A.
Autoignition Temp: N.A.

NFPA Classification:
LEL: N.A.
UEL: N.A.

Fire Extinguishing Methods: Use agent suitable for surrounding fire.

Special Fire Fighting Considerations: Structural fire fighters protective clothing will not provide adequate protection. Approach fire from upwind. Water spray should be used to keep closed containers cool. For large fires, if possible, withdraw and allow to burn. May explosively decompose when heated.

		Small Spill: 600 ft	Large Spill: 900 ft
DOT Recommended Isolation Zones:		Small Spill: 600 ft	Large Spill: 900 ft
DOT Recommended Down Wind Take Cover Distance:		Small Spill: 2 miles	Large Spill: 3 miles

TOXICOLOGY
QUESTIONABLE CARCINOGEN

Odor:
Physical Contact: Material is extremely destructive to human tissues.
TLV = 0.7 mg/m^3 **STEL** = N.D.

Odor Threshold: N.D.

IDLH = 28 mg/m^3

Routes of Entry and Relative LD$_{50}$ (or LC$_{50}$)
  Inhalation   N.D.
  Ingestion   6   (250 mg/kg)
  Skin Absorption   N.D.

RTECS # PB6300000

Symptoms of Exposure: Possible burns or irritation to skin, eyes and upper respiratory tract; headache; nausea. Lachrymation. Inhalation of vapors may be fatal by causing glottis to spasm and suffocation. Exposure may result in chemical pneumonitis or pulmonary edema.

Emergency/First Aid Treatment: Remove to ventilated area; immediately remove any contaminated clothing and wash contaminated areas for 15 minutes using water. Treat supportively and observe for possible shock. If ingested, seek immediate medical aid.

Recommended Clean-Up Procedures

Personal Protection: Level A Ensemble
RCRA Waste # None

Recommended Material Teflon (0)
Reportable Quantities: None

Spills: Dike to contain spill and collect liquid for disposal. Evacuate area. Stop leak if it can be safely be done. Take up with a non-combustible absorbent and dispose. Decontaminate spill area using soapy water. Treat all materials used as contaminated by a hazardous waste.

Special Emergency Information: May be fatal if inhaled, swallowed, or absorbed through skin. Extreme irritant, highly toxic. Exposure to 28 mg/m^3 will render most workers unable to function. Concentrations of 140 mg/m^3 for one or two minutes may result in lung lesions and death.

CHLOROPIVAVOYLCHLORIDE

Synonyms:

CAS Number: 4300-97-4   **Description:** Colorless Liquid
DOT Number: 9263   **DOT Classification:** N.D.

Molecular Weight: 155.0
Melting Point: N.D.   **Vapor Density:** N.D.   **Vapor Pressure:** N.D.
Boiling Point: N.D.   **Specific Gravity:** 2.00   **Water Solubility:** Decomposes

Chemical Incompatibilities or Instabilities: Water, alkali, alcohols, strong oxidizers.

FLAMMABILITY

NFPA Hazard Code: **NFPA Classification:** Class III A Combustible
Flash Point: 62°C (144°F)   **LEL:** N.D.
Autoignition Temp: N.D.   **UEL:** N.D.

Fire Extinguishing Methods: Carbon Dioxide, Dry Chemical, Foam.

Special Fire Fighting Considerations: Structural fire fighters protective clothing will not provide adequate protection. Fight fire from a distance or protected location, if possible. Treat all materials used or generated and equipment involved as contaminated by hazardous waste. May release toxic fumes upon heating (HCl, phosgene).

DOT Recommended Isolation Zones: Small Spill: 150 ft   Large Spill: 150 ft
DOT Recommended Down Wind Take Cover Distance Small Spill: 0.2 miles Large Spill: 0.2 miles

TOXICOLOGY

Odor: **Odor Threshold:** N.D.
Physical Contact: Material is extremely destructive to human tissues.
TLV = N.D.   **STEL** = N.D.   **IDLH** = N.D.

Routes of Entry and Relative LD$_{50}$ (or LC$_{50}$)   **RTECS #** N.D.
   Inhalation   N.D.
   Ingestion   N.D.
   Skin Absorption   N.D.

Symptoms of Exposure: Possible burns or irritation to skin, eyes and upper respiratory tract; headache; nausea. Lachrymation. Inhalation of vapors may be fatal by causing glottis to spasm and suffocation. Exposure may result in chemical pneumonitis or pulmonary edema.

Emergency/First Aid Treatment: Remove to ventilated area; immediately remove any contaminated clothing and wash contaminated areas for 15 minutes using water. Treat supportively and observe for possible shock. If ingested, seek immediate medical aid.

Recommended Clean-Up Procedures

Personal Protection: Level A Ensemble   **Recommended Material** N.D.

RCRA Waste #   None   **Reportable Quantities:**   None

Spills: Remove all potential ignition sources. Dike to contain spill, absorb with non-combustible absorbent and take up using non-sparking tools. Decontaminate spill area using dilute alkaline solution. Treat all materials used or generated and equipment involved as contaminated by hazardous waste.

Special Emergency Information: May be harmful is inhaled, swallowed, or absorbed through the skin.

CHLOROPRENE - $CH_2=CClCH=CH_2$

Synonyms: 2-chloro-1,2-butadiene; neoprene

CAS Number: 126-99-8

DOT Number: 1991

Description: Colorless Liquid

DOT Classification: Flammable Liquid

Molecular Weight: 88.6

Melting Point: -103°C (-153°F)

Boiling Point: 59°C (138°F)

Vapor Density: 3.0

Specific Gravity: 0.96

Vapor Pressure: N.D.

Water Solubility: < 1%

Chemical Incompatibilities or Instabilities: Oxygen, peroxides, halogens, oxidizers.

FLAMMABILITY

NFPA Hazard Code: 2-3-0

Flash Point: -20°C (4°F)

Autoignition Temp: °C (°F)

NFPA Classification: Class I A Flammable

LEL: 4.0%

UEL: 20.0%

Fire Extinguishing Methods: Carbon Dioxide, Dry Chemical, Foam, Water.

Special Fire Fighting Considerations: Structural fire fighters protective clothing will not provide adequate protection. Approach fire from upwind. Isolate for 1/2 mile if rail or tank truck is involved in at fire. Water spray should be used to keep closed containers cool. For large fires, if possible, withdraw and allow to burn. Immediately withdraw if rising sound from venting device is heard or if fire is causing discoloration to the tank.

TOXICOLOGY
SUSPECTED CARCINOGEN

Odor: Rubber

Physical Contact: Irritant.

TLV = 36 mg/m^3 (skin)   **STEL** = N.D.

Odor Threshold: 0.4 mg/m^3

IDLH = 1500 mg/m^3

Routes of Entry and Relative LD$_{50}$ (or LC$_{50}$)

Inhalation	4	(47,200 mg/m^3/H)
Ingestion	5	(450 mg/kg)
Skin Absorption	N.D.	

RTECS # EI9625000

Symptoms of Exposure: Possible irritation to skin, eyes and upper respiratory tract; headache; nausea.

Emergency/First Aid Treatment: Remove to ventilated area; immediately remove any contaminated clothing and wash contaminated areas for 15 minutes using water. Treat supportively and observe for possible shock. If ingested, seek immediate medical aid.

Recommended Clean-Up Procedures

Personal Protection: Level A Ensemble

RCRA Waste # None

Recommended Material Polyvinyl Alcohol (+), Viton (-)

Reportable Quantities: None

Spills: Remove all potential ignition sources. Dike to contain spill, absorb with non-combustible absorbent and take up using non-sparking tools. Decontaminate spill area using soapy water.

Special Emergency Information: Vapors are heavier than air and may travel some distance to an ignition source. May form unstable peroxides upon exposure to air. May be harmful is inhaled, swallowed, or absorbed through the skin. Vapors are more dense than air and may settle in low lying areas.

2-CHLOROPROPANOL - CH₃CH(Cl)CH₂OH

Synonyms: 2-chloro-1-propanol

CAS Number: 78-89-7	**Description:** Colorless Liquid
DOT Number: 2611	**DOT Classification:** Poison B

Molecular Weight: 94.6

Melting Point: °C (°F)	**Vapor Density:** 3.3	**Vapor Pressure:** N.D.
Boiling Point: 134°C (273°F)	**Specific Gravity:** 1.1	**Water Solubility:** Soluble

Chemical Incompatibilities or Instabilities: Strong oxidizers.

FLAMMABILITY

NFPA Hazard Code: 2-2-0	**NFPA Classification:** Class II Combustible Liquid
Flash Point: 44°C (111°F)	**LEL:** N.D.
Autoignition Temp: N.D.	**UEL:** N.D.

Fire Extinguishing Methods: Alcohol Resistant Foam, Carbon Dioxide, Dry Chemical, Water.

Special Fire Fighting Considerations: Structural fire fighters protective clothing will not provide adequate protection. Water spray from an unmanned device should be used to keep closed containers cool. Continue to cool container after fire is extinguished. Fight fire from a distance or protected location, if possible. Treat all materials used or generated and equipment involved as contaminated by hazardous waste.

TOXICOLOGY

Odor:	**Odor Threshold:** N.D.
Physical Contact: Irritant.	
TLV = N.D. **STEL** = N.D.	**IDLH** = N.D.

Routes of Entry and Relative LD₅₀ (or LC₅₀)   **RTECS # UA8925000**

Inhalation	N.D.	
Ingestion	6	(218 mg/kg)
Skin Absorption	7	(529 mg/kg)

Symptoms of Exposure: Possible irritation to skin, eyes and upper respiratory tract, headache, nausea.

Emergency/First Aid Treatment: Remove to ventilated area; immediately remove any contaminated clothing and wash contaminated areas for 15 minutes using water. Treat supportively and observe for possible shock. If ingested, seek immediate medical aid.

Recommended Clean-Up Procedures

Personal Protection: Level B Ensemble	**Recommended Material** Butyl Rubber (-), Viton (-)
RCRA Waste # None	**Reportable Quantities:** None

Spills: Remove all potential ignition sources. Dike to contain spill, absorb with non-combustible absorbent and take up using non-sparking tools. Decontaminate spill area using water.

Special Emergency Information: May be harmful is inhaled, swallowed, or absorbed through the skin.

1,3-dichloropropanol (CAS # 96-23-1, RTECS # UB1400000, DOT # 2750),
3-chloro-1-propanol (CAS # 627-10-5, RTECS # UA8930000), and
2,3 dichloropropanol (CAS # 616-23-9, RTECTS # UB1225000) have similar chemical, physical, and toxicological properties.

2-CHLOROPROPENE - $CH_2=CClCH_3$

Synonyms: 2-chloropropylene

CAS Number: 557-98-2   **Description:** Colorless Liquid
DOT Number: 2456   **DOT Classification:** Flammable Liquid

Molecular Weight:
Melting Point: -138°C (-216°F)   **Vapor Density:**   2.6   **Vapor Pressure:**   N.D.
Boiling Point: 23°C (73°F)   **Specific Gravity:** 0.9   **Water Solubility:**   Insoluble

Chemical Incompatibilities or Instabilities: Strong oxidizers.

FLAMMABILITY

NFPA Hazard Code: 2-4-0   **NFPA Classification:** Class 1A Flammable
Flash Point: -21°C   (-6°F)   **LEL:**   4.5%
Autoignition Temp: N.A.   **UEL:**   16%

Fire Extinguishing Methods: Alcohol Resistant Foam, Carbon Dioxide, Dry Chemical.

Special Fire Fighting Considerations: Isolate for 1/2 mile if rail or tank truck is involved in at fire. Water spray from an unmanned device should be used to keep closed containers cool. Continue to cool container after fire is extinguished. For large fires, if possible, withdraw and allow to burn. May release toxic fumes upon heating (HCl). Immediately withdraw if rising sound from venting device is heard or if fire is causing discoloration to the tank.

TOXICOLOGY

Odor: **Odor Threshold:** N.D.
Physical Contact: Material is extremely destructive to human tissues.
TLV = N.D.   **STEL** = N.D.   **IDLH** = N.D.

Routes of Entry and Relative LD$_{50}$ (or LC$_{50}$)   **RTECS # UC7200000**
    Inhalation   N.D.
    Ingestion   N.D.
    Skin Absorption   N.D.

Symptoms of Exposure: Possible irritation to skin, eyes and upper respiratory tract, headache, nausea. Laryngitis. Lachrymation.

Emergency/First Aid Treatment: Remove to ventilated area; immediately remove any contaminated clothing and wash contaminated areas for 15 minutes using water. Treat supportively and observe for possible shock. If ingested, seek immediate medical aid.

Recommended Clean-Up Procedures

Personal Protection: Level B Ensemble   **Recommended Material**   N.D.

RCRA Waste #   None   **Reportable Quantities:**   None

Spills: Remove all potential ignition sources. Dike to contain spill and collect liquid for disposal. Absorb with non-combustible absorbent and take up using non-sparking tools. Decontaminate spill area using soapy water.

Special Emergency Information: Vapors are heavier than air and may travel some distance to an ignition source. May be harmful is inhaled, swallowed, or absorbed through the skin. Vapors are more dense than air and may settle in low lying areas.

2-CHLOROPYRIDINE

Synonyms: α-chloropyridine; o-chloropyridine

CAS Number: 109-09-1   **Description:** Colorless, Viscous Liquid
DOT Number: 2822   **DOT Classification:** Poison B

Molecular Weight: 113.6
Melting Point: °C (°F)   **Vapor Density:** 3.9   **Vapor Pressure:** **2** (2 mm Hg)
Boiling Point: 170°C (338°F)   **Specific Gravity:** 1.2   **Water Solubility:** Soluble

Chemical Incompatibilities or Instabilities: Strong oxidizers.

FLAMMABILITY

NFPA Hazard Code: **NFPA Classification:** Class IIIA Combustible
Flash Point: 65°C (149°F)   **LEL:** N.D.
Autoignition Temp: N.D.   **UEL:** N.D.

Fire Extinguishing Methods: Carbon Dioxide, Dry Chemical, Foam, Water.

Special Fire Fighting Considerations: Water spray should be used to keep closed containers cool. May release toxic fumes upon heating (HCl, NO_x, phosgene).

TOXICOLOGY

Odor: **Odor Threshold:** N.D.
Physical Contact: Irritant.
TLV = N.D.   **STEL** = N.D.   **IDLH** = N.D.

Routes of Entry and Relative LD_{50} (or LC_{50})   **RTECS #** US5950000
     Inhalation   N.D.
     Ingestion   N.D.
     Skin Absorption   9   (64 mg/kg)

Symptoms of Exposure: Possible irritation to skin, eyes and upper respiratory tract, headache, nausea.

Emergency/First Aid Treatment: Remove to ventilated area; immediately remove any contaminated clothing and wash contaminated areas for 15 minutes using water. Treat supportively and observe for possible shock. If ingested, seek immediate medical aid.

Recommended Clean-Up Procedures

Personal Protection: Level A Ensemble   **Recommended Material** Butyl Rubber (-)

RCRA Waste # None   **Reportable Quantities:** None

Spills: Dike to contain spill and collect liquid for disposal. Decontaminate spill area using soapy water.

Special Emergency Information: May be fatal if inhaled, swallowed, or absorbed through skin.

CHLOROSULFONIC ACID - ClSO₃H

Synonyms: sulfuric chlorohydrin

CAS Number: 7790-94-5
DOT Number: 1754

Description: Colorless Liquid
DOT Classification: Corrosive Material

Molecular Weight: 116.5
Melting Point: -80°C (-112°F)
Boiling Point: 151°C (304°F)

Vapor Density: 4.0
Specific Gravity: 1.8

Vapor Pressure: 2 (1 mm Hg)
Water Solubility: Decomposes

Chemical Incompatibilities or Instabilities: Water, finely divided metals, alcohols, acids, organics, ordinary combustibles, nitrates, amines, peroxides.

FLAMMABILITY

NFPA Hazard Code:
Flash Point: N.A.
Autoignition Temp: N.A.

NFPA Classification:
LEL: N.D.
UEL: N.D.

Fire Extinguishing Methods: Carbon Dioxide, Dry Chemical.

Special Fire Fighting Considerations: Structural fire fighters protective clothing will not provide adequate protection. Water spray should be used to keep closed containers cool. DO NOT get water inside container.

DOT Recommended Isolation Zones: Small Spill: 150 ft   Large Spill: 150 ft
DOT Recommended Down Wind Take Cover Distance: Small Spill: 0.2 miles   Large Spill: 0.3 miles

TOXICOLOGY

Odor: Sharp
Physical Contact: Material is extremely destructive to human tissues.
TLV = N.D.   **STEL** = N.D.

Odor Threshold: N.D.

IDLH = N.D.

Routes of Entry and Relative LD₅₀ (or LC₅₀)
  Inhalation   N.D.
  Ingestion   N.D.
  Skin Absorption N.D.

RTECS # FX5730000

Symptoms of Exposure: Possible burns or irritation to skin, eyes and upper respiratory tract; headache; nausea. Laryngitis. Lachrymation. Inhalation of vapors may be fatal by causing glottis to spasm and suffocation. Exposure may result in chemical pneumonitis or pulmonary edema.

Emergency/First Aid Treatment: Remove to ventilated area; immediately remove any contaminated clothing and wash contaminated areas for 15 minutes using water. Treat supportively and observe for possible shock. If ingested, seek immediate medical aid.

Recommended Clean-Up Procedures

Personal Protection: Level A Ensemble

RCRA Waste # None

Recommended Material Teflon (0), Saranex (0)

Reportable Quantities: 1000 lb (1000 lb)

Spills: Dike to contain spill and collect liquid for disposal. Decontaminate spill area using flooding amounts of water.

Special Emergency Information: May be fatal if inhaled, swallowed, or absorbed through skin.

CHLORPYRIFOS

Synonyms: DOWCO 214; ENT 27520; reldan; dursban F; dursban, pyrinex

CAS Number: 2921-88-2	**Description:** White Powder
DOT Number: 2783	**DOT Classification:** ORM-A

Molecular Weight:

Melting Point:	41°C (106°F)	**Vapor Density:** N.D.	**Vapor Pressure:**	1 (< 1 mm Hg)
Boiling Point:	N.D.	**Specific Gravity:** N.D.	**Water Solubility:**	< 1%

Chemical Incompatibilities or Instabilities: Strong oxidizers.-

FLAMMABILITY

NFPA Hazard Code:		**NFPA Classification:** N.A.	
Flash Point:	N.A.	**LEL:** N.A.	
Autoignition Temp:	N.A.	**UEL:** N.A.	

Fire Extinguishing Methods: Dry Chemical, Foam, Water.

Special Fire Fighting Considerations: Structural fire fighters protective clothing will not provide adequate protection. Approach fire from upwind. Fight fire from a distance or protected location, if possible. Treat all materials used or generated and equipment involved as contaminated by hazardous waste.

TOXICOLOGY
QUESTIONABLE TETRAGEN

Odor:

Odor Threshold: N.D.

Physical Contact: Irritant.

$TLV = 0.2$ mg/m^3 (skin) $STEL =$ N.D. $IDLH =$ N.D.

Routes of Entry and Relative LD$_{50}$ (or LC$_{50}$)

RTECS # TF6300000

Inhalation	N.D.	
Ingestion	8	(82 mg/kg)
Skin Absorption	3	(2000 mg/kg)

Symptoms of Exposure: Possible irritation to skin, eyes and upper respiratory tract; headache; nausea. Paresthesia. Weakness.

Emergency/First Aid Treatment: Remove to ventilated area; immediately remove any contaminated clothing and wash contaminated areas for 15 minutes using water. Treat supportively and observe for possible shock. If ingested, seek immediate medical aid.

Recommended Clean-Up Procedures

Personal Protection:	Level A Ensemble	**Recommended Material** N.D.
RCRA Waste #	None	**Reportable Quantities:** 1 lb (1 lb)

Spills: Take up spill. Decontaminate spill area using soapy water. Treat all materials used or generated and equipment involved as contaminated by hazardous waste.

Special Emergency Information: May be fatal is inhaled, swallowed, or absorbed through the skin.

CHROMIC ACID (mixture) - CrO₃

Synonyms: chromium trioxide; chromium (VI) oxide; chromic anhydride

CAS Number: 1333-82-0   **Description:** Dark Red Powder
DOT Number: 1463   **DOT Classification:** Oxidizer

Molecular Weight: 100.6
Melting Point: 196°C (385°F)   **Vapor Density:**   N.A.   **Vapor Pressure:**   N.A.
Boiling Point: (D)   250°C (482°F)   **Specific Gravity:** 2.7   **Water Solubility:**   Soluble

Chemical Incompatibilities or Instabilities: Ordinary combustibles, halogens, reducing agents, organics, hydrogen sulfide, phosphorous, heat.

FLAMMABILITY

NFPA Hazard Code: **3-O-1-OX**   **NFPA Classification:**
Flash Point: N.A.   **LEL:** N.A.
Autoignition Temp: N.A.   **UEL:** N.A.

Fire Extinguishing Methods: Water ONLY.

Special Fire Fighting Considerations: Fight fire from a distance or protected location, if possible. Water spray from an unmanned device should be used to keep closed containers cool. Continue to cool container after fire is extinguished. For large fires, if possible, withdraw and allow to burn. Will react violently with many materials, especially at elevated temperatures. May explosively decompose at elevated temperature. Treat all materials used or generated and equipment involved as contaminated by hazardous waste.

TOXICOLOGY
CONFIRMED HUMAN CARCINOGEN

Odor: None   **Odor Threshold:** N.A.
Physical Contact: Material is extremely destructive to human tissues.
TLV = 0.05 mg (Cr)/m³   **STEL** = N.D.   **IDLH** = 30 mg (Cr)/m³

Routes of Entry and Relative LD₅₀ (or LC₅₀)   **RTECS # GB6650000**
   Inhalation   N.D.
   Ingestion   **8**   (80 mg/kg)
   Skin Absorption   N.D.

Symptoms of Exposure: Possible burns or irritation to skin, eyes and upper respiratory tract; headache; nausea. Inhalation of vapors may be fatal by causing glottis to spasm and suffocation. Exposure may result in chemical pneumonitis or pulmonary edema.

Emergency/First Aid Treatment: Remove to ventilated area; immediately remove any contaminated clothing and wash contaminated areas for 15 minutes using water. Treat supportively and observe for possible shock. If ingested, seek immediate medical aid.

Recommended Clean-Up Procedures

Personal Protection: Level A Ensemble   **Recommended Material**   Butyl Rubber (0), Polyvinyl Chloride (0)

RCRA Waste #   None   **Reportable Quantities:**   10 lb (1000 lb)

Spills: Take up in non-combustible absorbent and dispose. Treat all materials used or generated and equipment involved as contaminated by hazardous waste. May generate gases.

Special Emergency Information: May be fatal if inhaled, swallowed, or absorbed through skin.

Chromic acid solution (CAS # 1308-14-1, DOT # 1755, RTECS # GB2650000) has similar chemical, physical, and toxicological properties.

CHROMIC NITRATE - Cr(NO$_3$)$_3$

Synonyms: chromium (III) nitrate

CAS Number: 13548-38-4	**Description:** Blue to Purple Powder
DOT Number: 2720	**DOT Classification:** Oxidizer

Molecular Weight: 238.0

Melting Point: (D) 60°C (140°F)	**Vapor Density:** N.A.	**Vapor Pressure:** N.A.
Boiling Point: N.A.	**Specific Gravity:** N.D.	**Water Solubility:** Soluble

Chemical Incompatibilities or Instabilities: Organics, ordinary combustibles, reducing agents.

FLAMMABILITY

NFPA Hazard Code:	**NFPA Classification:**
Flash Point: N.A.	**LEL:** N.A.
Autoignition Temp: N.A.	**UEL:** N.A.

Fire Extinguishing Methods: Water ONLY.

Special Fire Fighting Considerations: Water spray should be used to keep closed containers cool. Treat all materials used or generated and equipment involved as contaminated by hazardous waste.

TOXICOLOGY
QUESTIONABLE CARCINOGEN

Odor: None | **Odor Threshold:** N.A.

Physical Contact: Irritant.

TLV = 0.5 mg (Cr)/m^3 **STEL** = N.D.   **IDLH** = N.D.

Routes of Entry and Relative LD$_{50}$ (or LC$_{50}$)   **RTECS # GB8630000**

Inhalation	N.D.	
Ingestion	3	(3250 mg/kg)
Skin Absorption	N.D.	

Symptoms of Exposure: Possible irritation to skin, eyes and upper respiratory tract, headache, nausea. May cause allergic reaction.

Emergency/First Aid Treatment: Remove to ventilated area; immediately remove any contaminated clothing and wash contaminated areas for 15 minutes using water. Treat supportively and observe for possible shock. If ingested, seek immediate medical aid.

Recommended Clean-Up Procedures

Personal Protection: Level B Ensemble	**Recommended Material** N.D.
RCRA Waste # None	**Reportable Quantities:** None (1 lb)

Spills: Take up spill. Decontaminate spill area using water. Treat all materials used or generated and equipment involved as contaminated by hazardous waste.

Special Emergency Information: May be harmful is inhaled, swallowed, or absorbed through the skin.

Chromium nitrate nonahydrate (CAS # 7789-02-8) has similar chemical, physical, and toxicological properties.

CHROMIC OXIDE - Cr_2O_3

Synonyms: chrome green; green cinnabar; chrome (III) oxide

CAS Number: 1308-38-9   **Description:** Green Powder
DOT Number: ----   **DOT Classification:** ----

Molecular Weight: 152.0
Melting Point: 2435°C (4415°F)   **Vapor Density:** N.A.   **Vapor Pressure:** N.A.
Boiling Point: 3000°C (5432°F)   **Specific Gravity:** 5.2   **Water Solubility:** Insoluble

Chemical Incompatibilities or Instabilities: ClF_3, glycerol, OF_2, lithium.

FLAMMABILITY

NFPA Hazard Code: **NFPA Classification:**
Flash Point: N.A.   **LEL:** N.A.
Autoignition Temp: N.A.   **UEL:** N.A.

Fire Extinguishing Methods: Use agent suitable for surrounding fire.

Special Fire Fighting Considerations: Treat all materials used or generated and equipment involved as contaminated by hazardous waste.

TOXICOLOGY
CONFIRMED CARCINOGEN

Odor: None   **Odor Threshold:** N.A.
Physical Contact: Irritant.
TLV = 0.5 mg (Cr)/m^3 **STEL** = N.D.   **IDLH** = N.D.

Routes of Entry and Relative LD$_{50}$ (or LC$_{50}$)   **RTECS # GB6475000**
    Inhalation   N.D.
    Ingestion   N.D.
    Skin Absorption   N.D.

Symptoms of Exposure: Possible irritation to skin, eyes and upper respiratory tract, headache, nausea. May cause an allergic reaction.

Emergency/First Aid Treatment: Remove to ventilated area; immediately remove any contaminated clothing and wash contaminated areas for 15 minutes using water. Treat supportively and observe for possible shock. If ingested, seek immediate medical aid.

Recommended Clean-Up Procedures

Personal Protection: Level A Ensemble   **Recommended Material** N.D.

RCRA Waste #   None   **Reportable Quantities:** None (1 lb)

Spills: Take up spill. Decontaminate spill area using soapy water. Treat all materials used or generated and equipment involved as contaminated by hazardous waste.

Special Emergency Information: May be harmful is inhaled, swallowed, or absorbed through the skin.

CHROMYLCHLORIDE - CrO_2Cl_2

Synonyms: chromic oxychloride; chromium (III) oxychloride

CAS Number: 14977-61-8   **Description:** Dark Red Liquid
DOT Number: 1758   **DOT Classification:** Corrosive Material

Molecular Weight: 154.9
Melting Point: -97°C (-143°F)   **Vapor Density:** N.D.   **Vapor Pressure:** N.D.
Boiling Point: 117°C (243°F)   **Specific Gravity:** 1.2   **Water Solubility:** Decomposes

Chemical Incompatibilities or Instabilities: Light water, organics, reducing agents, ordinary combustibles, non metal halides, hydrides, phosphorous, azides, sulfur.

FLAMMABILITY

NFPA Hazard Code: 3-0-2-W̶   **NFPA Classification:**
Flash Point: N.A.   **LEL:** N.A.
Autoignition Temp: N.A.   **UEL:** N.A.

Fire Extinguishing Methods: Use agent suitable for surrounding small fire. Use flooding quantities of water for large fires.

Special Fire Fighting Considerations: Water spray should be used to keep closed containers cool. Structural fire fighters protective clothing will not provide adequate protection. Treat all materials used or generated and equipment involved as contaminated by hazardous waste.

TOXICOLOGY
POTENTIAL CARCINOGEN

Odor: Musty, burning   **Odor Threshold:** N.D.
Physical Contact: Material is extremely destructive to human tissues.
TLV = 0.16 mg/m^3 **STEL** = N.D.   **IDLH** = N.D.

Routes of Entry and Relative LD$_{50}$ (or LC$_{50}$)   **RTECS #** GB5775000
  Inhalation   N.D.
  Ingestion   N.D.
  Skin Absorption   N.D.

Symptoms of Exposure: Possible burns or irritation to skin, eyes and upper respiratory tract; headache; nausea. Inhalation of vapors may be fatal by causing glottis to spasm and suffocation. Exposure may result in chemical pneumonitis or pulmonary edema.

Emergency/First Aid Treatment: Remove to ventilated area; immediately remove any contaminated clothing and wash contaminated areas for 15 minutes using water. Treat supportively and observe for possible shock. If ingested, seek immediate medical aid.

Recommended Clean-Up Procedures

Personal Protection: Level A Ensemble   **Recommended Material**   N.D.

RCRA Waste #   None   **Reportable Quantities:**   None (1 lb)

Spills: Take up with a non-combustible absorbent and dispose. Decontaminate spill area using water. Treat all materials used or generated and equipment involved as contaminated by hazardous waste.

Special Emergency Information: May be harmful is inhaled, swallowed, or absorbed through the skin. Reacts with water to yield chromic acid and hydrochloric acid.

COAL TAR

Synonyms: tar (liquid); asphalt

CAS Number: 8007-45-2

DOT Number: 1999

Description: Black Liquid

DOT Classification: Flammable or Combustible Liquid

Molecular Weight: N.A.

Melting Point: N.D.

Boiling Point: N.D.

Vapor Density: N.D.   **Vapor Pressure:** N.D.

Specific Gravity: N.D.   **Water Solubility:** Insoluble

Chemical Incompatibilities or Instabilities: Strong oxidizers.

FLAMMABILITY

NFPA Hazard Code:

Flash Point: N.D.

Autoignition Temp: N.D.

NFPA Classification:

LEL: N.D.

UEL: N.D.

Fire Extinguishing Methods: Carbon Dioxide, Dry Chemical, Foam, Water.

Special Fire Fighting Considerations: Isolate for 1/2 mile if rail or tank truck is involved in at fire. Water spray should be used to keep closed containers cool. Keep away from ends of tank. For large fires, if possible, withdraw and allow to burn. Immediately withdraw if rising sound from venting device is heard or if fire is causing discoloration to the tank. Treat all materials used or generated and equipment involved as contaminated by hazardous waste.

TOXICOLOGY
CONFIRMED CARCINOGEN

Odor: Kerosene

Physical Contact: Irritant.

TLV = 0.2 mg/m^3 **STEL** = N.D.

Odor Threshold: N.D.

IDLH = 700 mg/m^3

Routes of Entry and Relative LD$_{50}$ (or LC$_{50}$)

Inhalation   N.D.

Ingestion   N.D.

Skin Absorption   N.D.

RTECS # GF8600000

Symptoms of Exposure: Possible irritation to skin, eyes and upper respiratory tract, headache, nausea.

Emergency/First Aid Treatment: Remove to ventilated area; immediately remove any contaminated clothing and wash contaminated areas for 15 minutes using water. Treat supportively and observe for possible shock. If ingested, seek immediate medical aid.

Recommended Clean-Up Procedures

Personal Protection: Level A Ensemble   **Recommended Material** N.D.

RCRA Waste #   None   **Reportable Quantities:** None

Spills: Remove all potential ignition sources. Dike to contain spill, absorb with non-combustible absorbent and take up using non-sparking tools. Decontaminate spill area using soapy water. Treat all materials used or generated and equipment involved as contaminated by hazardous waste.

Special Emergency Information: May be harmful is inhaled, swallowed, or absorbed through the skin. Vapors are heavier than air and may travel some distance to an ignition source.

Coal tar distillate (CAS # 65996-92-1, DOT # 1136, RTECS # GB8617500) and
coal tar oil (CAS # 65996-91-0, DOT # 1137, RTECS # GF8615000) have similar chemical, physical, and toxicological properties.

COPPER (II) CYANIDE - Cu(CN)$_2$

Synonyms: cupric cyanide

CAS Number: 14765-77-0   **Description:** Yellow to Green Powder
DOT Number: 1587   **DOT Classification:** Poison B

Molecular Weight: 115.6
Melting Point: (D)   °C   (°F)   **Vapor Density:** N.A.   **Vapor Pressure:** N.A.
Boiling Point: N.A.   **Specific Gravity:** N.D. **Water Solubility:** Insoluble

Chemical Incompatibilities or Instabilities: Strong oxidizers, magnesium.

FLAMMABILITY

NFPA Hazard Code: N.D.   **NFPA Classification:**
Flash Point: N.A.   **LEL:** N.A.
Autoignition Temp: N.A.   **UEL:** N.A.

Fire Extinguishing Methods: Use agent suitable for surrounding fire.

Special Fire Fighting Considerations: Structural fire fighter protective clothing will not provide adequate protection. May release toxic fumes upon heating (NO$_x$, HCN). Treat all materials used or generated and equipment involved as contaminated by hazardous waste.

TOXICOLOGY

Odor: None   **Odor Threshold:** N.A.
Physical Contact: Irritant.
TLV = 1 mg (Cu)/m^3 **STEL** = N.D.   **IDLH** = 50 mg (Cu)/m^3

Routes of Entry and Relative LD$_{50}$ (or LC$_{50}$)   **RTECS #** GL7175000
     Inhalation   N.D.
     Ingestion   N.D.
     Skin Absorption   N.D.

Symptoms of Exposure: Possible irritation to skin, eyes and upper respiratory tract; headache; nausea. Cyanosis.

Emergency/First Aid Treatment: Remove to ventilated area; immediately remove any contaminated clothing and wash contaminated areas for 15 minutes using water. Treat supportively and observe for possible shock. If ingested, seek immediate medical aid.

Recommended Clean-Up Procedures

Personal Protection: Level A Ensemble   **Recommended Material**   N.D.

RCRA Waste #   P029   **Reportable Quantities:**   10 lb (1 lb)

Spills: Take up spill. Decontaminate spill area using dilute alkaline solution. Treat all materials used as contaminated by a hazardous waste.

Special Emergency Information: May be fatal if inhaled, swallowed, or absorbed through skin.

COPPER (II) NITRATE - Cu(NO$_3$)$_2$

Synonyms: cupric nitrate

CAS Number: 3251-23-8
DOT Number: 1479

Description: Blue Green Powder
DOT Classification: Oxidizer

Molecular Weight: 187.6
Melting Point: 255°C (491°F)
Boiling Point: °C (°F)

Vapor Density: N.A.
Specific Gravity: 2.0

Vapor Pressure: N.A.
Water Solubility: Soluble

Chemical Incompatibilities or Instabilities: Ammonia compounds, organics, ordinary combustibles, tin..

FLAMMABILITY

NFPA Hazard Code: N.D.
Flash Point: N.A.
Autoignition Temp: N.A.

NFPA Classification:
LEL: N.A.
UEL: N.A.

Fire Extinguishing Methods: WATER ONLY.

Special Fire Fighting Considerations: Water spray should be used to keep closed containers cool. For large fires, if possible, withdraw and allow to burn.

TOXICOLOGY

Odor: None
Physical Contact: Material is extremely destructive to human tissues.
TLV = 1 mg/m^3 (Cu)   **STEL** = N.D.

Odor Threshold: N.A.

IDLH = N.D.

Routes of Entry and Relative LD$_{50}$ (or LC$_{50}$)
  Inhalation   N.D.
  Ingestion   **4**   (940 mg/kg)
  Skin Absorption   N.D.

RTECS # QU7400000

Symptoms of Exposure: Possible burns or irritation to skin, eyes and upper respiratory tract; headache; nausea. Laryngitis. Inhalation of vapors may be fatal by causing glottis to spasm and suffocation. Exposure may result in chemical pneumonitis or pulmonary edema.

Emergency/First Aid Treatment: Remove to ventilated area; immediately remove any contaminated clothing and wash contaminated areas for 15 minutes using water. Treat supportively and observe for possible shock. If ingested, seek immediate medical aid.

Recommended Clean-Up Procedures

Personal Protection: Level B Ensemble   **Recommended Material**   N.D.

RCRA Waste #   None   **Reportable Quantities:**   None (1 lb)

Spills: Take up spill. Decontaminate spill area using water

Special Emergency Information: May be harmful is inhaled, swallowed, or absorbed through the skin.

Copper (II) chlorate (CAS # 26506-47-8, DOT # 2721) has similar chemical, physical, and toxicological properties.

m-CRESOL

Synonyms: m-methylphenol; m-hydroxytoluene

CAS Number: 108-39-4
DOT Number: 2076

Description: Colorless to Yellow Liquid or Solid
DOT Classification: Poison B

Molecular Weight: 108.2

Melting Point:	11°C (52°F)	**Vapor Density:**	3.7	**Vapor Pressure:**	0 (< 1 mm Hg)
Boiling Point:	203°C (397°F)	**Specific Gravity:**	> 1.0	**Water Solubility:**	Insoluble

Chemical Incompatibilities or Instabilities: Strong oxidizers.

FLAMMABILITY

NFPA Hazard Code: 3-2-0
Flash Point: 86°C (187°F)
Autoignition Temp: 620°C (1148°F)

NFPA Classification: Class IIIA Combustible
LEL: 1.0%
UEL: N.D.

Fire Extinguishing Methods: Carbon Dioxide, Dry Chemical, Foam, Water.

Special Fire Fighting Considerations: Structural fire fighters protective clothing will not provide adequate protection. Approach fire from upwind. Water spray should be used to keep closed containers cool. Keep away from ends of tank. Fight fire from a distance or protected location, if possible. Treat all materials used or generated and equipment involved as contaminated by hazardous waste.

TOXICOLOGY
QUESTIONABLE CARCINOGEN

Odor:

Odor Threshold:

Physical Contact: Material is extremely destructive to human tissues.

TLV = 23 mg/m^3 **STEL** = N.D.   **IDLH** = 1125 mg/m^3

Routes of Entry and Relative LD$_{50}$ (or LC$_{50}$)   **RTECS # GO6125000**

Inhalation	N.D.	
Ingestion	6	(242 mg/kg)
Skin Absorption	3	(2050 mg/kg)

Symptoms of Exposure: Possible burns or irritation to skin, eyes and upper respiratory tract; headache; nausea. Inhalation of vapors may be fatal by causing glottis to spasm and suffocation. Exposure may result in chemical pneumonitis or pulmonary edema.

Emergency/First Aid Treatment: Remove to ventilated area; immediately remove any contaminated clothing and wash contaminated areas for 15 minutes using water. Treat supportively and observe for possible shock. If ingested, seek immediate medical aid.

Recommended Clean-Up Procedures

Personal Protection: Level B Ensemble   **Recommended Material** Butyl Rubber (+), Viton (+)

RCRA Waste # U052   **Reportable Quantities:** 1000 lb (1000 lb)

Spills: Dike to contain spill and collect liquid for disposal. Decontaminate spill area using soapy water. Treat all materials used as contaminated by a hazardous waste.

Special Emergency Information: May be harmful if inhaled, swallowed or absorbed through the skin.

p-cresol (CAS # 106-44-5, DOT # 2076, RTECS # GO6475000),
o-cresol (CAS # 95-48-7, DOT # 2076, RTECS # GO6475000), and
cresol, mixture (CAS # 1319-77-3, DOT # 2076, RTECS # GO6950000) have the same RCRA waste number, the same reportable quantities and similar chemical, physical, and toxicological properties.

CROTONALDEHYDE - CH$_3$CH=CHCHO

Synonyms: 2-butenal

CAS Number: 4170-30-3
DOT Number: 1143

Description: Colorless to White Liquid
DOT Classification: Flammable Liquid

Molecular Weight: 70.1
Melting Point: -76°C (105°F)
Boiling Point: 104°C (219°F)

Vapor Density: 2.4
Specific Gravity: 0.9

Vapor Pressure: 5 (30 mm Hg)
Water Solubility: 18%

Chemical Incompatibilities or Instabilities: Nitric acid, 1,2-butadiene.

FLAMMABILITY

NFPA Hazard Code: 3-3-2
Flash Point: 13°C (55°F)
Autoignition Temp: 207°C (405°F)

NFPA Classification: Class 1B Flammable
LEL: 2.1%
UEL: 15.5%

Fire Extinguishing Methods: Alcohol Resistant Foam, Carbon Dioxide, Dry Chemical, Water.

Special Fire Fighting Considerations: Isolate for 1/2 mile if rail or tank truck is involved in at fire. Structural fire fighters protective clothing will not provide adequate protection. Water spray should be used to keep closed containers cool. Immediately withdraw if rising sound from venting device is heard or if fire is causing discoloration to the tank.

DOT Recommended Isolation Zones: Small Spill: 150 ft   Large Spill: 150 ft
DOT Recommended Down Wind Take Cover Distance: Small Spill: 0.2 miles   Large Spill: 0.4 miles

TOXICOLOGY
SUSPECTED CARCINOGEN, SENSITIZER

Odor: Pungent
Physical Contact: Material is extremely destructive to human tissues.
TLV = 6 mg/m^3 **STEL** = N.D.

Odor Threshold: 0.2 mg/m^3

IDLH = 1200 mg/m^3

Routes of Entry and Relative LD$_{50}$ (or LC$_{50}$)
    Inhalation   9   (400 mg/m^3/H)
    Ingestion   6   (206 mg/kg)
    Skin Absorption N.D.

RTECS # GP9625000

Symptoms of Exposure: Possible burns or irritation to skin, eyes and upper respiratory tract; headache; nausea. Lachrymation. Inhalation of vapors may be fatal by causing glottis to spasm and suffocation. Exposure may result in chemical pneumonitis or pulmonary edema.

Emergency/First Aid Treatment: Remove to ventilated area; immediately remove any contaminated clothing and wash contaminated areas for 15 minutes using water. Treat supportively and observe for possible shock. If ingested, seek immediate medical aid.

Recommended Clean-Up Procedures

Personal Protection: Level B Ensemble   **Recommended Material** Butyl Rubber (+), Teflon (0)

RCRA Waste #   U053   **Reportable Quantities:**   100 lb (100 lb)

Spills: Remove all potential ignition sources. Dike to contain spill, absorb with non-combustible absorbent and take up using non-sparking tools. Decontaminate spill area using water. Treat all materials used as contaminated by a hazardous waste.

Special Emergency Information: May be fatal if inhaled, swallowed, or absorbed through skin. May undergo hazardous polymerization. Vapors are heavier than air and may travel some distance to an ignition source. Vapors are more dense than air and may settle in low lying areas.

CROTONIC ACID - CH₃CH=CHCOOH

Synonyms: 2-butenoic acid, methyl acrylic acid

CAS Number: 3724-65-0   **Description:** White Powder
DOT Number: 2823   **DOT Classification:** Corrosive Material

Molecular Weight: 8.61
Melting Point: 72°C (162°F)   **Vapor Density:** 3.0   **Vapor Pressure:** 1 (< 1 mm Hg)
Boiling Point: 185°C (365°F)   **Specific Gravity:** 1.0 **Water Solubility:** 50%

Chemical Incompatibilities or Instabilities: Strong oxidizers.

FLAMMABILITY

NFPA Hazard Code: 3-2-0   **NFPA Classification:**
Flash Point: 88°C (190°F)   **LEL:** N.A.
Autoignition Temp: 396°C (745°F) **UEL:** N.A.

Fire Extinguishing Methods: Carbon Dioxide, Dry Chemical, Foam, Water.

Special Fire Fighting Considerations: Water spray should be used to keep closed containers cool.

TOXICOLOGY

Odor: **Odor Threshold:** N.D.
Physical Contact: Material is extremely destructive to human tissues.
TLV = N.D.   **STEL** = N.D.   **IDLH** = N.D.

Routes of Entry and Relative LD₅₀ (or LC₅₀)   **RTECS # GQ2800000**
   Inhalation   N.D.
   Ingestion   4   (1000 mg/kg)
   Skin Absorption N.D.

Symptoms of Exposure: Possible burns or irritation to skin, eyes and upper respiratory tract; headache; nausea. Inhalation of vapors may be fatal by causing spasms and suffocation. Exposure may result in chemical pneumonitis or pulmonary edema.

Emergency/First Aid Treatment: Remove to ventilated area; immediately remove any contaminated clothing and wash contaminated areas for 15 minutes using water. Treat supportively and observe for possible shock. If ingested, seek immediate medical aid.

Recommended Clean-Up Procedures

Personal Protection: Level B Ensemble   **Recommended Material** N.D.

RCRA Waste # None   **Reportable Quantities:** None

Spills: Take up spill. Decontaminate spill area using water.

Special Emergency Information: May be harmful is inhaled, swallowed, or absorbed through the skin.

CROTONYLENE - $CH_3C{\equiv}CCH_3$

Synonyms: 2-butyne

CAS Number: 503-17-3	**Description:** Colorless Liquid
DOT Number: 1144	**DOT Classification:** Flammable Liquid

Molecular Weight: 54.1

Melting Point:	-32°C (-26°F)	**Vapor Density:** 1.9		**Vapor Pressure:**	9 (400 mm Hg)
Boiling Point:	27°C (81°F)	**Specific Gravity:** 0.7		**Water Solubility:**	Insoluble

Chemical Incompatibilities or Instabilities: Strong oxidizers.

FLAMMABILITY

NFPA Hazard Code:	**NFPA Classification:** Class 1A Flammable Liquid
Flash Point: <-22°C (<-8°F)	**LEL:** 1.4%
Autoignition Temp: °C (°F)	**UEL:** N.D.

Fire Extinguishing Methods: Carbon Dioxide, Dry Chemical, Foam, Water.

Special Fire Fighting Considerations: Isolate for 1/2 mile if rail or tank truck is involved in at fire. Water spray should be used to keep closed containers cool. For large fires, if possible, withdraw and allow to burn. Immediately withdraw if rising sound from venting device is heard or if fire is causing discoloration to the tank.

TOXICOLOGY

Odor:	**Odor Threshold:**
Physical Contact: Irritant.	
TLV = N.D. **STEL** = N.D.	**IDLH** = N.D.

Routes of Entry and Relative LD_{50} (or LC_{50})   **RTECS #** GQ7210000

Inhalation	N.D.
Ingestion	N.D.
Skin Absorption	N.D.

Symptoms of Exposure: A simple asphyxiant.

Emergency/First Aid Treatment: Remove to ventilated area; immediately remove any contaminated clothing and wash contaminated areas for 15 minutes using water. Treat supportively and observe for possible shock. If ingested, seek immediate medical aid.

Recommended Clean-Up Procedures

Personal Protection: Level B Ensemble	**Recommended Material** N.D.
RCRA Waste # None	**Reportable Quantities:** None

Spills: Remove all potential ignition sources. Dike to contain spill and collect liquid for disposal. Absorb with non-combustible absorbent and take up using non-sparking tools. Decontaminate spill area using soapy water.

Special Emergency Information: May be harmful is inhaled, swallowed, or absorbed through the skin. Vapors are heavier than air and may travel some distance to an ignition source. Vapors are more dense than air and may settle in low lying areas.

CUMENE - C$_6$H$_5$CH(CH$_3$)$_2$

Synonyms: isopropyl benzene, 2-phenyl propane

CAS Number: 98-82-2

DOT Number: 1918

Description: Colorless Liquid

DOT Classification: Flammable Liquid

Molecular Weight: 120.2

Melting Point: -96°C (-141°F)	**Vapor Density:** 4.1	**Vapor Pressure:** 3 (8 mm Hg)
Boiling Point: 152°C (306°F)	**Specific Gravity:** 0.9	**Water Solubility:** Insoluble

Chemical Incompatibilities or Instabilities: Nitric acid, oleum, chlorosulfonic acid, strong oxidizers.

FLAMMABILITY

NFPA Hazard Code: 2-3-1

Flash Point: 44°C (111°F)

Autoignition Temp: 424°C (795°F)

NFPA Classification: Class II Combustible Liquid

LEL: 0.9%

UEL: 6.5%

Fire Extinguishing Methods: Alcohol Resistant Foam, Carbon Dioxide, Dry Chemical, Water.

Special Fire Fighting Considerations: Isolate for 1/2 mile if rail or tank truck is involved in at fire. Water spray from an unmanned device should be used to keep closed containers cool. Continue to cool container after fire is extinguished. For large fires, if possible, withdraw and allow to burn. Immediately withdraw if rising sound from venting device is heard or if fire is causing discoloration to the tank.

TOXICOLOGY

Odor: Sharp, aromatic

Physical Contact: Irritant.

TLV = 245 mg/m^3 (skin)   **STEL** = N.D.

Odor Threshold: 0.025 mg/m^3

IDLH = 40,000 mg/m^3

RTECS # GR8575000

Routes of Entry and Relative LD$_{50}$ (or LC$_{50}$)

Inhalation	3	(160,000 mg/m^3/H)
Ingestion	3	(1400 mg/kg)
Skin Absorption	1	(12,300 mg/kg)

Symptoms of Exposure: Possible irritation to skin, eyes and upper respiratory tract; headache; nausea. May act as a narcotic in high concentrations.

Emergency/First Aid Treatment: Remove to ventilated area; immediately remove any contaminated clothing and wash contaminated areas for 15 minutes using water. Treat supportively and observe for possible shock. If ingested, seek immediate medical aid.

Recommended Clean-Up Procedures

Personal Protection: Level B Ensemble   **Recommended Material** Viton (0)

RCRA Waste # V055   **Reportable Quantities:** 5000 lb (1 lb)

Spills: Remove all potential ignition sources. Dike to contain spill and collect liquid for disposal. Absorb with non-combustible absorbent and take up using non-sparking tools. Decontaminate spill area using soapy water.

Special Emergency Information: May be harmful is inhaled, swallowed, or absorbed through the skin. May form unstable peroxides upon exposure to air. Vapors are heavier than air and may travel some distance to an ignition source. Vapors are more dense than air and may settle in low lying areas.

CUMENE HYDROPEROXIDE - $C_6H_5C(CH_3)_2OOH$

Synonyms: cumyl hydroperoxide, isopropyl benzene hydroperoxide

CAS Number: 80-15-9   **Description:** Colorless to Yellow Liquid
DOT Number: 2116   **DOT Classification:** Organic Peroxide

Molecular Weight: 152.2
Melting Point: °C   (°F)   **Vapor Density:**   N.D.   **Vapor Pressure:**   1 (< 1 mm Hg)
Boiling Point: 153°C (307°F)   **Specific Gravity:**   1.1   **Water Solubility:**   < 1%

Chemical Incompatibilities or Instabilities: Heat, acids, reducing agents, finely divided metals.

FLAMMABILITY

NFPA Hazard Code: **1-2-4-OX**   **NFPA Classification:**   **Class IIIB Combustible**
Flash Point: 79°C (174°F)   **LEL:**   N.D.
Autoignition Temp: N.D.   **UEL:**   N.D.

Fire Extinguishing Methods: Carbon Dioxide, Dry Chemical, Foam, Water.

Special Fire Fighting Considerations: Fight fire from a distance or protected location, if possible. For large fires, if possible, withdraw and allow to burn. May explosively decompose at elevated temperature.

TOXICOLOGY
QUESTIONABLE CARCINOGEN, SENSITIZER

Odor: **Odor Threshold:** N.D.
Physical Contact: Material is extremely destructive to human tissues.
TLV = N.D.   **STEL** = N.D.   **IDLH** = N.D.

Routes of Entry and Relative LD$_{50}$ (or LC$_{50}$)   **RTECS # MX2450000**
   Inhalation   5   (5600 mg/m^3/H)
   Ingestion   5   (382 mg/kg)
   Skin Absorption   N.D.

Symptoms of Exposure: Possible burns or irritation to skin, eyes and upper respiratory tract; headache; nausea.

Emergency/First Aid Treatment: Remove to ventilated area; immediately remove any contaminated clothing and wash contaminated areas for 15 minutes using water. Treat supportively and observe for possible shock. If ingested, seek immediate medical aid.

Recommended Clean-Up Procedures

Personal Protection: Level B Ensemble   **Recommended Material**   Teflon (0)

RCRA Waste #   None   **Reportable Quantities:**   None

Spills: Take up with a non-combustible absorbent and dispose. Decontaminate spill area using water.

Special Emergency Information: May be harmful is inhaled, swallowed, or absorbed through the skin.

CYANOGEN - N≡C-C≡N

Synonyms: dicyanogen; ethanedinitrile; prussite

CAS Number: 460-19-5   **Description:** Colorless Gas
DOT Number: 1026   **DOT Classification:** Poison A

Molecular Weight: 52.0
Melting Point: -34°C (-29°F)   **Vapor Density:** 1.8   **Vapor Pressure:** GAS
Boiling Point: -21°C (-6°F)   **Specific Gravity:**   **Water Solubility:** Soluble

Chemical Incompatibilities or Instabilities: Strong oxidizers, acids, water.

FLAMMABILITY

NFPA Hazard Code: 4-4-2   **NFPA Classification:**
Flash Point: N.A.   **LEL:** 6.6%
Autoignition Temp: N.D.   **UEL:** 43%

Fire Extinguishing Methods: Foam, Water Fog.

Special Fire Fighting Considerations: Isolate for 1/2 mile if rail or tank truck is involved in at fire. Structural fire fighters protective clothing will not provide adequate protection. Water spray should be used to keep closed containers cool. For large fires, if possible, withdraw and allow to burn. Stop leak if it can be safely be done.

DOT Recommended Isolation Zones: Small Spill: 300 ft   Large Spill: 300 ft
DOT Recommended Down Wind Take Cover Distance: Small Spill: 1 mile   Large Spill: 1 mile

TOXICOLOGY

Odor: Pungent   **Odor Threshold:** N.D.
Physical Contact: Material is extremely destructive to human tissues.
TLV = 21 mg/m^3 **STEL** = N.D.   **IDLH** = N.D.

Routes of Entry and Relative LD$_{50}$ (or LC$_{50}$)   **RTECS #** GT1925000
   Inhalation   8   (750 mg/m^3/H)
   Ingestion   N.D.
   Skin Absorption   N.D.

Symptoms of Exposure: Possible irritation to skin, eyes and upper respiratory tract; headache; nausea. Cyanosis.

Emergency/First Aid Treatment: Remove to ventilated area; immediately remove any contaminated clothing and wash contaminated areas for 15 minutes using water. Treat supportively and observe for possible shock. If ingested, seek immediate medical aid. Treat for cyanide poisoning as necessary.

Recommended Clean-Up Procedures

Personal Protection: Level A Ensemble   **Recommended Material**   N.D.

RCRA Waste #   P031   **Reportable Quantities:**   100 lb (1 lb)

Spills: Remove all potential ignition sources. Stop leak if it can be safely be done. Keep area isolated until vapors have dissipated.

Special Emergency Information: May be fatal if inhaled, swallowed, or absorbed through skin. May form unstable peroxides upon exposure to air. Vapors are heavier than air and may travel some distance to an ignition source. Vapors are more dense than air and may settle in low lying areas.

CYANOGEN BROMIDE - BrC≡N

Synonyms: bromine cyanide

CAS Number: 506-68-3   **Description:** White Powder
DOT Number: 1889   **DOT Classification:** Poison B

Molecular Weight: 105.9
Melting Point: 52°C (126°F)   **Vapor Density:**   ---   **Vapor Pressure:**   7 (100 mm Hg)
Boiling Point: 61°C (142°F)   **Specific Gravity:** 2.0   **Water Solubility:**   Soluble

Chemical Incompatibilities or Instabilities: heat, strong oxidizers.

FLAMMABILITY

NFPA Hazard Code: 3-0-1   **NFPA Classification:**
Flash Point: N.D.   **LEL:** N.D.
Autoignition Temp: N.D.   **UEL:** N.D.

Fire Extinguishing Methods: Foam, Water Fog.

Special Fire Fighting Considerations: Structural fire fighters protective clothing will not provide adequate protection. Fight fire from a distance or protected location, if possible. May release toxic fumes upon heating (HCN, HBr). Fight fire from a distance or protected location, if possible. Treat all materials used or generated and equipment involved as contaminated by hazardous waste.

DOT Recommended Isolation Zones: Small Spill: 900 ft   Large Spill: 900 ft
DOT Recommended Down Wind Take Cover Distance: Small Spill: 3 miles   Large Spill: 3 miles

TOXICOLOGY

Odor: **Odor Threshold:** N.D.
Physical Contact: Irritant.
TLV = N.D.   **STEL** = N.D.   **IDLH** = N.D.

Routes of Entry and Relative LD$_{50}$ (or LC$_{50}$)   **RTECS # GT2100000**
   Inhalation   N.D.
   Ingestion   N.D.
   Skin Absorption   N.D.

Symptoms of Exposure: Possible irritation to skin, eyes and upper respiratory tract; headache; nausea. Exposure may result in chemical pneumonitis or pulmonary edema.

Emergency/First Aid Treatment: Remove to ventilated area; immediately remove any contaminated clothing and wash contaminated areas for 15 minutes using water. Treat supportively and observe for possible shock. If ingested, seek immediate medical aid.

Recommended Clean-Up Procedures

Personal Protection: Level B Ensemble   **Recommended Material**   N.D.

RCRA Waste #   U246   **Reportable Quantities:**   1000 lb (1 lb)

Spills: Take up spill. Decontaminate spill area using water. Keep area isolated until vapors have dissipated. Treat all materials used as contaminated by a hazardous waste.

Special Emergency Information: May be fatal if inhaled, swallowed, or absorbed through skin. May undergo hazardous polymerization. Vapors are heavier than air and may travel some distance to an ignition source. Vapors are more dense than air and may settle in low lying areas.

CYANOGEN CHLORIDE - ClC≡N

Synonyms: chlorocyanide

CAS Number: 506-77-4   **Description:** Colorless Liquid or Gas
DOT Number: 1589   **DOT Classification:** Poison A

Molecular Weight: 61.5
Melting Point: -7°C (19°F)   **Vapor Density:** 2.0   **Vapor Pressure:** GAS
Boiling Point: 13°C (55°F)   **Specific Gravity:** 1.2 **Water Solubility:** Soluble

Chemical Incompatibilities or Instabilities: Heat, strong oxidizers.

FLAMMABILITY

NFPA Hazard Code: N.D.   **NFPA Classification:**
Flash Point: N.A.   **LEL:** N.A.
Autoignition Temp: N.A.   **UEL:** N.A.

Fire Extinguishing Methods: Carbon Dioxide, Dry Chemical, Foam, Water.

Special Fire Fighting Considerations: Isolate for 150 ft if rail or tank truck is involved in at fire. Structural fire fighters protective clothing will not provide adequate protection. Water spray should be used to keep closed containers cool.

DOT Recommended Isolation Zones: Small Spill: 1500 ft   Large Spill: 1500 ft
DOT Recommended Down Wind Take Cover Distance: Small Spill: 5 miles   Large Spill: 5 miles

TOXICOLOGY

Odor: **Odor Threshold:** N.D.
Physical Contact: Irritant.
TLV = 0.75 mg/m^3 **STEL** = N.D.   **IDLH** = N.D.

Routes of Entry and Relative LD$_{50}$ (or LC$_{50}$)   **RTECS #** GT2275000
    Inhalation   **10**   (270 mg/m^3/H)
    Ingestion   N.D.
    Skin Absorption N.D.

Symptoms of Exposure: Possible irritation to skin, eyes and upper respiratory tract; headache; nausea. Lachrymation. Weakness. Exposure may result in chemical pneumonitis or pulmonary edema.

Emergency/First Aid Treatment: Remove to ventilated area; immediately remove any contaminated clothing and wash contaminated areas for 15 minutes using water. Treat supportively and observe for possible shock. If ingested, seek immediate medical aid. Treat for cyanide poisoning as necessary.

Recommended Clean-Up Procedures

Personal Protection: Level A Ensemble   **Recommended Material**   N.D.

RCRA Waste #   P033   **Reportable Quantities:**   10 lb (10 lb)

Spills: Remove all potential ignition sources. Stop leak if it can be safely be done. Take up spill. Decontaminate spill area using alkaline hypochlorite solution. Ventilate to dissipate vapors.

Special Emergency Information: May be fatal if inhaled, swallowed, or absorbed through skin. Vapors are heavier than air and may travel some distance to an ignition source. Vapors are more dense than air and may settle in low lying areas.

CYANURIC CHLORIDE

Synonyms: 2,4,6 trichlorotriazine

CAS Number: 108-77-0	**Description:** White Powder	
DOT Number: 2670	**DOT Classification:** N.D.	

Molecular Weight: 184.4

Melting Point: 146°C (295°F)	**Vapor Density:** 6.3	**Vapor Pressure:** 2 (2 mm Hg)
Boiling Point: 90°C (374°F)	**Specific Gravity:** 1.3	**Water Solubility:** Decomposes

Chemical Incompatibilities or Instabilities: Water, alcohols, strong acids, strong oxidizers, acetone.

FLAMMABILITY

NFPA Hazard Code: N.D.	**NFPA Classification:**	
Flash Point: N.A.	**LEL:** N.A.	
Autoignition Temp: N.A.	**UEL:** N.A.	

Fire Extinguishing Methods: Carbon Dioxide, Dry Chemical, Foam, Water.

Special Fire Fighting Considerations: DO NOT allow water to enter container. Water spray should be used to keep closed containers cool. Reaction with water above 30°C may be extremely violent.

TOXICOLOGY
QUESTIONABLE CARCINOGEN

Odor: Pungent   **Odor Threshold:**

Physical Contact: Material is extremely destructive to human tissues.

TLV = N.D.   **STEL** = N.D.   **IDLH** = N.D.

Routes of Entry and Relative LD50 (or LC50)   **RTECS # XZ1400000**

Inhalation	N.D.	
Ingestion	**5**	(485 mg/kg)
Skin Absorption	N.D.	

Symptoms of Exposure: Possible burns or irritation to skin, eyes and upper respiratory tract; headache; nausea. Lachrymation. Inhalation of vapors may be fatal by causing glottis to spasm and suffocation. Exposure may result in chemical pneumonitis or pulmonary edema.

Emergency/First Aid Treatment: Remove to ventilated area; immediately remove any contaminated clothing and wash contaminated areas for 15 minutes using water. Treat supportively and observe for possible shock. If ingested, seek immediate medical aid.

Recommended Clean-Up Procedures

Personal Protection: Level B Ensemble	**Recommended Material** N.D.	
RCRA Waste # None	**Reportable Quantities:** None	

Spills: Take up spill. Use dry decontamination techniques.

Special Emergency Information: May be harmful is inhaled, swallowed, or absorbed through the skin.

CYCLOHEXANE

Synonyms:

CAS Number: 110-82-7
DOT Number: 1145

Description: Colorless Liquid
DOT Classification: Flammable Liquid

Molecular Weight: 84.2
Melting Point: 7°C (45°F)
Boiling Point: 81°C (178°F)

Vapor Density: 2.9
Specific Gravity: 0.8

Vapor Pressure: 6 (77 mm Hg)
Water Solubility: Insoluble

Chemical Incompatibilities or Instabilities: Strong oxidizers.

FLAMMABILITY

NFPA Hazard Code: 1-3-0
Flash Point: -17°C (2°F)
Autoignition Temp: 245°C (473°F)

NFPA Classification: Class 1B Flammable
LEL: 1.3%
UEL: 8.4%

Fire Extinguishing Methods: Alcohol Resistant Foam, Carbon Dioxide, Dry Chemical, Water.

Special Fire Fighting Considerations: Isolate for 1/2 mile if rail or tank truck is involved in at fire. Water spray should be used to keep closed containers cool. For large fires, if possible, withdraw and allow to burn. Immediately withdraw if rising sound from venting device is heard or if fire is causing discoloration to the tank.

TOXICOLOGY

Odor: Pungent
Physical Contact: Irritant
TLV = 1030 mg/m^3 **STEL** = N.D.

Odor Threshold: 2 mg/m^3

IDLH = 35,000 mg/m^3

Routes of Entry and Relative LD50 (or LC50)
Inhalation   N.D.
Ingestion   1   (29,820 mg/kg)
Skin Absorption   N.D.

RTECS # GV6300000

Symptoms of Exposure: Possible irritation to skin, eyes and upper respiratory tract; headache; nausea. Exposure may result in chemical pneumonitis or pulmonary edema.

Emergency/First Aid Treatment: Remove to ventilated area; immediately remove any contaminated clothing and wash contaminated areas for 15 minutes using water. Treat supportively and observe for possible shock. If ingested, seek immediate medical aid.

Recommended Clean-Up Procedures

Personal Protection: Level C Ensemble   **Recommended Material** Nitrile Rubber (+), Viton (+), teflon (0)

RCRA Waste # U056   **Reportable Quantities:** 1000 lb (1000 lb)

Spills: Remove all potential ignition sources. Dike to contain spill and collect liquid for disposal. Absorb with non-combustible absorbant and take up using non-sparking tools. Decontaminate spill area using soapy water.

Special Emergency Information: May be harmful is inhaled, swallowed, or absorbed through the skin. Vapors are heavier than air and may travel sone distance to an ignition source.

Cycloheptane (CAS # 544-25-2, DOT # 2241, RTECS # GU3140000),
cycloheptene (CAS # 628-92-2, DOT # 2242, RTECS # GU4615000), and
cyclohexene (CAS # 110-83-8, DOT # 2256, RTECS # GW2500000), have similar chemical, physical, and toxicological properties.

CYCLOHEXANOL

Synonyms:

CAS Number: 108-93-0
DOT Number: ----

Description: Viscous Liquid
DOT Classification: ----

Molecular Weight: 100.2
Melting Point: 22°C (72°F)
Boiling Point: 162°C (324°F)

Vapor Density: 3.5
Specific Gravity: 0.95

Vapor Pressure: **2** (1 mm Hg)
Water Solubility: 3%

Chemical Incompatibilities or Instabilities: Strong oxidizers.

FLAMMABILITY

NFPA Hazard Code: 1-2-0
Flash Point: 68°C (154°F)
Autoignition Temp: 300°C (572°F)

NFPA Classification: Class IIIA Combustible
LEL: N.D.
UEL: N.D.

Fire Extinguishing Methods: Alcohol Resistant Foam, Carbon Dioxide, Dry Chemical, Water.

Special Fire Fighting Considerations: Water spray should be used to keep closed containers cool. Isolate for 1/2 mile if rail or tank truck is involved in at fire. Immediately withdraw if rising sound from venting device is heard or if fire is causing discoloration to the tank.

TOXICOLOGY

Odor: Comphor-like
Physical Contact: Irritant
TLV = 206 mg/m^3 (skin)   **STEL** = N.D.

Odor Threshold: 0.25 mg/kg

IDLH = N.D.

Routes of Entry and Relative LD50 (or LC50)
    Inhalation   N.D.
    Ingestion   **3**   (2060 mg/kg)
    Skin Absorption   N.D.

RTECS # GV7875000

Symptoms of Exposure: Possible irritation to skin, eyes and upper respiratory tract; headache; nausea.

Emergency/First Aid Treatment: Remove to ventilated area; immediately remove any contaminated clothing and wash contaminated areas for 15 minutes using water. Treat supportively and observe for possible shock. If ingested, seek immediate medical aid.

Recommended Clean-Up Procedures

Personal Protection: Level B Ensemble

RCRA Waste # None

Recommended Material Butyl Rubber (+), Nitrile Rubber (+), Polyvinyl Alcohol (+), Viton (+)

Reportable Quantities: None

Spills: Dike to contain spill and collect liquid for later disposal. Decontaminate spill area using soapy water.

Special Emergency Information: May be harmful is inhaled, swallowed, or absorbed through the skin.

CYCLOHEXANONE

Synonyms:

CAS Number: 108-94-1
DOT Number: 1915

Description: Colorless Liquid
DOT Classification: Flammable Liquid

Molecular Weight: 98.2
Melting Point: -45°C (-49°F)
Boiling Point: 116°C (241°F)

Vapor Density: 3.4
Specific Gravity: 0.9

Vapor Pressure: **3** (10 mm Hg)
Water Solubility: Insoluble

Chemical Incompatibilities or Instabilities: Strong oxidizers.

FLAMMABILITY

NFPA Hazard Code:
Flash Point: °C (°F)
Autoignition Temp: °C (°F)

NFPA Classification:
LEL:
UEL:

Fire Extinguishing Methods: Alcohol Resistant Foam, Carbon Dioxide, Dry Chemical,, Water.

Special Fire Fighting Considerations: Isolate for 1/2 mile if rail or tank truck is involved in a fire. Water spray from an unmanned device should be used to keep closed containers cool. Continue to cool container after fire is extinguished. For large fires, if possible, withdraw and allow to burn. Immediately withdraw if rising sound from venting device is heard or if fire is causing discoloration to the tank.

TOXICOLOGY

Odor: Nail polish remover
Physical Contact: Irritant
TLV = 100 mg/m^3 (skin)   **STEL** = N.D.

Odor Threshold: 0.2 mg/m^3

IDLH = 20,000 mg/m^3

RTECS # GW1050000

Routes of Entry and Relative LD$_{50}$ (or LC$_{50}$)
Inhalation	3	(144,000 mg/m^3/H)
Ingestion	3	(1535 mg/kg)
Skin Absorption	5	(948 mg/kg)

Symptoms of Exposure: Possible irritation to skin, eyes and upper respiratory tract; headache; nausea. Exposure may result in chemical pneumonitis or pulmonary edema.

Emergency/First Aid Treatment: Remove to ventilated area; immediately remove any contaminated clothing and wash contaminated areas for 15 minutes using water. Treat supportively and observe for possible shock. If ingested, seek immediate medical aid.

Recommended Clean-Up Procedures

Personal Protection: Level B Ensemble

RCRA Waste # U057

Recommended Material Butyl Rubber (+), Teflon (+)

Reportable Quantities: 5000 lb (1 lb)

Spills: Remove all potential ignition sources. Dike to contain spill, absorb with non-combustible absorbant and take up using non-sparking tools. Decontaminate spill area using soapy water.

Special Emergency Information: May be harmful if inhaled, swallowed or absorbed through the skin. Vapors are heavier than air and may travel some distance to an ignition source. Vapors are more dense than air and may settle in low lying areas.

CYCLOHEXYL AMINE

Synonyms: aminocyclohexane

CAS Number: 108-91-8	**Description:** Colorless Liquid	
DOT Number: 2357	**DOT Classification:** Flammable Liquid	

Molecular Weight: 99.2

Melting Point: -18°C (0°F)	**Vapor Density:** 3.4	**Vapor Pressure:** 3 (10 mm Hg)
Boiling Point: 135°C (275°F)	**Specific Gravity:** 0.9	**Water Solubility:** Soluble

Chemical Incompatibilities or Instabilities: Oxidizers, halgens.

FLAMMABILITY

NFPA Hazard Code: 2-3-0	**NFPA Classification:** Class IC Flammable
Flash Point: 28°C (83°F)	**LEL:** 1.5%
Autoignition Temp 293°C (560°F)	**UEL:** 9.4%

Fire Extinguishing Methods: Alcohol Resistant Foam, Carbon Dioxide, Dry Chemical, Water.

Special Fire Fighting Considerations: Isolate for 1/2 mile if rail or tank truck is involved in at fire. Water spray should be used to keep closed containers cool. For large fires, if possible, withdraw and allow to burn. Immediately withdraw if rising sound from venting device is heard or if fire is causing discoloration to the tank.

TOXICOLOGY
QUESTIONABLE CARCINOGEN

Odor: Fish

Odor Threshold: N.D.

Physical Contact: Material is extremely destructive to human tissues.

TLV = 41 mg/m^3 **STEL** = N.D.   **IDLH** = N.D.

Routes of Entry and Relative LD50 (or LC50)

			RTECS # GX0700000
Inhalation	5	(7500 mg/m^3)	
Ingestion	7	(156 mg/kg)	
Skin Absorption	8	(277 mg/kg)	

Symptoms of Exposure: Possible burns or irritation to skin, eyes and upper respiratory tract; headache; nausea. Inhalation of vapors may be fatal by causing glottis to spasm and suffocation. Exposure may result in chemical pneumonitis or pulmonary edema.

Emergency/First Aid Treatment: Remove to ventilated area; immediately remove any contaminated clothing and wash contaminated areas for 15 minutes using water. Treat supportively and observe for possible shock. If ingested, seek immediate medical aid.

Recommended Clean-Up Procedures

Personal Protection: Level B Ensemble	**Recommended Material** N.D.
RCRA Waste # None	**Reportable Quantities:** None

Spills: Remove all potential ignition sources. Dike to contain spill and collect liquid for disposal. Absorb with non-combustible absorbant and take up using non-sparking tools. Decontaminate spill area using water.

Special Emergency Information: May be harmful is inhaled, swallowed, or absorbed through the skin. Vapors are heavier than air and may travel sone distance to an ignition source. Vapors are more dense than air and may settle in low lying areas.

CYCLONITE

Synonyms: cyclotrimethylene nitramine

CAS Number: 121-82-4   **Description:** White Powder
DOT Number: 0072, 0118   **DOT Classification:** Class A Explosive

Molecular Weight: 222.2
Melting Point: 204°C (399°F)   **Vapor Density:** N.A.   **Vapor Pressure:** N.A.
Boiling Point: °C (°F)   **Specific Gravity:** 1.8   **Water Solubility:** Insoluble

Chemical Incompatibilities or Instabilities: Oxidizers, combustibles.

FLAMMABILITY

NFPA Hazard Code: **NFPA Classification:**
Flash Point: N.A.   **LEL:** N.A.
Autoignition Temp: N.A.   **UEL:** N.A.

Fire Extinguishing Methods: Carbon Dioxide, Foam, Water.

Special Fire Fighting Considerations: Fight fire from a distance or protected location, if possible. Water spray should be used to keep closed containers cool. For large fires, if possible, withdraw and allow to burn. Isolate for 1/2 mile if rail or tank truck is involved in at fire.

TOXICOLOGY
QUESTIONABLE TERATGEN

Odor: **Odor Threshold:** N.D.
Physical Contact: Material is extremely destructive to human tissues.
TLV = 1.5 mg/m^3 (skin)   **STEL** = N.D.   **IDLH** = N.D.

Routes of Entry and Relative LD50 (or LC50)   **RTECS # XY9450000**
    Inhalation   N.D.
    Ingestion   8   (100 mg/kg)
    Skin Absorption   N.D.

Symptoms of Exposure: Possible burns or irritation to skin, eyes and upper respiratory tract; headache; nausea. Epileptiform convulsions. Inhalation of vapors may be fatal by causing glottis to spasm and suffocation. Exposure may result in chemical pneumonitis or pulmonary edema.

Emergency/First Aid Treatment: Remove to ventilated area; immediately remove any contaminated clothing and wash contaminated areas for 15 minutes using water. Treat supportively and observe for possible shock. If ingested, seek immediate medical aid.

Recommended Clean-Up Procedures

Personal Protection: Level A Ensemble   **Recommended Material** N.D.

RCRA Waste # None   **Reportable Quantities:** None

Spills: Clean up only under expert supervision.

Special Emergency Information: May be harmful is inhaled, swallowed, or absorbed through the skin. Use extreme caution; this material is a powerful explosive.

CYCLOOCTATETRAENE

Synonyms: 1,3,5,7-cyclooctatetraene

CAS Number: 629-20-9	**Description:** Yellow Liquid	
DOT Number: 2538	**DOT Classification:** Flammable Liquid	

Molecular Weight: 104.2

Melting Point:	-27°C (-17°F)	**Vapor Density:** ---	**Vapor Pressure:**	3 (9 mm Hg)
Boiling Point:	142°C (288°F)	**Specific Gravity:** 0.9	**Water Solubility:**	Insoluble

Chemical Incompatibilities or Instabilities: Strong Oxidizers.

FLAMMABILITY

NFPA Hazard Code:

Flash Point:	22°C (72°F)	**NFPA Classification:** Class IB Flammable
Autoignition Temp:	°C (°F)	**LEL:** N.D.
		UEL: N.D.

Fire Extinguishing Methods: Carbon Dioxide, Dry Chemical, Foam, Water.

Special Fire Fighting Considerations: Isolate for 1/2 mile if rail or tank truck is involved in at fire. Water spray should be used to keep closed containers cool. Keep away from ends of tank. For large fires, if possible, withdraw and allow to burn.

TOXICOLOGY

Odor:

Physical Contact: A simple asphyxiant

Odor Threshold: N.D.

TLV = N.D.   **STEL** = N.D.   **IDLH** = N.D.

Routes of Entry and Relative LD50 (or LC50)

Inhalation	N.D.	**RTECS # GY0175600**
Ingestion	N.D.	
Skin Absorption	N.D.	

Symptoms of Exposure: A simple asphyxiant.

Emergency/First Aid Treatment: Remove to ventilated area; immediately remove any contaminated clothing and wash contaminated areas for 15 minutes using water. Treat supportively and observe for possible shock. If ingested, seek immediate medical aid.

Recommended Clean-Up Procedures

Personal Protection:	Level B Ensemble	**Recommended Material** N.D.
RCRA Waste #	None	**Reportable Quantities:** None

Spills: Remove all potential ignition sources. Dike to contain spill and collect liquid for disposal. Absorb with non-combustible absorbant and take up using non-sparking tools. Decontaminate spill area using soapy water.

Special Emergency Information: May be harmful is inhaled, swallowed, or absorbed through the skin. May form unstable peroxides upon exposure to air. Vapors are heavier than air and may travel sone distance to an ignition source. Vapors are more dense than air and may settle in low lying areas.

1,5-cyclooctadiene (CAS # 1552-12-1, DOT # 2520, RTECS # GX9620000) has similar chemical, physical, and toxicological properties.

CYCLOPENTADIENE

Synonyms: 1,3-cyclopentadiene

CAS Number: 542-92-7   **Description:** Colorless to Dark Liquid
DOT Number: ---   **DOT Classification:** ---

Molecular Weight: 66.1
Melting Point: -85°C (-121°F)   **Vapor Density:** N.D.   **Vapor Pressure:** N.D.
Boiling Point: 43°C (109°F)   **Specific Gravity:** 0.8   **Water Solubility:** Insoluble

Chemical Incompatibilities or Instabilities: Strong acids, nitrogen oxides, potassium hydroxide, oxidizers.

FLAMMABILITY

NFPA Hazard Code: N.D.   **NFPA Classification:** Class IB Flammable
Flash Point: 25°C   (77°F)   **LEL:** N.D.
Autoignition Temp: °C   (°F)   **UEL:** N.D.

Fire Extinguishing Methods: Carbon Dioxide, Dry Chemical, Foam, Water.

Special Fire Fighting Considerations: Isolate for 1/2 mile if rail or tank truck is involved in at fire. Water spray should be used to keep closed containers cool. For large fires, if possible, withdraw and allow to burn. Immediately withdraw if rising sound from venting device is heard or if fire is causing discoloration to the tank.

TOXICOLOGY

Odor: Pine   **Odor Threshold:** 5 mg/m^3
Physical Contact: Mild irritant
TLV = 200 mg/m^3 **STEL** = 400 mg/m^3 **IDLH** = 2000 mg/m^3

Routes of Entry and Relative LD50 (or LC50)   **RTECS # GY1000000**
    Inhalation   N.D.
    Ingestion   N.D.
    Skin Absorption   N.D.

Symptoms of Exposure: Possible irritation to skin, eyes and upper respiratory tract; headache; nausea. Depression of circulatory and respiratory systems.

Emergency/First Aid Treatment: Remove to ventilated area; immediately remove any contaminated clothing and wash contaminated areas for 15 minutes using water. Treat supportively and observe for possible shock. If ingested, seek immediate medical aid.

Recommended Clean-Up Procedures

Personal Protection: Level B Ensemble   **Recommended Material** N.D.

RCRA Waste #   None   **Reportable Quantities:**   None

Spills: Remove all potential ignition sources. Dike to contain spill and collect liquid for disposal. Absorb with non-combustible absorbant and take up using non-sparking tools. Decontaminate spill area using soapy water.

Special Emergency Information: May be harmful is inhaled, swallowed, or absorbed through the skin. May undergo hazardous polymerization. Vapors are heavier than air and may travel sone distance to an ignition source. Vapors are more dense than air and may settle in low lying areas. May undergo explosive decomposition at elevated temperatures.

CYCLOPENTENE

Synonyms:

CAS Number: 142-29-0

DOT Number: 2246

Description: Colorless Liquid

DOT Classification: Flammable Liquid

Molecular Weight: 63.1

Melting Point: -94°C (-137°F)

Boiling Point: 44°C (111°F)

Vapor Density: 2.4

Specific Gravity: 0.8

Vapor Pressure: N.D.

Water Solubility: Insoluble

Chemical Incompatibilities or Instabilities: Strong oxidizers.

FLAMMABILITY

NFPA Hazard Code: 1-3-1

Flash Point: -29°C (-20°F)

Autoignition Temp: 395°C (743°F)

NFPA Classification: Class 1A Flammable

LEL: N.D.

UEL: N.D.

Fire Extinguishing Methods: Carbon Dioxide, Dry Chemical, Foam, Water.

Special Fire Fighting Considerations: Isolate for 1/2 mile if rail or tank truck is involved in at fire. Water spray should be used to keep closed containers cool. For large fires, if possible, withdraw and allow to burn. Immediately withdraw if rising sound from venting device is heard or if fire is causing discoloration to the tank.

TOXICOLOGY

Odor:

Physical Contact: Irritant.

TLV = N.D.

STEL = N.D.

Odor Threshold: N.D.

IDLH = N.D.

RTECS # GY5950000

Routes of Entry and Relative LD50 (or LC50)

Inhalation	N.D.	
Ingestion	3	(1656 mg/kg)
Skin Absorption	4	(1231 mg/kg)

Symptoms of Exposure: Possible irritation to skin, eyes and upper respiratory tract; headache; nausea.

Emergency/First Aid Treatment: Remove to ventilated area; immediately remove any contaminated clothing and wash contaminated areas for 15 minutes using water. Treat supportively and observe for possible shock. If ingested, seek immediate medical aid.

Recommended Clean-Up Procedures

Personal Protection: Level B Ensemble

RCRA Waste # None

Recommended Material N.D.

Reportable Quantities: None

Spills: Remove all potential ignition sources. Dike to contain spill and collect liquid for disposal. Absorb with non-combustible absorbant and take up using non-sparking tools. Decontaminate spill area using soapy water.

Special Emergency Information: May be harmful is inhaled, swallowed, or absorbed through the skin. Vapors are heavier than air and may travel sone distance to an ignition source. Vapors are more dense than air and may settle in low lying areas.

Cyclopentane (CAS # 287-92-3, DOT # 1146, RTECS # BY2390000) has similar chemical, physical, and toxicological properties.

CYCLOPENTANONE

Synonyms:

CAS Number: 120-92-3

DOT Number: 2245

Description: Colorless Liquid

DOT Classification: Flammable Liquid

Molecular Weight: 84.1

Melting Point: -58°C (-72°F)

Boiling Point: 131°C (267°F)

Vapor Density: 2.3

Specific Gravity: 0.95

Vapor Pressure: N.D.

Water Solubility: < 1%

Chemical Incompatibilities or Instabilities: Nitric acid.

FLAMMABILITY

NFPA Hazard Code: 2-3-0

Flash Point: 26°C (79°F)

Autoignition Temp: N.D.

NFPA Classification: Class IC Flammable

LEL: N.D.

UEL: N.D.

Fire Extinguishing Methods: Alcohol Resistant Foam, Carbon Dioxide, Dry Chemical, Water.

Special Fire Fighting Considerations: Isolate for 1/2 mile if rail or tank truck is involved in at fire. Water spray should be used to keep closed containers cool. For large fires, if possible, withdraw and allow to burn. Immediately withdraw if rising sound from venting device is heard or if fire is causing discoloration to the tank.

TOXICOLOGY

Odor:

Physical Contact: Irritant

TLV = N.D.

Odor Threshold: N.D.

STEL = N.D.

IDLH = N.D.

Routes of Entry and Relative LD50 (or LC50)

Inhalation   N.D.

Ingestion   N.D.

Skin Absorption   N.D.

RTECS # GY4725000

Symptoms of Exposure: Possible irritation to skin, eyes and upper respiratory tract, headache, nausea.

Emergency/First Aid Treatment: Remove to ventilated area; immediately remove any contaminated clothing and wash contaminated areas for 15 minutes using water. Treat supportively and observe for possible shock. If ingested, seek immediate medical aid.

Recommended Clean-Up Procedures

Personal Protection: Level B Ensemble

RCRA Waste # None

Recommended Material N.D.

Reportable Quantities: None

Spills: Remove all potential ignition sources. Dike to contain spill and collect liquid for disposal. Absorb with non-combustible absorbant and take up using non-sparking tools. Decontaminate spill area using soapy water.

Special Emergency Information: May be harmful is inhaled, swallowed, or absorbed through the skin. Vapors are heavier than air and may travel sone distance to an ignition source.

Cyclopentanol (CAS # 96-41-3, DOT # 2244, RTECS # GY4702000) has similar chemical, physical, and toxicological properties.

CYCLOPROPANE

Synonyms:

CAS Number: 75-19-4   **Description:** Colorless Gas
DOT Number: 1027   **DOT Classification:** Flammable Gas

Molecular Weight: 42.1
Melting Point: -127°C (-197°F)   **Vapor Density:** 1.5   **Vapor Pressure:** GAS
Boiling Point: -34°C (-29°F)   **Specific Gravity:** --- **Water Solubility:** Insoluble

Chemical Incompatibilities or Instabilities: Strong oxidizers.

FLAMMABILITY

NFPA Hazard Code: 1-4-0   **NFPA Classification:**
Flash Point: °C (°F)   **LEL:** 2.4%
Autoignition Temp: 500°C (932°F) **UEL:** 10.4%

Fire Extinguishing Methods: Carbon Dioxide, Dry Chemical, Foam, Water Spray or Fog..

Special Fire Fighting Considerations: Isolate for 1/2 mile if rail or tank truck is involved in at fire. Water spray should be used to keep closed containers cool. For large fires, if possible, withdraw and allow to burn. Immediately withdraw if rising sound from venting device is heard or if fire is causing discoloration to the tank.

TOXICOLOGY
QUESTIONABLE CARCINOGEN

Odor: **Odor Threshold:** N.D.
Physical Contact: A simple asphyxiant.
TLV = N.D.   **STEL** = N.D.   **IDLH** = N.D.

Routes of Entry and Relative LD50 (or LC50)   **RTECS # GZ0690000**
   Inhalation   N.D.
   Ingestion   N.D.
   Skin Absorption   N.D.

Symptoms of Exposure: A simple asphysiant. May act as a narcotic in high concentrations.

Emergency/First Aid Treatment: Remove to ventilated area; immediately remove any contaminated clothing and wash contaminated areas for 15 minutes using water. Treat supportively and observe for possible shock. If ingested, seek immediate medical aid.

Recommended Clean-Up Procedures

Personal Protection: Level B Ensemble   **Recommended Material** N.D.

RCRA Waste # None   **Reportable Quantities:** None

Spills: Remove all potential ignition sources. Stop leak if it can safely be done. Ventilate to dissipate vapors.

Special Emergency Information: May be harmful is inhaled, swallowed, or absorbed through the skin. Vapors are heavier than air and may travel sone distance to an ignition source. Vapors are more dense than air and may settle in low lying areas.

2,4-D

Synonyms: 2,4-dichlorophenoxyacetic acid; weed b-gon

CAS Number: 94-75-7	**Description:** White Powder
DOT Number: 2765	**DOT Classification:** ORM-A

Molecular Weight: 221.0

Melting Point:	141°C (286°F)	**Vapor Density:**	7.6	**Vapor Pressure:**	1 (< 1 mm Hg)
Boiling Point:	°C (°F)	**Specific Gravity:**		**Water Solubility:**	Insoluble

Chemical Incompatibilities or Instabilities: Strong oxidizers.

FLAMMABILITY

NFPA Hazard Code:		**NFPA Classification:**	
Flash Point:	N.A.	**LEL:**	N.A.
Autoignition Temp:	N.A.	**UEL:**	N.A.

Fire Extinguishing Methods: Carbon Dioxide, Dry Chemical, Foam, Water.

Special Fire Fighting Considerations: Structural fire fighters protective clothing will not provide adequate protection. Fight fire from a distance or protected location, if possible. Treat all materials used or generated and equipment involved as contaminated by hazardous waste.

TOXICOLOGY
SUSPECTED CARCINOGEN

Odor: **Odor Threshold:** N.D.

Physical Contact: Irritant

TLV = 10 mg/m^3 **STEL** = N.D.   **IDLH** = 500 mg/m^3

Routes of Entry and Relative LD50 (or LC50)   **RTECS # AG6825000**

Inhalation	N.D.	
Ingestion	5	(370 mg/kg)
Skin Absorption	4	(1400 mg/kg)

Symptoms of Exposure: Skin and eye irritation, weakness,nausea, convulsions.

Emergency/First Aid Treatment: Remove to ventilated area; immediately remove any contaminated clothing and wash contaminated areas for 15 minutes using water. Treat supportively and observe for possible shock. If ingested, seek immediate medical aid.

Recommended Clean-Up Procedures

Personal Protection: Level B Ensemble	**Recommended Material** Natural Rubber (+), Neoprene (+), Natural Rubber (+), Polyvinyl Chloride (+)
RCRA Waste # U240	**Reportable Quantities:** 100 lb (100 lb)

Spills: Take up in non-combustible absorbant and dispose. Decontaminate spill area using soapy water.

Special Emergency Information: May be harmful is inhaled, swallowed, or absorbed through the skin.

DDT

Synonyms: dichlorodiphenyltrichloroethane

CAS Number: 50-29-3		**Description:** White Powder	
DOT Number: 2761		**DOT Classification:** ORM-A	

Molecular Weight: 354.5

Melting Point: 109°C (228°F)	**Vapor Density:** N.A.	**Vapor Pressure:** N.A.
Boiling Point: °C (°F)	**Specific Gravity:** N.D.	**Water Solubility:** Insoluble

Chemical Incompatibilities or Instabilities: Strong oxidizers.

FLAMMABILITY

NFPA Hazard Code:	**NFPA Classification:**
Flash Point: N.A.	**LEL:** N.A.
Autoignition Temp: N.A.	**UEL:** N.A.

Fire Extinguishing Methods: Carbon Dioxide, Dry Chemical, Foam, Water.

Special Fire Fighting Considerations: Structural fire fighters protective clothing will not provide adequate protection. Fight fire from a distance or protected location, if possible. Treat all materials used or generated and equipment involved as contaminated by hazardous waste.

TOXICOLOGY
CARCINOGEN

Odor:	**Odor Threshold:** N.D.
Physical Contact: Mild irritant.	
TLV = N.D. **STEL** = N.D.	**IDLH** = N.D.

Routes of Entry and Relative LD50 (or LC50)   **RTECS # KJ3325000**

Inhalation	N.D.	
Ingestion	8	(87 mg/kg)
Skin Absorption	8	(300 mg/kg)

Symptoms of Exposure: Headache, nausea, convulsions, numbness in the extremities, sweating.

Emergency/First Aid Treatment: Remove to ventilated area; immediately remove any contaminated clothing and wash contaminated areas for 15 minutes using water. Treat supportively and observe for possible shock. If ingested, seek immediate medical aid.

Recommended Clean-Up Procedures

Personal Protection: Level B Ensemble	**Recommended Material** N.D.
RCRA Waste # U061	**Reportable Quantities:** 1 lb (1 lb)

Spills: Take up spill. Decontaminate spill area using soapy water. Treat all materials used as contaminated by a hazardous waste.

Special Emergency Information: May be harmful is inhaled, swallowed, or absorbed through the skin. Toxicity of DDT may be enhanced if dissolved in an organic solvent.

DECABORANE - B$_{10}$H$_{14}$

Synonyms:

CAS Number: 17702-41-9   **Description:** White Powder
DOT Number: 1868   **DOT Classification:** Flammable Solid

Molecular Weight: 122.2
Melting Point: 99°C (210°F)   **Vapor Density:** N.A.   **Vapor Pressure:** N.A.
Boiling Point: 213°C (415°F)   **Specific Gravity:** 0.9   **Water Solubility:** Insoluble

Chemical Incompatibilities or Instabilities: Oxidizers, halogenated compounds, ethers, oxygen.

FLAMMABILITY

NFPA Hazard Code: 3-2-1   **NFPA Classification:**
Flash Point: N.A.   **LEL:** N.A.
Autoignition Temp: 149°C (300°F)   **UEL:** N.A.

Fire Extinguishing Methods: Dry sand, water.

Special Fire Fighting Considerations: Water spray should be used to keep closed containers cool. Approach fire from upwind. Structural fire fighters protective clothing will not provide adequate protection. For large fires, if possible, withdraw and allow to burn. Treat all materials used or generated and equipment involved as contaminated by hazardous waste.

TOXICOLOGY

Odor: Pungent   **Odor Threshold:** 0.3 mg/m^3
Physical Contact: Material is extremely destructive to human tissues.
TLV = 0.3 mg/m^3 **STEL** = 0.9 mg/m^3(skin)   **IDLH** = 100 mg/m^3

Routes of Entry and Relative LD50 (or LC50)   **RTECS #**   **HD1400000**
   Inhalation   7   (1100 mg/m^3/H)
   Ingestion   8   (64 mg/kg)
   Skin Absorption   9   (71 mg/kg)

Symptoms of Exposure: Possible burns or irritation to skin, eyes and upper respiratory tract; headache; nausea. Muscular tremors.

Emergency/First Aid Treatment: Remove to ventilated area; immediately remove any contaminated clothing and wash contaminated areas for 15 minutes using water. Treat supportively and observe for possible shock. If ingested, seek immediate medical aid.

Recommended Clean-Up Procedures

Personal Protection: Level B Ensemble   **Recommended Material**   N.D.

RCRA Waste #   None   **Reportable Quantities:**   None

Spills: Remove all potential ignition sources. Dike to contain spill and collect liquid for disposal. Absorb with non-combustible absorbant and take up using non-sparking tools. Decontaminate spill area using soapy water. Treat all materials used or generated and equipment involved as contaminated by hazardous waste.

Special Emergency Information: May be fatal if inhaled, swallowed, or absorbed through skin.

DECAHYDRONAPHTHALENE

Synonyms: decalin

CAS Number: 91-17-8**Description:** Colorless Liquid
DOT Number: 1147**DOT Classification:** Combustible Liquid

Molecular Weight: 138.3
Melting Point: -31°C (-24°F)   **Vapor Density:**   4.8   **Vapor Pressure:**   2 (5 mm Hg)
Boiling Point: 187°C (369°F)   **Specific Gravity:** 0.9   **Water Solubility:** Insoluble

Chemical Incompatibilities or Instabilities: Strong oxidizers.

FLAMMABILITY

NFPA Hazard Code: 2-2-0   **NFPA Classification:**   Class II Combustible
Flash Point: 58°C (136°F)   **LEL:**   0.7%
Autoignition Temp: 250°C (482°F)   **UEL:**   4.9%

Fire Extinguishing Methods: Carbon Dioxide, Dry Chemical, Foam, Water.

Special Fire Fighting Considerations: Isolate for 1/2 mile if rail or tank truck is involved in at fire. Water spray should be used to keep closed containers cool. For large fires, if possible, withdraw and allow to burn. Immediately withdraw if rising sound from venting device is heard or if fire is causing discoloration to the tank.

TOXICOLOGY
QUESTIONABLE CARCINOGEN

Odor: **Odor Threshold:** N.D.
Physical Contact: Irritant.
TLV = N.D.   **STEL** = N.D.   **IDLH** = N.D.

Routes of Entry and Relative LD50 (or LC50)   **RTECS # QJ3150000**
    Inhalation   N.D.
    Ingestion   3   (4170 mg/kg)
    Skin Absorption   1   (5900 mg/kg)

Symptoms of Exposure: Possible irritation to skin, eyes and upper respiratory tract; headache; nausea.

Emergency/First Aid Treatment: Remove to ventilated area; immediately remove any contaminated clothing and wash contaminated areas for 15 minutes using water. Treat supportively and observe for possible shock. If ingested, seek immediate medical aid.

Recommended Clean-Up Procedures

Personal Protection: Level B Ensemble   **Recommended Material**   N.D.

RCRA Waste #   None   **Reportable Quantities:**   None

Spills: Remove all potential ignition sources. Dike to contain spill, absorb with non-combustible absorbant and take up using non-sparking tools. Decontaminate spill area using soapy water.

Special Emergency Information: May be harmful is inhaled, swallowed, or absorbed through the skin.

DIACETONE ALCOHOL - CH3COCH2C(CH3)2OH

Synonyms: 4-hydroxy-4 methyl pentanone

CAS Number: 123-42-2 **Description:** Colorless Liquid
DOT Number: 1148 **DOT Classification:** Flammable Liquid

Molecular Weight: 116.2
Melting Point: -50°C (-58°F) **Vapor Density:** 4.0 **Vapor Pressure:** 1 (< 1 mm Hg)
Boiling Point: 168°C (334°F) **Specific Gravity:** 0.9 **Water Solubility:** Soluble

Chemical Incompatibilities or Instabilities: Strong oxidizers.

FLAMMABILITY

NFPA Hazard Code: 1-2-0 **NFPA Classification:** Class IIA Combustible
Flash Point: 61°C (143°F) **LEL:** 1.8%
Autoignition Temp: 604°C (1118°F) **UEL:** 6.9%

Fire Extinguishing Methods: Carbon Dioxide, Dry Chemical, Foam, Water.

Special Fire Fighting Considerations: Isolate for 1/2 mile if rail or tank truck is involved in at fire. Water spray should be used to keep closed containers cool. For large fires, if possible, withdraw and allow to burn. Immediately withdraw if rising sound from venting device is heard or if fire is causing discoloration to the tank.

TOXICOLOGY

Odor: Sweet **Odor Threshold:** 1.3 mg/m^3
Physical Contact: Irritant.
TLV = 238 mg/m^3 **STEL** = N.D. **IDLH** = 10,000 mg/m^3

Routes of Entry and Relative LD50 (or LC50) **RTECS # SA9100000**
　　Inhalation　　　N.D.
　　Ingestion　　　3　　(4000 mg/kg)
　　Skin Absorption　1　　(13,500 mg/kg)

Symptoms of Exposure: Possible irritation to skin, eyes and upper respiratory tract, headache, nausea.

Emergency/First Aid Treatment: Remove to ventilated area; immediately remove any contaminated clothing and wash contaminated areas for 15 minutes using water. Treat supportively and observe for possible shock. If ingested, seek immediate medical aid.

Recommended Clean-Up Procedures

Personal Protection: Level B Ensemble **Recommended Material** Neoprene (0)

RCRA Waste # None **Reportable Quantities:** None

Spills: Remove all potential ignition sources. Dike to contain spill, absorb with non-combustible absorbant and take up using non-sparking tools. Decontaminate spill area using water.

Special Emergency Information: May be harmful is inhaled, swallowed, or absorbed through the skin. Vapors are heavier than air and may travel some distance to an ignition source. Vapors are more dense than air and may settle in low lying areas.

DIACETYL - CH3C(O)C(O)CH3

Synonyms: 2,3-butanedione, dimethylglyoxal

CAS Number: 431-03-8   **Description:** Yellow to Green Liquid
DOT Number: 2346   **DOT Classification:** Flammable Liquid

Molecular Weight: 86.1
Melting Point: °C   (°F)   **Vapor Density:**   3.0   **Vapor Pressure:**
Boiling Point: 88°C (190°F)   **Specific Gravity:** 0.99   **Water Solubility:**   25%

Chemical Incompatibilities or Instabilities: Strong oxidizers.

FLAMMABILITY

NFPA Hazard Code: 1-3-0   **NFPA Classification:**   Class IB Flammable
Flash Point: 26°C   (80°F)   **LEL:** N.D.
Autoignition Temp: °C   (°F)   **UEL:** N.D.

Fire Extinguishing Methods: Alcohol Resistant Foam, Carbon Dioxide, Dry Chemical, Water.

Special Fire Fighting Considerations: Isolate for 1/2 mile if rail or tank truck is involved in at fire. Water spray should be used to keep closed containers cool. For large fires, if possible, withdraw and allow to burn. Immediately withdraw if rising sound from venting device is heard or if fire is causing discoloration to the tank.

TOXICOLOGY

Odor: **Odor Threshold:** N.D.
Physical Contact: Iirritant.
TLV = N.D.   **STEL** = N.D.   **IDLH** = N.D.

Routes of Entry and Relative LD50 (or LC50)   **RTECS # EK2625000**
   Inhalation   N.D.
   Ingestion   3   (1580 mg/kg)
   Skin Absorption   N.D.

Symptoms of Exposure: Possible irritation to skin, eyes and upper respiratory tract, headache, nausea.

Emergency/First Aid Treatment: Remove to ventilated area; immediately remove any contaminated clothing and wash contaminated areas for 15 minutes using water. Treat supportively and observe for possible shock. If ingested, seek immediate medical aid.

Recommended Clean-Up Procedures

Personal Protection: Level B Ensemble   **Recommended Material**   N.D.

RCRA Waste #   None   **Reportable Quantities:**   None

Spills: Remove all potential ignition sources. Dike to contain spill and collect liquid for disposal. Absorb with non-combustible absorbant and take up using non-sparking tools. Decontaminate spill area using water.

Special Emergency Information: May be harmful is inhaled, swallowed, or absorbed through the skin.

DIALLYLAMINE - $(CH_2=CHCH_2)_2NH$

Synonyms: D1-2-propenylamine

CAS Number: 124-02-7   **Description:** Colorless Liquid
DOT Number: 2359   **DOT Classification:** Flammable Liquid

Molecular Weight: 97.2
Melting Point: -88°C (-126°F)   **Vapor Density:** 3.4   **Vapor Pressure:** 4 (18 mm Hg)
Boiling Point: 112°C (234°F)   **Specific Gravity:** 0.8   **Water Solubility:** Soluble

Chemical Incompatibilities or Instabilities: Acid chlorides, acid anydrides, strong oxidizers.

FLAMMABILITY

NFPA Hazard Code: **NFPA Classification:**
Flash Point: 21°C (70°F)   **LEL:** N.D.
Autoignition Temp: N.D.   **UEL:** N.D.

Fire Extinguishing Methods: Carbon Dioxide, Dry Chemical, Foam, Water.

Special Fire Fighting Considerations: Isolate for 1/2 mile if rail or tank truck is involved in at fire. Water spray should be used to keep closed containers cool. Immediately withdraw if rising sound from venting device is heard or if fire is causing discoloration to the tank.

TOXICOLOGY

Odor: **Odor Threshold:** N.D.
Physical Contact: Material is extremely destructive to human tissues.
TLV = N.D.   **STEL** = N.D.   **IDLH** = N.D.

Routes of Entry and Relative LD50 (or LC50)   **RTECS # UC6650000**
     Inhalation   N.D.
     Ingestion   4   (650 mg/m^3)
     Skin Absorption   8   (280 mg/m^3)

Symptoms of Exposure: Possible irritation to skin, eyes and upper respiratory tract, headache, nausea. Lachrymation. Exposure may result in chemical pneumonitis or pulmonary edema. Inhalation of vapors may be fatal by causing spasms and suffocation.

Emergency/First Aid Treatment: Remove to ventilated area; immediately remove any contaminated clothing and wash contaminated areas for 15 minutes using water. Treat supportively and observe for possible shock. If ingested, seek immediate medical aid.

Recommended Clean-Up Procedures

Personal Protection: Level B Ensemble   **Recommended Material** Polyvinyl Alcohol (0), Viton (0)

RCRA Waste #   None   **Reportable Quantities:** None

Spills: Remove all potential ignition sources. Dike to contain spill and collect liquid for disposal. Absorb with non-combustible absorbant and take up using non-sparking tools. Decontaminate spill area using water.

Special Emergency Information: May be harmful is inhaled, swallowed, or absorbed through the skin. Vapors are heavier than air and may travel sone distance to an ignition source.

Diamylamine (CAS # 2050-92-2, DOT # 2841, RTECS # RZ9100000, has similar chemical, physical, and toxicological properties.

DIALLYL ETHER - $(CH_2=CHCH_2)O$

Synonyms: allyl ether ; propenyl ether

CAS Number: 557-40-4   **Description:** Colorless Liquid
DOT Number: 2360   **DOT Classification:** Flammable Liquid

Molecular Weight: 98.2
Melting Point: °C   (°F)   **Vapor Density:** 3.4   **Vapor Pressure:**
Boiling Point: 94°C (205°F)   **Specific Gravity:** 0.8   **Water Solubility:** Insoluble

Chemical Incompatibilities or Instabilities: Strong oxidizers.

FLAMMABILITY

NFPA Hazard Code: 3-3-2   **NFPA Classification:** Class IB Flammable
Flash Point: -7°C   (20°F)   **LEL:** N.D.
Autoignition Temp: N.D.   **UEL:** N.D.

Fire Extinguishing Methods: Alcohol Resistant Foam, Carbon Dioxide, Dry Chemical, Water.

Special Fire Fighting Considerations: Isolate for 1/2 mile if rail or tank truck is involved in at fire. Water spray should be used to keep closed containers cool. Immediately withdraw if rising sound from venting device is heard or if fire is causing discoloration to the tank.

TOXICOLOGY

Odor: Radishes   **Odor Threshold:** N.D.
Physical Contact: Irritant.
TLV = N.D.   **STEL** = N.D.   **IDLH** = N.D.

Routes of Entry and Relative LD50 (or LC50)   **RTECS # KN7525000**
    Inhalation   N.D.
    Ingestion   **5**   (320 mg/kg)
    Skin Absorption   **7**   (600 mg/kg)

Symptoms of Exposure: Possible irritation to skin, eyes and upper respiratory tract, headache, nausea.

Emergency/First Aid Treatment: Remove to ventilated area; immediately remove any contaminated clothing and wash contaminated areas for 15 minutes using water. Treat supportively and observe for possible shock. If ingested, seek immediate medical aid.

Recommended Clean-Up Procedures

Personal Protection: Level B Ensemble   **Recommended Material**   N.D.

RCRA Waste #   None   **Reportable Quantities:**   None

Spills: Remove all potential ignition sources. Dike to contain spill and collect liquid for disposal. Absorb with non-combustible absorbant and take up using non-sparking tools. Decontaminate spill area using soapy water.

Special Emergency Information: May be fatal if inhaled, swallowed, or absorbed through skin. Vapors are heavier than air and may travel sone distance to an ignition source. Vapors are more dense than air and may settle in low lying areas. May form unstable peroxides upon prolonged exposure to air.

Allyl ethyl ether (CAS # 557-31-3, DOT # 2335, RTECS # KM8120000) has similar chemical, physical, and toxicological properties.

DIAZINON

Synonyms:

CAS Number: 333-41-5   **Description:** Colorless Liquid
DOT Number: 2783   **DOT Classification:** ORM-A

Molecular Weight: 304.4
Melting Point: (D) 120°C (248°F)   **Vapor Density:** N.A.   **Vapor Pressure:** 1 (< 1 mm Hg)
Boiling Point: °C (°F)   **Specific Gravity:** 1.1   **Water Solubility:** Insoluble

Chemical Incompatibilities or Instabilities: Strong oxidizers.

FLAMMABILITY

NFPA Hazard Code: **NFPA Classification:**
Flash Point: N.A.   **LEL:** N.A.
Autoignition Temp: N.A.   **UEL:** N.A.

Fire Extinguishing Methods: Dry Chemical, Foam, Water.

Special Fire Fighting Considerations: Structural fire fighters protective clothing will not provide adequate protection. Fight fire from a distance or protected location, if possible. Treat all materials used or generated and equipment involved as contaminated by hazardous waste.

TOXICOLOGY
POSSIBLE MUTAGEN

Odor: Sweet   **Odor Threshold:** N.D.
Physical Contact: Irritant.
TLV = 0.1 mg/m^3(skin)   **STEL** = N.D.   **IDLH** = N.D.

Routes of Entry and Relative LD50 (or LC50)   **RTECS # TF3325000**
   Inhalation   4   (14,000 mg/m^3/H)
   Ingestion   8   (66 mg/kg)
   Skin Absorption   9   (180 mg/kg)

Symptoms of Exposure: Possible irritation to skin, eyes and upper respiratory tract, headache, nausea, muscle weakness, sweating.

Emergency/First Aid Treatment: Remove to ventilated area; immediately remove any contaminated clothing and wash contaminated areas for 15 minutes using water. Treat supportively and observe for possible shock. If ingested, seek immediate medical aid.

Recommended Clean-Up Procedures

Personal Protection: Level A Ensemble   **Recommended Material** N.D.

RCRA Waste # None   **Reportable Quantities:** 1 lb (1 lb)

Large Spills: Take up spill. Decontaminate spill area using soapy water.

Special Emergency Information: May be fatal if inhaled, swallowed, or absorbed through skin.

DIBORANE - B₂H₆

Synonyms: boron hydride

CAS Number: 19287-45-7	**Description:** Colorless Gas
DOT Number: 1911	**DOT Classification:** Flammable Gas

Molecular Weight: 27.7

Melting Point: -166°C (-267°F)	**Vapor Density:** 0.95	**Vapor Pressure:** GAS
Boiling Point: -93°C (-135°F)	**Specific Gravity:** ---	**Water Solubility:** Decomposes

Chemical Incompatibilities or Instabilities: Water, halogens, halogenated hydrocarbons, acids, oxidized surface, strong oxidizers, aluminum, lithium.

FLAMMABILITY

NFPA Hazard Code: 3-4-3-W̶	**NFPA Classification:**
Flash Point: -68°C (-90°F)	**LEL:** 0.9%
Autoignition Temp: 40°C (104°F)	**UEL:** 98%

Fire Extinguishing Methods: Water Spray.

Special Fire Fighting Considerations: Structural fire fighters protective clothing will not provide adequate protection. Isolate for 1/2 mile if rail or tank truck is involved in at fire. For large fires, if possible, withdraw and allow to burn. Fight fire from a distance or protected location, if possible. Approach fire from upwind. Water spray from an unmanned device should be used to keep closed containers cool. Continue to cool container after fire is extinguished. Immediately withdraw if rising sound from venting device is heard or if fire is causing discoloration to the tank.

DOT Recommended Isolation Zones:	Small Spill: 1500 ft	Large Spill: 1500 ft
DOT Recommended Down Wind Take Cover Distance:	Small Spill: 5 miles	Large Spill: 5 miles

TOXICOLOGY

Odor: Replusive	**Odor Threshold:** 2 mg/m³

Physical Contact: Material is extremely destructive to human tissues.

TLV = 0.1 mg/m³	**STEL** = N.D.	**IDLH** = 44 mg/m³

Routes of Entry and Relative LD50 (or LC50)   **RTECS # HQ9275000**

Inhalation	9	(44 mg/m³/4 HR)
Ingestion	N.D.	
Skin Absorption	N.D.	

Symptoms of Exposure: Possible burns or irritation to skin, eyes and upper respiratory tract; headache; nausea. Vertigo. Muscle spasm.

Emergency/First Aid Treatment: Remove to ventilated area; immediately remove any contaminated clothing and wash contaminated areas for 15 minutes using water. Treat supportively and observe for possible shock. If ingested, seek immediate medical aid.

Recommended Clean-Up Procedures

Personal Protection: Level A Ensemble	**Recommended Material**	N.D.
RCRA Waste # None	**Reportable Quantities:**	None

Spills: Remove all potential ignition sources. Stop leak if it can be safely be done. Keep area isolated until vapors have dissipated.

Special Emergency Information: May be fatal if inhaled, swallowed, or absorbed through skin. Reacts with water to yield hydrogen gas.

1,2-DIBROMO-3-CHLOROPROPANE - BrCH$_2$CH(Br)CH$_2$Cl

Synonyms: nematox

CAS Number: 96-12-8   **Description:** Yellow-Brown Liquid
DOT Number: 2872   **DOT Classification:** Poison B

Molecular Weight: 236.4
Melting Point: 6°C (43°F)   **Vapor Density:** N.D.   **Vapor Pressure:** N.D.
Boiling Point: 196°C (385°F)   **Specific Gravity:** 2.1   **Water Solubility:** N.D.

Chemical Incompatibilities or Instabilities: Strong oxidizers.

FLAMMABILITY

NFPA Hazard Code: **NFPA Classification:** Class IIIA Combustible
Flash Point: 77°C (170°F)   **LEL:** N.D.
Autoignition Temp: N.D.   **UEL:** N.D.

Fire Extinguishing Methods: Carbon Dioxide, Dry Chemical, Foam, Water.

Special Fire Fighting Considerations: Structural fire fighters protective clothing will not provide adequate protection.

TOXICOLOGY
CARCINOGEN

Odor: **Odor Threshold:** N.D.
Physical Contact: Irritant.
TLV = 0.01 mg/m^3 **STEL** = N.D.   **IDLH** = N.D.

Routes of Entry and Relative LD50 (or LC50)   **RTECS # TX8750000**

Route		
Inhalation	5	(7600 mg/m^3/H)
Ingestion	7	(170 mg/kg)
Skin Absorption	4	(1400 mg/kg)

Symptoms of Exposure: Possible irritation to skin, eyes and upper respiratory tract, headache, nausea. May act as a narcotic in high concentrations.

Emergency/First Aid Treatment: Remove to ventilated area; immediately remove any contaminated clothing and wash contaminated areas for 15 minutes using water. Treat supportively and observe for possible shock. If ingested, seek immediate medical aid.

Recommended Clean-Up Procedures

Personal Protection: Level B Ensemble   **Recommended Material** N.D.

RCRA Waste # U066   **Reportable Quantities:** 1 lb (1 lb)

Spills: Dike to contain spill and collect liquid for disposal. Decontaminate spill area using soapy water. Treat all materials used as contaminated by a hazardous waste.

Special Emergency Information: May be harmful is inhaled, swallowed, or absorbed through the skin.

1,2-DIBROMOETHANE - BrCH₂CH₂Br

Synonyms: 1,2-ethylene dibromide

CAS Number: 106-93-4   **Description:** Colorless Liquid
DOT Number: 1605   **DOT Classification:** ORM-A

Molecular Weight: 187.9
Melting Point: 10°C   (50°F)   **Vapor Density:**   6.5   **Vapor Pressure:**   4 (17 mm Hg)
Boiling Point: 132°C (270°F)   **Specific Gravity:** 2.2   **Water Solubility:**   Insoluble

Chemical Incompatibilities or Instabilities:

FLAMMABILITY

NFPA Hazard Code: **NFPA Classification:**
Flash Point: None   **LEL:** N.A.
Autoignition Temp: N.D.   **UEL:** N.A.

Fire Extinguishing Methods: Dry Chemical, Foam, Water.

Special Fire Fighting Considerations: Structural fire fighters protective clothing will not provide adequate protection. Fight fire from a distance or protected location, if possible. Approach fire from upwind. Water spray should be used to keep closed containers cool. Continue to cool container after fire is extinguished.

DOT Recommended Isolation Zones: Small Spill: 150 ft   Large Spill: 150 ft
DOT Recommended Down Wind Take Cover Distance: Small Spill: 0.2 miles   Large Spill: 0.2 miles

TOXICOLOGY
CARCINOGEN, POSSIBLE MUTAGEN

Odor: **Odor Threshold:** N.D.
Physical Contact: Irritant.
TLV = 150 mg/m³   **STEL** = N.D.   **IDLH** = 3200 mg/m³

Routes of Entry and Relative LD50 (or LC50)   **RTECS # KH9275000**
   Inhalation   5   (7,150 mg/kg/H)
   Ingestion   7   (108 mg/kg)
   Skin Absorption   8   (300 mg/kg)

Symptoms of Exposure: Possible irritation to skin, eyes and upper respiratory tract, headache, nausea.

Emergency/First Aid Treatment: Remove to ventilated area; immediately remove any contaminated clothing and wash contaminated areas for 15 minutes using water. Treat supportively and observe for possible shock. If ingested, seek immediate medical aid.

Recommended Clean-Up Procedures

Personal Protection: Level B Ensemble   **Recommended Material**   N.D.

RCRA Waste #   None   **Reportable Quantities:**   None

Spills: Dike to contain spill and collect liquid for disposal. Decontaminate spill area using soapy water. Treat all materials used as contaminated by a hazardous waste.

Special Emergency Information: May be harmful is inhaled, swallowed, or absorbed through the skin.

DIBROMOMETHANE - CH$_2$Br$_2$

Synonyms: methylene bromide

CAS Number: 74-95-3	**Description:** Colorless Liquid	
DOT Number: 2664	**DOT Classification:** Poison B	

Molecular Weight: 173.9

Melting Point:	-52°C (-62°F)	**Vapor Density:** 6.1	**Vapor Pressure:**	**5** (35 mm Hg)
Boiling Point:	96°C (205°F)	**Specific Gravity:** 2.5	**Water Solubility:**	Insoluble

Chemical Incompatibilities or Instabilities: Potassium, strong oxidizers.

FLAMMABILITY

NFPA Hazard Code:	**NFPA Classification:**
Flash Point: None	**LEL:** N.A.
Autoignition Temp: N.A.	**UEL:** N.A.

Fire Extinguishing Methods: Carbon Dioxide, Dry Chemical, Foam, Water.

Special Fire Fighting Considerations: Isolate for 1/2 mile if rail or tank truck is involved in at fire. Water spray should be used to keep closed containers cool.

TOXICOLOGY

Odor:	**Odor Threshold:** N.D.
Physical Contact: Irritant.	
TLV = N.D. **STEL** = N.D.	**IDLH** = N.D.

Routes of Entry and Relative LD50 (or LC50)   **RTECS #** PA7350000

Inhalation	2	(80,0000 mg/m^3/H)
Ingestion	7	(108 mg/kg)
Skin Absorption	N.D.	

Symptoms of Exposure: Possible irritation to skin, eyes and upper respiratory tract; headache; nausea. May act as a narcotic in high concentrations.

Emergency/First Aid Treatment: Remove to ventilated area; immediately remove any contaminated clothing and wash contaminated areas for 15 minutes using water. Treat supportively and observe for possible shock. If ingested, seek immediate medical aid.

Recommended Clean-Up Procedures

Personal Protection: Level B Ensemble	**Recommended Material** Polyvinyl Alcohol (0)
RCRA Waste # None	**Reportable Quantities:** None

Spills: Dike to contain spill and collect liquid for disposal. Decontaminate spill area using soapy water. Keep area isolated until vapors have dissipated.

Special Emergency Information: May be harmful is inhaled, swallowed, or absorbed through the skin.

n-DIBUTYLAMINE - (CH$_2$CH$_2$CH$_2$CH$_2$)$_2$NH

Synonyms: di-n-butylamine

CAS Number: 111-92-2	**Description:** Colorless Liquid	
DOT Number: 2248	**DOT Classification:** IMO	

Molecular Weight: 129.3

Melting Point: -59°C (-74°F)	**Vapor Density:** 4.5	**Vapor Pressure:** **2** (2 mm Hg)
Boiling Point: °C (°F)	**Specific Gravity:** 0.8	**Water Solubility:** Insoluble

Chemical Incompatibilities or Instabilities: Acid chlorides, carbon dioxide, acid anhydrides strong oxidizers.

FLAMMABILITY

NFPA Hazard Code: 3-2-0	**NFPA Classification:** Class II Combustible
Flash Point: 47°C (117°F)	**LEL:** 1.1%
Autoignition Temp: N.D.	**UEL:** N.D.

Fire Extinguishing Methods: Dry Chemical, Foam, Water.

Special Fire Fighting Considerations: Isolate for 1/2 mile if rail or tank truck is involved in at fire. Water spray should be used to keep closed containers cool. Immediately withdraw if rising sound from venting device is heard or if fire is causing discoloration to the tank.

TOXICOLOGY

Odor: **Odor Threshold:** N.D.
Physical Contact: Material is extremely destructive to human tissues.
TLV = N.D.   **STEL** = N.D.   **IDLH** = N.D.

Routes of Entry and Relative LD50 (or LC50)   **RTECS #** HR7780000

Inhalation	N.D.	
Ingestion	6	(220 mg/kg)
Skin Absorption	4	(1010 mg/kg)

Symptoms of Exposure: Possible burns or irritation to skin, eyes and upper respiratory tract; headache; nausea. Lachrymation. Inhalation of vapors may be fatal by causing spasms and suffocation. Exposure may result in chemical pneumonitis or pulmonary edema.

Emergency/First Aid Treatment: Remove to ventilated area; immediately remove any contaminated clothing and wash contaminated areas for 15 minutes using water. Treat supportively and observe for possible shock. If ingested, seek immediate medical aid.

Recommended Clean-Up Procedures

Personal Protection:	Level B Ensemble	**Recommended Material** Nitrile Rubber (+), Polyvinyl Alcohol (+), Viton (+)
RCRA Waste #	None	**Reportable Quantities:** None

Large Spills: Remove all potential ignition sources. Dike to contain spill and collect liquid for disposal. Absorb with non-combustible absorbant and take up using non-sparking tools. Decontaminate spill area using soapy water.

Special Emergency Information: May be fatal if inhaled, swallowed, or absorbed through skin. Vapors are heavier than air and may travel some distance to an ignition source. Vapors are more dense than air and may settle in low lying areas.

DICHLOROACETIC ACID - Cl₂CHCOOH

Synonyms: dichloroethnoic acid, urner's liquid, dichloracetic acid

CAS Number: 79-43-6

DOT Number: 1764

Description: Colorless Liquid

DOT Classification: Corrosive Material

Molecular Weight: 128.9

Melting Point: 10°C (50°F)	**Vapor Density:** 4.5	**Vapor Pressure:** **2** (1 mm Hg)
Boiling Point: 194°C (381°F)	**Specific Gravity:** 1.6	**Water Solubility:** Soluble

Chemical Incompatibilities or Instabilities: Strong oxidizers.

FLAMMABILITY

NFPA Hazard Code:

Flash Point: 110°C (230°F)

Autoignition Temp: N.D.

NFPA Classification: Class IIIB Combustible

LEL: N.D.

UEL: N.D.

Fire Extinguishing Methods: Carbon Dioxide, Dry Chemical, Foam, Water.

Special Fire Fighting Considerations: Water spray should be used to keep closed containers cool. Keep away from ends of tank.

TOXICOLOGY
QUESTIONABLE CARCINOGEN

Odor:

Odor Threshold: N.D.

Physical Contact: Material is extremely destructive to human tissues.

TLV = N.D.   **STEL** = N.D.   **IDLH** = N.D.

Routes of Entry and Relative LD50 (or LC50)   **RTECS # AG6125000**

Inhalation	N.D.	
Ingestion	3	(2820 mg/kg)
Skin Absorption	7	(510 mg/kg)

Symptoms of Exposure: Possible burns or irritation to skin, eyes and upper respiratory tract; headache; nausea. Exposure may result in chemical pneumonitis or pulmonary edema. Inhalation of vapors may be fatal by causing spasms and suffocation.

Emergency/First Aid Treatment: Remove to ventilated area; immediately remove any contaminated clothing and wash contaminated areas for 15 minutes using water. Treat supportively and observe for possible shock. If ingested, seek immediate medical aid.

Recommended Clean-Up Procedures

Personal Protection: Level B Ensemble   **Recommended Material** N.D.

RCRA Waste # None   **Reportable Quantities:** None

Spills: Dike to contain spill and collect liquid for disposal. Decontaminate spill area using dilute alkaline solution.

Special Emergency Information: May be harmful is inhaled, swallowed, or absorbed through the skin.

DICHLOROACETYLE CHLORIDE - Cl₂CHCOCl

Synonyms: 2,2-dichloroacetyl chloride

CAS Number: 79-36-7	**Description:** Fuming Liquid
DOT Number: 1765	**DOT Classification:** Corrosive Material

Molecular Weight: 147.4

Melting Point: °C (°F)	**Vapor Density:** 5.8	**Vapor Pressure:**
Boiling Point: 107°C (225°F)	**Specific Gravity:** 1.5	**Water Solubility:** Decomposes

Chemical Incompatibilities or Instabilities: Water, alcohols, strong oxidizers.

FLAMMABILITY

NFPA Hazard Code: 3-2-1-W	**NFPA Classification:** Class IIIA Combustible
Flash Point: 66°C (151°F)	**LEL:** N.D.
Autoignition Temp: N.D.	**UEL:** N.D.

Fire Extinguishing Methods: Carbon Dioxide, Dry Chemical (Foam or Water with caution).

Special Fire Fighting Considerations: Water spray should be used to keep closed containers cool. Keep away from ends of tank. Do not allow water to enter container. Reacts violently with water to yield hydrochloric acid and dichloroacetic acid.

TOXICOLOGY
QUESTIONABLE CARCINOGEN

Odor: **Odor Threshold:** N.D.

Physical Contact: Material is extremely destructive to human tissues.

TLV = N.D.   **STEL** = N.D.   **IDLH** ≈ N.D.

Routes of Entry and Relative LD50 (or LC50)   **RTECS #** AO6650000

Inhalation	N.D.	
Ingestion	3	(2460 mg/kg)
Skin Absorption	6	(650 mg/kg)

Symptoms of Exposure: Possible burns or irritation to skin, eyes and upper respiratory tract; headache; nausea. Lachrymation. Laryngitis. Inhalation of vapors may be fatal by causing spasms and suffocation. Exposure may result in chemical pneumonitis or pulmonary edema.

Emergency/First Aid Treatment: Remove to ventilated area; immediately remove any contaminated clothing and wash contaminated areas for 15 minutes using water. Treat supportively and observe for possible shock. If ingested, seek immediate medical aid.

Recommended Clean-Up Procedures

Personal Protection: Level B Ensemble	**Recommended Material** Viton (-)
RCRA Waste # None	**Reportable Quantities:** None

Spills: Dike to contain spill and collect liquid for disposal. Decontaminate spill area using dilute alkaline solution.

Special Emergency Information: May be harmful is inhaled, swallowed, or absorbed through the skin.

3,4-DICHLOROANILINE - $Cl_2(C_6H_4)NH_2$

Synonyms: 3,4-dichlorobenzamine

CAS Number: 95-76-1	**Description:** Brown Powder or Solid
DOT Number: 1590	**DOT Classification:** ---

Molecular Weight: 162

Melting Point:	71°C (160°F)	**Vapor Density:** 5.6	**Vapor Pressure:**	1 (< 1 mm Hg)
Boiling Point:	272°C (522°F)	**Specific Gravity:** ---	**Water Solubility:**	Insoluble

Chemical Incompatibilities or Instabilities: Acids, oxidizers.

FLAMMABILITY

NFPA Hazard Code:	3-1-0	**NFPA Classification:**	
Flash Point:	166°C (331°F)	**LEL:**	2.8%
Autoignition Temp:	265°C (509°F)	**UEL:**	7.2%

Fire Extinguishing Methods: Dry Chemical, Foam, Water.

Special Fire Fighting Considerations: Structural fire fighters protective clothing will not provide adequate protection. Approach fire from upwind. Fight fire from a distance or protected location, if possible. Water spray should be used to keep closed containers cool.

TOXICOLOGY

Odor:	**Odor Threshold:** N.D.

Physical Contact: Material is extremely destructive to human tissues.

TLV = N.D.	**STEL** = N.D.	**IDLH** = N.D.

Routes of Entry and Relative LD50 (or LC50)   **RTECS # BX2625000**

Inhalation	N.D.	
Ingestion	4	(648 mg/kg)
Skin Absorption	N.D.	

Symptoms of Exposure: Possible burns or irritation to skin, eyes and upper respiratory tract; headache; nausea. Exposure may result in chemical pneumonitis or pulmonary edema. Inhalation of vapors may be fatal by causing glottis to spasm and suffocation. Absorption may lead to the formation of methemoglobin resulting in cyanosis several hours after exposure. May cause an allergic reaction.

Emergency/First Aid Treatment: Remove to ventilated area; immediately remove any contaminated clothing and wash contaminated areas for 15 minutes using water. Treat supportively and observe for possible shock. If ingested, seek immediate medical aid.

Recommended Clean-Up Procedures

Personal Protection: Level A Ensemble	**Recommended Material** N.D.
RCRA Waste # None	**Reportable Quantities:** None

Spills: Take up with a non-combustible absorbant and dispose. Decontaminate spill area using soapy water. Treat all materials used as contaminated by a hazardous waste.

Special Emergency Information: May be fatal in inhaled, ingested or absorbed through the skin.

2,3-dichloroaniline (CAS # 608-27-5),
2,4-dichloroaniline (CAS # 554-00-7, RTECS # BX2600000),
2,6-dichloroaniline (CAS # 608-31-1),
3,4-dichloroaniline (CAS # 95-76-1, RTECS # BX2625000), and
3,5-dichloroanile (CAS # 626-43-7) have similar chemical, physical and toxicological properties.

o-DICHLOROBENZENE - $Cl_2C_6H_4$

Synonyms: ortho dichlorobenzene, 1,2-dichlorobenzene

CAS Number: 95-50-1	**Description:** Colorless Liquid	
DOT Number: 1591	**DOT Classification:** ORM-A	

Molecular Weight: 147.0

Melting Point: -18°C (0°F)	**Vapor Density:** 5.1	**Vapor Pressure:** **2** (1 mm Hg)
Boiling Point: 180°C (357°F)	**Specific Gravity:** 1.3	**Water Solubility:** Insoluble

Chemical Incompatibilities or Instabilities: Strong oxidizers, aluminum and alloys.

FLAMMABILITY

NFPA Hazard Code: 2-2-0	**NFPA Classification:** Class IIA Combustible	
Flash Point: 66°C (151°F)	**LEL:** 2.2%	
Autoignition Temp: 648°C (1198°F)	**UEL:** 9.2%	

Fire Extinguishing Methods: Carbon Dioxide, Dry Chemical, Foam, Water.

Special Fire Fighting Considerations: Structural fire fighters protective clothing will not provide adequate protection. Approach fire from upwind. Water spray should be used to keep closed containers cool.

TOXICOLOGY
QUESTIONABLE CARCINOGEN

Odor: Camphor

Physical Contact: Irritant

TLV = 150 mg/m^3 **STEL** = 300 mg/m^3

Odor Threshold: 0.1 mg/m^3

IDLH = 6000 mg/m^3

Routes of Entry and Relative LD50 (or LC50)

Inhalation	N.D.
Ingestion	**5** (500 mg/kg)
Skin Absorption	N.D.

RTECS # CZ4500000

Symptoms of Exposure: Possible irritation to skin, eyes and upper respiratory tract, headache, nausea. May cause an allergic reaction.

Emergency/First Aid Treatment: Remove to ventilated area; immediately remove any contaminated clothing and wash contaminated areas for 15 minutes using water. Treat supportively and observe for possible shock. If ingested, seek immediate medical aid.

Recommended Clean-Up Procedures

Personal Protection: Level B Ensemble	**Recommended Material** Viton (0)	
RCRA Waste # U070	**Reportable Quantities:** 100 lb (100 lb)	

Spills: Take up in non-combustible absorbant and dispose. Decontaminate spill area using soapy water. Treat all materials used as contaminated by a hazardous waste.

Special Emergency Information: May be harmful is inhaled, swallowed, or absorbed through the skin.

m-dichlorobenzene (Synonym: 1,3-dichlorobenzene, CAS # 541-73-1, RTECS # CZ4499000, RCRA Waste # U071, RQ: 100 lb [1 lb]) has similar chemical, physical, and toxicological properties.

p-dichlorobenzene (Synonym: 1,4-dichlorobenzene, CAS # 106-46-7, DOT # 1592, RTECS # CZ4550000, RCRA Waste # U072, RQ: 100 lb [100 lb]) is a solid at room temperature and a confirmed CARCENOGEN. Otherwise, it has similar chemical, physical, and toxicological properties.

DICHLORODIETHYL ETHER - $(ClCH_2CH_3)_2O$

Synonyms: dichloroethyl ether

CAS Number: 111-44-4
DOT Number: 1916

Description: Colorless Liquid
DOT Classification: Poison B

Molecular Weight: 143.0

Melting Point:	-52°C (-62°F)	**Vapor Density:**	4.9	**Vapor Pressure:**	2 (1 mm Hg)
Boiling Point:	178°C (352°F)	**Specific Gravity:**	1.2	**Water Solubility:**	Insoluble

Chemical Incompatibilities or Instabilities: Sulfuric acid, chlorosulfonic acid, oxidizers.

FLAMMABILITY

NFPA Hazard Code: 3-2-1
Flash Point: 55°C (131°F)
Autoignition Temp: 396°C (696°F)

NFPA Classification: Class II Combustible
LEL: 2.7%
UEL: N.D.

Fire Extinguishing Methods: Carbon Dioxide, Dry Chemical, Foam, Water.

Special Fire Fighting Considerations: Structural fire fighters protective clothing will not provide adequate protection. Fight fire from a distance or protected location, if possible. Water spray should be used to keep closed containers cool. Immediately withdraw if rising sound from venting device is heard or if fire is causing discoloration to the tank.

DOT Recommended Isolation Zones:	Small Spill: 150 ft	Large Spill: 150 ft
DOT Recommended Down Wind Take Cover Distance:	Small Spill: 0.2 miles	Large Spill: 0.2 miles

TOXICOLOGY
CARCINOGEN

Odor:
Physical Contact: Material is extremely destructive to human tissues.
TLV = 29 mg/m^3

STEL = 58 mg/m^3

Odor Threshold: N.D.

IDLH = 1500 mg/m^3

RTECS # KN0875000

Routes of Entry and Relative LD50 (or LC50)

Inhalation	4	(330 mg/m^3/4 HR)
Ingestion	8	(75 mg/kg)
Skin Absorption	6	(720 mg/kg)

Symptoms of Exposure: Possible burns or irritation to skin, eyes and upper respiratory tract; headache; nausea. Inhalation of vapors may be fatal by causing glottis to spasm and suffocation. Exposure may result in chemical pneumonitis or pulmonary edema.

Emergency/First Aid Treatment: Remove to ventilated area; immediately remove any contaminated clothing and wash contaminated areas for 15 minutes using water. Treat supportively and observe for possible shock. If ingested, seek immediate medical aid.

Recommended Clean-Up Procedures

Personal Protection: Level B Ensemble

RCRA Waste # None

Recommended Material N.D.

Reportable Quantities: None

Spills: Remove all potential ignition sources. Dike to contain spill, absorb with non-combustible absorbant and take up using non-sparking tools. Decontaminate spill area using dilute alkaline solution.

Special Emergency Information: May be harmful is inhaled, swallowed, or absorbed through the skin. Decomposes in the presence of water to yield hydrochloric acid. May form unstable peroxides upon exposure to air.

DICHLORODIFLUOROMETHANE - Cl_2F_2C

Synonyms: freon F-12, propellant 12, refrigerant 12, halon 122

CAS Number: 75-71-8	**Description:** Odorless Gas	
DOT Number: 1028	**DOT Classification:** Nonflammable Gas	

Molecular Weight: 120.9

Melting Point: -158°C (-252°F)	**Vapor Density:** 4.2	**Vapor Pressure:** Gas
Boiling Point: -30°C (-22°F)	**Specific Gravity:** ---	**Water Solubility:** Insoluble

Chemical Incompatibilities or Instabilities: Aluminum, magnesium, silver, copper and their alloys.

FLAMMABILITY

NFPA Hazard Code:	**NFPA Classification:**
Flash Point: N.A.	**LEL:** N.A.
Autoignition Temp: N.A.	**UEL:** N.A.

Fire Extinguishing Methods: Carbon Dioxide, Dry Chemical, Foam, Water.

Special Fire Fighting Considerations: Isolate for 1/2 mile if rail or tank truck is involved in at fire. Water spray should be used to keep closed containers cool. May release toxic fumes upon heating (phosgene).

TOXICOLOGY

Odor: **Odor Threshold:** N.D.

Physical Contact: A simple asphyxiant.

TLV = 4950 mg/m^3 **STEL** = N.D.   **IDLH** = 250,000 mg/m^3

Routes of Entry and Relative LD50 (or LC50)   **RTECS #** PA8200000

    Inhalation   1(80,000,000 mg/m^3/H)
    Ingestion   N.D.
    Skin Absorption   N.D.

Symptoms of Exposure: May act as a narcotic in high concentrations.

Emergency/First Aid Treatment: Remove to ventilated area; immediately remove any contaminated clothing and wash contaminated areas for 15 minutes using water. Treat supportively and observe for possible shock. If ingested, seek immediate medical aid.

Recommended Clean-Up Procedures

Personal Protection: Level B Ensemble	**Recommended Material** N.D.
RCRA Waste # U075	**Reportable Quantities:** 5000 lb (1 lb)

Spills: Stop leak if it can safely be done. Keep area isolated until vapors have dissipated.

Special Emergency Information: May be harmful if inhaled, swallowed or absorbed through the skin.

1,1-DICHLOROETHANE - Cl$_2$CHCH$_3$

Synonyms: ethylidene dichloride

CAS Number: 75-34-5   **Description:** Colorless Liquid
DOT Number: 2362   **DOT Classification:** Flammable Liquid

Molecular Weight: 99.0
Melting Point: -98°C (-144°F)   **Vapor Density:** 3.4   **Vapor Pressure:** 8 (230 mm Hg)
Boiling Point: 57°C (135°F)   **Specific Gravity:** 1.2   **Water Solubility:** Insoluble

Chemical Incompatibilities or Instabilities: Strong Oxidizers

FLAMMABILITY

NFPA Hazard Code: 2-3-0   **NFPA Classification:** Class 1B Flammable
Flash Point: -17°C (2°F)   **LEL:** 5.4%
Autoignition Temp: °C (856°F)   **UEL:** 11.2%

Fire Extinguishing Methods: Carbon Dioxide, Dry Chemical, Foam, Water Spray or Fog.

Special Fire Fighting Considerations: Isolate for 1/2 mile if rail or tank truck is involved in a fire. Water spray should be used to keep closed containers cool. For large fires, if possible, withdraw and allow to burn. Immediately withdraw if rising sound from venting device is heard or if fire is causing discoloration to the tank. Immediately withdraw if rising sound from venting device is heard or if fire is causing discoloration to the tank.

TOXICOLOGY
QUESTIONABLE CARCINOGEN

Odor: Ethereal, chloroform-like   **Odor Threshold:** 200mg/m^3
Physical Contact: Irritant
TLV = 405 mg^3 **STEL** = N.D.   **IDLH** = 16,000 mg/m^3

Routes of Entry and Relative LD$_{50}$ (or LC$_{50}$)   **RTECS # KI0175000**
   Inhalation   N.D.
   Ingestion   4   (725 mg/kg)
   Skin Absorption   N.D.

Symptoms of Exposure: Possible irritation to skin, eyes and upper respiratory tract; headache; nausea. May act as a narcotic in high concentrations.

Emergency/First Aid Treatment: Remove to ventilated area; immediately remove any contaminated clothing and wash contaminated areas for 15 minutes using water. Treat supportively and observe for possible shock. If ingested, seek immediate medical aid.

Recommended Clean-Up Procedures

Personal Protection: Level B Ensemble   **Recommended Material** N.D.

RCRA Waste # U076   **Reportable Quantities:** 1000 lb (1 lb)

Spills: Remove all potential ignition sources. Decontaminate spill area using soapy water. Keep area isolated until vapors have dissipated. Treat all materials used or generated and equipment involved as contaminated by hazardous waste.

Special Emergency Information: Vapors are heavier than air and may travel some distance to an ignition source. May be harmful if inhaled, swallowed or absorbed through the skin. Vapors are more dense than air and may settle in low lying areas.

1,2-DICHLOROETHYLENE - CLCH=CHCL

Synonyms: acetylene dichloride

CAS Number: 540-59-0
DOT Number: 1150

Description: Colorless liquid
DOT Classification: Flammable Liquid

Molecular Weight: 97.0
Melting Point: -57°C   (°F)
Boiling Point: 48°C   (119°F)

Vapor Density: 3.3
Specific Gravity: 1.3

Vapor Pressure: 8 (250mm Hg)
Water Solubility: Insoluble

Chemical Incompatibilities or Instabilities: Strong oxidizers, strong caustics, sodium, N_2O_4

FLAMMABILITY

NFPA Hazard Code: 2-3-2
Flash Point: 6°C   (43°F)
Autoignition Temp: 458°C (855°F)

NFPA Classification: Class 1B Flammable
LEL: 5.6%
UEL: 12.8%

Fire Extinguishing Methods: Carbon Dioxide, Dry Chemical, Foam, Water Spray or Fog.

Special Fire Fighting Considerations: Water spray should be used to keep closed containers cool. Isolate for 1/2 mile if rail or tank truck is involved in a fire. For large fires, if possible, withdraw and allow to burn. Immediately withdraw if rising sound from venting device is heard or if fire is causing discoloration to the tank.

TOXICOLOGY

Odor: Pleasant
Physical Contact: Iirritant
TLV = 790 mg/m^3

STEL = N.D.

Odor Threshold: N.D.

IDLH = 16,000 mg/m^3

Routes of Entry and Relative LD$_{50}$ (or LC$_{50}$)
   Inhalation   N.D.
   Ingestion   4   (770 mg/kg)
   Skin Absorption N.D.

RTECS # KV9360000

Symptoms of Exposure: Possible irritation to skin, eyes and upper respiratory tract; headache; nausea. May act as a narcotic in high concentrations.

Emergency/First Aid Treatment: Remove to ventilated area; immediately remove any contaminated clothing and wash contaminated areas for 15 minutes using water. Treat supportively and observe for possible shock. If ingested, seek immediate medical aid.

Recommended Clean-Up Procedures

Personal Protection: Level B Ensemble
RCRA Waste #   U079

Recommended Material   N.D.
Reportable Quantities: 1000 lb (1 lb)

Spills: Remove all potential ignition sources. Absorb with noncombustible absorbant and take up using nonsparking tools. Decontaminate spill area using soapy water. Keep area isolated until vapors have dissipated. Treat all materials used or generated and equipment involved as contaminated by hazardous waste.

Special Emergency Information: Vapors are heavier than air and may travel some distance to an ignition source. Vapors are more dense than air and may settle in low lying areas. May be harmful if inhaled, swallowed or absorbed through the skin.

cis-Dichloroethylene (CAS # 156-59-2, RTECS # KV9420000),
trans-dichloroethylene (CAS # 156-60-5, RTECS # KV9400000), and
dichloroethylene (CAS # 25323-30-2, RTECS # KV925000) have similar chemical, physical and toxicological properties.

DICHLOROISOPROPYL ETHER - $(CLCH_2CH(CH_3))_2O$

Synonyms: bis(2-chloroisopropyl) ether; bis(2-chloro-1-methylethyl)ether; dichlorodiisopropyl ether

CAS Number: 108-60-1
DOT Number: 2490

Description: Colorless liquid
DOT Classification: Corrosive Material

Molecular Weight: 171.1
Melting Point: -20°C (°F)
Boiling Point: 188°C (369°F)

Vapor Density: 6.0
Specific Gravity: 1.1

Vapor Pressure: 1 (0.1mm Hg)
Water Solubility: Insoluble

Chemical Incompatibilities or Instabilities: Strong oxidizers.

FLAMMABILITY

NFPA Hazard Code: 2-2-0
Flash Point: 85°C (185°F)
Autoignition Temp: N.D.

NFPA Classification: Class IIIA Combustible
LEL: N.D.
UEL: N.D.

Fire Extinguishing Methods: Carbon Dioxide, Dry Chemical, Foam, Water Spray or Fog.

Special Fire Fighting Considerations: Structural fire fighter protective clothing will not provide adequate protection. Structural fire fighter protective clothing will not provide adequate protection. Treat all materials used or generated and equipment involved as contaminated by hazardous waste.

TOXICOLOGY
QUESTIONABLE CARCINOGEN

Odor:
Physical Contact: Material is extremely destructive to human tissues.
TLV = N.D.   **STEL** = N.D.

Odor Threshold: N.D.

IDLH = N.D.

Routes of Entry and Relative LD50 (or LC50)

Inhalation	N.D.	
Ingestion	6	(240 mg/kg)
Skin Absorption	2	(3000 mg/kg)

RTECS # KN1750000

Symptoms of Exposure: Possible irritation and burns to skin, eyes and upper respiratory tract; headache; nausea. Inhalation of vapors may be fatal by causing glottis to spasm and suffocation. Exposure may result in chemical pneumonitis or pulmonary edema.

Emergency/First Aid Treatment: Remove to ventilated area; immediately remove any contaminated clothing and wash contaminated areas for 15 minutes using water. Treat supportively and observe for possible shock. If ingested, seek immediate medical aid.

Recommended Clean-Up Procedures

Personal Protection: Level B Ensemble   **Recommended Material** N.D.

RCRA Waste # U027   **Reportable Quantities:** 1000 lb (1 lb)

Spills: Dike to contain spill and collect liquid for later disposal. Decontaminate spill area using soapy water. Treat all materials used or generated and equipment involved as contaminated by hazardous waste.

Special Emergency Information: May be fatal in inhaled, ingested or absorbed through the skin. May form unstable peroxides upon exposure to air.

1,1-DICHLORO-1-NITROETHANE - $CL_2C(NO_2)CH_3$

Synonyms: dichloronitroethane; ethide

CAS Number: 594-72-9   **Description:** Colorless liquid
DOT Number: 2650   **DOT Classification:** Poison B

Molecular Weight: 144.0
Melting Point: °C   (°F)   **Vapor Density:**   5.0   **Vapor Pressure:**   **4** (15 mm Hg)
Boiling Point: 124°C (255°F)   **Specific Gravity:** 1.4   **Water Solubility:**   Insoluble

Chemical Incompatibilities or Instabilities: Oxidizers.

FLAMMABILITY

NFPA Hazard Code: 2-2-3   **NFPA Classification:** Class II Combustible
Flash Point: 76°C (168°F)   **LEL:** N.D.
Autoignition Temp: °C   (°F)   **UEL:** N.D.

Fire Extinguishing Methods: Carbon Dioxide, Dry Chemical, Foam, Water Spray or Fog.

Special Fire Fighting Considerations: Structural fire fighter protective clothing will not provide adequate protection. ight fire from a distance or protected location, if possible. Water spray from an unmanned device should be used to keep closed containers cool. Continue to cool container after fire is extinguished. Treat all materials used or generated and equipment involved as contaminated by hazardous waste.

TOXICOLOGY

Odor: Unpleasant   **Odor Threshold:** N.D.
Physical Contact: Irritant
TLV = 12 mg/m^3.   **STEL** = N.D.   **IDLH** = 900 mg/m^3

Routes of Entry and Relative LD$_{50}$ (or LC$_{50}$)   **RTECS # KI1050000**
    Inhalation   N.D.
    Ingestion   **5**   (410 mg/kg)
    Skin Absorption   N.D.

Symptoms of Exposure: Possible irritation to skin, eyes and upper respiratory tract; headache; nausea. Inhalation of vapors may cause pulmonary edema. Inhalation of vapors may be fatal by causing glottis to spasm and suffocation.

Emergency/First Aid Treatment: Remove to ventilated area; immediately remove any contaminated clothing and wash contaminated areas for 15 minutes using water. Treat supportively and observe for possible shock. If ingested, seek immediate medical aid.

Recommended Clean-Up Procedures

Personal Protection: Level B Ensembled   **Recommended Material**   N.D.

RCRA Waste #   None   **Reportable Quantities:**   None

Spills: Dike to contain spill and collect liquid for later disposal. Decontaminate spill area using soapy water. Treat all materials used or generated and equipment involved as contaminated by hazardous waste.

Special Emergency Information: May be harmful if inhaled, swallowed or absorbed through the skin. Vapors may collect in low lying areas. Vapors are more dense than air and may settle in low lying areas.

DICHLOROPENTANE - ClCH$_2$CH$_2$CH$_2$CH$_2$CH$_2$Cl

Synonyms: amylene chloride

CAS Number: 628-76-2
DOT Number:

Description: Colorless liquid
DOT Classification: Flammable liquid

Molecular Weight: 141.1
Melting Point: -72°C (-98°F)
Boiling Point: 180°C (356°F)

Vapor Density: 4.9
Specific Gravity: 1.1

Vapor Pressure:
Water Solubility: Insoluble

Chemical Incompatibilities or Instabilities: Strong oxidizers.

FLAMMABILITY

NFPA Hazard Code: 2-3-2
Flash Point: 26°C (80°F)
Autoignition Temp: °C (°F)

NFPA Classification: Class 1C Flammable
LEL: N.D.
UEL: N.D.

Fire Extinguishing Methods: Carbon Dioxide, Dry Chemical, Foam, Water Spray or Fog.

Special Fire Fighting Considerations: Isolate for 1/2 mile if rail or tank truck is involved in a fire. Water spray should be used to keep closed containers cool until well after fire is out. For large fires, if possible, withdraw and allow to burn. Immediately withdraw if rising sound from venting device is heard or if fire is causing discoloration to the tank.

TOXICOLOGY

Odor: Sweet
Physical Contact: Irritant
TLV = N.D.

STEL = N.D.

Odor Threshold: N.D.

IDLH = N.D.

RTECS # SA0350000

Routes of Entry and Relative LD50 (or LC50)
Inhalation   N.D.
Ingestion   N.D.
Skin Absorption   N.D.

Symptoms of Exposure: Possible irritation to skin, eyes and upper respiratory tract; headache; nausea.

Emergency/First Aid Treatment: Remove to ventilated area; immediately remove any contaminated clothing and wash contaminated areas for 15 minutes using water. Treat supportively and observe for possible shock. If ingested, seek immediate medical aid.

Recommended Clean-Up Procedures

Personal Protection: Level B Ensemble

RCRA Waste # None

Recommended Material N.D.

Reportable Quantities: None

Spills: Remove all potential ignition sources. Dike to contain spill, absorb with non-combustible absorbant and take up using non-sparking tools. Decontaminate spill area using soapy water.

Special Emergency Information: May be harmful if inhaled, swallowed or absorbed through the skin. Vapors are heavier than air and may travel some distance to an ignition source. Vapors are more dense than air and may settle in low lying areas.

DICHLOROPHENYL TRICHLOROSILANE

Synonyms:

CAS Number: 27137-85-5   **Description:** Clear to tan colored liquid
DOT Number: 1766   **DOT Classification:** Corrosive Material

Molecular Weight: 280.4

Melting Point:	N. A.°C	(°F)	**Vapor Density:** N.D.	**Vapor Pressure:**	N.D.
Boiling Point:	260°C	(°F)	**Specific Gravity:** 1.6	**Water Solubility:**	Decomposes

Chemical Incompatibilities or Instabilities: Oxidizers.

FLAMMABILITY

NFPA Hazard Code: N.D.   **NFPA Classification:** Class IIIB Combustible
Flash Point: 286°C (°F)   **LEL:** N.D..
Autoignition Temp: °C (°F)   **UEL:** N.D.

Fire Extinguishing Methods: Carbon Dioxide, Dry Chemical, Foam, Water Spray or Fog.

Special Fire Fighting Considerations: Keep away from ends of tank. Water spray should be used to keep closed containers cool. Continue to cool container after fire is extinguished.

TOXICOLOGY

Odor: **Odor Threshold:** N.D.

Physical Contact: Material is extremely destructive to human tissues.

TLV = N.D.   **STEL** = N.D.   **IDLH** = N.D.

Routes of Entry and Relative LD$_{50}$ (or LC$_{50}$)   **RTECS #** VV3540000
    Inhalation   N.D.
    Ingestion   N.D.
    Skin Absorption   N.D.

Symptoms of Exposure: Irritation and burns to skin, eyes and upper respiratory tract; headache; nausea. Inhalation of vapors may be fatal by causing glottis to spasm and suffocation. Exposure may result in chemical pneumonitis or pulmonary edema.

Emergency/First Aid Treatment: Remove to ventilated area; immediately remove any contaminated clothing and wash contaminated areas for 15 minutes using water. Treat supportively and observe for possible shock. If ingested, seek immediate medical aid.

Recommended Clean-Up Procedures

Personal Protection: Level A Ensemble   **Recommended Material**   N.D.

RCRA Waste #   None   **Reportable Quantities:**   None

Spills: Dike to contain spill and collect liquid for later disposal. Decontaminate spill area using dilute alkaline solution. Keep area isolated until vapors have dissipated.

Special Emergency Information: May be fatal in inhaled, ingested or absorbed through the skin. Reacts with water to yield hydrochloric acid.

1,3-DICHLOROPROPENE

Synonyms: 1,3-dichloropropylene; telone; vidden D

CAS Number: 542-75-6

DOT Number: 2047

Description: Colorless to yellow liquid

DOT Classification: Flammable Liquid

Molecular Weight: 111.0

Melting Point:	°C (°F)	**Vapor Density:** 3.8	**Vapor Pressure:**	**5** (48 mm Hg)
Boiling Point:	104°C (219°F)	**Specific Gravity:** 1.2	**Water Solubility:**	Insoluble

Chemical Incompatibilities or Instabilities: Oxidizers, acids, active metals

FLAMMABILITY

NFPA Hazard Code: 3-3-0

Flash Point: 35°C (95°F)

Autoignition Temp: °C (°F)

NFPA Classification: Class IC Flammable

LEL: 2.6%

UEL: 7.8%

Fire Extinguishing Methods: Carbon Dioxide, Dry Chemical, Foam, Water Spray or Fog.

Special Fire Fighting Considerations: Isolate for 1/2 mile if rail or tank truck is involved in a fire. Water spray should be used to keep closed containers cool. Continue to cool container after fire is extinguished. Immediately withdraw if rising sound from venting device is heard or if fire is causing discoloration to the tank.

TOXICOLOGY
CONFIRMED CARCINOGEN

Odor: Chloroform-like

Physical Contact: Material is extremely destructive to human tissues.

TLV = 4.5 mg/m^3 **STEL** = N.D.

Odor Threshold: N.D.

IDLH = N.D.

TECS #   UC8310000

Routes of Entry and Relative LD$_{50}$ (or LC$_{50}$)

Inhalation	N.D.	
Ingestion	6	(250 mg/kg)
Skin Absorption	7	(500 mg/m^3)

Symptoms of Exposure: Irritation and burns to skin, eyes and upper respiratory tract; headache; nausea. Lachrymation. Inhalation of vapors may be fatal by causing glottis to spasm and suffocation. Exposure may result in chemical pneumonitis or pulmonary edema.

Emergency/First Aid Treatment: Remove to ventilated area; immediately remove any contaminated clothing and wash contaminated areas for 15 minutes using water. Treat supportively and observe for possible shock. If ingested, seek immediate medical aid.

Recommended Clean-Up Procedures

Personal Protection: Level A Ensemble

RCRA Waste #   U084

Recommended Material   Polyvinyl Alcohol (+), Viton (+)

Reportable Quantities: 100 lb (5000 lb)

Spills: Remove all potential ignition sources. Dike to contain spill, absorb with noncombustible absorbant and take up using nonsparking tools. Decontaminate spill area using soapy water. Keep area isolated until vapors have dissipated.

Special Emergency Information: Vapors are heavier than air and may travel some distance to an ignition source. Vapors may collect in low lying areas. May be fatal in inhaled, ingested or absorbed through the skin.

1,2-dichloropropene (CAS # 563-54-2, RTECS # UC8300000),
1,1-dichloropropene (CAS # 563-58-6, RTECS # UC8290000),
2,3-dichloropropene (CAS # 78-88-6, RTECS # UC8400000),
trans-1,3-dichloropropene (CAS # 10061-02-6, RTECS # UC8320000),
cis-1,3-dichloropropene (CAS # 10061-01-5, RTECS # UB8325000), and
dichloropropane-dichloropropene mixtures (CAS # 8003-19-8, RTECS # TX9800000, DOT #2047) have similar chemical, physical and toxicological properties.

2,2-DICHLOROPROPIONIC ACID. - CH$_3$C(Cl$_2$)COOH

Synonyms: basfapon; basinex; BH dalapon; crisapon; Ded-Weed; dowpon; gramevin; kenapon; liropon; proprop; radapon; revenge; unipon

CAS Number: 75-99-0		**Description:** Colorless liquid	
DOT Number: 1760		**DOT Classification:** Corrosive Material	

Molecular Weight: 143.0

Melting Point:	°C (°F)	**Vapor Density:** N.D.	**Vapor Pressure:** N.D.
Boiling Point:	190°C (374°F)	**Specific Gravity:** 1.4	**Water Solubility:** Soluble

Chemical Incompatibilities or Instabilities: Strong oxidizers.

FLAMMABILITY

NFPA Hazard Code:	**NFPA Classification:** Class IIIB Combustible
Flash Point: 110°C (230°F)	**LEL:** N.D.
Autoignition Temp: °C (°F)	**UEL:** N.D.

Fire Extinguishing Methods: Carbon Dioxide, Dry Chemical, Foam, Water Spray or Fog.

Special Fire Fighting Considerations: Water spray should be used to keep closed containers cool. Continue to cool container after fire is extinguished.

TOXICOLOGY

Odor: N.D.	**Odor Threshold:** N.D.

Physical Contact: Material is extremely destructive to human tissues.

TLV = 5.8 mg/m^3	**STEL** = N.D.	**IDLH** = N.D.

Routes of Entry and Relative LD$_{50}$ (or LC$_{50}$)   **RTECS # UF0690000**

Inhalation	N.D.	
Ingestion	4	(970 mg/kg)
Skin Absorption	N.D.	

Symptoms of Exposure: Irritation and burns to skin, eyes and upper respiratory tract; headache; nausea. Inhalation of vapors may be fatal by causing glottis to spasm and suffocation. Exposure may result in chemical pneumonitis or pulmonary edema.

Emergency/First Aid Treatment: Remove to ventilated area; immediately remove any contaminated clothing and wash contaminated areas for 15 minutes using water. Treat supportively and observe for possible shock. If ingested, seek immediate medical aid.

Recommended Clean-Up Procedures

Personal Protection:	Level B Ensemble	**Recommended Material**	N.D.
RCRA Waste #	None	**Reportable Quantities:**	5000 lb (5000 lb)

Spills: Dike to contain spill and collect liquid for later disposal. Wash contaminated areas with larges volumes of water.

Special Emergency Information: May be fatal in inhaled, ingested or absorbed through the skin.

This material may be transported as the sodium salt (CAS # 127-20-8, RTECS # UF1225000) which is present as a white solid. The sodium salt has similar toxicological properties.

DICHLOROSILANE - SiH$_2$Cl$_2$

Synonyms:

CAS Number: 4109-96-0
DOT Number: 2189

Description: Colorless gas
DOT Classification: Poison A

Molecular Weight: 101.0
Melting Point: -122°C (-188°F)
Boiling Point: 8°C (46°F)

Vapor Density: 3.5
Specific Gravity: N. A.

Vapor Pressure: Gas
Water Solubility: Decomposes

Chemical Incompatibilities or Instabilities: Water, air, oxidizers.

FLAMMABILITY

NFPA Hazard Code: 3-4-2-W
Flash Point: -52°C (-62°F)
Autoignition Temp: 44°C (111°F)

NFPA Classification:
LEL: 4.7%
UEL: 96.0%

Fire Extinguishing Methods: Water Spray or Fog.

Special Fire Fighting Considerations: Isolate for 1/2 mile if rail or tank truck is involved in a fire. Keep away from ends of tank. Water spray should be used to keep closed containers cool. For large fires, if possible, withdraw and allow to burn. Immediately withdraw if rising sound from venting device is heard or if fire is causing discoloration to the tank. Pyrophoric. May spontaneously reignite in air. May accumulate aa static electric electric charge. Low ignition energy.

DOT Recommended Isolation Zones: Small Spill: 300 ft   Large Spill: 1200 ft
DOT Recommended Down Wind Take Cover Distance: Small Spill: 1 mile   Large Spill: 4 miles

TOXICOLOGY

Odor: Acrid
Physical Contact: Material is extremely destructive to human tissues.
TLV = N.D.   **STEL** = N.D.

Odor Threshold: N.D.

IDLH = N.D.

Routes of Entry and Relative LD$_{50}$ (or LC$_{50}$)
    Inhalation   N.D.
    Ingestion   N.D.
    Skin Absorption N.D.

RTECS # VV304000

Symptoms of Exposure: Irritation and burns to skin, eyes and upper respiratory tract; headache; nausea. Inhalation of vapors may be fatal by causing glottis to spasm and suffocation. Exposure may result in chemical pneumonitis or pulmonary edema.

Emergency/First Aid Treatment: Remove to ventilated area; immediately remove any contaminated clothing and wash contaminated areas for 15 minutes using water. Treat supportively and observe for possible shock. If ingested, seek immediate medical aid.

Recommended Clean-Up Procedures

Personal Protection: Level A Ensemble   **Recommended Material** N.D.

RCRA Waste # None   **Reportable Quantities:** None

Spills: Remove all potential ignition sources. Use water fog and ventillation to disperse vapors. Stop leak if it can safely be done. Keep area isolated until vapors have dissipated.

Special Emergency Information: May be fatal in inhaled, ingested or absorbed through the skin. Vapors may collect in low lying areas and reignite. Vapors are heavier than air and may travel some distance to an ignition source. Reacts with water to yield hydrochloric acid.

3,5-DICHLORO-2,4,6-TRIFLUOROPYRIDINE

Synonyms:

CAS Number: 1737-93-5   **Description:** Colorless liquid
DOT Number: 9264   **DOT Classification:**

Molecular Weight: 202.0
Melting Point: °C   (°F)   **Vapor Density:**   N.D.   **Vapor Pressure:**   N.D.
Boiling Point: 160°C (320°F)   **Specific Gravity:** 1.6   **Water Solubility:**   N.D.

Chemical Incompatibilities or Instabilities: Strong oxidizers.

FLAMMABILITY

NFPA Hazard Code: **NFPA Classification:**   Class IIIB Combustible
Flash Point: 110°C (230°F)   **LEL:**   N.D.
Autoignition Temp: °C   (°F)   **UEL:**   N.D.

Fire Extinguishing Methods: Carbon Dioxide, Dry Chemical, Foam, Water Spray or Fog.

Special Fire Fighting Considerations: Structural fire fighter protective clothing will not provide adequate protection. Fight fire from a distance or protected location, if possible. Treat all materials used as contaminated by hazardous waste.

DOT Recommended Isolation Zones: Small Spill: 150 ft   Large Spill: 150 ft
DOT Recommended Down Wind Take Cover Distance: Small Spill: 0.2 miles   Large Spill: 0.2 miles

TOXICOLOGY

Odor: N.D.   **Odor Threshold:** N.D.
Physical Contact: Irritant
TLV = N.D.   **STEL** = N.D.   **IDLH** = N.D.

Routes of Entry and Relative LD50 (or LC50)   **RTECS #** N.D.
    Inhalation   N.D.
    Ingestion   N.D.
    Skin Absorption   N.D.

Symptoms of Exposure: Possible irritation to skin, eyes and upper respiratory tract; headache; nausea.

Emergency/First Aid Treatment: Remove to ventilated area; immediately remove any contaminated clothing and wash contaminated areas for 15 minutes using water. Treat supportively and observe for possible shock. If ingested, seek immediate medical aid.

Recommended Clean-Up Procedures

Personal Protection: Level B Ensemble   **Recommended Material**   N.D.

RCRA Waste #   None   **Reportable Quantities:**   None

Spills: Dike to contain spill and collect liquid for later disposal. Decontaminate spill area using soapy water. Treat all materials used as contaminated by hazardous waste.

Special Emergency Information: May be harmful if inhaled, swallowed or absorbed through the skin.

DICHLORVOS

Synonyms: apavap; astrobot; DDVP; SD 1750; canogard; dedevap; dichlorman; divipan; equiguard; equigel; esterol; herkol; nogos, nuvan; task; vapona; verdisol; Fly-Die; lindan; vinylophos

CAS Number: 62-73-7	**Description:** Liquid
DOT Number: 2783	**DOT Classification:** Poison B

Molecular Weight: 221.0

Melting Point: °C (°F)	**Vapor Density:**	**Vapor Pressure:**	1 (0.001 mm Hg)
Boiling Point: Decomposes	**Specific Gravity:** 1.4	**Water Solubility:**	1%

Chemical Incompatibilities or Instabilities: Strong acids, strong alkalis.

FLAMMABILITY

NFPA Hazard Code:	**NFPA Classification:**
Flash Point: °C (°F)	**LEL:** N.D.
Autoignition Temp: °C (°F)	**UEL:** N.D.

Fire Extinguishing Methods: Carbon Dioxide, Dry Chemical, Foam, Water Spray or Fog.

Special Fire Fighting Considerations: Structural fire fighter protective clothing will not provide adequate protection. Fight fire from a distance or protected location, if possible. Treat all materials used as contaminated by hazardous waste.

TOXICOLOGY

Odor: N.D.	**Odor Threshold:** N.D.
Physical Contact: Irritation	
TLV = 0.9 mg/m^3 **STEL** = N.D.	**IDLH** = 200 mg/m^3
	RTECS # TC0350000

Routes of Entry and Relative LD$_{50}$ (or LC$_{50}$)

Inhalation	10	(60 mg/m^3/H)
Ingestion	9	(25mg/kg)
Skin Absorption	9	(107 mg/kg)

Symptoms of Exposure: Possible irritation to skin, eyes and upper respiratory tract; headache; nausea; ataxia; convulsions; cardiac irregularities. Inhalation of vapors may be fatal by causing glottis to spasm and suffocation.

mergency/First Aid Treatment: Remove to ventilated area; immediately remove any contaminated clothing and wash contaminated areas for 15 minutes using water. Treat supportively and observe for possible shock. If ingested, seek immediate medical aid.

Recommended Clean-Up Procedures

Personal Protection: Level A Ensemble	**Recommended Material** N.D.
RCRA Waste # None	**Reportable Quantities:** 10 lb (10 lb)

Spills: Take up in non-combustible absorbant and dispose. Decontaminate spill area using soapy water. Treat all materials used as contaminated by hazardous waste.

Special Emergency Information: May be fatal in inhaled, ingested or absorbed through the skin.

DICYCLOHEXYLAMINE

Synonyms: CDHA; N-cyclohexylhexanamine; N,N-dicyclohexylamine

CAS Number: 101-83-7
DOT Number: 2565

Description: Colorless liquid
DOT Classification: Corrosive Material

Molecular Weight: 181.4
Melting Point: -1°C (30°F)
Boiling Point: 256°C (492°F)

Vapor Density: 6.0
Specific Gravity: 0.9

Vapor Pressure: 3 (8 mm HG)
Water Solubility: <1%

Chemical Incompatibilities or Instabilities: Acid chlorides, acid anhydrides, chloroformates.

FLAMMABILITY

NFPA Hazard Code: 3-1-0
Flash Point: 96°C (205°F)
Autoignition Temp: °C (°F)

NFPA Classification: Class IIIB Combustible
LEL: N.D.
UEL: N.D.

Fire Extinguishing Methods: Carbon Dioxide, Dry Chemical, Foam, Water Spray or Fog.

Special Fire Fighting Considerations: Water spray should be used to keep closed containers cool. Continue to cool container after fire is extinguished.

TOXICOLOGY
QUESTIONABLE CARCINOGEN

Odor: Faint fish-like odor
Physical Contact: Material is extremely destructive to human tissues.
TLV = N.D.   **STEL** = N.D.

Odor Threshold: N.D.

IDLH = N.D.

Routes of Entry and Relative LD50 (or LC50)
Inhalation	N.D.	
Ingestion	5	(373 mg/kg)
Skin Absorption	N.D.	

RTECS # HY4025000

Symptoms of Exposure: Possible irritation to skin, eyes and upper respiratory tract; headache; nausea. Inhalation of vapors may be fatal by causing glottis to spasm and suffocation. Exposure may result in chemical pneumonitis or pulmonary edema.

Emergency/First Aid Treatment: Remove to ventilated area; immediately remove any contaminated clothing and wash contaminated areas for 15 minutes using water. Treat supportively and observe for possible shock. If ingested, seek immediate medical aid.

Recommended Clean-Up Procedures

Personal Protection: Level B Ensemble   **Recommended Material** N.D.

RCRA Waste #   None   **Reportable Quantities:** None

Spills: Dike to contain spill and collect liquid for later disposal. Decontaminate spill area using soapy water.

Special Emergency Information: May be harmful if inhaled, swallowed or absorbed through the skin.

DICYCLOHEXYLAMMONIUM NITRITE

Synonyms: dechan; dicyclohexylamine nitrite; dusitan

CAS Number: 3129-91-7	**Description:** White powder
DOT Number: 2687	**DOT Classification:** N.D.

Molecular Weight: 228.4

Melting Point: °C (°F)	**Vapor Density:** N. A.	**Vapor Pressure:** N. A.
Boiling Point: 180°C (356°F)	**Specific Gravity:** N.D.	**Water Solubility:** N.D.

Chemical Incompatibilities or Instabilities: Oxidizers, acid anhydrides, acid chlorides, chloroformates.

FLAMMABILITY

NFPA Hazard Code:	**NFPA Classification:** N. A.
Flash Point: °C (°F)	**LEL:** N. A.
Autoignition Temp: °C (°F)	**UEL:** N. A.

Fire Extinguishing Methods: Carbon Dioxide, Dry Chemical, Foam, Water Spray or Fog.

Special Fire Fighting Considerations: Water spray should be used to keep closed containers cool. Continue to cool container after fire is extinguished.

TOXICOLOGY
QUESTIONABLE CARCINOGEN

Odor: N. A.	**Odor Threshold:** N. A.

Physical Contact: Irritant

TLV = N.D.	**STEL** = N.D.	**IDLH** = N.D.

Routes of Entry and Relative LD50 (or LC50)   **RTECS # HY4200000**

Inhalation	N.D.	
Ingestion	6	(284 mg/kg)
Skin Absorption	N.D.	

Symptoms of Exposure: Possible irritation to skin, eyes and upper respiratory tract; headache; nausea. Absorption into the body may yield methemoglobin formation. Symptons may be delayed by several hours.

Emergency/First Aid Treatment: Remove to ventilated area; immediately remove any contaminated clothing and wash contaminated areas for 15 minutes using water. Treat supportively and observe for possible shock. If ingested, seek immediate medical aid.

Recommended Clean-Up Procedures

Personal Protection: Level B Ensemble	**Recommended Material** N.D.
RCRA Waste # None	**Reportable Quantities:** None

Spills: Sweep up spill and dispose. Decontaminate spill area using soapy water.

Special Emergency Information: May be harmful if inhaled, swallowed or absorbed through the skin.

DICYCLOPENTADIENE

Synonyms: bicyclopentadiene; biscyclopentadiene

CAS Number: 77-73-6   **Description:** Colorless liquid or white powder
DOT Number: 2048   **DOT Classification:** IMO

Molecular Weight: 132.2
Melting Point: 33°C (91°F)   **Vapor Density:** 4.6   **Vapor Pressure:** 2 (6mm Hg)
Boiling Point: 172°C (342°F)   **Specific Gravity:** 1.1   **Water Solubility:** Insoluble

Chemical Incompatibilities or Instabilities: Oxidizers.

FLAMMABILITY

NFPA Hazard Code: 1-3-1   **NFPA Classification:** Class 1C Flammable
Flash Point: 32°C (90°F)   **LEL:**   1.0%
Autoignition Temp: 503°C (937°F)   **UEL:** 10.0%

Fire Extinguishing Methods: Alcohol Resistant Foam, Carbon Dioxide, Dry Chemical, Water Spray or Fog.

Special Fire Fighting Considerations: Isolate for 1/2 mile if rail or tank truck is involved in a fire. Water spray should be used to keep closed containers cool. Continue to cool container after fire is extinguished. For large fires, if possible, withdraw and allow to burn. Immediately withdraw if rising sound from venting device is heard or if fire is causing discoloration to the tank.

TOXICOLOGY

Odor: Sweet or sharp   **Odor Threshold:** 0.02 mg/m^3
Physical Contact: Irritant
TLV = 27 mg/m^3 **STEL** = N.D.   **IDLH** = N.D.

Routes of Entry and Relative LD$_{50}$ (or LC$_{50}$)   **RTECS # PC1050000**
   Inhalation   7   (1500 ppm)
   Ingestion   5   (353mg/kg)
   Skin Absorption   1   (5080 mg/kg)

Symptoms of Exposure: Possible irritation to skin, eyes and upper respiratory tract; headache; nausea.

Emergency/First Aid Treatment: Remove to ventilated area; immediately remove any contaminated clothing and wash contaminated areas for 15 minutes using water. Treat supportively and observe for possible shock. If ingested, seek immediate medical aid.

Recommended Clean-Up Procedures

Personal Protection: Level B Ensemble   **Recommended Material**   N.D.

RCRA Waste #   None   **Reportable Quantities:**   None

Spills: Remove all potential ignition sources. Dike to contain spill, absorb with non-combustible absorbant and take up using non-sparking tools. Decontaminate spill area using soapy water.

Special Emergency Information: May be harmful if inhaled, swallowed or absorbed through the skin.

Dicycloheptadiene (DOT # 2251) has similar chemical, physical and toxicological properties.

DIELDRIN

Synonyms: alvit; Compound 497; dieldrex; heod; illoxol; octalox; quintox

CAS Number: 60-57-1
DOT Number: 2761

Description: White to tan powder
DOT Classification: ORM-A

Molecular Weight: 380.9
Melting Point: 144°C (°F)
Boiling Point: °C (°F)

Vapor Density: 13.2
Specific Gravity: 1.8

Vapor Pressure: N. D.
Water Solubility: Insoluble

Chemical Incompatibilities or Instabilities: Copper, iron, oxidizer.

FLAMMABILITY

NFPA Hazard Code:
Flash Point: °C (°F)
Autoignition Temp: °C (°F)

NFPA Classification:
LEL: N. A.
UEL: N. A.

Fire Extinguishing Methods: Carbon Dioxide, Dry Chemical, Foam, Water Spray or Fog.

Special Fire Fighting Considerations: Structural fire fighter protective clothing will not provide adequate protection. Fight fire from a distance or protected location, if possible. Treat all materials used as contaminated by hazardous waste.

TOXICOLOGY
POSSIBLE CARCINOGEN

Odor: N.D.
Physical Contact: Irritant
TLV = 0.25 mg/m^3 **STEL** = N.D.

Odor Threshold: N.D.

IDLH = 450 mg/m^3

Routes of Entry and Relative LD50 (or LC50)

Inhalation	10	(3 ppm)
Ingestion	9	(38 mg/kg)
Skin Absorption	8	(250 mg/kg)

RTECS # IO1750000

Symptoms of Exposure: Possible irritation to skin, eyes and upper respiratory tract; headache; nausea, cunvulsions.

Emergency/First Aid Treatment: Remove to ventilated area; immediately remove any contaminated clothing and wash contaminated areas for 15 minutes using water. Treat supportively and observe for possible shock. If ingested, seek immediate medical aid.

Recommended Clean-Up Procedures

Personal Protection: Level A Ensemble
RCRA Waste # P037

Recommended Material N.D.
Reportable Quantities: 1 lb (1 lb)

Spills: Sweep up spill and dispose. Decontaminate spill area using soapy water. Treat all materials used as contaminated by hazardous waste.

Special Emergency Information: May be fatal in inhaled, ingested or absorbed through the skin.

DIETHANOLAMINE - (HOCH$_2$CH$_2$)$_2$NH

Synonyms: bis(2-hydroxyethyl)amine; DEA; 2,2'-dihydroxydiethylamine; diolamine

CAS Number: 111-42-2   **Description:** White solid or colorless liquid
DOT Number: N.D.   **DOT Classification:** N.D.

Molecular Weight: 105.1
Melting Point: 28°C   (82°F)   **Vapor Density:**   3.7   **Vapor Pressure:**   1 (<1 mm Hg)
Boiling Point: 269°C (516°F)   **Specific Gravity:** 1.1   **Water Solubility:**   Soluble

Chemical Incompatibilities or Instabilities: Acid, oxidizers, copper, copper alloys, zinc.

FLAMMABILITY

NFPA Hazard Code: 1-1-0   **NFPA Classification:** Class IIIB Combustible
Flash Point: 172°C (342°F)   **LEL:**   N.D.
Autoignition Temp: 662°C (1224°F)   **UEL:**   N.D.

Fire Extinguishing Methods: Alcohol Resistant Foam, Carbon Dioxide, Dry Chemical, Water Spray or Fog.

Special Fire Fighting Considerations: Water spray should be used to keep closed containers cool. Immediately withdraw if rising sound from venting device is heard or if fire is causing discoloration to the tank.

TOXICOLOGY

Odor: **Odor Threshold:** N.D.
Physical Contact: Material is extremely destructive to human tissues.
TLV = 13 mg/m^3 **STEL** = N.D.   **IDLH** = N.D.

Routes of Entry and Relative LD$_{50}$ (or LC$_{50}$)   **RTECS # KL2975000**
    Inhalation   N.D.
    Ingestion   4   (710 mg/kg)
    Skin Absorption   1   (12,200 mg/kg)

Symptoms of Exposure: Possible irritation to skin, eyes and upper respiratory tract; headache; nausea. Inhalation of vapors may be fatal by causing glottis to spasm and suffocation. Inhalation of vapors may be fatal by causing glottis to spasm and suffocation. May cause an allergic reaction.

Emergency/First Aid Treatment: Remove to ventilated area; immediately remove any contaminated clothing and wash contaminated areas for 15 minutes using water. Treat supportively and observe for possible shock. If ingested, seek immediate medical aid.

Recommended Clean-Up Procedures

Personal Protection: Level B Ensemble   **Recommended Material**   Butyl Rubber (+), Viton (+),
                                                                          Polyvinyl Alcohol (+), Neoprene,
                                                                          Nitrile Rubber

RCRA Waste #   None   **Reportable Quantities:**   None

Spills: Dike to contain spill and collect liquid for later disposal. decon Water spray should be used to keep closed containers cool. Continue to cool container after fire is extinguished.

Special Emergency Information: May be harmful if inhaled, swallowed or absorbed through the skin.

DIETHOXYDIMETHYLSILANE - (CH₃CH₂O)₂Si(CH₃)

Synonyms: dimethyldiethoxysilane

CAS Number: 78-62-6	**Description:** Colorless liquid	
DOT Number: 2380	**DOT Classification:** Flammable Liquid	

Molecular Weight: 148.3

Melting Point: °C (°F)	**Vapor Density:** 5.1	**Vapor Pressure:** 4 (15 mm Hg)	
Boiling Point: 114°C (°F)	**Specific Gravity:** 0.8	**Water Solubility:** Decomposes	

Chemical Incompatibilities or Instabilities: Water, Strong oxidizers.

FLAMMABILITY

NFPA Hazard Code: N.D.	**NFPA Classification:** Class IB Flammable
Flash Point: 11°C (53°F)	**LEL:** N.D.
Autoignition Temp: °C (°F)	**UEL:** N.D.

Fire Extinguishing Methods: Alcohol Resistant Foam, Carbon Dioxide, Dry Chemical, Water Spray or Fog.

Special Fire Fighting Considerations: Structural fire fighter protective clothing will not provide adequate protection. Water spray should be used to keep closed containers cool. Continue to cool container after fire is extinguished. For large fires, if possible, withdraw and allow to burn. Immediately withdraw if rising sound from venting device is heard or if fire is causing discoloration to the tank.

TOXICOLOGY

Odor:	**Odor Threshold:** N.D.
Physical Contact: Irritant	
TLV = N.D. **STEL** = N.D.	**IDLH** = N.D.

Routes of Entry and Relative LD₅₀ (or LC₅₀) **RTECS # VV359000**

Inhalation	N.D.	
Ingestion	2	(9289 mg/kg)
Skin Absorption	N.D.	

Symptoms of Exposure: Possible irritation to skin, eyes and upper respiratory tract; headache; nausea; dizziness.

Emergency/First Aid Treatment: Remove to ventilated area; immediately remove any contaminated clothing and wash contaminated areas for 15 minutes using water. Treat supportively and observe for possible shock. If ingested, seek immediate medical aid.

Recommended Clean-Up Procedures

Personal Protection: Level B Ensemble	**Recommended Material** N.D.
RCRA Waste # None	**Reportable Quantities:** None

Spills: Remove all potential ignition sources. Dike to contain spill, absorb with non-combustible absorbant and take up using non-sparking tools. Decontaminate spill area using soapy water.

Special Emergency Information: May be harmful if inhaled, swallowed or absorbed through the skin. Vapors are heavier than air and may travel some distance to an ignition source. Vapors are more dense than air and may settle in low lying areas.

DIETHYLAMINE

Synonyms: n-ethyl-ethane amine

CAS Number: 109-89-7	**Description:** Colorless liquid
DOT Number: 1154	**DOT Classification:** Flammable Liquid

Molecular Weight: 73.2

Melting Point: -50°C (-58°F)	**Vapor Density:** 2.5	**Vapor Pressure:** 7 (189 mm Hg)
Boiling Point: 56°C (132°F)	**Specific Gravity:** 0.7	**Water Solubility:** Soluble

Chemical Incompatibilities or Instabilities: Oxidizers, halogens, hypochlorite, sulfuric acid.

FLAMMABILITY

NFPA Hazard Code: 3-3-0	**NFPA Classification:** Class IB Flammable
Flash Point: -28°C (-18°F)	**LEL:** 1.8%
Autoignition Temp: 312°C (594°F)	**UEL:** 10.1%

Fire Extinguishing Methods: Alcohol Resistant Foam, Carbon Dioxide, Dry Chemical, Water Spray or Fog.

Special Fire Fighting Considerations: Isolate for 1/2 mile if rail or tank truck is involved in a fire. Water spray should be used to keep closed containers cool. Continue to cool container after fire is extinguished. Immediately withdraw if rising sound from venting device is heard or if fire is causing discoloration to the tank.

TOXICOLOGY

Odor: Fishy or ammonia-like

Physical Contact: Material is extremely destructive to human tissues.

Odor Threshold: 0.05 mg/m^3

$TLV = 30 \text{ mg/m}^3$ $STEL = 75 \text{ mg/m}^3$ $IDLH = 6000 \text{ mg/m}^3$

Routes of Entry and Relative LD_{50} (or LC_{50})

Inhalation	4	(16,000 ppm)	
Ingestion	4	(540 mg/kg)	
Skin Absorption	5	(820 mg/kg)	

RTECS # HZ8750000

Symptoms of Exposure: Possible irritation to skin, eyes and upper respiratory tract; headache; nausea. Inhalation of vapors may be fatal by causing glottis to spasm and suffocation. Exposure may result in chemical pneumonitis or pulmonary edema. May cause an allergic reaction.

Emergency/First Aid Treatment: Remove to ventilated area; immediately remove any contaminated clothing and wash contaminated areas for 15 minutes using water. Treat supportively and observe for possible shock. If ingested, seek immediate medical aid.

Recommended Clean-Up Procedures

Personal Protection: Level B Ensemble	**Recommended Material** Teflon (-)
RCRA Waste # None	**Reportable Quantities:** 100 lb (1000 lb)

Spills: Remove all potential ignition sources. Dike to contain spill and take up using nonsparking tools. Decontaminate spill area using water.

Special Emergency Information: May be harmful if inhaled, swallowed or absorbed through the skin. Vapors are heavier than air and may travel some distance to an ignition source. Vapors may collect in low lying areas.

Dimethylamine (CAS # 124-40-3, DOT # 1032, 1160, RTECS # IP8750000, RCRA Waste # U092, RQ 1000 lb [1000 lb]), dipropylamine (CAS # 142-84-7, DOT # 2383, RTECS # JL9200000), diisobutylamine (CAS # 110-96-3, DOT # 2361, RTECS # TX1750000) and dibutylamine (CAS # 111-92-2, DOT # 2248, RTECS # HR7780000) have similar chemical, physical and toxicological properties.

DIETHYLAMINOETHANOL - $HOCH_2CH_2N(CH_2CH_3)_2$

Synonyms: DEAE; 2-diethylaminoethanol; n,n-diethylethanolamine; n-diethylaminoethanol

CAS Number: 100-37-8
DOT Number: 2686

Description: Colorless liquid
DOT Classification: Flammable Liquid

Molecular Weight: 117.2

Melting Point:	°C (°F)	**Vapor Density:** 4.0	**Vapor Pressure:**	**2** (2 mm Hg)
Boiling Point:	162°C (324°F)	**Specific Gravity:** 0.9	**Water Solubility:**	Soluble

Chemical Incompatibilities or Instabilities: Strong oxidizers.

FLAMMABILITY

NFPA Hazard Code: 3-2-0
Flash Point: 48°C (120°F)
Autoignition Temp: 320°C (608°F)

NFPA Classification: Class II Combustible
LEL: N.D.
UEL: N.D.

Fire Extinguishing Methods: Carbon Dioxide, Dry Chemical, Foam, Water Spray or Fog.

Special Fire Fighting Considerations: Isolate for 1/2 mile if rail or tank truck is involved in a fire. Water spray should be used to keep closed containers cool. Continue to cool container after fire is extinguished. Immediately withdraw if rising sound from venting device is heard or if fire is causing discoloration to the tank.

TOXICOLOGY

Odor: Sharp ammonia-like
Physical Contact: Material is extremely destructive to human tissues.
TLV = 48 mg/m^3 **STEL** = N.D.

Odor Threshold: 0.05 mg/m^3

IDLH = 2500 mg/m^3

Routes of Entry and Relative LD$_{50}$ (or LC$_{50}$)

Inhalation	N.D.	
Ingestion	**3**	1300 mg/kg)
Skin Absorption	N.D.	

RTECS # KK5075000

Symptoms of Exposure: Possible irritation to skin, eyes and upper respiratory tract; headache; nausea. Inhalation of vapors may be fatal by causing glottis to spasm and suffocation. Exposure may result in chemical pneumonitis or pulmonary edema. May cause an allergic reaction. Vapors may cause eye damage.

Emergency/First Aid Treatment: Remove to ventilated area; immediately remove any contaminated clothing and wash contaminated areas for 15 minutes using water. Treat supportively and observe for possible shock. If ingested, seek immediate medical aid.

Recommended Clean-Up Procedures

Personal Protection: Level B Ensemble

Recommended Material Butyl Rubber (+), Viton (+), Polyvinyl Alcohol (+), Nitrile Rubber (+)

RCRA Waste # None

Reportable Quantities: None

Spills: Remove all potential ignition sources. Dike to contain spill, absorb with non-combustible absorbant and take up using non-sparking tools. Decontaminate spill area using water.

Special Emergency Information: May be harmful if inhaled, swallowed or absorbed through the skin. Vapors are heavier than air and may travel some distance to an ignition source. Vapors are more dense than air and may settle in low lying areas.

DIETHYLALUMINUM CHLORIDE - $(CH_2CH_3)_2AlCl$

Synonyms:

CAS Number: 96-10-6
DOT Number: 1101

Description: Colorless liquid
DOT Classification: N.D.

Molecular Weight: 120.5
Melting Point: -50°C (-58°F)
Boiling Point: 214°C (417°F)

Vapor Density:
Specific Gravity: 0.97

Vapor Pressure: 1 (<1 mm Hg)
Water Solubility: Reacts violently

Chemical Incompatibilities or Instabilities: Water, air, oxidizers, acids, alcohols, azides.

FLAMMABILITY

NFPA Hazard Code: 3-4-3-W
Flash Point: °C (°F)
Autoignition Temp: °C (°F)

NFPA Classification:
LEL: N. A.
UEL: N. A.

Fire Extinguishing Methods: Graphite powder, soda ash; **DO NOT USE WATER.**

Special Fire Fighting Considerations: Approach fire from upwind. Fight fire from a distance or protected location, if possible. Structural fire fighter protective clothing will not provide adequate protection. Water spray should be used to keep closed containers cool. Do not allow water to enter container.

TOXICOLOGY

Odor: N.D.
Odor Threshold: N.D.
Physical Contact: Thermal burns, material is extremely destructive to human tissues.
TLV = 5 mg (Al)/m^3 **STEL** = N.D.   **IDLH** = N.D.

Routes of Entry and Relative LD50 (or LC50)
Inhalation   N.D.
Ingestion   N.D.
Skin Absorption   N.D.

RTECS # BD0558000

Symptoms of Exposure: Possible burns or irritation to skin, eyes and upper respiratory tract; headache; nausea. Inhalation of vapors may be fatal by causing glottis to spasm and suffocation. Exposure may result in chemical pneumonitis or pulmonary edema.

Emergency/First Aid Treatment: Remove to ventilated area; immediately remove any contaminated clothing and wash contaminated areas for 15 minutes using water. Treat supportively and observe for possible shock. If ingested, seek immediate medical aid.

Recommended Clean-Up Procedures

Personal Protection: Level A Ensemble
Recommended Material N.D.

RCRA Waste # None
Reportable Quantities: None

Spills: Clean up only under expert supervision.

Special Emergency Information: May be fatal in inhaled, ingested or absorbed through the skin. Triethylaluminum is likely to be present as a solution. The solvent used to disolve it may be flammable and will evaporate allowing the triethylaluminum to come in contact with air and ignite.

DIETHYLAMINOPROPYLAMINE - (CH$_3$CH$_2$)$_2$NCH$_2$CH$_2$CH$_3$

Synonyms: n,n-Diethyl-1,3-diaminopropane; 3-(diethylamino)propylamine; diethylaminotriethyleneamine

CAS Number: 104-78-9
DOT Number: 2684

Description: Colorless liquid
DOT Classification: Corrosive Material

Molecular Weight: 130.3
Melting Point: 58°C (138°F)
Boiling Point: 169°C (337°F)

Vapor Density: 4.5
Specific Gravity: 0.8

Vapor Pressure:
Water Solubility: Soluble

Chemical Incompatibilities or Instabilities: Acid anhydrides, acid chlorides, oxidizers.

FLAMMABILITY

NFPA Classification:
Flash Point: 59°C (138°F)
Autoignition Temp: °C (°F)

NFPA Hazard Code: 2-2-0
LEL: N.D.
UEL: N.D.

Fire Extinguishing Methods: Carbon Dioxide, Dry Chemical, Foam, Water Spray or Fog.

Special Fire Fighting Considerations: Isolate for 1/2 mile if rail or tank truck is involved in a fire. Water spray should be used to keep closed containers cool. Continue to cool container after fire is extinguished. Immediately withdraw if rising sound from venting device is heard or if fire is causing discoloration to the tank.

TOXICOLOGY

Odor:
Physical Contact: Material is extremely destructive to human tissues.
TLV = N.D.   **STEL** = N.D.

Odor Threshold: N.D.

IDLH = N.D.

Routes of Entry and Relative LD50 (or LC50)
Inhalation	N.D.	
Ingestion	3	(1410 mg/kg)
Skin Absorption	6	(750mg/kg)

RTECS # TX7350000

Symptoms of Exposure: Possible burns or irritation to skin, eyes and upper respiratory tract; headache; nausea. Inhalation of vapors may be fatal by causing glottis to spasm and suffocation. Exposure may result in chemical pneumonitis or pulmonary edema.

Emergency/First Aid Treatment: Remove to ventilated area; immediately remove any contaminated clothing and wash contaminated areas for 15 minutes using water. Treat supportively and observe for possible shock. If ingested, seek immediate medical aid.

Recommended Clean-Up Procedures

Personal Protection: Level A Ensemble
RCRA Waste # None

Recommended Material N.D.
Reportable Quantities: None

Spills: Remove all potential ignition sources. Dike to contain spill, absorb with non-combustible absorbant and take up using non-sparking tools. Decontaminate spill area using water.

Special Emergency Information: May be fatal in inhaled, ingested or absorbed through the skin.

DIETHYLANILINE

Synonyms: n,n-diethylaminobenzene; n,n-diethylaniline

CAS Number: 91-66-7   **Description:** Pale yellow liquid
DOT Number: 2432   **DOT Classification:** Poison B

Molecular Weight: 149.3
Melting Point: -38°C (-36°F)   **Vapor Density:** 5.0   **Vapor Pressure:** 1 (<1 mm Hg)
Boiling Point: 215°C (419°F)   **Specific Gravity:** 0.9   **Water Solubility:** 15%

Chemical Incompatibilities or Instabilities: Strong oxidizers.

FLAMMABILITY

NFPA Hazard Code: 3-2-2   **NFPA Classification:** Class IIA Combustible \
Flash Point: 85°C (185°F)   **LEL:** N.D.
Autoignition Temp: 630°C (1166°F)   **UEL:** N.D.

Fire Extinguishing Methods: Carbon Dioxide, Dry Chemical, Foam, Water Spray or Fog.

Special Fire Fighting Considerations: Structural fire fighter protective clothing will not provide adequate protection. Water spray should be used to keep closed containers cool. Continue to cool container after fire is extinguished. Fight fire from a distance or protected location, if possible.

TOXICOLOGY

Odor: **Odor Threshold:** N.D.
Physical Contact: Irritant
TLV = N.D.   **STEL** = N.D.   **IDLH** = N.D.

Routes of Entry and Relative LD$_{50}$ (or LC$_{50}$)   **RTECS # BX3400000**
    Inhalation   7   (1300 ppm/hr)
    Ingestion   4   (780 mg/kg)
    Skin Absorption   N.D.

Symptoms of Exposure: Possible irritation to skin, eyes and upper respiratory tract; headache; nausea. Absorption may lead to the formation of methemoglobin resulting in cyanosis several hours after exposure.

Emergency/First Aid Treatment: Remove to ventilated area; immediately remove any contaminated clothing and wash contaminated areas for 15 minutes using water. Treat supportively and observe for possible shock. If ingested, seek immediate medical aid.

Recommended Clean-Up Procedures

Personal Protection: Level A Ensemble   **Recommended Material**   N.D.

RCRA Waste #   None   **Reportable Quantities:**   None

Spills: Dike to contain spill and collect liquid for later disposal. Decontaminate spill area using water.

Special Emergency Information: May be harmful if inhaled, swallowed or absorbed through the skin.

DIETHYLBENZENE

Synonyms:

CAS Number: 25340-17-4   **Description:** Colorless liquid
DOT Number: 2049   **DOT Classification:** Flammable Liquid

Molecular Weight: 134.2
Melting Point: -30 C (F)   **Vapor Density:** 4.6   **Vapor Pressure:** 2 (2 mm Hg)
Boiling Point: 181 C (358 F)   **Specific Gravity:** 0.9   **Water Solubility:** Insoluble

Chemical Incompatibilities or Instabilities: Strong oxidizers

FLAMMABILITY

NFPA Hazard Code: 2-2-0   **NFPA Classification:** Class II Combustible
Flash Point: 55 C (132 F)   **LEL:** N.D.
Autoignition Temp: 430 C (806 F)   **UEL:** N.D.

Fire Extinguishing Methods: Carbon Dioxide, Dry Chemical, Foam, Water Spray or Fog.

Special Fire Fighting Considerations: Isolate for 1/2 mile if rail or tank truck is involved in a fire. Water spray should be used to keep closed containers cool. Continue to cool container after fire is extinguished. Immediately withdraw if rising sound from venting device is heard or if fire is causing discoloration to the tank.

TOXICOLOGY

Odor: N.D.   **Odor Threshold:** N.D.
Physical Contact: Irritant
TLV = N.D.   **STEL** = N.D.   **IDLH** = N.D.

Routes of Entry and Relative LD50 (or LC50)   **RTECS # CZ5600000**
   Inhalation   N.D.
   Ingestion   2   (5000 mg/kg)
   Skin Absorption   N.D.

Symptoms of Exposure: Possible irritation to skin, eyes and upper respiratory tract; headache; nausea.

Emergency/First Aid Treatment: Remove to ventilated area; immediately remove any contaminated clothing and wash contaminated areas for 15 minutes using water. Treat supportively and observe for possible shock. If ingested, seek immediate medical aid.

Recommended Clean-Up Procedures

Personal Protection: Level B Ensemble   **Recommended Material** N.D.

RCRA Waste # None   **Reportable Quantities:** None

Spills: Remove all potential ignition sources. Dike to contain spill, absorb with non-combustible absorbant and take up using non-sparking tools. Decontaminate spill area using soapy water.

Special Emergency Information: May be harmful if inhaled, swallowed or absorbed through the skin.

1,2-diethylbenzene (CAS # 3135-01-3, RTECS # CZ564000),
1,3-diethylbenzene (CAS # 141-93-5, RTECS # CZ5620000), and
1,4-diethylbenzene (CAS # 105-05-5) have similar chemical, physical and toxicological properties.

2-DIETHYLDICHLOROSILANE - (CH$_3$CH$_2$)$_2$SiCl$_2$

Synonyms: dichlorodiethylsilane

CAS Number: 1719-53-5
DOT Number: 1767

Description: Colorless liquid
DOT Classification: Flammable Liquid

Molecular Weight: 157.1
Melting Point: -96°C (-141°F)
Boiling Point: 131°C (268°F)

Vapor Density: 5.4
Specific Gravity: 1.1

Vapor Pressure: N.D.
Water Solubility: Decomposes

Chemical Incompatibilities or Instabilities: Water, strong oxidizers.

FLAMMABILITY

NFPA Hazard Code:
Flash Point: 28°C (83°F)
Autoignition Temp: 320°C (608°F)

NFPA Classification: Class IB Flammable
LEL: N.D.
UEL: N.D.

Fire Extinguishing Methods: Carbon Dioxide, Dry Chemical, Foam, Water Spray or Fog.

Special Fire Fighting Considerations: Isolate for 1/2 mile if rail or tank truck is involved in a fire. Water spray should be used to keep closed containers cool. Continue to cool container after fire is extinguished. Do not allow water to enter container. Immediately withdraw if rising sound from venting device is heard or if fire is causing discoloration to the tank.

TOXICOLOGY

Odor: N.D.   **Odor Threshold:** N.D.
Physical Contact: Material is extremely destructive to human tissues.
TLV =N.D.   **STEL** = N.D.   **IDLH** =N.D.

Routes of Entry and Relative LD50 (or LC50)   **RTECS # VV3060000**
  Inhalation   N.D.
  Ingestion   N.D.
  Skin Absorption   N.D.

Symptoms of Exposure: Possible burns or irritation to skin, eyes and upper respiratory tract; headache; nausea. Laryngitis. Inhalation of vapors may be fatal by causing glottis to spasm and suffocation. Exposure may result in chemical pneumonitis or pulmonary edema.

Emergency/First Aid Treatment: Remove to ventilated area; immediately remove any contaminated clothing and wash contaminated areas for 15 minutes using water. Treat supportively and observe for possible shock. If ingested, seek immediate medical aid.

Recommended Clean-Up Procedures

Personal Protection: Level B Ensemble   **Recommended Material** Viton (+), Nitrile Rubber (+)

RCRA Waste #   None   **Reportable Quantities:**   None

Spills: Remove all potential ignition sources. Dike to contain spill and collect liquid for later disposal. Absorb with noncombustible absorbant and take up using nonsparking tools. Decontaminate spill area using dilute alkaline solution.

Special Emergency Information: May be harmful if inhaled, swallowed or absorbed through the skin. Vapors are heavier than air and may travel some distance to an ignition source. Vapors are more dense than air and may settle in low lying areas.

Dimethyldichlorosilane (CAS # 75-78-5, DOT # 1162, RETCS # VV3150000) has similar chemical, physical and toxicological properties.

DIETHYLCARBONATE - (CH$_3$CH$_2$)$_2$CO$_3$

Synonyms: DEC; ethoxyformic anhyride; ethyl carbonate, eufin

CAS Number: 105-58-8   **Description:** Colorless liquid
DOT Number: 2366   **DOT Classification:** Flammable Liquid

Molecular Weight: 118.2

Melting Point:	-43°C (-45°F)	**Vapor Density:** 4.1	**Vapor Pressure:**	3 (10 mm Hg)
Boiling Point:	126°C (259°F)	**Specific Gravity:** 0.9	**Water Solubility:**	Insoluble

Chemical Incompatibilities or Instabilities: Oxidizers.

FLAMMABILITY

NFPA Hazard Code: 2-3-1   **NFPA Classification:** Class IC Flammable
Flash Point: 25°C (77°F)   **LEL:** N.D.
Autoignition Temp: °C (°F)   **UEL:** N.D.

Fire Extinguishing Methods: Carbon Dioxide, Dry Chemical, Foam, Water Spray or Fog.

Special Fire Fighting Considerations: Isolate for 1/2 mile if rail or tank truck is involved in a fire. Water spray should be used to keep closed containers cool. Continue to cool container after fire is extinguished. For large fires, if possible, withdraw and allow to burn. Immediately withdraw if rising sound from venting device is heard or if fire is causing discoloration to the tank.

TOXICOLOGY
QUESTIONABLE CARCINOGEN

Odor: N.D.   **Odor Threshold:** N.D.
Physical Contact: Irritant
TLV = N.D.   **STEL** = N.D.   **IDLH** = N.D.

Routes of Entry and Relative LD50 (or LC50)   **RTECS # FF9800000**
Inhalation   N.D.
    Ingestion   N.D.
    Skin Absorption   N.D.

Symptoms of Exposure: Possible irritation to skin, eyes and upper respiratory tract; headache; nausea.

Emergency/First Aid Treatment: Remove to ventilated area; immediately remove any contaminated clothing and wash contaminated areas for 15 minutes using water. Treat supportively and observe for possible shock. If ingested, seek immediate medical aid.

Recommended Clean-Up Procedures

Personal Protection: Level B Ensemble   **Recommended Material** N.D.

RCRA Waste # None   **Reportable Quantities:** None

Spills: Remove all potential ignition sources. Dike to contain spill and collect liquid for later disposal. Absorb with noncombustible absorbant and take up using nonsparking tools. Decontaminate spill area using soapy water.

Special Emergency Information: May be harmful if inhaled, swallowed or absorbed through the skin. Vapors are heavier than air and may travel some distance to an ignition source. Vapors are more dense than air and may settle in low lying areas.

DIETHYLENETRIAMINE - $H_2NCH_2CH_2NHCH_2CH_2NH_2$

Synonyms: aminoethylendiamine; n-(2-aminoethyl)ethylendiamine

CAS Number: 111-40-0	**Description:** Pale yellow liquid			
DOT Number: 2079	**DOT Classification:** Corrosive Material			

Molecular Weight: 103.2			
Melting Point: -35°C (-31°F)	**Vapor Density:** 3.5	**Vapor Pressure:** 1 (<1 mm Hg)
Boiling Point: 207°C (405°F)	**Specific Gravity:** 0.9	**Water Solubility:** Soluble

Chemical Incompatibilities or Instabilities: Oxidizers, chlorine, acids, halogenated compounds, copper and its alloys.

FLAMMABILITY

NFPA Hazard Code: 3-1-0	**NFPA Classification:** Class IIIB Combustible
Flash Point: 102°C (215°F)	**LEL:** N.D.
Autoignition Temp: 385°C (676°F)	**UEL:** N.D.

Fire Extinguishing Methods: Alcohol Resistant Foam, Carbon Dioxide, Dry Chemical, Water Spray or Fog.

Special Fire Fighting Considerations: Isolate for 1/2 mile if rail or tank truck is involved in a fire. Water spray should be used to keep closed containers cool. Immediately withdraw if rising sound from venting device is heard or if fire is causing discoloration to the tank.

TOXICOLOGY

Odor: Ammonia   **Odor Threshold:** N.D.

Physical Contact: Material is extremely destructive to human tissues.

$TLV = 4.2 \ mg/m^3$ **STEL** = N.D.   **IDLH** = N.D.

Routes of Entry and Relative LD50 (or LC50)   **RTECS # IE2250000**

Inhalation	N.D.	
Ingestion	3	(1080mg/kg)
Skin Absorption	4	(1090 mg/kg)

Symptoms of Exposure: Possible burns or irritation to skin, eyes and upper respiratory tract; headache; nausea. Inhalation of vapors may be fatal by causing glottis to spasm and suffocation. Exposure may result in chemical pneumonitis or pulmonary edema.

Emergency/First Aid Treatment: Remove to ventilated area; immediately remove any contaminated clothing and wash contaminated areas for 15 minutes using water. Treat supportively and observe for possible shock. If ingested, seek immediate medical aid.

Recommended Clean-Up Procedures

Personal Protection: Level B Ensemble	**Recommended Material** Butyl Rubber (+)
RCRA Waste # None	**Reportable Quantities:** None

Spills: Dike to contain spill and collect liquid for later disposal. Decontaminate spill area using water.

Special Emergency Information: May be harmful if inhaled, swallowed or absorbed through the skin.

Diethylethylenediamine (CAS # 100-36-7, DOT # 2685, RTECS # KV3500000) has similar chemical, physical and toxicological properties.

DIETHYL ETHER - CH$_3$CH$_2$OCH$_2$CH$_3$

Synonyms: ethyl ether; ether; diethyl oxide

CAS Number: 60-29-7 **Description:** Colorless liquid
DOT Number: 1155 **DOT Classification:** Flammable Liquid

Molecular Weight: 74.1
Melting Point: -116°C (-177°F) **Vapor Density:** 2.6 **Vapor Pressure:** 9 (450 mm Hg)
Boiling Point: 35°C (94°F) **Specific Gravity:** 0.7 **Water Solubility:** Insoluble

Chemical Incompatibilities or Instabilities: Strong oxidizers.

FLAMMABILITY

NFPA Hazard Code: 2-4-1 **NFPA Classification:** Class IA Flammable
Flash Point: -45°C (-49°F) **LEL:** 1.9%
Autoignition Temp: 160°C (320°F) **UEL:** 36%

Fire Extinguishing Methods: Alcohol Resistant Foam, Carbon Dioxide, Dry Chemical, Water Spray or Fog.

Special Fire Fighting Considerations: Isolate for 1/2 mile if rail or tank truck is involved in a fire. Water spray should be used to keep closed containers cool. Continue to cool container after fire is extinguished. For large fires, if possible, withdraw and allow to burn. Immediately withdraw if rising sound from venting device is heard or if fire is causing discoloration to the tank.

TOXICOLOGY

Odor: Sweet, pungent **Odor Threshold:** N.D.
Physical Contact: Irritant
TLV = 1210 mg/m^3 **STEL** = 1520 mg/m^3 **IDLH** = 57,000 mg/m^3

Routes of Entry and Relative LD50 (or LC50) **RTECS # KI5775000**
  Inhalation 1 (145,000 ppm)
  Ingestion 3 (1215 mg/kg)
  Skin Absorption N.D.

Symptoms of Exposure: Possible irritation to skin, eyes and upper respiratory tract; headache; nausea.

Emergency/First Aid Treatment: Remove to ventilated area; immediately remove any contaminated clothing and wash contaminated areas for 15 minutes using water. Treat supportively and observe for possible shock. If ingested, seek immediate medical aid.

Recommended Clean-Up Procedures

Personal Protection: Level B Ensemble **Recommended Material** Teflon (+), Polyvinyl Ether (+)

RCRA Waste # None **Reportable Quantities:** None

Spills: Remove all potential ignition sources. Dike to contain spill and collect liquid for later disposal. Absorb with noncombustible absorbant and take up using nonsparking tools. Decontaminate spill area using soapy water.

Special Emergency Information: May be harmful if inhaled, swallowed or absorbed through the skin. Vapors are heavier than air and may travel some distance to an ignition source. Vapors are more dense than air and may settle in low lying areas. Use extreme caution, diethyl ether is extremely flammable and has a low ignition energy. May form unstable peroxides upon prolonged exposure to air.

Isopropyl ether (CAS # 108-20-3, DOT # 1159, RTECS # TZ5425000) and
methyl ethyl ether (CAS # 115-10-6, DOT # 1039, RETCS # KO026000) have similar chemical, physical and toxicological properties.

DIETHYL KETONE - CH₃CH₂C(O)CH₂CH₃

Synonyms: 3-pentanone

CAS Number: 96-22-0   **Description:** Colorless liquid
DOT Number: 1156   **DOT Classification:** Flammable Liquid

Molecular Weight: 86.2
Melting Point: -42°C (-44°F)   **Vapor Density:**   3.0   **Vapor Pressure:**   4 (20 mm Hg)
Boiling Point: 101°C (214°F)   **Specific Gravity:** 0.8   **Water Solubility:** 2%

Chemical Incompatibilities or Instabilities: Strong oxidizers.

FLAMMABILITY

NFPA Hazard Code: 1-3-0   **NFPA Classification:** Class IB Flammable
Flash Point: 12°C   (55°F)   **LEL:**   1.6%
Autoignition Temp: 452°C (845°F)   **UEL:**   N.D.

Fire Extinguishing Methods: Alcohol Resistant Foam, Carbon Dioxide, Dry Chemical, Water Spray or Fog.

Special Fire Fighting Considerations: Isolate for 1/2 mile if rail or tank truck is involved in a fire. Water spray should be used to keep closed containers cool. Continue to cool container after fire is extinguished. For large fires, if possible, withdraw and allow to burn. Immediately withdraw if rising sound from venting device is heard or if fire is causing discoloration to the tank.

TOXICOLOGY

Odor: Acetone-like   **Odor Threshold:** 3 mg/m³
Physical Contact: Irritant
TLV = 700 mg/m³   **STEL** = N.D.   **IDLH** = N.D.

Routes of Entry and Relative LD50 (or LC50)   **RTECS # SA8050000**
    Inhalation   N.D.
    Ingestion   3   (2140 mg/kg)
    Skin Absorption 1

Symptoms of Exposure: Possible irritation to skin, eyes and upper respiratory tract; headache; nausea. May act as a narcotic in high concentrations.

Emergency/First Aid Treatment: Remove to ventilated area; immediately remove any contaminated clothing and wash contaminated areas for 15 minutes using water. Treat supportively and observe for possible shock. If ingested, seek immediate medical aid.

Recommended Clean-Up Procedures

Personal Protection: Level B Ensemble

Small Spills: Remove all potential ignition sources. Absorb with noncombustible absorbant and take up using nonsparking tools.

Large Spills: Remove all potential ignition sources. Dike to contain spill and collect liquid for later disposal. Absorb with noncombustible absorbant and take up using nonsparking tools.

Special Emergency Information: May be harmful if inhaled, swallowed or absorbed through the skin. Vapors are heavier than air and may travel some distance to an ignition source. Vapors are more dense than air and may settle in low lying areas.

Methyl ethyl ketone CAS # 78-98-3; DOT # 1193, 1232; RTECS # EL6475000) has similar chemical, physical and toxicological properties.

DIETHYL SULFATE - $(CH_3CH_2)_2SO_4$

Synonyms: diethyl ester sulfuric acid; ethyl sulfate; sulfuric acid diethyl ester

CAS Number: 64-67-5	**Description:** Colorless liquid
DOT Number: 1594	**DOT Classification:** Poison B

Molecular Weight: 154.2

Melting Point:	-25°C (-13°F)	**Vapor Density:** 5.3	**Vapor Pressure:**	1 (<1 mm Hg)
Boiling Point:	210°C (410°F)	**Specific Gravity:** 1.2	**Water Solubility:**	Insoluble

Chemical Incompatibilities or Instabilities: Heat, moisture, iron, aquesou alkaline solutions.

FLAMMABILITY

NFPA Hazard Code:	3-1-1	**NFPA Classification:** Class IIIB Combustible
Flash Point:	104°C (220°F)	**LEL:** 4.7%
Autoignition Temp:	436°C (817°F)	**UEL:** N.D.

Fire Extinguishing Methods: Carbon Dioxide, Dry Chemical, Foam, Water Spray or Fog.

Special Fire Fighting Considerations: Structural fire fighter protective clothing will not provide adequate protection. Fight fire from a distance or protected location, if possible. Treat all materials used as contaminated by hazardous waste. Water spray should be used to keep closed containers cool.

TOXICOLOGY
CONFIRMED CARCINOGEN

Odor: Faint ether-like **Odor Threshold:** N.D.

Physical Contact: Material is extremely destructive to human tissues.

TLV = N.D. **STEL** = N.D. **IDLH** = N.D.

Routes of Entry and Relative LD50 (or LC50) **RTECS # WS7875000**

Inhalation	N.D.	
Ingestion	4	(880 mg/kg)
Skin Absorption	7	(600 mg/kg)

Symptoms of Exposure: Possible burns and irritation to skin, eyes and upper respiratory tract; headache; nausea. Laryngitis. Exposure may result in chemical pneumonitis or pulmonary edema. Inhalation of vapors may be fatal by causing glottis to spasm and suffocation.

Emergency/First Aid Treatment: Remove to ventilated area; immediately remove any contaminated clothing and wash contaminated areas for 15 minutes using water. Treat supportively and observe for possible shock. If ingested, seek immediate medical aid.

Recommended Clean-Up Procedures

Personal Protection:	Level A Ensemble	**Recommended Material** N.D.
RCRA Waste #	None	**Reportable Quantities:** None

Spills: Dike to contain spill and collect liquid for later disposal. Take up in non-combustible absorbant and dispose. Decontaminate spill area using soapy water. Treat all materials used as contaminated by hazardous waste.

Special Emergency Information: May be fatal in inhaled, ingested or absorbed through the skin. May explosively decompose over 100°C.

DIETHYLZINC - (CH3CH2)2Zn

Synonyms: zinc ethide; zinc ethyl

CAS Number: 557-20-0   **Description:** Colorless liquid
DOT Number: 1366, 2845   **DOT Classification:** Flammable Liquid

Molecular Weight: 123.5
Melting Point: -28°C   (-18°F)   **Vapor Density:**   N.D.   **Vapor Pressure:**   N.D.
Boiling Point: 118°C   (243°F)   **Specific Gravity:**   1.2   **Water Solubility:**   Decomposes

Chemical Incompatibilities or Instabilities: Halogens, halogenated compounds, water, air, alcohols, nitro compounds, hydrazine, sulfur dioxide, arsenic trichloride, phosphorous trichloride.

FLAMMABILITY

NFPA Hazard Code: 1-4-3-~~W~~   **NFPA Classification:**
Flash Point: N. A.   **LEL:** N. A.
Autoignition Temp: N. A.   **UEL:** N. A.

Fire Extinguishing Methods: Dry Chemical, Dry sand, Graphite.

Special Fire Fighting Considerations: For large fires, if possible, withdraw and allow to burn. Fight fire from a distance or protected location, if possible. Explodes when heated above 120°C (248°F).

TOXICOLOGY

Odor: **Odor Threshold:** N.D.
Physical Contact: Material is extremely destructive to human tissues.
TLV = N.D.   **STEL** = N.D.   **IDLH** = N.D.

Routes of Entry and Relative LD50 (or LC50)   **RTECS # ZH2070000**
    Inhalation   N.D.
    Ingestion   N.D.
    Skin Absorption   N.D.

Symptoms of Exposure: Possible burns and irritation to skin, eyes and upper respiratory tract; headache; nausea. Laryngitis. Exposure may result in chemical pneumonitis or pulmonary edema. Inhalation of vapors may be fatal by causing glottis to spasm and suffocation.

Emergency/First Aid Treatment: Remove to ventilated area; immediately remove any contaminated clothing and wash contaminated areas for 15 minutes using water. Treat supportively and observe for possible shock. If ingested, seek immediate medical aid.

Recommended Clean-Up Procedures

Personal Protection: Level B Ensemble   **Recommended Material**   N.D.

RCRA Waste #   None   **Reportable Quantities:**   None

Spills: Dike to contain spill and collect liquid for later disposal. Dike to contain spill, absorb with non-combustible absorbant and take up using non-sparking tools. Decontaminate spill area using dry decon techniques.

Special Emergency Information: May be fatal in inhaled, ingested or absorbed through the skin. Spontaneously ignites or explodes in air. Reaction with water yields flammable and explosive gases. Caution: Product may be packaged as a solution with a flammable solvent.

Dimethyl zinc (CAS # 544-97-8, DOT # 1370), other than being a solid, has similar chemical, physical and toxicological properties.

DIHYDROPYRAN

Synonyms: 3,4-dihydropyran

CAS Number: 110-87-2	**Description:** Colorless liquid
DOT Number: 2376	**DOT Classification:** Flammable Liquid

Molecular Weight: 84.1

Melting Point: -70°C (-94°F)	**Vapor Density:** 2.9	**Vapor Pressure:** N.D.
Boiling Point: 86°C (186°F)	**Specific Gravity:** 0.9	**Water Solubility:** Insoluble

Chemical Incompatibilities or Instabilities: Strong oxidizers.

FLAMMABILITY

NFPA Hazard Code: 2-3-0	**NFPA Classification:** Class 1B Flammable
Flash Point: -18°C (0°F)	**LEL:** N.D.
Autoignition Temp: °C (°F)	**UEL:** N.D.

Fire Extinguishing Methods: Alcohol Resistant Foam, Carbon Dioxide, Dry Chemical, Water Spray or Fog.

Special Fire Fighting Considerations: Isolate for 1/2 mile if rail or tank truck is involved in a fire. Water spray should be used to keep closed containers cool. Continue to cool container after fire is extinguished. For large fires, if possible, withdraw and allow to burn. Immediately withdraw if rising sound from venting device is heard or if fire is causing discoloration to the tank.

TOXICOLOGY

Odor:	**Odor Threshold:** N.D.
Physical Contact: Irritant	
TLV = N.D. **STEL** = N.D.	**IDLH** = N.D.

Routes of Entry and Relative LD50 (or LC50)   **RTECS # UP7700000**

Inhalation	N.D.
Ingestion	N.D.
Skin Absorption	N.D.

Symptoms of Exposure: Possible irritation to skin, eyes and upper respiratory tract; headache; nausea.

Emergency/First Aid Treatment: Remove to ventilated area; immediately remove any contaminated clothing and wash contaminated areas for 15 minutes using water. Treat supportively and observe for possible shock. If ingested, seek immediate medical aid.

Recommended Clean-Up Procedures

Personal Protection: Level B Ensemble	**Recommended Material** N.D.
RCRA Waste # None	**Reportable Quantities:** None

Spills: Remove all potential ignition sources. Dike to contain spill, absorb with non-combustible absorbant and take up using non-sparking tools.

Special Emergency Information: May be harmful if inhaled, swallowed or absorbed through the skin. Vapors are heavier than air and may travel some distance to an ignition source. Vapors are more dense than air and may settle in low lying areas. May form unstable peroxides upon prolonged exposure to air.

DIISOBUTYL KETONE - [(CH$_3$)$_2$CHCH$_2$]$_2$CO

Synonyms: 2,6-dimethyl-4-heptanone; isovalerone; valerone

CAS Number: 108-83-8	**Description:** Colorless liquid	
DOT Number: 1157	**DOT Classification:** Flammable Liquid	

Molecular Weight: 142.3

Melting Point:	-42°C (-43°F)	**Vapor Density:** 4.9	**Vapor Pressure:**	**2** (2 mm Hg)
Boiling Point:	168°C (335°F)	**Specific Gravity:** 0.8	**Water Solubility:**	Insoluble

Chemical Incompatibilities or Instabilities: Strong oxidizers.

FLAMMABILITY

NFPA Hazard Code:	**1-2-0**	**NFPA Classification:** Class II Combustible	
Flash Point:	49°C (120°F)	**LEL:** N.D.	
Autoignition Temp:	396°C (745°F)	**UEL:** N.D.	

Fire Extinguishing Methods: Alcohol Resistant Foam, Carbon Dioxide, Dry Chemical, Water Spray or Fog.

Special Fire Fighting Considerations: Isolate for 1/2 mile if rail or tank truck is involved in a fire. Water spray should be used to keep closed containers cool. Continue to cool container after fire is extinguished. For large fires, if possible, withdraw and allow to burn. Immediately withdraw if rising sound from venting device is heard or if fire is causing discoloration to the tank.

TOXICOLOGY

Odor: Mild, sweet	**Odor Threshold:** N.D.

Physical Contact:

TLV = 145 mg/m^3	**STEL** = N.D.	**IDLH** = 12,000 mg/m^3

Routes of Entry and Relative LD50 (or LC50)   **RTECS # MJ5775000**

Inhalation	N.D.	
Ingestion	2	(5720 mg/kg)
Skin Absorption	1	(16,000 mg/kg)

Symptoms of Exposure: Possible irritation to skin, eyes and upper respiratory tract; headache; nausea and possible visual disturbances.

Emergency/First Aid Treatment: Remove to ventilated area; immediately remove any contaminated clothing and wash contaminated areas for 15 minutes using water. Treat supportively and observe for possible shock. If ingested, seek immediate medical aid.

Recommended Clean-Up Procedures

Personal Protection:	Level B Ensemble	**Recommended Material** Polyvinyl Alcohol (0)
RCRA Waste #	None	**Reportable Quantities:** None

Spills: Remove all potential ignition sources. Dike to contain spill, absorb with noncombustible absorbant and take up using nonsparking tools. Decontaminate spill area using soapy water.

Special Emergency Information: May be harmful if inhaled, swallowed or absorbed through the skin. Vapors are heavier than air and may travel some distance to an ignition source. Vapors are more dense than air and may settle in low lying areas.

DIISOPROPYLAMINE - [(CH$_3$)$_2$CH]$_2$NH

Synonyms: DIPA; n-(1-methylethyl)-2-propanamine

CAS Number: 108-18-9	**Description:** Colorless liquid
DOT Number: 1158	**DOT Classification:** Flammable Liquid

Molecular Weight: 101.2

Melting Point: -96°C (-141°F)	**Vapor Density:** 3.5	**Vapor Pressure:** 6 (70 mm Hg)
Boiling Point: 84°C (183°F)	**Specific Gravity:** 0.7	**Water Solubility:** < 1%

Chemical Incompatibilities or Instabilities: Strong oxidizers.

FLAMMABILITY

NFPA Hazard Code: 3-3-0	**NFPA Classification:** Class IB Flammable
Flash Point: -6°C (21°F)	**LEL:** 1.1%
Autoignition Temp: 314°C (599°F)	**UEL:** 7.1%

Fire Extinguishing Methods: Carbon Dioxide, Dry Chemical, Foam, Water Spray or Fog.

Special Fire Fighting Considerations: Isolate for 1/2 mile if rail or tank truck is involved in a fire. Water spray should be used to keep closed containers cool. Continue to cool container after fire is extinguished. Immediately withdraw if rising sound from venting device is heard or if fire is causing discoloration to the tank.

TOXICOLOGY

Odor: Ammonia or possibly fish-like   **Odor Threshold:** 0.08 mg/m^3
Physical Contact: Material is extremely destructive to human tissues.
TLV = 21 mg/m^3 **STEL** = N.D.   **IDLH** = 4,000 mg/m^3

Routes of Entry and Relative LD$_{50}$ (or LC$_{50}$)   **RTECS # IM4025000**
    Inhalation   7   (9600 mg/m^3/h)
    Ingestion   5   (770 mg/kg)
    Skin Absorption   N.D.

Symptoms of Exposure: Possible burns or irritation to skin, eyes and upper respiratory tract; headache; nausea and difficulty breathing. Laryngitis. Inhalation of vapors may be fatal by causing glottis to spasm and suffocation. Exposure may result in chemical pneumonitis or pulmonary edema.

Emergency/First Aid Treatment: Remove to ventilated area; immediately remove any contaminated clothing and wash contaminated areas for 15 minutes using water. Treat supportively and observe for possible shock. If ingested, seek immediate medical aid.

Recommended Clean-Up Procedures

Personal Protection: Level B Ensemble	**Recommended Material** Teflon (+), Viton (+)
RCRA Waste # None	**Reportable Quantities:** None

Spills: Remove all potential ignition sources. Dike to contain spill, absorb with noncombustible absorbant and take up using nonsparking tools. Decontaminate spill area using soapy water. Keep area isolated until vapors have dissipated.

Special Emergency Information: May be harmful if inhaled, swallowed or absorbed through the skin. Vapors are heavier than air and may travel some distance to an ignition source. Vapors are more dense than air and may settle in low lying areas.

DIISOPROPYL PEROXYDICARBONATE - [(CH$_3$)$_2$CHOC(O)O-]$_2$

Synonyms: diisopropyl perdicarbonate; isopropyl percarbonate; isopropyl peroxydicarbonate

CAS Number: 105-64-6
DOT Number: 2133, 2134

Description: White powder
DOT Classification: Organic Peroxide

Molecular Weight: 206.2
Melting Point: 9°C (48°F)
Boiling Point: Decomposes

Vapor Density: N. A.
Specific Gravity: 1.1

Vapor Pressure: N. A.
Water Solubility: Insoluble

Chemical Incompatibilities or Instabilities: Amines; potassium iodide; organic matter.

FLAMMABILITY

NFPA Hazard Code: N. A.
Flash Point: °C (°F)
Autoignition Temp: °C (°F)

NFPA Classification: N. A.
LEL: N. A.
UEL: N. A.

Fire Extinguishing Methods: Carbon Dioxide, Dry Chemical, Foam, Water Spray or Fog.

Special Fire Fighting Considerations: Cooling of containers must be maintained using ice, dry ice or liquid nitrogen. If cooling is lost, evacuate area. For large fires, use flooding quantities of water using unmanned hose holder. Continue to apply water until well after fire is extinguished. If fire cannot be controlled, evacuate area and allow to burn.

TOXICOLOGY

Odor: N. A.
Physical Contact: Irritant
TLV = N.D.   **STEL** = N.D.

Odor Threshold: N. A.

IDLH = N.D.

Routes of Entry and Relative LD$_{50}$ (or LC$_{50}$)
Inhalation	N.D.	
Ingestion	3	(2140 mg/kg)
Skin Absorption	3	(2025 mg/kg)

RTECS # SD9800000

Symptoms of Exposure: Possible irritation to skin, eyes and upper respiratory tract; headache; nausea.

Emergency/First Aid Treatment: Remove to ventilated area; immediately remove any contaminated clothing and wash contaminated areas for 15 minutes using water. Treat supportively and observe for possible shock. If ingested, seek immediate medical aid.

Recommended Clean-Up Procedures

Personal Protection: Level B Ensemble
RCRA Waste # None

Recommended Material N.D.
Reportable Quantities: None

Spills: Remove all potential ignition sources. Carefully take up material and place in an appropriate container. Decontaminate spill area using water.

Special Emergency Information: May be harmful if inhaled, swallowed or absorbed through the skin.

DIKETENE

Synonyms: 3-buteno-beta-lactone; ketene dimer; 4-methylene-2-oxetanone

CAS Number: 674-82-8

DOT Number: 2521

Description: Colorless to orange liquid

DOT Classification: Flammable Liquid

Molecular Weight: 84.1

Melting Point:	-8°C (18°F)	**Vapor Density:** 2.9	**Vapor Pressure:**	3 (8 mm Hg)
Boiling Point:	127°C (261°F)	**Specific Gravity:** 1.1	**Water Solubility:**	Decomposes

Chemical Incompatibilities or Instabilities: Acids; bases, oxidizers; halogenated materials; reactive metals.

FLAMMABILITY

NFPA Hazard Code: 3-2-2

Flash Point: 34°C (93°F)

Autoignition Temp: 310°C (590°F)

NFPA Classification: Class 1C Flammable

LEL: N.D.

UEL: N.D.

Fire Extinguishing Methods: Carbon Dioxide, Dry Chemical, Foam, Water Spray or Fog.

Special Fire Fighting Considerations: Structural fire fighter protective clothing will not provide adequate protection. Fight fire from a distance or protected location, if possible. Water spray should be used to keep closed containers cool. Continue to cool container after fire is extinguished.

DOT Recommended Isolation Zones:	Small Spill: 150 ft	Large Spill: 600 ft
DOT Recommended Down Wind Take Cover Distance:	Small Spill: 0.8 mile	Large Spill: 2 miles

TOXICOLOGY

Odor: Pungent

Odor Threshold: N.D.

Physical Contact: Material is extremely destructive to human tissues.

TLV = N.D.　　　　**STEL** = N.D.　　　　**IDLH** = N.D.

Routes of Entry and Relative LD50 (or LC50)　　　　**RTECS # RQ8225000**

Inhalation	N.D.	
Ingestion	4	(560 mg/kg)
Skin Absorption	2	(2830 mg/kg)

Symptoms of Exposure: Possible burns or irritation to skin, eyes and upper respiratory tract; headache; nausea. Laryngitis. Exposure may result in chemical pneumonitis or pulmonary edema. Inhalation of vapors may be fatal by causing glottis to spasm and suffocation.

Emergency/First Aid Treatment: Remove to ventilated area; immediately remove any contaminated clothing and wash contaminated areas for 15 minutes using water. Treat supportively and observe for possible shock. If ingested, seek immediate medical aid.

Recommended Clean-Up Procedures

Personal Protection: Level B Ensemble　　　**Recommended Material** N.D.

RCRA Waste # None　　　**Reportable Quantities:** None

Spills: Remove all potential ignition sources. Dike to contain spill, absorb with non-combustible absorbant and take up using non-sparking tools. Decontaminate spill area using water.

Special Emergency Information: May be fatal in inhaled, ingested or absorbed through the skin. May undergo hazardous polymerization. Material is thermially unstable. Vapors are heavier than air and may travel some distance to an ignition source. Vapors are more dense than air and may settle in low lying areas.

1,2-DIMETHOXYETHANE - CH₃OCH₂CH₂OCH₃

Synonyms: dimethylcellusolve; ethylene dimethy ether; ethylene glycol dimethyl ether; glyme

CAS Number: 110-71-4   **Description:** Colorless liquid
DOT Number: 2252   **DOT Classification:** Flammable Liquid

Molecular Weight: 90.1
Melting Point: -58°C (-72°F)   **Vapor Density:**   **Vapor Pressure:**   **6** (60 mm Hg)
Boiling Point: 82°C (180°F)   **Specific Gravity:** 0.9   **Water Solubility:**   Insoluble

Chemical Incompatibilities or Instabilities: Strong oxidizers.

FLAMMABILITY

NFPA Hazard Code: 2-2-0   **NFPA Classification:**   Class 1B Flammable
Flash Point: -2°C   (29°F)   **LEL:** N.D.
Autoignition Temp: 202°C (395°F)   **UEL:** N.D.

Fire Extinguishing Methods: Carbon Dioxide, Dry Chemical, Foam, Water Spray or Fog.

Special Fire Fighting Considerations: Isolate for 1/2 mile if rail or tank truck is involved in a fire. Water spray should be used to keep closed containers cool. Continue to cool container after fire is extinguished. Fight fire from a distance or protected location, if possible. For large fires, if possible, withdraw and allow to burn. Immediately withdraw if rising sound from venting device is heard or if fire is causing discoloration to the tank.

TOXICOLOGY

Odor: Sharp, ether-like   **Odor Threshold:** N.D.
Physical Contact: Irritant
TLV = N.D.   **STEL** = N.D.   **IDLH** = N.D.

Routes of Entry and Relative LD₅₀ (or LC₅₀)   **RTECS # KI1451000**
    Inhalation   N.D.
    Ingestion   N.D.
    Skin Absorption   N.D.

Symptoms of Exposure: Possible irritation to skin, eyes and upper respiratory tract; headache; nausea.

Emergency/First Aid Treatment: Remove to ventilated area; immediately remove any contaminated clothing and wash contaminated areas for 15 minutes using water. Treat supportively and observe for possible shock. If ingested, seek immediate medical aid.

Recommended Clean-Up Procedures

Personal Protection: Level B Ensemble   **Recommended Material**   Butyl Rubber (0),

RCRA Waste #   None   **Reportable Quantities:**   None

Spills: Remove all potential ignition sources. Dike to contain spill, absorb with noncombustible absorbant and take up using nonsparking tools. Decontaminate spill area using soapy water.

Special Emergency Information: May be harmful if inhaled, swallowed or absorbed through the skin. May form unstable peroxides upon exposure to air. Vapors are heavier than air and may travel some distance to an ignition source. Vapors are more dense than air and may settle in low lying areas.

DIMETHYLAMINE - CH_3NHCH_3

Synonyms: DMA

CAS Number: 124-40-3
DOT Number: 1032, 1160

Description: Colorless gas
DOT Classification: Flammable Gas, Flammable liquid (solutions)

Molecular Weight: 45.1
Melting Point: -92°C (-134°F)
Boiling Point: 7°C (44°F)

Vapor Density: 1.6
Specific Gravity: N. A.

Vapor Pressure: Gas
Water Solubility: Soluble

Chemical Incompatibilities or Instabilities: Acids; halogens; hypochlorite; acid chlorides; zinc; tin; aluminum; mercury, acetaldehyde, oxides of nitrogen.

FLAMMABILITY

NFPA Hazard Code: 3-4-0
Flash Point: -7°C (20°F)
Autoignition Temp: 430°C (806°F)

NFPA Classification: Class 1A Flammable (solutions)
LEL: 2.8%
UEL: 14.4%

Fire Extinguishing Methods: Alcohol Resistant Foam, Carbon Dioxide, Dry Chemical, Water Spray or Fog.

Special Fire Fighting Considerations: Isolate for 1/2 mile if rail or tank truck is involved in a fire. Water spray should be used to keep closed containers cool. Continue to cool container after fire is extinguished. For large fires, if possible, withdraw and allow to burn. Immediately withdraw if rising sound from venting device is heard or if fire is causing discoloration to the tank.

TOXICOLOGY

Odor: Rotten fish
Physical Contact: Material is extremely destructive to human tissues.
TLV = 9 mg/m³　　　　　　　　**STEL** = 28 mg/m³

Odor Threshold: 0.002 mg/m³

IDLH = 4,600 mg/m³

RTECS # IP8750000

Routes of Entry and Relative LD_{50} (or LC_{50})
Inhalation　　　2 (51,750 mg/m³/h)
Ingestion　　　4 (700 mg/kg)
Skin Absorption　N.D.

Symptoms of Exposure: Possible burns or irritation to skin, eyes and upper respiratory tract; headache; nausea. Laryngitis. Exposure may result in chemical pneumonitis or pulmonary edema. Inhalation of vapors may be fatal by causing glottis to spasm and suffocation.

Emergency/First Aid Treatment: Remove to ventilated area; immediately remove any contaminated clothing and wash contaminated areas for 15 minutes using water. Treat supportively and observe for possible shock. If ingested, seek immediate medical aid.

Recommended Clean-Up Procedures

Personal Protection: Level B Ensemble

RCRA Waste Number: U092

Recommended Material Butyl Rubber (+), Neoprene (+)

Reportable Quantities: 1000 lbs (1000 lb)

Spills: Remove all potential ignition sources. For aqueous solutions, absorb with non-combustible absorbant and take up using non-sparking tools. Decontaminate spill area using water. For gases, turn off source if it can be safely done and allow vapors to dissipate.

Special Emergency Information: May be harmful if inhaled, swallowed or absorbed through the skin. Vapors are heavier than air and may travel some distance to an ignition source. Vapors are more dense than air and may settle in low lying areas.

DIMETHYLAMINOETHANOL - (CH₃)₂NCH₂CH₂OH

Synonyms: deanol; n,n-dimethylethanolamine; DMAE

CAS Number: 108-01-0   **Description:** Colorless liquid
DOT Number: 2051   **DOT Classification:** Flammable Liquid

Molecular Weight: 89.2
Melting Point: °C   (°F)   **Vapor Density:** 3.1   **Vapor Pressure:** **2** (4 mm Hg)
Boiling Point: 133°C (272°F)   **Specific Gravity:** 0.9   **Water Solubility:** Soluble

Chemical Incompatibilities or Instabilities: Acid; Zn; Cu; Cu alloys; Cellulose nitrate; Oxidizers

FLAMMABILITY

NFPA Hazard Code: 2-2-0   **NFPA Classification:** Class II Combustible
Flash Point: 41°C (105°F)   **LEL:**   1.9%
Autoignition Temp: 295°C (563°F)   **UEL:**   10.0%

Fire Extinguishing Methods: Alcohol Resistant Foam, Carbon Dioxide, Dry Chemical, Water Spray or Fog.

Special Fire Fighting Considerations: Isolate for 1/2 mile if rail or tank truck is involved in a fire. Water spray should be used to keep closed containers cool. Continue to cool container after fire is extinguished. Immediately withdraw if rising sound from venting device is heard or if fire is causing discoloration to the tank.

TOXICOLOGY
POSSIBLE ALLERGEN

Odor: **Odor Threshold:** N.D.
Physical Contact: Material is extremely destructive to human tissues.
TLV = N.D.   **STEL** = N.D.   **IDLH** = N.D.

Routes of Entry and Relative LD₅₀ (or LC₅₀)   **RTECS # KK6125000**
     Inhalation   N.D.
     Ingestion   3   2,000 mg/kg)
     Skin Absorption   4   1370 mg/kg)

Symptoms of Exposure: Possible burns or irritation to skin, eyes and upper respiratory tract; headache; nausea. Inhalation of vapors may be fatal by causing glottis to spasm and suffocation. Exposure may result in chemical pneumonitis or pulmonary edema. Vapors mays cause damage specific to the eyes.

Emergency/First Aid Treatment: Remove to ventilated area; immediately remove any contaminated clothing and wash contaminated areas for 15 minutes using water. Treat supportively and observe for possible shock. If ingested, seek immediate medical aid.

Recommended Clean-Up Procedures

Personal Protection: Level B Ensemble   **Recommended Material**   Butyl Rubber (+), Nitrile Rubber (+)

RCRA Waste #   None   **Reportable Quantities:**   None

Spills: Remove all potential ignition sources. Dike to contain spill, absorb with noncombustible absorbant and take up using nonsparking tools. Decontaminate spill area using water.

Special Emergency Information: May be harmful if inhaled, swallowed or absorbed through the skin. Vapors are heavier than air and may travel some distance to an ignition source. Vapors are more dense than air and may settle in low lying areas.

N,N-DIMETHYLANILINE

Synonyms: dimethylaminobenzene; dimethylphenylamine

CAS Number: 121-69-7   **Description:** Yellow liquid
DOT Number: 2253   **DOT Classification:** Poison B

Molecular Weight: 121.2
Melting Point: 2°C (36°F)   **Vapor Density:** 4.2   **Vapor Pressure:** 1 (<1 mm Hg)
Boiling Point: 193°C (379°F)   **Specific Gravity:** 0.9   **Water Solubility:** 2%

Chemical Incompatibilities or Instabilities: Acid chloride, acid anhydrides, chloroformates, halogens, peroxides.

FLAMMABILITY

NFPA Hazard Code: 3-2-0   **NFPA Classification:** Class IIIA Combustible
Flash Point: 63°C (145°F)   **LEL:** 1.0%
Autoignition Temp: 371°C (700°F)   **UEL:** N.D.

Fire Extinguishing Methods: Carbon Dioxide, Dry Chemical, Foam, Water Spray or Fog.

Special Fire Fighting Considerations: Structural fire fighter protective clothing will not provide adequate protection. Water spray from an unmanned device should be used to keep closed containers cool. Treat all materials used as contaminated by hazardous waste.

TOXICOLOGY

Odor: Oily, amine-like   **Odor Threshold:** 0.005 mg/m^3
Physical Contact: Irritant
TLV = 25 mg/m^3 **STEL** = 50 mg/m^3 **IDLH** = 500 mg/m^3

Routes of Entry and Relative LD$_{50}$ (or LC$_{50}$)   **RTECS #** BX4725000
     Inhalation   N.D.
     Ingestion   3   (1,410 mg/kg)
     Skin Absorption   3   (1,770 mg/kg)

Symptoms of Exposure: Possible irritation to skin, eyes and upper respiratory tract; headache; nausea. Absorption may lead to the formation of methemoglobin resulting in cyanosis several hours after exposure.

Emergency/First Aid Treatment: Remove to ventilated area; immediately remove any contaminated clothing and wash contaminated areas for 15 minutes using water. Treat supportively and observe for possible shock. If ingested, seek immediate medical aid.

Recommended Clean-Up Procedures

Personal Protection: Level B Ensemble   **Recommended Material** N.D.

RCRA Waste # None   **Reportable Quantities:** None

Spills: Remove all potential ignition sources. Dike to contain spill, absorb with non-combustible absorbant and take up using non-sparking tools. Decontaminate spill area using water. Treat all materials used or generated and equipment involved as contaminated by hazardous waste.

Special Emergency Information: May be fatal in inhaled, ingested or absorbed through the skin.

1,3-DIMETHYL BUTYLAMINE - $(CH_3)_2CHCH_2CH(CH_3)NH_2$

Synonyms: 2-amino-4-methylpentane

CAS Number: 108-09-8	**Description:** Colorless liquid	
DOT Number: 2379	**DOT Classification:** Flammable Liquid	

Molecular Weight: 101.2

Melting Point:	°C (°F)	**Vapor Density:** 3.5	**Vapor Pressure:** N.D.	
Boiling Point:	109°C (228°F)	**Specific Gravity:** 0.7	**Water Solubility:** Insoluble	

Chemical Incompatibilities or Instabilities: Oxidizers, halogens

FLAMMABILITY

NFPA Hazard Code: 2-3-0	**NFPA Classification:** Class IB Flammable	
Flash Point: 13°C (55°F)	**LEL:** N.D.	
Autoignition Temp: °C (°F)	**UEL:** N.D.	

Fire Extinguishing Methods: Carbon Dioxide, Dry Chemical, Foam, Water Spray or Fog.

Special Fire Fighting Considerations: Isolate for 1/2 mile if rail or tank truck is involved in a fire. Water spray should be used to keep closed containers cool. Continue to cool container after fire is extinguished. For large fires, if possible, withdraw and allow to burn. Immediately withdraw if rising sound from venting device is heard or if fire is causing discoloration to the tank.

TOXICOLOGY

Odor:

Odor Threshold: N.D.

Physical Contact: Irritant

TLV = N.D.	**STEL** = N.D.	**IDLH** = N.D.

Routes of Entry and Relative LD$_{50}$ (or LC$_{50}$)

RTECS # ED4460000

Inhalation	N.D.	
Ingestion	N.D.	
Skin Absorption	7	(600 mg/kg)

Symptoms of Exposure: Possible burns or irritation to skin, eyes and upper respiratory tract; headache; nausea. Laryngitis. Exposure may result in chemical pneumonitis or pulmonary edema. Inhalation of vapors may be fatal by causing glottis to spasm and suffocation.

Emergency/First Aid Treatment: Remove to ventilated area; immediately remove any contaminated clothing and wash contaminated areas for 15 minutes using water. Treat supportively and observe for possible shock. If ingested, seek immediate medical aid.

Recommended Clean-Up Procedures

Personal Protection: Level B Ensemble	**Recommended Material**	Nitrile Rubber (+), Polyvinyl Alcohol (+), Viton (+)
RCRA Waste # None	**Reportable Quantities:**	None

Spills: Remove all potential ignition sources. Dike to contain spill, absorb with non-combustible absorbant and take up using non-sparking tools. Decontaminate spill area using soapy water.

Special Emergency Information: May be harmful if inhaled, swallowed or absorbed through the skin. Vapors are heavier than air and may travel some distance to an ignition source. Vapors are more dense than air and may settle in low lying areas.

DIMETHYLCARBAMOYL CHLORIDE - (CH$_3$)$_2$NC(O)Cl

Synonyms: chloroformic acid dimethylamide; DDC; dimethylcarbamic acid chloride; DMCC

CAS Number: 79-44-7	**Description:** Colorless liquid	
DOT Number: 2262	**DOT Classification:** Corrosive Material	

Molecular Weight: 107.6

Melting Point:	-33°C (-27°F)	**Vapor Density:** 3.7	**Vapor Pressure:**	N.D.
Boiling Point:	166°C (331°F)	**Specific Gravity:** 1.2	**Water Solubility:**	Decomposes

Chemical Incompatibilities or Instabilities: Oxidizers, water.

FLAMMABILITY

NFPA Hazard Code:	**NFPA Classification:** Class IIIA Combustible
Flash Point: 68°C (155°F)	**LEL:** N.D.
Autoignition Temp: °C (°F)	**UEL:** N.D.

Fire Extinguishing Methods: Carbon Dioxide, Dry Chemical, Foam, Water Spray or Fog.

Special Fire Fighting Considerations: Water spray should be used to keep closed containers cool. Continue to cool container after fire is extinguished.

TOXICOLOGY
CONFIRMED CARCINOGEN

Odor:	**Odor Threshold:** N.D.
Physical Contact: Material is extremely destructive to human tissues.	
TLV = N.D. **STEL** = N.D.	**IDLH** = N.D.

Routes of Entry and Relative LD$_{50}$ (or LC$_{50}$)   **RTECS # FD4200000**

Inhalation	7	(4,700 mg/m^3/h)
Ingestion	4	(1,000 mg/kg)
Skin Absorption	N.D.	

Symptoms of Exposure: Possible burns or irritation to skin, eyes and upper respiratory tract; headache; nausea. Lachrymation. Exposure may result in chemical pneumonitis or pulmonary edema. Inhalation of vapors may be fatal by causing glottis to spasm and suffocation.

Emergency/First Aid Treatment: Remove to ventilated area; immediately remove any contaminated clothing and wash contaminated areas for 15 minutes using water. Treat supportively and observe for possible shock. If ingested, seek immediate medical aid.

Recommended Clean-Up Procedures

Personal Protection: Level B Ensemble	**Recommended Material** N.D.
RCRA Waste #: U097	**Reportable Quantity:** 1 lb (1 lb)

Spills: Dike to contain spill and collect liquid for later disposal. Decontaminate spill area using water.

Special Emergency Information: May be harmful if inhaled, swallowed or absorbed through the skin

DIMETHYL CHLOROTHIOPHOSPHATE

Synonyms: dimethyl phosphorochloridothioate; o,o-dimethylphosphorochloridothionate; methyl PCT

CAS Number: 2524-03-0   **Description:** White powder
DOT Number: 2267   **DOT Classification:** Corrosive Material

Molecular Weight: 160.6
Melting Point: °C (°F)   **Vapor Density:** N.D.   **Vapor Pressure:** 1 (<1 mm Hg)
Boiling Point: °C (°F)   **Specific Gravity:** 1.3   **Water Solubility:** Decomposes

Chemical Incompatibilities or Instabilities: Strong oxidizers.

FLAMMABILITY

NFPA Hazard Code: N.D.   **NFPA Classification:** N. A.
Flash Point: 105°C (221°F)   **LEL:** N. A.
Autoignition Temp: 105°C (221°F)   **UEL:** N. A.

Fire Extinguishing Methods: Carbon Dioxide, Dry Chemical, Foam, Water Spray or Fog.

Special Fire Fighting Considerations: Structural fire fighter protective clothing will not provide adequate protection. Water spray should be used to keep closed containers cool. Continue to cool container after fire is extinguished.

DOT Recommended Isolation Zones: Small Spill: 150 ft   Large Spill: 150 ft
DOT Recommended Down Wind Take Cover Distance: Small Spill: 0.2 mile   Large Spill: 0.2 mile

TOXICOLOGY

Odor: **Odor Threshold:** N.D.
Physical Contact: Material is extremely destructive to human tissues.
TLV = N.D.   **STEL** = N.D.   **IDLH** = N.D.

Routes of Entry and Relative LD$_{50}$ (or LC$_{50}$)   **RTECS # TD1830000**
   Inhalation   **8** (1350 mg/m^3/H)
   Ingestion   N.D.
   Skin Absorption   N.D.

Symptoms of Exposure: Possible burns or irritation to skin, eyes and upper respiratory tract; headache; nausea. Inhalation of vapors may be fatal by causing glottis to spasm and suffocation. Exposure may result in chemical pneumonitis or pulmonary edema.

Emergency/First Aid Treatment: Remove to ventilated area; immediately remove any contaminated clothing and wash contaminated areas for 15 minutes using water. Treat supportively and observe for possible shock. If ingested, seek immediate medical aid.

Recommended Clean-Up Procedures

Personal Protection: Level A Ensemble   **Recommended Material** N.D.

RCRA Waste # None   **Reportable Quantities:** None

Spills: Take up in non-combustible absorbant and dispose. Decontaminate spill area using soapy water. Treat all materials used as contaminated by hazardous waste.

Special Emergency Information: May be fatal in inhaled, ingested or absorbed through the skin. May undergo thermal decomposition at temperatures above 120°C to generate oxides of phosphorous and phosphine gas.

DIMETHYLDISULFIDE - CH3SSCH3

Synonyms: methyl disulfide

CAS Number: 624-92-0
DOT Number: 2381

Description: Colorless liquid
DOT Classification: Flammable Liquid

Molecular Weight: 94.2
Melting Point: -85°C (-117°F)
Boiling Point: 110°C (230°F)

Vapor Density: 3.2
Specific Gravity: 1.1

Vapor Pressure: 5 (29 mm Hg)
Water Solubility:

Chemical Incompatibilities or Instabilities: Oxidizers.

FLAMMABILITY

NFPA Hazard Code:
Flash Point: 24°C (76°F)
Autoignition Temp: °C (°F)

NFPA Classification: Class IC Flammable
LEL: 1.1%
UEL: 16.0%

Fire Extinguishing Methods: Carbon Dioxide, Dry Chemical, Foam, Water Spray or Fog.

Special Fire Fighting Considerations: Isolate for 1/2 mile if rail or tank truck is involved in a fire. Water spray should be used to keep closed containers cool. Continue to cool container after fire is extinguished. For large fires, if possible, withdraw and allow to burn. Immediately withdraw if rising sound from venting device is heard or if fire is causing discoloration to the tank.

TOXICOLOGY

Odor:
Physical Contact: Irritant
TLV = N.D.

STEL = N.D.

Odor Threshold: N.D.

IDLH = N.D.

Routes of Entry and Relative LD$_{50}$ (or LC$_{50}$)
  Inhalation   10 (16 mg/m^3/h)
  Ingestion   N.D.
  Skin Absorption N.D.

RTECS # JO1927500

Symptoms of Exposure: Possible irritation to skin, eyes and upper respiratory tract; headache; nausea.

Emergency/First Aid Treatment: Remove to ventilated area; immediately remove any contaminated clothing and wash contaminated areas for 15 minutes using water. Treat supportively and observe for possible shock. If ingested, seek immediate medical aid.

Recommended Clean-Up Procedures

Personal Protection: Level A Ensemble
RCRA Waste # None

Recommended Material N.D.
Reportable Quantities: None

Spills: Remove all potential ignition sources. Dike to contain spill, absorb with non-combustible absorbant and take up using non-sparking tools. Decontaminate spill area using soapy water.

Special Emergency Information: May be fatal in inhaled, ingested or absorbed through the skin. Vapors are heavier than air and may travel some distance to an ignition source. Vapors are more dense than air and may settle in low lying areas.

DIMETHYL ETHER - CH3OCH3

Synonyms: methyl ether; wood ether

CAS Number: 115-10-6
DOT Number: 1033

Description: Colorless Gas
DOT Classification: Flammable Gas

Molecular Weight: 46.1
Melting Point: -139°C (-217°F)
Boiling Point: -24°C (-11°F)

Vapor Density: 1.6
Specific Gravity: N. A.

Vapor Pressure: Gas
Water Solubility: Soluble

Chemical Incompatibilities or Instabilities: Strong oxidizers; halogens, aluminum hydride, lithium aluminum hydride.

FLAMMABILITY

NFPA Hazard Code: 2-4-1
Flash Point: °C (°F)
Autoignition Temp: 350°C (662°F)

NFPA Classification:
LEL: 3.4%
UEL: 27.0%

Fire Extinguishing Methods: Alcohol Resistant Foam, Carbon Dioxide, Dry Chemical, Water Spray or Fog.

Special Fire Fighting Considerations: Isolate for 1/2 mile if rail or tank truck is involved in a fire. Water spray should be used to keep closed containers cool. Continue to cool container after fire is extinguished. For large fires, if possible, withdraw and allow to burn. Immediately withdraw if rising sound from venting device is heard or if fire is causing discoloration to the tank.

TOXICOLOGY

Odor: Ether-like
Physical Contact:
TLV = N.D.   **STEL** = N.D.

Odor Threshold: N.D.

IDLH = N.D.

Routes of Entry and Relative LD$_{50}$ (or LC$_{50}$)
Inhalation   1   (308,000 mg/m^3/h)
Ingestion   N.D.
Skin Absorption   N.D.

RTECS # PM4780000

Symptoms of Exposure: A simple asphyxiant. Disorientation, headache, depression of the central nervous system, unconsciousness.

Emergency/First Aid Treatment: Remove to ventilated area; immediately remove any contaminated clothing and wash contaminated areas for 15 minutes using water. Treat supportively and observe for possible shock. If ingested, seek immediate medical aid.

Recommended Clean-Up Procedures

Personal Protection: Level B Ensemble
RCRA Waste # None

Recommended Material Butyl Rubber (+), Neoprene (+)
Reportable Quantities: None

Spills: Remove all potential ignition sources. Stop leak if it can safely be done. Keep area isolated until vapors have dissipated.

Special Emergency Information: May be harmful if inhaled, swallowed or absorbed through the skin. May form unstable peroxides upon prolonged exposure to air. Vapors are heavier than air and may travel some distance to an ignition source. Vapors are more dense than air and may settle in low lying areas.

DIMETHYLFORMAMIDE - (CH3)2NCHO

Synonyms: n,n-dimethylformamide, DMF; n-formyldimethylamine

CAS Number: 68-12-2
DOT Number: 2265

Description: Colorless liquid
DOT Classification: Flammable Liquid

Molecular Weight: 73.1
Melting Point: -61°C (-78°F)
Boiling Point: 153°C (307°F)

Vapor Density: 2.5
Specific Gravity: 0.9

Vapor Pressure: 2 (4 mm Hg)
Water Solubility: Soluble

Chemical Incompatibilities or Instabilities: Halogens; halogenated hydrocarbons; nitrates; strong oxidizers; strong reducers.

FLAMMABILITY

NFPA Hazard Code: 1-2-0
Flash Point: 58°C (136°F)
Autoignition Temp: 445°C (833°F)

NFPA Classification: Class IIIA Combustible
LEL: N.D.
UEL: 15.2%

Fire Extinguishing Methods: Alcohol Resistant Foam, Carbon Dioxide, Dry Chemical, Water Spray or Fog.

Special Fire Fighting Considerations: Isolate for 1/2 mile if rail or tank truck is involved in a fire. Water spray should be used to keep closed containers cool. Continue to cool container after fire is extinguished. For large fires, if possible, withdraw and allow to burn. Immediately withdraw if rising sound from venting device is heard or if fire is causing discoloration to the tank.

TOXICOLOGY
SUSPECTED CARCINOGEN

Odor: Fish
Physical Contact:
TLV = 30 mg/m^3 **STEL** = N.D.

Odor Threshold: 1.5 mg/m^3

IDLH = 10,500 mg/m^3

Routes of Entry and Relative LD$_{50}$ (or LC$_{50}$)

Inhalation	N.D.	
Ingestion	3	(2,800 mg/kg)
Skin Absorption	2	(4,720 mg/kg)

RTECS # LQ2100000

Symptoms of Exposure: Possible irritation to skin, eyes and upper respiratory tract; headache; nausea. Alcohol intolerance may be observed for several days after exposure.

Emergency/First Aid Treatment: Remove to ventilated area; immediately remove any contaminated clothing and wash contaminated areas for 15 minutes using water. Treat supportively and observe for possible shock. If ingested, seek immediate medical aid.

Recommended Clean-Up Procedures

Personal Protection: Level B Ensemble

RCRA Waste # None

Recommended Material Butyl Rubber (+), Teflon (+)

Reportable Quantities: None

Spills: Remove all potential ignition sources. Dike to contain spill, absorb with non-combustible absorbant and take up using non-sparking tools. Decontaminate spill area using water. Treat all materials used as contaminated by hazardous waste.

Special Emergency Information: May be harmful if inhaled, swallowed or absorbed through the skin.

1,1-DIMETHYLHYDRAZINE - $(CH_3)_2NNH_2$

Synonyms: dimazine; asym-dimethylhydrazine; DMH, UDMH

CAS Number: 57-14-7	**Description:** Colorless liquid, fumes in air	
DOT Number: 1163	**DOT Classification:** Flammable Liquid	

Molecular Weight:	60.1			
Melting Point:	-72°C (-58°F)	**Vapor Density:** 1.9	**Vapor Pressure:**	7 (157 mm Hg)
Boiling Point:	64°C (147°F)	**Specific Gravity:** 0.8	**Water Solubility:**	Soluble

Chemical Incompatibilities or Instabilities: Oxidizers, halogens, nitric acid, peroxides, mercury.

FLAMMABILITY

NFPA Hazard Code:	3-3-1	**NFPA Classification:** Class IB Flammable	
Flash Point:	-15°C (5°F)	**LEL:**	2.0%
Autoignition Temp:	249°C (480°F)	**UEL:**	95.0%

Fire Extinguishing Methods: Alcohol Resistant Foam, Carbon Dioxide, Dry Chemical, Water Spray or Fog.

Special Fire Fighting Considerations: Structural fire fighter protective clothing will not provide adequate protection. Fight fire from a distance or protected location, if possible. Fight fire until using unmanned devices until fire is out.

DOT Recommended Isolation Zones:	Small Spill: 1200 ft	Large Spill: 1500 ft
DOT Recommended Down Wind Take Cover Distance:	Small Spill: 4 miles	Large Spill: 5 miles

TOXICOLOGY
CARCINOGEN

Odor: Ammonia-like or fishy

Physical Contact: Material is extremely destructive to human tissues.

Odor Threshold: 1 mg/m^3

$TLV = 1.2$ mg/m^3　　　　　　$STEL = N.D.$　　　　　　$IDLH = 150$ mg/m^3

Routes of Entry and Relative LD$_{50}$ (or LC$_{50}$)

Inhalation	8	3,000 mg/m^3/H)
Ingestion	7	122 mg/kg)
Skin Absorption	N.D.	

RTECS # V2450000

Symptoms of Exposure: Possible burns or irritation to skin, eyes and upper respiratory tract; headache; nausea. Inhalation of vapors may be fatal by causing glottis to spasm and suffocation. Exposure may result in chemical pneumonitis or pulmonary edema. May cause convulsions leading to death.

Emergency/First Aid Treatment: Remove to ventilated area; immediately remove any contaminated clothing and wash contaminated areas for 15 minutes using water. Treat supportively and observe for possible shock. If ingested, seek immediate medical aid.

Recommended Clean-Up Procedures

Personal Protection:	Level A Ensemble	**Recommended Material** Butyl Rubber (+)
RCRA Waste #	U098	**Reportable Quantity:** 10 lb (1 lb)

Spills: Remove all potential ignition sources. Dike to contain spill, absorb with non-combustible absorbant and take up using non-sparking tools. Decontaminate spill area using water. Treat all materials used or generated and equipment involved as contaminated by hazardous waste.

Special Emergency Information: May be fatal in inhaled, ingested or absorbed through the skin. Vapors are heavier than air and may travel some distance to an ignition source. Vapors are more dense than air and may settle in low lying areas.

1,2-Dimethylhydrazine (CAS # 540-73-8, DOT # 2382, RTECS # MV2625000, RCRA Waste # U099, RQ: 1 lb [1 lb]) has similar chemical, physical and toxicological properties.

DIMETHYL SULFATE - $(CH_3)_2SO_4$

Synonyms: dimethyl monosulfate; DMS; methyl sulfate; sulfuric acid, dimethyl ester

CAS Number: 77-78-1

DOT Number: 1595

Description: Colorless liquid

DOT Classification: Corrosive Material

Molecular Weight: 126.1

Melting Point: -32°C (-25°F)

Boiling Point: 188°C (370°F)

Vapor Density: 4.4

Specific Gravity: 1.3

Vapor Pressure: 1 (<1 mmHg)

Water Solubility: Insoluble

Chemical Incompatibilities or Instabilities: Strong oxidizers, strong bases, ammonia.

FLAMMABILITY

NFPA Hazard Code: 4-2-0

Flash Point: 83°C (182°F)

Autoignition Temp: 188°C (370°F)

NFPA Classification: Class IIIA Combustible

LEL: N.D.

UEL: N.D.

Fire Extinguishing Methods: Carbon Dioxide, Dry Chemical, Foam, Water Spray or Fog.

Special Fire Fighting Considerations: Structural fire fighter protective clothing will not provide adequate protection. Water spray should be used to keep closed containers cool. Continue to cool container after fire is extinguished. Fight fire from a distance or protected location, if possible. Treat all materials used as contaminated by hazardous waste.

DOT Recommended Isolation Zones:　　　　　　　　Small Spill: 150 ft　　Large Spill: 150 ft
DOT Recommended Down Wind Take Cover Distance:　Small Spill: 0.4 mile　Large Spill: 0.4 mile

TOXICOLOGY
CONFIRMED CARCINOGEN

Odor: Faint onion-like

Physical Contact: Material is extremely destructive to human tissues.

TLV = 0.5 mg/m^3　　　　**STEL** = N.D.

Odor Threshold: N.D.

IDLH = 52 mg/m^3

Routes of Entry and Relative LD$_{50}$ (or LC$_{50}$)

Inhalation	10	(180 mg/m^3/h)
Ingestion	6	(205 mg/kg)
Skin Absorption	N.D.	

RTECS # WS8225000

Symptoms of Exposure: Possible burns or irritation to skin, eyes and upper respiratory tract; headache; nausea. Laryngitis. Inhalation of vapors may be fatal by causing glottis to spasm and suffocation. Exposure may result in chemical pneumonitis or pulmonary edema.

Emergency/First Aid Treatment: Remove to ventilated area; immediately remove any contaminated clothing and wash contaminated areas for 15 minutes using water. Treat supportively and observe for possible shock. If ingested, seek immediate medical aid.

Recommended Clean-Up Procedures

Personal Protection: Level A Ensemble

RCRA Waste # U103

Recommended Material N.D.

Reportable Quantity: 100 lbs (1 lb)

Spills: Dike to contain spill and collect liquid for later disposal. Decontaminate spill area using water. Treat all materials used as contaminated by hazardous waste.

Special Emergency Information: May be fatal in inhaled, ingested or absorbed through the skin.

DIMETHYL SULFIDE - CH₃SCH₃

Synonyms: methyl sulfide; methylthiomethane; 2-thiapropane; 2-thiopropane

CAS Number: 75-18-3   **Description:** Colorless liquid
DOT Number: 1164   **DOT Classification:** Flammable Liquid

Molecular Weight: 62.1
Melting Point: -83°C (-117°F)   **Vapor Density:** 2.1   **Vapor Pressure:** 8 (388 mm Hg)
Boiling Point: 38°C (100°F)   **Specific Gravity:** 0.8   **Water Solubility:** Insoluble

Chemical Incompatibilities or Instabilities: Oxidizers.

FLAMMABILITY

NFPA Hazard Code: 2-4-0   **NFPA Classification:** Class IA Flammable
Flash Point: -36°C (-34°F)   **LEL:** 2.2%
Autoignition Temp: 206°C (402°F) **UEL:** 19.7%

Fire Extinguishing Methods: Carbon Dioxide, Dry Chemical, Foam, Water Spray or Fog.

Special Fire Fighting Considerations: Isolate for 1/2 mile if rail or tank truck is involved in a fire. Water spray should be used to keep closed containers cool. Continue to cool container after fire is extinguished. For large fires, if possible, withdraw and allow to burn. Immediately withdraw if rising sound from venting device is heard or if fire is causing discoloration to the tank. May release toxic fumes upon heating (SO_2).

TOXICOLOGY

Odor: Cabbage or radish   **Odor Threshold:** N.D.
Physical Contact: Irritant
TLV = N.D.   **STEL** = N.D.   **IDLH** = N.D.

Routes of Entry and Relative LD₅₀ (or LC₅₀)   **RTECS #** PV5075000
   Inhalation   1   (100,000 mg/m³/h)
   Ingestion   4   (535 mg/kg)
   Skin Absorption   N.D.

Symptoms of Exposure: Possible irritation to skin, eyes and upper respiratory tract; headache; nausea.

Emergency/First Aid Treatment: Remove to ventilated area; immediately remove any contaminated clothing and wash contaminated areas for 15 minutes using water. Treat supportively and observe for possible shock. If ingested, seek immediate medical aid.

Recommended Clean-Up Procedures

Personal Protection: Level B Ensemble   **Recommended Material** N.D.

RCRA Waste # None   **Reportable Quantities:** None

Spills: Remove all potential ignition sources. Dike to contain spill, absorb with non-combustible absorbant and take up using non-sparking tools. Decontaminate spill area using soapy water.

Special Emergency Information: May be harmful if inhaled, swallowed or absorbed through the skin. Vapors are heavier than air and may travel some distance to an ignition source. Vapors are more dense than air and may settle in low lying areas.

DIMINODIPHENYL METHANE - $(NH_2C_6H_4)_2CH_2$

Synonyms: 4,4-methylenedianiline

CAS Number: 101-77-9   **Description:** White Powder
DOT Number: 2651   **DOT Classification:** Poison B

Molecular Weight: 198.3
Melting Point: 90°C (194°F)   **Vapor Density:** N.A.   **Vapor Pressure:** N.A.
Boiling Point: 399°C (750°F)   **Specific Gravity:** N.D.   **Water Solubility:** < 1%

Chemical Incompatibilities or Instabilities: Strong oxidizers.

FLAMMABILITY

NFPA Hazard Code: **NFPA Classification:**
Flash Point: 221°C (430°F)   **LEL:** N.A.
Autoignition Temp: °C   (°F)   **UEL:** N.A.

Fire Extinguishing Methods: Carbon Dioxide, Dry Chemical, Foam, Water.

Special Fire Fighting Considerations: Water spray should be used to keep closed containers cool. Continue to cool container after fire is extinguished.

TOXICOLOGY
CARCINOGEN

Odor: **Odor Threshold:** N.D.
Physical Contact: Irritant.
TLV = N.D.   **STEL** = N.D.   **IDLH** = N.D.

Routes of Entry and Relative LD50 (or LC50)   **RTECS # BY5425000**
    Inhalation   N.D.
    Ingestion   5   (347 mg/kg)
    Skin Absorption N.D.

Symptoms of Exposure: Possible irritation to skin, eyes and upper respiratory tract, headache, nausea. May cause cyanosis several hours after exposure.

Emergency/First Aid Treatment: Remove to ventilated area; immediately remove any contaminated clothing and wash contaminated areas for 15 minutes using water. Treat supportively and observe for possible shock. If ingested, seek immediate medical aid.

Recommended Clean-Up Procedures

Personal Protection: Level B Ensemble   **Recommended Material**   N.D.

RCRA Waste #   None   **Reportable Quantities:**   None

Spills: Take up spill. Decontaminate spill area using soapy water. Treat all materials used as contaminated by a hazardous waste.

Special Emergency Information: May be harmful is inhaled, swallowed, or absorbed through the skin.

2,4-DINITROANILINE

Synonyms: 2,4-dinitro-1-aminobenzene; 2,4-dinitrobenzamine

CAS Number: 97-02-9	**Description:** Yellow Powder	
DOT Number: 1596	**DOT Classification:** N. A.	

Molecular Weight: 183.1

Melting Point: 188°C (370°F)	**Vapor Density:** 6.3	**Vapor Pressure:**
Boiling Point: °C (°F)	**Specific Gravity:** 1.6	**Water Solubility:** Insoluble

Chemical Incompatibilities or Instabilities: Halogens, oxidizers, acid chlorides.

FLAMMABILITY

NFPA Hazard Code: 3-1-3	**NFPA Classification:** N. A.
Flash Point: 224°C (435°F)	**LEL:** N. A.
Autoignition Temp: °C (°F)	**UEL:** N. A.

Fire Extinguishing Methods: Carbon Dioxide, Dry Chemical, Foam, Water Spray or Fog.

Special Fire Fighting Considerations: Structural fire fighter protective clothing will not provide adequate protection. Water spray should be used to keep closed containers cool. Continue to cool container after fire is extinguished. For large fires, if possible, withdraw and allow to burn. Fires involving this material should be considered an explosion hazard.

TOXICOLOGY

Odor: N.D.	**Odor Threshold:** N.D.

Physical Contact:

TLV = N.D.	**STEL** = N.D.	**IDLH** = N.D.

Routes of Entry and Relative LD$_{50}$ (or LC$_{50}$)   **RTECS # BX9100000**

Inhalation	N.D.	
Ingestion	5	(370 mg/kg)
Skin Absorption	N.D.	

Symptoms of Exposure: Possible irritation to skin, eyes and upper respiratory tract; headache; nausea. Absorption may lead to the formation of methemoglobin resulting in cyanosis several hours after exposure.

Emergency/First Aid Treatment: Remove to ventilated area; immediately remove any contaminated clothing and wash contaminated areas for 15 minutes using water. Treat supportively and observe for possible shock. If ingested, seek immediate medical aid.

Recommended Clean-Up Procedures

Personal Protection: Level B Ensemble	**Recommended Material** N.D.	
RCRA Waste # None	**Reportable Quantities:** None	

Spills: Take up in non-combustible absorbant and dispose. Decontaminate spill area using soapy water. Treat all materials used as contaminated by hazardous waste.

Special Emergency Information: May be fatal in inhaled, ingested or absorbed through the skin.

2,6-dinitroaniline (CAS # 3606-22-4, RTECS # BX920000) and
3,5-dinitroaniline (CAS # 618-87-1, RTECS # BX9200100) have similar chemical, physical and toxicological properties.

1,3-DINITROBENZENE

Synonyms: m-dinitrobenzene

CAS Number: 99-65-0
DOT Number: 1597

Description: Yellow to orange powder
DOT Classification: Poison B

Molecular Weight: 168.1
Melting Point: 89°C (192°F)
Boiling Point: 301°C (574°F)

Vapor Density: N. A.
Specific Gravity: 1.4

Vapor Pressure: N.D.
Water Solubility: 0.02%

Chemical Incompatibilities or Instabilities: Oxidizers, heat, shock, friction.

FLAMMABILITY

NNFPA Hazard Code: 3-1-4
Flash Point: °C (°F)
Autoignition Temp: °C (°F)

FPA Classification: N. A.
LEL: N.D.
UEL: N.D.

Fire Extinguishing Methods: Carbon Dioxide, Dry Chemical, Foam, Water Spray or Fog.

Special Fire Fighting Considerations: Structural fire fighter protective clothing will not provide adequate protection. Water spray should be used to keep closed containers cool. Continue to cool container after fire is extinguished. For large fires, if possible, withdraw and allow to burn. Fires involving this material should be considered an explosion hazard.

TOXICOLOGY
SUSPECTED CARCINOGEN

Odor: N.D.
Physical Contact: Irritant
TLV = 1 mg/m^3

STEL = N.D.

Odor Threshold: N.D.

IDLH = 200 mg/m^3

RTECS # CZ7350000

Routes of Entry and Relative LD$_{50}$ (or LC$_{50}$)
Inhalation N.D.
Ingestion 8 (83 mg/kg)
Skin Absorption N.D.

Symptoms of Exposure: Possible irritation to skin, eyes and upper respiratory tract; headache; nausea. Absorption may lead to the formation of methemoglobin resulting in cyanosis several hours after exposure.

Emergency/First Aid Treatment: Remove to ventilated area; immediately remove any contaminated clothing and wash contaminated areas for 15 minutes using water. Treat supportively and observe for possible shock. If ingested, seek immediate medical aid.

Recommended Clean-Up Procedures

Personal Protection: Level B Ensemble

RCRA Waste # None

Recommended Material N.D.

Reportable Quantities: 1000 lb

Spills: Take up in non-combustible absorbant and dispose. Decontaminate spill area using soapy water. Treat all materials used as contaminated by hazardous waste.

Special Emergency Information: May be fatal in inhaled, ingested or absorbed through the skin.

1.2-dinitrobenzene (CAS # 528-29-6, RTECS # CZ7450000),
1,4-dinitrobenzene (CAS # 100-25-4, RTECS # CZ7525000), and
chlorodinitrobenzene (CAS # 25567-67-3, DOT # 1577, RTECS # CZ0490000) have similar chemical, physical and toxicological properties.

DINITRO-o-CRESOL

Synonyms: 2,4-dinitro-o-cresol; 4,6-dinitro-o-cresol.; 2,4-dinitro-6-methylphenol.

CAS Number: 534-52-1   **Description:** Yellow Powder
DOT Number: 1598   **DOT Classification:** N.D.

Molecular Weight: 198.1
Melting Point: 88°C   (190°F)   **Vapor Density:**   6.8   **Vapor Pressure:**   1 (<1 mm Hg)
Boiling Point: 312°C   (594°F)   **Specific Gravity:** N.D.   **Water Solubility:** Insoluble

Chemical Incompatibilities or Instabilities: Oxidizers.

FLAMMABILITY

NFPA Hazard Code: **NFPA Classification:**
Flash Point: °C   (°F)   **LEL:** N. A.
Autoignition Temp: °C   (°F)   **UEL:** N. A.

Fire Extinguishing Methods: Carbon Dioxide, Dry Chemical, Foam, Water Spray or Fog.

Special Fire Fighting Considerations: Structural fire fighter protective clothing will not provide adequate protection. Fight fire from a distance or protected location, if possible. Water spray from an unmanned device should be used to keep closed containers cool. Continue to cool container after fire is extinguished. May explode when heated, may explode from shock or friction.

TOXICOLOGY

Odor: None   **Odor Threshold:** N. A.
Physical Contact: Irritant
TLV = 0.2 mg/m^3 **STEL** = N.D.   **IDLH** = 5 mg/m^3

Routes of Entry and Relative LD$_{50}$ (or LC$_{50}$)   **RTECS # GO9625000**
  Inhalation   N.D.
  Ingestion   9   (10 mg/kg)
  Skin Absorption   N.D.

Symptoms of Exposure: Possible irritation to skin, eyes and upper respiratory tract; headache; nausea. Absorption may lead to the formation of methemoglobin resulting in cyanosis several hours after exposure.

Emergency/First Aid Treatment: Remove to ventilated area; immediately remove any contaminated clothing and wash contaminated areas for 15 minutes using water. Treat supportively and observe for possible shock. If ingested, seek immediate medical aid.

Recommended Clean-Up Procedures

Personal Protection: Level B Ensemble   **Recommended Material**   N.D.

RCRA Waste #   P047   **Reportable Quantity:**   10 lb (1 lb)

Spills: Take up in non-combustible absorbent and dispose. Decontaminate spill area using soapy water. Treat all materials used as contaminated by hazardous waste.

Special Emergency Information: May be fatal in inhaled, ingested or absorbed through the skin.

3,5-dinitro-o-cresol (CAS # 497-56-3, RTECS # GO9500000),
3,5-dinitro-p-cresol (CAS # 63989-82-2, RTECS # GO9900000) and
2,6-dinitro-p-cresol (CAS # 609-93-8, RTECS # GO9800000) have similar chemical, physical and toxicological properties.

2,4-DINITROPHENOL

Synonyms: 2,4-DNP; 1-hydroxy-2,4-dinitrobenzene

CAS Number: 51-28-5   **Description:** Yellow Powder
DOT Number: 1320, 1321, 1599 (Solution)   **DOT Classification:** Poison B (1599), Flammable Solid (1320, 1321)

Molecular Weight: 184.1
Melting Point: 112°C (234°F)   **Vapor Density:** 6.4   **Vapor Pressure:** N.D.
Boiling Point: °C (°F)   **Specific Gravity:** 1.7   **Water Solubility:** <1%

Chemical Incompatibilities or Instabilities: Alkali; ammonia; oxidizers.

FLAMMABILITY

NFPA Hazard Code: **NFPA Classification:**
Flash Point: °C (°F)   **LEL:** N.D.
Autoignition Temp: °C (°F)   **UEL:** N.D.

Fire Extinguishing Methods: Carbon Dioxide, Dry Chemical, Foam Water Spray or Fog.

Special Fire Fighting Considerations: Structural fire fighter protective clothing will not provide adequate protection. Water spray should be used to keep closed containers cool. Continue to cool container after fire is extinguished. For large fires, if possible, withdraw and allow to burn. Fires involving this material should be considered an explosion hazard.

TOXICOLOGY

Odor: N.D.   **Odor Threshold:** N.D.
Physical Contact: Irritant
TLV = N.D.   **STEL** = N.D.   **IDLH** = N.D.

Routes of Entry and Relative LD_{50} (or LC_{50})   **RTECS # SL2800000**

    Inhalation   N.D.
    Ingestion   9   (30 mg/kg)
    Skin Absorption   N.D.

Symptoms of Exposure: Possible irritation to skin, eyes and upper respiratory tract; headache; nausea.

Emergency/First Aid Treatment: Remove to ventilated area; immediately remove any contaminated clothing and wash contaminated areas for 15 minutes using water. Treat supportively and observe for possible shock. If ingested, seek immediate medical aid.

Recommended Clean-Up Procedures

Personal Protection: Level B Ensemble   **Recommended Material** N.D.

RCRA Waste #   P048   **Reportable Quantity:**   10 lb (1000 lb)

Spills: Take up in non-combustible absorbent and dispose. Decontaminate spill area using soapy water. Treat all materials used as contaminated by hazardous waste.

Special Emergency Information: May be fatal in inhaled, ingested or absorbed through the skin. May form shock, heat, or friction sensitive explosive salts with ammonia or alkali.

2,3-dinitrophenol (CAS # 66-56-8, RTECS # SL2700000),
2,5-dinitrophenol (CAS # 329-71-5, RTECS # SL2900000),
2,6-dinitrophenol (CAS # 573-56-8, RTECS # SL2975000),
3,4-dinitrophenol (CAS # 577-71-9, RTECS # SL3000000), and
3,5-dinitrophenol (CAS # 586-11-8, RTECS # SL3050000) have similar chemical, physical and toxicological properties.

2,4-DINITROTOLUENE

Synonyms: DNT; 2,4-DNT; 1-methyl-2,4-dinitrobenzene

CAS Number: 121-14-2	**Description:** Yellow Powder		
DOT Number: 1600 (molten), 2038	**DOT Classification:** Poison B		

Molecular Weight: 182.1

Melting Point: 70°C (158°F)	**Vapor Density:** 6.3	**Vapor Pressure:** 1 (<1 mm Hg)
Boiling Point: 300°C (572°F)	**Specific Gravity:** 1.5	**Water Solubility:** Insoluble

Chemical Incompatibilities or Instabilities: Strong oxidizers.

FLAMMABILITY

NFPA Hazard Code: 3-1-3	**NFPA Classification:**	
Flash Point: 207°C (404°F)	**LEL:** N. A.	
Autoignition Temp: °C (°F)	**UEL:** N. A.	

Fire Extinguishing Methods: Carbon Dioxide, Dry Chemical, Foam Water Spray or Fog.

Special Fire Fighting Considerations: Structural fire fighter protective clothing will not provide adequate protection. Water spray should be used to keep closed containers cool. Continue to cool container after fire is extinguished. For large fires, if possible, withdraw and allow to burn. Fires involving this material should be considered an explosion hazard.

TOXICOLOGY
SUSPECTED CARCINOGEN

Odor: N.D.   **Odor Threshold:** N.D.
Physical Contact: Irritant
$TLV = 0.15$ mg/m^3 $STEL$ = N.D. $IDLH = 200$ mg/m^3

Routes of Entry and Relative LD$_{50}$ (or LC$_{50}$)   **RTECS # XT1575000**
    Inhalation   N.D.
    Ingestion   6   (268 mg/kg)
    Skin Absorption N.D.

Symptoms of Exposure: Possible irritation to skin, eyes and upper respiratory tract; headache; nausea. Absorption may lead to the formation of methemoglobin resulting in cyanosis several hours after exposure.

Emergency/First Aid Treatment: Remove to ventilated area; immediately remove any contaminated clothing and wash contaminated areas for 15 minutes using water. Treat supportively and observe for possible shock. If ingested, seek immediate medical aid.

Recommended Clean-Up Procedures

Personal Protection: Level A Ensemble	**Recommended Material** Saranex	
RCRA Waste # U105	**Reportable Quantity:** 10 lb (1000 lb)	

Spills: Take up in non-combustible absorbent and dispose. Decontaminate spill area using soapy water. Treat all materials used as contaminated by hazardous waste.

Special Emergency Information: Use extreme caution, may be shock sensitive. May be fatal in inhaled, ingested or absorbed through the skin.

2,3-dinitrotoluene (CAS # 602-01-7, RTECS # XT1400000),
2,5-dinitrotoluene (CAS # 619-15-8, RTECS # XT1750000),
2,6-dinitrotoluene (CAS # 606-20-2, RTECS # XT1925000, RCRA Waste # U106, RQ 100 lb [1000l lb]),
3,4-dinitrotoluene CAS # 610-39-9, RTECS # XT2100000), and
3,5-dinitrotoluene (CAS # 618-85-9, RTECS # XT2150000) have similar chemical, physical and toxicological properties.

DIOXANE

Synonyms: diethylene dioxide; diethylene ether; 1,4-dioxane

CAS Number: 123-91-1   **Description:** Colorless liquid
DOT Number: 1165   **DOT Classification:** Flammable Liquid

Molecular Weight: 88.1
Melting Point: 12°C (53°F)   **Vapor Density:** 3.0   **Vapor Pressure:** **5** (40 mm Hg)
Boiling Point: 101°C (214°F)   **Specific Gravity:** 1.0   **Water Solubility:** Soluble

Chemical Incompatibilities or Instabilities: Oxidizers, halogens, sulfur trioxide

FLAMMABILITY

NFPA Hazard Code: 2-3-1   **NFPA Classification:** Class IB Flammable
Flash Point: 12°C (54°F)   **LEL:** 2.0%
Autoignition Temp: 180°C (356°F)   **UEL:** 22.0%

Fire Extinguishing Methods: Alcohol Resistant Foam, Carbon Dioxide, Dry Chemical, Water Spray or Fog.

Special Fire Fighting Considerations: Isolate for 1/2 mile if rail or tank truck is involved in a fire. Water spray should be used to keep closed containers cool. Continue to cool container after fire is extinguished. For large fires, if possible, withdraw and allow to burn. Immediately withdraw if rising sound from venting device is heard or if fire is causing discoloration to the tank. Treat all materials used or generated and equipment involved as contaminated by hazardous waste.

TOXICOLOGY
CARCINOGEN

Odor: Faint ether-like, sweet   **Odor Threshold:** 0.08 mg/m^3
Physical Contact: Irritant
TLV = 90 mg/m^3 **STEL** = N.D.   **IDLH** = 7000 mg/m^3

Routes of Entry and Relative LD$_{50}$ (or LC$_{50}$)   **RTECS # JG8225000**
     Inhalation   2   (96,000 mg/m^3/H)
     Ingestion   N.D.
     Skin Absorption   1   (7600 mg/kg)

Symptoms of Exposure: Possible irritation to skin, eyes and upper respiratory tract; headache; nausea.

Emergency/First Aid Treatment: Remove to ventilated area; immediately remove any contaminated clothing and wash contaminated areas for 15 minutes using water. Treat supportively and observe for possible shock. If ingested, seek immediate medical aid.

Recommended Clean-Up Procedures

Personal Protection: Level B Ensemble   **Recommended Material**   Butyl Rubber (0), Polyvinyl Alcohol (0), Teflon (0)

RCRA Waste #   U108   **Reportable Quantity:**   100 lb (1 lb)

Spills: Remove all potential ignition sources. Dike to contain spill and collect liquid for later disposal. Dike to contain spill, absorb with non-combustible absorbent and take up using non-sparking tools. Treat all materials used as contaminated by hazardous waste.

Special Emergency Information: May be harmful if inhaled, swallowed or absorbed through the skin. Vapors are heavier than air and may travel some distance to an ignition source. Vapors are more dense than air and may settle in low lying areas. May form unstable peroxides upon prolonged exposure to air.

DIVINYL ETHER - (CH₂=CH)₂O

Synonyms: divinyl oxide; ethenyloxyethene; vinyl ether

CAS Number: 109-93-3	**Description:** Colorless liquid
DOT Number: 1167	**DOT Classification:** Flammable Liquid

Molecular Weight: 70.1

Melting Point:	-101°C (-150°F)	**Vapor Density:** 2.4	**Vapor Pressure:**	N.D.
Boiling Point:	28°C (83°F)	**Specific Gravity:** 0.8	**Water Solubility:**	Insoluble

Chemical Incompatibilities or Instabilities: Oxidizers.

FLAMMABILITY

NFPA Hazard Code: 2-3-2	**NFPA Classification:** Class IA Flammable
Flash Point: $< -30°C (< -22°F)$	**LEL:** 1.7%
Autoignition Temp: 360°C (680°F)	**UEL:** 27.0%

Fire Extinguishing Methods: Carbon Dioxide, Dry Chemical, Foam, Water Spray or Fog.

Special Fire Fighting Considerations: Structural fire fighter protective clothing will not provide adequate protection. Isolate for 1/2 mile if rail or tank truck is involved in a fire. Water spray should be used to keep closed containers cool. Continue to cool container after fire is extinguished. For large fires, if possible, withdraw and allow to burn. Immediately withdraw if rising sound from venting device is heard or if fire is causing discoloration to the tank.

TOXICOLOGY

Odor: Ether-like   **Odor Threshold:** N.D.
Physical Contact: Irritant
TLV = N.D.   **STEL** = N.D.   **IDLH** = N.D.

Routes of Entry and Relative LD₅₀ (or LC₅₀)   **RTECS # YZ6700000**
  Inhalation   N.D.
  Ingestion   N.D.
  Skin Absorption   N.D.

Symptoms of Exposure: Possible irritation to skin, eyes and upper respiratory tract; headache; nausea.

Emergency/First Aid Treatment: Remove to ventilated area; immediately remove any contaminated clothing and wash contaminated areas for 15 minutes using water. Treat supportively and observe for possible shock. If ingested, seek immediate medical aid.

Recommended Clean-Up Procedures

Personal Protection: Level B Ensemble	**Recommended Material** N.D.
RCRA Waste # None	**Reportable Quantities:** None

Spills: Remove all potential ignition sources. Dike to contain spill and collect liquid for later disposal. Dike to contain spill, absorb with non-combustible absorbent and take up using non-sparking tools. Decontamination will occur via evaporation

Special Emergency Information: May be harmful if inhaled, swallowed or absorbed through the skin. May form unstable peroxides upon exposure to air. May undergo hazardous polymerization. Vapors are heavier than air and may travel some distance to an ignition source. Vapors are more dense than air and may settle in low lying areas. May form unstable peroxides upon prolonged exposure to air.

ENDRIN

Synonyms: endrex; hexadrin; mendrin; nendrin

CAS Number: 72-20-8	**Description:** White powder	
DOT Number: 2761	**DOT Classification:** Poison B	

Molecular Weight:

Melting Point: 200°C (d)(392°F)	**Vapor Density:** N.D.	**Vapor Pressure:** 1 (< 1 mm Hg)
Boiling Point: °C (°F)	**Specific Gravity:** 1.7	**Water Solubility:** Insoluble

Chemical Incompatibilities or Instabilities: Strong oxidizers, parathion in organic solvents.

FLAMMABILITY

NFPA Hazard Code:	**NFPA Classification:** N. A.
Flash Point: °C (°F)	**LEL:** N. A.
Autoignition Temp: °C (°F)	**UEL:** N. A.

Fire Extinguishing Methods: Carbon Dioxide, Dry Chemical, Foam Water Spray or Fog.

Special Fire Fighting Considerations: Structural fire fighter protective clothing will not provide adequate protection. Fight fire from a distance or protected location, if possible. Treat all materials used as contaminated by hazardous waste.

TOXICOLOGY
QUESTIONABLE CARCINOGEN

Odor: Chemical (insecticide)	**Odor Threshold:** N.D.
Physical Contact: Irritant	
TLV = 0.1 mg/m^3 **STEL** = N.D.	**IDLH** = 2000 mg/m^3

Routes of Entry and Relative LD$_{50}$ (or LC$_{50}$)   **RTECS #** IO1575000

Inhalation	N.D.	
Ingestion	10	(3 mg/kg)
Skin Absorption	9	(60 mg/kg)

Symptoms of Exposure: Possible irritation to skin, eyes and upper respiratory tract; headache; nausea; weakness; convulsions.

Emergency/First Aid Treatment: Remove to ventilated area; immediately remove any contaminated clothing and wash contaminated areas for 15 minutes using water. Treat supportively and observe for possible shock. If ingested, seek immediate medical aid.

Recommended Clean-Up Procedures

Personal Protection: Level A Ensemble	**Recommended Material** N.D.
RCRA Waste # P051	**Reportable Quantity:** 1 lb (1 lb)

Spills: Take up in non-combustible absorbent and dispose. Avoid raising dust. Decontaminate spill area using soapy water. Treat all materials used as contaminated by hazardous waste.

Special Emergency Information: May be fatal in inhaled, ingested or absorbed through the skin.

EPICHLOROHYDRIN

Synonyms: 1-chloro-2,3-epoxypropane; (chloromethyl)ethylene oxide; 2,3-epoxypropyl chloride; glycerol epichlorohydrin

CAS Number: 106-89-8   **Description:** Colorless liquid
DOT Number: 2023   **DOT Classification:** Flammable Liquid

Molecular Weight: 92.5
Melting Point: -57°C (-71°F)   **Vapor Density:** 3.3   **Vapor Pressure:** **3** (10 mm Hg)
Boiling Point: 118°C (244°F)   **Specific Gravity:** 1.2   **Water Solubility:** 7%

Chemical Incompatibilities or Instabilities: Strong acids, strong alkali, aniline, trichloroethylene, isopropylamine, aluminum, aluminum chloride, ferric chloride, zinc, potassium, tert-butoxide.

FLAMMABILITY

NFPA Hazard Code: **3-3-2**   **NFPA Classification:** Class IC Flammable
Flash Point: 33°C   (91°F)   **LEL:**   3.8%
Autoignition Temp: 411°C (772°F)   **UEL:**   21.0%

Fire Extinguishing Methods: Alcohol Resistant Foam, Carbon Dioxide, Dry Chemical, Water Spray or Fog.

Special Fire Fighting Considerations: Isolate for 1/2 mile if rail or tank truck is involved in a fire. Structural fire fighter protective clothing will not provide adequate protection. Approach fire from upwind. Fight fire from a distance or protected location, if possible. Water spray should be used to keep closed containers cool. Continue to cool container after fire is extinguished. Immediately withdraw if rising sound from venting device is heard or if fire is causing discoloration to the tank.

TOXICOLOGY
CARCINOGEN

Odor: Chloroform-like   **Odor Threshold:** 0.3 mg/m^3
Physical Contact: Material is extremely destructive to human tissues.
TLV = 8 mg/m^3 **STEL** = N.D.   **IDLH** = 1,000 mg/m^3

Routes of Entry and Relative LD$_{50}$ (or LC$_{50}$)   **RTECS # TX4900000**
   Inhalation   7   (7,700 mg/m^3/h)
   Ingestion   8   (90 mg/kg)
   Skin Absorption   7   (515 mg/kg)

Symptoms of Exposure: Possible irritation to skin, eyes and upper respiratory tract; headache; nausea. Repeated exposures may cause allergic reactions or clouding of the cornea.

Emergency/First Aid Treatment: Remove to ventilated area; immediately remove any contaminated clothing and wash contaminated areas for 15 minutes using water. Treat supportively and observe for possible shock. If ingested, seek immediate medical aid.

Recommended Clean-Up Procedures

Personal Protection: Level B Ensemble   **Recommended Material** Butyl Rubber (+), Teflon (0)

RCRA Waste #   U041   **Reportable Quantity:**   100 lb (1000 lb)

Spills: Remove all potential ignition sources. Dike to contain spill and collect liquid for later disposal. Dike to contain spill, absorb with non-combustible absorbent and take up using non-sparking tools. Treat all materials used as contaminated by hazardous waste.

Special Emergency Information: May be fatal in inhaled, ingested or absorbed through the skin. Vapors are heavier than air and may travel some distance to an ignition source. Vapors are more dense than air and may settle in low lying areas. May undergo hazardous polymerization.

Epibromohydrin (CAS # 3132-64-7, DOT # 2558, RTECS # TX4115000) has similar chemical, physical and toxicological properties.

ETHANOLAMINE - $CH_3CH_2NH_2$

Synonyms: beta-aminoethyl alcohol; beta-ethanolamine; 2-hydroxyethylamine

CAS Number: 141-43-5   **Description:** Colorless liquid
DOT Number: 2491   **DOT Classification:** Corrosive Material

Molecular Weight: 61.1
Melting Point: 11°C (52°F)   **Vapor Density:** 2.1   **Vapor Pressure:** 1 (< 1 mm Hg)
Boiling Point: 172°C (342°F)   **Specific Gravity:** 1.0   **Water Solubility:** Soluble

Chemical Incompatibilities or Instabilities: Acetic acid, acetic anhydride, acids, acrolein, acrylonitrile, cellulose, chlorosulfonic acid, epichlorhydrin, beta-propiolactone, vinyl chloride.

FLAMMABILITY

NFPA Hazard Code: 2-2-0   **NFPA Classification:** Class IIIA Combustible
Flash Point: 86°C (186°F)   **LEL:** N.D.
Autoignition Temp: 410°C (770°F)   **UEL:** N.D.

Fire Extinguishing Methods: Alcohol Resistant Foam, Carbon Dioxide, Dry Chemical, Water Spray or Fog.

Special Fire Fighting Considerations: Water spray should be used to keep closed containers cool. Continue to cool container after fire is extinguished.

TOXICOLOGY

Odor: Unpleasant Ammonia-like   **Odor Threshold:** N.D.
Physical Contact: Material is extremely destructive to human tissues.
$TLV = 7.5$ mg/m^3 $STEL = 15$ mg/m^3 $IDLH = 2,500$ mg/m^3

Routes of Entry and Relative LD$_{50}$ (or LC$_{50}$)   **RTECS # KJ5775000**
    Inhalation   N.D.
    Ingestion   3   (1720 mg/kg)
    Skin Absorption   5   (100 mg/kg)

Symptoms of Exposure: Possible irritation to skin, eyes and upper respiratory tract; headache; nausea. Inhalation of vapors may be fatal by causing glottis to spasm and suffocation. Exposure may result in chemical pneumonitis or pulmonary edema.

Emergency/First Aid Treatment: Remove to ventilated area; immediately remove any contaminated clothing and wash contaminated areas for 15 minutes using water. Treat supportively and observe for possible shock. If ingested, seek immediate medical aid.

Recommended Clean-Up Procedures

Personal Protection: Level B Ensemble   **Recommended Material**   Butyl Rubber (+), Neoprene (+), Nitrile Rubber (+), Viton (+)

RCRA Waste #   None   **Reportable Quantities:**   None

Spills: Dike to contain spill and collect liquid for later disposal. Decontaminate spill area using water.

Special Emergency Information: May be harmful if inhaled, swallowed or absorbed through the skin.

ETHION

Synonyms: bis(s-(diethoxyphosphinothioyl)mercapto)methane; ethyl methylene phosphorodithiolate; O,O,O',O',-Tetraethyl-S,S'-Methylenediphosphorodithioate

CAS Number: 563-12-2		**Description:** Liquid	
DOT Number: 2783		**DOT Classification:** Poison B	

Molecular Weight: 384.5

Melting Point: -13°C (9°F)	**Vapor Density:** N.D.	**Vapor Pressure:** 1 (<1 mm Hg)
Boiling Point: °C (°F)	**Specific Gravity:** 1.2	**Water Solubility:** < 1%

Chemical Incompatibilities or Instabilities: Oxidizers.

FLAMMABILITY

NFPA Hazard Code: N.D.	**NFPA Classification:** N.D.
Flash Point: °C (°F)	**LEL:**
Autoignition Temp: °C (°F)	**UEL:**

Fire Extinguishing Methods: Carbon Dioxide, Dry Chemical, Foam, Water Spray or Fog.

Special Fire Fighting Considerations: Structural fire fighter protective clothing will not provide adequate protection. Fight fire from a distance or protected location, if possible. Treat all materials used as contaminated by hazardous waste.

TOXICOLOGY

Odor: **Odor Threshold:** N.D.

Physical Contact: Irritant

TLV = 0.4 mg/m^3 **STEL** = N.D.   **IDLH** = N.D.

Routes of Entry and Relative LD$_{50}$ (or LC$_{50}$)   **RTECS # TE4550000**

Inhalation	N.D.	
Ingestion	9	(13 mg/kg)
Skin Absorption	5	(890 mg/kg)

Symptoms of Exposure: Headache, nausea, blurred vision, lachrymation, muscle tremors or cramps, convulsions.

Emergency/First Aid Treatment: Remove to ventilated area; immediately remove any contaminated clothing and wash contaminated areas for 15 minutes using water. Treat supportively and observe for possible shock. If ingested, seek immediate medical aid. Treat for anticholinesterase poisoning, as necessary.

Recommended Clean-Up Procedures

Personal Protection: Level A Ensemble	**Recommended Material** Teflon (0)
RCRA Waste # N.D.	**Reportable Quantity:** 10 lb (10 lb)

Spills: Dike to contain spill and collect liquid for later disposal. Treat all materials used as contaminated by hazardous waste.

Special Emergency Information: May be fatal in inhaled, ingested or absorbed through the skin.

ETHYL ACETATE - CH$_3$CH$_2$OC(O)CH$_3$

Synonyms:

CAS Number: 141-78-6		**Description:** Colorless Liquid		
DOT Number: 1173		**DOT Classification:** Flammable Liquid		

Molecular Weight: 88.1

Melting Point: -84°C (-119°F)	**Vapor Density:** 3.0	**Vapor Pressure:** 6 (73 mm Hg)
Boiling Point: 77°C (171°F)	**Specific Gravity:** 0.9	**Water Solubility:** < 1%

Chemical Incompatibilities or Instabilities: Chlorosulfonic acid, lithium aluminum hydride, fuming sulfuric acid (oleum)

FLAMMABILITY

NFPA Hazard Code: 1-3-0	**NFPA Classification:** Class IB Flammable	
Flash Point: -4°C (24°F)	**LEL:** 2.0%	
Autoignition Temp: 426°C (800°F)	**UEL:** 11.5%	

Fire Extinguishing Methods: Alcohol Resistant Foam, Carbon Dioxide, Dry Chemical, Water Spray or Fog.

Special Fire Fighting Considerations: Isolate for 1/2 mile if rail or tank truck is involved in a fire. Water spray should be used to keep closed containers cool. Continue to cool container after fire is extinguished. For large fires, if possible, withdraw and allow to burn. Immediately withdraw if rising sound from venting device is heard or if fire is causing discoloration to the tank.

TOXICOLOGY

Odor: Fingernail Polish

Physical Contact: Irritant

Odor Threshold: 0.6 mg/m^3

TLV = 1440 mg/m^3	**STEL** = N.D.	**IDLH** = 35,000 mg/m^3

Routes of Entry and Relative LD$_{50}$ (or LC$_{50}$)

Inhalation	4	(45,000 mg/m^3)
Ingestion	2	(5620 mg/kg)
Skin Absorption	N.D.	

RTECS # AH5425000

Symptoms of Exposure: Possible irritation to skin, eyes and upper respiratory tract; headache; nausea. May act as a narcotic in high concentrations.

Emergency/First Aid Treatment: Remove to ventilated area; immediately remove any contaminated clothing and wash contaminated areas for 15 minutes using water. Treat supportively and observe for possible shock. If ingested, seek immediate medical aid.

Recommended Clean-Up Procedures

Personal Protection: Level B Ensemble	**Recommended Material**	Butyl Rubber (0), Polyvinyl Alcohol (0), Teflon (0)
RCRA Waste # U112	**Reportable Quantity:**	5000 lb (1 lb)

Spills: Dike to contain spill, absorb with non-combustible absorbent and take up using non-sparking tools. Decontaminate spill area using soapy water.

Special Emergency Information: May be harmful if inhaled, swallowed or absorbed through the skin. Vapors are heavier than air and may travel some distance to an ignition source. Vapors are more dense than air and may settle in low lying areas.

Ethoxyethyl acetate (CAS # 111-15-9, DOT # 1172, RTECS # KK8225000) and
propyl acetate (CAS # 109-60-4, DOT # 1276, RTECS # AJ3675000) have similar chemical, physical and toxicological properties.

ETHYL ACRYLATE - $CH_3CH_2OC(O)CH=CH_2$

Synonyms: ethyl propenoate

CAS Number: 140-88-5   **Description:** Colorless Liquid
DOT Number: 1917   **DOT Classification:** Flammable Liquid

Molecular Weight: 100.1
Melting Point: -72°C (-98°F)   **Vapor Density:** 3.5   **Vapor Pressure:** **5** (31 mm Hg)
Boiling Point: 100°C (212°F)   **Specific Gravity:** 0.9   **Water Solubility:** Insoluble

Chemical Incompatibilities or Instabilities: Strong acids, strong bases, oxidizers, chlorosulfonic acid.

FLAMMABILITY

NFPA Hazard Code: **2-3-2**   **NFPA Classification:** Class IB Flammable
Flash Point: 9°C (48°F)   **LEL:** 1.4%
Autoignition Temp: 372°C (702°F)   **UEL:** 14.0%

Fire Extinguishing Methods: Alcohol Resistant Foam, Carbon Dioxide, Dry Chemical, Water Spray or Fog.

Special Fire Fighting Considerations: Isolate for 1/2 mile if rail or tank truck is involved in a fire. Water spray should be used to keep closed containers cool. Continue to cool container after fire is extinguished. For large fires, if possible, withdraw and allow to burn. Immediately withdraw if rising sound from venting device is heard or if fire is causing discoloration to the tank. Treat all materials used or generated and equipment involved as contaminated by hazardous waste.

TOXICOLOGY
CARCINOGEN

Odor: Sweet, possibly acrid - penetrating   **Odor Threshold:** 0.001 mg/m^3
Physical Contact: Material is extremely destructive to human tissues.
TLV = 20 mg/m^3 **STEL** = 61 mg/m^3 **IDLH** = 8,000 mg/m^3

Routes of Entry and Relative LD$_{50}$ (or LC$_{50}$)   **RTECS # AT0700000**
  Inhalation   4   (73,000 mg/m^3/H)
  Ingestion   4   (800 mg/kg)
  Skin Absorption   N.D.

Symptoms of Exposure: Possible irritation to skin, eyes and upper respiratory tract; headache; nausea; lachrymation; shortness of breath; laryngitis. Inhalation of vapors may be fatal by causing glottis to spasm and suffocation. Exposure may result in chemical pneumonitis or pulmonary edema.

Emergency/First Aid Treatment: Remove to ventilated area; immediately remove any contaminated clothing and wash contaminated areas for 15 minutes using water. Treat supportively and observe for possible shock. If ingested, seek immediate medical aid.

Recommended Clean-Up Procedures

Personal Protection: Level B Ensemble   **Recommended Material** Butyl Rubber (+), Polyvinyl Alcohol (+), Teflon (+)

RCRA Waste # U113   **Reportable Quantities:** 1000 lb (1 lb)

Spills: Remove all potential ignition sources. Dike to contain spill, absorb with non-combustible absorbent and take up using non-sparking tools. Decontaminate spill area using soapy water. Treat all materials used or generated and equipment involved as contaminated by hazardous waste.

Special Emergency Information: May be harmful if inhaled, swallowed or absorbed through the skin. May undergo hazardous polymerization. Vapors are heavier than air and may travel some distance to an ignition source. Vapors are more dense than air and may settle in low lying areas.

ETHYLAMINE - CH3CH2NH2

Synonyms: 1-aminoethane; ethanamine; monoethylamine

CAS Number: 75-04-7	**Description:** Colorless Liquid
DOT Number: 1036, 2270	**DOT Classification:** Flammable Liquid

Molecular Weight: 45.1

Melting Point:	-81°C (-114°F)	**Vapor Density:** 1.6	**Vapor Pressure:**	8 (400 mg Hg)
Boiling Point:	17°C (62°F)	**Specific Gravity:** 0.7	**Water Solubility:**	Soluble

Chemical Incompatibilities or Instabilities: Oxidizers, brass, cellulose nitrate, mercury, silver

FLAMMABILITY

NFPA Hazard Code:	3-4-0	**NFPA Classification:**	Class IA Flammable
Flash Point:	-18°C (0°F)	**LEL:**	3.5%
Autoignition Temp:	385°C (725°F)	**UEL:**	14.0%

Fire Extinguishing Methods: Carbon Dioxide, Dry Chemical, Foam, Water Spray or Fog.

Special Fire Fighting Considerations: Isolate for 1/2 mile if rail or tank truck is involved in a fire. Water spray should be used to keep closed containers cool. Continue to cool container after fire is extinguished. Immediately withdraw if rising sound from venting device is heard or if fire is causing discoloration to the tank.

TOXICOLOGY

Odor: Sharp Ammonia-like	**Odor Threshold:** 0.06 mg/m^3

Physical Contact: Material is extremely destructive to human tissues.

TLV = 18 mg/m^3	**STEL** = N.D.	**IDLH** = 30,000 mg/m^3

Routes of Entry and Relative LD$_{50}$ (or LC$_{50}$)   **RTECS # KH2100000**

Inhalation	N.D.	
Ingestion	5	(400 mg/kg)
Skin Absorption	8	(390 mg/kg)

Symptoms of Exposure: Possible irritation to skin, eyes and upper respiratory tract; headache; nausea; laryngitis. Inhalation of vapors may be fatal by causing glottis to spasm and suffocation. Exposure may result in chemical pneumonitis or pulmonary edema.

Emergency/First Aid Treatment: Remove to ventilated area; immediately remove any contaminated clothing and wash contaminated areas for 15 minutes using water. Treat supportively and observe for possible shock. If ingested, seek immediate medical aid.

Recommended Clean-Up Procedures

Personal Protection:	Level B Ensemble	**Recommended Material** Butyl Rubber (+), Teflon (0)
RCRA Waste #	None	**Reportable Quantities:** None

Spills: Remove all potential ignition sources. Dike to contain spill, absorb with non-combustible absorbent and take up using non-sparking tools. Decontaminate spill area using water.

Special Emergency Information: May be harmful if inhaled, swallowed or absorbed through the skin. Vapors are heavier than air and may travel some distance to an ignition source. Vapors are more dense than air and may settle in low lying areas.

n-propylamine (CAS # 107-10-8, DOT # 1277, RTECS # UH9100000) has similar chemical, physical and toxicological properties.

N-ETHYLANILINE

Synonyms: n-ethylaminobenzene

CAS Number: 103-69-5	**Description:** Pale Yellow Liquid		
DOT Number: 2272	**DOT Classification:** Poison B		

Molecular Weight: 121.2

Melting Point: -64°C (-83°F)	**Vapor Density:**	**Vapor Pressure:** 1 (<1 mm Hg)	
Boiling Point: 204°C (400°F)	**Specific Gravity:** 0.9	**Water Solubility:** Insoluble	

Chemical Incompatibilities or Instabilities: Fuming nitric acid

FLAMMABILITY

NFPA Hazard Code: 3-2-0	**NFPA Classification:** Class IIIA Combustible
Flash Point: 85°C (185°F)	**LEL:** 1.6%
Autoignition Temp: °C (°F)	**UEL:** 9.5%

Fire Extinguishing Methods: Carbon Dioxide, Dry Chemical, Foam, Water Spray or Fog.

Special Fire Fighting Considerations: Structural fire fighter protective clothing will not provide adequate protection. Fight fire from a distance or protected location, if possible. Treat all materials used as contaminated by hazardous waste.

TOXICOLOGY

Odor:	**Odor Threshold:** N.D.
Physical Contact: Irritant	
TLV = N.D. **STEL** = N.D.	**IDLH** = N.D.

Routes of Entry and Relative LD$_{50}$ (or LC$_{50}$)   **RTECS # BX9780000**

Inhalation	N.D.
Ingestion	5 (334 mg/kg)
Skin Absorption	N.D.

Symptoms of Exposure: Possible irritation to skin, eyes and upper respiratory tract; headache; nausea.

Emergency/First Aid Treatment: Remove to ventilated area; immediately remove any contaminated clothing and wash contaminated areas for 15 minutes using water. Treat supportively and observe for possible shock. If ingested, seek immediate medical aid.

Recommended Clean-Up Procedures

Personal Protection: Level B Ensemble	**Recommended Material** N.D.	
RCRA Waste # None	**Reportable Quantities:** None	

Spills: Dike to contain spill and collect liquid for later disposal. Take up in non-combustible absorbent and dispose. Decontaminate spill area using soapy water. Treat all materials used or generated and equipment involved as contaminated by hazardous waste.

Special Emergency Information: May be harmful if inhaled, swallowed or absorbed through the skin.

2-ETHYLANILINE

Synonyms: o-ethylaniline; 2-ethylbenzamine

CAS Number: 578-54-1   **Description:** Yellow to Red Liquid
DOT Number: 2273   **DOT Classification:** Poison B

Molecular Weight: 121.2
Melting Point: -44°C (-47°F)   **Vapor Density:** 4.2   **Vapor Pressure:**   1 (< 1 mm Hg)
Boiling Point: 215°C (419°F)   **Specific Gravity:** 1.0   **Water Solubility:** Soluble

Chemical Incompatibilities or Instabilities: Chloroformates, strong oxidizers.

FLAMMABILITY

NFPA Hazard Code: N.D.   **NFPA Classification:** Class IIIA Combustible
Flash Point: 91°C (196°F)   **LEL:** N.D.
Autoignition Temp: °C   (°F)   **UEL:** N.D.

Fire Extinguishing Methods: Carbon Dioxide, Dry Chemical, Foam, Water Spray or Fog.

Special Fire Fighting Considerations: Structural fire fighter protective clothing will not provide adequate protection. Fight fire from a distance or protected location, if possible. Treat all materials used as contaminated by hazardous waste.

TOXICOLOGY

Odor: **Odor Threshold:** N.D.
Physical Contact: Irritant
TLV = N.D.   **STEL** = N.D.   **IDLH** = N.D.

Routes of Entry and Relative LD$_{50}$ (or LC$_{50}$)   **RTECS #** BX9800000
    Inhalation   N.D.
    Ingestion   **3**   (1260 mg/kg)
    Skin Absorption   N.D.

Symptoms of Exposure: Possible irritation to skin, eyes and upper respiratory tract; headache; nausea. Absorption may lead to the formation of methemoglobin resulting in cyanosis several hours after exposure.

Emergency/First Aid Treatment: Remove to ventilated area; immediately remove any contaminated clothing and wash contaminated areas for 15 minutes using water. Treat supportively and observe for possible shock. If ingested, seek immediate medical aid.

Recommended Clean-Up Procedures

Personal Protection: Level B Ensemble   **Recommended Material** N.D.

RCRA Waste #   None   **Reportable Quantities:**   None

Spills: Dike to contain spill and collect liquid for later disposal. Take up in non-combustible absorbent and dispose. Decontaminate spill area using soapy water. Treat all materials used or generated and equipment involved as contaminated by hazardous waste.

Special Emergency Information: May be harmful if inhaled, swallowed or absorbed through the skin.

3-ethylaniline (CAS # 587-02-0, RTECS # BX9770000) and
4-ethylaniline (CAS # 589-16-2, RTECS # BX9900000) have similar chemical, physical and toxicological properties.

ETHYLBENZENE - $CH_3CH_2C_6H_5$

Synonyms: phenylethane

CAS Number: 100-41-4   **Description:** Colorless Liquid
DOT Number: 1175   **DOT Classification:** Flammable Liquid

Molecular Weight: 106.2
Melting Point: -95°C (-139°F)   **Vapor Density:** 3.7   **Vapor Pressure:** 3 (7 mm Hg)
Boiling Point: 136°C (277°F)   **Specific Gravity:** 0.9   **Water Solubility:** Insoluble

Chemical Incompatibilities or Instabilities: Strong oxidizers.

FLAMMABILITY

NFPA Hazard Code: 2-3-0   **NFPA Classification:** Class IB Flammable
Flash Point: 21°C (70°F)   **LEL:** 1.0%
Autoignition Temp: 432°C (810°F)   **UEL:** 6.7%

Fire Extinguishing Methods: Alcohol Resistant Foam, Carbon Dioxide, Dry Chemical, Water Spray or Fog.

Special Fire Fighting Considerations: Isolate for 1/2 mile if rail or tank truck is involved in a fire. Water spray should be used to keep closed containers cool. Continue to cool container after fire is extinguished. For large fires, if possible, withdraw and allow to burn. Immediately withdraw if rising sound from venting device is heard or if fire is causing discoloration to the tank.

TOXICOLOGY

Odor: Aromatic   **Odor Threshold:** 0.2 mg/m^3
Physical Contact: Irritant
TLV = 435 mg/m^3 **STEL** = 545 mg/m^3 **IDLH** = 8,800 mg/m^3

Routes of Entry and Relative LD$_{50}$ (or LC$_{50}$)   **RTECS # DA0700000**
    Inhalation   N.D.
    Ingestion   3   (3500 mg/kg)
    Skin Absorption   1   (17800 mg/kg)

Symptoms of Exposure: Possible irritation to skin, eyes and upper respiratory tract; headache; nausea.

Emergency/First Aid Treatment: Remove to ventilated area; immediately remove any contaminated clothing and wash contaminated areas for 15 minutes using water. Treat supportively and observe for possible shock. If ingested, seek immediate medical aid.

Recommended Clean-Up Procedures

Personal Protection: Level B Ensemble   **Recommended Material**   Viton (+), Teflon (0)

RCRA Waste #   None   **Reportable Quantities:**   1000 lb (1000 lb)

Spills: Remove all potential ignition sources. Dike to contain spill, absorb with non-combustible absorbent and take up using non-sparking tools. Decontaminate spill area using soapy water.

Special Emergency Information: May be harmful if inhaled, swallowed or absorbed through the skin.

Propylbenzene (CAS # 103-65-1, DOT # 2364, RTECS # DA8750000) has similar chemical, physical and toxicological properties.

ETHYL CHLOROACETATE - ClCH$_2$C(O)OCH$_2$CH$_3$

Synonyms: chloroacetic acid, ethyl ester; ethyl monochloroacetate

CAS Number: 105-39-5   **Description:** Colorless Liquid
DOT Number: 1181   **DOT Classification:** Flammable Liquid

Molecular Weight: 122.6
Melting Point: -27°C (-17°F)   **Vapor Density:** 4.3   **Vapor Pressure:** **3** (9 mm Hg)
Boiling Point: 146°C (295°F)   **Specific Gravity:** 1.2   **Water Solubility:** Insoluble

Chemical Incompatibilities or Instabilities: Sodium cyanide, oxidizers.

FLAMMABILITY

NFPA Hazard Code: ?-3-0   **NFPA Classification:** Class IIIA Combustible
Flash Point: 64°C (147°F)   **LEL:** N.D.
Autoignition Temp: °C (°F)   **UEL:** N.D.

Fire Extinguishing Methods: Carbon Dioxide, Dry Chemical, Foam, Water Spray or Fog.

Special Fire Fighting Considerations: Structural fire fighter protective clothing will not provide adequate protection. Fight fire from a distance or protected location, if possible.

TOXICOLOGY
QUESTIONABLE CARCINOGEN

Odor: Pungent, Fruity   **Odor Threshold:** N.D.
Physical Contact: Irritant
TLV = N.D.   **STEL** = N.D.   **IDLH** = N.D.

Routes of Entry and Relative LD$_{50}$ (or LC$_{50}$)   **RTECS # AF9110000**
   Inhalation   N.D.
   Ingestion   N.D.
   Skin Absorption   **8**   (230 mg/kg)

Symptoms of Exposure: Possible irritation to skin, eyes and upper respiratory tract; headache; nausea. Lachrymation.

Emergency/First Aid Treatment: Remove to ventilated area; immediately remove any contaminated clothing and wash contaminated areas for 15 minutes using water. Treat supportively and observe for possible shock. If ingested, seek immediate medical aid.

Recommended Clean-Up Procedures

Personal Protection: Level B Ensemble   **Recommended Material** N.D.

RCRA Waste # None   **Reportable Quantities:** None

Spills: Remove all potential ignition sources. Dike to contain spill, absorb with non-combustible absorbant and take up using non-sparking tools. Decontaminate spill area using soapy water.

Special Emergency Information: May be fatal in inhaled, ingested or absorbed through the skin.

ETHYL CHLORIDE - CH$_3$CH$_2$Cl

Synonyms: chloroethane; monochloroethane

CAS Number: 75-00-3   **Description:** Colorless Liquid
DOT Number: 1037   **DOT Classification:** Flammable Liquid

Molecular Weight: 64.5
Melting Point: -136°C (-213°F)   **Vapor Density:** 2.2   **Vapor Pressure:** GAS
Boiling Point: 12°C (54°F)   **Specific Gravity:** 0.9 **Water Solubility:** Insoluble

Chemical Incompatibilities or Instabilities: Aluminum, copper, magnesium, zinc, alkali metals.

FLAMMABILITY

NFPA Hazard Code: 2-4-0   **NFPA Classification:** Class IA Flammable
Flash Point: -50°C (-58°F)   **LEL:**   3.8%
Autoignition Temp: 519°C (966°F)   **UEL:**   15.4%

Fire Extinguishing Methods: Carbon Dioxide, Dry Chemical, Foam, Water Spray or Fog.

Special Fire Fighting Considerations: Isolate for 1/2 mile if rail or tank truck is involved in a fire. Water spray should be used to keep closed containers cool. Continue to cool container after fire is extinguished. For large fires, if possible, withdraw and allow to burn. Immediately withdraw if rising sound from venting device is heard or if fire is causing discoloration to the tank.

TOXICOLOGY

Odor: Ethereal, Pungent   **Odor Threshold:** N.D.
Physical Contact: Irritant
TLV = 2640 mg/m^3 **STEL** = N.D.   **IDLH** = 50,000 mg/m^3

Routes of Entry and Relative LD$_{50}$ (or LC$_{50}$)   **RTECS # KH7525000**
  Inhalation   1   (320,000 mg/m^3/H)
  Ingestion   N.D.
  Skin Absorption   N.D.

Symptoms of Exposure: Possible irritation to skin, eyes and upper respiratory tract; headache; nausea. May act as a simple asphyxiant.

Emergency/First Aid Treatment: Remove to ventilated area; immediately remove any contaminated clothing and wash contaminated areas for 15 minutes using water. Treat supportively and observe for possible shock. If ingested, seek immediate medical aid.

Recommended Clean-Up Procedures

Personal Protection: Level B Ensemble   **Recommended Material**   N.D.

RCRA Waste #   None   **Reportable Quantities:**   None

Spills: Remove all potential ignition sources. Stop leak if it can safely be done. Dike to contain spill, absorb with non-combustible absorbant and take up using non-sparking tools. Decontaminate spill area using soapy water.

Special Emergency Information: May be harmful if inhaled, swallowed or absorbed through the skin. Vapors are heavier than air and may travel some distance to an ignition source. Vapors are more dense than air and may settle in low lying areas. Rapid release of vapors may cause frostbite.

Ethyl bromide (CAS # 74-96-4, DOT # 1891, RTECS # KH6475000) has similar chemical, physical and toxicological properties.

ETHYL CHLOROFORMATE - CH₃CH₂OC(O)Cl

Synonyms: chloroformic acid, ethyl ester; ethyl chlorocarbonate

CAS Number: 541-41-3

DOT Number: 1182

Description: Colorless Liquid

DOT Classification: Flammable Liquid

Molecular Weight: 108.5

Melting Point: -81°C (-113°F) **Vapor Density:** 3.7 **Vapor Pressure:** 6 (53 mm Hg)

Boiling Point: 95°C (203°F) **Specific Gravity:** 1.1 **Water Solubility:** Decomposes

Chemical Incompatibilities or Instabilities: Oxidizers, rust.

FLAMMABILITY

NFPA Hazard Code: 3-3-1 **NFPA Classification:** Class IB Flammable

Flash Point: 16°C (61°F) **LEL:** N.D.

Autoignition Temp: 500°C (932°F) **UEL:** N.D.

Fire Extinguishing Methods: Alcohol Resistant Foam, Carbon Dioxide, Dry Chemical, Water Spray or Fog.

Special Fire Fighting Considerations: Structural fire fighter protective clothing will not provide adequate protection. Isolate for 1/2 mile if rail or tank truck is involved in a fire. Water spray should be used to keep closed containers cool. Continue to cool container after fire is extinguished. Immediately withdraw if rising sound from venting device is heard or if fire is causing discoloration to the tank.

DOT Recommended Isolation Zones: Small Spill: 150 ft Large Spill: 150 ft

DOT Recommended Down Wind Take Cover Distance Small Spill: 0.2 miles Large Spill: 0.2 miles

TOXICOLOGY

Odor: **Odor Threshold:** N.D.

Physical Contact: Material is extremely destructive to human tissues.

TLV = N.D. **STEL** = N.D. **IDLH** = N.D.

Routes of Entry and Relative LD₅₀ (or LC₅₀) **RTECS #** LQ6125000

Inhalation	10	(645 mg/m³/H)
Ingestion	6	(270 mg/kg)
Skin Absorption	1	(7120 mg/kg)

Symptoms of Exposure: Possible irritation to skin, eyes and upper respiratory tract; headache; nausea; laryngitis. Inhalation of vapors may be fatal by causing glottis to spasm and suffocation. Exposure may result in chemical pneumonitis or pulmonary edema.

Emergency/First Aid Treatment: Remove to ventilated area; immediately remove any contaminated clothing and wash contaminated areas for 15 minutes using water. Treat supportively and observe for possible shock. If ingested, seek immediate medical aid.

Recommended Clean-Up Procedures

Personal Protection: Level B Ensemble **Recommended Material** N.D.

RCRA Waste # None **Reportable Quantities:** None

Spills: Remove all potential ignition sources. Dike to contain spill, absorb with non-combustible absorbent and take up using non-sparking tools. Decontaminate spill area using water. Keep area isolated until vapors have dissipated.

Special Emergency Information: May be fatal in inhaled, ingested or absorbed through the skin. Vapors are heavier than air and may travel some distance to an ignition source. Vapors are more dense than air and may settle in low lying areas.

ETHYLENE - $CH_2=CH_2$

Synonyms: acetene; ethene

CAS Number: 75-85-1
DOT Number: 1038 (liq), 1962

Description: Colorless Gas
DOT Classification: Flammable Gas

Molecular Weight: 28.1
Melting Point: -169°C (-272°F)
Boiling Point: -104°C (-155°F)

Vapor Density: 0.9
Specific Gravity: N. A.

Vapor Pressure: Gas
Water Solubility: Insoluble

Chemical Incompatibilities or Instabilities: Halogens, oxidizers, copper, copper compounds, aluminum trichloride, mercury oxides, silver oxide, nickel, ozone, halogenated hydrocarbons.

FLAMMABILITY

NFPA Hazard Code: 1-4-2
Flash Point: °C (°F)
Autoignition Temp: 490°C (914°F)

NFPA Classification:
LEL: 2.7%
UEL: 36.0%

Fire Extinguishing Methods: Carbon Dioxide, Dry Chemical, Foam, Water Spray or Fog.

Special Fire Fighting Considerations: Isolate for 1/2 mile if rail or tank truck is involved in a fire. Water spray should be used to keep closed containers cool. Continue to cool container after fire is extinguished. For large fires, if possible, withdraw and allow to burn. Immediately withdraw if rising sound from venting device is heard or if fire is causing discoloration to the tank.

TOXICOLOGY

Odor: Sweet, Grass-like
Physical Contact: Possible frostbite from liquefied ethylene
TLV = N.D.   **STEL** = N.D.

Odor Threshold: 20 mg/m^3

IDLH = N.D.

Routes of Entry and Relative LD$_{50}$ (or LC$_{50}$)
    Inhalation   N.D.
    Ingestion   N.D.
    Skin Absorption   N.D.

RTECS # KU5340000

Symptoms of Exposure: A simple asphyxiant.

Emergency/First Aid Treatment: Remove to ventilated area; immediately remove any contaminated clothing and wash contaminated areas for 15 minutes using water. Treat supportively and observe for possible shock. If ingested, seek immediate medical aid.

Recommended Clean-Up Procedures

Personal Protection: Level b Ensemble
RCRA Waste # None

Recommended Material N.D.
Reportable Quantities: None

Spills: Remove all potential ignition sources. Stop leak if it can safely be done. Keep area isolated until vapors have dissipated.

Special Emergency Information: May be harmful if inhaled, swallowed or absorbed through the skin. Vapors are less dense than air and will tend to rise. Vapors may still find an ignition source and flash back.

ETHYLENE CHLOROHYDRIN - $ClCH_2CH_2OH$

Synonyms: 2-chloroethyl alcohol; 2-chloroethanol

CAS Number: 107-07-3
DOT Number: 1135

Description: Colorless Liquid
DOT Classification: Poison B

Molecular Weight: 80.5
Melting Point: -68°C (-90°F)
Boiling Point: 130°C (266°F)

Vapor Density: 2.8
Specific Gravity: 1.2

Vapor Pressure: **2** (5 mm Hg)
Water Solubility: Soluble

Chemical Incompatibilities or Instabilities: Chlorosulfonic acid; ethylene diamine; sodium hydroxide; steam, strong oxidizers.

FLAMMABILITY

NFPA Hazard Code: 4-2-0
Flash Point: 59°C (139°F)
Autoignition Temp: 425°C (797°F)

NFPA Classification: Class IIIA Combustible
LEL: 4.9%
UEL: 15.9%

Fire Extinguishing Methods: Carbon Dioxide, Dry Chemical, Foam, Water Spray or Fog.

Special Fire Fighting Considerations: Structural fire fighter protective clothing will not provide adequate protection. Fight fire from a distance or protected location, if possible. May release toxic fumes upon heating (phosgene). Treat all materials used as contaminated by hazardous waste.

DOT Recommended Isolation Zones: Small Spill: 150 ft   Large Spill: 0.8 miles
DOT Recommended Down Wind Take Cover Distance   Small Spill: 150 ft   Large Spill: 0.8 miles

TOXICOLOGY

Odor: Ether-like
Physical Contact: Irritant
TLV = 3.3 mg/m^3 **STEL** = N.D.

Odor Threshold: N.D.

IDLH = 33 mg/m^3

RTECS # KK0875000

Routes of Entry and Relative LD$_{50}$ (or LC$_{50}$)

Inhalation	10	(290 mg/m^3/H)
Ingestion	8	(71 mg/kg)
Skin Absorption	9	(67 mg/kg)

Symptoms of Exposure: Possible irritation to skin, eyes and upper respiratory tract; headache; nausea.

Emergency/First Aid Treatment: Remove to ventilated area; immediately remove any contaminated clothing and wash contaminated areas for 15 minutes using water. Treat supportively and observe for possible shock. If ingested, seek immediate medical aid.

Recommended Clean-Up Procedures

Personal Protection: Level A Ensemble

Recommended Material Butyl Rubber (+), Polyvinyl Alcohol (+), Viton (+)

RCRA Waste # None

Reportable Quantities: None

Spills: Remove all potential ignition sources. Dike to contain spill, absorb with non-combustible absorbant and take up using non-sparking tools. Decontaminate spill area using water. Treat all materials used as contaminated by hazardous waste.

Special Emergency Information: May be fatal in inhaled, ingested or absorbed through the skin. Vapors are more dense than air and may settle in low lying areas. Vapors are heavier than air and may travel some distance to an ignition source.

ETHYLENEDIAMINE - NH₂CH₂CH₂NH₂

Synonyms: 1,2-ethanediamine

CAS Number: 107-15-3   **Description:** Colorless Liquid
DOT Number: 1604   **DOT Classification:** Corrosive Material

Molecular Weight: 60.1
Melting Point: 9°C (47°F)   **Vapor Density:** 2.1   **Vapor Pressure:** **3** (10 mm Hg)
Boiling Point: 115°C (239°F)   **Specific Gravity:** 0.9   **Water Solubility:** Soluble

Chemical Incompatibilities or Instabilities: Acetic acid; acetic anhydride; acrolein, acrylic acid; acrylonitrile; allyl chloride; carbon disulfide, chlorosulfonic acid; epichlorohydrin; ethylene chlorohydrin, strong acids, vinyl chloride.

FLAMMABILITY

NFPA Hazard Code: 3-3-0   **NFPA Classification:** Class IC Flammable
Flash Point: 33°C (91°F)   **LEL:**   1.1%
Autoignition Temp: 385°C (725°F) **UEL:**   5.8%

Fire Extinguishing Methods: Carbon Dioxide, Dry Chemical, Foam, Water Spray or Fog.

Special Fire Fighting Considerations: Isolate for 1/2 mile if rail or tank truck is involved in a fire. Water spray should be used to keep closed containers cool. Continue to cool container after fire is extinguished. Immediately withdraw if rising sound from venting device is heard or if fire is causing discoloration to the tank.

TOXICOLOGY
ALLERGEN AND SENSITIZER

Odor: Ammonia-like   **Odor Threshold:** N.D.
Physical Contact: Material is extremely destructive to human tissues.
TLV = 25 mg/m³   **STEL** = N.D.   **IDLH** = 5,000 mg/m³

Routes of Entry and Relative LD₅₀ (or LC₅₀)   **RTECS # KH8575000**
   Inhalation   N.D.
   Ingestion   5   (500 mg/kg)
   Skin Absorption   6   (730 mg/kg)

Symptoms of Exposure: Possible irritation to skin, eyes and upper respiratory tract; headache; nausea. Inhalation of vapors may be fatal by causing glottis to spasm and suffocation. Exposure may result in chemical pneumonitis or pulmonary edema.

Emergency/First Aid Treatment: Remove to ventilated area; immediately remove any contaminated clothing and wash contaminated areas for 15 minutes using water. Treat supportively and observe for possible shock. If ingested, seek immediate medical aid.

Recommended Clean-Up Procedures

Personal Protection: Level B Ensemble   **Recommended Material**   Butyl Rubber (+), Saranex (+)

RCRA Waste #   None   **Reportable Quantities:**   5000 lb (1000 lb)

Spills: Remove all potential ignition sources. Dike to contain spill, absorb with non-combustible absorbent and take up using non-sparking tools. Decontaminate spill area using water. Treat all materials used or generated and equipment involved as contaminated by hazardous waste.

Special Emergency Information: May be harmful if inhaled, swallowed or absorbed through the skin. Vapors are heavier than air and may travel some distance to an ignition source. Vapors are more dense than air and may settle in low lying areas.

ETHYLENE DICHLORIDE - ClCH₂CH₂Cl

Synonyms: 1,2-dichloroethane; ethylene chloride

CAS Number: 107-06-2 **Description:** Colorless Liquid
DOT Number: 1184 **DOT Classification:** Flammable Liquid

Molecular Weight: 99.0
Melting Point: -35°C (-32°F) **Vapor Density:** 3.4 **Vapor Pressure:** 6 (100 mm Hg)
Boiling Point: 84°C (183°F) **Specific Gravity:** 1.3 **Water Solubility:** Insoluble

Chemical Incompatibilities or Instabilities: Aluminum, ammonia, reaction with oxidizers may form vinyl chloride

FLAMMABILITY

NFPA Hazard Code: 2-3-0 **NFPA Classification:** Class IB Flammable
Flash Point: 13°C (56°F) **LEL:** 6.2%
Autoignition Temp: 413°C (775°F) **UEL:** 15.9%

Fire Extinguishing Methods: Alcohol Resistant Foam, Carbon Dioxide, Dry Chemical, Water Spray or Fog.

Special Fire Fighting Considerations: Isolate for 1/2 mile if rail or tank truck is involved in a fire. Water spray should be used to keep closed containers cool. Continue to cool container after fire is extinguished. For large fires, if possible, withdraw and allow to burn. Immediately withdraw if rising sound from venting device is heard or if fire is causing discoloration to the tank. May release toxic fumes upon heating (phosgene).

TOXICOLOGY
CARCINOGEN

Odor: Pleasant, chloroform-like **Odor Threshold:** 16 mg/m³
Physical Contact: Irritant
$TLV = 40$ mg/m³ $STEL = $ N.D. $IDLH = 4000$ mg/m³

Routes of Entry and Relative LD₅₀ (or LC₅₀) **RTECS # KI0525000**
  Inhalation 5 (28,000 mg/m³/H)
  Ingestion 4 (670 mg/kg)
  Skin Absorption N.D.

Symptoms of Exposure: Possible irritation to skin, eyes and upper respiratory tract; headache; nausea; CNS depression. May act as a narcotic in high concentrations.

Emergency/First Aid Treatment: Remove to ventilated area; immediately remove any contaminated clothing and wash contaminated areas for 15 minutes using water. Treat supportively and observe for possible shock. If ingested, seek immediate medical aid.

Recommended Clean-Up Procedures

Personal Protection: Level B Ensemble **Recommended Material** Teflon (+), Viton (+)

RCRA Waste # U077 **Reportable Quantities:** 100 lb (500 lb)

Spills: Remove all potential ignition sources. Dike to contain spill, absorb with non-combustible absorbent and take up using non-sparking tools. Decontaminate spill area using soapy water. Treat all materials used as contaminated by hazardous waste.

Special Emergency Information: May be harmful if inhaled, swallowed or absorbed through the skin. Vapors are heavier than air and may travel some distance to an ignition source. Vapors are more dense than air and may settle in low lying areas.

ETHYLENE GLYCOL DIETHYL ETHER - CH$_3$CH$_2$OCH$_2$CH$_2$OCH$_2$CH$_3$

Synonyms: 1,2-diethoxyethane; diethyl cellusolve

CAS Number: 629-14-1   **Description:** Colorless Liquid
DOT Number: 1153   **DOT Classification:** Combustible Liquid

Molecular Weight: 118.2
Melting Point: -74°C (-101°F)   **Vapor Density:** 4.1   **Vapor Pressure:** 3 (9 mm Hg)
Boiling Point: 122°C (251°F)   **Specific Gravity:** 0.8   **Water Solubility:** Insoluble

Chemical Incompatibilities or Instabilities: Strong oxidizers

FLAMMABILITY

NFPA Hazard Code: 1-3-0   **NFPA Classification:** Class IC Flammable
Flash Point: 35°C (95°F)   **LEL:** N.D.
Autoignition Temp: 208°C (406°F)   **UEL:** N.D.

Fire Extinguishing Methods: Alcohol Resistant Foam, Carbon Dioxide, Dry Chemical, Water Spray or Fog.

Special Fire Fighting Considerations: Isolate for 1/2 mile if rail or tank truck is involved in a fire. Water spray should be used to keep closed containers cool. Continue to cool container after fire is extinguished. For large fires, if possible, withdraw and allow to burn. Immediately withdraw if rising sound from venting device is heard or if fire is causing discoloration to the tank.

TOXICOLOGY

Odor: Ethereal   **Odor Threshold:** N.D.
Physical Contact: Irritant
TLV = N.D.   **STEL** = N.D.   **IDLH** = N.D.

Routes of Entry and Relative LD$_{50}$ (or LC$_{50}$)   **RTECS # KI1225000**
   Inhalation   N.D.
   Ingestion   3   (4390 mg/kg)
   Skin Absorption   N.D.

Symptoms of Exposure: Possible irritation to skin, eyes and upper respiratory tract; headache; nausea.

Emergency/First Aid Treatment: Remove to ventilated area; immediately remove any contaminated clothing and wash contaminated areas for 15 minutes using water. Treat supportively and observe for possible shock. If ingested, seek immediate medical aid.

Recommended Clean-Up Procedures

Personal Protection: Level B Ensemble   **Recommended Material** Butyl Rubber (-)

RCRA Waste #   None   **Reportable Quantities:**   None

Spills: Remove all potential ignition sources. Dike to contain spill, absorb with non-combustible absorbent and take up using non-sparking tools. Decontaminate spill area using soapy water.

Special Emergency Information: May be harmful if inhaled, swallowed or absorbed through the skin. May form unstable peroxides upon exposure to air. Vapors are heavier than air and may travel some distance to an ignition source. Vapors are more dense than air and may settle in low lying areas.

Ethylene glycol methyl ether (CAS # 109-86-4, DOT # 1188, RTECS # KL5775000),
ethylene glycol monomethyl ether acetate (CAS # 110-49-6, DOT # 1189, RTECS # KL5950000)
have similar chemical, physical and toxicological properties.

ETHYLENEIMINE

Synonyms: aziridine; azirane; dimethyleneinie

CAS Number: 151-56-4   **Description:** Colorless Liquid
DOT Number: 1185   **DOT Classification:** Flammable Liquid

Molecular Weight: 43.1
Melting Point: -72°C (-981.5°F)   **Vapor Density:** 1.5   **Vapor Pressure:** 7 (160 mm Hg)
Boiling Point: 57°C (135°F)   **Specific Gravity:** 0.8 **Water Solubility:** Soluble

Chemical Incompatibilities or Instabilities: Acids, acrolein, carbon disulfide, chloride salts, epichlorohydrin, glyoxal, halogens, silver, vinyl acetate

FLAMMABILITY

NFPA Hazard Code: 3-3-3   **NFPA Classification:** Class IB Flammable
Flash Point: -11°C (12°F)   **LEL:** 3.3%
Autoignition Temp: 320°C (608°F)   **UEL:** 46.0%

Fire Extinguishing Methods: Alcohol Resistant Foam, Carbon Dioxide, Dry Chemical, Foam, Water Spray or Fog.

Special Fire Fighting Considerations: Isolate for 1/2 mile if rail or tank truck is involved in a fire. Structural fire fighter protective clothing will not provide adequate protection. Water spray should be used to keep closed containers cool. Continue to cool container after fire is extinguished. For large fires, if possible, withdraw and allow to burn. Immediately withdraw if rising sound from venting device is heard or if fire is causing discoloration to the tank. Treat all materials used or generated and equipment involved as contaminated by hazardous waste.

DOT Recommended Isolation Zones: Small Spill: 600 ft   Large Spill: 900 ft
DOT Recommended Down Wind Take Cover Distance   Small Spill: 2 miles   Large Spill: 3 miles

TOXICOLOGY
CARCINOGEN

Odor: Ammonia-like   **Odor Threshold:** 1.3 mg/m^3
Physical Contact: Material is extremely destructive to human tissues.
TLV = See 29 CFR 1910.1012   **STEL** = N.D.   **IDLH** = 180 mg/m^3

Routes of Entry and Relative LD$_{50}$ (or LC$_{50}$)   **RTECS # KX5075000**
    Inhalation   10   (200 mg/m^3/H)
    Ingestion   9   (15 mg/kg)
    Skin Absorption   N.D.

Symptoms of Exposure: Possible irritation to skin, eyes and upper respiratory tract; headache; nausea; eye damage. Inhalation of vapors may be fatal by causing glottis to spasm and suffocation. Exposure may result in chemical pneumonitis or pulmonary edema.

Emergency/First Aid Treatment: Remove to ventilated area; immediately remove any contaminated clothing and wash contaminated areas for 15 minutes using water. Treat supportively and observe for possible shock. If ingested, seek immediate medical aid. If ingested, carbonated beverages is recommended as an antidote.

Recommended Clean-Up Procedures

Personal Protection: Level A Ensemble   **Recommended Material** Butyl Rubber (+)

RCRA Waste # P054   **Reportable Quantity:** 1 lb (1 lb)

Spills: Remove all potential ignition sources. Dike to contain spill, absorb with non-combustible absorbent and take up using non-sparking tools. Decontaminate spill area using water. Treat all materials used as contaminated by hazardous waste.

Special Emergency Information: May be fatal in inhaled, ingested or absorbed through the skin. Vapors are heavier than air and may travel some distance to an ignition source. Vapors are more dense than air and may settle in low lying areas.

ETHYLENE OXIDE

Synonyms: dimethylene oxide; epoxyethane; oxirane

CAS Number: 75-21-8   **Description:** Colorless Gas (Colorless Liquid below 50 F)
DOT Number: 1040   **DOT Classification:** Flammable Gas

Molecular Weight: 44.1
Melting Point: -112°C (-170°F)   **Vapor Density:** 1.5   **Vapor Pressure:**   GAS
Boiling Point: 11°C   (51°F)   **Specific Gravity:** 0.9   **Water Solubility:** Soluble

Chemical Incompatibilities or Instabilities: Acids, ammonia, amines, alkali hydroxides, aluminum and iron metals and oxides.

FLAMMABILITY

NFPA Hazard Code: 2-4-3   **NFPA Classification:** Class IA Flammable
Flash Point: -20°C   (-4°F)   **LEL:**   3.0%
Autoignition Temp: 429°C (804°F)   **UEL:**   100%

Fire Extinguishing Methods: Alcohol Resistant Foam, Carbon Dioxide, Dry Chemical, Water Spray or Fog.

Special Fire Fighting Considerations: Structural fire fighter protective clothing will not provide adequate protection. Isolate for 1 mile if rail or tank truck is involved in a fire. Stop leak if it can safely be done. Fight fire from a distance or protected location, if possible. For large fires, if possible, withdraw and allow to burn. Immediately withdraw if rising sound from venting device is heard or if fire is causing discoloration to the tank.

DOT Recommended Isolation Zones: Small Spill: 150 ft   Large Spill: 600 ft
DOT Recommended Down Wind Take Cover Distance   Small Spill: 0.8 miles   Large Spill: 2 miles

TOXICOLOGY
CARCINOGEN

Odor: Sweet, Ether-like   **Odor Threshold:** 1.5 mg/m^3
Physical Contact: Irritant
TLV = 1.8 mg/m^3 **STEL** = N.D.   **IDLH** = 1500 mg/m^3

Routes of Entry and Relative LD$_{50}$ (or LC$_{50}$)   **RTECS # KX2450000**
    Inhalation   6   (5800 mg/m^3/H)
    Ingestion   N.D.
    Skin Absorption   N.D.

Symptoms of Exposure: Possible irritation to skin, eyes and upper respiratory tract; headache; nausea; convulsions. Exposure may result in chemical pneumonitis or pulmonary edema.

Emergency/First Aid Treatment: Remove to ventilated area; immediately remove any contaminated clothing and wash contaminated areas for 15 minutes using water. Treat supportively and observe for possible shock. If ingested, seek immediate medical aid.

Recommended Clean-Up Procedures

Personal Protection: Level A Ensemble   **Recommended Material**   N.D.

RCRA Waste #   U115   **Reportable Quantity:**   10 lb (1 lb)

Spills: Remove all potential ignition sources. Stop leak if it can safely be done. Water spray May be used to keep vapors down. Vent to dissipate vapors. Treat all materials used as contaminated by hazardous waste.

Special Emergency Information: May be fatal in inhaled, ingested or absorbed through the skin. Vapors are heavier than air and may travel some distance to an ignition source. Vapors are more dense than air and may settle in low lying areas. May undergo hazardous polymerization.

ETHYL FORMATE - CH₃CH₂OC(O)H

Synonyms: ethyl formic ester; ethyl methanoate; formic acid, ethyl ester

CAS Number: 109-94-4	**Description:** Colorless Liquid	
DOT Number: 1190	**DOT Classification:** Flammable Liquid	

Molecular Weight: 74.1

Melting Point: °C (°F)	**Vapor Density:** 2.6	**Vapor Pressure:** 7 (200 mm Hg)
Boiling Point: 54°C (130°F)	**Specific Gravity:** 0.9	**Water Solubility:** Insoluble

Chemical Incompatibilities or Instabilities: Strong oxidizers.

FLAMMABILITY

NFPA Hazard Code: 2-3-0	**NFPA Classification:** Class IB Flammable
Flash Point: -20°C (-4°F)	**LEL:** 2.8%
Autoignition Temp: 455°C (851°F)	**UEL:** 16.0%

Fire Extinguishing Methods: Alcohol Resistant Foam, Carbon Dioxide, Dry Chemical, Water Spray or Fog.

Special Fire Fighting Considerations: Isolate for 1/2 mile if rail or tank truck is involved in a fire. Water spray should be used to keep closed containers cool. Continue to cool container after fire is extinguished. For large fires, if possible, withdraw and allow to burn. Immediately withdraw if rising sound from venting device is heard or if fire is causing discoloration to the tank.

TOXICOLOGY

Odor: Rum-like or Fruity   **Odor Threshold:** 60 mg/m³

Physical Contact: Material is extremely destructive to human tissues.

TLV = 300 mg/m³   **STEL** = N.D.   **IDLH** = 2500 mg/m³

Routes of Entry and Relative LD₅₀ (or LC₅₀)   **RTECS #** LQ8400000

Inhalation	N.D.	
Ingestion	3	(1850 mg/kg)
Skin Absorption	1	(20,000 mg/kg)

Symptoms of Exposure: Possible irritation to skin, eyes and upper respiratory tract; headache; nausea; laryngitis. May act as a narcotic.

Emergency/First Aid Treatment: Remove to ventilated area; immediately remove any contaminated clothing and wash contaminated areas for 15 minutes using water. Treat supportively and observe for possible shock. If ingested, seek immediate medical aid.

Recommended Clean-Up Procedures

Personal Protection: Level B Ensemble	**Recommended Material** N.D.
RCRA Waste # None	**Reportable Quantities:** None

Spills: Remove all potential ignition sources. Dike to contain spill, absorb with non-combustible absorbent and take up using non-sparking tools. Decontaminate spill area using soapy water.

Special Emergency Information: May be harmful if inhaled, swallowed or absorbed through the skin. Vapors are heavier than air and may travel some distance to an ignition source. Vapors are more dense than air and may settle in low lying areas.

Methyl formate (CAS # 107-31-3, DOT # 1243, RTECS # LQ8925000) has similar chemical, physical and toxicological properties.

2-ETHYLHEXYLAMINE - $CH_3CH_2CH_2CH(CH_3CH_2)CH_2NH_2$

Synonyms:

CAS Number: 104-75-6   **Description:** Colorless Liquid
DOT Number: 2276   **DOT Classification:** Corrosive Material

Molecular Weight: 129.3
Melting Point: °C   (°F)   **Vapor Density:** 4.5   **Vapor Pressure:**   **2** (2 mm Hg)
Boiling Point: 337°C (169°F)   **Specific Gravity:** 0.8   **Water Solubility:**   Soluble

Chemical Incompatibilities or Instabilities: Oxidizers.

FLAMMABILITY

NFPA Hazard Code: **2-2-0**   **NFPA Classification:** Class IIIA Combustible
Flash Point: 60°C (140°F)   **LEL:** N.D.
Autoignition Temp: 295°C (563°F)   **UEL:** N.D.

Fire Extinguishing Methods: Carbon Dioxide, Dry Chemical, Foam, Water Spray or Fog.

Special Fire Fighting Considerations: Isolate for 1/2 mile if rail or tank truck is involved in a fire. Water spray should be used to keep closed containers cool. Continue to cool container after fire is extinguished. Immediately withdraw if rising sound from venting device is heard or if fire is causing discoloration to the tank.

TOXICOLOGY

Odor: **Odor Threshold:** N.D.
Physical Contact: Material is extremely destructive to human tissues.
TLV = N.D.   **STEL** = N.D.   **IDLH** = N.D.

Routes of Entry and Relative LD$_{50}$ (or LC$_{50}$)   **RTECS #** MQ5250000
   Inhalation   N.D.
   Ingestion   5   (450 mg/kg)
   Skin Absorption   7   (600 mg/kg)

Symptoms of Exposure: Possible irritation to skin, eyes and upper respiratory tract; headache; nausea; laryngitis. Inhalation of vapors may be fatal by causing glottis to spasm and suffocation. Exposure may result in chemical pneumonitis or pulmonary edema.

Emergency/First Aid Treatment: Remove to ventilated area; immediately remove any contaminated clothing and wash contaminated areas for 15 minutes using water. Treat supportively and observe for possible shock. If ingested, seek immediate medical aid.

Recommended Clean-Up Procedures

Personal Protection: Level B Ensemble   **Recommended Material**   N.D.

RCRA Waste #   None   **Reportable Quantities:**   None

Spills: Remove all potential ignition sources. Dike to contain spill, absorb with non-combustible absorbant and take up using non-sparking tools. Decontaminate spill area using water.

Special Emergency Information: May be harmful if inhaled, swallowed or absorbed through the skin.

ETHYL ISOCYANATE - CH₃CH₂N=C=O

Synonyms: isocyanic acid, ethyl ester

CAS Number: 109-90-0
DOT Number: 2841

Description: Colorless Liquid
DOT Classification: Flammable Liquid

Molecular Weight: 71.1
Melting Point: °C   (°F)
Boiling Point: 60°C   (140°F)

Vapor Density: 2.5
Specific Gravity: 0.9

Vapor Pressure: 4 (13 mm Hg)
Water Solubility: Insoluble

Chemical Incompatibilities or Instabilities: Water, heat, oxidizers.

FLAMMABILITY

NFPA Hazard Code: N.D.
Flash Point: -6°C   (20°F)
Autoignition Temp: °C   (°F)

NFPA Classification: Class IB Flammable
LEL: N.D.
UEL: N.D.

Fire Extinguishing Methods: Alcohol Resistant Foam, Carbon Dioxide, Dry Chemical, Water Spray or Fog.

Special Fire Fighting Considerations: Structural fire fighter protective clothing will not provide adequate protection. Isolate for 1/2 mile if rail or tank truck is involved in a fire. Water spray should be used to keep closed containers cool. Continue to cool container after fire is extinguished. Immediately withdraw if rising sound from venting device is heard or if fire is causing discoloration to the tank. Treat all materials used as contaminated by hazardous waste.

DOT Recommended Isolation Zones: Small Spill: 150 ft   Large Spill: 150 ft
DOT Recommended Down Wind Take Cover Distance   Small Spill: 0.8 miles   Large Spill: 0.8 miles

TOXICOLOGY

Odor:
Physical Contact: Material is extremely destructive to human tissues.
TLV = N.D.   **STEL** = N.D.

Odor Threshold: N.D.

IDLH = N.D.

Routes of Entry and Relative LD₅₀ (or LC₅₀)
Inhalation   N.D.
Ingestion   N.D.
Skin Absorption   N.D.

RTECS # NQ8825000

Symptoms of Exposure: Possible irritation to skin, eyes and upper respiratory tract; headache; nausea; burns. Lachrymation. Inhalation of vapors may be fatal by causing glottis to spasm and suffocation. May cause an allergic reaction.Exposure may result in chemical pneumonitis or pulmonary edema.

Emergency/First Aid Treatment: Remove to ventilated area; immediately remove any contaminated clothing and wash contaminated areas for 15 minutes using water. Treat supportively and observe for possible shock. If ingested, seek immediate medical aid.

Recommended Clean-Up Procedures

Personal Protection: Level B Ensemble   **Recommended Material**   N.D.

RCRA Waste #   None   **Reportable Quantities:**   None

Spills: Remove all potential ignition sources. Dike to contain spill, absorb with non-combustible absorbent and take up using non-sparking tools. Decontaminate spill area using soapy water. Treat all materials used as contaminated by hazardous waste.

Special Emergency Information: May be fatal in inhaled, ingested or absorbed through the skin. Vapors are heavier than air and may travel some distance to an ignition source. Vapors are more dense than air and may settle in low lying areas.

ETHYL MERCAPTAN - CH_3CH_2SH

Synonyms: ethanethiol; ethyl hydrosulfide; mercaptoethane; thioethanol

CAS Number: 75-08-1   **Description:** Colorless Liquid
DOT Number: 2363   **DOT Classification:** Flammable Liquid

Molecular Weight: 62.1
Melting Point: -144°C (-228°F)   **Vapor Density:** 2.1   **Vapor Pressure:** 9 (450 mm Hg)
Boiling Point: 35°C (95°F)   **Specific Gravity:** 0.8   **Water Solubility:** 1%

Chemical Incompatibilities or Instabilities: Calcium hypochlorite, oxidizers, steam.

FLAMMABILITY

NFPA Hazard Code: 2-4-0   **NFPA Classification:** Class IA Flammable
Flash Point: -17°C (1°F)   **LEL:** 2.8%
Autoignition Temp: 300°C (572°F)   **UEL:** 18.0%

Fire Extinguishing Methods: Carbon Dioxide, Dry Chemical, Foam, Water Spray or Fog.

Special Fire Fighting Considerations: Isolate for 1/2 mile if rail or tank truck is involved in a fire. Water spray should be used to keep closed containers cool. Continue to cool container after fire is extinguished. For large fires, if possible, withdraw and allow to burn. Immediately withdraw if rising sound from venting device is heard or if fire is causing discoloration to the tank.

TOXICOLOGY

Odor: Skunk, Garlic, Penetrating Stench   **Odor Threshold:** 0.00025 mg/m^3
Physical Contact: Irritant
TLV = 1 mg/m^3 **STEL** = N.D.   **IDLH** = 6,250 mg/m^3

Routes of Entry and Relative LD$_{50}$ (or LC$_{50}$)   **RTECS #** KI9625000
   Inhalation   4   (44,000 mg/m^3/H)
   Ingestion   3   (1960 mg/kg)
   Skin Absorption   N.D.

Symptoms of Exposure: Possible irritation to skin, eyes and upper respiratory tract; headache; nausea. May act as a narcotic in high concentrations.

Emergency/First Aid Treatment: Remove to ventilated area; immediately remove any contaminated clothing and wash contaminated areas for 15 minutes using water. Treat supportively and observe for possible shock. If ingested, seek immediate medical aid.

Recommended Clean-Up Procedures

Personal Protection: Level B Ensemble   **Recommended Material** N.D.

RCRA Waste # None   **Reportable Quantities:** None

Spills: Remove all potential ignition sources. Dike to contain spill, absorb with non-combustible absorbent and take up using non-sparking tools. Decontaminate spill area using water.

Special Emergency Information: May be harmful if inhaled, swallowed or absorbed through the skin. Vapors are heavier than air and may travel some distance to an ignition source. Vapors are more dense than air and may settle in low lying areas.

ETHYL NITRITE - CH₃CH₂ONO

Synonyms: nitrosyl ethoxide; nitrous acid, ethyl ester

CAS Number: 109-95-5	**Description:** Colorless Liquid	
DOT Number: 1194	**DOT Classification:** Flammable Liquid	

Molecular Weight: 75.1

Melting Point:	°C (°F)	**Vapor Density:** 2.6	**Vapor Pressure:**	Gas above 63°F
Boiling Point:	17°C (63°F)	**Specific Gravity:** 0.9	**Water Solubility:**	Decomposes

Chemical Incompatibilities or Instabilities: Acids, reducing agents, explodes above 194°F (90°C).

FLAMMABILITY

NFPA Hazard Code: 2-4-4	**NFPA Classification:** Class IA Flammable	
Flash Point: -31°C (-31°F)	**LEL:** 3.0%	
Autoignition Temp: 90°C (194°F)	**UEL:** 50.0%	

Fire Extinguishing Methods: Carbon Dioxide, Dry Chemical, Foam, Water Spray or Fog.

Special Fire Fighting Considerations: Structural fire fighter protective clothing will not provide adequate protection. Isolate for 1/2 mile if rail or tank truck is involved in a fire. Water spray should be used to keep closed containers cool. Continue to cool container after fire is extinguished. For large fires, if possible, withdraw and allow to burn. Immediately withdraw if rising sound from venting device is heard or if fire is causing discoloration to the tank.

TOXICOLOGY

Odor: Aromatic, Ethereal	**Odor Threshold:** N.D.

Physical Contact: Material is extremely destructive to human tissues.

TLV = N.D.	**STEL** = N.D.	**IDLH** = N.D.

Routes of Entry and Relative LD₅₀ (or LC₅₀)   **RTECS # RA0810000**

Inhalation	N.D.
Ingestion	N.D.
Skin Absorption	N.D.

Symptoms of Exposure: Possible burns or irritation to skin, eyes and upper respiratory tract; headache; nausea. May act as a narcotic in high concentrations. Absorption may lead to the formation of methemoglobin resulting in cyanosis several hours after exposure. Inhalation of vapors may be fatal by causing glottis to spasm and suffocation. Exposure may result in chemical pneumonitis or pulmonary edema.

Emergency/First Aid Treatment: Remove to ventilated area; immediately remove any contaminated clothing and wash contaminated areas for 15 minutes using water. Treat supportively and observe for possible shock. If ingested, seek immediate medical aid.

Recommended Clean-Up Procedures

Personal Protection:	Level B Ensemble	**Recommended Material** N.D.
RCRA Waste #	None	**Reportable Quantities:** None

Spills: Remove all potential ignition sources. Dike to contain spill, absorb with non-combustible absorbent and take up using non-sparking tools. Decontaminate spill area using dilute alkaline solution.

Special Emergency Information: May be fatal in inhaled, ingested or absorbed through the skin. Vapors are heavier than air and may travel some distance to an ignition source. Vapors are more dense than air and may settle in low lying areas.

ETHYL PROPIONATE - CH$_3$CH$_2$C(O)OCH$_2$CH$_3$

Synonyms: propionic acid, ethyl ether

CAS Number: 105-37-3
DOT Number: 1195

Description: Colorless Liquid
DOT Classification: Flammable Liquid

Molecular Weight: 102.2
Melting Point: °C (°F)
Boiling Point: 99°C (210°F)

Vapor Density: 3.5
Specific Gravity: 0.9

Vapor Pressure: **5** (40 mm Hg)
Water Solubility: Insoluble

Chemical Incompatibilities or Instabilities: Oxidizers

FLAMMABILITY

NFPA Hazard Code: ?-3-0
Flash Point: 12°C (54°F)
Autoignition Temp: 440°C (824°F)

NFPA Classification: Class IB Flammable
LEL: 1.9%
UEL: 11.0%

Fire Extinguishing Methods: Alcohol Resistant Foam, Carbon Dioxide, Dry Chemical, Water Spray or Fog.

Special Fire Fighting Considerations: Isolate for 1/2 mile if rail or tank truck is involved in a fire. Water spray should be used to keep closed containers cool. Continue to cool container after fire is extinguished. For large fires, if possible, withdraw and allow to burn. Immediately withdraw if rising sound from venting device is heard or if fire is causing discoloration to the tank.

TOXICOLOGY

Odor: Fruity, Rum
Physical Contact: Irritant
TLV = N.D.   **STEL** = N.D.

Odor Threshold: N.D.

IDLH = N.D.

Routes of Entry and Relative LD$_{50}$ (or LC$_{50}$)
  Inhalation   N.D.
  Ingestion   **3**   (3500 mg/kg)
  Skin Absorption   N.D.

RTECS # UF3675000

Symptoms of Exposure: Possible irritation to skin, eyes and upper respiratory tract; headache; nausea. May act as a narcotic in high concentrations.

Emergency/First Aid Treatment: Remove to ventilated area; immediately remove any contaminated clothing and wash contaminated areas for 15 minutes using water. Treat supportively and observe for possible shock. If ingested, seek immediate medical aid.

Recommended Clean-Up Procedures

Personal Protection: Level B Ensemble
RCRA Waste # None

Recommended Material Teflon (-)
Reportable Quantities: None

Spills: Remove all potential ignition sources. Dike to contain spill, absorb with non-combustible absorbent and take up using non-sparking tools. Decontaminate spill area using soapy water.

Special Emergency Information: May be harmful if inhaled, swallowed or absorbed through the skin. Vapors are heavier than air and may travel some distance to an ignition source. Vapors are more dense than air and may settle in low lying areas.

ETHYL SILICATE - (CH3CH2O)4Si

Synonyms: ethyl orthosilicate; tetraethoxysilane; tetraethyl silicate

CAS Number: 78-10-4	**Description:** Colorless Liquid		
DOT Number: 1292	**DOT Classification:** Flammable Liquid		

Molecular Weight: 208.4

Melting Point:	-77°C (-107°F)	**Vapor Density:** 7.2	**Vapor Pressure:**	1 (< 1 mm Hg)
Boiling Point:	168°C (334°F)	**Specific Gravity:** 0.9	**Water Solubility:**	Decomposes

Chemical Incompatibilities or Instabilities: Oxidizers.

FLAMMABILITY

NFPA Hazard Code:	2-2-0	**NFPA Classification:** Class IC Flammable
Flash Point:	52°C (125°F)	**LEL:** N.D.
Autoignition Temp:	°C (°F)	**UEL:** N.D.

Fire Extinguishing Methods: Carbon Dioxide, Dry Chemical, Foam, Water Spray or Fog.

Special Fire Fighting Considerations: Isolate for 1/2 mile if rail or tank truck is involved in a fire. Water spray should be used to keep closed containers cool. Continue to cool container after fire is extinguished. Immediately withdraw if rising sound from venting device is heard or if fire is causing discoloration to the tank.

TOXICOLOGY

Odor: Sweet, Alcohol   **Odor Threshold:** 31 mg/m^3
Physical Contact:
TLV = 85 mg/m^3 **STEL** = N.D.   **IDLH** = 8500 mg/m^3

Routes of Entry and Relative LD$_{50}$ (or LC$_{50}$)   **RTECS # VV9450000**

Inhalation	N.D.	
Ingestion	2	(6270 mg/kg)
Skin Absorption	1	(5878 mg/kg)

Symptoms of Exposure: Possible irritation to skin, eyes and upper respiratory tract; headache; nausea. May act as a narcotic in high concentrations.

Emergency/First Aid Treatment: Remove to ventilated area; immediately remove any contaminated clothing and wash contaminated areas for 15 minutes using water. Treat supportively and observe for possible shock. If ingested, seek immediate medical aid.

Recommended Clean-Up Procedures

Personal Protection:	Level B Ensemble	**Recommended Material**	N.D.
RCRA Waste #	None	**Reportable Quantities:**	None

Spills: Remove all potential ignition sources. Dike to contain spill, absorb with non-combustible absorbent and take up using non-sparking tools. Decontaminate spill area using soapy water.

Special Emergency Information: May be harmful if inhaled, swallowed or absorbed through the skin.

ETHYL TRICHLOROSILANE - CH3CH2SiCl3

Synonyms: trichloroethylsilane

CAS Number: 115-21-9
DOT Number: 1196

Description: Colorless Liquid
DOT Classification: Flammable Liquid

Molecular Weight: 163.5
Melting Point: -106°C (-158°F)
Boiling Point: 98°C (208°F)

Vapor Density: 5.6
Specific Gravity: 1.2

Vapor Pressure:
Water Solubility: Decomposes

Chemical Incompatibilities or Instabilities: Water, oxidizers.

FLAMMABILITY

NFPA Hazard Code: 3-3-2-~~W~~
Flash Point: 18°C (64°F)
Autoignition Temp: °C (°F)

NFPA Classification:
LEL: N.D.
UEL: N.D.

Fire Extinguishing Methods: Carbon Dioxide, Dry Chemical, Foam, Water Spray or Fog.

Special Fire Fighting Considerations: Isolate for 1/2 mile if rail or tank truck is involved in a fire. Do not get water inside container. Water spray should be used to keep closed containers cool. Continue to cool container after fire is extinguished. Immediately withdraw if rising sound from venting device is heard or if fire is causing discoloration to the tank.

TOXICOLOGY

Odor:
Physical Contact: Material is extremely destructive to human tissues.
TLV = N.D.   **STEL** = N.D.

Odor Threshold: N.D.

IDLH = N.D.

Routes of Entry and Relative LD50 (or LC50)
Inhalation	N.D.	
Ingestion	3	(1330 mg/kg)
Skin Absorption	N.D.	

RTECS # VV4200000

Symptoms of Exposure: Possible burns to skin, eyes and upper respiratory tract; headache; nausea. Inhalation of vapors may be fatal by causing glottis to spasm and suffocation. Inhalation of vapors may cause chemical pneumonitis and pulmonary edema.

Emergency/First Aid Treatment: Remove to ventilated area; immediately remove any contaminated clothing and wash contaminated areas for 15 minutes using water. Treat supportively and observe for possible shock. If ingested, seek immediate medical aid.

Recommended Clean-Up Procedures

Personal Protection: Level B Ensemble
RCRA Waste # None

Recommended Material N.D.
Reportable Quantities: None

Spills: Remove all potential ignition sources. Dike to contain spill, absorb with non-combustible absorbant and take up using non-sparking tools. Decontaminate spill area using dilute alkaline solution.

Special Emergency Information: May be fatal in inhaled, ingested or absorbed through the skin. Vapors are heavier than air and may travel some distance to an ignition source. Vapors are more dense than air and may settle in low lying areas.

FERRIC CHLORIDE - FeCl₃

Synonyms: iron (III) chloride

CAS Number: 7705-08-0
DOT Number: 1773, 2582

Description: Black Solid
DOT Classification: ORM-B (DOT #1773),
Corrosive Material DOT #2582)

Molecular Weight:
Melting Point: 292°C (558°F)
Boiling Point: 319°C (606°F)

Vapor Density: N. A.
Specific Gravity: 2.9

Vapor Pressure: N. A.
Water Solubility: Soluble

Chemical Incompatibilities or Instabilities: Alkali metals, allyl alcohol, ethylene oxide.

FLAMMABILITY

NFPA Hazard Code:
Flash Point: N. A.
Autoignition Temp: N. A.

NFPA Classification: N. A.
LEL: N. A.
UEL: N. A.

Fire Extinguishing Methods: Use agent suitable for surrounding fire.

Special Fire Fighting Considerations: Water spray should be used to keep closed containers cool. Continue to cool container after fire is extinguished.

TOXICOLOGY

Odor: N. A.
Physical Contact: Material is extremely destructive to human tissues.
TLV = 1 mg/m³   **STEL** = N.D.

Odor Threshold: N.D.

IDLH = N.D.

RTECS # LJ9100000

Routes of Entry and Relative LD₅₀ (or LC₅₀)
Inhalation   N.D.
Ingestion   **3**   (1872 mg/kg)
Skin Absorption   N.D.

Symptoms of Exposure: Possible irritation to skin, eyes and upper respiratory tract; headache; nausea. Inhalation of vapors may be fatal by causing glottis to spasm and suffocation. Inhalation of vapors may cause chemical pneumonitis and pulmonary edema.

Emergency/First Aid Treatment: Remove to ventilated area; immediately remove any contaminated clothing and wash contaminated areas for 15 minutes using water. Treat supportively and observe for possible shock. If ingested, seek immediate medical aid.

Recommended Clean-Up Procedures

Personal Protection: Level B Ensemble
RCRA Waste # None

Recommended Material N.D.
Reportable Quantities: 1000 lb (1000 lb)

Spills: Take up in non-combustible absorbent and dispose.

Special Emergency Information: May be harmful if inhaled, swallowed or absorbed through the skin.

Ferrous chloride (CAS # 7758-94-3, DOT # 1759, 1760, RTECS # NO5400000, RCRZ Waste # N.D., RQ: 100 lb [100 lb]) has similar chemical, physical and toxicological properties.

FLUORINE - F$_2$

Synonyms:

CAS Number: 7782-41-4	**Description:** Pale Yellow Gas
DOT Number: 1045, 9192	**DOT Classification:** Nonflammable Gas

Molecular Weight: 38

Melting Point: -219°C (-365°F)	**Vapor Density:** 1.3	**Vapor Pressure:** GAS
Boiling Point: -188°C (-307°F)	**Specific Gravity:** N. A.	**Water Solubility:** Decomposes

Chemical Incompatibilities or Instabilities: All oxidizable materials, water.

FLAMMABILITY

NFPA Hazard Code: 4-0-4-W	**NFPA Classification:** N. A.
Flash Point: N. A.	**LEL:** N. A.
Autoignition Temp: N. A.	**UEL:** N. A.

Fire Extinguishing Methods: Use agent suitable for surrounding fire.

Special Fire Fighting Considerations: Isolate for 1/2 mile if rail or tank truck is involved in a fire. Stop leak before extinguishing fire if it can safely be done. Water spray should be used to keep closed containers cool. Continue to cool container after fire is extinguished. Fight fire from a distance or protected location, if possible.

TOXICOLOGY

Odor: Pungent   **Odor Threshold:** 0.15 mg/m^3

Physical Contact: Material is extremely destructive to human tissues.

TLV = 1.6 mg/m^3 **STEL** = 3.1 mg/m^3 **IDLH** = 40 mg/m^3

Routes of Entry and Relative LD$_{50}$ (or LC$_{50}$)   **RTECS # LM6475000**

Inhalation	**10**	(300 mg/m^3/H)
Ingestion	N.D.	
Skin Absorption	N.D.	

Symptoms of Exposure: Possible burns or irritation to skin, eyes and upper respiratory tract; headache; nausea. Inhalation of vapors may be fatal by causing glottis to spasm and suffocation. Inhalation of vapors may cause chemical pneumonitis and pulmonary edema.

Emergency/First Aid Treatment: Remove to ventilated area; immediately remove any contaminated clothing and wash contaminated areas for 15 minutes using water. Treat supportively and observe for possible shock. If ingested, seek immediate medical aid.

Recommended Clean-Up Procedures

Personal Protection: Level A Ensemble	**Recommended Material** N.D.
RCRA Waste # P056	**Reportable Quantities:** 10 lb (1 lb)

Spills: Stop leak if it can safely be done. Keep area isolated until vapors have dissipated. Water spray may be use to keep vapors down. If this is done, then treat all materials used or generated and equipment involved as contaminated by hazardous waste.

Special Emergency Information: May be fatal in inhaled, ingested or absorbed through the skin.

4-FLUOROANILINE

Synonyms: p-fluoroaniline

CAS Number: 371-40-4   **Description:** Pale Yellow Liquid
DOT Number: 2941, 2944   **DOT Classification:** Poison B

Molecular Weight: 111.3
Molecular Weight:
Melting Point: °C   (°F)   **Vapor Density:**   **Vapor Pressure:**
Boiling Point: °C   (°F)   **Specific Gravity:**   **Water Solubility:**

Chemical Incompatibilities or Instabilities: Acid chlorides, chloroformates, oxidizers

FLAMMABILITY

NFPA Hazard Code: **NFPA Classification:**
Flash Point: °C   (°F)   **LEL:**   N.D.
Autoignition Temp: °C   (°F)   **UEL:**   N.D.

Fire Extinguishing Methods: Carbon Dioxide, Dry Chemical, Foam, Water Spray or Fog.

Special Fire Fighting Considerations: Structural fire fighter protective clothing will not provide adequate protection. Fight fire from a distance or protected location, if possible. Water spray should be used to keep closed containers cool. Continue to cool container after fire is extinguished. Treat all materials used as contaminated by hazardous waste.

TOXICOLOGY

Odor: **Odor Threshold:** N.D.
Physical Contact: Irritant
TLV = N.D.   **STEL** = N.D.   **IDLH** = N.D.

Routes of Entry and Relative LD$_{50}$ (or LC$_{50}$)   **RTECS #** BY1575000
    Inhalation   N.D.
    Ingestion   5   (417 mg/kg)
    Skin Absorption   N.D.

Symptoms of Exposure: Possible irritation to skin, eyes and upper respiratory tract; headache; nausea. Absorption may lead to the formation of methemoglobin resulting in cyanosis several hours after exposure.

Emergency/First Aid Treatment: Remove to ventilated area; immediately remove any contaminated clothing and wash contaminated areas for 15 minutes using water. Treat supportively and observe for possible shock. If ingested, seek immediate medical aid.

Recommended Clean-Up Procedures

Personal Protection: Level B Ensemble   **Recommended Material**   N.D.

RCRA Waste #   None   **Reportable Quantities:**   None

Spills: Take up in non-combustible absorbant and dispose. Decontaminate spill area using soapy water. Treat all materials used as contaminated by hazardous waste.

Special Emergency Information: May be fatal in inhaled, ingested or absorbed through the skin.

2-Fluoroaniline (CAS # 348-54-9, RTECS # BY1390000) and
3-Flouroaniline (CAS # 372-19-0, RTECS # BY1400000) have similar chemical, physical and toxicological properties.

FLUOROBENZENE - C$_6$H$_5$F

Synonyms: phenyl fluoride

CAS Number: 462-06-6   **Description:** Colorless Liquid
DOT Number: 2387   **DOT Classification:** Flammable Liquid

Molecular Weight: 96.1
Melting Point: -42°C (-44°F)   **Vapor Density:** 3.3   **Vapor Pressure:** 6 (100 mm Hg)
Boiling Point: 85°C (185°F)   **Specific Gravity:** 1.0   **Water Solubility:** Insoluble

Chemical Incompatibilities or Instabilities: Strong oxidizers.

FLAMMABILITY

NFPA Hazard Code: ?-3-0   **NFPA Classification:** Class IB Flammable
Flash Point: -15°C (5°F)   **LEL:** N.D.
Autoignition Temp: °C (°F)   **UEL:** N.D.

Fire Extinguishing Methods: Carbon Dioxide, Dry Chemical, Foam, Water Spray or Fog.

Special Fire Fighting Considerations: Isolate for 1/2 mile if rail or tank truck is involved in a fire. Water spray should be used to keep closed containers cool. Continue to cool container after fire is extinguished. For large fires, if possible, withdraw and allow to burn. Immediately withdraw if rising sound from venting device is heard or if fire is causing discoloration to the tank.

TOXICOLOGY

Odor: **Odor Threshold:** N.D.
Physical Contact: Irritant
TLV = N.D.   **STEL** = N.D.   **IDLH** = N.D.

Routes of Entry and Relative LD$_{50}$ (or LC$_{50}$)   **RTECS # DA0800000**
   Inhalation   5   (26908 mg/m^3/H)
   Ingestion   3   (4399 mg/kg)
   Skin Absorption   N.D.

Symptoms of Exposure: Possible irritation to skin, eyes and upper respiratory tract; headache; nausea.

Emergency/First Aid Treatment: Remove to ventilated area; immediately remove any contaminated clothing and wash contaminated areas for 15 minutes using water. Treat supportively and observe for possible shock. If ingested, seek immediate medical aid.

Recommended Clean-Up Procedures

Personal Protection: Level B Ensemble   **Recommended Material** N.D.

RCRA Waste # None   **Reportable Quantities:** None

Spills: Remove all potential ignition sources. Dike to contain spill, absorb with non-combustible absorbant and take up using non-sparking tools. Decontaminate spill area using soapy water.

Special Emergency Information: May be harmful if inhaled, swallowed or absorbed through the skin.

FLUOROSULFONIC ACID - FO₂SOH

Synonyms: fluorosulfuric acid

CAS Number: 7789-21-1   **Description:** Colorless, Fuming Liquid
DOT Number: 1777   **DOT Classification:** Corrosive Material

Molecular Weight: 100.1
Melting Point: -87°C (-125°F)   **Vapor Density:**   **Vapor Pressure:**
Boiling Point: 166°C (331°F)   **Specific Gravity:** 1.8   **Water Solubility:** Decomposes

Chemical Incompatibilities or Instabilities: Alcohols, water.

FLAMMABILITY

NFPA Hazard Code: **NFPA Classification:**
Flash Point: N.A.   **LEL:** N.A.
Autoignition Temp: N.A.   **UEL:** N.A.

Fire Extinguishing Methods: Carbon Dioxide, Dry Chemical, or, for large fires, flooding quantities of water.

Special Fire Fighting Considerations: Fight fire from a distance or protected location, if possible. Water spray should be used to keep closed containers cool. Continue to cool container after fire is extinguished. Do not put water directly on spilled material.

TOXICOLOGY

Odor: **Odor Threshold:** N.D.
Physical Contact: Material is extremely destructive to human tissues.
TLV = 2.5 mg (F)/m³   **STEL** = N.D.   **IDLH** = 500 mg (F)/m³

Routes of Entry and Relative LD₅₀ (or LC₅₀)   **RTECS #** LP0715000
   Inhalation   N.D.
   Ingestion   N.D.
   Skin Absorption   N.D.

Symptoms of Exposure: Possible burns or irritation to skin, eyes and upper respiratory tract; headache; nausea. Inhalation of vapors may be fatal by causing glottis to spasm and suffocation. Exposure may result in chemical pneumonitis or pulmonary edema.

Emergency/First Aid Treatment: Remove to ventilated area; immediately remove any contaminated clothing and wash contaminated areas for 15 minutes using water. Treat supportively and observe for possible shock. If ingested, seek immediate medical aid.

Recommended Clean-Up Procedures

Personal Protection: Level A Ensemble   **Recommended Material** Saranex (0)

RCRA Waste #   None   **Reportable Quantities:** None

Spills: Dike to contain spill and collect liquid for later disposal. Take up in non-combustible absorbant and dispose. Decontaminate spill area using dilute alkaline solution.

Special Emergency Information: May be fatal in inhaled, ingested or absorbed through the skin.

4-FLUOROTOLUENE

Synonyms: p-fluorotoluene

CAS Number: 352-32-9   **Description:** Colorless Liquid
DOT Number: 2388   **DOT Classification:** Flammable Liquid

Molecular Weight: 110.1
Melting Point: -56°C (-69°F)   **Vapor Density:**   **Vapor Pressure:**   4 (20 mm Hg)
Boiling Point: 116°C (241°F)   **Specific Gravity:** 1.0   **Water Solubility:** Insoluble

Chemical Incompatibilities or Instabilities: Strong oxidizers

FLAMMABILITY

NFPA Hazard Code: **NFPA Classification:**   Class II Combustoble
Flash Point: 40°C (105°F)   **LEL:** N.D.
Autoignition Temp: °C   (°F)   **UEL:** N.D.

Fire Extinguishing Methods: Carbon Dioxide, Dry Chemical, Foam, Water Spray or Fog.

Special Fire Fighting Considerations: Isolate for 1/2 mile if rail or tank truck is involved in a fire. Water spray should be used to keep closed containers cool. Continue to cool container after fire is extinguished. For large fires, if possible, withdraw and allow to burn. Immediately withdraw if rising sound from venting device is heard or if fire is causing discoloration to the tank.

TOXICOLOGY

Odor: **Odor Threshold:** N.D.
Physical Contact: Irritant
TLV = N.D.   **STEL** = N.D.   **IDLH** = N.D.

Routes of Entry and Relative LD$_{50}$ (or LC$_{50}$)   **RTECS # XT2580000**
    Inhalation   N.D.
    Ingestion   N.D.
    Skin Absorption   N.D.

Symptoms of Exposure: Possible irritation to skin, eyes and upper respiratory tract; headache; nausea.

Emergency/First Aid Treatment: Remove to ventilated area; immediately remove any contaminated clothing and wash contaminated areas for 15 minutes using water. Treat supportively and observe for possible shock. If ingested, seek immediate medical aid.

Recommended Clean-Up Procedures

Personal Protection: Level B Ensemble   **Recommended Material**   Polyvinyl Alcohol (-), Teflon (-), Viton (-).

RCRA Waste #   None   **Reportable Quantities:**   None

Spills: Remove all potential ignition sources. Dike to contain spill, absorb with non-combustible absorbant and take up using non-sparking tools. Decontaminate spill area using soapy water.

Special Emergency Information: May be harmful if inhaled, swallowed or absorbed through the skin.

2-Fluorotoluene (CAS # 95-52-3, RTECS # XT2579000) and
3-Fluorotoluene (CAS # 352-70-5, RTECS # XT2578000) have similar chemical, physical and toxicological properties.

FORMALDEHYDE - $H_2C=O$

Synonyms: formalin; methanal; methylene oxide

CAS Number: 50-00-0   **Description:** Colorless 37% - 50% aqueous solution or colorless gas with methanol

DOT Number: 1198, 2209   **DOT Classification:** Colorless Liquid

Molecular Weight: 30.0

Melting Point: (gas) °C (°F)	**Vapor Density:** 1.0	**Vapor Pressure:** Gas
Boiling Point: (gas) -20°C (-3°F)	**Specific Gravity:** 1.1	**Water Solubility:** Soluble

Chemical Incompatibilities or Instabilities: Strong oxidizers.

FLAMMABILITY

NFPA Hazard Code: 3-2-0 (solution), 3-4-0 (gas)   **NFPA Classification:** Class II Combustible (solution)

Flash Point: 56°C (133°F)   **LEL:** 7.0%

Autoignition Temp: 430°C (806°F)   **UEL:** 73.0%

Fire Extinguishing Methods: Carbon Dioxide, Dry Chemical, Foam, Water Spray or Fog.

Special Fire Fighting Considerations: Isolate for 1/2 mile if rail or tank truck is involved in a fire. Water spray should be used to keep closed containers cool. Continue to cool container after fire is extinguished. Immediately withdraw if rising sound from venting device is heard or if fire is causing discoloration to the tank.

TOXICOLOGY
CARCINOGEN

Odor: Pungent   **Odor Threshold:** 0.03 mg/m^3

Physical Contact: Irritant

TLV = 1 mg/m^3 **STEL** = 2 mg/m^3.   **IDLH** = 30 mg/m^3

Routes of Entry and Relative LD$_{50}$ (or LC$_{50}$)   **RTECS #** LP8925000

Inhalation	8	(590 mg/m^3/H)
Ingestion	4	(800 mg/kg)
Skin Absorption	8	(270 mg/kg)

Symptoms of Exposure: Possible irritation to skin, eyes and upper respiratory tract; headache; nausea. May cause an allergic reaction. Exposure may result in chemical pneumonitis or pulmonary edema.

Emergency/First Aid Treatment: Remove to ventilated area; immediately remove any contaminated clothing and wash contaminated areas for 15 minutes using water. Treat supportively and observe for possible shock. If ingested, seek immediate medical aid.

Recommended Clean-Up Procedures

Personal Protection: Level B Ensemble   **Recommended Material**   Butyl Rubber (+), Nitrile Rubber (+), Polyethylene (+), Viton (+)

RCRA Waste # U122   **Reportable Quantities:** 100 lbs (1000 lbs)

Spills: Remove all potential ignition sources. Dike to contain spill, absorb with non-combustible absorbant and take up using non-sparking tools. Decontaminate spill area using water.

Special Emergency Information: May be fatal in inhaled, ingested or absorbed through the skin.

FORMIC ACID - HCO₂H

Synonyms: methanoic acid

CAS Number: 64-18-6
DOT Number: 1779

Description: Colorless Liquid
DOT Classification: Corrosive Material

Molecular Weight: 46.0
Melting Point: 9°C   (47°F)
Boiling Point: 101°C (213°F)

Vapor Density: 1.6
Specific Gravity: 1.22

Vapor Pressure: **5** (22 mm Hg)
Water Solubility: Soluble

Chemical Incompatibilities or Instabilities: Furfuryl alcohol, hydrogen peroxide, nitromethane, phosphorous pentoxide, finely divided metals.

FLAMMABILITY

NFPA Hazard Code: 3-2-0
Flash Point: 50°C (122°F)
Autoignition Temp: 601°C (813°F)

NFPA Classification: Class IIIB Flammable
LEL: 18%
UEL: 57%

Fire Extinguishing Methods: Carbon Dioxide, Dry Chemical, Foam, Water Spray or Fog.

Special Fire Fighting Considerations: Water spray should be used to keep closed containers cool. Continue to cool container after fire is extinguished.

TOXICOLOGY

Odor: Sharp, pungent
Physical Contact: Material is extremely destructive to human tissues.
$TLV = 9.4 \ mg/m^3$ $STEL = 19 \ mg/m^3$

Odor Threshold: $3 \ mg/m^3$

$IDLH = 57 \ mg/m^3$

Routes of Entry and Relative LD₅₀ (or LC₅₀)
    Inhalation   7 $(3750 \ mg/m^3/H)$
    Ingestion   3   (1100 mg/kg)
    Skin Absorption   N.D.

RTECS # LQ4900000

Symptoms of Exposure: Possible burns or irritation to skin, eyes and upper respiratory tract; headache; nausea. Inhalation of vapors may be fatal by causing glottis to spasm and suffocation. Exposure may result in chemical pneumonitis or pulmonary edema.

Emergency/First Aid Treatment: Remove to ventilated area; immediately remove any contaminated clothing and wash contaminated areas for 15 minutes using water. Treat supportively and observe for possible shock. If ingested, seek immediate medical aid.

Recommended Clean-Up Procedures

Personal Protection: Level B Ensemble

RCRAWaste #   U123

Recommended Material   Butyl Rubber (+), Saranex (+)

Reportable Quantities: 5000 lbs (5000lbs)

Spills: Take up in non-combustible absorbant and dispose. Decontaminate spill area using water.

Special Emergency Information: May be fatal in inhaled, ingested or absorbed through the skin.

FUMARYL CHLORIDE - ClC(O)CH=CHC(O)Cl

Synonyms: fumaroyl chloride

CAS Number: 627-63-4
DOT Number: 1780

Description: Clear Amber Liquid
DOT Classification: Corrosive Material

Molecular Weight: 153.0
Melting Point: °C (°F)
Boiling Point: 160°C (320°F)

Vapor Density:
Specific Gravity: 1.4

Vapor Pressure:
Water Solubility: Decomposes

Chemical Incompatibilities or Instabilities: Alcohols, water.

FLAMMABILITY

NFPA Hazard Code: N.D.
Flash Point: 73°C (165°F)
Autoignition Temp: °C (°F)

NFPA Classification: Class IIIB Combustible
LEL: N.D.
UEL: N.D.

Fire Extinguishing Methods: Carbon Dioxide, Dry Chemical, Foam, Water Spray or Fog.

Special Fire Fighting Considerations: Water spray should be used to keep closed containers cool. Continue to cool container after fire is extinguished. May release toxic fumes upon heating (phosgene). Will react with water to produce corrosive and toxic fumes.

TOXICOLOGY

Odor: Sharp, pungent
Physical Contact: Material is extremely destructive to human tissues.
TLV = N.D.   **STEL** = N.D.

Odor Threshold: N.D.

IDLH = N.D.

Routes of Entry and Relative LD$_{50}$ (or LC$_{50}$)
Inhalation	N.D.	
Ingestion	4	(810 mg/kg)
Skin Absorption	4	(1410 mg/kg)

RTECS # LT2800000

Symptoms of Exposure: Possible burns or irritation to skin, eyes and upper respiratory tract; headache; nausea; lachymation. Inhalation of vapors may be fatal by causing glottis to spasm and suffocation. Exposure may result in chemical pneumonitis or pulmonary edema.

Emergency/First Aid Treatment: Remove to ventilated area; immediately remove any contaminated clothing and wash contaminated areas for 15 minutes using water. Treat supportively and observe for possible shock. If ingested, seek immediate medical aid.

Recommended Clean-Up Procedures

Personal Protection: Level B Ensemble   **Recommended Material** N.D.

RCRA Waste # None   **Reportable Quantities:** None

Spills: Dike to contain spill and collect liquid for later disposal. Decontaminate spill area using a dilute alkaline solution.

Special Emergency Information: May be fatal in inhaled, ingested or absorbed through the skin.

FURAN

Synonyms: divinylene oxide; furfuran

CAS Number:	110-00-9	**Description:**	Colorless Liquid
DOT Number:	2389	**DOT Classification:**	Flammable Liquid

Molecular Weight: 68.1

Melting Point:	-86°C (-123°F)	**Vapor Density:**	2.3	**Vapor Pressure:**	9 (493 mm Hg)
Boiling Point:	31°C (88°F)	**Specific Gravity:**	0.9	**Water Solubility:**	Insoluble

Chemical Incompatibilities or Instabilities: Acids, strong oxidizers.

FLAMMABILITY

NFPA Hazard Code:	1-4-1	**NFPA Classification:**	Class IA Flammable
Flash Point:	-35°C (-32°F)	**LEL:**	2.3%
Autoignition Temp:	°C (°F)	**UEL:**	14.3%

Fire Extinguishing Methods: Alcohol Resistant Foam, Carbon Dioxide, Dry Chemical, Water Spray or Fog.

Special Fire Fighting Considerations: Isolate for 1/2 mile if rail or tank truck is involved in a fire. Water spray should be used to keep closed containers cool. Continue to cool container after fire is extinguished. For large fires, if possible, withdraw and allow to burn. Immediately withdraw if rising sound from venting device is heard or if fire is causing discoloration to the tank. Treat all materials used or generated and equipment involved as contaminated by hazardous waste.

TOXICOLOGY

Odor: Ethereal   **Odor Threshold:** N.D.
Physical Contact: Irritant
TLV = N.D.   **STEL** = N.D.   **IDLH** = N.D.

Routes of Entry and Relative LD_{50} (or LC_{50})   **RTECS # LT8524000**
 Inhalation   N.D.
 Ingestion   N.D.
 Skin Absorption   N.D.

Symptoms of Exposure: Possible irritation to skin, eyes and upper respiratory tract; headache; nausea. May act as a narcotic in high concentrations.

Emergency/First Aid Treatment: Remove to ventilated area; immediately remove any contaminated clothing and wash contaminated areas for 15 minutes using water. Treat supportively and observe for possible shock. If ingested, seek immediate medical aid.

Recommended Clean-Up Procedures

Personal Protection:	Level B Ensemble	**Recommended Material**	N.D.
RCRA Waste #	U124	**Reportable Quantities:**	100 lbs (1 lb)

Spills: Remove all potential ignition sources. Dike to contain spill, absorb with non-combustible absorbant and take up using non-sparking tools. Decontaminate spill area using soapy water.

Special Emergency Information: May be harmful if inhaled, swallowed or absorbed through the skin. Vapors are heavier than air and may travel some distance to an ignition source. Vapors are more dense than air and may settle in low lying areas. May form unstable peroxides upon prolonged exposure to air. Treat all materials used or generated and equipment involved as contaminated by hazardous waste.

FURFURAL

Synonyms: furaldehyde

CAS Number: 98-01-1   **Description:** Colorless Liquid
DOT Number: 1199   **DOT Classification:** Combustible Liquid

Molecular Weight: 96.1
Melting Point: -39°C (-38°F)   **Vapor Density:** 3.3   **Vapor Pressure:** 2 (2 mm Hg)
Boiling Point: 162°C (323°F)   **Specific Gravity:** 1.2 **Water Solubility:** Soluble

Chemical Incompatibilities or Instabilities: Acids, alkalies, oxidizers.

FLAMMABILITY

NFPA Hazard Code: 2-2-0   **NFPA Classification:** Class IIIA Combustible
Flash Point: 60°C (140°F) **LEL:**   2.1%
Autoignition Temp: 316°C (601°F) **UEL:**   19.3%

Fire Extinguishing Methods: Alcohol Resistant Foam, Carbon Dioxide, Dry Chemical, Water Spray or Fog.

Special Fire Fighting Considerations: Isolate for 1/2 mile if rail or tank truck is involved in a fire. Water spray should be used to keep closed containers cool. Continue to cool container after fire is extinguished. Immediately withdraw if rising sound from venting device is heard or if fire is causing discoloration to the tank.

TOXICOLOGY

Odor: Almond or bread-like   **Odor Threshold:** 0.004 mg/m^3
Physical Contact: Material is extremely destructive to human tissues.
TLV = 7.9 mg/m^3 **STEL** = N.D.   **IDLH** = 1000 mg/m^3

Routes of Entry and Relative LD$_{50}$ (or LC$_{50}$)   **RTECS # LT7000000**
  Inhalation   8   (2440 mg/m^3/H)
  Ingestion   8   (65 mg/kg)
  Skin Absorption N.D.

Symptoms of Exposure: Possible burns or irritation to skin, eyes and upper respiratory tract; headache; nausea. Inhalation of vapors may be fatal by causing glottis to spasm and suffocation. Exposure may result in chemical pneumonitis or pulmonary edema. May cause an allergic reaction.

Emergency/First Aid Treatment: Remove to ventilated area; immediately remove any contaminated clothing and wash contaminated areas for 15 minutes using water. Treat supportively and observe for possible shock. If ingested, seek immediate medical aid.

Recommended Clean-Up Procedures

Personal Protection: Level B Ensemble   **Recommended Material**   Butyl Rubber

RCRA Waste #   U125   **Reportable Quantities:**   5000 lbs (1000 lbs)

Spills: Remove all potential ignition sources. Dike to contain spill, absorb with non-combustible absorbant and take up using non-sparking tools. Decontaminate spill area using water. Treat all materials used as contaminated by hazardous waste.

Special Emergency Information: May be fatal in inhaled, ingested or absorbed through the skin. May undergo hazardous polymerization.

FURFURYL ALCOHOL

Synonyms: 2-furancarbinol; 2-furanmethanol; furyl alcohol

CAS Number: 98-00-0   **Description:** Pale Yellow Liquid
DOT Number: 2874   **DOT Classification:** Poison B

Molecular Weight: 98.1
Melting Point: -29°C (-20°F)   **Vapor Density:** 3.4   **Vapor Pressure:**   1 (<1 mm Hg)
Boiling Point: 171°C (340°F)   **Specific Gravity:** 1.1   **Water Solubility:**   Soluble

Chemical Incompatibilities or Instabilities: Acids, strong oxidizers.

FLAMMABILITY

NFPA Hazard Code: **1-2-1**   **NFPA Classification:** Class IIIA Combustible
Flash Point: 75°C (167°F)   **LEL:**   1.8%
Autoignition Temp: 491°C (915°F)   **UEL:**   16.3%

Fire Extinguishing Methods: Carbon Dioxide, Dry Chemical, Foam, Water Spray or Fog.

Special Fire Fighting Considerations: Structural fire fighter protective clothing will not provide adequate protection. Fight fire from a distance or protected location, if possible. Treat all materials used as contaminated by hazardous waste. Water spray should be used to keep closed containers cool. Continue to cool container after fire is extinguished.

TOXICOLOGY

Odor: Sweet, ethereal   **Odor Threshold:** 32 mg/m^3
Physical Contact: Irritant
TLV = 40 mg/m^3 **STEL** = 60 mg/m^3 **IDLH** = 1000 mg/m^3

Routes of Entry and Relative LD$_{50}$ (or LC$_{50}$)   **RTECS # LU9100000**
    Inhalation   8   (3800 mg/m^3/H)
    Ingestion   8   (88 mg/kg)
    Skin Absorption   8   (400 mg/kg)

Symptoms of Exposure: Possible irritation to skin, eyes and upper respiratory tract; headache; nausea.

Emergency/First Aid Treatment: Remove to ventilated area; immediately remove any contaminated clothing and wash contaminated areas for 15 minutes using water. Treat supportively and observe for possible shock. If ingested, seek immediate medical aid.

Recommended Clean-Up Procedures

Personal Protection: Level B Ensemble   **Recommended Material**   N.D.

RCRA Waste #   None   **Reportable Quantities:**   None

Spills: Dike to contain spill and collect liquid for later disposal. Decontaminate spill area using water.

Special Emergency Information: May be harmful if inhaled, swallowed or absorbed through the skin..

GASOLINE

Synonyms: motor fuel; petrol

CAS Number: 8006-61-9
DOT Number: 1203, 1257

Description: Colorless Liquid
DOT Classification: Flammable Liquid

Molecular Weight: N.A.

Melting Point:	°C (°F)	**Vapor Density:** 3-4	**Vapor Pressure:**
Boiling Point:	204°C (399°F)	**Specific Gravity:** 0.8	**Water Solubility:** Insoluble

Chemical Incompatibilities or Instabilities: Strong oxidizers.

FLAMMABILITY

NFPA Hazard Code: 1-3-0
Flash Point: -46°C (-50°F)
Autoignition Temp: 280°C (536°F)

NFPA Classification: Class IB Flammable
LEL:
UEL:

Fire Extinguishing Methods: Alcohol Resistant Foam, Carbon Dioxide, Dry Chemical, Foam, Water Spray or Fog.

Special Fire Fighting Considerations: Isolate for 1/2 mile if rail or tank truck is involved in a fire. Water spray from an unmanned device should be used to keep closed containers cool. Continue to cool container after fire is extinguished. For large fires, if possible, withdraw and allow to burn. Immediately withdraw if rising sound from venting device is heard or if fire is causing discoloration to the tank. Treat all materials used or generated and equipment involved as contaminated by hazardous waste.

TOXICOLOGY
SUSPECTED CARCINOGEN

Odor: Gasoline
Physical Contact: Irritant
TLV = 890 mg/m^3 **STEL** = 1480 mg/m^3

Odor Threshold: N.D.

IDLH = N.D.

Routes of Entry and Relative LD$_{50}$ (or LC$_{50}$)
  Inhalation   3   (25,000 mg/m^3/H)
  Ingestion   N.D.
  Skin Absorption   N.D.

RTECS # LX3300000

Symptoms of Exposure: Possible irritation to skin, eyes and upper respiratory tract; headache; nausea. Exposure may result in chemical pneumonitis or pulmonary edema.

Emergency/First Aid Treatment: Remove to ventilated area; immediately remove any contaminated clothing and wash contaminated areas for 15 minutes using water. Treat supportively and observe for possible shock. If ingested, seek immediate medical aid.

Recommended Clean-Up Procedures

Personal Protection: Level A Ensemble
RCRA Waste #   None

Recommended Material Nitrile Rubber (+), Teflon (+), Viton (+)
Reportable Quantities: None

Spills: Remove all potential ignition sources. Dike to contain spill, absorb with non-combustible absorbant and take up using non-sparking tools. Decontaminate spill area using soapy water. Treat all materials used or generated and equipment involved as contaminated by hazardous waste.

Special Emergency Information: May be harmful if inhaled, swallowed or absorbed through the skin. Vapors are heavier than air and may travel some distance to an ignition source. Vapors are more dense than air and may settle in low lying areas.

GERMANE - GeH₄

Synonyms: germanium tetrahydride

CAS Number: 7782-65-2   **Description:** Colorless Gas
DOT Number: 2192   **DOT Classification:** Poison A

Molecular Weight:
Melting Point: -165°C (-265°F)   **Vapor Density:** 2.6   **Vapor Pressure:**   6 (76 mm Hg)
Boiling Point: -89°C (-127°F)   **Specific Gravity:** 1.5   **Water Solubility:** Decomposes

Chemical Incompatibilities or Instabilities: Pyrophoric, Reacts violently with numerous chemical classes.

FLAMMABILITY

NFPA Hazard Code: 4-4-3   **NFPA Classification:**
Flash Point: °C   (°F)   **LEL:**
Autoignition Temp: °C   (°F)   **UEL:**

Fire Extinguishing Methods: Carbon Dioxide, Dry Chemical, Foam, Water Spray or Fog.

Special Fire Fighting Considerations: Stop leak if it can safely be done, otherwise, let small fires burn out. Water spray should be used to keep closed containers cool. Continue to cool container after fire is extinguished. Do not allow water to enter container. For large fires, if possible, withdraw and allow to burn. Immediately withdraw if rising sound from venting device is heard or if fire is causing discoloration to the tank.

DOT Recommended Isolation Zones: Small Spill: 1500 ft   Large Spill: 1500 ft
DOT Recommended Down Wind Take Cover Distance   Small Spill: 5 miles   Large Spill: 5 miles

TOXICOLOGY

Odor: **Odor Threshold:** N.D.
Physical Contact: Material is extremely destructive to human tissues.
TLV = 0.6 mg/m³   **STEL** = N.D.   **IDLH** = N.D.

Routes of Entry and Relative LD₅₀ (or LC₅₀)   **RTECS # LY4900000**
  Inhalation   N.D.
  Ingestion   N.D.
  Skin Absorption   N.D.

Symptoms of Exposure: Possible burns or irritation to skin, eyes and upper respiratory tract; headache; nausea.

Emergency/First Aid Treatment: Remove to ventilated area; immediately remove any contaminated clothing and wash contaminated areas for 15 minutes using water. Treat supportively and observe for possible shock. If ingested, seek immediate medical aid.

Recommended Clean-Up Procedures

Personal Protection: Level A Ensemble   **Recommended Material**   N.D.

RCRA Waste #   None   **Reportable Quantities:**   None

Spills: Remove all potential ignition sources. Stop leak if it can safely be done. Keep area isolated until vapors have dissipated.

Special Emergency Information: May be fatal in inhaled, ingested or absorbed through the skin. Spontaneous explosion may occur without any ignition source.

GUANIDINE NITRATE

Synonyms:

CAS Number: 506-93-4   **Description:** White to Grey Powder
DOT Number: 1467   **DOT Classification:** Oxidizer

Molecular Weight: 122.1
Melting Point: 214°C (417°F)   **Vapor Density:** N.D.   **Vapor Pressure:** N.D.
Boiling Point: °C (°F)   **Specific Gravity:**   **Water Solubility:** Soluble

Chemical Incompatibilities or Instabilities: Metals, oxidizer, heat, shock..

FLAMMABILITY

NFPA Hazard Code: **NFPA Classification:**
Flash Point: N.D.   **LEL:** N.D.
Autoignition Temp: N.D.   **UEL:** N.D.

Fire Extinguishing Methods: WATER ONLY.

Special Fire Fighting Considerations: Structural fire fighter protective clothing will not provide adequate protection. Isolate for 1/2 mile if rail or tank truck is involved in a fire. Water spray should be used to keep closed containers cool. Continue to cool container after fire is extinguished. Fight fire from a distance or protected location, if possible. Use flooding quantities of water. For large fires, if possible, withdraw and allow to burn.

TOXICOLOGY

Odor: **Odor Threshold:** N.D.
Physical Contact: Irritant
TLV = N.D.   **STEL** = N.D.   **IDLH** = N.D.

Routes of Entry and Relative LD$_{50}$ (or LC$_{50}$)   **RTECS #** MF4350000
    Inhalation   N.D.
    Ingestion   N.D.
    Skin Absorption   N.D.

Symptoms of Exposure: Possible irritation to skin, eyes and upper respiratory tract; headache; nausea.

Emergency/First Aid Treatment: Remove to ventilated area; immediately remove any contaminated clothing and wash contaminated areas for 15 minutes using water. Treat supportively and observe for possible shock. If ingested, seek immediate medical aid.

Recommended Clean-Up Procedures

Personal Protection: Level B Ensemble   **Recommended Material** N.D.

RCRA Waste # None   **Reportable Quantities:** None

Spills: Take up and dispose. Decontaminate spill area using water.

Special Emergency Information: May be harmful if inhaled, swallowed or absorbed through the skin. Treat as an explosion hazard.

HAFNIUM - Hf

Synonyms:

CAS Number: 7440-58-6   **Description:** Grey Powder
DOT Number: 2545(dry), 1326(wet)   **DOT Classification:** Flammable Solid

Molecular Weight: 178.5
Melting Point: 2227°C (4041°F)   **Vapor Density:**   N.D.   **Vapor Pressure:**   1 (< 1 mm Hg)
Boiling Point: °C   (°F)   **Specific Gravity:** 13.3   **Water Solubility:**   Insoluble

Chemical Incompatibilities or Instabilities: Halogens, nitrogen, oxidizers, oxygen, phosphorous, sulfur, heat, shock.

FLAMMABILITY

NFPA Hazard Code: N.D.   **NFPA Classification:**
Flash Point: N.A.   **LEL:** N.A.
Autoignition Temp: N.A.   **UEL:** N.A.

Fire Extinguishing Methods: Wet: Carbon Dioxide, Dry Chemical, Foam, Water Spray or Fog. Dry: Dry Chemical, Soda Ash, Lime, Sand.

Special Fire Fighting Considerations: For large fires, if possible, withdraw and allow to burn. If wet, water spray should be used to keep closed containers cool.

TOXICOLOGY

Odor: **Odor Threshold:** N.D.
Physical Contact: Irritant
TLV = 0.5 mg/m^3 **STEL** = N.D.   **IDLH** = N.D.

Routes of Entry and Relative LD$_{50}$ (or LC$_{50}$)   **RTECS # MG4600000**
   Inhalation   N.D.
   Ingestion   N.D.
   Skin Absorption   N.D.

Symptoms of Exposure: Possible irritation to skin, eyes and upper respiratory tract; headache; nausea.

Emergency/First Aid Treatment: Remove to ventilated area; immediately remove any contaminated clothing and wash contaminated areas for 15 minutes using water. Treat supportively and observe for possible shock. If ingested, seek immediate medical aid.

Recommended Clean-Up Procedures

Personal Protection: Level B Ensemble   **Recommended Material**   N.D.

RCRA Waste #   None   **Reportable Quantities:**   None

Spills: Remove all potential ignition sources. Absorb with non-combustible absorbant and take up using non-sparking tools. Decontaminate spill area using soapy water.

Special Emergency Information: May be harmful if inhaled, swallowed or absorbed through the skin. Heat, shock, or friction may cause powdered metal to explode.

HEXACHLOROACETONE - Cl₃CC(O)CCl₃

Synonyms: hexachloro-2-pentanone; perchloroacetone

CAS Number: 116-16-5
DOT Number: 2661

Description: Colorless Liquid
DOT Classification: Poison B

Molecular Weight: 264.7
Melting Point: -2°C (28°F)
Boiling Point: 203°C (397°F)

Vapor Density: 9.2
Specific Gravity: 1.7

Vapor Pressure:
Water Solubility: 1%

Chemical Incompatibilities or Instabilities: Strong oxidizers.

FLAMMABILITY

NFPA Hazard Code:
Flash Point: N.A.
Autoignition Temp: N.A.

NFPA Classification:
LEL: N.A.
UEL: N.A.

Fire Extinguishing Methods: Use agent suitable for surrounding fire.

Special Fire Fighting Considerations: Water spray should be used to keep closed containers cool. Continue to cool container after fire is extinguished.

TOXICOLOGY

Odor:
Physical Contact: Irritant
TLV = N.D.

STEL = N.D.

Odor Threshold: N.D.

IDLH = N.D.

RTECS # UC2100000

Routes of Entry and Relative LD$_{50}$ (or LC$_{50}$)

Inhalation	6	(22,800 mg/m^3/H)
Ingestion	3	(1290 mg/kg)
Skin Absorption	2	(2980 mg/kg)

Symptoms of Exposure: Possible irritation to skin, eyes and upper respiratory tract; headache; nausea.

Emergency/First Aid Treatment: Remove to ventilated area; immediately remove any contaminated clothing and wash contaminated areas for 15 minutes using water. Treat supportively and observe for possible shock. If ingested, seek immediate medical aid.

Recommended Clean-Up Procedures

Personal Protection: Level B Ensemble

Recommended Material N.D.

RCRA Waste # None

Reportable Quantities: None

Spills: Dike to contain spill and collect liquid for later disposal. Decontaminate spill area using water.

Special Emergency Information: May be harmful if inhaled, swallowed or absorbed through the skin.

HEXACHLOROBENZENE - C6Cl6

Synonyms:

CAS Number: 118-74-1   **Description:** White powder
DOT Number: 2729   **DOT Classification:** Poison B

Molecular Weight: 284.8
Melting Point: 231°C (448°F)   **Vapor Density:**   9.8   **Vapor Pressure:**   1(< 1 mm Hg)
Boiling Point: 326°C (619°F)   **Specific Gravity:** 2.4   **Water Solubility:**   Insoluble

Chemical Incompatibilities or Instabilities: Dimethylformamide.

FLAMMABILITY

NFPA Hazard Code: N.D.   **NFPA Classification:**
Flash Point: 242°C (468°F)   **LEL:**   N.D.
Autoignition Temp: °C   (°F)   **UEL:**   N.D.

Fire Extinguishing Methods: Carbon Dioxide, Dry Chemical, Foam, Water Spray or Fog.

Special Fire Fighting Considerations: Water spray should be used to keep closed containers cool. Continue to cool container after fire is extinguished.

TOXICOLOGY
CARCINOGEN

Odor: N.A.   **Odor Threshold:** N.D.
Physical Contact: Irritant
TLV = N.D.   **STEL** = N.D.   **IDLH** = N.D.

Routes of Entry and Relative LD$_{50}$ (or LC$_{50}$)   **RTECS #** DA2975000
    Inhalation   9   (3600 mg/m^3/H)
    Ingestion   1   (10,000 mg/kg)
    Skin Absorption   N.D.

Symptoms of Exposure: Possible irritation to skin, eyes and upper respiratory tract; headache; nausea.

Emergency/First Aid Treatment: Remove to ventilated area; immediately remove any contaminated clothing and wash contaminated areas for 15 minutes using water. Treat supportively and observe for possible shock. If ingested, seek immediate medical aid.

Recommended Clean-Up Procedures

Personal Protection: Level B Ensemble   **Recommended Material**   N.D.

RCRA Waste #   U127   **Reportable Quantities:**   10 lbs (1 lb)

Spills: Take up in non-combustible absorbant and dispose. Decontaminate spill area using soapy water.

Special Emergency Information: May be harmful if inhaled, swallowed or absorbed through the skin.

HEXACHLOROBUTADIENE -$Cl_2C=CClClC=CCl_2$

Synonyms:

CAS Number: 87-68-3
DOT Number: 2279

Description: Colorless Liquid
DOT Classification: Poison B

Molecular Weight: 260.1
Melting Point: -19°C (-2°F)
Boiling Point: 220°C (428°F)

Vapor Density: 9.0
Specific Gravity: 1.7

Vapor Pressure: N.D.
Water Solubility: Insoluble

Chemical Incompatibilities or Instabilities: Perchlorates, strong oxidizers.

FLAMMABILITY

NFPA Hazard Code: 2-1-1
Flash Point: N.A.
Autoignition Temp: 620°C (1130°F)

NFPA Classification:
LEL: N.D.
UEL: N.D.

Fire Extinguishing Methods: Carbon Dioxide, Dry Chemical, Foam, Water Spray or Fog.

Special Fire Fighting Considerations: Structural fire fighter protective clothing will not provide adequate protection. Fight fire from a distance or protected location, if possible. Treat all materials used as contaminated by hazardous waste.

TOXICOLOGY
SUSPECTED CARCINOGEN

Odor:
Physical Contact: Material is extremely destructive to human tissues.
TLV = 0.21 mg/m^3 **STEL** = N.D.

Odor Threshold: N.D.

IDLH = N.D.

Routes of Entry and Relative LD$_{50}$ (or LC$_{50}$)

Inhalation	N.D.	
Ingestion	8	(90 mg/kg)
Skin Absorption	4	(1210 mg/kg)

RTECS # EJ0700000

Symptoms of Exposure: Possible burns or irritation to skin, eyes and upper respiratory tract; headache; nausea. Laryngitis. Inhalation of vapors may be fatal by causing glottis to spasm and suffocation. Exposure may result in chemical pneumonitis or pulmonary edema.

Emergency/First Aid Treatment: Remove to ventilated area; immediately remove any contaminated clothing and wash contaminated areas for 15 minutes using water. Treat supportively and observe for possible shock. If ingested, seek immediate medical aid.

Recommended Clean-Up Procedures

Personal Protection: Level A Ensemble   **Recommended Material** N.D.

RCRA Waste #   U128   **Reportable Quantities :**   1 lb (1 lb)

Large Spills: Dike to contain spill and collect liquid for later disposal. Take up in non-combustible absorbant and dispose. Decontaminate spill area using soapy water. Treat all materials used as contaminated by hazardous waste.

Special Emergency Information: May be harmful if inhaled, swallowed or absorbed through the skin.

HEXACHLOROCYCLOPENTADIENE

Synonyms:

CAS Number: 77-47-4
DOT Number: 2646

Description: Amber Liquid
DOT Classification: Corrosive Material

Molecular Weight: 272.8
Melting Point: 10°C (50°F)
Boiling Point: 239°C (462°F)

Vapor Density: 9.4
Specific Gravity: 1.7

Vapor Pressure: 1 (< 1 mm Hg)
Water Solubility: Insoluble

Chemical Incompatibilities or Instabilities: Strong oxidizers, alkali metals.

FLAMMABILITY

NFPA Hazard Code:
Flash Point: N.A.
Autoignition Temp: N.D.

NFPA Classification:
LEL: N.A.
UEL: N.A.

Fire Extinguishing Methods: Carbon Dioxide, Dry Chemical, Foam, Water Spray or Fog.

Special Fire Fighting Considerations: Structural fire fighter protective clothing will not provide adequate protection. Fight fire from a distance or protected location, if possible. Treat all materials used as contaminated by hazardous waste.

DOT Recommended Isolation Zones: Small Spill: 150 ft   Large Spill: 150 ft
DOT Recommended Down Wind Take Cover Distance   Small Spill: 0.4 miles   Large Spill: 0.4 miles

TOXICOLOGY

Odor:
Physical Contact: Material is extremely destructive to human tissues.
$TLV = 0.11$ mg/m^3 **STEL** = N.D.

Odor Threshold: N.D.

$IDLH = $ N.D.

Routes of Entry and Relative LD$_{50}$ (or LC$_{50}$)
Inhalation   10   (69 mg/m^3/H)
Ingestion   3   (1300 mg/kg)
Skin Absorption   7   (430 mg/kg)

RTECS # GY1225000

Symptoms of Exposure: Possible burns or irritation to skin, eyes and upper respiratory tract; headache; nausea; lachrymation. Laryngitis. Inhalation of vapors may be fatal by causing glottis to spasm and suffocation. Exposure may result in chemical pneumonitis or pulmonary edema.

Emergency/First Aid Treatment: Remove to ventilated area; immediately remove any contaminated clothing and wash contaminated areas for 15 minutes using water. Treat supportively and observe for possible shock. If ingested, seek immediate medical aid.

Recommended Clean-Up Procedures

Personal Protection: Level A Ensemble

Recommended Material Butyl Rubber (+), Polyvinyl Alcohol (+), Viton (+)

RCRA Waste # U130

Reportable Quantities: 10 lbs (1 lb)

Spills: Dike to contain spill and collect liquid for later disposal. Take up in non-combustible absorbant and dispose. Decontaminate spill area using soapy water. Treat all materials used as contaminated by hazardous waste.

Special Emergency Information: May be fatal in inhaled, ingested or absorbed through the skin.

HEXACHLOROETHANE - Cl₃CCCl₃

Synonyms: ethane hexachloride; perchloroethane

CAS Number: 67-72-1

DOT Number: 9037

Description: White powder

DOT Classification: ORM-A

Molecular Weight: 236.7

Melting Point: (s)   187°C   (369°F)

Boiling Point: °C   (°F)

Vapor Density: 6.3

Specific Gravity: 2.1

Vapor Pressure: 1 (< 1 mm Hg)

Water Solubility: Insoluble

Chemical Incompatibilities or Instabilities: Strong oxidizers.

FLAMMABILITY

NFPA Hazard Code:

Flash Point: N.A.

Autoignition Temp: N.D.

NFPA Classification:

LEL: N.A.

UEL: N.A.

Fire Extinguishing Methods: Use agent suitable for surrounding fire.

Special Fire Fighting Considerations: Water spray should be used to keep closed containers cool. Continue to cool container after fire is extinguished.

TOXICOLOGY
SUSPECTED CARCINOGEN

Odor: Camphor-like

Physical Contact: Irritant

TLV = 9.7   **STEL** = N.D.

Odor Threshold: N.D.

IDLH = 2850 mg/m³

Routes of Entry and Relative LD₅₀ (or LC₅₀)

Inhalation	N.D.	
Ingestion	3	(4460 mg/kg)
Skin Absorption	1	(32,000 mg/kg)

RTECS # KI4025000

Symptoms of Exposure: Possible irritation to skin, eyes and upper respiratory tract; headache; nausea.

Emergency/First Aid Treatment: Remove to ventilated area; immediately remove any contaminated clothing and wash contaminated areas for 15 minutes using water. Treat supportively and observe for possible shock. If ingested, seek immediate medical aid.

Recommended Clean-Up Procedures

Personal Protection: Level B Ensemble

RCRA Waste #   U131

Recommended Material   N.D.

Reportable Quantities: 100 lbs (1 lb)

Spills: Take up in non-combustible absorbant and dispose. Decontaminate spill area using soapy water. Treat all materials used as contaminated by hazardous waste.

Special Emergency Information: May be harmful if inhaled, swallowed or absorbed through the skin.

HEXAFLUOROACETONE - F$_3$CC(O)CF$_3$

Synonyms:

CAS Number: 684-16-2   **Description:** Colorless Liquid
DOT Number: 2420   **DOT Classification:** Poison B

Molecular Weight: 166.0
Melting Point: -129°C (-200°F)   **Vapor Density:** 1.7   **Vapor Pressure:** Gas
Boiling Point: °C   (°F)   **Specific Gravity:** N.A.   **Water Solubility:** N.A.

Chemical Incompatibilities or Instabilities: Strong oxidizers.

FLAMMABILITY

NFPA Hazard Code: **NFPA Classification:**
Flash Point: N.A.   **LEL:** N.A.
Autoignition Temp: N.D.   **UEL:** N.A.

Fire Extinguishing Methods: Use agent suitable for surrounding fire.

Special Fire Fighting Considerations: Structural fire fighter protective clothing will not provide adequate protection. Water spray should be used to keep closed containers cool. Continue to cool container after fire is extinguished. Keep area isolated until gas has dispersed.

DOT Recommended Isolation Zones: Small Spill: 1500 ft   Large Spill: 1500 ft
DOT Recommended Down Wind Take Cover Distance   Small Spill: 5 miles   Large Spill: 5 miles

TOXICOLOGY

Odor: **Odor Threshold:** N.D.
Physical Contact: Material is extremely destructive to human tissues.
TLV = 0.68 mg/m^3 **STEL** = N.D.   **IDLH** = N.D.

Routes of Entry and Relative LD$_{50}$ (or LC$_{50}$)   **RTECS # UC2450000**
   Inhalation   **8**   (5500 mg/m^3/H)
   Ingestion   N.D.
   Skin Absorption   N.D.

Symptoms of Exposure: Possible irritation to skin, eyes and upper respiratory tract; headache; nausea. Laryngitis. Inhalation of vapors may be fatal by causing glottis to spasm and suffocation. Exposure may result in chemical pneumonitis or pulmonary edema.

Emergency/First Aid Treatment: Remove to ventilated area; immediately remove any contaminated clothing and wash contaminated areas for 15 minutes using water. Treat supportively and observe for possible shock. If ingested, seek immediate medical aid.

Recommended Clean-Up Procedures

Personal Protection: Level A Ensemble   **Recommended Material**   N.D.

RCRA Waste #   None   **Reportable Quantities:**   None

Spills: Stop leak if it can safely be done. Keep area isolated until vapors have dissipated.

Special Emergency Information: May be fatal in inhaled, ingested or absorbed through the skin. Vapors are more dense than air and may settle in low lying areas.

Hexafluoroacetone hydrate (CAS # 34202-69-2, DOT # 2552, RTECS # UC2700000), a liquid, has similar toxicological properties.

HEXAFLUOROPHOSPHORIC ACID - HPF$_6$

Synonyms:

CAS Number: 16940-81-1   **Description:** Colorless Liquid
DOT Number: 1782   **DOT Classification:** Corrosive Material

Molecular Weight: 146
Melting Point: 31°C (88°F)   **Vapor Density:** N.D.   **Vapor Pressure:** N.D.
Boiling Point: °C (°F)   **Specific Gravity:** 1.7   **Water Solubility:** Soluble

Chemical Incompatibilities or Instabilities: Alkalis.

FLAMMABILITY

NFPA Hazard Code: **NFPA Classification:**
Flash Point: N.A.   **LEL:** N.A.
Autoignition Temp: N.A.   **UEL:** N.A.

Fire Extinguishing Methods: Carbon Dioxide, Dry Chemical, Foam, Water Spray or Fog.

Special Fire Fighting Considerations: Structural fire fighter protective clothing will not provide adequate protection. Water spray should be used to keep closed containers cool. Continue to cool container after fire is extinguished. May release toxic fumes upon heating (phosphorous oxides, phosphine).

TOXICOLOGY

Odor: **Odor Threshold:** N.D.
Physical Contact: Material is extremely destructive to human tissues.
TLV = 2.5 mg (F)/m^3 **STEL** = N.D.   **IDLH** = 500 mg (F)/m^3

Routes of Entry and Relative LD$_{50}$ (or LC$_{50}$)   **RTECS # SY7115000**
    Inhalation   N.D.
    Ingestion   N.D.
    Skin Absorption   N.D.

Symptoms of Exposure: Possible burns or irritation to skin, eyes and upper respiratory tract; headache; nausea. Laryngitis. Inhalation of vapors may be fatal by causing glottis to spasm and suffocation. Exposure may result in chemical pneumonitis or pulmonary edema.

Emergency/First Aid Treatment: Remove to ventilated area; immediately remove any contaminated clothing and wash contaminated areas for 15 minutes using water. Treat supportively and observe for possible shock. If ingested, seek immediate medical aid.

Recommended Clean-Up Procedures

Personal Protection: Level B Ensemble   **Recommended Material** N.D.

RCRA Waste # None   **Reportable Quantities:** None

Spills: Dike to contain spill and collect liquid for later disposal. Decontaminate spill area using water.

Special Emergency Information: May be fatal in inhaled, ingested or absorbed through the skin.

HEXALDEHYDE - CH$_3$(CH$_2$)$_4$CHO

Synonyms: hexanal; caproaldehyde

CAS Number: 66-25-1	**Description:** Colorless Liquid
DOT Number: 1207	**DOT Classification:** Flammable Liquid

Molecular Weight: 100.2

Melting Point: -56°C (-69°F)	**Vapor Density:** 3.5	**Vapor Pressure:** **3** (8 mm Hg)
Boiling Point: 131°C (268°F)	**Specific Gravity:** 0.8	**Water Solubility:** < 1%

Chemical Incompatibilities or Instabilities: Oxidizers.

FLAMMABILITY

NFPA Hazard Code: **2-3-1**	**NFPA Classification:** Class IC Flammable
Flash Point: 32°C (90°F)	**LEL:** N.D.
Autoignition Temp: N.D.	**UEL:** N.D.

Fire Extinguishing Methods: Alcohol Resistant Foam, Carbon Dioxide, Dry Chemical, Water Spray or Fog.

Special Fire Fighting Considerations: Isolate for 1/2 mile if rail or tank truck is involved in a fire. Water spray should be used to keep closed containers cool. Continue to cool container after fire is extinguished. For large fires, if possible, withdraw and allow to burn. Immediately withdraw if rising sound from venting device is heard or if fire is causing discoloration to the tank.

TOXICOLOGY

Odor:	**Odor Threshold:** N.D.
Physical Contact: Irritant	
TLV = N.D. **STEL** = N.D.	**IDLH** = N.D.

Routes of Entry and Relative LD$_{50}$ (or LC$_{50}$)   **RTECS # MN7175000**

Inhalation	N.D.	
Ingestion	3	(4890 mg/kg)
Skin Absorption	N.D.	

Symptoms of Exposure: Possible irritation to skin, eyes and upper respiratory tract; headache; nausea.

Emergency/First Aid Treatment: Remove to ventilated area; immediately remove any contaminated clothing and wash contaminated areas for 15 minutes using water. Treat supportively and observe for possible shock. If ingested, seek immediate medical aid.

Recommended Clean-Up Procedures

Personal Protection: Level B Ensemble	**Recommended Material** N.D.
RCRA Waste # None	**Reportable Quantities:** None

Spills: Remove all potential ignition sources. Dike to contain spill, absorb with non-combustible absorbant and take up using non-sparking tools. Decontaminate spill area using soapy water.

Special Emergency Information: May be harmful if inhaled, swallowed or absorbed through the skin. Vapors are heavier than air and may travel some distance to an ignition source. Vapors are more dense than air and may settle in low lying areas.

n-Heptaldehyde (CAS # 111-71-7, DOT # 3056, RTECS # MI6900000) has similar chemical, physical and toxicological properties.

HEXAMINE

Synonyms: hexamethylenetetramine; formamine; hexamethyleneamine

CAS Number: 100-97-0	**Description:** White powder
DOT Number: 1328	**DOT Classification:** IMO

Molecular Weight:

Melting Point: (s) 280°C (536°F)	**Vapor Density:** N.D.	**Vapor Pressure:** N.D.
Boiling Point: °C (°F)	**Specific Gravity:** 1.3	**Water Solubility:** Soluble

Chemical Incompatibilities or Instabilities: Formaldehyde released upon contact with acids.

FLAMMABILITY

NFPA Hazard Code:	**NFPA Classification:**
Flash Point: 250°C (482°F)	**LEL:** N.A.
Autoignition Temp: N.A.	**UEL:** N.A.

Fire Extinguishing Methods: Carbon Dioxide, Dry Chemical, Foam, Water Spray or Fog.

Special Fire Fighting Considerations: Water spray should be used to keep closed containers cool. Continue to cool container after fire is extinguished. For large fires, if possible, withdraw and allow to burn.

TOXICOLOGY
QUESTIONABLE CARCINOGEN

Odor:	**Odor Threshold:** N.D.
Physical Contact: Irritant	
TLV = N.D. **STEL** = N.D.	**IDLH** = N.D.

Routes of Entry and Relative LD$_{50}$ (or LC$_{50}$)   **RTECS # MN4725000**

Inhalation	N.D.
Ingestion	N.D.
Skin Absorption	N.D.

Symptoms of Exposure: Possible irritation to skin, eyes and upper respiratory tract; headache; nausea.

Emergency/First Aid Treatment: Remove to ventilated area; immediately remove any contaminated clothing and wash contaminated areas for 15 minutes using water. Treat supportively and observe for possible shock. If ingested, seek immediate medical aid.

Recommended Clean-Up Procedures

Personal Protection: Level B Ensemble	**Recommended Material** N.D.
RCRA Waste # None	**Reportable Quantities:** None

Spills: Take up in non-combustible absorbant and dispose. Decontaminate spill area using water.

Special Emergency Information: May be harmful if inhaled, swallowed or absorbed through the skin.

HEXANE - $CH_3(CH_2)_4CH_3$

Synonyms: n-hexane

CAS Number: 110-54-3   **Description:** Colorless Liquid
DOT Number: 1208   **DOT Classification:** Flammable Liquid

Molecular Weight: 86.2
Melting Point: -95°C (139°F)   **Vapor Density:** 3.0   **Vapor Pressure:** 7 (125 mm Hg)
Boiling Point: 69°C (156°F)   **Specific Gravity:** 0.7   **Water Solubility:** Insoluble

Chemical Incompatibilities or Instabilities: Strong oxidizers.

FLAMMABILITY

NFPA Hazard Code: 1-3-0   **NFPA Classification:** Class IB Flammable
Flash Point: -22°C   (-7°F)   **LEL:**   1.1%
Autoignition Temp: 225°C (437°F)   **UEL:**   7.5%

Fire Extinguishing Methods: Carbon Dioxide, Dry Chemical, Foam, Water Spray or Fog.

Special Fire Fighting Considerations: Isolate for 1/2 mile if rail or tank truck is involved in a fire. Water spray should be used to keep closed containers cool. Continue to cool container after fire is extinguished. For large fires, if possible, withdraw and allow to burn. Immediately withdraw if rising sound from venting device is heard or if fire is causing discoloration to the tank.

TOXICOLOGY

Odor: Gasoline-like   **Odor Threshold:** 230 mg/m^3
Physical Contact: Irritant
TLV = 180 mg/m^3 **STEL** = N.D.   **IDLH** = 18,000 mg/m^3

Routes of Entry and Relative LD$_{50}$ (or LC$_{50}$)   **RTECS # MN9275000**
    Inhalation   N.D.
    Ingestion   1   (28,710 mg/kg)
    Skin Absorption   N.D.

Symptoms of Exposure: Possible irritation to skin, eyes and upper respiratory tract; headache; nausea. May act as a narcotic or neurotoxin in high concentrations.

Emergency/First Aid Treatment: Remove to ventilated area; immediately remove any contaminated clothing and wash contaminated areas for 15 minutes using water. Treat supportively and observe for possible shock. If ingested, seek immediate medical aid.

Recommended Clean-Up Procedures

Personal Protection: Level B Ensemble   **Recommended Material** Nitrile Rubber (+), Polyvinyl Alcohol (+),
                                                                         Teflon (+), Viton (+)

RCRA Waste #   None   **Reportable Quantities:** None

Spills: Remove all potential ignition sources. Dike to contain spill, absorb with non-combustible absorbant and take up using non-sparking tools. Vent to dissipate vapors.

Special Emergency Information: May be harmful if inhaled, swallowed or absorbed through the skin. Vapors are heavier than air and may travel some distance to an ignition source. Vapors are more dense than air and may settle in low lying areas.

Pentane (CAS # 109-66-0, DOT # 1265, RTECS # RZ9450000) and
heptane (CAS # 142-82-5, DOT # 1206, RTECS # MI7700000) have similar chemical, physical and toxicological properties.

HEXANOIC ACID - CH$_3$(CH$_2$)$_3$CH$_2$COOH

Synonyms: caproic acid

CAS Number: 142-62-1   **Description:** Colorless Liquid
DOT Number: 1706   **DOT Classification:** Corrosive Material

Molecular Weight: 116.2
Melting Point: -4°C   (25°F)   **Vapor Density:** 4.0   **Vapor Pressure:** 1 (<1 mm Hg)
Boiling Point: 204°C   (400°F)   **Specific Gravity:** 0.9   **Water Solubility:** Insoluble

Chemical Incompatibilities or Instabilities: Strong oxidizers.

FLAMMABILITY

NFPA Hazard Code: 2-1-0   **NFPA Classification:** Class IIIB Combustible
Flash Point: 104°C (220°F)   **LEL:** N.A.
Autoignition Temp: 380°C (716°F)   **UEL:** N.A.

Fire Extinguishing Methods: Carbon Dioxide, Dry Chemical, Foam, Water Spray or Fog.

Special Fire Fighting Considerations: Water spray should be used to keep closed containers cool. Continue to cool container after fire is extinguished.

TOXICOLOGY

Odor: Limburger Cheese   **Odor Threshold:** N.D.
Physical Contact: Material is extremely destructive to human tissues.
TLV = N.D.   **STEL** = N.D.   **IDLH** = N.D.

Routes of Entry and Relative LD$_{50}$ (or LC$_{50}$)   **RTECS #** MO5250000
   Inhalation   N.D.
   Ingestion   3   (3000 mg/kg)
   Skin Absorption   6   (630 mg/kg)

Symptoms of Exposure: Possible burns or irritation to skin, eyes and upper respiratory tract; headache; nausea. Laryngitis. Inhalation of vapors may be fatal by causing glottis to spasm and suffocation. Exposure may result in chemical pneumonitis or pulmonary edema.

Emergency/First Aid Treatment: Remove to ventilated area; immediately remove any contaminated clothing and wash contaminated areas for 15 minutes using water. Treat supportively and observe for possible shock. If ingested, seek immediate medical aid.

Recommended Clean-Up Procedures

Personal Protection: Level B Ensemble   **Recommended Material**   N.D.

RCRA Waste #   None   **Reportable Quantities:**   None

Spills: Dike to contain spill and collect liquid for later disposal. Decontaminate spill area using soapy water.

Special Emergency Information: May be harmful if inhaled, swallowed or absorbed through the skin.

Pentanoic acid (CAS # 109-52-4, DOT # 1760, RTECS # YV6100000) and
heptanoic acid (CAS # 111-14-8, RTECS # MJ1575000) have similar chemical, physical and toxicological properties.

n-HEXANOL - $CH_3(CH_2)_4CH_2OH$

Synonyms: n-hexyl alcohol; 1-hexanol; amylcarbinol

CAS Number: 111-27-3   **Description:** Colorless Liquid
DOT Number: 2282   **DOT Classification:** Flammable Liquid

Molecular Weight: 102.2
Melting Point: -45°C (-49°F)   **Vapor Density:** 3.5   **Vapor Pressure:**   1 (1 mm Hg)
Boiling Point: 155°C (311°F)   **Specific Gravity:** 0.8   **Water Solubility:**   1%

Chemical Incompatibilities or Instabilities: Strong oxidizers.

FLAMMABILITY

NFPA Hazard Code: 1-2-0   **NFPA Classification:**   Class IIIA Combustible
Flash Point: 63°C (145°F)   **LEL:** N.D.
Autoignition Temp: N.D.   **UEL:** N.D.

Fire Extinguishing Methods: Alcohol Resistant Foam, Carbon Dioxide, Dry Chemical, Water Spray or Fog.

Special Fire Fighting Considerations: Isolate for 1/2 mile if rail or tank truck is involved in a fire. Water spray should be used to keep closed containers cool. Continue to cool container after fire is extinguished. For large fires, if possible, withdraw and allow to burn. Immediately withdraw if rising sound from venting device is heard or if fire is causing discoloration to the tank.

TOXICOLOGY

Odor: Sweet   **Odor Threshold:** N.A.
Physical Contact: Irritant
TLV = N.D.   **STEL** = N.D.   **IDLH** = N.D.

Routes of Entry and Relative LD$_{50}$ (or LC$_{50}$)   **RTECS # MQ4025000**
   Inhalation   N.D.
   Ingestion   4   (740 mg/kg)
   Skin Absorption   2   (3100 mg/kg)

Symptoms of Exposure: Possible irritation to skin, eyes and upper respiratory tract; headache; nausea. Eye specific damage has been reported.

Emergency/First Aid Treatment: Remove to ventilated area; immediately remove any contaminated clothing and wash contaminated areas for 15 minutes using water. Treat supportively and observe for possible shock. If ingested, seek immediate medical aid.

Recommended Clean-Up Procedures

Personal Protection: Level B Ensemble   **Recommended Material**   N.D.

RCRA Waste #   None   **Reportable Quantities:**   None

Spills: Dike to contain spill and collect liquid for later disposal. Take up in non-combustible absorbant and dispose. Decontaminate spill area using water.

Special Emergency Information: May be harmful if inhaled, swallowed or absorbed through the skin.

1-HEXENE - CH$_2$=CH$_2$CH$_2$CH$_2$CH$_2$CH$_3$

Synonyms:

CAS Number: 592-41-6	**Description:** Colorless Liquid	
DOT Number: 2370	**DOT Classification:** Flammable Liquid	

Molecular Weight:

Melting Point:	°C	(°F)	**Vapor Density:** 3.0	**Vapor Pressure:**	**8** (310 mm Hg)
Boiling Point:	63°C	(146°F)	**Specific Gravity:** 0.7	**Water Solubility:**	Insoluble

Chemical Incompatibilities or Instabilities: Oxidizers.

FLAMMABILITY

NFPA Hazard Code: 1-3-0	**NFPA Classification:** Class IB Flammable	
Flash Point: -26°C (-15°F)	**LEL:** 1.2%	
Autoignition Temp: 253°C (487°F)	**UEL:** 6.9%	

Fire Extinguishing Methods: Carbon Dioxide, Dry Chemical, Foam, Water Spray or Fog.

Special Fire Fighting Considerations: Isolate for 1/2 mile if rail or tank truck is involved in a fire. Water spray should be used to keep closed containers cool. Continue to cool container after fire is extinguished. For large fires, if possible, withdraw and allow to burn. Immediately withdraw if rising sound from venting device is heard or if fire is causing discoloration to the tank.

TOXICOLOGY

Odor:	**Odor Threshold:** N.D.
Physical Contact: Irritant	
TLV = N.D. **STEL** = N.D.	**IDLH** = N.D.

Routes of Entry and Relative LD$_{50}$ (or LC$_{50}$)		**RTECS # MP6600100**
Inhalation	N.D.	
Ingestion	N.D.	
Skin Absorption	N.D.	

Symptoms of Exposure: Possible irritation to skin, eyes and upper respiratory tract; headache; nausea.

Emergency/First Aid Treatment: Remove to ventilated area; immediately remove any contaminated clothing and wash contaminated areas for 15 minutes using water. Treat supportively and observe for possible shock. If ingested, seek immediate medical aid.

Recommended Clean-Up Procedures

Personal Protection: Level B Ensemble	**Recommended Material** N.D.
RCRA Waste # None	**Reportable Quantities:** None

Spills: Remove all potential ignition sources. Dike to contain spill, absorb with non-combustible absorbant and take up using non-sparking tools. Decontaminate spill area using soapy water.

Special Emergency Information: May be harmful if inhaled, swallowed or absorbed through the skin.

Heptene (CAS # 25339-56-4, DOT # 2278, RTECS # MJ8850000) has similar chemical, physical and toxicological properties.

HYDRAZINE - H$_2$NNH$_2$

Synonyms: diamide; diamine

CAS Number: 302-01-2   **Description:** Colorless Fuming Liquid or White powder
DOT Number: 2029, 2030   **DOT Classification:** Flammable Liquid (2029), Corrosive Material (2030)

Molecular Weight: 32.0
Melting Point: 2°C (36°F)   **Vapor Density:** 1.1   **Vapor Pressure:** 4 (10 mm Hg)
Boiling Point: 113°C (236°F)   **Specific Gravity:** 1.0   **Water Solubility:** Soluble

Chemical Incompatibilities or Instabilities: Oxidizers, copper, zinz, organics, ordinary combustibles.

FLAMMABILITY

NFPA Hazard Code: 3-3-3   **NFPA Classification:** Class IC Flammable
Flash Point: 38°C (100°F)   **LEL:**   4.7%
Autoignition Temp: 270°C (518°F)   **UEL:**   100%

Fire Extinguishing Methods: Alcohol Resistant Foam, Carbon Dioxide, Dry Chemical, Water Spray or Fog.

Special Fire Fighting Considerations: Structural fire fighter protective clothing will not provide adequate protection. Isolate for 1/2 mile if rail or tank truck is involved in a fire. Water spray should be used to keep closed containers cool. Continue to cool container after fire is extinguished. Immediately withdraw if rising sound from venting device is heard or if fire is causing discoloration to the tank. Treat all materials used as contaminated by hazardous waste.

TOXICOLOGY
CARCINOGEN

Odor: Ammonia   **Odor Threshold:** 4 mg/m^3
Physical Contact: Material is extremely destructive to human tissues.
TLV = 0.13 mg/m^3 **STEL** = N.D.   **IDLH** = 105 mg/m^3

Routes of Entry and Relative LD$_{50}$ (or LC$_{50}$)   **RTECS # MU7175000**
   Inhalation   6   (3030 mg/m^3/H)
   Ingestion   8   (60 mg/kg)
   Skin Absorption   9   (91 mg/kg)

Symptoms of Exposure: Possible burns or irritation to skin, eyes and upper respiratory tract; headache; nausea. Laryngitis. Inhalation of vapors may be fatal by causing glottis to spasm and suffocation. Exposure may result in chemical pneumonitis or pulmonary edema.

Emergency/First Aid Treatment: Remove to ventilated area; immediately remove any contaminated clothing and wash contaminated areas for 15 minutes using water. Treat supportively and observe for possible shock. If ingested, seek immediate medical aid.

Recommended Clean-Up Procedures

Personal Protection: Level A Ensemble   **Recommended Material**   Butyl Rubber (+), Nitrile Rubber (+),
                                                                               Polyvinyl Alcohol (+), Teflon (+), Saranex (+)

RCRA Waste #   U133   **Reportable Quantities:**   1 lb (1lb)

Spills: Clean up only under expert supervision.

Special Emergency Information: May be fatal in inhaled, ingested or absorbed through the skin.

HYDROBROMIC ACID SOLUTIONS (> 49%) - HBr

Synonyms:

CAS Number: 10035-10-6	**Description:** Colorless Liquid	
DOT Number: 1788	**DOT Classification:** Corrosive Material	

Molecular Weight: 80.9

Melting Point: -89°C (-127°F)	**Vapor Density:**	**Vapor Pressure:** Gas
Boiling Point: 74°C (165°F)	**Specific Gravity:**	**Water Solubility:** Soluble

Chemical Incompatibilities or Instabilities: Halogens, iron (II) oxide, ammonia, ozone

FLAMMABILITY

NFPA Hazard Code: 3-0-0	**NFPA Classification:**	
Flash Point: N.A.	**LEL:** N.A.	
Autoignition Temp: N.A.	**UEL:** N.A.	

Fire Extinguishing Methods: Use agent suitable for surrounding fire.

Special Fire Fighting Considerations: Water spray should be used to keep closed containers cool. Continue to cool container after fire is extinguished.

TOXICOLOGY

Odor: Pungent, Sharp   **Odor Threshold:** N.D.

Physical Contact: Material is extremely destructive to human tissues.

TLV = 9.9 mg/m^3 **STEL** = N.D.   **IDLH** = 150 mg/m^3

Routes of Entry and Relative LD$_{50}$ (or LC$_{50}$)   **RTECS # MW3850000**

Inhalation	6	(9700 mg/m^3/H)
Ingestion	N.D.	
Skin Absorption	N.D.	

Symptoms of Exposure: Possible burns or irritation to skin, eyes and upper respiratory tract; headache; nausea. Laryngitis. Inhalation of vapors may be fatal by causing glottis to spasm and suffocation. Exposure may result in chemical pneumonitis or pulmonary edema.

Emergency/First Aid Treatment: Remove to ventilated area; immediately remove any contaminated clothing and wash contaminated areas for 15 minutes using water. Treat supportively and observe for possible shock. If ingested, seek immediate medical aid.

Recommended Clean-Up Procedures

Personal Protection: Level B Ensemble	**Recommended Material** Butyl Rubber (+), Neoprene (+), Saranex (+)
RCRA Waste # None	**Reportable Quantities:** None

Spills: Dike to contain spill and collect liquid for later disposal. Decontaminate spill area using a dilute alkaline solution.

Special Emergency Information: May be fatal in inhaled, ingested or absorbed through the skin. Vapors are more dense than air and may settle in low lying areas.

Hydrochloric acid (CAS # 7647-01-0, DOT # 1789, RTECS # MW4025000) has similar chemical, physical and toxicological properties.

HYDROGEN - H$_2$

Synonyms:

CAS Number: 1333-74-0   **Description:** Colorless Gas
DOT Number: 1049 (gas), 1966 (cryogenic liquid)   **DOT Classification:** Flammable Gas

Molecular Weight: 2.0
Melting Point: -259°C (-434°F)   **Vapor Density:**   0.07   **Vapor Pressure:**   Gas
Boiling Point: -253°C (-423°F)   **Specific Gravity:** N.A.   **Water Solubility:**   Insoluble

Chemical Incompatibilities or Instabilities: Oxidizers, halogens, reactive metals.

FLAMMABILITY

NFPA Hazard Code: 0-4-0 (Gas) 3-4-0 (Liquid)   **NFPA Classification:**
Flash Point: Gas   **LEL:**   4.0%
Autoignition Temp: 500°C (932°F)   **UEL:**   75.0%

Fire Extinguishing Methods: Carbon Dioxide, Dry Chemical, Water Spray or Fog.

Special Fire Fighting Considerations: Isolate for 1/2 mile if rail or tank truck is involved in a fire. Stop leak if it can safely be done. If this cannot be done, then isolate from other combustible materials and allow to burn. Water spray should be used to keep closed containers cool. Continue to cool container after fire is extinguished. For large fires, if possible, withdraw and allow to burn. Immediately withdraw if rising sound from venting device is heard or if fire is causing discoloration to the tank.

TOXICOLOGY

Odor: None   **Odor Threshold:** N.D.
Physical Contact: Frostbite from contact with cryogenic liquid.
TLV = N.D.   **STEL** = N.D.   **IDLH** = N.D.

Routes of Entry and Relative LD$_{50}$ (or LC$_{50}$)   **RTECS # MW8900000**
    Inhalation   N.D.
    Ingestion   N.D.
    Skin Absorption   N.D.

Symptoms of Exposure: A simple asphyxiant.

Emergency/First Aid Treatment: Remove to ventilated area; immediately remove any contaminated clothing and wash contaminated areas for 15 minutes using water. Treat supportively and observe for possible shock. If ingested, seek immediate medical aid.

Recommended Clean-Up Procedures

Personal Protection: Level B Ensemble   **Recommended Material**   N.D.

RCRA Waste #   None   **Reportable Quantities:**   None

Spills: Remove all potential ignition sources. Stop leak if it can safely be done. Keep area isolated until vapors have dissipated.

Special Emergency Information: May be harmful if inhaled, swallowed or absorbed through the skin.

HYDROGEN BROMIDE - HBr

Synonyms:

CAS Number: 10035-10-6	**Description:** Colorless Gas	
DOT Number: 1048	**DOT Classification:** Corrosive Material	

Molecular Weight: 80.9

Melting Point: -86°C (-124°F)	**Vapor Density:**	**Vapor Pressure:** Gas	
Boiling Point: -66°C (-88°F)	**Specific Gravity:**	**Water Solubility:** Soluble	

Chemical Incompatibilities or Instabilities: Halogens, iron (II) oxide, ammonia, ozone.

FLAMMABILITY

NFPA Hazard Code: 0-3-0	**NFPA Classification:**	
Flash Point: N.A.	**LEL:** N.A.	
Autoignition Temp: N.A.	**UEL:** N.A.	

Fire Extinguishing Methods: Use agent suitable for surrounding fire.

Special Fire Fighting Considerations: Structural fire fighter protective clothing will not provide adequate protection. Water spray should be used to keep closed containers cool. Continue to cool container after fire is extinguished. Keep area isolated until gas has dispersed.

DOT Recommended Isolation Zones:	Small Spill: 1500 ft	Large Spill: 1500 ft
DOT Recommended Down Wind Take Cover Distance	Small Spill: 5 miles	Large Spill: 5 miles

TOXICOLOGY

Odor: Sharp, Irritating   **Odor Threshold:** N.D.

Physical Contact: Material is extremely destructive to human tissues.

TLV = N.D.   **STEL** = N.D.   **IDLH** = N.D.

Routes of Entry and Relative LD$_{50}$ (or LC$_{50}$)   **RTECS # MW3850000**

Inhalation	6	(9700 mg/m^3/H)
Ingestion	N.D.	
Skin Absorption	N.D.	

Symptoms of Exposure: Possible burns or irritation to skin, eyes and upper respiratory tract; headache; nausea. Laryngitis. Inhalation of vapors may be fatal by causing glottis to spasm and suffocation. Exposure may result in chemical pneumonitis or pulmonary edema.

Emergency/First Aid Treatment: Remove to ventilated area; immediately remove any contaminated clothing and wash contaminated areas for 15 minutes using water. Treat supportively and observe for possible shock. If ingested, seek immediate medical aid.

Recommended Clean-Up Procedures

Personal Protection: Level B Ensemble	**Recommended Material** Butyl Rubber (+), Neoprene (+), Sranex (+)
RCRA Waste # None	**Reportable Quantities:** None

Spills: Stop leak if it can safely be done. Keep area isolated until vapors have dispersed.

Special Emergency Information: May be fatal in inhaled, ingested or absorbed through the skin.

HYDROGEN CHLORIDE - HCl

Synonyms:

CAS Number:	7647-01-0	**Description:** Colorless Gas
DOT Number:	1050 (gas), 2186 (cryogenic liquid)	**DOT Classification:** Corrosive Material

Molecular Weight: 36.5

Melting Point:	-114°C (-174°F)	**Vapor Density:** 1.3	**Vapor Pressure:**	Gas
Boiling Point:	-85°C (-121°F)	**Specific Gravity:**	**Water Solubility:**	Soluble

Chemical Incompatibilities or Instabilities: Halogens, insaturated hydrocarbons, carbides, ammonium hydroxide, aluminum, reactive metals, strong oxidizers.

FLAMMABILITY

NFPA Hazard Code:	3-0-0	**NFPA Classification:**	
Flash Point:	N.A.	**LEL:**	N.A.
Autoignition Temp:	N.A.	**UEL:**	N.A.

Fire Extinguishing Methods: Use agent suitable for surrounding fire.

Special Fire Fighting Considerations: Structural fire fighter protective clothing will not provide adequate protection. Water spray should be used to keep closed containers cool. Continue to cool container after fire is extinguished. Keep area isolated until vapors have dispersed.

DOT Recommended Isolation Zones:		Small Spill: 1200 ft	Large Spill: 1500 ft
DOT Recommended Down Wind Take Cover Distance		Small Spill: 4 miles	Large Spill: 5 miles

TOXICOLOGY

Odor: Sharp, Pungent	**Odor Threshold:** 0.4 mg/m^3

Physical Contact: Material is extremely destructive to human tissues.

TLV = 7.5 mg/m^3	**STEL** = N.D.	**IDLH** = 150 mg/m^3

Routes of Entry and Relative LD$_{50}$ (or LC$_{50}$)   **RTECS # MW4025000**

Inhalation	6	(4750 mg/m^3/H)
Ingestion	N.D.	
Skin Absorption	N.D.	

Symptoms of Exposure: Possible burns or irritation to skin, eyes and upper respiratory tract; headache; nausea. Laryngitis. Inhalation of vapors may be fatal by causing glottis to spasm and suffocation. Exposure may result in chemical pneumonitis or pulmonary edema.

Emergency/First Aid Treatment: Remove to ventilated area; immediately remove any contaminated clothing and wash contaminated areas for 15 minutes using water. Treat supportively and observe for possible shock. If ingested, seek immediate medical aid.

Recommended Clean-Up Procedures

Personal Protection:	Level B Ensemble	**Recommended Material**	Butyl Rubber (+), Neoprene (+), Saranex (+)
RCRA Waste #	None	**Reportable Quantities:**	5000 lb (5000 lb)

Spills: Stop leak if it can safely be done. Keep area isolated until vapors have dispersed.

Special Emergency Information: May be fatal in inhaled, ingested or absorbed through the skin.

HYDROCYANIC ACID - HC≡N

Synonyms: prussic acid

CAS Number: 74-90-8
DOT Number: 1051, 1613, 1614

Description: Colorless Liquid
DOT Classification: Poison A

Molecular Weight: 27.0
Melting Point: -14°C (7°F)
Boiling Point: 26°C (78°F)

Vapor Density: 0.9
Specific Gravity: 0.7

Vapor Pressure: 8 (400 mm Hg)
Water Solubility: Soluble

Chemical Incompatibilities or Instabilities: Acetaldehyde.

FLAMMABILITY

NFPA Hazard Code: 4-4-2
Flash Point: -18°C (0°F)
Autoignition Temp: 540°C (1004°F)

NFPA Classification: Class IA Flammable
LEL: 6.0%
UEL: 41.0%

Fire Extinguishing Methods: Carbon Dioxide, Dry Chemical, Foam, Water Spray or Fog.

Special Fire Fighting Considerations: Structural fire fighter protective clothing will not provide adequate protection. Isolate for 1/2 mile if rail or tank truck is involved in a fire. Water spray from an unmanned device should be used to keep closed containers cool. Continue to cool container after fire is extinguished. Keep area isolated until vapors have dispersed.

DOT Recommended Isolation Zones:	Small Spill: 600 ft	Large Spill: 600 ft
DOT Recommended Down Wind Take Cover Distance	Small Spill: 2 miles	Large Spill: 2 miles

TOXICOLOGY

Odor: Almonds
Physical Contact:
TLV = 11 mg/m^3

STEL = N.D.

Odor Threshold: N.D.

IDLH = 60 mg/m^3

Routes of Entry and Relative LD$_{50}$ (or LC$_{50}$)
Inhalation 10 (44 mg/m^3/H)
Ingestion N.D.
Skin Absorption N.D.

RTECS # MW6825000

Symptoms of Exposure: Headache; nausea; weakness; cyanosis.

Emergency/First Aid Treatment: Remove to ventilated area; immediately remove any contaminated clothing and wash contaminated areas for 15 minutes using water. Treat supportively and observe for possible shock. If ingested, seek immediate medical aid. Treat for cyanide poisoning as necessary.

Recommended Clean-Up Procedures

Personal Protection: Level A Ensemble
RCRA Waste # P063

Recommended Material Teflon (+)
Reportable Quantities: 10 lb (10 lb)

Spills: Remove all potential ignition sources. Stop leak if it can safely be done. Take up in non-combustible absorbant and dispose. Decontaminate spill area using a dilute alkaline hypochlorite solution.

Special Emergency Information: May be fatal in inhaled, ingested or absorbed through the skin. May undergo hazardous polymerization.

HYDROFLUORIC ACID - HF

Synonyms:

CAS Number: 7664-39-3	**Description:** Colorless, Fuming Liquid		
DOT Number: 1052, 1790	**DOT Classification:** Corrosive Material		

Molecular Weight: 20.0

Melting Point: -83°C (117°F)	**Vapor Density:** 1.3	**Vapor Pressure:** 8 (400 mm Hg)
Boiling Point: 19°C (66°F)	**Specific Gravity:** 1.2	**Water Solubility:** Soluble

Chemical Incompatibilities or Instabilities: Reacts with numerous classes of materials.

FLAMMABILITY

NFPA Hazard Code: 4-0-1   **NFPA Classification:**

Flash Point: None	**LEL:** N.A.
Autoignition Temp: None	**UEL:** N.A.

Fire Extinguishing Methods: Use agent suitable for surrounding fire.

Special Fire Fighting Considerations: Structural fire fighter protective clothing will not provide adequate protection. Water spray should be used to keep closed containers cool. Continue to cool container after fire is extinguished. Keep area isolated until vapors are dispersed.

DOT Recommended Isolation Zones:	Small Spill: 300 ft	Large Spill: 900 ft
DOT Recommended Down Wind Take Cover Distance	Small Spill: 1 mile	Large Spill: 3 miles

---

TOXICOLOGY

Odor: Sharp   **Odor Threshold:** 0.04 mg/m^3

Physical Contact: Material is extremely destructive to human tissues.

TLV = 2.6 mg/m^3 **STEL** = N.D.   **IDLH** = 25 mg/m^3

Routes of Entry and Relative LD$_{50}$ (or LC$_{50}$)   **RTECS # MW7875000**

Inhalation	**8**	(800 mg/m^3/H)
Ingestion	N.D.	
Skin Absorption	N.D.	

Symptoms of Exposure: Possible burns or irritation to skin, eyes and upper respiratory tract; headache; nausea. Burns may take several days to manifest themselves.

Emergency/First Aid Treatment: Remove to ventilated area; immediately remove any contaminated clothing and wash contaminated areas for 15 minutes using water. Treat supportively and observe for possible shock. If ingested, seek immediate medical aid.

Recommended Clean-Up Procedures

Personal Protection: Level A Ensemble	**Recommended Material** Teflon (0)
RCRA Waste # U134	**Reportable Quantities:** 100 lb (5000 lb)

Spills: Stop leak if it can safely be done. Take up liquid spills in a non-combustible absorbant and dispose. Decontaminate spill area using a dilute sodium bicarbonate solution. Keep area isolated until vapors disipate.

Special Emergency Information: May be fatal in inhaled, ingested or absorbed through the skin.

HYDROGEN PEROXIDE (35% -52%) - H$_2$O$_2$

Synonyms:

CAS Number: 7722-84-1   **Description:** Colorless Liquid
DOT Number: 2014   **DOT Classification:** Oxidizer

Molecular Weight: 34.0

Melting Point:	°C	(°F)	**Vapor Density:** 1.0	**Vapor Pressure:**	5 (5 mm Hg)
Boiling Point:	114°C	(237°F)	**Specific Gravity:** 1.1	**Water Solubility:**	Soluble

Chemical Incompatibilities or Instabilities: Alkalis, oxidizable materials, powdered metals, alcohols.

FLAMMABILITY

NFPA Hazard Code: 2-0-1-OX   **NFPA Classification:**
Flash Point: N.A.   **LEL:** N.A.
Autoignition Temp: N.A.   **UEL:** N.A.

Fire Extinguishing Methods: Alcohol Resistant Foam, Carbon Dioxide, Dry Chemical, Water Spray or Fog.

Special Fire Fighting Considerations: Water spray should be used to keep closed containers cool. Continue to cool container after fire is extinguished. For large fires, if possible, withdraw and allow to burn. Flood burning areas with water.

TOXICOLOGY

Odor: Irritating, Penetrating   **Odor Threshold:** N.D.
Physical Contact: Material is extremely destructive to human tissues.
TLV = 1.4 mg/m^3 **STEL** = N.D.   **IDLH** = 105 mg/m^3

Routes of Entry and Relative LD$_{50}$ (or LC$_{50}$)   **RTECS # MX0899000**
   Inhalation   N.D.
   Ingestion   N.D.
   Skin Absorption   N.D.

Symptoms of Exposure: Possible burns or irritation to skin, eyes and upper respiratory tract; headache; nausea. Inhalation of vapors may be fatal by causing glottis to spasm and suffocation. Exposure may result in chemical pneumonitis or pulmonary edema.

Emergency/First Aid Treatment: Remove to ventilated area; immediately remove any contaminated clothing and wash contaminated areas for 15 minutes using water. Treat supportively and observe for possible shock. If ingested, seek immediate medical aid.

Recommended Clean-Up Procedures

Personal Protection: Level B Ensemble   **Recommended Material**   N.D.

RCRA Waste #   None   **Reportable Quantities:**   None

Spills: Dike to contain spill and collect liquid for later disposal. Decontaminate spill area using water.

Special Emergency Information: May be harmful if inhaled, swallowed or absorbed through the skin.

HYDROGEN SELENIDE - H$_2$Se

Synonyms: selenium hydride

CAS Number: 7783-07-5	**Description:** Colorless Gas	
DOT Number: 2202	**DOT Classification:** Flammable Gas	

Molecular Weight: 81.0

Melting Point:	-64°C (-87°F)	**Vapor Density:**	**Vapor Pressure:**	Gas
Boiling Point:	-41°C (-42°F)	**Specific Gravity:**	**Water Solubility:**	1%

Chemical Incompatibilities or Instabilities: Oxidizers.

FLAMMABILITY

NFPA Hazard Code:		**NFPA Classification:**
Flash Point:	GAS	**LEL:** N.A.
Autoignition Temp:	N.A.	**UEL:** N.A.

Fire Extinguishing Methods: Dry Chemical, Foam, Water Spray or Fog.

Special Fire Fighting Considerations: Structural fire fighter protective clothing will not provide adequate protection. Stop leak if it can safely be done. If not, then isolate from combustibles and allow to burn. Immediately withdraw if rising sound from venting device is heard or if fire is causing discoloration to the tank.

DOT Recommended Isolation Zones:	Small Spill: 1500 ft	Large Spill: 1500 ft
DOT Recommended Down Wind Take Cover Distance	Small Spill: 5 miles	Large Spill: 5 miles

TOXICOLOGY

Odor: Garlic	**Odor Threshold:** 1 mg/m^3
Physical Contact: Irritant	
TLV = 0.16 mg/m^3 **STEL** = N.D.	**IDLH** = 6.8 mg/m^3

Routes of Entry and Relative LD$_{50}$ (or LC$_{50}$)   **RTECS # MX1050000**

Inhalation	N.D.
Ingestion	N.D.
Skin Absorption	N.D.

Symptoms of Exposure: Possible irritation to skin, eyes and upper respiratory tract; headache; nausea. Exposure may result in chemical pneumonitis or pulmonary edema. May cause an allergic reaction.

Emergency/First Aid Treatment: Remove to ventilated area; immediately remove any contaminated clothing and wash contaminated areas for 15 minutes using water. Treat supportively and observe for possible shock. If ingested, seek immediate medical aid.

Recommended Clean-Up Procedures

Personal Protection:	Level A Ensemble	**Recommended Material** N.D.
RCRA Waste #	None	**Reportable Quantities:** None

Spills: Remove all potential ignition sources. Stop leak if it can safely be done. Keep area isolated until vapors dissipate.

Special Emergency Information: May be fatal in inhaled, ingested or absorbed through the skin. Material is extremely toxic. Olfactory fatigue can easily cause one to lose the ability to smell the gas.

HYDROGEN SULFIDE - H₂S

Synonyms: hydrosulfuric acid

CAS Number: 7783-06-4	**Description:** Colorless Gas	
DOT Number: 1053	**DOT Classification:** Flammable Gas	

Molecular Weight: 34.1

Melting Point:	86°C (122°F)	**Vapor Density:** 1.2	**Vapor Pressure:** **5** (20 mm Hg)
Boiling Point:	60°C (76°F)	**Specific Gravity:**	**Water Solubility:** Soluble

Chemical Incompatibilities or Instabilities: Metal oxides, powdered metals, halogens.

FLAMMABILITY

NFPA Hazard Code: 3-4-0	**NFPA Classification:**	
Flash Point: Gas	**LEL:** 4.3%	
Autoignition Temp: 260°C (500°F)	**UEL:** 46.0%	

Fire Extinguishing Methods: Foam, Water Spray or Fog.

Special Fire Fighting Considerations: Structural fire fighter protective clothing will not provide adequate protection. Isolate for 1/2 mile if rail or tank truck is involved in a fire. Stop leak if it can safely be done. If leak cannot be stopped, then isolate from other combustibles and allow to burn. Immediately withdraw if rising sound from venting device is heard or if fire is causing discoloration to the tank. Water spray from an unmanned device should be used to keep closed containers cool. Continue to cool container after fire is extinguished. Keep area isolated until after vapors have dispersed.

DOT Recommended Isolation Zones:	Small Spill: 1500 ft	Large Spill: 1500 ft
DOT Recommended Down Wind Take Cover Distance	Small Spill: 5 miles	Large Spill: 5 miles

---

TOXICOLOGY

Odor: Rotten Eggs	**Odor Threshold:** 0.0001 mg/m³
Physical Contact: Irritant	
TLV = 14 mg/m³ **STEL** = 21 mg/m³	**IDLH** = 450 mg/m³

Routes of Entry and Relative LD₅₀ (or LC₅₀)   **RTECS # MX1225000**

Inhalation	9	(630 mg/m³/H)
Ingestion	N.D.	
Skin Absorption	N.D.	

Symptoms of Exposure: Possible irritation to skin, eyes and upper respiratory tract; headache; nausea. Inhalation of vapors may be fatal by causing glottis to spasm and suffocation. Exposure may result in chemical pneumonitis or pulmonary edema.

Emergency/First Aid Treatment: Remove to ventilated area; immediately remove any contaminated clothing and wash contaminated areas for 15 minutes using water. Treat supportively and observe for possible shock. If ingested, seek immediate medical aid.

Recommended Clean-Up Procedures

Personal Protection: Level B Ensemble	**Recommended Material** N.D.
RCRA Waste # None	**Reportable Quantities:** None

Spills: Remove all potential ignition sources. Stop leak if it can safely be done. Keep area isolated until vapors disipate.

Special Emergency Information: May be fatal in inhaled, ingested or absorbed through the skin. Material is extremely toxic. Olfactory fatigue can easily cause one to lose the ability to smell the gas.

HYDROIODIC ACID - HI

Synonyms:

CAS Number: 10034-85-2   **Description:** Colorless to Amber Liquid
DOT Number: 1787, 2197   **DOT Classification:** Corrosive Material

Molecular Weight: 127.9
Melting Point: °C   (°F)   **Vapor Density:**   **Vapor Pressure:**
Boiling Point: 127°C (261°F)   **Specific Gravity:** 1.7   **Water Solubility:**   Soluble

Chemical Incompatibilities or Instabilities: Ozone, metals, phosphorous.

FLAMMABILITY

NFPA Hazard Code: 3-0-0   **NFPA Classification:**
Flash Point: N.A.   **LEL:**   N.A.
Autoignition Temp: N.A.   **UEL:**   N.A.

Fire Extinguishing Methods: Use agent suitable for surrounding fire.

Special Fire Fighting Considerations: Water spray should be used to keep closed containers cool. Continue to cool container after fire is extinguished.

TOXICOLOGY

Odor: Pungent   **Odor Threshold:** N.D.
Physical Contact: Material is extremely destructive to human tissues.
TLV = N.D.   **STEL** = N.D.   **IDLH** = N.D.

Routes of Entry and Relative LD$_{50}$ (or LC$_{50}$)   **RTECS # MW3760000**
    Inhalation   N.D.
    Ingestion   N.D.
    Skin Absorption   N.D.

Symptoms of Exposure: Possible burns or irritation to skin, eyes and upper respiratory tract; headache; nausea. Laryngitis. Inhalation of vapors may be fatal by causing glottis to spasm and suffocation. Exposure may result in chemical pneumonitis or pulmonary edema.

Emergency/First Aid Treatment: Remove to ventilated area; immediately remove any contaminated clothing and wash contaminated areas for 15 minutes using water. Treat supportively and observe for possible shock. If ingested, seek immediate medical aid.

Recommended Clean-Up Procedures

Personal Protection: Level B Ensemble   **Recommended Material**   N.D.

RCRA Waste #   None   **Reportable Quantities:**   None

Spills: Dike to contain spill and collect liquid for later disposal. Take up in non-combustible absorbant and dispose. Decontaminate spill area using a dilute alkaline solution.

Special Emergency Information: May be harmful if inhaled, swallowed or absorbed through the skin.

HYDROXYLAMINE - H$_2$NOH

Synonyms: oxammonium

CAS Number: 7803-49-8   **Description:** White powder or Colorless Liquid
DOT Number: N.D.   **DOT Classification:** N.D.

Molecular Weight: 164.2
Melting Point: 34°C   (93°F)   **Vapor Density:**   **Vapor Pressure:**   **3** (6 mm Hg)
Boiling Point: 70°C (158°F)   **Specific Gravity:** 1.2   **Water Solubility:**   Soluble

Chemical Incompatibilities or Instabilities: Acids, oxidizers, reactive metals, alkali metals, phosphorous chlorides, pyridine, copper sulfate, air..

FLAMMABILITY

NFPA Hazard Code: 2-0-3   **NFPA Classification:**
Flash Point: N.A.   **LEL:** N.A.
Autoignition Temp: N.A.   **UEL:** N.A.

Fire Extinguishing Methods: Use agent suitable for surrounding fire.

Special Fire Fighting Considerations: Isolate for 1/2 mile if rail or tank truck is involved in a fire. Water spray should be used to keep closed containers cool. Continue to cool container after fire is extinguished. Fight fire from a distance or protected location, if possible. Explosive decomposition may occur at temperatures as low as 130°C (265°F).

TOXICOLOGY

Odor: **Odor Threshold:** N.D.
Physical Contact: Material is extremely destructive to human tissues.
TLV = N.D.   **STEL** = N.D.   **IDLH** = N.D.

Routes of Entry and Relative LD$_{50}$ (or LC$_{50}$)   **RTECS # NC2975000**
          Inhalation   N.D.
          Ingestion   N.D.
          Skin Absorption   N.D.

Symptoms of Exposure: Possible burns or irritation to skin, eyes and upper respiratory tract; headache; nausea. Laryngitis. Inhalation of vapors may be fatal by causing glottis to spasm and suffocation. Exposure may result in chemical pneumonitis or pulmonary edema.

Emergency/First Aid Treatment: Remove to ventilated area; immediately remove any contaminated clothing and wash contaminated areas for 15 minutes using water. Treat supportively and observe for possible shock. If ingested, seek immediate medical aid.

Recommended Clean-Up Procedures

Personal Protection: Level B Ensemble   **Recommended Material**   N.D.

RCRA Waste #   None   **Reportable Quantities:**   None

Spills: Dike to contain spill and collect liquid for later disposal. Decontaminate spill area using water.

Special Emergency Information: May be fatal in inhaled, ingested or absorbed through the skin. May spontaneously ignite if large surface area is present.

IMINOBISPROPYLAMINE - $(H_2NCH_2CH_2CH_2)_2NH$

Synonyms: aminobis(propylamine); 3,3'-iminobispropylamine

CAS Number: 56-18-8	**Description:** Colorless Liquid	
DOT Number: 2269	**DOT Classification:** Corrosive Material	

Molecular Weight: 131.2

Melting Point: -14°C (7°F)	**Vapor Density:** N.D.	**Vapor Pressure:** N.D.
Boiling Point: °C (°F)	**Specific Gravity:** 0.9	**Water Solubility:** Soluble

Chemical Incompatibilities or Instabilities: Oxidizers, halogens, chloroformates, acid chlorides.

FLAMMABILITY

NFPA Hazard Code:	**NFPA Classification:** Class IIIB Combustible
Flash Point: 118°C (245°F)	**LEL:** N.A.
Autoignition Temp: N.A.	**UEL:** N.A.

Fire Extinguishing Methods: Carbon Dioxide, Dry Chemical, Foam, Water Spray or Fog.

Special Fire Fighting Considerations: Water spray should be used to keep closed containers cool. Continue to cool container after fire is extinguished.

TOXICOLOGY

Odor:	**Odor Threshold:** N.D.

Physical Contact: Material is extremely destructive to human tissues.

TLV = N.D.	**STEL** = N.D.	**IDLH** = N.D.

Routes of Entry and Relative LD$_{50}$ (or LC$_{50}$)   **RTECS # JL945000**

Inhalation	N.D.	
Ingestion	4	(810 mg/kg)
Skin Absorption	9	(110 mg/kg)

Symptoms of Exposure: Possible burns or irritation to skin, eyes and upper respiratory tract; headache; nausea. Laryngitis. Inhalation of vapors may be fatal by causing glottis to spasm and suffocation. Exposure may result in chemical pneumonitis or pulmonary edema.

Emergency/First Aid Treatment: Remove to ventilated area; immediately remove any contaminated clothing and wash contaminated areas for 15 minutes using water. Treat supportively and observe for possible shock. If ingested, seek immediate medical aid.

Recommended Clean-Up Procedures

Personal Protection: Level B Ensemble	**Recommended Material** N.D.	
RCRA Waste # None	**Reportable Quantities:** None	

Spills: Dike to contain spill and collect liquid for later disposal. Decontaminate spill area using water.

Special Emergency Information: May be fatal in inhaled, ingested or absorbed through the skin. This materail may be considered an explosive.

IODINE MONOCHLORIDE - ICl

Synonyms:

CAS Number: 7790-99-0
DOT Number: 1792

Description: Red-Brown Oily Liquid or Solid
DOT Classification: Corrosive Material

Molecular Weight: 162.4

Melting Point:	27°C (81°F)	**Vapor Density:**	N.A.	**Vapor Pressure:**	N.A.
Boiling Point: (d)	97°C (207°F)	**Specific Gravity:**	3.1	**Water Solubility:**	Soluble

Chemical Incompatibilities or Instabilities: Reactive metals, metal sulfides, organic matter, phosphorous, alkali metals.

FLAMMABILITY

NFPA Hazard Code:

Flash Point: N.A.
Autoignition Temp: N.A.

NFPA Classification:

LEL: N.A.
UEL: N.A.

Fire Extinguishing Methods: Carbon Dioxide, Dry Chemical, Foam, Water Spray or Fog.

Special Fire Fighting Considerations: Structural fire fighter protective clothing will not provide adequate protection. Water spray should be used to keep closed containers cool. Continue to cool container after fire is extinguished. May explode when heated.

TOXICOLOGY

Odor:

Physical Contact: Material is extremely destructive to human tissues.

TLV = N.D.   **STEL** = N.D.

Odor Threshold: N.D.

IDLH = N.D.

Routes of Entry and Relative LD$_{50}$ (or LC$_{50}$)

Inhalation	N.D.
Ingestion	N.D.
Skin Absorption	N.D.

RTECS # NN1650000

Symptoms of Exposure: Possible burns or irritation to skin, eyes and upper respiratory tract; headache; nausea. Laryngitis. Inhalation of vapors may be fatal by causing glottis to spasm and suffocation. Exposure may result in chemical pneumonitis or pulmonary edema.

Emergency/First Aid Treatment: Remove to fresh air, remove contaminated clothing, and immediately wash contacted areas with 20% hydrochloric acid. Continue washing for 15 minutes with water. Treat supportively and observe for possible shock. If ingested, seek immediate medical aid.

Recommended Clean-Up Procedures

Personal Protection: Level B Ensemble   **Recommended Material**   N.D.

RCRA Waste #   None   **Reportable Quantities:**   None

Spills: Take up in non-combustible absorbant and dispose. Decontaminate spill area using water.

Special Emergency Information: May be fatal in inhaled, ingested or absorbed through the skin.

2-IODOBUTANE - CH3CHICH2CH3

Synonyms:

CAS Number: 513-48-4   **Description:** Pale Orange Liquid
DOT Number: 2390   **DOT Classification:** Flammable Liquid

Molecular Weight: 184.0
Melting Point: -14°C (7°F)   **Vapor Density:**   **Vapor Pressure:** N.D.
Boiling Point: 120°C (248°F)   **Specific Gravity:** 1.6   **Water Solubility:** Insoluble

Chemical Incompatibilities or Instabilities: Oxidizers.

FLAMMABILITY

NFPA Hazard Code: **NFPA Classification:** Class IC Flammable
Flash Point: 23°C (75°F)   **LEL:** N.D.
Autoignition Temp: N.D.   **UEL:** N.D.

Fire Extinguishing Methods: Alcohol Resistant Foam, Carbon Dioxide, Dry Chemical, Water Spray or Fog.

Special Fire Fighting Considerations: Isolate for 1/2 mile if rail or tank truck is involved in a fire. Water spray should be used to keep closed containers cool. Continue to cool container after fire is extinguished. For large fires, if possible, withdraw and allow to burn. Immediately withdraw if rising sound from venting device is heard or if fire is causing discoloration to the tank.

TOXICOLOGY
QUESTIONABLE CARCINOGEN

Odor: **Odor Threshold:** N.D.
Physical Contact: Irritant
TLV = N.D.   **STEL** = N.D.   **IDLH** = N.D.

Routes of Entry and Relative LD$_{50}$ (or LC$_{50}$)   **RTECS # EK4410000**
    Inhalation   N.D.
    Ingestion   N.D.
    Skin Absorption N.D.

Symptoms of Exposure: Possible irritation to skin, eyes and upper respiratory tract; headache; nausea.

Emergency/First Aid Treatment: Remove to ventilated area; immediately remove any contaminated clothing and wash contaminated areas for 15 minutes using water. Treat supportively and observe for possible shock. If ingested, seek immediate medical aid.

Recommended Clean-Up Procedures

Personal Protection: Level B Ensemble   **Recommended Material** N.D.

RCRA Waste # None   **Reportable Quantities:** None

Spills: Dike to contain spill and collect liquid for later disposal. Dike to contain spill, absorb with non-combustible absorbant and take up using non-sparking tools. Decontaminate spill area using soapy water.

Special Emergency Information: May be harmful if inhaled, swallowed or absorbed through the skin.

Iodopropane (CAS # 107-08-4, DOT # 2392, RTECS # TZ4100000) have similar chemical, physical and toxicological properties.

IRON PENTACARBONYL - Fe(CO)₅

Synonyms: iron carbonyl

CAS Number: 13463-40-6	**Description:** Orange Liquid	
DOT Number: 1994	**DOT Classification:** Poison B	

Molecular Weight: 195.9

Melting Point: -20°C (-4°F)	**Vapor Density:** 6.7	**Vapor Pressure:** 5 (40 mm Hg)
Boiling Point: 105°C (221°F)	**Specific Gravity:** 1.5	**Water Solubility:** Decomposes

Chemical Incompatibilities or Instabilities: Pyrophoric in air, halogens, oxidizers, amines, acetic acid.

FLAMMABILITY

NFPA Hazard Code: 2-3-1-~~W~~	**NFPA Classification:**
Flash Point: -15°C (5°F)	**LEL:** N.A.
Autoignition Temp: °C (°F)	**UEL:** N.A.

Fire Extinguishing Methods: Carbon Dioxide, Dry Chemical, Foam, Water Spray or Fog.

Special Fire Fighting Considerations: Structural fire fighter protective clothing will not provide adequate protection. Water spray should be used to keep closed containers cool. Continue to cool container after fire is extinguished. Treat all materials used as contaminated by hazardous waste.

DOT Recommended Isolation Zones:	Small Spill: 1500 ft	Large Spill: 1500 ft
DOT Recommended Down Wind Take Cover Distance	Small Spill: 5 miles	Large Spill: 5 miles

TOXICOLOGY

Odor:	**Odor Threshold:** N.D.
Physical Contact: Irritant	
TLV = 0.23 mg/m³ **STEL** = 0.45 mg/m³	**IDLH** = N.D.

Routes of Entry and Relative LD₅₀ (or LC₅₀)   **RTECS # NO4900000**

Inhalation	N.D.	
Ingestion	9	(12 mg/kg)
Skin Absorption	8	(240 mg/kg)

Symptoms of Exposure: Possible irritation to skin, eyes and upper respiratory tract; headache; nausea. Laryngitis. Inhalation of vapors may be fatal by causing glottis to spasm and suffocation. Exposure may result in chemical pneumonitis or pulmonary edema. Absorption may lead to the formation of methemoglobin resulting in cyanosis several hours after exposure.

Emergency/First Aid Treatment: Remove to ventilated area; immediately remove any contaminated clothing and wash contaminated areas for 15 minutes using water. Treat supportively and observe for possible shock. If ingested, seek immediate medical aid.

Recommended Clean-Up Procedures

Personal Protection: Level A Ensemble	**Recommended Material** N.D.
RCRA Waste # None	**Reportable Quantities:** None

Spills: Remove all potential ignition sources. Dike to contain spill, absorb with non-combustible absorbant and take up using non-sparking tools. Decontaminate spill area using water.

Special Emergency Information: May be fatal in inhaled, ingested or absorbed through the skin.

ISOBUTYL FORMATE - (CH₃)₂CHCH₂OCHO

Synonyms: formic acid, ethyl ester

CAS Number: 542-55-2
DOT Number: 2393

Description: Colorless Liquid
DOT Classification: Flammable Liquid

Molecular Weight: 102.2
Melting Point: -95°C (-139°F)
Boiling Point: 98°C (208°F)

Vapor Density: 3.5
Specific Gravity: 0.9

Vapor Pressure: 5 (40 mm Hg)
Water Solubility: 1%

Chemical Incompatibilities or Instabilities: Oxidizers.

FLAMMABILITY

NFPA Hazard Code: ?-3-?
Flash Point: 10°C (50°F)
Autoignition Temp: 320°C (608°F)

NFPA Classification: Class IB Flammable
LEL: 1.7%
UEL: 8.0%

Fire Extinguishing Methods: Carbon Dioxide, Dry Chemical, Foam, Water Spray or Fog.

Special Fire Fighting Considerations: Isolate for 1/2 mile if rail or tank truck is involved in a fire. Water spray should be used to keep closed containers cool. Continue to cool container after fire is extinguished. Immediately withdraw if rising sound from venting device is heard or if fire is causing discoloration to the tank.

TOXICOLOGY

Odor:
Physical Contact: Material is extremely destructive to human tissues.
TLV = N.D.   **STEL** = N.D.

Odor Threshold: N.D.

IDLH = N.D.

Routes of Entry and Relative LD₅₀ (or LC₅₀)
Inhalation   N.D.
Ingestion   N.D.
Skin Absorption   N.D.

RTECS # LQ8650000

Symptoms of Exposure: Possible burns or irritation to skin, eyes and upper respiratory tract; headache; nausea. Laryngitis. Inhalation of vapors may be fatal by causing glottis to spasm and suffocation. Exposure may result in chemical pneumonitis or pulmonary edema.

Emergency/First Aid Treatment: Remove to ventilated area; immediately remove any contaminated clothing and wash contaminated areas for 15 minutes using water. Treat supportively and observe for possible shock. If ingested, seek immediate medical aid.

Recommended Clean-Up Procedures

Personal Protection: Level B Ensemble
RCRA Waste # None

Recommended Material Butyl Rubber (-), Polyvinyl Alcohol (-)
Reportable Quantities: None

Spills: Remove all potential ignition sources. Dike to contain spill, absorb with non-combustible absorbant and take up using non-sparking tools. Decontaminate spill area using soapy water.

Special Emergency Information: May be harmful if inhaled, swallowed or absorbed through the skin. Vapors are heavier than air and may travel some distance to an ignition source. Vapors are more dense than air and may settle in low lying areas.

ISOBUTYRONITRILE - (CH₃)₂CHCN

Synonyms: isopropyl cyanide; 2-methylpropionitrile

CAS Number: 78-82-0   **Description:** Colorless Liquid
DOT Number: 2284   **DOT Classification:** Flammable Liquid

Molecular Weight: 69.1

Melting Point:	°C (°F)	**Vapor Density:** 2.4	**Vapor Pressure:**	3 (10 mm Hg)
Boiling Point:	102°C (216°F)	**Specific Gravity:** 0.8	**Water Solubility:**	1%

Chemical Incompatibilities or Instabilities: Strong oxidizers.

FLAMMABILITY

NFPA Hazard Code: 3-3-0   **NFPA Classification:** Class IB Flammable
Flash Point: 8°C (47°F)   **LEL:** N.D.
Autoignition Temp: 482°C (900°F)   **UEL:** N.D.

Fire Extinguishing Methods: Alcohol Resistant Foam, Carbon Dioxide, Dry Chemical, Water Spray or Fog.

Special Fire Fighting Considerations: Structural fire fighter protective clothing will not provide adequate protection. Isolate for 1/2 mile if rail or tank truck is involved in a fire. Water spray should be used to keep closed containers cool. Continue to cool container after fire is extinguished. Immediately withdraw if rising sound from venting device is heard or if fire is causing discoloration to the tank. Treat all materials used or generated and equipment used involved as contaminated by hazardous waste.

TOXICOLOGY

Odor: **Odor Threshold:** N.D.
Physical Contact: Irritant
TLV = N.D.   **STEL** = N.D.   **IDLH** = N.D.

Routes of Entry and Relative LD₅₀ (or LC₅₀)   **RTECS # TZ4900000**

Inhalation	N.D.	
Ingestion	7	(102 mg/kg)
Skin Absorption	8	(310 mg/kg)

Symptoms of Exposure: Possible irritation to skin, eyes and upper respiratory tract; headache; nausea. Absorption may lead to the formation of methemoglobin resulting in cyanosis several hours after exposure.

Emergency/First Aid Treatment: Remove to ventilated area; immediately remove any contaminated clothing and wash contaminated areas for 15 minutes using water. Treat supportively and observe for possible shock. If ingested, seek immediate medical aid.

Recommended Clean-Up Procedures

Personal Protection: Level B Ensemble   **Recommended Material** N.D.

RCRA Waste # None   **Reportable Quantities:** None

Spills: Remove all potential ignition sources. Dike to contain spill, absorb with non-combustible absorbant and take up using non-sparking tools. Decontaminate spill area using dilute alkaline hypochlorite solution. Treat all materials used or generated and equipment used involved as contaminated by hazardous waste.

Special Emergency Information: May be fatal in inhaled, ingested or absorbed through the skin. Vapors are heavier than air and may travel some distance to an ignition source. Vapors are more dense than air and may settle in low lying areas.

ISOPHORONE DIISOCYANATE

Synonyms: IPDI

CAS Number: 4098-71-9
DOT Number: 2290

Description: Colorless Liquid
DOT Classification: N.A.

Molecular Weight: 222.3
Melting Point: °C (°F)
Boiling Point: °C (°F)

Vapor Density:
Specific Gravity: 1.1

Vapor Pressure: 1 (<1 mm Hg)
Water Solubility: N.A.

Chemical Incompatibilities or Instabilities: Alcohols, oxidizers, alkali, amines.

FLAMMABILITY

NFPA Hazard Code:
Flash Point: 162°C (325°F)
Autoignition Temp: N.D.

NFPA Classification:
LEL: N.A.
UEL: N.A.

Fire Extinguishing Methods: Carbon Dioxide, Dry Chemical, Foam, Water Spray or Fog.

Special Fire Fighting Considerations: Structural fire fighter protective clothing will not provide adequate protection. Fight fire from a distance or protected location, if possible. Treat all materials used or generated and equipment used involved as contaminated by hazardous waste.

DOT Recommended Isolation Zones:	Small Spill: 900 ft	Large Spill: 900 ft
DOT Recommended Down Wind Take Cover Distance	Small Spill: 3 miles	Large Spill: 3 miles

TOXICOLOGY

Odor:
Physical Contact: Irritant
TLV = 0.05 mg/m^3

STEL = N.D.

Odor Threshold: N.D.

IDLH = N.D.

Routes of Entry and Relative LD$_{50}$ (or LC$_{50}$)
Inhalation 10 (1040 mg/m^3/H)
Ingestion N.D.
Skin Absorption N.D.

RTECS # NQ9370000

Symptoms of Exposure: Possible irritation to skin, eyes and upper respiratory tract; headache; nausea; lachrymation.

Emergency/First Aid Treatment: Remove to ventilated area; immediately remove any contaminated clothing and wash contaminated areas for 15 minutes using water. Treat supportively and observe for possible shock. If ingested, seek immediate medical aid.

Recommended Clean-Up Procedures

Personal Protection: Level B Ensemble

Recommended Material Butyl Rubber (+), Polyvinyl Alcohol (+), Viton (+)

RCRA Waste # None

Reportable Quantities: None

Spills: Dike to contain spill and collect liquid for later disposal. Treat all materials used or generated and equipment used involved as contaminated by hazardous waste.

Special Emergency Information: May be harmful if inhaled, swallowed or absorbed through the skin.

ISOPRENE - $CH_2=C(CH_3)CH=CH_2$

Synonyms: 2-methyl-1,3-butadiene

CAS Number: 78-79-5
DOT Number: 1218

Description: Colorless Liquid
DOT Classification: Flammable Liquid

Molecular Weight: 68.1
Melting Point: -146°C (-231°F)
Boiling Point: 34°C (93°F)

Vapor Density: 2.4
Specific Gravity: 0.7

Vapor Pressure: 9 (400 mg Hg)
Water Solubility: Insoluble

Chemical Incompatibilities or Instabilities: Oxidizers, peroxides, acids, chlorosulfonic acid, reducing agents

FLAMMABILITY

NFPA Hazard Code: 2-4-2
Flash Point: -65°C (-54°F)
Autoignition Temp: 220°C (428°F)

NFPA Classification: Class IA Flammable
LEL: 2.0%
UEL: 9.0%

Fire Extinguishing Methods: Carbon Dioxide, Dry Chemical, Foam, Water Spray or Fog.

Special Fire Fighting Considerations: Isolate for 1/2 mile if rail or tank truck is involved in a fire. Water spray should be used to keep closed containers cool. Continue to cool container after fire is extinguished. For large fires, if possible, withdraw and allow to burn. Immediately withdraw if rising sound from venting device is heard or if fire is causing discoloration to the tank. Heat or sunlight may cause explosive polymerization.

TOXICOLOGY

Odor:
Physical Contact: Irritant
TLV = N.D.   **STEL** = N.D.

Odor Threshold: N.D.

IDLH = N.D.

Routes of Entry and Relative LD_{50} (or LC_{50})
   Inhalation   1   (720,000 mg/m^3/H)
   Ingestion   N.D.
   Skin Absorption   N.D.

RTECS # NT4037000

Symptoms of Exposure: Possible irritation to skin, eyes and upper respiratory tract; headache; nausea. May act as a narcotic in high concentrations.

Emergency/First Aid Treatment: Remove to ventilated area; immediately remove any contaminated clothing and wash contaminated areas for 15 minutes using water. Treat supportively and observe for possible shock. If ingested, seek immediate medical aid.

Recommended Clean-Up Procedures

Personal Protection: Level B Ensemble

RCRA Waste # None

Recommended Material Polyvinyl Alcohol (+), Viton (0)

Reportable Quantities: 100 lb (1000 lb)

Spills: Remove all potential ignition sources. Dike to contain spill, absorb with non-combustible absorbant and take up using non-sparking tools. Decontaminate spill area using soapy water.

Special Emergency Information: May be harmful if inhaled, swallowed or absorbed through the skin. May undergo hazardous polymerization.

ISOPROPANOL - (CH3)2CHOH

Synonyms: 2-propanol

CAS Number: 67-63-0	**Description:** Colorless Liquid	
DOT Number: 1219	**DOT Classification:** Flammable Liquid	

Molecular Weight: 60.1

Melting Point: -89°C (-128°F)	**Vapor Density:** 2.1	**Vapor Pressure:** 5 (33 mm Hg)
Boiling Point: 83°C (181°F)	**Specific Gravity:** 0.8	**Water Solubility:** Soluble

Chemical Incompatibilities or Instabilities: Oxidizers, aluminum, perchlorates, phosgene.

FLAMMABILITY

NFPA Hazard Code: 1-3-0	**NFPA Classification:** Class IB Flammable	
Flash Point: 22°C (72°F)	**LEL:** 2.5%	
Autoignition Temp: 460°C (860°F)	**UEL:** 12.0%	

Fire Extinguishing Methods: Alcohol Resistant Foam, Carbon Dioxide, Dry Chemical, Water Spray or Fog.

Special Fire Fighting Considerations: Isolate for 1/2 mile if rail or tank truck is involved in a fire. Water spray should be used to keep closed containers cool. Continue to cool container after fire is extinguished. For large fires, if possible, withdraw and allow to burn. Immediately withdraw if rising sound from venting device is heard or if fire is causing discoloration to the tank.

TOXICOLOGY

Odor: Rubbing Alcohol

Physical Contact: Irritant

TLV = 983 mg/m^3 **STEL** = 1230 mg/m^3

Odor Threshold: 2.5 mg/m^3

IDLH = 30,000 mg/m^3

Routes of Entry and Relative LD$_{50}$ (or LC$_{50}$)

Inhalation	N.D.	
Ingestion	2	(5045 mg/kg)
Skin Absorption	1	(12,800 mg/kg)

RTECS # NT8050000

Symptoms of Exposure: Possible irritation to skin, eyes and upper respiratory tract; headache; nausea. May act as a narcotic in high concentrations.

Emergency/First Aid Treatment: Remove to ventilated area; immediately remove any contaminated clothing and wash contaminated areas for 15 minutes using water. Treat supportively and observe for possible shock. If ingested, seek immediate medical aid.

Recommended Clean-Up Procedures

Personal Protection: Level B Ensemble	**Recommended Material**	Nitrile Rubber (+), Neoprene (0), Teflon (0)
RCRA Waste # None	**Reportable Quantities:**	None

Spills: Remove all potential ignition sources. Dike to contain spill, absorb with non-combustible absorbant and take up using non-sparking tools. Decontaminate spill area using water.

Special Emergency Information: May be harmful if inhaled, swallowed or absorbed through the skin. Vapors are heavier than air and may travel some distance to an ignition source. Vapors are more dense than air and may settle in low lying areas.

1-Propanol (CAS # 71-23-8, DOT # 1274, RTECS # UH8225000) has similar chemical, physical and toxicological properties.

ISOPROPYL ACETATE - (CH₃)₂CHOC(O)CH₃

Synonyms: acetic acid, isopropyl ester

CAS Number: 108-21-4	**Description:** Colorless Liquid
DOT Number: 1220	**DOT Classification:** Flammable Liquid

Molecular Weight: 102.1

Melting Point:	°C	(°F)	**Vapor Density:** 3.5	**Vapor Pressure:**	5 (47 mm Hg)
Boiling Point:	85°C	(185°F)	**Specific Gravity:** 0.9	**Water Solubility:**	Insoluble

Chemical Incompatibilities or Instabilities: Oxidizers.

FLAMMABILITY

NFPA Hazard Code: 1-3-0	**NFPA Classification:**	
Flash Point: 16°C (62°F)	**LEL:**	1.8%
Autoignition Temp: 479°C (894°F)	**UEL:**	8.0%

Fire Extinguishing Methods: Alcohol Resistant Foam, Carbon Dioxide, Dry Chemical, Water Spray or Fog.

Special Fire Fighting Considerations: Isolate for 1/2 mile if rail or tank truck is involved in a fire. Water spray should be used to keep closed containers cool. Continue to cool container after fire is extinguished. For large fires, if possible, withdraw and allow to burn. Immediately withdraw if rising sound from venting device is heard or if fire is causing discoloration to the tank.

TOXICOLOGY

Odor: Pleasant, Fruity	**Odor Threshold:** 1.5 mg/m³
Physical Contact: Irritant	
TLV = 1040 mg/m³ **STEL** = 1290 mg/m³	**IDLH** = 68,000 mg/m³

Routes of Entry and Relative LD₅₀ (or LC₅₀)   **RTECS # AI4930000**

Inhalation	N.D.	
Ingestion	3	(3000 mg/kg)
Skin Absorption	N.D.	

Symptoms of Exposure: Possible irritation to skin, eyes and upper respiratory tract; headache; nausea. May act as a narcotic in high concentrations.

Emergency/First Aid Treatment: Remove to ventilated area; immediately remove any contaminated clothing and wash contaminated areas for 15 minutes using water. Treat supportively and observe for possible shock. If ingested, seek immediate medical aid.

Recommended Clean-Up Procedures

Personal Protection: Level B Ensemble	**Recommended Material** N.D.
RCRA Waste # None	**Reportable Quantities:** None

Spills: Remove all potential ignition sources. Dike to contain spill, absorb with non-combustible absorbant and take up using non-sparking tools. Decontamination will occur via evaporation.

Special Emergency Information: May be harmful if inhaled, swallowed or absorbed through the skin. Vapors are heavier than air and may travel some distance to an ignition source. Vapors are more dense than air and may settle in low lying areas.

n-Propyl acetate (CAS # 109-60-4, DOT # 1276, RTECS # AJ3675000) has similar chemical, physical and toxicological properties.

ISOPROPENYL ACETATE - $CH_2=C(CH_3)OC(O)CH_3$

Synonyms: methylvinyl acetate

CAS Number: 108-22-5	**Description:** Colorless Liquid
DOT Number: 2403	**DOT Classification:** Flammable Liquid

Molecular Weight: 100.1

Melting Point:	°C (°F)	**Vapor Density:** 3.5	**Vapor Pressure:** N.A.
Boiling Point:	97°C (207°F)	**Specific Gravity:** 0.9	**Water Solubility:** Insoluble

Chemical Incompatibilities or Instabilities: Oxidizers, peroxides.

FLAMMABILITY

NFPA Hazard Code:	2-3-0	**NFPA Classification:** Class IB Flammable
Flash Point:	16°C (60°F)	**LEL:** N.A.
Autoignition Temp:	431°C (800°F)	**UEL:** N.A.

Fire Extinguishing Methods: Alcohol Resistant Foam, Carbon Dioxide, Dry Chemical, Water Spray or Fog.

Special Fire Fighting Considerations: Isolate for 1/2 mile if rail or tank truck is involved in a fire. Water spray should be used to keep closed containers cool. Continue to cool container after fire is extinguished. For large fires, if possible, withdraw and allow to burn. Immediately withdraw if rising sound from venting device is heard or if fire is causing discoloration to the tank.

TOXICOLOGY

Odor:	**Odor Threshold:** N.D.
Physical Contact: Irritant	
TLV = N.D.　　　　　　**STEL** = N.D.	**IDLH** = N.D.

Routes of Entry and Relative LD$_{50}$ (or LC$_{50}$)　　　　**RTECS # UD4200000**

Inhalation	N.D.	
Ingestion	3	(3000 mg/kg)
Skin Absorption	N.D.	

Symptoms of Exposure: Possible irritation to skin, eyes and upper respiratory tract; headache; nausea. May act as a narcotic in high concentrations.

Emergency/First Aid Treatment: Remove to ventilated area; immediately remove any contaminated clothing and wash contaminated areas for 15 minutes using water. Treat supportively and observe for possible shock. If ingested, seek immediate medical aid.

Recommended Clean-Up Procedures

Personal Protection: Level B Ensemble	**Recommended Material** N.D.
RCRA Waste # None	**Reportable Quantities:** None

Spills: Remove all potential ignition sources. Dike to contain spill, absorb with non-combustible absorbant and take up using non-sparking tools. Decontaminate spill area using soapy water.

Special Emergency Information: May be harmful if inhaled, swallowed or absorbed through the skin. Vapors are heavier than air and may travel some distance to an ignition source. Vapors are more dense than air and may settle in low lying areas. May form unstable peroxides upon prolonged exposure to air.

ISOPROPENYL BENZENE

Synonyms: alpha-methyl styrene

CAS Number: 98-83-9
DOT Number: 2303

Description: Colorless Liquid
DOT Classification: Flammable Liquid

Molecular Weight: 118.2
Melting Point: -24°C (-11°F)
Boiling Point: 169°C (336°F)

Vapor Density: 4.1
Specific Gravity: 0.9

Vapor Pressure: 2 (2 mm Hg)
Water Solubility: Insoluble

Chemical Incompatibilities or Instabilities: Oxidizers.

FLAMMABILITY

NFPA Hazard Code: 1-2-1
Flash Point: 54°C (129°F)
Autoignition Temp: 574°C (1066°F)

NFPA Classification: Class II Combustible
LEL: 1.9%
UEL: 6.1%

Fire Extinguishing Methods: Carbon Dioxide, Dry Chemical, Foam, Water Spray or Fog.

Special Fire Fighting Considerations: Isolate for 1/2 mile if rail or tank truck is involved in a fire. Water spray should be used to keep closed containers cool. Continue to cool container after fire is extinguished. For large fires, if possible, withdraw and allow to burn. Immediately withdraw if rising sound from venting device is heard or if fire is causing discoloration to the tank.

TOXICOLOGY

Odor:
Physical Contact: Irritant
TLV = 242 mg/m^3

STEL = 483 mg/m^3

Odor Threshold: N.D.

IDLH = 2500 mg/m^3

Routes of Entry and Relative LD$_{50}$ (or LC$_{50}$)
    Inhalation   N.D.
    Ingestion   N.D.
    Skin Absorption   N.D.

RTECS # DA8400000

Symptoms of Exposure: Possible irritation to skin, eyes and upper respiratory tract; headache; nausea; laryngitis.

Emergency/First Aid Treatment: Remove to ventilated area; immediately remove any contaminated clothing and wash contaminated areas for 15 minutes using water. Treat supportively and observe for possible shock. If ingested, seek immediate medical aid.

Recommended Clean-Up Procedures

Personal Protection: Level B Ensemble

RCRA Waste # None

Recommended Material N.D.

Reportable Quantities: None

Spills: Remove all potential ignition sources. Dike to contain spill, absorb with non-combustible absorbent and take up using non-sparking tools. Decontaminate spill area using soapy water.

Special Emergency Information: May be harmful if inhaled, swallowed or absorbed through the skin.

ISOPROPYLAMINE - $(CH_3)_2CHNH_2$

Synonyms: 2-aminopropane; 2-propanamine

CAS Number: 75-31-0	**Description:** Colorless Liquid
DOT Number: 1221	**DOT Classification:** Flammable Liquid

Molecular Weight: 59.1

Melting Point:	-101°C (-150°F)	**Vapor Density:** 2.0	**Vapor Pressure:**	9 (480 mm Hg)
Boiling Point:	34°C (93°F)	**Specific Gravity:** 0.7	**Water Solubility:**	Soluble

Chemical Incompatibilities or Instabilities: Oxidizers, epoxides.

FLAMMABILITY

NFPA Hazard Code:	3-4-0	**NFPA Classification:**	Class IA Flammable
Flash Point:	-26°C (-15°F)	**LEL:**	2.0%
Autoignition Temp:	402°C (756°F)	**UEL:**	10.4%

Fire Extinguishing Methods: Alcohol Resistant Foam, Carbon Dioxide, Dry Chemical, Water Spray or Fog.

Special Fire Fighting Considerations: Isolate for 1/2 mile if rail or tank truck is involved in a fire. Water spray should be used to keep closed containers cool. Continue to cool container after fire is extinguished. Immediately withdraw if rising sound from venting device is heard or if fire is causing discoloration to the tank.

TOXICOLOGY

Odor: Ammonia
Physical Contact: Material is extremely destructive to human tissues.
TLV = 12 mg/m^3 **STEL** = 24 mg/m^3

Odor Threshold: 0.5 mg/m^3

IDLH = 10,000 mg/m^3

Routes of Entry and Relative LD$_{50}$ (or LC$_{50}$)

Inhalation	N.D.	
Ingestion	4	(820 mg/kg)
Skin Absorption	8	(380 mg/m^3)

RTECS # NT8400000

Symptoms of Exposure: Possible burns or irritation to skin, eyes and upper respiratory tract; headache; nausea. Inhalation of vapors may be fatal by causing glottis to spasm and suffocation. Exposure may result in chemical pneumonitis or pulmonary edema.

Emergency/First Aid Treatment: Remove to ventilated area; immediately remove any contaminated clothing and wash contaminated areas for 15 minutes using water. Treat supportively and observe for possible shock. If ingested, seek immediate medical aid.

Recommended Clean-Up Procedures

Personal Protection:	Level B Ensemble	**Recommended Material**	Teflon (0)
RCRA Waste #	None	**Reportable Quantities:**	None

Spills: Remove all potential ignition sources. Dike to contain spill, absorb with non-combustible absorbent and take up using non-sparking tools. Decontaminate spill area using water.

Special Emergency Information: May be harmful if inhaled, swallowed or absorbed through the skin. Vapors are heavier than air and may travel some distance to an ignition source. Vapors are more dense than air and may settle in low lying areas.

n-Propylamine (CAS # 107-10-8, DOT # 1277, RTECS # UH9100000, RCRA Waste # U194, RQ : 5000 lb [1 lb]) has similar chemical, physical and toxicological properties.

ISOPROPYLCHLOROFORMATE - (CH3)2CHOC(O)Cl

Synonyms: isopropyl chlorocarbonate

CAS Number: 108-23-6
DOT Number: 2407

Description: Colorless Liquid
DOT Classification: Flammable Liquid

Molecular Weight: 122.6
Melting Point: -80°C (-112°F)
Boiling Point: 105°C (221°F)

Vapor Density: 4.2
Specific Gravity: 1.1

Vapor Pressure: **5** (21 mm Hg)
Water Solubility: Insoluble

Chemical Incompatibilities or Instabilities: Iron salts, acids, amines, alkalis, alcohols, water.

FLAMMABILITY

NFPA Hazard Code:
Flash Point: -11°C (11°F)
Autoignition Temp: N.D.

NFPA Classification: Class IB Flammable
LEL: 4.0%
UEL: 15.0%

Fire Extinguishing Methods: Carbon Dioxide, Dry Chemical, Foam, Water.

Special Fire Fighting Considerations: Keep area isolated until vapors have dissipated. Fight fire from a distance or protected location, if possible. Water spray from an unmanned device should be used to keep closed containers cool. Continue to cool container after fire is extinguished. For large fires, if possible, withdraw and allow to burn. Immediately withdraw if rising sound from venting device is heard or if fire is causing discoloration to the tank.

DOT Recommended Isolation Zones:
DOT Recommended Down Wind Take CoverDistance

Small Spill: 0.2 miles
Small Spill: 150 ft

Large Spill: 0.2 miles
Large Spill: 150 ft

TOXICOLOGY

Odor:
Physical Contact: Material is extremely destructive to human tissues.
TLV = N.D.   **STEL** = N.D.

Odor Threshold: N.D.

IDLH = N.D.

Routes of Entry and Relative LD$_{50}$ (or LC$_{50}$)
   Inhalation   N.D.
   Ingestion   3   (1070 mg/kg)
   Skin Absorption 1   (11,300 mg/kg)

RTECS # LQ6475000

Symptoms of Exposure: Possible burns or irritation to skin, eyes and upper respiratory tract; headache; nausea. Laryngitis. Inhalation of vapors may be fatal by causing glottis to spasm and suffocation. Exposure may result in chemical pneumonitis or pulmonary edema.

Emergency/First Aid Treatment: Remove to ventilated area; immediately remove any contaminated clothing and wash contaminated areas for 15 minutes using water. Treat supportively and observe for possible shock. If ingested, seek immediate medical aid.

Recommended Clean-Up Procedures

Personal Protection: Level B Ensemble

RCRA Waste # None

Recommended Material N.D.

Reportable Quantities: None

Spills: Dike to contain spill and collect liquid for later disposal. Dike to contain spill, absorb with non-combustible absorbent and take up using non-sparking tools. Decontaminate spill area using soapy water.

Special Emergency Information: May be fatal in inhaled, ingested or absorbed through the skin.

ISOPROPYL FORMATE - (CH$_3$)$_2$CHOCHO

Synonyms: formic acid, isopropyl ester

CAS Number: 625-55-8   **Description:** Colorless Liquid
DOT Number: 1281   **DOT Classification:** Flammable Liquid

Molecular Weight: 88.1
Melting Point: °C   (°F)   **Vapor Density:** 3.0   **Vapor Pressure:** 7 (100 mm Hg)
Boiling Point: 67°C (153°F)   **Specific Gravity:** 0.9   **Water Solubility:** Decomposes

Chemical Incompatibilities or Instabilities: Oxidizers; causes generation of formic acid upon contact with water.

FLAMMABILITY

NFPA Hazard Code: 2-3-0   **NFPA Classification:** Class IB Flammable
Flash Point: -6°C   (22°F)   **LEL:** N.D.
Autoignition Temp: 485°C (905°F)   **UEL:** N.D.

Fire Extinguishing Methods: Carbon Dioxide, Dry Chemical, Foam, Water Spray or Fog.

Special Fire Fighting Considerations: Isolate for 1/2 mile if rail or tank truck is involved in a fire. Water spray should be used to keep closed containers cool. Continue to cool container after fire is extinguished. For large fires, if possible, withdraw and allow to burn. Immediately withdraw if rising sound from venting device is heard or if fire is causing discoloration to the tank.

TOXICOLOGY

Odor: **Odor Threshold:** N.D.
Physical Contact: Irritant
TLV = N.D.   **STEL** = N.D.   **IDLH** = N.D.

Routes of Entry and Relative LD$_{50}$ (or LC$_{50}$)   **RTECS #** LQ8750000
   Inhalation   N.D.
   Ingestion   N.D.
   Skin Absorption   N.D.

Symptoms of Exposure: Possible irritation to skin, eyes and upper respiratory tract; headache; nausea.

Emergency/First Aid Treatment: Remove to ventilated area; immediately remove any contaminated clothing and wash contaminated areas for 15 minutes using water. Treat supportively and observe for possible shock. If ingested, seek immediate medical aid.

Recommended Clean-Up Procedures

Personal Protection: Level B Ensemble   **Recommended Material** N.D.

RCRA Waste #   None   **Reportable Quantities:** None

Spills: Dike to contain spill and collect liquid for later disposal. Dike to contain spill, absorb with non-combustible absorbent and take up using non-sparking tools. Decontaminate spill area using dilute alkaline solution.

Special Emergency Information: May be harmful if inhaled, swallowed or absorbed through the skin. Vapors are heavier than air and may travel some distance to an ignition source. Vapors are more dense than air and may settle in low lying areas.

ISOPROPYLMERCAPTAN - (CH3)2CHSH

Synonyms: isopropyl mercaptan; 2-propanethiol

CAS Number: 75-33-3
DOT Number: 2402, 2703

Description: Colorless Liquid
DOT Classification: Flammable Liquid

Molecular Weight: 72.2
Melting Point: -131°C (-204°F) **Vapor Density:** 2.6 **Vapor Pressure:** 9 (455 mm Hg)
Boiling Point: 60°C (140°F) **Specific Gravity:** 0.8 **Water Solubility:** 1%

Chemical Incompatibilities or Instabilities: Oxidizers.

FLAMMABILITY

NFPA Hazard Code:
Flash Point: -34°C (-30°F)
Autoignition Temp: °C (°F)

NFPA Classification: Class IB Flammable
LEL: N.D.
UEL: N.D.

Fire Extinguishing Methods: Carbon Dioxide, Dry Chemical, Foam, Water Spray or Fog.

Special Fire Fighting Considerations: Isolate for 1/2 mile if rail or tank truck is involved in a fire. Water spray should be used to keep closed containers cool. Continue to cool container after fire is extinguished. For large fires, if possible, withdraw and allow to burn. Immediately withdraw if rising sound from venting device is heard or if fire is causing discoloration to the tank.

TOXICOLOGY

Odor: Strong, unpleasant
Physical Contact: Irritant
TLV = N.D.   **STEL** = N.D.

Odor Threshold: N.A.

IDLH = N.D.

Routes of Entry and Relative LD$_{50}$ (or LC$_{50}$)
Inhalation   N.D.
Ingestion   N.D.
Skin Absorption   N.D.

RTECS # TZ7302000

Symptoms of Exposure: Possible irritation to skin, eyes and upper respiratory tract; headache; nausea.

Emergency/First Aid Treatment: Remove to ventilated area; immediately remove any contaminated clothing and wash contaminated areas for 15 minutes using water. Treat supportively and observe for possible shock. If ingested, seek immediate medical aid.

Recommended Clean-Up Procedures

Personal Protection: Level B Ensemble   **Recommended Material** N.D.

RCRA Waste # None   **Reportable Quantities:** None

Spills: Remove all potential ignition sources. Dike to contain spill, absorb with non-combustible absorbent and take up using non-sparking tools. Decontaminate spill area using water.

Special Emergency Information: May be harmful if inhaled, swallowed or absorbed through the skin. Vapors are heavier than air and may travel some distance to an ignition source. Vapors are more dense than air and may settle in low lying areas.

ISOPROPYL NITRATE - (CH3)2CHONO2

Synonyms: nitric acid, isopropyl ester

CAS Number: 1712-64-7   **Description:** Colorless Liquid
DOT Number: 1222   **DOT Classification:** Flammable Liquid

Molecular Weight: 105.1
Melting Point: °C   (°F)   **Vapor Density:** N.A.   **Vapor Pressure:** N.A.
Boiling Point: 102°C (216°F)   **Specific Gravity:** 1.0   **Water Solubility:** Insoluble

Chemical Incompatibilities or Instabilities: Pure vapor may spontaneously ignite. Powdered metals.

FLAMMABILITY

NFPA Hazard Code: **NFPA Classification:** Class IB Flammable
Flash Point: 11°C   (52°F)   **LEL:** N.A.
Autoignition Temp: 12°C   (54°F)   **UEL:** 100%

Fire Extinguishing Methods: Carbon Dioxide, Dry Chemical, Foam, Water Spray or Fog.

Special Fire Fighting Considerations: Structural fire fighter protective clothing will not provide adequate protection. Isolate for 1/2 mile if rail or tank truck is involved in a fire. Water spray should be used to keep closed containers cool. Continue to cool container after fire is extinguished. For large fires, if possible, withdraw and allow to burn. Immediately withdraw if rising sound from venting device is heard or if fire is causing discoloration to the tank.

TOXICOLOGY

Odor: **Odor Threshold:** N.D.
Physical Contact: I rritant
TLV = N.D.   **STEL** = N.D.   **IDLH** = N.D.

Routes of Entry and Relative LD50 (or LC50)   **RTECS #** QU8930000
    Inhalation   N.D.
    Ingestion   N.D.
    Skin Absorption   N.D.

Symptoms of Exposure: Possible irritation to skin, eyes and upper respiratory tract; headache; nausea. Absorption may lead to the formation of methemoglobin resulting in cyanosis several hours after exposure.

Emergency/First Aid Treatment: Remove to ventilated area; immediately remove any contaminated clothing and wash contaminated areas for 15 minutes using water. Treat supportively and observe for possible shock. If ingested, seek immediate medical aid.

Recommended Clean-Up Procedures

Personal Protection: Level B Ensemble   **Recommended Material** N.D.

RCRA Waste #   None   **Reportable Quantities:** None

Spills: Remove all potential ignition sources. Dike to contain spill, absorb with non-combustible absorbent and take up using non-sparking tools. Decontamination will occur via evaporation

Special Emergency Information: May be harmful if inhaled, swallowed or absorbed through the skin. Vapors are heavier than air and may travel some distance to an ignition source. Vapors are more dense than air and may settle in low lying areas.

n-Propyl nitrate (CAS # 108-03-2, DOT # 1865, RTECS # TZ5075000) has similar chemical, physical and toxicological properties.

KEROSENE

Synonyms: fuel oil no. 1; coal oil

CAS Number: 8008-20-6
DOT Number: 1223

Description: Pale Amber Liquid
DOT Classification: Combustible Liquid

Molecular Weight: N.A.
Melting Point: N.A.
Boiling Point: 151°C-301°C(304°F-574°F)

Vapor Density: N.A.
Specific Gravity: <1.0

Vapor Pressure: N.A.
Water Solubility: Insoluble

Chemical Incompatibilities or Instabilities: Oxidizers.

FLAMMABILITY

NFPA Hazard Code: 0-2-0
Flash Point: 43°C-72°C (100°F-162°F)
Autoignition Temp: 210°C (410°F)

NFPA Classification: Class II Combustible
LEL: 0.7%
UEL: 5%

Fire Extinguishing Methods: Carbon Dioxide, Dry Chemical, Foam, Water Spray or Fog.

Special Fire Fighting Considerations: Isolate for 1/2 mile if rail or tank truck is involved in a fire. Water spray should be used to keep closed containers cool. Continue to cool container after fire is extinguished. For large fires, if possible, withdraw and allow to burn. Immediately withdraw if rising sound from venting device is heard or if fire is causing discoloration to the tank.

TOXICOLOGY
SUSPECTED CARCINOGEN

Odor:
Physical Contact: Irritant
TLV = N.D.

STEL = N.D.

Odor Threshold: N.D.

IDLH = N.D.

Routes of Entry and Relative LD$_{50}$ (or LC$_{50}$)
Inhalation   N.D.
Ingestion   N.D.
Skin Absorption N.D.

RTECS # OA5500000

Symptoms of Exposure: Possible irritation to skin, eyes and upper respiratory tract; headache; nausea.

Emergency/First Aid Treatment: Remove to ventilated area; immediately remove any contaminated clothing and wash contaminated areas for 15 minutes using water. Treat supportively and observe for possible shock. If ingested, seek immediate medical aid.

Recommended Clean-Up Procedures

Personal Protection: Level B Ensemble

Recommended Material Nitrile Rubber (+), Polyvinyl Alcohol (+), Viton (0)

RCRA Waste # None

Reportable Quantities: None

Spills: Remove all potential ignition sources. Dike to contain spill, absorb with non-combustible absorbent and take up using non-sparking tools. Decontaminate spill area using soapy water.

Special Emergency Information: May be harmful if inhaled, swallowed or absorbed through the skin. Vapors are heavier than air and may travel some distance to an ignition source. Vapors are more dense than air and may settle in low lying areas.

LEAD ARSENATE - PbAsHO$_4$

Synonyms:

CAS Number: 7784-40-9   **Description:** White powder
DOT Number: 1617   **DOT Classification:** Poison B

Molecular Weight: 347.1
Melting Point: °C   (°F)   **Vapor Density:**   N.A.   **Vapor Pressure:**   1 (0 mm Hg)
Boiling Point: (d)   720°C (1328°F)   **Specific Gravity:**   5.9   **Water Solubility:**   Insoluble

Chemical Incompatibilities or Instabilities:

FLAMMABILITY

NFPA Hazard Code: 2-0-0   **NFPA Classification:**
Flash Point: N.A.   **LEL:**   N.A.
Autoignition Temp: N.A.   **UEL:**   N.A.

Fire Extinguishing Methods: Use agent suitable for surrounding fire.

Special Fire Fighting Considerations: Treat all materials used or generated and equipment involved as contaminated by hazardous waste.

TOXICOLOGY
CARCINOGEN

Odor: None   **Odor Threshold:** N.D.
Physical Contact: Irritant
TLV = 0.15 mg/m^3 **STEL** = N.D.   **IDLH** = 700 mg/m^3

Routes of Entry and Relative LD$_{50}$ (or LC$_{50}$)   **RTECS # CG0980000**
   Inhalation   N.D.
   Ingestion   **8**   (100 mg/kg)
   Skin Absorption   N.D.

Symptoms of Exposure: Possible irritation to skin, eyes and upper respiratory tract; headache; nausea.

Emergency/First Aid Treatment: Remove to ventilated area; immediately remove any contaminated clothing and wash contaminated areas for 15 minutes using water. Treat supportively and observe for possible shock. If ingested, seek immediate medical aid.

Recommended Clean-Up Procedures

Personal Protection: Level B Ensemble   **Recommended Material**   N.D.

RCRA Waste #   None   **Reportable Quantities:**   1 lb (5000 lb)

Spills: Take up in non-combustible absorbent and dispose. Decontaminate spill area using soapy water. Treat all materials used or generated and equipment used involved as contaminated by hazardous waste.

Special Emergency Information: May be fatal if inhaled, swallowed or absorbed through the skin.

Lead arsenite (CAS # 10031-13-7, DOT # 1618, RTECS # OF8600000) has similar chemical, physical and toxicological properties.

LEAD NITRATE - Pb(NO$_3$)$_2$

Synonyms:

CAS Number: 10099-74-8	**Description:** White powder
DOT Number: 1469	**DOT Classification:** Oxidizer

Molecular Weight: 331.2

Melting Point: (d) 470°C (878°F)	**Vapor Density:** N.A.	**Vapor Pressure:**	1 (0 mm Hg)
Boiling Point: °C (°F)	**Specific Gravity:** 4.5	**Water Solubility:**	50%

Chemical Incompatibilities or Instabilities: Powdered metals, ammonium thiocyanate, carbon, lead hypophosphite.

FLAMMABILITY

NFPA Hazard Code:		**NFPA Classification:**
Flash Point:	N.A.	**LEL:** N.A.
Autoignition Temp:	N.A.	**UEL:** N.A.

Fire Extinguishing Methods: WATER ONLY

Special Fire Fighting Considerations: Water spray should be used to keep closed containers cool. Continue to cool container after fire is extinguished. For large fires, use unmanned devices. Fight fire from a distance or protected location, if possible.

TOXICOLOGY
QUESTIONABLE CARCINOGEN

Odor: None   **Odor Threshold:** N.D.

Physical Contact: Irritant

TLV = 0.05 mg (Pb)/m   **STEL** = N.D.   **IDLH** = 700 mg (Pb)/m^3

Routes of Entry and Relative LD$_{50}$ (or LC$_{50}$)   **RTECS # OG2100000**

Inhalation	N.D.
Ingestion	N.D.
Skin Absorption	N.D.

Symptoms of Exposure: Possible irritation to skin, eyes and upper respiratory tract; headache; nausea. Repeated exposures may cause a blue-green line to form on the gums.

Emergency/First Aid Treatment: Remove to ventilated area; immediately remove any contaminated clothing and wash contaminated areas for 15 minutes using water. Treat supportively and observe for possible shock. If ingested, seek immediate medical aid.

Recommended Clean-Up Procedures

Personal Protection: Level B Ensemble	**Recommended Material** N.D.
RCRA Waste # None	**Reportable Quantities:** 100 lb (5000 lb)

Spills: Take up in non-combustible absorbent and dispose. Decontaminate spill area using water. Treat all materials used or generated and equipment used involved as contaminated by hazardous waste.

Special Emergency Information: May be harmful if inhaled, swallowed or absorbed through the skin.

LEAD PERCHLORATE - Pb(ClO$_4$)$_2$

Synonyms:

CAS Number: 13637-76-8   **Description:** White powder
DOT Number: 1470   **DOT Classification:** Oxidizer

Molecular Weight: 406.1
Melting Point: N.D.   **Vapor Density:** N.A.   **Vapor Pressure:** 1 (0 mm Hg)
Boiling Point: N.D.   **Specific Gravity:**   **Water Solubility:** Soluble

Chemical Incompatibilities or Instabilities: Reacts with alcohols to form shock sensitive compounds.

FLAMMABILITY

NFPA Hazard Code: **NFPA Classification:**
Flash Point: N.A.   **LEL:** N.A.
Autoignition Temp: N.A.   **UEL:** N.A.

Fire Extinguishing Methods: WATER ONLY

Special Fire Fighting Considerations: Water spray should be used to keep closed containers cool. Continue to cool container after fire is extinguished. For large fires, use unmanned devices. Fight fire from a distance or protected location, if possible.

TOXICOLOGY

Odor: **Odor Threshold:** N.D.
Physical Contact: Irritant
TLV = 0.05 mg (Pb)/m^3 **STEL** = N.D.   **IDLH** = 700 mg (Pb)/m^3

Routes of Entry and Relative LD$_{50}$ (or LC$_{50}$)   **RTECS #** OG3160000
   Inhalation   N.D.
   Ingestion   N.D.
   Skin Absorption N.D.

Symptoms of Exposure: Possible irritation to skin, eyes and upper respiratory tract; headache; nausea.

Emergency/First Aid Treatment: Remove to ventilated area; immediately remove any contaminated clothing and wash contaminated areas for 15 minutes using water. Treat supportively and observe for possible shock. If ingested, seek immediate medical aid.

Recommended Clean-Up Procedures

Personal Protection: Level B Ensemble   **Recommended Material**   N.D.

RCRA Waste #   None   **Reportable Quantities:**   None

Spills: Take up in non-combustible absorbent and dispose. Decontaminate spill area using water. Treat all materials used or generated and equipment used involved as contaminated by hazardous waste.

Special Emergency Information: May be harmful if inhaled, swallowed or absorbed through the skin.

LEAD, TETRAETHYL - Pb(CH$_2$CH$_3$)$_4$

Synonyms: tetraethyl lead

CAS Number: 78-00-2
DOT Number: 1649

Description: Colorless Liquid
DOT Classification: Poison B

Molecular Weight: 323.5
Melting Point: N.D.
Boiling Point: (d) 110°C (230°F)

Vapor Density: 8.6
Specific Gravity: 1.6

Vapor Pressure: 1 (<1 mm Hg)
Water Solubility: Insoluble

Chemical Incompatibilities or Instabilities: Oxidizers.

FLAMMABILITY

NFPA Hazard Code: 3-2-3
Flash Point: 93°C (200°F)
Autoignition Temp: N.D.

NFPA Classification: Class IIIA Combustible
LEL: 1.8%
UEL: N.D.

Fire Extinguishing Methods: Carbon Dioxide, Dry Chemical, Foam, Water Spray or Fog.

Special Fire Fighting Considerations: Structural fire fighter protective clothing will not provide adequate protection. Water spray should be used to keep closed containers cool. Continue to cool container after fire is extinguished. For large fires, if possible, withdraw and allow to burn.

TOXICOLOGY
QUESTIONABLE CARCINOGEN

Odor: Pleasant
Physical Contact: Irritant
TLV = 0.075 mg/m^3 **STEL** = N.D.

Odor Threshold: N.D.

IDLH = 40 mg/m^3

Routes of Entry and Relative LD$_{50}$ (or LC$_{50}$)
Inhalation	10	(850 mg/m^3/H)
Ingestion	9	(12 mg/kg)
Skin Absorption	N.D.	

RTECS # TP4550000

Symptoms of Exposure: Possible irritation to skin, eyes and upper respiratory tract; headache; nausea.

Emergency/First Aid Treatment: Remove to ventilated area; immediately remove any contaminated clothing and wash contaminated areas for 15 minutes using water. Treat supportively and observe for possible shock. If ingested, seek immediate medical aid.

Recommended Clean-Up Procedures

Personal Protection: Level B Ensemble
RCRA Waste # P110

Recommended Material N.D.
Reportable Quantities: 10 lb (100 lb)

Spills: Remove all potential ignition sources. Dike to contain spill, absorb with non-combustible absorbent and take up using non-sparking tools. Decontaminate spill area using soapy water. Treat all materials used or generated and equipment used involved as contaminated by hazardous waste.

Special Emergency Information: May be fatal in inhaled, ingested or absorbed through the skin.

Tetramethyl lead (CAS # 75-74-1, DOT # 1649, RTECS # TP4725000) has similar chemical, physical and toxicological properties.

LITHIUM - Li

Synonyms:

CAS Number:　7439-93-2　　　　　　**Description:** Silver-White Metal
DOT Number:　1415　　　　　　　　**DOT Classification:** Flammable Solid

Molecular Weight:　6.9
Melting Point:　　179°C　(354°F)　**Vapor Density:**　N.A.　**Vapor Pressure:**　**1** (0 mm Hg)
Boiling Point:　　1317°C (2502°F)　**Specific Gravity:** 0.5　**Water Solubility:**　Decomposes

Chemical Incompatibilities or Instabilities: Reacts violently with numerous classes of compounds including water.

FLAMMABILITY

NFPA Hazard Code:　　**3-2-2-W**　　　　**NFPA Classification:**
Flash Point:　　　　°C　　(°F)　　　　**LEL:**　N.A.
Autoignition Temp:　179°C　(354°F)　　**UEL:**　N.A.

Fire Extinguishing Methods: Dry Sand, Lith-X® powder, Dry Sand, Dry Limestone.

Special Fire Fighting Considerations: Violent reaction with water yields hydrogen gas.

TOXICOLOGY

Odor: None　　　　　　　　　　　　　　　　　**Odor Threshold:** N.A.
Physical Contact: Material is extremely destructive to human tissues.
TLV = N.D.　　　　　　　　**STEL** = N.D.　　　　　**IDLH** = N.D.

Routes of Entry and Relative LD$_{50}$ (or LC$_{50}$)　　　**RTECS # OJ5540000**
　　Inhalation　　　N.D.
　　Ingestion　　　N.D.
　　Skin Absorption　N.D.

Symptoms of Exposure: Possible burns or irritation to skin, eyes and upper respiratory tract; headache; nausea.

Emergency/First Aid Treatment: Remove to ventilated area; immediately remove any contaminated clothing and wash contaminated areas for 15 minutes using water. Treat supportively and observe for possible shock. If ingested, seek immediate medical aid.

Recommended Clean-Up Procedures

Personal Protection:　Level B Ensemble　　**Recommended Material**　N.D.

RCRA Waste #　　　None　　　　　　**Reportable Quantities:**　None

Spills: Remove all potential ignition sources. Cover with dry sand, limestone or Lith-X® and place in metal containers for disposal.

Special Emergency Information: May be fatal in inhaled, ingested or absorbed through the skin.

LITHIUM ALUMINUM HYDRIDE - LiAlH$_4$

Synonyms: lithium tetrahydroaluminate

CAS Number: 16853-85-3
DOT Number: 1410

Description: Light Gray Powder
DOT Classification: Flammable Solid

Molecular Weight: 38.0
Melting Point: (D) 125°C (257°F)
Boiling Point: N.A.

Vapor Density: N.A.
Specific Gravity: 0.9

Vapor Pressure: 1 (0 mm Hg)
Water Solubility: Decomposes

Chemical Incompatibilities or Instabilities: Reacts violently with numerous classes of compounds including water.

FLAMMABILITY

NFPA Hazard Code: 3-2-2-W
Flash Point: N.A.
Autoignition Temp: N.A.

NFPA Classification:
LEL: N.A.
UEL: N.A.

Fire Extinguishing Methods: Class D Fire Extinguishers, Dry Sand, Dry Limestone.

Special Fire Fighting Considerations: Will react with water to yields hydrogen gas.

TOXICOLOGY

Odor: None
Physical Contact: Material is extremely destructive to human tissues.
TLV = N.D.   **STEL** = N.D.

Odor Threshold: N.A.

IDLH = N.D.

Routes of Entry and Relative LD$_{50}$ (or LC$_{50}$)
   Inhalation   N.D.
   Ingestion   N.D.
   Skin Absorption   N.D.

RTECS # BD0100000

Symptoms of Exposure: Possible burns or irritation to skin, eyes and upper respiratory tract; headache; nausea.

Emergency/First Aid Treatment: Remove to ventilated area; immediately remove any contaminated clothing and wash contaminated areas for 15 minutes using water. Treat supportively and observe for possible shock. If ingested, seek immediate medical aid.

Recommended Clean-Up Procedures

Personal Protection: Level B Ensemble   **Recommended Material**   N.D.

RCRA Waste #   None   **Reportable Quantities:**   None

Spills: Remove all potential ignition sources. Take up in non-combustible absorbent and place in a loosely covered metal can.

Special Emergency Information: May be harmful if inhaled, swallowed or absorbed through the skin.

Lithium aluminum hydride in an ether solution (CAS # 16853-85-3, DOT # 1411, RTECS # BD0100000) will also have the characteristics of diethyl ether. Events involving this material should be handled with extreme care. See Diethyl Ether.

LITHIUM AMIDE - LiNH$_2$

Synonyms:

CAS Number: 7782-89-0		**Description:** White powder		
DOT Number: 1412		**DOT Classification:** Flammable Solid		

Molecular Weight: 23.0

Melting Point: 400°C (752°F)	**Vapor Density:** N.A.	**Vapor Pressure:** 1 (0 mm Hg)	
Boiling Point: °C (°F)	**Specific Gravity:** 1.2	**Water Solubility:** Decomposes	

Chemical Incompatibilities or Instabilities: Oxidizers, acids, water, alcohols.

FLAMMABILITY

NFPA Hazard Code:		**NFPA Classification:**
Flash Point: N.D.		**LEL:** N.A.
Autoignition Temp: N.D.		**UEL:** N.A.

Fire Extinguishing Methods: Class D Fire Extinguisher, Dry Sand, Dry Limestone.

Special Fire Fighting Considerations: Reacts violently with water to yield ammonia.

TOXICOLOGY

Odor: None   **Odor Threshold:** N.A.

Physical Contact: Material is extremely destructive to human tissues.

TLV = N.D.   **STEL** = N.D.   **IDLH** = N.D.

Routes of Entry and Relative LD$_{50}$ (or LC$_{50}$)   **RTECS # OJ5590000**

Inhalation	N.D.
Ingestion	N.D.
Skin Absorption	N.D.

Symptoms of Exposure: Possible burns or irritation to skin, eyes and upper respiratory tract; headache; nausea. Laryngitis. . Inhalation of vapors may be fatal by causing glottis to spasm and suffocation. Exposure may result in chemical pneumonitis or pulmonary edema.

Emergency/First Aid Treatment: Remove to ventilated area; immediately remove any contaminated clothing and wash contaminated areas for 15 minutes using water. Treat supportively and observe for possible shock. If ingested, seek immediate medical aid.

Recommended Clean-Up Procedures

Personal Protection: Level B Ensemble	**Recommended Material** N.D.
RCRA Waste # None	**Reportable Quantities:** None

Spills: Remove all potential ignition sources. Take up in non-combustible absorbent and place in a loosely covered metal can.

Special Emergency Information: May be harmful if inhaled, swallowed or absorbed through the skin.

LITHIUM BOROHYDRIDE - LiBH$_4$

Synonyms: lithium tetrahydroborate

CAS Number: 16949-15-8 **Description:** White powder
DOT Number: 1413 **DOT Classification:** Flammable Solid

Molecular Weight: 21.8
Melting Point: 268°C (514°F) **Vapor Density:** N.A. **Vapor Pressure:** 1 (0 mm Hg)
Boiling Point: (D) 380°C (716°F) **Specific Gravity:** 0.6 **Water Solubility:** Decomposes

Chemical Incompatibilities or Instabilities: Acids, Alcohols, Water.

FLAMMABILITY

NFPA Hazard Code: **NFPA Classification:**
Flash Point: N.D. **LEL:** N.A.
Autoignition Temp: N.D. **UEL:** N.A.

Fire Extinguishing Methods: Class D Fire Extinguisher, Dry Sand, Dry Limestone.

Special Fire Fighting Considerations: May react with water to yield hydrogen gas.

TOXICOLOGY

Odor: **Odor Threshold:** N.D.
Physical Contact: Material is extremely destructive to human tissues.
TLV = N.D. **STEL** = N.D. **IDLH** = N.D.

Routes of Entry and Relative LD$_{50}$ (or LC$_{50}$) **RTECS # OJ6427000**
  Inhalation N.D.
  Ingestion N.D.
  Skin Absorption N.D.

Symptoms of Exposure: Possible burns or irritation to skin, eyes and upper respiratory tract; headache; nausea. Laryngitis. Inhalation of vapors may be fatal by causing glottis to spasm and suffocation. Exposure may result in chemical pneumonitis or pulmonary edema.

Emergency/First Aid Treatment: Remove to ventilated area; immediately remove any contaminated clothing and wash contaminated areas for 15 minutes using water. Treat supportively and observe for possible shock. If ingested, seek immediate medical aid.

Recommended Clean-Up Procedures

Personal Protection: Level B Ensemble **Recommended Material** N.D.

RCRA Waste # None **Reportable Quantities:** None

Spills: Remove all potential ignition sources. Take up in non-combustible absorbent and place in a loosely covered metal can.

Special Emergency Information: May be harmful if inhaled, swallowed or absorbed through the skin.

LITHIUM HYDRIDE - LiH

Synonyms:

CAS Number: 7580-67-8   **Description:** White to Pale Blue Powder
DOT Number: 1414, 2805   **DOT Classification:** Flammable Solid

Molecular Weight: 8.0
Melting Point: 680°C (1256°F)   **Vapor Density:** N.A.   **Vapor Pressure:** 1 (0 mm Hg)
Boiling Point: °C (°F)   **Specific Gravity:** 0.8   **Water Solubility:** Decomposes

Chemical Incompatibilities or Instabilities: Acids, alcohols, oxidizers, water, may be pyrophoric in moist air.

FLAMMABILITY

NFPA Hazard Code: **NFPA Classification:**
Flash Point: N.A.   **LEL:** N.A.
Autoignition Temp: N.A.   **UEL:** N.A.

Fire Extinguishing Methods: Class D Fire Extinguisher, Dry Sand, Dry Limestone.

Special Fire Fighting Considerations: Reacts with many materials, including water, to release hydrogen gas.

TOXICOLOGY

Odor: **Odor Threshold:** N.D.
Physical Contact: Material is extremely destructive to human tissues.
TLV = 0.025 mg/m^3 **STEL** = N.D.   **IDLH** = 55 mg/m^3

Routes of Entry and Relative LD$_{50}$ (or LC$_{50}$)   **RTECS #** OJ6300000
  Inhalation   N.D.
  Ingestion   N.D.
  Skin Absorption   N.D.

Symptoms of Exposure: Possible burns or irritation to skin, eyes and upper respiratory tract; headache; nausea. Laryngitis. Inhalation of vapors may be fatal by causing glottis to spasm and suffocation. Exposure may result in chemical pneumonitis or pulmonary edema.

Emergency/First Aid Treatment: Remove to ventilated area; immediately remove any contaminated clothing and wash contaminated areas for 15 minutes using water. Treat supportively and observe for possible shock. If ingested, seek immediate medical aid.

Recommended Clean-Up Procedures

Personal Protection: Level B Ensemble   **Recommended Material** N.D.

RCRA Waste #   None   **Reportable Quantities:** None

Spills: Remove all potential ignition sources. Take up in non-combustible absorbent and place in a loosely covered metal container.

Special Emergency Information: May be harmful if inhaled, swallowed or absorbed through the skin.

LITHIUM NITRIDE - Li3N

Synonyms:

CAS Number: 26134-62-3

DOT Number: 2806

Description: Red-Brown Powder

DOT Classification: Flammable Solid

Molecular Weight: 34.8

Melting Point: 845°C (1553°F)

Boiling Point: °C (°F)

Vapor Density: N.A.

Specific Gravity: 1.3

Vapor Pressure: 1 (0 mm Hg)

Water Solubility: Decomposes

Chemical Incompatibilities or Instabilities: May be pyrophoric in moist air.

FLAMMABILITY

NFPA Hazard Code:

Flash Point: N.D.

Autoignition Temp: N.D.

NFPA Classification:

LEL: N.A.

UEL: N.A.

Fire Extinguishing Methods: Class D Fire Extinguisher, Dry Sand, Dry Limestone.

Special Fire Fighting Considerations: Reacts with water to yield ammonia.

TOXICOLOGY

Odor:

Physical Contact: Material is extremely destructive to human tissues.

TLV = N.D.   **STEL** = N.D.

Odor Threshold: N.D.

IDLH = N.D.

Routes of Entry and Relative LD$_{50}$ (or LC$_{50}$)

  Inhalation   N.D.
  Ingestion   N.D.
  Skin Absorption   N.D.

RTECS # OJ6325000

Symptoms of Exposure: Possible burns or irritation to skin, eyes and upper respiratory tract; headache; nausea.

Emergency/First Aid Treatment: Remove to ventilated area; immediately remove any contaminated clothing and wash contaminated areas for 15 minutes using water. Treat supportively and observe for possible shock. If ingested, seek immediate medical aid.

Recommended Clean-Up Procedures

Personal Protection: Level B Ensemble

RCRA Waste # None

Recommended Material N.D.

Reportable Quantities: None

Spills: Remove all potential ignition sources. Take up in non-combustible absorbent and place in a loosely covered metal can.

Special Emergency Information: May be harmful if inhaled, swallowed or absorbed through the skin.

MAGNESIUM - Mg

Synonyms:

CAS Number: 7439-95-4   **Description:** Silver-White Metal
DOT Number: 1418, 1869, 2950   **DOT Classification:** Flammable Solid

Molecular Weight: 24.3
Melting Point: 650°C (1202°F)   **Vapor Density:** N.A.   **Vapor Pressure:**   1 (0 mm Hg)
Boiling Point: 1107°C (2025°F)   **Specific Gravity:** 1.7   **Water Solubility:**   Insoluble

Chemical Incompatibilities or Instabilities: Halogens, halocarbons, metal oxides, ammonium nitrate, phosphates.

FLAMMABILITY

NFPA Hazard Code: 0-1-1   **NFPA Classification:**
Flash Point: N.A.   **LEL:** N.A.
Autoignition Temp: 473°C (883°F)   **UEL:** N.A.

Fire Extinguishing Methods: Dry Sand, Met-L-X®, G-1 graphite powder.

Special Fire Fighting Considerations: For large fires, if possible, withdraw and allow to burn. Powdered magnesium can form explosive mixtures with numerous materials including air and water. Reactions with acids and water yield hydrogen gas.

TOXICOLOGY

Odor: None   **Odor Threshold:** N.A.
Physical Contact: Irritant
TLV = 10 mg/m^3 (dust)   **STEL** = N.D.   **IDLH** = N.D.

Routes of Entry and Relative LD$_{50}$ (or LC$_{50}$)   **RTECS # OM2100000**
    Inhalation   N.D.
    Ingestion   N.D.
    Skin Absorption   N.D.

Symptoms of Exposure: Possible irritation to skin, eyes and upper respiratory tract; headache; nausea.

Emergency/First Aid Treatment: Remove to ventilated area; immediately remove any contaminated clothing and wash contaminated areas for 15 minutes using water. Treat supportively and observe for possible shock. If ingested, seek immediate medical aid.

Recommended Clean-Up Procedures

Personal Protection: Level C Ensemble   **Recommended Material**   N.D.

RCRA Waste #   None   **Reportable Quantities:**   None

Spills: Remove all potential ignition sources. Carefully sweep up spill and place in an appropriate container. Avoid raising any dust.

Special Emergency Information: May be harmful if inhaled, swallowed or absorbed through the skin.

MAGNESIUM CHLORATE - Mg(ClO₃)₂

Synonyms:

CAS Number: 10326-21-3   **Description:** White powder
DOT Number: 2723   **DOT Classification:** Oxidizer

Molecular Weight: 191.2
Melting Point: 35°C (95°F)   **Vapor Density:** N.A.   **Vapor Pressure:** 1 (0 mm Hg)
Boiling Point: (d) 120°C (248°F)   **Specific Gravity:** 1.8   **Water Solubility:** Soluble

Chemical Incompatibilities or Instabilities: Metal sulfides, aluminum, arsenic, prganic matter.

FLAMMABILITY

NFPA Hazard Code: **NFPA Classification:**
Flash Point: N.A.   **LEL:** N.A.
Autoignition Temp: N.A.   **UEL:** N.A.

Fire Extinguishing Methods: WATER ONLY

Special Fire Fighting Considerations: Fight fire from a distance or protected location, if possible. Use flooding quantities of water. Water spray should be used to keep closed containers cool. Continue to cool container after fire is extinguished. For large fires, if possible, withdraw and allow to burn.

TOXICOLOGY

Odor: **Odor Threshold:** N.D.
Physical Contact: Material is extremely destructive to human tissues.
TLV = N.D.   **STEL** = N.D.   **IDLH** = N.D.

Routes of Entry and Relative LD₅₀ (or LC₅₀)   **RTECS # FO0175000**
    Inhalation   N.D.
    Ingestion   **2**   (6348 mg/kg)
    Skin Absorption   N.D.

Symptoms of Exposure: Possible burns or irritation to skin, eyes and upper respiratory tract; headache; nausea.

Emergency/First Aid Treatment: Remove to ventilated area; immediately remove any contaminated clothing and wash contaminated areas for 15 minutes using water. Treat supportively and observe for possible shock. If ingested, seek immediate medical aid.

Recommended Clean-Up Procedures

Personal Protection: Level B Ensemble   **Recommended Material**   N.D.

RCRA Waste #   None   **Reportable Quantities:**   None

Spills: Take up in non-combustible absorbent and dispose.

Special Emergency Information: May be harmful if inhaled, swallowed or absorbed through the skin.

Magnesium bromate (CAS # 7789-36-8, DOT # 1473) has similar chemical, physical and toxicological properties.

MAGNESIUM HYDRIDE - MgH$_2$

Synonyms:

CAS Number: 60616-74-2	**Description:** White powder	
DOT Number: 2010	**DOT Classification:** Flammable Solid	

Molecular Weight: 26.3

Melting Point: (d) 200°C (392°F)	**Vapor Density:** N.A.	**Vapor Pressure:** 1 (0 mm Hg)
Boiling Point: N.A.	**Specific Gravity:** 1.4	**Water Solubility:** Decomposes

Chemical Incompatibilities or Instabilities: Pyrophoric, reacts violently with water to yield hydrogen.

FLAMMABILITY

NFPA Hazard Code:	**NFPA Classification:**	
Flash Point: N.A.	**LEL:** N.A.	
Autoignition Temp: N.A.	**UEL:** N.A.	

Fire Extinguishing Methods: Class D Fire Extinguisher, Dry Sand, Dry limestone.

Special Fire Fighting Considerations: For large fires, if possible, withdraw and allow to burn.

TOXICOLOGY

Odor: **Odor Threshold:** N.D.

Physical Contact: Material is extremely destructive to human tissues.

TLV = N.D.   **STEL** = N.D.   **IDLH** = N.D.

Routes of Entry and Relative LD$_{50}$ (or LC$_{50}$)   **RTECS #** None

Inhalation	N.D.
Ingestion	N.D.
Skin Absorption	N.D.

Symptoms of Exposure: Possible burns or irritation to skin, eyes and upper respiratory tract; headache; nausea.

Emergency/First Aid Treatment: Remove to ventilated area; immediately remove any contaminated clothing and wash contaminated areas for 15 minutes using water. Treat supportively and observe for possible shock. If ingested, seek immediate medical aid.

Recommended Clean-Up Procedures

Personal Protection: Level B Ensemble	**Recommended Material** N.D.	
RCRA Waste # None	**Reportable Quantities:** None	

Spills: Remove all potential ignition sources. Take up in non-combustible absorbent and place in a loosely covered metal can.

Special Emergency Information: May be harmful if inhaled, swallowed or absorbed through the skin.

MAGNESIUM NITRATE HEXAHYDRATE - Mg(NO$_3$)$_2$·6H$_2$O

Synonyms:

CAS Number: 13446-18-9   **Description:** White powder
DOT Number: 1474   **DOT Classification:** Oxidizer

Molecular Weight: 256.4
Melting Point: 89°C (192°F)   **Vapor Density:** N.A.   **Vapor Pressure:** 1 (0 mm Hg)
Boiling Point: (d) 330°C (626°F)   **Specific Gravity:** 1.6   **Water Solubility:** Soluble

Chemical Incompatibilities or Instabilities: Dimethylformamide, reducing agents, organics, ordinary combustibles.

FLAMMABILITY

NFPA Hazard Code: **NFPA Classification:**
Flash Point: N.A.   **LEL:** N.A.
Autoignition Temp: N.A.   **UEL:** N.A.

Fire Extinguishing Methods: WATER ONLY

Special Fire Fighting Considerations: Fight fire from a distance or protected location, if possible. Use flooding quantities of water. Water spray should be used to keep closed containers cool. Continue to cool container after fire is extinguished. For large fires, if possible, withdraw and allow to burn.

TOXICOLOGY

Odor: None   **Odor Threshold:** N.A.
Physical Contact: Irritant
TLV = N.D.   **STEL** = N.D.   **IDLH** = N.D.

Routes of Entry and Relative LD$_{50}$ (or LC$_{50}$)   **RTECS #** OM3756000
   Inhalation   N.D.
   Ingestion   3   (5440 mg/kg)
   Skin Absorption   N.D.

Symptoms of Exposure: Possible irritation to skin, eyes and upper respiratory tract; headache; nausea.

Emergency/First Aid Treatment: Remove to ventilated area; immediately remove any contaminated clothing and wash contaminated areas for 15 minutes using water. Treat supportively and observe for possible shock. If ingested, seek immediate medical aid.

Recommended Clean-Up Procedures

Personal Protection: Level B Ensemble   **Recommended Material** N.D.

RCRA Waste # None   **Reportable Quantities:** None

Spills: Remove all potential ignition sources. Take up in non-combustible absorbent and dispose. Decontaminate spill area using water.

Special Emergency Information: May be harmful if inhaled, swallowed or absorbed through the skin.

Magnesium peroxide (CAS # 14452-57-4, DOT # 1476, RTECS # OM4100000) has similar chemical, physical and toxicological properties.

MALATHION

Synonyms:

CAS Number: 121-75-5   **Description:** Brown to Amber Liquid
DOT Number: 2783   **DOT Classification:** ORM-A

Molecular Weight: 330.4
Melting Point: 3°C (37°F)   **Vapor Density:** N.A.   **Vapor Pressure:** N.A.
Boiling Point: N.D.   **Specific Gravity:** 1.2   **Water Solubility:** 1%

Chemical Incompatibilities or Instabilities: Oxidizers.

FLAMMABILITY

NFPA Hazard Code: **NFPA Classification:**
Flash Point: N.A.   **LEL:** N.A.
Autoignition Temp: N.A.   **UEL:** N.A.

Fire Extinguishing Methods: Carbon Dioxide, Dry Chemical, Foam, Water Spray or Fog.

Special Fire Fighting Considerations: Structural fire fighter protective clothing will not provide adequate protection. Fight fire from a distance or protected location, if possible. Treat all materials used or generated and equipment used involved as contaminated by hazardous waste.

TOXICOLOGY
QUESTIONABLE CARCINOGEN

Odor: Garlic-like   **Odor Threshold:** N.D.
Physical Contact: Irritant
$TLV = 10$ mg/m^3 (skin)   **STEL** = N.D. $IDLH = 5000$ mg/m^3

Routes of Entry and Relative LD$_{50}$ (or LC$_{50}$)   **RTECS # WM8400000**
   Inhalation   10   (340 mg/m^3/H)
   Ingestion   6   (290 mg/kg)
   Skin Absorption   2   (4100 mg/kg)

Symptoms of Exposure: Possible irritation to skin, eyes and upper respiratory tract; headache; nausea; blood pressure depression; breathing difficulties; coma.

Emergency/First Aid Treatment: Remove to ventilated area; immediately remove any contaminated clothing and wash contaminated areas for 15 minutes using water. Treat supportively and observe for possible shock. If ingested, seek immediate medical aid. Treat for anticholinesterase poisoning.

Recommended Clean-Up Procedures

Personal Protection: Level A Ensemble   **Recommended Material**   Nitrile Rubber (+), Teflon (+),

RCRA Waste #   None   **Reportable Quantities:**   10 lb (100 lb)

Spills: Dike to contain spill and collect liquid for later disposal. Decontaminate spill area using soapy water. Treat all materials used or generated and equipment involved as contaminated by hazardous waste.

Special Emergency Information: May be fatal in inhaled, ingested or absorbed through the skin. Exposures may cause allergic sensitization.

MALEIC ANHYDRIDE

Synonyms: cis-butenedioic anhydride; toxilic anhydride

CAS Number: 108-31-6	**Description:** White powder	
DOT Number: 2215	**DOT Classification:** ORM-A	

Molecular Weight: 98.1

Melting Point: 53°C (127°F)	**Vapor Density:** 3.4	**Vapor Pressure:** 1 (<1 mm Hg)	
Boiling Point: 202°C (395°F)	**Specific Gravity:** 1.5	**Water Solubility:** Decomposes	

Chemical Incompatibilities or Instabilities: Oxidizers.

FLAMMABILITY

NFPA Hazard Code: 2-3-1	**NFPA Classification:**
Flash Point: 102°C (215°F)	**LEL:** 1.4%
Autoignition Temp: 890°C (477°F)	**UEL:** 7.1%

Fire Extinguishing Methods: Carbon Dioxide, Dry Chemical, Foam, Water Spray or Fog.

Special Fire Fighting Considerations: Water spray should be used to keep closed containers cool. Continue to cool container after fire is extinguished. Treat all materials used or generated and equipment involved as contaminated by hazardous waste.

TOXICOLOGY
QUESTIONABLE CARCINOGEN

Odor: Acrid   **Odor Threshold:** 1 mg/m^3

Physical Contact: Material is extremely destructive to human tissues.

TLV = 1 mg/m^3 **STEL** = N.D.   **IDLH** = N.D.

Routes of Entry and Relative LD$_{50}$ (or LC$_{50}$)   **RTECS # ON3675000**

Inhalation	N.D.	
Ingestion	5	(400 mg/kg)
Skin Absorption	2	(2620 mg/kg)

Symptoms of Exposure: Possible burns or irritation to skin, eyes and upper respiratory tract; headache; nausea. Laryngitis. Inhalation of vapors may be fatal by causing glottis to spasm and suffocation. Exposure may result in chemical pneumonitis or pulmonary edema.

Emergency/First Aid Treatment: Remove to ventilated area; immediately remove any contaminated clothing and wash contaminated areas for 15 minutes using water. Treat supportively and observe for possible shock. If ingested, seek immediate medical aid.

Recommended Clean-Up Procedures

Personal Protection:	Level B Ensemble	**Recommended Material** N.D.
RCRA Waste #	U147	**Reportable Quantities:** 500 lb (5000 lb)

Spills: Take up in non-combustible absorbent and dispose. Decontaminate spill area using soapy water. Treat all materials used or generated and equipment involved as contaminated by hazardous waste.

Special Emergency Information: May be harmful if inhaled, swallowed or absorbed through the skin. Small quantities or concentrations of alkali metals or amines may cause explosive decarboxylation and hazardous polymerization at temperatures above 150°C.

MALONONITRILE - NCCH$_2$CN

Synonyms: cyanoacetonitrile; malonic dinitrile

CAS Number: 109-77-3   **Description:** White powder
DOT Number: 2647   **DOT Classification:** Poison B

Molecular Weight: 66.1
Melting Point: 31°C (88°F)   **Vapor Density:** N.D.   **Vapor Pressure:** N.D..
Boiling Point: 220°C (428°F) **Specific Gravity:** 1.0   **Water Solubility:** Soluble

Chemical Incompatibilities or Instabilities: Heating to 70°C may cause an explosion, strong alkali.

FLAMMABILITY

NFPA Hazard Code: **NFPA Classification:**
Flash Point: 112°C (234°F)   **LEL:** N.A.
Autoignition Temp: N.D.   **UEL:** N.A.

Fire Extinguishing Methods: Carbon Dioxide, Dry Chemical, Foam, Water Spray or Fog.

Special Fire Fighting Considerations: Water spray should be used to keep closed containers cool. Continue to cool container after fire is extinguished.

TOXICOLOGY

Odor: **Odor Threshold:** N.D.
Physical Contact: Irritant
TLV = N.D.   **STEL** = N.D.   **IDLH** = N.D.

Routes of Entry and Relative LD$_{50}$ (or LC$_{50}$)   **RTECS # OO3150000**
    Inhalation   N.D.
    Ingestion   **8**   (61 mg/kg)
    Skin Absorption   N.D.

Symptoms of Exposure: Possible irritation to skin, eyes and upper respiratory tract; headache; nausea. Absorption may lead to the formation of methemoglobin resulting in cyanosis several hours after exposure.

Emergency/First Aid Treatment: Remove to ventilated area; immediately remove any contaminated clothing and wash contaminated areas for 15 minutes using water. Treat supportively and observe for possible shock. If ingested, seek immediate medical aid.

Recommended Clean-Up Procedures

Personal Protection: Level B Ensemble   **Recommended Material** N.D.

RCRA Waste # U149   **Reportable Quantities:** 1000 lb (1 lb)

Spills: Take up in non-combustible absorbent and dispose. Decontaminate spill area using dilute alkaline hypochlorite solution.

Special Emergency Information: May be fatal in inhaled, ingested or absorbed through the skin. May undergo hazardous polymerization at elevated temperatures or in the presence of strong alkali.

MERCURIC ACETATE - Hg(OOCCH₃)₂

Synonyms: mercury (II) acetate; mercury diacetate

CAS Number: 1600-27-7   **Description:** White powder
DOT Number: 1629   **DOT Classification:** Poison B

Molecular Weight: 318.7
Melting Point: 182°C (360°F)   **Vapor Density:** N.A.   **Vapor Pressure:** 1 (<1 mm Hg)
Boiling Point: °C (°F)   **Specific Gravity:** 3.3   **Water Solubility:** Soluble

Chemical Incompatibilities or Instabilities: Oxidizers.

FLAMMABILITY

NFPA Hazard Code: **NFPA Classification:**
Flash Point: N.A.   **LEL:** N.A.
Autoignition Temp: N.A.   **UEL:** N.A.

Fire Extinguishing Methods: Use agent suitable for surrounding fire.

Special Fire Fighting Considerations: Structural fire fighter protective clothing will not provide adequate protection. Treat all materials used or generated and equipment involved as contaminated by hazardous waste.

TOXICOLOGY

Odor: **Odor Threshold:** N.D.
Physical Contact: Irritant
TLV = 0.1 mg/m³   **STEL** = N.D.   **IDLH** = 10 mg/m³

Routes of Entry and Relative LD₅₀ (or LC₅₀)   **RTECS #** AI8575000

Inhalation	N.D.	
Ingestion	9	(41 mg/kg)
Skin Absorption	N.D.	

Symptoms of Exposure: Possible irritation to skin, eyes and upper respiratory tract; headache; nausea.

Emergency/First Aid Treatment: Remove to ventilated area; immediately remove any contaminated clothing and wash contaminated areas for 15 minutes using water. Treat supportively and observe for possible shock. If ingested, seek immediate medical aid.

Recommended Clean-Up Procedures

Personal Protection: Level B Ensemble   **Recommended Material** N.D.

RCRA Waste # None   **Reportable Quantities:** 1 lb

Spills: Take up in non-combustible absorbent and dispose. Decontaminate spill area using water. Treat all materials used or generated and equipment used involved as contaminated by hazardous waste.

Special Emergency Information: May be fatal in inhaled, ingested or absorbed through the skin.

Mercury (II) bromide (CAS # 7789-47-1, DOT # 1634, RTECS # OV7415000),
mercury (II) chloride (CAS # 7487-94-7, DOT # 1624, RTECS # OV9100000),
mercury (I) acetate (CAS # 631-60-7, DOT # 1629, RTECS # AI8570000),
mercury (I) bromide (CAS # 10031-18-2, DOT # 1634, RTECS # OV7410000),
mercury (I) chloride (CAS # 7546-30-7, DOT # 1624, RTECS # OV8750000),
mercury (I) sulfate (CAS # 7783-36-0, DOT # 1645, RTECS # OX0480000, RCRA Waste # None, RQ: 10 lb [10 lb]), and
mercury (II) sulfate (CAS # 7783-35-9, DOT # 1628, RTECS # OX0500000) have similar chemical, physical and toxicological properties.

MERCURY - Hg

Synonyms:

CAS Number: 7439-97-6	**Description:** Silver colored liquid
DOT Number: 2809	**DOT Classification:** Corrosive Material

Molecular Weight: 200.6

Melting Point:	-39°C (-38°F)	**Vapor Density:** 7.0	**Vapor Pressure:**	1 (<1 mm Hg)
Boiling Point:	357°C (675°F)	**Specific Gravity:** 13.5	**Water Solubility:**	Insoluble

Chemical Incompatibilities or Instabilities: Halogens, strong oxidizers, acetylene, ammonia.

FLAMMABILITY

NFPA Hazard Code:		**NFPA Classification:**	
Flash Point:	N.A.	**LEL:**	N.A.
Autoignition Temp:	N.A.	**UEL:**	N.A.

Fire Extinguishing Methods: Alcohol Resistant Foam, Carbon Dioxide, Dry Chemical, Water Spray or Fog.

Special Fire Fighting Considerations: Water spray should be used to keep closed containers cool. Continue to cool container after fire is extinguished. Treat all materials used or generated and equipment involved as contaminated by hazardous waste.

TOXICOLOGY

Odor: None   **Odor Threshold:** N.A.

Physical Contact: Irritant

TLV = 0.05 mg/m^3 **STEL** = N.D.   **IDLH** = 28 mg/m^3

Routes of Entry and Relative LD$_{50}$ (or LC$_{50}$)   **RTECS #** OV4550000

Inhalation	N.D.
Ingestion	N.D.
Skin Absorption	N.D.

Symptoms of Exposure: Possible irritation to skin, eyes and upper respiratory tract; headache; nausea. Chronic exposures may cause nervous system disturbances.

Emergency/First Aid Treatment: Remove to ventilated area; immediately remove any contaminated clothing and wash contaminated areas for 15 minutes using water. Treat supportively and observe for possible shock. If ingested, seek immediate medical aid.

Recommended Clean-Up Procedures

Personal Protection:	Level B Ensemble	**Recommended Material**	N.D.
RCRA Waste #	U151	**Reportable Quantities:**	1 lb (1 lb)

Spills: Dike to contain spill and collect liquid for later disposal. Take up in non-combustible absorbent and dispose. Treat all materials used or generated and equipment used involved as contaminated by hazardous waste.

Special Emergency Information: May be fatal in inhaled, ingested or absorbed through the skin.

MERCURIC CYANIDE - Hg(CN)$_2$

Synonyms: mercury (II) cyanide

CAS Number: 592-04-1
DOT Number: 1636

Description: White powder
DOT Classification: Poison B

Molecular Weight: 252.6
Melting Point: (d) 320°C (608°F)
Boiling Point: N.A.

Vapor Density: N.A.
Specific Gravity: 4.0

Vapor Pressure: 1 (<1 mm Hg)
Water Solubility: 7%

Chemical Incompatibilities or Instabilities: Contact with strong acids may liberate hydrogen cyanide gas.

FLAMMABILITY

NFPA Hazard Code: 3-0-0
Flash Point: N.A.
Autoignition Temp: N.A.

NFPA Classification:
LEL: N.A.
UEL: N.A.

Fire Extinguishing Methods: Carbon Dioxide, Dry Chemical, Foam, Water Spray or Fog.

Special Fire Fighting Considerations: Treat all materials used or generated and equipment used involved as contaminated by hazardous waste.

TOXICOLOGY

Odor:
Physical Contact: Irritant
TLV = 0.1 mg/m^3 **STEL** = N.D.

Odor Threshold: N.D.

IDLH = 10 mg/m^3

Routes of Entry and Relative LD$_{50}$ (or LC$_{50}$)
   Inhalation   N.D.
   Ingestion   N.D.
   Skin Absorption N.D.

RTECS # OW1515000

Symptoms of Exposure: Possible irritation to skin, eyes and upper respiratory tract; headache; nausea. Absorption may lead to the formation of methemoglobin resulting in cyanosis several hours after exposure.

Emergency/First Aid Treatment: Remove to ventilated area; immediately remove any contaminated clothing and wash contaminated areas for 15 minutes using water. Treat supportively and observe for possible shock. If ingested, seek immediate medical aid. Treat for cyanide poisoning as necessary.

Recommended Clean-Up Procedures

Personal Protection: Level B Ensemble

RCRA Waste #   None

Recommended Material   N.D.

Reportable Quantities: 1 lb (1 lb)

Spills: Take up in non-combustible absorbent and dispose. Decontaminate spill area using a dilute alkaline hypochlorite solution. Treat all materials used or generated and equipment used involved as contaminated by hazardous waste.

Special Emergency Information: May be fatal in inhaled, ingested or absorbed through the skin.

MERCURIC NITRATE - Hg(NO3)2

Synonyms: mercury (II) nitrate

CAS Number: 10045-94-0
DOT Number: 1625

Description: White powder
DOT Classification: Oxidizer

Molecular Weight: 324.6
Melting Point: 79°C   (174°F)
Boiling Point: °C   (°F)

Vapor Density: N.A.
Specific Gravity: 4.4

Vapor Pressure: **1** (<1 mm Hg)
Water Solubility:

Chemical Incompatibilities or Instabilities: Acetylene, ethanol, phosphine, cyanides, unsaturated hydrocarbons, sulfur, [hosphinic acid, hypophosphoric acid.

FLAMMABILITY

NFPA Hazard Code:
Flash Point: N.A.
Autoignition Temp: N.A.

NFPA Classification:
LEL: N.A.
UEL: N.A.

Fire Extinguishing Methods: WATER ONLY

Special Fire Fighting Considerations: Fight fire from a distance or protected location, if possible. Flooding quantities of water should be used to extinguish the fire. Water spray should be used to keep closed containers cool. Continue to cool container after fire is extinguished. For large fires, if possible, withdraw and allow to burn.

TOXICOLOGY

Odor:
Physical Contact: Material is extremely destructive to human tissues.
TLV = 0.1 mg/m^3 **STEL** = N.D.

Odor Threshold: N.D.

IDLH = 10 mg/m^3

Routes of Entry and Relative LD$_{50}$ (or LC$_{50}$)
   Inhalation   N.D.
   Ingestion   **9**   (26 mg/m^3)
   Skin Absorption   N.D.

RTECS # OW8225000

Symptoms of Exposure: Possible burns or irritation to skin, eyes and upper respiratory tract; headache; nausea. Inhalation of vapors may be fatal by causing glottis to spasm and suffocation. Exposure may result in chemical pneumonitis or pulmonary edema.

Emergency/First Aid Treatment: Remove to ventilated area; immediately remove any contaminated clothing and wash contaminated areas for 15 minutes using water. Treat supportively and observe for possible shock. If ingested, seek immediate medical aid.

Recommended Clean-Up Procedures

Personal Protection: Level B Ensemble   **Recommended Material**   N.D.

RCRA Waste #   None   **Reportable Quantities:**   10 lb (10 lb)

Spills: Take up in non-combustible absorbent and dispose. Decontaminate spill area using water. Treat all materials used or generated and equipment used involved as contaminated by hazardous waste.

Special Emergency Information: May be fatal in inhaled, ingested or absorbed through the skin.

Mercury (I) nitrate (CAS # 10415-75-5, DOT # 1627, RTECS # OW8000000, RCRA Waste # None, RQ 10 lb [10 lb]) has similar chemical, physical and toxicological properties.

MESITYLENE

Synonyms: 1,3,5 trimethyl benzene

CAS Number: 1060-67-8	**Description:** Colorless Liquid	
DOT Number: 2325	**DOT Classification:** Flammable Liquid	

Molecular Weight: 120.2

Melting Point: -45°C (-49°F)	**Vapor Density:** 4.1	**Vapor Pressure:** 2 (2 mm Hg)	
Boiling Point: 165°C (328°F)	**Specific Gravity:** 0.8	**Water Solubility:** Insoluble	

Chemical Incompatibilities or Instabilities: Oxidizers, nitric acid.

FLAMMABILITY

NFPA Hazard Code: 0-2-0	**NFPA Classification:** Class II Combustible
Flash Point: 50°C (122°F)	**LEL:** N.A.
Autoignition Temp: 559°C (1039°F)	**UEL:** N.A.

Fire Extinguishing Methods: Alcohol Resistant Foam, Carbon Dioxide, Dry Chemical, Water Spray or Fog.

Special Fire Fighting Considerations: Isolate for 1/2 mile if rail or tank truck is involved in a fire. Water spray should be used to keep closed containers cool. Continue to cool container after fire is extinguished. For large fires, if possible, withdraw and allow to burn. Immediately withdraw if rising sound from venting device is heard or if fire is causing discoloration to the tank.

TOXICOLOGY

Odor:

Physical Contact: Irritant

Odor Threshold: N.D.

TLV = N.D.	**STEL** = N.D.	**IDLH** = N.D.

RTECS # OX6825000

Routes of Entry and Relative LD_{50} (or LC_{50})

Inhalation	10	(96 mg/m^3/H)
Ingestion	N.D.	
Skin Absorption	N.D.	

Symptoms of Exposure: Possible irritation to skin, eyes and upper respiratory tract; headache; nausea.

Emergency/First Aid Treatment: Remove to ventilated area; immediately remove any contaminated clothing and wash contaminated areas for 15 minutes using water. Treat supportively and observe for possible shock. If ingested, seek immediate medical aid.

Recommended Clean-Up Procedures

Personal Protection: Level A Ensemble	**Recommended Material** N.D.	
RCRA Waste # None	**Reportable Quantities:** None	

Spills: Remove all potential ignition sources. Dike to contain spill, absorb with non-combustible absorbent and take up using non-sparking tools. Decontaminate spill area using soapy water.

Special Emergency Information: May be fatal in inhaled, ingested or absorbed through the skin. Vapors are heavier than air and may travel some distance to an ignition source. Vapors are more dense than air and may settle in low lying areas.

MESITYL OXIDE - $(CH_3)_2C=CHC(O)CH_3$

Synonyms: isobutenyl methyl ketone; methyl isobutenyl ketone; 4-methyl-3-penten-2-one

CAS Number: 141-79-7	**Description:** Colorless Liquid	
DOT Number: 1229	**DOT Classification:** Flammable Liquid	

Molecular Weight: 98.2

Melting Point:	-59°C (-74°F)	**Vapor Density:** 3.4	**Vapor Pressure:**	3 (9 mm Hg)
Boiling Point:	130°C (266°F)	**Specific Gravity:** 0.86	**Water Solubility:**	Insoluble

Chemical Incompatibilities or Instabilities: Acids, alkalis, halogens, oxidizers, reducing agents.

FLAMMABILITY

NFPA Hazard Code:	3-3-1	**NFPA Classification:**	Class IC Flammable
Flash Point:	31°C (87°F)	**LEL:**	1.4%
Autoignition Temp:	344°C (652°F)	**UEL:**	7.2%

Fire Extinguishing Methods: Alcohol Resistant Foam, Carbon Dioxide, Dry Chemical, Water Spray or Fog.

Special Fire Fighting Considerations: Isolate for 1/2 mile if rail or tank truck is involved in a fire. Water spray should be used to keep closed containers cool. Continue to cool container after fire is extinguished. For large fires, if possible, withdraw and allow to burn. Immediately withdraw if rising sound from venting device is heard or if fire is causing discoloration to the tank.

TOXICOLOGY

Odor: Sweet, strong

Odor Threshold: 0.08 mg/m^3

Physical Contact: Material is extremely destructive to human tissues.

TLV = 60 mg/m^3 **STEL** = 100 mg/m^3

IDLH = 20,000 mg/m^3

Routes of Entry and Relative LD$_{50}$ (or LC$_{50}$)

Inhalation	1	(36,000 mg/m^3/H)
Ingestion	3	(1120 mg/kg)
Skin Absorption	1	(5150 mg/kg)

RTECS # SB4200000

Symptoms of Exposure: Possible burns or irritation to skin, eyes and upper respiratory tract; headache; nausea; laryngitis; lachrymation. Exposure may result in chemical pneumonitis or pulmonary edema.

Emergency/First Aid Treatment: Remove to ventilated area; immediately remove any contaminated clothing and wash contaminated areas for 15 minutes using water. Treat supportively and observe for possible shock. If ingested, seek immediate medical aid.

Recommended Clean-Up Procedures

Personal Protection:	Level B Ensemble	**Recommended Material** N.D.
RCRA Waste #	None	**Reportable Quantities:** None

Spills: Remove all potential ignition sources. Dike to contain spill, absorb with non-combustible absorbent and take up using non-sparking tools.

Special Emergency Information: May be harmful if inhaled, swallowed or absorbed through the skin. May undergo hazardous polymerization. Vapors are heavier than air and may travel some distance to an ignition source. Vapors are more dense than air and may settle in low lying areas.

METHACRYLALDEHYDE - $CH_2=C(CH_3)CHO$

Synonyms: isobutenal; methacrolein; methylacrylaldehyde; 2-methylpropenal

CAS Number: 78-85-3

DOT Number: 2396

Description: Colorless Liquid

DOT Classification: Flammable Liquid

Molecular Weight: 70.1

Melting Point: -81°C (-114°F)	**Vapor Density:** 2.4	**Vapor Pressure:** 7 (120 mm Hg)
Boiling Point: 68°C (154°F)	**Specific Gravity:** 0.8	**Water Solubility:** Soluble

Chemical Incompatibilities or Instabilities: Alkali, oxidizers, reducing agents.

FLAMMABILITY

NFPA Hazard Code: 3-3-2

Flash Point: 2°C (35°F)

Autoignition Temp: N.D.

NFPA Classification: Class IB Flammable

LEL: N.D.

UEL: N.D.

Fire Extinguishing Methods: Alcohol Resistant Foam, Carbon Dioxide, Dry Chemical, Water Spray or Fog.

Special Fire Fighting Considerations: Structural fire fighter protective clothing will not provide adequate protection. Isolate for 1/2 mile if rail or tank truck is involved in a fire. Water spray should be used to keep closed containers cool. Continue to cool container after fire is extinguished. Immediately withdraw if rising sound from venting device is heard or if fire is causing discoloration to the tank. Treat all materials used or generated and equipment used involved as contaminated by hazardous waste.

TOXICOLOGY

Odor:

Physical Contact: Material is extremely destructive to human tissues.

TLV = N.D.   **STEL** = N.D.

Odor Threshold: N.D.

IDLH = N.D.

Routes of Entry and Relative LD$_{50}$ (or LC$_{50}$)

Inhalation	N.D.	
Ingestion	7	(111 mg/kg)
Skin Absorption	8	(364 mg/kg)

RTECS # OZ2625000

Symptoms of Exposure: Possible burns or irritation to skin, eyes and upper respiratory tract; laryngitis; headache; nausea. Inhalation of vapors may be fatal by causing glottis to spasm and suffocation. Exposure may result in chemical pneumonitis or pulmonary edema.

Emergency/First Aid Treatment: Remove to ventilated area; immediately remove any contaminated clothing and wash contaminated areas for 15 minutes using water. Treat supportively and observe for possible shock. If ingested, seek immediate medical aid.

Recommended Clean-Up Procedures

Personal Protection: Level B Ensemble   **Recommended Material** N.D.

RCRA Waste # None   **Reportable Quantities:** None

Spills: Remove all potential ignition sources. Dike to contain spill, absorb with non-combustible absorbent and take up using non-sparking tools. Decontaminate spill area using waster. Treat all materials used or generated and equipment involved as contaminated by hazardous waste.

Special Emergency Information: May be harmful if inhaled, swallowed or absorbed through the skin. May undergo hazardous polymerization. Vapors are heavier than air and may travel some distance to an ignition source. Vapors are more dense than air and may settle in low lying areas.

METHACRYLIC ACID - $CH_2=C(CH_3)COOH$

Synonyms: 2-methylpropenoic acid

CAS Number: 79-41-4
DOT Number: 2531

Description: Colorless Liquid
DOT Classification: Corrosive Material

Molecular Weight: 86.1
Melting Point: 16°C (61°F)
Boiling Point: 163°C (325°F)

Vapor Density:
Specific Gravity: 1.0

Vapor Pressure: 1 (1 mm Hg)
Water Solubility: Soluble

Chemical Incompatibilities or Instabilities: Oxidizers.

FLAMMABILITY

NFPA Hazard Code: 3-2-2
Flash Point: 68°C (154°F)
Autoignition Temp: 400°C (752°F)

NFPA Classification: Class IIIA Combustible
LEL: 1.6%
UEL: 8.8%

Fire Extinguishing Methods: Carbon Dioxide, Dry Chemical, Foam, Water Spray or Fog.

Special Fire Fighting Considerations: Water spray should be used to keep closed containers cool. Continue to cool container after fire is extinguished.

TOXICOLOGY

Odor: Repulsive
Physical Contact: Material is extremely destructive to human tissues.
$TLV = 70$ mg/m^3 **STEL** = N.D.

Odor Threshold: N.D.

IDLH = N.D.

Routes of Entry and Relative LD$_{50}$ (or LC$_{50}$)
  Inhalation   N.D.
  Ingestion   N.D.
  Skin Absorption   7   (500 mg/kg)

RTECS # OZ2975000

Symptoms of Exposure: Possible irritation to skin, eyes and upper respiratory tract; headache; nausea; laryngitis. Inhalation of vapors may be fatal by causing glottis to spasm and suffocation. Exposure may result in chemical pneumonitis or pulmonary edema.

Emergency/First Aid Treatment: Remove to ventilated area; immediately remove any contaminated clothing and wash contaminated areas for 15 minutes using water. Treat supportively and observe for possible shock. If ingested, seek immediate medical aid.

Recommended Clean-Up Procedures

Personal Protection: Level B Ensemble
RCRA Waste # None

Recommended Material Butyl Rubber (+), Viton (+)
Reportable Quantities: None

Spills: Dike to contain spill and collect liquid for later disposal. Decontaminate spill area using water.

Special Emergency Information: May be harmful if inhaled, swallowed or absorbed through the skin. May undergo hazardous polymerization.

METHACRYLONITRILE - $CH_2=C(CH_3)CN$

Synonyms: 2-cyanopropene

CAS Number: 126-98-7

DOT Number: 3079

Description: Colorless Liquid

DOT Classification: N.D.

Molecular Weight: 67.1

Melting Point: -36°C (-33°F)

Boiling Point: 90°C (194°F)

Vapor Density: 2.3

Specific Gravity: 0.8

Vapor Pressure: 6 (64 mm Hg)

Water Solubility: 1%

Chemical Incompatibilities or Instabilities: Oxidizers.

FLAMMABILITY

NFPA Hazard Code: 2-3-2

Flash Point: 1°C (34°F)

Autoignition Temp: N.D.

NFPA Classification: Class IB Flammable

LEL: 2.0%

UEL: 6.8%

Fire Extinguishing Methods: Alcohol Resistant Foam, Carbon Dioxide, Dry Chemical, Water Spray or Fog.

Special Fire Fighting Considerations: Structural fire fighter protective clothing will not provide adequate protection. Isolate for 1/2 mile if rail or tank truck is involved in a fire. Water spray should be used to keep closed containers cool. Continue to cool container after fire is extinguished. Immediately withdraw if rising sound from venting device is heard or if fire is causing discoloration to the tank. Treat all materials used or generated and equipment involved as contaminated by hazardous waste.

TOXICOLOGY

Odor:

Physical Contact: Material is extremely destructive to human tissues.

TLV = 2.7 mg/m^3 **STEL** = N.D.

Odor Threshold: 19 mg/m^3

IDLH = N.D.

RTECS # UD1400000

Routes of Entry and Relative LD$_{50}$ (or LC$_{50}$)

Inhalation	7	(3600 mg/m^3/H)
Ingestion	6	(250 mg/kg)
Skin Absorption	8	(320 mg/kg)

Symptoms of Exposure: Possible burns or irritation to skin, eyes and upper respiratory tract; headache; nausea. Absorption may lead to the formation of methemoglobin resulting in cyanosis several hours after exposure.

Emergency/First Aid Treatment: Remove to ventilated area; immediately remove any contaminated clothing and wash contaminated areas for 15 minutes using water. Treat supportively and observe for possible shock. If ingested, seek immediate medical aid.

Recommended Clean-Up Procedures

Personal Protection: Level A Ensemble

RCRA Waste # U152

Recommended Material Butyl Rubber (+)

Reportable Quantities: 1000 lb (1 lb)

Spills: Remove all potential ignition sources. Dike to contain spill, absorb with non-combustible absorbent and take up using non-sparking tools. Decontaminate spill area using soapy water. Treat all materials used or generated and equipment involved as contaminated by hazardous waste.

Special Emergency Information: May be fatal in inhaled, ingested or absorbed through the skin. Vapors are heavier than air and may travel some distance to an ignition source. Vapors are more dense than air and may settle in low lying areas. May undergo hazardous polymerization.

METHALLYL ALCOHOL - $CH_2=C(CH_3)CH_2OH$

Synonyms: isopropenyl alcohol; 2-methyl-2-propen-1-ol

CAS Number: 513-42-8	**Description:** Colorless Liquid
DOT Number: 2614	**DOT Classification:** Flammable Liquid

Molecular Weight: 72.1

Melting Point: °C (°F)	**Vapor Density:** 2.5	**Vapor Pressure:** N.D.
Boiling Point: 114°C (237°F)	**Specific Gravity:** 0.9	**Water Solubility:** Soluble

Chemical Incompatibilities or Instabilities: Oxidizers.

FLAMMABILITY

NFPA Hazard Code: 2-3-0	**NFPA Classification:** Class IC Flammable
Flash Point: 33°C (92°F)	**LEL:** N.D.
Autoignition Temp: N.D.	**UEL:** N.D.

Fire Extinguishing Methods: Alcohol Resistant Foam, Carbon Dioxide, Dry Chemical, Water Spray or Fog.

Special Fire Fighting Considerations: Isolate for 1/2 mile if rail or tank truck is involved in a fire. Water spray should be used to keep closed containers cool. Continue to cool container after fire is extinguished. For large fires, if possible, withdraw and allow to burn. Immediately withdraw if rising sound from venting device is heard or if fire is causing discoloration to the tank.

TOXICOLOGY

Odor: **Odor Threshold:** N.D.
Physical Contact: Irritant
TLV = N.D.   **STEL** = N.D.   **IDLH** = N.D.

Routes of Entry and Relative LD$_{50}$ (or LC$_{50}$)   **RTECS # UD5250000**
   Inhalation   N.D.
   Ingestion   N.D.
   Skin Absorption   N.D.

Symptoms of Exposure: Possible irritation to skin, eyes and upper respiratory tract; headache; nausea.

Emergency/First Aid Treatment: Remove to ventilated area; immediately remove any contaminated clothing and wash contaminated areas for 15 minutes using water. Treat supportively and observe for possible shock. If ingested, seek immediate medical aid.

Recommended Clean-Up Procedures

Personal Protection: Level B Ensemble	**Recommended Material** N.D.
RCRA Waste # None	**Reportable Quantities:** None

Spills: Remove all potential ignition sources. Dike to contain spill, absorb with non-combustible absorbent and take up using non-sparking tools. Decontaminate spill area using water.

Special Emergency Information: May be harmful if inhaled, swallowed or absorbed through the skin. Vapors are heavier than air and may travel some distance to an ignition source. Vapors are more dense than air and may settle in low lying areas.

METHANESULFONYL CHLORIDE

Synonyms:

CAS Number: 124-63-0
DOT Number: 9265

Description: Colorless Liquid
DOT Classification: N.D.

Molecular Weight: 114.6
Melting Point: N.D.
Boiling Point: N.D.

Vapor Density: 3.9
Specific Gravity: 1.5

Vapor Pressure: N.D.
Water Solubility: Decomposes

Chemical Incompatibilities or Instabilities: Alcohols, oxidizers, water.

FLAMMABILITY

NFPA Hazard Code:
Flash Point: 110°C (230°F)
Autoignition Temp: N.D.

NFPA Classification: Class IIIB Combustible
LEL: N.D.
UEL: N.D.

Fire Extinguishing Methods: Carbon Dioxide, Dry Chemical, Foam, Water Spray or Fog.

Special Fire Fighting Considerations: Structural fire fighter protective clothing will not provide adequate protection. Water spray should be used to keep closed containers cool. Continue to cool container after fire is extinguished.

		Small Spill:	Large Spill:
DOT Recommended Isolation Zones:		150 ft	150 ft
DOT Recommended Down Wind Take Cover Distance		.2 miles	0.2 miles

TOXICOLOGY

Odor: Sharp, suffocating
Physical Contact: Material is extremely destructive to human tissues.
TLV = N.D.   **STEL** = N.D.

Odor Threshold: N.D.

IDLH = N.D.

RTECS # None

Routes of Entry and Relative LD$_{50}$ (or LC$_{50}$)
   Inhalation   N.D.
   Ingestion   N.D.
   Skin Absorption   N.D.

Symptoms of Exposure: Possible burns or irritation to skin, eyes and upper respiratory tract; headache; nausea; lachrymation. Inhalation of vapors may be fatal by causing glottis to spasm and suffocation. Exposure may result in chemical pneumonitis or pulmonary edema.

Emergency/First Aid Treatment: Remove to ventilated area; immediately remove any contaminated clothing and wash contaminated areas for 15 minutes using water. Treat supportively and observe for possible shock. If ingested, seek immediate medical aid.

Recommended Clean-Up Procedures

Personal Protection: Level A Ensemble

RCRA Waste # None

Recommended Material N.D.

Reportable Quantities: None

Spills: Dike to contain spill and collect liquid for later disposal.

Special Emergency Information: May be fatal in inhaled, ingested or absorbed through the skin. This material is extremely toxic. Methane sulfonyl fluoride is an analogous toxin. The LD$_{50}$ (ingestion) for this material is 2 mg/kg which would rate a relative score of **10**.

METHANOL - CH3OH

Synonyms: carbinol; methyl alcohol; wood alcohol

CAS Number: 67-56-1
DOT Number: 1230

Description: Colorless Liquid
DOT Classification: Flammable Liquid

Molecular Weight: 32.1
Melting Point: -98°C (-144°F)
Boiling Point: 64°C (147°F)

Vapor Density: 1.1
Specific Gravity: 0.8

Vapor Pressure: 7 (100 mm Hg)
Water Solubility: Soluble

Chemical Incompatibilities or Instabilities: Strong oxidizers, perchlorates; reactive metals, hydrides.

FLAMMABILITY

NFPA Hazard Code: 1-3-0
Flash Point: 11°C (52°F)
Autoignition Temp: 464°C (867°F)

NFPA Classification: Class IB Flammable
LEL: 6.0%
UEL: 36.0%

Fire Extinguishing Methods: Alcohol Resistant Foam, Carbon Dioxide, Dry Chemical, Water Spray or Fog.

Special Fire Fighting Considerations: Structural fire fighter protective clothing will not provide adequate protection. Isolate for 1/2 mile if rail or tank truck is involved in a fire. Water spray should be used to keep closed containers cool. Continue to cool container after fire is extinguished. Immediately withdraw if rising sound from venting device is heard or if fire is causing discoloration to the tank.

TOXICOLOGY

Odor: Sweet
Physical Contact: Irritant
TLV = 260 mg/m^3

Odor Threshold: 4 mg/m^3

STEL = 420 mg/m^3

IDLH = 33,250 mg/m^3

Routes of Entry and Relative LD$_{50}$ (or LC$_{50}$)

Inhalation	1	(330,000 mg/m^3/H)
Ingestion	2	(5628 mg/kg)
Skin Absorption	N.D.	

RTECS # PC1400000

Symptoms of Exposure: Possible irritation to skin, eyes and upper respiratory tract; headache; nausea; acidosis, visual impairment or blindness; convulsions.

Emergency/First Aid Treatment: Remove to ventilated area; immediately remove any contaminated clothing and wash contaminated areas for 15 minutes using water. Treat supportively and observe for possible shock. If ingested, seek immediate medical aid.

Recommended Clean-Up Procedures

Personal Protection: Level B Ensemble

Recommended Material Butyl Rubber (+), Saranex (+), Teflon (+), Viton (+)

RCRA Waste # U154

Reportable Quantities: 5000 lb (1 lb)

Spills: Remove all potential ignition sources. Dike to contain spill, absorb with non-combustible absorbent and take up using non-sparking tools. Decontaminate spill area using water.

Special Emergency Information: May be harmful if inhaled, swallowed or absorbed through the skin. Vapors are heavier than air and may travel some distance to an ignition source. Vapors are more dense than air and may settle in low lying areas.

METHYL ACETATE - CH3OC(O)CH3

Synonyms: acetic acid, methyl ester

CAS Number: 79-20-9

DOT Number: 1231

Description: Colorless Liquid

DOT Classification: Flammable Liquid

Molecular Weight: 74.1

Melting Point: -98°C (-144°F)

Boiling Point: 60°C (140°F)

Vapor Density: 2.8

Specific Gravity: 0.9

Vapor Pressure: 7 (170 mm Hg)

Water Solubility: Soluble

Chemical Incompatibilities or Instabilities: Oxidizers.

FLAMMABILITY

NFPA Hazard Code: 1-3-0

Flash Point: -10°C (14°F)

Autoignition Temp: 454°C (850°F)

NFPA Classification: Class IB Flammable

LEL: 3.1%

UEL: 16.0%

Fire Extinguishing Methods: Alcohol Resistant Foam, Carbon Dioxide, Dry Chemical, Water Spray or Fog.

Special Fire Fighting Considerations: Isolate for 1/2 mile if rail or tank truck is involved in a fire. Water spray should be used to keep closed containers cool. Continue to cool container after fire is extinguished. For large fires, if possible, withdraw and allow to burn. Immediately withdraw if rising sound from venting device is heard or if fire is causing discoloration to the tank.

TOXICOLOGY

Odor: Sweet, fruity

Physical Contact: Irritant

TLV = 606 mg/m^3

Odor Threshold: 0.6 mg/m^3

STEL = 757 mg/m^3

IDLH = 30,000 mg/m^3

Routes of Entry and Relative LD$_{50}$ (or LC$_{50}$)

Inhalation	N.D.
Ingestion	N.D.
Skin Absorption	N.D.

RTECS # AI9100000

Symptoms of Exposure: Possible irritation to skin, eyes and upper respiratory tract; headache; nausea. May act as a narcotic in high concentrations.

Emergency/First Aid Treatment: Remove to ventilated area; immediately remove any contaminated clothing and wash contaminated areas for 15 minutes using water. Treat supportively and observe for possible shock. If ingested, seek immediate medical aid.

Recommended Clean-Up Procedures

Personal Protection: Level B Ensemble

RCRA Waste # None

Recommended Material Butyl Rubber (+)

Reportable Quantities: None

Spills: Remove all potential ignition sources. Dike to contain spill, absorb with non-combustible absorbent and take up using non-sparking tools. Decontaminate spill area using soapy water.

Special Emergency Information: May be harmful if inhaled, swallowed or absorbed through the skin. Vapors are heavier than air and may travel some distance to an ignition source. Vapors are more dense than air and may settle in low lying areas.

METHYL ACRYLATE - CH$_2$=CHC(O)OCH$_3$

Synonyms: acrylic acid, methyl ester; methyl propenate; 2-propenic acid, methyl ester

CAS Number: 96-33-3
DOT Number: 1919

Description: Colorless Liquid
DOT Classification: Flammable Liquid

Molecular Weight: 86.1
Melting Point: -77°C (-106°F)
Boiling Point: 81°C (177°F)

Vapor Density:
Specific Gravity: <1.0

Vapor Pressure: 6 (68 mm Hg)
Water Solubility: Insoluble

Chemical Incompatibilities or Instabilities: Oxidizers, peroxides.

FLAMMABILITY

NFPA Hazard Code: 2-3-2
Flash Point: -3°C (27°F)
Autoignition Temp: 468°C (875°F)

NFPA Classification: Class IB Flammable
LEL: 2.8%
UEL: 25.0%

Fire Extinguishing Methods: Alcohol Resistant Foam, Carbon Dioxide, Dry Chemical, Water Spray or Fog.

Special Fire Fighting Considerations: Isolate for 1/2 mile if rail or tank truck is involved in a fire. Water spray should be used to keep closed containers cool. Continue to cool container after fire is extinguished. For large fires, if possible, withdraw and allow to burn. Immediately withdraw if rising sound from venting device is heard or if fire is causing discoloration to the tank.

TOXICOLOGY

Odor: Airplane glue
Physical Contact: Material is extremely destructive to human tissues.
TLV = 35 mg/m^3 **STEL** = N.D.

Odor Threshold: 0.01 mg/m^3

IDLH = 3500 mg/m^3

Routes of Entry and Relative LD$_{50}$ (or LC$_{50}$)

Inhalation	5	(19,000 mg/m^3/H)
Ingestion	6	(300 mg/kg)
Skin Absorption	4	(1243 mg/kg)

RTECS # AT2800000

Symptoms of Exposure: Possible burns or irritation to skin, eyes and upper respiratory tract; headache; nausea. Exposure may result in chemical pneumonitis or pulmonary edema.

Emergency/First Aid Treatment: Remove to ventilated area; immediately remove any contaminated clothing and wash contaminated areas for 15 minutes using water. Treat supportively and observe for possible shock. If ingested, seek immediate medical aid.

Recommended Clean-Up Procedures

Personal Protection: Level B Ensemble

Recommended Material Butyl Rubber (+), Teflon (0)

RCRA Waste # None

Reportable Quantities: None

Spills: Remove all potential ignition sources. Dike to contain spill, absorb with non-combustible absorbent and take up using non-sparking tools. Decontaminate spill area using soapy water.

Special Emergency Information: May be harmful if inhaled, swallowed or absorbed through the skin. Vapors are heavier than air and may travel some distance to an ignition source. Vapors are more dense than air and may settle in low lying areas. May undergo hazardous polymerization. May form unstable peroxides upon exposure to air. Peroxides can initiate hazardous polymerization.

METHYLAL - CH3OCH3OCH3

Synonyms: dimethoxymethane; formal; methylene dimethyl ether

CAS Number: 109-87-5
DOT Number: 1234

Description: Colorless Liquid
DOT Classification: Flammable Liquid

Molecular Weight: 76.1
Melting Point: N.D.
Boiling Point: 44°C (111°F)

Vapor Density: 2.6
Specific Gravity: 0.9

Vapor Pressure: 8 (330 mm Hg)
Water Solubility: Soluble

Chemical Incompatibilities or Instabilities: Oxidizers.

FLAMMABILITY

NFPA Hazard Code: 2-3-2
Flash Point: -32°C (-26°F)
Autoignition Temp: 237°C (459°F)

NFPA Classification: Class IB Flammable
LEL: 2.2%
UEL: 13.8%

Fire Extinguishing Methods: Alcohol Resistant Foam, Carbon Dioxide, Dry Chemical, Water Spray or Fog.

Special Fire Fighting Considerations: Structural fire fighter protective clothing will not provide adequate protection. Isolate for 1/2 mile if rail or tank truck is involved in a fire. Water spray should be used to keep closed containers cool. Continue to cool container after fire is extinguished. Immediately withdraw if rising sound from venting device is heard or if fire is causing discoloration to the tank.

TOXICOLOGY

Odor: Chloroform-like
Physical Contact: Irritant
$TLV = 3110$ mg/m^3 **STEL** = N.D.

Odor Threshold: N.D.

$IDLH = 47,000$ mg/m^3

Routes of Entry and Relative LD$_{50}$ (or LC$_{50}$)
Inhalation   4   (47,000 mg/m^3/H)
Ingestion   N.D.
Skin Absorption   N.D.

RTECS # PA8750000

Symptoms of Exposure: Possible irritation to skin, eyes and upper respiratory tract; headache; nausea. May act as a narcotic in high concentrations.

Emergency/First Aid Treatment: Remove to ventilated area; immediately remove any contaminated clothing and wash contaminated areas for 15 minutes using water. Treat supportively and observe for possible shock. If ingested, seek immediate medical aid.

Recommended Clean-Up Procedures

Personal Protection: Level B Ensemble
Recommended Material N.D.

RCRA Waste # None
Reportable Quantities: None

Spills: Remove all potential ignition sources. Dike to contain spill, absorb with non-combustible absorbent and take up using non-sparking tools. Decontaminate spill area using soapy water.

Special Emergency Information: May be harmful if inhaled, swallowed or absorbed through the skin. Vapors are heavier than air and may travel some distance to an ignition source. Vapors are more dense than air and may settle in low lying areas. May form unstable peroxides upon exposure to air.

METHYLALLYL CHLORIDE - $CH_2=C(CH_3)CH_2Cl$

Synonyms: 3-chloro-2-methylpropene

CAS Number: 563-47-3	**Description:** Colorless Liquid	
DOT Number: 2554	**DOT Classification:** IMO	

Molecular Weight: 90.6

Melting Point: <-80°C (<-62°F)	**Vapor Density:** 3.1	**Vapor Pressure:** 7 (102 mm Hg)
Boiling Point: 72°C (160°F)	**Specific Gravity:** 0.9	**Water Solubility:** Insoluble

Chemical Incompatibilities or Instabilities: Oxidizers.

FLAMMABILITY

NFPA Hazard Code:	**NFPA Classification:** Class IB Flammable
Flash Point: -12°C (9°F)	**LEL:** 3.2%
Autoignition Temp: 482°C (899°F)	**UEL:** 8.1%

Fire Extinguishing Methods: Alcohol Resistant Foam, Carbon Dioxide, Dry Chemical, Water Spray or Fog.

Special Fire Fighting Considerations: Isolate for 1/2 mile if rail or tank truck is involved in a fire. Water spray should be used to keep closed containers cool. Continue to cool container after fire is extinguished. For large fires, if possible, withdraw and allow to burn. Immediately withdraw if rising sound from venting device is heard or if fire is causing discoloration to the tank.

TOXICOLOGY
CARCINOGEN

Odor: Disagreeable   **Odor Threshold:** N.D.
Physical Contact: Irritant
TLV = N.D.   **STEL** = N.D.   **IDLH** = N.D.

Routes of Entry and Relative LD$_{50}$ (or LC$_{50}$)   **RTECS # UC8050000**
  Inhalation   6   (17,000 mg/m^3/H)
  Ingestion   N.D.
  Skin Absorption   N.D.

Symptoms of Exposure: Possible irritation to skin, eyes and upper respiratory tract; headache; nausea; laryngitis; lachrymation.

Emergency/First Aid Treatment: Remove to ventilated area; immediately remove any contaminated clothing and wash contaminated areas for 15 minutes using water. Treat supportively and observe for possible shock. If ingested, seek immediate medical aid.

Recommended Clean-Up Procedures

Personal Protection: Level B Ensemble	**Recommended Material** Viton (+)
RCRA Waste # None	**Reportable Quantities:** None

Spills: Remove all potential ignition sources. Dike to contain spill, absorb with non-combustible absorbent and take up using non-sparking tools. Decontaminate spill area using soapy water.

Special Emergency Information: May be harmful if inhaled, swallowed or absorbed through the skin. Vapors are heavier than air and may travel some distance to an ignition source. Vapors are more dense than air and may settle in low lying areas.

METHYL ALUMINUM SESQUICHLORIDE - $(CH_3)_2AlCl\cdot Cl_2AlCH_3$

Synonyms: trichlorotrimethyldialuminum

CAS Number: 12542-85-7   **Description:** Colorless Liquid (above 23°C)
DOT Number: 1927   **DOT Classification:** Flammable Solid

Molecular Weight: 205.4

Melting Point:	23°C (73°F)	**Vapor Density:** N.A.	**Vapor Pressure:** N.A.
Boiling Point:	144°C (291°F)	**Specific Gravity:** 1.2	**Water Solubility:** Decomposes

Chemical Incompatibilities or Instabilities: Alcohols, oxidizers, oxygen, water.

FLAMMABILITY

NFPA Hazard Code: ?-3-3-~~W~~   **NFPA Classification:**
Flash Point: -18°C (1°F)   **LEL:** N.A.
Autoignition Temp: N.D.   **UEL:** N.A.

Fire Extinguishing Methods: Dry Chemical, Dry Sand, Dry Limestone.

Special Fire Fighting Considerations: For large fires, if possible, withdraw and allow to burn. Pyrophoric. Reacts violently with water to yield flammable or explosive gas.

TOXICOLOGY

Odor: **Odor Threshold:** N.D.
Physical Contact: Material is extremely destructive to human tissues.
TLV = 2 mg (Al)/m^3 **STEL** = N.D.   **IDLH** = N.D.

Routes of Entry and Relative LD$_{50}$ (or LC$_{50}$)   **RTECS # BD1970000**
  Inhalation   N.D.
  Ingestion   N.D.
  Skin Absorption   N.D.

Symptoms of Exposure: Possible burns or irritation to skin, eyes and upper respiratory tract; headache; nausea; laryngitis; shortness of breath. Inhalation of vapors may be fatal by causing glottis to spasm and suffocation. Exposure may result in chemical pneumonitis or pulmonary edema.

Emergency/First Aid Treatment: Remove to ventilated area; immediately remove any contaminated clothing and wash contaminated areas for 15 minutes using water. Treat supportively and observe for possible shock. If ingested, seek immediate medical aid.

Recommended Clean-Up Procedures

Personal Protection: Level B Ensemble   **Recommended Material** N.D.

RCRA Waste # None   **Reportable Quantities:** None

Spills: Remove all potential ignition sources. Dike to contain spill, absorb with non-combustible absorbent and take up using non-sparking tools. Place in loosely covered metal container. Do not allow water inside container.

Special Emergency Information: May be fatal in inhaled, ingested or absorbed through the skin. Material should be stored under nitrogen to minimize risk of explosion.

Methyl aluminum sesquibromide (CAS # 12263-85-3, DOT # 1296) has similar chemical, physical and toxicological properties.

METHYLAMINE - CH3NH2

Synonyms: aminomethane; monomethylamine

CAS Number: 74-89-5   **Description:** Colorless Gas
DOT Number: 1061   **DOT Classification:** Flammable Gas

Molecular Weight: 31.1
Melting Point: -94°C (-136°F)   **Vapor Density:**   1.1   **Vapor Pressure:**   Gas
Boiling Point: -6°C   (21°F)   **Specific Gravity:**   **Water Solubility:**   Soluble

Chemical Incompatibilities or Instabilities: Nitrated alkanes, oxidizers, halogens, nitric acid.

FLAMMABILITY

NFPA Hazard Code: 3-4-0   **NFPA Classification:**
Flash Point: GAS   **LEL:**   4.9%
Autoignition Temp: 430°C (806°F) **UEL:**   20.7%

Fire Extinguishing Methods: Carbon Dioxide, Dry Chemical, Foam, Water Spray or Fog.

Special Fire Fighting Considerations: Allow to burn unless the leak can safely be stopped. Water spray should be used to keep closed containers cool. Continue to cool container after fire is extinguished. For large fires, if possible, withdraw and allow to burn. Immediately withdraw if rising sound from venting device is heard or if fire is causing discoloration to the tank.

DOT Recommended Isolation Zones: Small Spill: 150 ft   Large Spill: 900 ft
DOT Recommended Down Wind Take Cover Distance   Small Spill: 0.8 miles   Large Spill: 3 miles

TOXICOLOGY

Odor: Ammonia, fishy   **Odor Threshold:** 0.001 mg/m^3
Physical Contact: Material is extremely destructive to human tissues.
TLV = 6.4 mg/m^3 **STEL** = 19 mg/m^3 **IDLH** = 130 mg/m^3

Routes of Entry and Relative LD$_{50}$ (or LC$_{50}$)   **RTECS # PF6300000**
    Inhalation   N.D.
    Ingestion   N.D.
    Skin Absorption N.D.

Symptoms of Exposure: Possible burns or irritation to skin, eyes and upper respiratory tract; headache; nausea; breathing difficulties. Inhalation of vapors may be fatal by causing glottis to spasm and suffocation. Exposure may result in chemical pneumonitis or pulmonary edema.

Emergency/First Aid Treatment: Remove to ventilated area; immediately remove any contaminated clothing and wash contaminated areas for 15 minutes using water. Treat supportively and observe for possible shock. If ingested, seek immediate medical aid.

Recommended Clean-Up Procedures

Personal Protection: Level B Ensemble   **Recommended Material**   Butyl Rubber (+), Nitrile Rubber (+),
                                                                          Viton (+)

RCRA Waste #   None   **Reportable Quantities:**   None

Spills: Remove all potential ignition sources. Stop leak if it can safely be done. Keep area isolated until vapors dissipate. A water fog may be used to keep vapors down.

Special Emergency Information: May be fatal in inhaled, ingested or absorbed through the skin.

Aqueous solutions of methylamine (CAS # 74-89-5, DOT # 1235) are common, but no less dangerous.

n-METHYLANILINE

Synonyms: n-methylaminobenzene; n-methylphenylamine

CAS Number: 100-61-8   **Description:** Colorless to amber liquid
DOT Number: 2294   **DOT Classification:** Poison B

Molecular Weight: 107.1
Melting Point: -57°C (-71°F)   **Vapor Density:**   N.D.   **Vapor Pressure:**   1 (<1 mm Hg)
Boiling Point: 197°C (387°F)   **Specific Gravity:** 1.0   **Water Solubility:**   <1%

Chemical Incompatibilities or Instabilities: Strong oxidizers.

FLAMMABILITY

NFPA Hazard Code: **NFPA Classification:**
Flash Point: 78°C (174°F)   **LEL:** N.D.
Autoignition Temp: N.D.   **UEL:** N.D.

Fire Extinguishing Methods: Carbon Dioxide, Dry Chemical, Foam, Water Spray or Fog.

Special Fire Fighting Considerations: Structural fire fighter protective clothing will not provide adequate protection. Fight fire from a distance or protected location, if possible. Water spray should be used to keep closed containers cool. Continue to cool container after fire is extinguished. Treat all materials used or generated and equipment used involved as contaminated by hazardous waste.

TOXICOLOGY

Odor: **Odor Threshold:** 7 mg/m^3
Physical Contact: Irritant
TLV = 2.2 mg/m^3 **STEL** = N.D.   **IDLH** = N.D.

Routes of Entry and Relative LD$_{50}$ (or LC$_{50}$)   **RTECS # BY4550000**
    Inhalation   N.D.
    Ingestion   N.D.
    Skin Absorption   N.D.

Symptoms of Exposure: Possible irritation to skin, eyes and upper respiratory tract; headache; nausea. Absorption may lead to the formation of methemoglobin resulting in cyanosis several hours after exposure.

Emergency/First Aid Treatment: Remove to ventilated area; immediately remove any contaminated clothing and wash contaminated areas for 15 minutes using water. Treat supportively and observe for possible shock. If ingested, seek immediate medical aid.

Recommended Clean-Up Procedures

Personal Protection: Level B Ensemble   **Recommended Material**   N.D.

RCRA Waste #   None   **Reportable Quantities:**   None

Spills: Dike to contain spill and collect liquid for later disposal. Decontaminate spill area using soapy water.

Special Emergency Information: May be harmful if inhaled, swallowed or absorbed through the skin.

METHYLBROMIDE - CH₃Br

Synonyms: bromomethane

CAS Number: 74-83-9   **Description:** Colorless Gas
DOT Number: 1062   **DOT Classification:** Poison A

Molecular Weight: 95.5
Melting Point: -93°C (-135°F)   **Vapor Density:**   3.3   **Vapor Pressure:**   GAS above 39°F
Boiling Point: 4°C   (39°F)   **Specific Gravity:** 1.7 @ 0°F **Water Solubility:**   Insoluble

Chemical Incompatibilities or Instabilities: Reactive metals, dimethylsulfoxide, ethylene oxide.

FLAMMABILITY

NFPA Hazard Code: 3-1-0   **NFPA Classification:**
Flash Point: Gas   **LEL:**   10.0%
Autoignition Temp: 537°C (999°F)   **UEL:**   16.0%

Fire Extinguishing Methods: Carbon Dioxide, Dry Chemical, Foam, Water Spray or Fog.

Special Fire Fighting Considerations: Structural fire fighter protective clothing will not provide adequate protection. Fight fire from a distance or protected location, if possible. Treat all materials used or generated and equipment used involved as contaminated by hazardous waste.

DOT Recommended Isolation Zones: Small Spill: 600 ft   Large Spill: 900 ft
DOT Recommended Down Wind Take Cover Distance   Small Spill: 2 miles   Large Spill: 3 miles

TOXICOLOGY
SUSPECTED CARCINOGEN

Odor: Chloroform-like   **Odor Threshold:** N.D.
Physical Contact: Material is extremely destructive to human tissues.
TLV = 19 mg/m³ (Skin)   **STEL** = N.D.   **IDLH** = 8,000 mg/m³

Routes of Entry and Relative LD₅₀ (or LC₅₀)   **RTECS # PA4900000**
   Inhalation   6   (9500 mg/m³/H)
   Ingestion   6   (214 mg/kg)
   Skin Absorption   N.D.

Symptoms of Exposure: Possible burns or irritation to skin, eyes and upper respiratory tract; headache; nausea. Inhalation of vapors may be fatal by causing glottis to spasm and suffocation. Exposure may result in chemical pneumonitis or pulmonary edema.

Emergency/First Aid Treatment: Remove to ventilated area; immediately remove any contaminated clothing and wash contaminated areas for 15 minutes using water. Treat supportively and observe for possible shock. If ingested, seek immediate medical aid.

Recommended Clean-Up Procedures

Personal Protection: Level B Ensemble   **Recommended Material**   Polyvinyl Alcohol (0)

RCRA Waste #   U029   **Reportable Quantities:**   1000 lb (1 lb)

Spills: Remove all potential ignition sources. Stop leak if it can safely be done. Keep area isolated until vapors dissipate.

Special Emergency Information: May be fatal in inhaled, ingested or absorbed through the skin. Vapors are heavier than air and may travel some distance to an ignition source. Vapors are more dense than air and may settle in low lying areas.

METHYL BUTANONE - $(CH_3)_2CHC(O)CH_3$

Synonyms: methyl isopropyl ketone; 3-methyl butan-2-one

CAS Number: 563-80-4   **Description:** Colorless Liquid
DOT Number: 2397   **DOT Classification:** Flammable Liquid

Molecular Weight: 86.2
Melting Point: -92°C (-134°F)   **Vapor Density:** N.A.   **Vapor Pressure:** 5 (40 mm Hg)
Boiling Point: 95°C (203°F)   **Specific Gravity:** 0.8   **Water Solubility:** Insoluble

Chemical Incompatibilities or Instabilities: Strong oxidizers.

FLAMMABILITY

NFPA Hazard Code: **NFPA Classification:**
Flash Point: 6°C (43°F)   **LEL:** N.D.
Autoignition Temp: N.D.   **UEL:** N.D.

Fire Extinguishing Methods: Alcohol Resistant Foam, Carbon Dioxide, Dry Chemical, Water Spray or Fog.

Special Fire Fighting Considerations: Isolate for 1/2 mile if rail or tank truck is involved in a fire. Water spray should be used to keep closed containers cool. Continue to cool container after fire is extinguished. For large fires, if possible, withdraw and allow to burn. Immediately withdraw if rising sound from venting device is heard or if fire is causing discoloration to the tank.

TOXICOLOGY

Odor: Sweet   **Odor Threshold:** 16 mg/m^3
Physical Contact: Irritant
TLV = 705 mg/m^3 **STEL** = N.D.   **IDLH** = N.D.

Routes of Entry and Relative LD$_{50}$ (or LC$_{50}$)   **RTECS # EL9100000**

Route		
Inhalation	N.D.	
Ingestion	8	(148 mg/kg)
Skin Absorption	1	(6350 mg/kg)

Symptoms of Exposure: Possible irritation to skin, eyes and upper respiratory tract; headache; nausea.

Emergency/First Aid Treatment: Remove to ventilated area; immediately remove any contaminated clothing and wash contaminated areas for 15 minutes using water. Treat supportively and observe for possible shock. If ingested, seek immediate medical aid.

Recommended Clean-Up Procedures

Personal Protection: Level B Ensemble   **Recommended Material** N.D.

RCRA Waste # None   **Reportable Quantities:** None

Spills: Remove all potential ignition sources. Dike to contain spill, absorb with non-combustible absorbent and take up using non-sparking tools. Decontaminate spill area using soapy water.

Special Emergency Information: May be harmful if inhaled, swallowed or absorbed through the skin. Vapors are heavier than air and may travel some distance to an ignition source. Vapors are more dense than air and may settle in low lying areas.

3-METHYL-1-BUTENE - (CH₃)₂CHCH=CH₂

Synonyms:

CAS Number: 563-45-1   **Description:** Colorless Liquid
DOT Number: 2561   **DOT Classification:** Flammable Liquid

Molecular Weight: 70.1
Melting Point: -138°C (-216°F)   **Vapor Density:** 2.4   **Vapor Pressure:**   7 (200 mm Hg)
Boiling Point: 31°C (88°F)   **Specific Gravity:** 0.7   **Water Solubility:**   Insoluble

Chemical Incompatibilities or Instabilities: Strong oxidizers.

FLAMMABILITY

NFPA Hazard Code: 2-4-0   **NFPA Classification:**
Flash Point: -56°C (-70°F)   **LEL:**   1.5%
Autoignition Temp: N.D.   **UEL:**   9.1%

Fire Extinguishing Methods: Alcohol Resistant Foam, Carbon Dioxide, Dry Chemical, Water Spray or Fog.

Special Fire Fighting Considerations: Isolate for 1/2 mile if rail or tank truck is involved in a fire. Water spray should be used to keep closed containers cool. Continue to cool container after fire is extinguished. For large fires, if possible, withdraw and allow to burn. Immediately withdraw if rising sound from venting device is heard or if fire is causing discoloration to the tank.

TOXICOLOGY

Odor: **Odor Threshold:** N.D.
Physical Contact: Irritant
TLV = N.D.   **STEL** = N.D.   **IDLH** = N.D.

Routes of Entry and Relative LD₅₀ (or LC₅₀)   **RTECS # EM7600000**
    Inhalation   N.D.
    Ingestion   N.D.
    Skin Absorption   N.D.

Symptoms of Exposure: Possible irritation to skin, eyes and upper respiratory tract; headache; nausea.

Emergency/First Aid Treatment: Remove to ventilated area; immediately remove any contaminated clothing and wash contaminated areas for 15 minutes using water. Treat supportively and observe for possible shock. If ingested, seek immediate medical aid.

Recommended Clean-Up Procedures

Personal Protection: Level B Ensemble   **Recommended Material**   N.D.

RCRA Waste #   None   **Reportable Quantities:**   None

Spills: Remove all potential ignition sources. Dike to contain spill, absorb with non-combustible absorbent and take up using non-sparking tools. Decontaminate spill area using soapy water.

Special Emergency Information: May be harmful if inhaled, swallowed or absorbed through the skin. Vapors are heavier than air and may travel some distance to an ignition source. Vapors are more dense than air and may settle in low lying areas.

2-Methyl-1-butene (CAS # 563-46-2, DOT # 2459, RTECS # EM7550000) and
2-methyl-2-butene (CAS # 513-35-9, DOT # 2460, RTECS # EM7650000) have similar chemical, physical and toxicological properties.

N-METHYLBUTYLAMINE - CH3NH CH2(CH2)2CH3

CAS Number: 110-68-9
DOT Number: 2945

Description: Colorless Liquid
DOT Classification: Flammable Liquid

Molecular Weight: 87.2
Melting Point: °C (°F)
Boiling Point: 91°C (196°F)

Vapor Density: 3.0
Specific Gravity: 0.7

Vapor Pressure: N.D.
Water Solubility: Soluble

Chemical Incompatibilities or Instabilities: Oxidizers.

FLAMMABILITY

NFPA Hazard Code: 3-3-0
Flash Point: 13°C (55°F)
Autoignition Temp: N.D.

NFPA Classification: Class IB Flammable
LEL: N.D.
UEL: N.D.

Fire Extinguishing Methods: Carbon Dioxide, Dry Chemical, Foam, Water Spray or Fog.

Special Fire Fighting Considerations: Isolate for 1/2 mile if rail or tank truck is involved in a fire. Water spray should be used to keep closed containers cool. Continue to cool container after fire is extinguished. Immediately withdraw if rising sound from venting device is heard or if fire is causing discoloration to the tank.

TOXICOLOGY

Odor:
Physical Contact: Material is extremely destructive to human tissues.
TLV = N.D.   **STEL** = N.D.

Odor Threshold: N.D.

IDLH = N.D.

Routes of Entry and Relative LD$_{50}$ (or LC$_{50}$)
Inhalation   N.D.
Ingestion   N.D.
Skin Absorption   **4**   (1260 mg/kg)

RTECS # EO5250000

Symptoms of Exposure: Possible burns or irritation to skin, eyes and upper respiratory tract; headache; nausea. Inhalation of vapors may be fatal by causing glottis to spasm and suffocation. Exposure may result in chemical pneumonitis or pulmonary edema.

Emergency/First Aid Treatment: Remove to ventilated area; immediately remove any contaminated clothing and wash contaminated areas for 15 minutes using water. Treat supportively and observe for possible shock. If ingested, seek immediate medical aid.

Recommended Clean-Up Procedures

Personal Protection: Level B Ensemble   **Recommended Material** N.D.

RCRA Waste #   None   **Reportable Quantities:** None

Spills: Remove all potential ignition sources. Dike to contain spill, absorb with non-combustible absorbent and take up using non-sparking tools. Decontaminate spill area using soapy water.

Special Emergency Information: May be harmful if inhaled, swallowed or absorbed through the skin. Vapors are heavier than air and may travel some distance to an ignition source. Vapors are more dense than air and may settle in low lying areas.

METHYL tert-BUTYL ETHER - $(CH_3)_3COCH_3$

Synonyms: tert-butyl methyl ether

CAS Number: 1634-04-4　　　　　　　　**Description:** Colorless Liquid
DOT Number: 2398　　　　　　　　　　**DOT Classification:** 2398

Molecular Weight: 88.2
Melting Point: 　　°C　　(°F)　　　**Vapor Density:** 3.1　　**Vapor Pressure:** 8 (240 mm Hg)
Boiling Point: 56°C (133°F)　　**Specific Gravity:** 0.8　　**Water Solubility:** Insoluble

Chemical Incompatibilities or Instabilities: Strong oxidizers.

FLAMMABILITY

NFPA Hazard Code:　　　　　　　　　　**NFPA Classification:** Class IB Flammable
Flash Point: -10°C (14°F)　　　　　**LEL:** 1.6%
Autoignition Temp: 191°C (377°F)　　**UEL:** 15.1%

Fire Extinguishing Methods: Alcohol Resistant Foam, Carbon Dioxide, Dry Chemical, Water Spray or Fog.

Special Fire Fighting Considerations: Isolate for 1/2 mile if rail or tank truck is involved in a fire. Water spray should be used to keep closed containers cool. Continue to cool container after fire is extinguished. For large fires, if possible, withdraw and allow to burn. Immediately withdraw if rising sound from venting device is heard or if fire is causing discoloration to the tank.

TOXICOLOGY

Odor:　　　　　　　　　　　　　　　　　　**Odor Threshold:** N.D.
Physical Contact: Irritant
TLV = N.D.　　　　　　　**STEL** = N.D.　　　　　　**IDLH** = N.D.

Routes of Entry and Relative LD$_{50}$ (or LC$_{50}$)　　　**RTECS # KN5250000**
　Inhalation　　　　N.D.
　Ingestion　　　　N.D.
　Skin Absorption　N.D.

Symptoms of Exposure: Possible irritation to skin, eyes and upper respiratory tract; headache; nausea. Exposure may result in chemical pneumonitis or pulmonary edema.

Emergency/First Aid Treatment: Remove to ventilated area; immediately remove any contaminated clothing and wash contaminated areas for 15 minutes using water. Treat supportively and observe for possible shock. If ingested, seek immediate medical aid.

Recommended Clean-Up Procedures

Personal Protection: Level B Ensemble　　**Recommended Material** N.D.

RCRA Waste #　　None　　　　　　　**Reportable Quantities:** None

Spills: Remove all potential ignition sources. Dike to contain spill, absorb with non-combustible absorbent and take up using non-sparking tools. Decontaminate spill area using soapy water.

Special Emergency Information: May be harmful if inhaled, swallowed or absorbed through the skin. Vapors are heavier than air and may travel some distance to an ignition source. Vapors are more dense than air and may settle in low lying areas. May form unstable peroxides upon exposure to air.

METHYL n-BUTYRATE - CH₃CH₂CH₂C(O)OCH₃

Synonyms: n-butyric acid, methyl ester

CAS Number: 623-42-7   **Description:** Colorless Liquid
DOT Number: 1237   **DOT Classification:** Flammable Liquid

Molecular Weight: 102.1
Melting Point: -97°C (-143°F)   **Vapor Density:** 3.5   **Vapor Pressure:** **5** (40 mm Hg)
Boiling Point: 102°C (215°F)   **Specific Gravity:** 0.9   **Water Solubility:** Insoluble

Chemical Incompatibilities or Instabilities: Strong oxidizers.

FLAMMABILITY

NFPA Hazard Code: 2-3-0   **NFPA Classification:** Class IB Flammable
Flash Point: 14°C   (57°F)   **LEL:**   0.9%
Autoignition Temp: N.D.   **UEL:**   3.5%

Fire Extinguishing Methods: Alcohol Resistant Foam, Carbon Dioxide, Dry Chemical, Water Spray or Fog.

Special Fire Fighting Considerations: Isolate for 1/2 mile if rail or tank truck is involved in a fire. Water spray should be used to keep closed containers cool. Continue to cool container after fire is extinguished. For large fires, if possible, withdraw and allow to burn. Immediately withdraw if rising sound from venting device is heard or if fire is causing discoloration to the tank.

TOXICOLOGY

Odor: **Odor Threshold:** N.D.
Physical Contact: Irritant
TLV = N.D.   **STEL** = N.D.   **IDLH** = N.D.

Routes of Entry and Relative LD₅₀ (or LC₅₀)   **RTECS # ET5500000**
    Inhalation   N.D.
    Ingestion   3   (3830 mg/kg)
    Skin Absorption   2   (3560 mg/kg)

Symptoms of Exposure: Possible irritation to skin, eyes and upper respiratory tract; headache; nausea.

Emergency/First Aid Treatment: Remove to ventilated area; immediately remove any contaminated clothing and wash contaminated areas for 15 minutes using water. Treat supportively and observe for possible shock. If ingested, seek immediate medical aid.

Recommended Clean-Up Procedures

Personal Protection: Level B Ensemble   **Recommended Material**   N.D.

RCRA Waste #   None   **Reportable Quantities:**   None

Spills: Remove all potential ignition sources. Dike to contain spill, absorb with non-combustible absorbent and take up using non-sparking tools. Decontaminate spill area using soapy water.

Special Emergency Information: May be harmful if inhaled, swallowed or absorbed through the skin. Vapors are heavier than air and may travel some distance to an ignition source. Vapors are more dense than air and may settle in low lying areas.

METHYL CHLORIDE - CH₃Cl

Synonyms: chloromethane

CAS Number: 74-87-3
DOT Number: 1063

Description: Colorless Gas
DOT Classification: Flammable Gas

Molecular Weight: 50.5
Melting Point: -98°C (-144°F)
Boiling Point: -24°C (-12°F)

Vapor Density: 1.7
Specific Gravity:

Vapor Pressure: Gas
Water Solubility: Insoluble

Chemical Incompatibilities or Instabilities: Oxidizers.

FLAMMABILITY

NFPA Hazard Code: 2-4-0
Flash Point: -46°C (-50°F)
Autoignition Temp: 632°C (1170°F)

NFPA Classification:
LEL: 8.1%
UEL: 17.4%

Fire Extinguishing Methods: Foam, Water Spray or Fog.

Special Fire Fighting Considerations: Structural fire fighter protective clothing will not provide adequate protection. Isolate for 1/2 mile if rail or tank truck is involved in a fire. Stop leak if it can safely be done. If not, then allow to burn. Water spray should be used to keep closed containers cool. Continue to cool container after fire is extinguished. For large fires, if possible, withdraw and allow to burn. Immediately withdraw if rising sound from venting device is heard or if fire is causing discoloration to the tank.

TOXICOLOGY

Odor: Sweet, ethereal
Physical Contact: Irritant
TLV = 103 mg/m³

STEL = 207 mg/m³

Odor Threshold: 20 mg/m³

IDLH = 21,000 mg/m³

Routes of Entry and Relative LD₅₀ (or LC₅₀)
Inhalation   5   (21,000 mg/m³/H)
Ingestion   3   (1800 mg/kg)
Skin Absorption   N.D.

RTECS # PA6300000

Symptoms of Exposure: Possible irritation to skin, eyes and upper respiratory tract; headache; nausea.

Emergency/First Aid Treatment: Remove to ventilated area; immediately remove any contaminated clothing and wash contaminated areas for 15 minutes using water. Treat supportively and observe for possible shock. If ingested, seek immediate medical aid.

Recommended Clean-Up Procedures

Personal Protection: Level B Ensemble

RCRA Waste # U045

Recommended Material N.D.

Reportable Quantities: 100 lb (1 lb)

Spills: Remove all potential ignition sources. Stop leak if it can safely be done. Keep area isolated until vapors have dissipated.

Special Emergency Information: May be harmful if inhaled, swallowed or absorbed through the skin. Vapors are heavier than air and may travel some distance to an ignition source. Vapors are more dense than air and may settle in low lying areas.

METHYL CHLOROACETATE - ClCH$_2$C(O)OCH$_3$

Synonyms: methyl monochloroacetate; monochloroacetic acid, ethylester

CAS Number: 96-34-4   **Description:** Colorless Liquid
DOT Number: 2295   **DOT Classification:** Flammable Liquid

Molecular Weight: 108.5
Melting Point: -33°C (-27°F)   **Vapor Density:** 3.8   **Vapor Pressure:** 4 (6 mm Hg)
Boiling Point: 130°C (266°F)   **Specific Gravity:** 1.2   **Water Solubility:** Insoluble

Chemical Incompatibilities or Instabilities: Strong oxidizers.

FLAMMABILITY

NFPA Hazard Code: 2-2-1   **NFPA Classification:** Class II Combustible
Flash Point: 57°C (135°F)   **LEL:** 7.5%
Autoignition Temp: 465°C (869°F)   **UEL:** 18.5%

Fire Extinguishing Methods: Carbon Dioxide, Dry Chemical, Foam, Water Spray or Fog.

Special Fire Fighting Considerations: Structural fire fighter protective clothing will not provide adequate protection. Fight fire from a distance or protected location, if possible. Water spray from an unmanned device should be used to keep closed containers cool. Treat all materials used or generated and equipment used involved as contaminated by hazardous waste.

TOXICOLOGY

Odor: **Odor Threshold:** N.D.
Physical Contact: Material is extremely destructive to human tissues.
TLV = N.D.   **STEL** = N.D.   **IDLH** = N.D.

Routes of Entry and Relative LD$_{50}$ (or LC$_{50}$)   **RTECS # AF9500000**
     Inhalation   N.D.
     Ingestion   N.D.
     Skin Absorption   N.D.

Symptoms of Exposure: Possible irritation to skin, eyes and upper respiratory tract; headache; nausea, lachrymation. Inhalation of vapors may be fatal by causing glottis to spasm and suffocation. Exposure may result in chemical pneumonitis or pulmonary edema.

Emergency/First Aid Treatment: Remove to ventilated area; immediately remove any contaminated clothing and wash contaminated areas for 15 minutes using water. Treat supportively and observe for possible shock. If ingested, seek immediate medical aid.

Recommended Clean-Up Procedures

Personal Protection: Level B Ensemble   **Recommended Material** Saranex (+)

RCRA Waste # None   **Reportable Quantities:** None

Spills: Dike to contain spill and collect liquid for later disposal. Decontaminate spill area using soapy water. Treat all materials used or generated and equipment used involved as contaminated by hazardous waste.

Special Emergency Information: May be harmful if inhaled, swallowed or absorbed through the skin.

METHYL CHLOROCARBONATE - ClC(O)OCH₃

Synonyms: methyl chloroformate

CAS Number: 79-22-1	**Description:** Colorless Liquid
DOT Number: 1238	**DOT Classification:** Flammable Liquid

Molecular Weight: 94.5

Melting Point: °C (°F)	**Vapor Density:** 3.3	**Vapor Pressure:** 7 (127 mm Hg)
Boiling Point: 72°C (162°F)	**Specific Gravity:** 1.2	**Water Solubility:** Decomposes

Chemical Incompatibilities or Instabilities: Water, alcohols.

FLAMMABILITY

NFPA Hazard Code:	**NFPA Classification:** Class IB Flammable
Flash Point: 17°C (64°F)	**LEL:** N.D.
Autoignition Temp: 540°C (940°F)	**UEL:** N.D.

Fire Extinguishing Methods: Alcohol Resistant Foam, Carbon Dioxide, Dry Chemical, Water Spray or Fog.

Special Fire Fighting Considerations: Structural fire fighter protective clothing will not provide adequate protection. Isolate for 1/2 mile if rail or tank truck is involved in a fire. Water spray should be used to keep closed containers cool. Continue to cool container after fire is extinguished. Immediately withdraw if rising sound from venting device is heard or if fire is causing discoloration to the tank. May release toxic fumes upon heating (phosgene). Treat all materials used or generated and equipment used involved as contaminated by hazardous waste.

DOT Recommended Isolation Zones:	Small Spill: 150 ft	Large Spill: 600 ft
DOT Recommended Down Wind Take Cover Distance	Small Spill: 0.8 miles	Large Spill: 2 miles

TOXICOLOGY

Odor: Sharp, penetrating	**Odor Threshold:** N.D.

Physical Contact: Material is extremely destructive to human tissues.

TLV = N.D.	**STEL** = N.D.	**IDLH** = N.D.

Routes of Entry and Relative LD₅₀ (or LC₅₀) **RTECS #** FG3675000

Inhalation	10	(340 mg/m³/H)
Ingestion	8	(60 mg/kg)
Skin Absorption	1	(7120 mg/kg)

Symptoms of Exposure: Possible burns or irritation to skin, eyes and upper respiratory tract; headache; nausea; laryngitis. Inhalation of vapors may be fatal by causing glottis to spasm and suffocation. Exposure may result in chemical pneumonitis or pulmonary edema.

Emergency/First Aid Treatment: Remove to ventilated area; immediately remove any contaminated clothing and wash contaminated areas for 15 minutes using water. Treat supportively and observe for possible shock. If ingested, seek immediate medical aid.

Recommended Clean-Up Procedures

Personal Protection: Level B Ensemble	**Recommended Material** N.D.
RCRA Waste # U156	**Reportable Quantities:** 1000 lbs (1 lb)

Spills: Remove all potential ignition sources. Dike to contain spill, absorb with non-combustible absorbent and take up using non-sparking tools. Decontaminate spill area using dilute alkaline solution. Treat all materials used or generated and equipment involved as contaminated by hazardous waste.

Special Emergency Information: May be fatal in inhaled, ingested or absorbed through the skin. Vapors are heavier than air and may travel some distance to an ignition source. Vapors are more dense than air and may settle in low lying areas.

METHYL CHLOROFORM

Synonyms: 1,1,1-trichloroethane

CAS Number: 71-55-6
DOT Number: 2831

Description: Colorless Liquid
DOT Classification: ORM-A

Molecular Weight: 133.4
Melting Point: -30°C (-22°F)
Boiling Point: 74°C (165°F)

Vapor Density:
Specific Gravity: 1.3

Vapor Pressure: 7 (100 mm Hg)
Water Solubility: Insoluble

Chemical Incompatibilities or Instabilities: Acetone, nitrites, oxygen, reactive metals and alloys.

FLAMMABILITY

NFPA Hazard Code: 2-1-0
Flash Point: None
Autoignition Temp: 500°C (932°F)

NFPA Classification:
LEL: 7.0%
UEL: 16.0%

Fire Extinguishing Methods: Carbon Dioxide, Dry Chemical, Foam, Water Spray or Fog.

Special Fire Fighting Considerations: Isolate for 1/2 mile if rail or tank truck is involved in a fire. Water spray should be used to keep closed containers cool. Continue to cool container after fire is extinguished. May release toxic fumes upon heating (phosgene).

TOXICOLOGY
QUESTIONABLE CARCINOGEN

Odor: Sweet, ethereal
Physical Contact: Irritant
TLV = 1910 mg/m^3 **STEL** = 2460 mg/m^3

Odor Threshold: 90 mg/m^3

IDLH = 5550 mg/m^3

RTECS # KJ2975000

Routes of Entry and Relative LD$_{50}$ (or LC$_{50}$)
  Inhalation   2   (400,000 mg/m^3/H)
  Ingestion   1   (10,300 mg/kg)
  Skin Absorption   N.D.

Symptoms of Exposure: Possible irritation to skin, eyes and upper respiratory tract; headache; nausea. May act as a narcotic in high concentrations.

Emergency/First Aid Treatment: Remove to ventilated area; immediately remove any contaminated clothing and wash contaminated areas for 15 minutes using water. Treat supportively and observe for possible shock. If ingested, seek immediate medical aid.

Recommended Clean-Up Procedures

Personal Protection: Level B Ensemble
RCRA Waste # U226

Recommended Material Polyvinyl Alcohol (+), Viton (+), Teflon (0)
Reportable Quantities: 1000 lbs (1 lb)

Spills: Remove all potential ignition sources. Dike to contain spill, absorb with non-combustible absorbent and take up using non-sparking tools. Decontaminate spill area using soapy water. Treat all materials used or generated and equipment used involved as contaminated by hazardous waste.

Special Emergency Information: May be harmful if inhaled, swallowed or absorbed through the skin. Vapors are heavier than air and may travel some distance to an ignition source. Vapors are more dense than air and may settle in low lying areas.

METHYL CHLOROMETHYL ETHER - ClCH$_2$OCH$_3$

Synonyms: chloromethyl methyl ether

CAS Number: 107-30-2   **Description:** Colorless Liquid
DOT Number: 1239   **DOT Classification:** Flammable Liquid

Molecular Weight: 80.5
Melting Point: °C (°F)   **Vapor Density:** N.A.   **Vapor Pressure:** 8 (260 mm Hg)
Boiling Point: 57°C (135°F)   **Specific Gravity:** 1.1   **Water Solubility:** Insoluble

Chemical Incompatibilities or Instabilities: Oxidizers, heat.

FLAMMABILITY

NFPA Hazard Code: **NFPA Classification:**
Flash Point: 15°C (60°F)   **LEL:** N.D.
Autoignition Temp: N.D.   **UEL:** N.D.

Fire Extinguishing Methods: Carbon Dioxide, Dry Chemical, Foam, Water Spray or Fog.

Special Fire Fighting Considerations: Structural fire fighter protective clothing will not provide adequate protection. Water spray should be used to keep closed containers cool. Continue to cool container after fire is extinguished. Fight fire from a distance or protected location, if possible. Treat all materials used or generated and equipment used involved as contaminated by hazardous waste.

DOT Recommended Isolation Zones: Small Spill: 150 ft   Large Spill: 150 ft
DOT Recommended Down Wind Take Cover Distance   Small Spill: 0.2 miles   Large Spill: 0.4 miles

TOXICOLOGY
CARCINOGEN

Odor: Sharp, penetrating   **Odor Threshold:** N.D.
Physical Contact: Irritant
TLV = See 29 CFR 1910.1006   **STEL** = N.D.   **IDLH** = N.D.

Routes of Entry and Relative LD$_{50}$ (or LC$_{50}$)   **RTECS # KN6650000**
  Inhalation   9   (1265 mg/m^3/H)
  Ingestion   4   (817 mg/kg)
  Skin Absorption   N.D.

Symptoms of Exposure: Possible irritation to skin, eyes and upper respiratory tract; headache; nausea; lachrymation. Inhalation of vapors may be fatal by causing glottis to spasm and suffocation. Exposure may result in chemical pneumonitis or pulmonary edema.

Emergency/First Aid Treatment: Remove to ventilated area; immediately remove any contaminated clothing and wash contaminated areas for 15 minutes using water. Treat supportively and observe for possible shock. If ingested, seek immediate medical aid.

Recommended Clean-Up Procedures

Personal Protection: Level A Ensemble   **Recommended Material** N.D.
RCRA Waste # None   **Reportable Quantities:** None

Spills: Remove all potential ignition sources. Dike to contain spill, absorb with non-combustible absorbent and take up using non-sparking tools. Decontaminate spill area using soapy water. Treat all materials used or generated and equipment used involved as contaminated by hazardous waste.

Special Emergency Information: May be harmful if inhaled, swallowed or absorbed through the skin. Vapors are heavier than air and may travel some distance to an ignition source. Vapors are more dense than air and may settle in low lying areas. May form unstable peroxides upon exposure to air.

METHYLCYCLOHEXANE

Synonyms:

CAS Number: 108-87-2

DOT Number: 2296

Description: Colorless Liquid

DOT Classification: Flammable Liquid

Molecular Weight: 98.2

Melting Point:	-126°C (-195°F)	**Vapor Density:** 3.4	**Vapor Pressure:**	5 (40 mm Hg)
Boiling Point:	101°C (214°F)	**Specific Gravity:** 0.8	**Water Solubility:**	Insoluble

Chemical Incompatibilities or Instabilities: Strong oxidizers.

FLAMMABILITY

NFPA Hazard Code: 2-3-0

Flash Point: -4°C (25°F)

Autoignition Temp: 250°C (482°F)

NFPA Classification: Class IB Flammable

LEL: 1.2%

UEL: 6.7%

Fire Extinguishing Methods: Carbon Dioxide, Dry Chemical, Foam, Water Spray or Fog.

Special Fire Fighting Considerations: Isolate for 1/2 mile if rail or tank truck is involved in a fire. Water spray should be used to keep closed containers cool. Continue to cool container after fire is extinguished. For large fires, if possible, withdraw and allow to burn. Immediately withdraw if rising sound from venting device is heard or if fire is causing discoloration to the tank.

TOXICOLOGY

Odor: Faint ethereal

Physical Contact: None

TLV = 1610 mg/m^3 **STEL** = N.D.

Odor Threshold: 2000 mg/m^3

IDLH = 40,000 mg/m^3

Routes of Entry and Relative LD$_{50}$ (or LC$_{50}$)

Inhalation	3	(83,000 mg/m^3/H)
Ingestion	N.D.	
Skin Absorption	N.D.	

RTECS # GV6125000

Symptoms of Exposure: May act as a narcotic in high concentrations.

Emergency/First Aid Treatment: Remove to ventilated area; immediately remove any contaminated clothing and wash contaminated areas for 15 minutes using water. Treat supportively and observe for possible shock. If ingested, seek immediate medical aid.

Recommended Clean-Up Procedures

Personal Protection: Level B Ensemble

RCRA Waste # None

Recommended Material N.D.

Reportable Quantities: None

Spills: Remove all potential ignition sources. Dike to contain spill, absorb with non-combustible absorbent and take up using non-sparking tools. Decontaminate spill area using soapy water.

Special Emergency Information: May be harmful if inhaled, swallowed or absorbed through the skin. Vapors are heavier than air and may travel some distance to an ignition source. Vapors are more dense than air and may settle in low lying areas.

2-METHYLCYCLOHEXANONE

Synonyms: o-cyclohexanone

CAS Number: 583-60-8   **Description:** Colorless Liquid
DOT Number: N.D.   **DOT Classification:** Flammable Liquid

Molecular Weight: 112.2
Melting Point: -14°C   (7°F)   **Vapor Density:**   3.9   **Vapor Pressure:**   1 (1 mm Hg)
Boiling Point: 163°C (325°F)   **Specific Gravity:** 0.9   **Water Solubility:** Insoluble

Chemical Incompatibilities or Instabilities: Strong oxidizers.

FLAMMABILITY

NFPA Hazard Code: ?-2-0   **NFPA Classification:** Class II Combustible
Flash Point: 48°C (118°F)   **LEL:** N.D.
Autoignition Temp: N.D.   **UEL:** N.D.

Fire Extinguishing Methods: Alcohol Resistant Foam, Carbon Dioxide, Dry Chemical, Water Spray or Fog.

Special Fire Fighting Considerations: Isolate for 1/2 mile if rail or tank truck is involved in a fire. Water spray should be used to keep closed containers cool. Continue to cool container after fire is extinguished. For large fires, if possible, withdraw and allow to burn. Immediately withdraw if rising sound from venting device is heard or if fire is causing discoloration to the tank.

TOXICOLOGY

Odor: Acetone-like   **Odor Threshold:** N.D.
Physical Contact: Irritant
$TLV = 229$ mg/m^3 $STEL = 344$ mg/m^3 $IDLH = 11,650$ mg/m^3

Routes of Entry and Relative LD$_{50}$ (or LC$_{50}$)   **RTECS # GW1750000**
    Inhalation   N.D.
    Ingestion   4   (2140 mg/kg)
    Skin Absorption   3   (1635 mg/kg)

Symptoms of Exposure: Possible irritation to skin, eyes and upper respiratory tract; headache; nausea.

Emergency/First Aid Treatment: Remove to ventilated area; immediately remove any contaminated clothing and wash contaminated areas for 15 minutes using water. Treat supportively and observe for possible shock. If ingested, seek immediate medical aid.

Recommended Clean-Up Procedures

Personal Protection: Level B Ensemble   **Recommended Material**   N.D.

RCRA Waste #   None   **Reportable Quantities:**   None

Spills: Remove all potential ignition sources. Dike to contain spill, absorb with non-combustible absorbent and take up using non-sparking tools. Decontaminate spill area using soapy water.

Special Emergency Information: May be harmful if inhaled, swallowed or absorbed through the skin. Vapors are heavier than air and may travel some distance to an ignition source. Vapors are more dense than air and may settle in low lying areas.

Methylcyclohexanone (CAS # 1331-22-2, DOT # 2297, RTECS # GW1575000),
3-methylcyclohexanone (CAS # 591-24-2, RTECS # GW1750100), and
4-cyclohexanone (CAS # 589-92-4, RTECS # GW1750200) have similar chemical, physical and toxicological properties.

METHYLCYCLOPENTANE

Synonyms:

CAS Number: 96-37-7 **Description:** Colorless Liquid
DOT Number: 2298 **DOT Classification:** Flammable Liquid

Molecular Weight: 84.7
Melting Point: -142°C (-224°F) **Vapor Density:** 2.9 **Vapor Pressure:** 7 (101 mm Hg)
Boiling Point: 72°C (161°F) **Specific Gravity:** 0.75 **Water Solubility:** Insoluble

Chemical Incompatibilities or Instabilities: Strong oxidizers.

FLAMMABILITY

NFPA Hazard Code: 2-3-0 **NFPA Classification:** Class IB Flammable
Flash Point: -23°C (-11°F) **LEL:** 1.1%
Autoignition Temp: 329°C (624°F) **UEL:** 8.7%

Fire Extinguishing Methods: Alcohol Resistant Foam, Carbon Dioxide, Dry Chemical, Water Spray or Fog.

Special Fire Fighting Considerations: Isolate for 1/2 mile if rail or tank truck is involved in a fire. Water spray should be used to keep closed containers cool. Continue to cool container after fire is extinguished. For large fires, if possible, withdraw and allow to burn. Immediately withdraw if rising sound from venting device is heard or if fire is causing discoloration to the tank.

TOXICOLOGY

Odor: **Odor Threshold:** N.D.
Physical Contact: Irritant
TLV = N.D. **STEL** = N.D. **IDLH** = N.D.

Routes of Entry and Relative LD$_{50}$ (or LC$_{50}$) **RTECS # GY4640000**
    Inhalation N.D.
    Ingestion N.D.
    Skin Absorption N.D.

Symptoms of Exposure: Possible irritation to skin, eyes and upper respiratory tract; headache; nausea. May act as a narcotic in high concentrations.

Emergency/First Aid Treatment: Remove to ventilated area; immediately remove any contaminated clothing and wash contaminated areas for 15 minutes using water. Treat supportively and observe for possible shock. If ingested, seek immediate medical aid.

Recommended Clean-Up Procedures

Personal Protection: Level B Ensemble **Recommended Material** N.D.

RCRA Waste # None **Reportable Quantities:** None

Spills: Remove all potential ignition sources. Dike to contain spill, absorb with non-combustible absorbent and take up using non-sparking tools. Decontaminate spill area using soapy water.

Special Emergency Information: May be harmful if inhaled, swallowed or absorbed through the skin. Vapors are heavier than air and may travel some distance to an ignition source. Vapors are more dense than air and may settle in low lying areas.

METHYL DICHLOROACETATE - Cl$_2$CHC(O)OCH$_3$

Synonyms: dichloroacetic acid, methyl ester; methyl dichloroethanoate

CAS Number: 116-54-1	**Description:** Colorless Liquid	
DOT Number: 2299	**DOT Classification:** Corrosive Material	

Molecular Weight: 143.0

Melting Point: -52°C (-62°F)	**Vapor Density:** 4.9	**Vapor Pressure:** N.D.
Boiling Point: 143°C (289°F)	**Specific Gravity:** 1.4	**Water Solubility:** Insoluble

Chemical Incompatibilities or Instabilities: Acids, oxidizers.

FLAMMABILITY

NFPA Hazard Code:	**NFPA Classification:** Class IIIA Combustible
Flash Point: 80°C (176°F)	**LEL:** N.D.
Autoignition Temp: N.D.	**UEL:** N.D.

Fire Extinguishing Methods: Carbon Dioxide, Dry Chemical, Foam, Water Spray or Fog.

Special Fire Fighting Considerations: Water spray should be used to keep closed containers cool. Continue to cool container after fire is extinguished.

TOXICOLOGY

Odor: Ethereal

Odor Threshold: N.A.

Physical Contact: Material is extremely destructive to human tissues.

TLV = N.D.   **STEL** = N.D.   **IDLH** = N.D.

Routes of Entry and Relative LD$_{50}$ (or LC$_{50}$)   **RTECS # AG6625000**

Inhalation	N.D.
Ingestion	N.D.
Skin Absorption	N.D.

Symptoms of Exposure: Possible burns or irritation to skin, eyes and upper respiratory tract; headache; nausea. Inhalation of vapors may be fatal by causing glottis to spasm and suffocation. Exposure may result in chemical pneumonitis or pulmonary edema.

Emergency/First Aid Treatment: Remove to ventilated area; immediately remove any contaminated clothing and wash contaminated areas for 15 minutes using water. Treat supportively and observe for possible shock. If ingested, seek immediate medical aid.

Recommended Clean-Up Procedures

Personal Protection: Level B Ensemble	**Recommended Material** N.D.
RCRA Waste # None	**Reportable Quantities:** None

Spills: Dike to contain spill and collect liquid for later disposal. Decontaminate spill area using dilute alkaline solution.

Special Emergency Information: May be harmful if inhaled, swallowed or absorbed through the skin. May hydrolyze upon contact with water to yield dichloroacetic acid.

METHYL DICHLOROSILANE - CH$_3$SiCl$_2$H

Synonyms: dichloromethylsilane

CAS Number: 75-54-7	**Description:** Colorless Liquid	
DOT Number: 1242	**DOT Classification:** Flammable Liquid	

Molecular Weight: 115.0

Melting Point: -93°C (-135°F)	**Vapor Density:** N.D.	**Vapor Pressure:** 9 (400 mm Hg)
Boiling Point: 42°C (107°F)	**Specific Gravity:** 1.1	**Water Solubility:** Decomposes

Chemical Incompatibilities or Instabilities: Oxidizers, lead oxides, copper oxides, silver oxide, water.

FLAMMABILIT

NFPA Hazard Code: 3-3-2-W	**NFPA Classification:** Class IB Flammable
Flash Point: -26°C (-14°F)	**LEL:** 6.0%
Autoignition Temp: 316°C (600°F)	**UEL:** 55.0%

Fire Extinguishing Methods: Carbon Dioxide, Dry Chemical, Foam, Water Spray or Fog.

Special Fire Fighting Considerations: Isolate for 1/2 mile if rail or tank truck is involved in a fire. Water spray should be used to keep closed containers cool. Continue to cool container after fire is extinguished. Do not get water inside container. Immediately withdraw if rising sound from venting device is heard or if fire is causing discoloration to the tank.

TOXICOLOGY

Odor: Sharp, penetrating   **Odor Threshold:** N.D.

Physical Contact: Material is extremely destructive to human tissues.

TLV = N.D.   **STEL** = N.D.   **IDLH** = N.D.

Routes of Entry and Relative LD$_{50}$ (or LC$_{50}$)   **RTECS # VV3500000**

Inhalation	N.D.
Ingestion	N.D.
Skin Absorption	N.D.

Symptoms of Exposure: Possible irritation to skin, eyes and upper respiratory tract; headache; nausea. Inhalation of vapors may be fatal by causing glottis to spasm and suffocation. Exposure may result in chemical pneumonitis or pulmonary edema.

Emergency/First Aid Treatment: Remove to ventilated area; immediately remove any contaminated clothing and wash contaminated areas for 15 minutes using water. Treat supportively and observe for possible shock. If ingested, seek immediate medical aid.

Recommended Clean-Up Procedures

Personal Protection: Level B Ensemble	**Recommended Material** N.D.
RCRA Waste # None	**Reportable Quantities:** None

Spills: Remove all potential ignition sources. Dike to contain spill, absorb with non-combustible absorbent and take up using non-sparking tools. Decontaminate spill area using dilute alkaline solution.

Special Emergency Information: May be fatal in inhaled, ingested or absorbed through the skin. Vapors are heavier than air and may travel some distance to an ignition source. Vapors are more dense than air and may settle in low lying areas. Methyl dichlorosilane may react violently to release hydrogen chloride gas.

METHYLENE BISPHENYL ISOCYANATE

Synonyms: 4,4'-diphenylmethane diisocyanate; MDI

CAS Number: 101-68-8	**Description:** White to yellow flakes		
DOT Number: 2489	**DOT Classification:** Poison B		

Molecular Weight: 250.3

Melting Point:	37°C (99°F)	**Vapor Density:** N.D.	**Vapor Pressure:**	1 (<1 mm Hg)	
Boiling Point:	172°C (342°F)	**Specific Gravity:** 1.2	**Water Solubility:**	0.2%	

Chemical Incompatibilities or Instabilities: Alkalis, acids, alcohols.

FLAMMABILITY

NFPA Hazard Code:		**NFPA Classification:**
Flash Point: 202°C (396°F)		**LEL:** N.D.
Autoignition Temp: N.D.		**UEL:** N.D.

Fire Extinguishing Methods: Carbon Dioxide, Dry Chemical, Foam, Water Spray or Fog.

Special Fire Fighting Considerations: Structural fire fighter protective clothing will not provide adequate protection. Fight fire from a distance or protected location, if possible. Treat all materials used or generated and equipment used involved as contaminated by hazardous waste.

TOXICOLOGY
QUESTIONABLE CARCINOGEN

Odor:

Odor Threshold: N.D.

Physical Contact: Irritant

$TLV = 0.2$ mg/m^3 $STEL =$ N.D. $IDLH = 100$ mg/m^3

Routes of Entry and Relative LD$_{50}$ (or LC$_{50}$)

Inhalation	10	(178 mg/m^3/H)	**RTECS # NQ9350000**
Ingestion	N.D.		
Skin Absorption	N.D.		

Symptoms of Exposure: Possible irritation to skin, eyes and upper respiratory tract; headache; nausea. An allergic sensitizer.

Emergency/First Aid Treatment: Remove to ventilated area; immediately remove any contaminated clothing and wash contaminated areas for 15 minutes using water. Treat supportively and observe for possible shock. If ingested, seek immediate medical aid.

Recommended Clean-Up Procedures

Personal Protection: Level A Ensemble	**Recommended Material**	N.D.
RCRA Waste # None	**Reportable Quantities:**	None

Spills: Take up in non-combustible absorbent and dispose. Decontaminate spill area using soapy water.

Special Emergency Information: May be fatal in inhaled, ingested or absorbed through the skin.

2-METHYL-5-ETHYLPYRIDINE

Synonyms: 5-ethyl-alpha-picoline

CAS Number: 104-90-5
DOT Number: 2300

Description: Colorless Liquid
DOT Classification: Corrosive Material

Molecular Weight: 121.2
Melting Point: °C (°F)
Boiling Point: 178°C (353°F)

Vapor Density: 4.2
Specific Gravity: 0.9

Vapor Pressure: N.D.
Water Solubility: <1%

Chemical Incompatibilities or Instabilities: Oxidizers, chloroformates.

FLAMMABILITY

NFPA Hazard Code: 3-3-2
Flash Point: 68°C (155°F)
Autoignition Temp: N.D.

NFPA Classification: Class IIIA Combustible
LEL: 1.1%
UEL: 6.6%

Fire Extinguishing Methods: Carbon Dioxide, Dry Chemical, Foam, Water Spray or Fog.

Special Fire Fighting Considerations: Water spray should be used to keep closed containers cool. Continue to cool container after fire is extinguished.

TOXICOLOGY

Odor: Aromatic
Physical Contact: Material is extremely destructive to human tissues.
TLV = N.D. **STEL** = N.D.

Odor Threshold: N.D.

IDLH = N.D.

RTECS # TJ6825000

Routes of Entry and Relative LD$_{50}$ (or LC$_{50}$)

Inhalation	N.D.	
Ingestion	5	(365 mg/kg)
Skin Absorption	5	(1000 mg/kg)

Symptoms of Exposure: Possible burns or irritation to skin, eyes and upper respiratory tract; headache; nausea; laryngitis. Inhalation of vapors may be fatal by causing glottis to spasm and suffocation. Exposure may result in chemical pneumonitis or pulmonary edema.

Emergency/First Aid Treatment: Remove to ventilated area; immediately remove any contaminated clothing and wash contaminated areas for 15 minutes using water. Treat supportively and observe for possible shock. If ingested, seek immediate medical aid.

Recommended Clean-Up Procedures

Personal Protection: Level B Ensemble
RCRA Waste # None

Recommended Material N.D.
Reportable Quantities: None

Spills: Dike to contain spill and collect liquid for later disposal. Decontaminate spill area using soapy water.

Special Emergency Information: May be harmful if inhaled, swallowed or absorbed through the skin.

METHYL HYDRAZINE - CH₃NHNH₂

Synonyms:

CAS Number: 60-34-4
DOT Number: 1244

Description: Colorless Liquid
DOT Classification: Flammable Liquid

Molecular Weight: 46.1
Melting Point: -52°C (-62°F)
Boiling Point: 88°C (190°F)

Vapor Density: 1.6
Specific Gravity: 0.9

Vapor Pressure: 5 (38 mm Hg)
Water Solubility: Soluble

Chemical Incompatibilities or Instabilities: Oxidizers, oxygen.

FLAMMABILITY

NFPA Hazard Code: 3-3-2
Flash Point: 21°C (70°F)
Autoignition Temp: 196°C (385°F)

NFPA Classification: Class IB Flammable
LEL: 2.5%
UEL: 97%

Fire Extinguishing Methods: Carbon Dioxide, Dry Chemical, Foam, Water Spray or Fog.

Special Fire Fighting Considerations: Structural fire fighter protective clothing will not provide adequate protection. Water spray from an unmanned device should be used to keep closed containers cool. Continue to cool container after fire is extinguished. Fight fire from a distance or protected location, if possible. Treat all materials used or generated and equipment involved as contaminated by hazardous waste.

DOT Recommended Isolation Zones: Small Spill: 1500 ft   Large Spill: 1500 ft
DOT Recommended Down Wind Take Cover Distance   Small Spill: 5 miles   Large Spill: 5 miles

TOXICOLOGY
SUSPECTED CARCINOGEN

Odor: Ammonia
Physical Contact: Material is extremely destructive to human tissues.
TLV = 0.38 mg/m³   **STEL** = N.D.

Odor Threshold: N.D.

IDLH = 96 mg/m³

Routes of Entry and Relative LD₅₀ (or LC₅₀)
Inhalation	10	(260 mg/m³/H)
Ingestion	9	(37 mg/kg)
Skin Absorption	9	(95 mg/kg)

RTECS # MV5600000

Symptoms of Exposure: Possible burns or irritation to skin, eyes and upper respiratory tract; headache; nausea: laryngitis. Inhalation of vapors may be fatal by causing glottis to spasm and suffocation. Exposure may result in chemical pneumonitis or pulmonary edema.

Emergency/First Aid Treatment: Remove to ventilated area; immediately remove any contaminated clothing and wash contaminated areas for 15 minutes using water. Treat supportively and observe for possible shock. If ingested, seek immediate medical aid.

Recommended Clean-Up Procedures

Personal Protection: Level A Ensemble

RCRA Waste # P068

Recommended Material N.D.

Reportable Quantities: 10 lbs (1 lb)

Spills: Remove all potential ignition sources. Dike to contain spill, absorb with non-combustible absorbent and take up using non-sparking tools. Decontaminate spill area using water. Material may ignite in air.

Special Emergency Information: May be fatal in inhaled, ingested or absorbed through the skin. Vapors are heavier than air and may travel some distance to an ignition source. Vapors are more dense than air and may settle in low lying areas.

METHYL IODIDE - CH3I

Synonyms: iodomethane

CAS Number: 78-88-4
DOT Number: 2644

Description: Colorless Liquid
DOT Classification: Poison B

Molecular Weight: 141.9
Melting Point: -66°C (-88°F)
Boiling Point: 43°C (109°F)

Vapor Density: 4.9
Specific Gravity: 2.3

Vapor Pressure: 8 (400mm Hg)
Water Solubility: Soluble

Chemical Incompatibilities or Instabilities: Strong oxidizers, trialkyl phosphines.

FLAMMABILITY

NFPA Hazard Code:
Flash Point: N.A.
Autoignition Temp: (d) 270°C (518°F)

NFPA Classification:
LEL: N.A.
UEL: N.A.

Fire Extinguishing Methods: Carbon Dioxide, Dry Chemical, Foam, Water Spray or Fog.

Special Fire Fighting Considerations: Structural fire fighter protective clothing will not provide adequate protection. Fight fire from a distance or protected location, if possible. Treat all materials used or generated and equipment used involved as contaminated by hazardous waste.

TOXICOLOGY
SUSPECTED CARCINOGEN

Odor: Pungent, ether-like
Physical Contact: Irritant
TLV = 12 mg/m^3

STEL = N.D.

Odor Threshold: N.D.

IDLH = 4720 mg/m^3

RTECS # PA9450000

Routes of Entry and Relative LD$_{50}$ (or LC$_{50}$)
Inhalation   5   (5200 mg/m^3/H)
Ingestion   N.D.
Skin Absorption   N.D.

Symptoms of Exposure: Possible burns, blisters or irritation to skin, eyes and upper respiratory tract; headache; nausea. Exposure may result in chemical pneumonitis or pulmonary edema.

Emergency/First Aid Treatment: Remove to ventilated area; immediately remove any contaminated clothing and wash contaminated areas for 15 minutes using water. Treat supportively and observe for possible shock. If ingested, seek immediate medical aid.

Recommended Clean-Up Procedures

Personal Protection: Level B Ensemble
RCRA Waste # U138

Recommended Material Viton (0)
Reportable Quantities: 100 lbs (1 lb)

Spills: Dike to contain spill and collect liquid for later disposal. Decontaminate spill area using soapy water. Treat all materials used or generated and equipment used involved as contaminated by hazardous waste.

Special Emergency Information: May be fatal in inhaled, ingested or absorbed through the skin.

METHYL ISOBUTYL CARBINOL - $CH_3CH(OH)CH_2CH(CH_3)_2$

Synonyms: 4-methyl-2-pentanol; methyl amylalcohol

CAS Number: 108-11-2	**Description:** Colorless Liquid		
DOT Number: 2053	**DOT Classification:** Flammable Liquid		

Molecular Weight: 102.2

Melting Point: -90°C (-130°F)	**Vapor Density:** 3.5	**Vapor Pressure:** **2** (3 mm Hg)
Boiling Point: 133°C (271°F)	**Specific Gravity:** 0.8	**Water Solubility:** 2%

Chemical Incompatibilities or Instabilities: Strong oxidizers.

FLAMMABILITY

NFPA Hazard Code: 2-2-0	**NFPA Classification:** Class II Combustible
Flash Point: 41°C (106°F)	**LEL:** 1.0%
Autoignition Temp: N.D.	**UEL:** 5.5%

Fire Extinguishing Methods: Alcohol Resistant Foam, Carbon Dioxide, Dry Chemical, Water Spray or Fog.

Special Fire Fighting Considerations: Isolate for 1/2 mile if rail or tank truck is involved in a fire. Water spray should be used to keep closed containers cool, Continue to cool container after fire is extinguished. For large fires, if possible, withdraw and allow to burn. Immediately withdraw if rising sound from venting device is heard or if fire is causing discoloration to the tank.

TOXICOLOGY

Odor: Mild   **Odor Threshold:** N.D.

Physical Contact: Irritant

$TLV = 104$ mg/m^3 $STEL = 167$ mg/m^3 $IDLH = 8500$ mg/m^3

Routes of Entry and Relative LD$_{50}$ (or LC$_{50}$)   **RTECS # SA3500000**

Inhalation	N.D.	
Ingestion	3	(2590 mg/kg)
Skin Absorption	2	(3560 mg/kg)

Symptoms of Exposure: Possible irritation to skin, eyes and upper respiratory tract; headache; nausea. May cause anesthesia.

Emergency/First Aid Treatment: Remove to ventilated area; immediately remove any contaminated clothing and wash contaminated areas for 15 minutes using water. Treat supportively and observe for possible shock. If ingested, seek immediate medical aid.

Recommended Clean-Up Procedures

Personal Protection: Level B Ensemble	**Recommended Material** N.D.
RCRA Waste # None	**Reportable Quantities:** None

Spills: Remove all potential ignition sources. Dike to contain spill, absorb with non-combustible absorbent and take up using non-sparking tools. Decontaminate spill area using water.

Special Emergency Information: May be harmful if inhaled, swallowed or absorbed through the skin. Vapors are heavier than air and may travel some distance to an ignition source. Vapors are more dense than air and may settle in low lying areas.

METHYL ISOCYANATE - CH3N=C=O

Synonyms: isocyanic acid, methyl ester

CAS Number: 624-83-9
DOT Number: 2480

Description: Colorless Liquid
DOT Classification: Flammable Liquid

Molecular Weight: 57.1
Melting Point: -80°C (-112°F)
Boiling Point: 39°C (102°F)

Vapor Density:
Specific Gravity: <1.0

Vapor Pressure: **8** (348 mm Hg)
Water Solubility: Decomposes

Chemical Incompatibilities or Instabilities: Alcohols, acids, amines, alkali, heat, oxidizers, steel.

FLAMMABILITY

NFPA Hazard Code: 4-3-2
Flash Point: -19°C (-7°F)
Autoignition Temp: 534°C (994°F)

NFPA Classification: Class IB Flammable
LEL: 5.3%
UEL: 26.0%

Fire Extinguishing Methods: Carbon Dioxide, Dry Chemical, Foam, Water Spray or Fog.

Special Fire Fighting Considerations: Structural fire fighter protective clothing will not provide adequate protection. Isolate for 1/2 mile if rail or tank truck is involved in a fire. Water spray should be used to keep closed containers cool. Continue to cool container after fire is extinguished. For large fires, if possible, withdraw and allow to burn. Immediately withdraw if rising sound from venting device is heard or if fire is causing discoloration to the tank.

DOT Recommended Isolation Zones: Small Spill: 1500 ft Large Spill: 1500 ft
DOT Recommended Down Wind Take Cover Distance Small Spill: 5 miles Large Spill: 5 miles

TOXICOLOGY

Odor:
Physical Contact: Material is extremely destructive to human tissues.
TLV = 0.047 mg/m^3 **STEL** = N.D.

Odor Threshold: 5 mg/m^3

IDLH = 47 mg/m^3

RTECS # NQ9450000

Routes of Entry and Relative LD$_{50}$ (or LC$_{50}$)
Inhalation	10	(86 mg/m^3/H)
Ingestion	8	(51 mg/kg)
Skin Absorption	8	(213 mg/kg)

Symptoms of Exposure: Possible burns or irritation to skin, eyes and upper respiratory tract; headache; nausea; laryngitis; lachrymation. Inhalation of vapors may be fatal by causing glottis to spasm and suffocation. Exposure may result in chemical pneumonitis or pulmonary edema. Repeated exposures may cause allergic reactions or asthma..cyanide poisoning may occur from impurities present.

Emergency/First Aid Treatment: Remove to ventilated area; immediately remove any contaminated clothing and wash contaminated areas for 15 minutes using water. Treat supportively and observe for possible shock. If ingested, seek immediate medical aid.

Recommended Clean-Up Procedures

Personal Protection: Level A Ensemble

RCRA Waste # P064

Recommended Material Polyvinyl Alcohol (+)

Reportable Quantities: 1 lb (1 lb)

Spills: Remove all potential ignition sources. Dike to contain spill, absorb with non-combustible absorbent and take up using non-sparking tools. Decontaminate spill area using dry decontamination techniques.

Special Emergency Information: May be fatal in inhaled, ingested or absorbed through the skin. Vapors are heavier than air and may travel some distance to an ignition source. Vapors are more dense than air and may settle in low lying areas. May undergo hazardous polymerization.

METHYL ISOTHIOCYANATE - CH₃N=C=S

Synonyms: isothiocyanatomethane

CAS Number: 556-61-6	**Description:** White to orange solid (below 36°C)
DOT Number: 2477	**DOT Classification:** Flammable Liquid

Molecular Weight: 73.1

Melting Point: 36°C (97°F)	**Vapor Density:** N.A.	**Vapor Pressure:** 5 (21 mm Hg)
Boiling Point: 119°C (246°F)	**Specific Gravity:** 1.1	**Water Solubility:** <1%

Chemical Incompatibilities or Instabilities: Oxidizers, water.

FLAMMABILITY

NFPA Hazard Code:	**NFPA Classification:** Class 1C Flammable
Flash Point: 32°C (90°F)	**LEL:** N.D.
Autoignition Temp: N.D.	**UEL:** N.D.

Fire Extinguishing Methods: Alcohol Resistant Foam, Carbon Dioxide, Dry Chemical, Water Spray or Fog.

Special Fire Fighting Considerations: Structural fire fighter protective clothing will not provide adequate protection. Isolate for 1/2 mile if rail or tank truck is involved in a fire. Water spray should be used to keep closed containers cool. Continue to cool container after fire is extinguished. Immediately withdraw if rising sound from venting device is heard or if fire is causing discoloration to the tank. Treat all materials used or generated and equipment used involved as contaminated by hazardous waste.

TOXICOLOGY

Odor: **Odor Threshold:** N.D.

Physical Contact: Material is extremely destructive to human tissues.

TLV = N.D.   **STEL** = N.D.   **IDLH** = N.D.

Routes of Entry and Relative LD₅₀ (or LC₅₀)   **RTECS # PA9625000**

Inhalation	N.D.	
Ingestion	8	(97 mg/kg)
Skin Absorption	10	(33 mg/kg)

Symptoms of Exposure: Possible burns or irritation to skin, eyes and upper respiratory tract; headache; nausea; lachrymation. Inhalation of vapors may be fatal by causing glottis to spasm and suffocation. Exposure may result in chemical pneumonitis or pulmonary edema. Repeated exposures may cause allergic reactions and asthma.

Emergency/First Aid Treatment: Remove to ventilated area; immediately remove any contaminated clothing and wash contaminated areas for 15 minutes using water. Treat supportively and observe for possible shock. If ingested, seek immediate medical aid.

Recommended Clean-Up Procedures

Personal Protection: Level A Ensemble	**Recommended Material**	N.D.
RCRA Waste # None	**Reportable Quantities:**	None

Spills: Remove all potential ignition sources. Dike to contain spill, absorb with non-combustible absorbent and take up using non-sparking tools. Decontaminate spill area using soapy water. Treat all materials used or generated and equipment used involved as contaminated by hazardous waste.

Special Emergency Information: May be fatal in inhaled, ingested or absorbed through the skin. Vapors are heavier than air and may travel some distance to an ignition source. Vapors are more dense than air and may settle in low lying areas.

METHYL MERCAPTAN - CH₃SH

Synonyms: methanethiol

CAS Number: 74-93-1   **Description:** Colorless Gas
DOT Number: 1064   **DOT Classification:** Flammable Gas

Molecular Weight: 48.1
Melting Point: -123°C (-189°F)   **Vapor Density:** 1.7   **Vapor Pressure:** Gas
Boiling Point: 6°C (43°F)   **Specific Gravity:** N.A.   **Water Solubility:** Soluble

Chemical Incompatibilities or Instabilities: Mercury (II) oxide, strong oxidizers.

FLAMMABILITY

NFPA Hazard Code: 2-4-0   **NFPA Classification:**
Flash Point: N.A.   **LEL:** 3.9%
Autoignition Temp: N.D.   **UEL:** 21.8%

Fire Extinguishing Methods: Foam, Water Spray or Fog.

Special Fire Fighting Considerations: Structural fire fighter protective clothing will not provide adequate protection. Isolate for 1/2 mile if rail or tank truck is involved in a fire. If leak cannot be safely stopped, then allow to burn. Water spray from an unmanned device should be used to keep closed containers cool. Continue to cool container after fire is extinguished. Immediately withdraw if rising sound from venting device is heard or if fire is causing discoloration to the tank.

DOT Recommended Isolation Zones:	Small Spill: 150 ft	Large Spill: 600 ft
DOT Recommended Down Wind Take Cover Distance	Small Spill: 0.8 miles	Large Spill: 2 miles

TOXICOLOGY

Odor: Rotten cabbage   **Odor Threshold:** 0.0000004 mg/m³
Physical Contact: Material is extremely destructive to human tissues.
TLV = 0.98 mg/m³   **STEL** = N.D.   **IDLH** = 800 mg/m³

Routes of Entry and Relative LD₅₀ (or LC₅₀)   **RTECS #** PB4375000
   Inhalation   6   (1350 mg/m³/H)
   Ingestion   N.D.
   Skin Absorption   N.D.

Symptoms of Exposure: Possible burns or irritation to skin, eyes and upper respiratory tract; headache; nausea. Inhalation of vapors may be fatal by causing glottis to spasm and suffocation. Exposure may result in chemical pneumonitis or pulmonary edema.

Emergency/First Aid Treatment: Remove to ventilated area; immediately remove any contaminated clothing and wash contaminated areas for 15 minutes using water. Treat supportively and observe for possible shock. If ingested, seek immediate medical aid.

Recommended Clean-Up Procedures

Personal Protection: Level B Ensemble   **Recommended Material** N.D.

RCRA Waste # U153   **Reportable Quantities:** 100 lbs (100 lbs)

Spills: Remove all potential ignition sources. Stop leak if it can safely be done. Vent to dissipate vapors.

Special Emergency Information: May be harmful if inhaled, swallowed or absorbed through the skin. Vapors are heavier than air and may travel some distance to an ignition source. Vapors are more dense than air and may settle in low lying areas.

METHYL METHACRYLATE - $CH_2=C(CH_3)C(O)CH_3$

Synonyms: methyl-2-methyl-2-propenoate; 2-methyl-2-propenoic acid, methyl ester; methacrylic acid; methyl ester

CAS Number: 80-62-6	**Description:** Colorless Liquid	
DOT Number: 1247	**DOT Classification:** Flammable Liquid	

Molecular Weight: 100.1

Melting Point:	-48°C (-55°F)	**Vapor Density:** 3.5	**Vapor Pressure:**	5 (29 mm Hg)
Boiling Point:	100°C (213°F)	**Specific Gravity:** 0.94	**Water Solubility:**	Insoluble

Chemical Incompatibilities or Instabilities: Amines, halogens, oxidizers, peroxides.

FLAMMABILITY

NFPA Hazard Code:	**2-3-2**	**NFPA Classification:** Class IB Flammable
Flash Point:	10°C (50°F)	**LEL:** 1.8%
Autoignition Temp:	435°C (815°F)	**UEL:** 8.2%

Fire Extinguishing Methods: Carbon Dioxide, Dry Chemical, Foam, Water Spray or Fog.

Special Fire Fighting Considerations: Isolate for 1/2 mile if rail or tank truck is involved in a fire. Water spray should be used to keep closed containers cool. Continue to cool container after fire is extinguished. For large fires, if possible, withdraw and allow to burn. Immediately withdraw if rising sound from venting device is heard or if fire is causing discoloration to the tank. Treat all materials used or generated and equipment involved as contaminated by hazardous waste.

TOXICOLOGY
QUESTIONABLE CARCINOGEN

Odor: Acrid, fruity

Physical Contact: Irritant

Odor Threshold: 0.06 mg/m^3

$TLV = 410$ mg/m^3 $STEL = $ N.D. $IDLH = 16,500$ mg/m^3

Routes of Entry and Relative LD$_{50}$ (or LC$_{50}$)

Inhalation	2	(15,600 mg/m^3/H)
Ingestion	2	(7872 mg/kg)
Skin Absorption	N.D.	

RTECS # OZ5075000

Symptoms of Exposure: Possible irritation to skin, eyes and upper respiratory tract; headache; nausea; laryngitis. May act as a narcotic in high concentrations. Repeated exposures may cause allergic reactions.

Emergency/First Aid Treatment: Remove to ventilated area; immediately remove any contaminated clothing and wash contaminated areas for 15 minutes using water. Treat supportively and observe for possible shock. If ingested, seek immediate medical aid.

Recommended Clean-Up Procedures

Personal Protection:	Level A Ensemble	**Recommended Material** Polyvinyl Alcohol (+), Teflon (0)
RCRA Waste #	U162	**Reportable Quantities:** 1000 lbs (5000 lbs)

Spills: Remove all potential ignition sources. Dike to contain spill, absorb with non-combustible absorbent and take up using non-sparking tools. Decontaminate spill area using soapy water. Treat all materials used or generated and equipment involved as contaminated by hazardous waste.

Special Emergency Information: May be harmful if inhaled, swallowed or absorbed through the skin. Vapors are heavier than air and may travel some distance to an ignition source. Vapors are more dense than air and may settle in low lying areas. May undergo hazardous polymerization.

METHYLMORPHOLINE

Synonyms: N-methylmorpholine; 4-methylmorpholine

CAS Number: 109-02-4	**Description:** Colorless Liquid	
DOT Number: 2535	**DOT Classification:** Flammable Liquid	

Molecular Weight: 101.2

Melting Point: -66°C (-87°F)	**Vapor Density:** 3.5	**Vapor Pressure:** N.A.
Boiling Point: 115°C (239°F)	**Specific Gravity:** 0.9	**Water Solubility:** Soluble

Chemical Incompatibilities or Instabilities: Oxidizers.

FLAMMABILITY

NFPA Hazard Code: 2-3-0	**NFPA Classification:** Class IC Flammable
Flash Point: 24°C (75°F)	**LEL:** N.D.
Autoignition Temp: N.D.	**UEL:** N.D.

Fire Extinguishing Methods: Carbon Dioxide, Dry Chemical, Foam, Water Spray or Fog.

Special Fire Fighting Considerations: Isolate for 1/2 mile if rail or tank truck is involved in a fire. Water spray should be used to keep closed containers cool. Continue to cool container after fire is extinguished. Immediately withdraw if rising sound from venting device is heard or if fire is causing discoloration to the tank.

TOXICOLOGY

Odor: **Odor Threshold:** N.D.

Physical Contact: Material is extremely destructive to human tissues.

TLV = N.D.   **STEL** = N.D.   **IDLH** = N.D.

Routes of Entry and Relative LD$_{50}$ (or LC$_{50}$)   **RTECS # QE5775000**

Inhalation	N.D.	
Ingestion	3	(2720 mg/kg)
Skin Absorption	4	(1242 mg/kg)

Symptoms of Exposure: Possible burns or irritation to skin, eyes and upper respiratory tract; headache; nausea; laryngitis. Inhalation of vapors may be fatal by causing glottis to spasm and suffocation. Exposure may result in chemical pneumonitis or pulmonary edema.

Emergency/First Aid Treatment: Remove to ventilated area; immediately remove any contaminated clothing and wash contaminated areas for 15 minutes using water. Treat supportively and observe for possible shock. If ingested, seek immediate medical aid.

Recommended Clean-Up Procedures

Personal Protection: Level B Ensemble	**Recommended Material** N.D.
RCRA Waste # None	**Reportable Quantities:** None

Spills: Remove all potential ignition sources. Decontaminate spill area using water. Dike to contain spill, absorb with non-combustible absorbent and take up using non-sparking tools.

Special Emergency Information: May be harmful if inhaled, swallowed or absorbed through the skin. Vapors are heavier than air and may travel some distance to an ignition source. Vapors are more dense than air and may settle in low lying areas.

METHYL PARATHION

Synonyms:

CAS Number: 298-00-0   **Description:** White powder
DOT Number: 2783   **DOT Classification:** Poison B

Molecular Weight: 263.2
Melting Point: 38°C (100°F)   **Vapor Density:**   9.1   **Vapor Pressure:**   N.D.
Boiling Point: °C   (°F)   **Specific Gravity:** 1.4   **Water Solubility:** <1%

Chemical Incompatibilities or Instabilities: Oxidizers.

FLAMMABILITY

NFPA Hazard Code: **NFPA Classification:**
Flash Point: N.A.   **LEL:** N.A.
Autoignition Temp: N.A.   **UEL:** N.A.

Fire Extinguishing Methods: Carbon Dioxide, Dry Chemical, Foam, Water Spray or Fog.

Special Fire Fighting Considerations: Structural fire fighter protective clothing will not provide adequate protection. Fight fire from a distance or protected location, if possible. Treat all materials used or generated and equipment used involved as contaminated by hazardous waste.

TOXICOLOGY

Odor: Pungent   **Odor Threshold:** 0.01 mg/m^3
Physical Contact: Irritant
TLV = 0.2 mg/m^3 (skin)   **STEL** = N.D.   **IDLH** = N.D.

Routes of Entry and Relative LD$_{50}$ (or LC$_{50}$)   **RTECS # TG0175000**
   Inhalation   10   (136 mg/m^3/H)
   Ingestion   9   (6 mg/kg)
   Skin Absorption   8   (300 mg/kg)

Symptoms of Exposure: Possible irritation to skin, eyes and upper respiratory tract; headache; nausea; blurred vision; cramps, convulsions.

Emergency/First Aid Treatment: Remove to ventilated area; immediately remove any contaminated clothing and wash contaminated areas for 15 minutes using water. Treat supportively and observe for possible shock. If ingested, seek immediate medical aid. Treat for anticholinesterase poisoning.

Recommended Clean-Up Procedures

Personal Protection: Level A Ensemble   **Recommended Material**   N.D.

RCRA Waste #   P071   **Reportable Quantities:**   100 lbs (100lbs)

Spills: Take up in non-combustible absorbent and dispose. Avoid raising dust. Decontaminate spill area using a dilute hypochlorite solution. Treat all materials used or generated and equipment involved as contaminated by hazardous waste.

Special Emergency Information: May be fatal in inhaled, ingested or absorbed through the skin.

2-METHYL-1,3-PENTADIENE - $CH_3CH=CHC(CH_3)=CH_2$

Synonyms:

CAS Number: 1118-58-7

DOT Number: None

Description: Colorless Liquid

DOT Classification: Flammable Liquid

Molecular Weight: 82.2

Melting Point: °C (°F)

Boiling Point: 76°C (169°F)

Vapor Density: 2.8

Specific Gravity: 0.7

Vapor Pressure: N.A.

Water Solubility: Insoluble

Chemical Incompatibilities or Instabilities: Oxidizers.

FLAMMABILITY

NFPA Hazard Code: 0-3-0

Flash Point: -20°C (-4°F)

Autoignition Temp: N.D.

NFPA Classification: Class IB Flammable

LEL: N.D.

UEL: N.D.

Fire Extinguishing Methods: Alcohol Resistant Foam, Carbon Dioxide, Dry Chemical, Water Spray or Fog.

Special Fire Fighting Considerations: Isolate for 1/2 mile if rail or tank truck is involved in a fire. Water spray should be used to keep closed containers cool. Continue to cool container after fire is extinguished. For large fires, if possible, withdraw and allow to burn. Immediately withdraw if rising sound from venting device is heard or if fire is causing discoloration to the tank.

TOXICOLOGY

Odor:

Physical Contact: Irritant

TLV = N.D.

Odor Threshold: N.D.

STEL = N.D.

IDLH = N.D.

Routes of Entry and Relative LD_{50} (or LC_{50})

Inhalation N.D.

Ingestion N.D.

Skin Absorption N.D.

RTECS # None

Symptoms of Exposure: Possible irritation to skin, eyes and upper respiratory tract; headache; nausea. May act as a narcotic in high concentrations.

Emergency/First Aid Treatment: Remove to ventilated area; immediately remove any contaminated clothing and wash contaminated areas for 15 minutes using water. Treat supportively and observe for possible shock. If ingested, seek immediate medical aid.

Recommended Clean-Up Procedures

Personal Protection: Level B Ensemble

RCRA Waste # None

Recommended Material N.D.

Reportable Quantities: None

Spills: Remove all potential ignition sources. Dike to contain spill, absorb with non-combustible absorbent and take up using non-sparking tools.

Special Emergency Information: May be harmful if inhaled, swallowed or absorbed through the skin. Vapors are heavier than air and may travel some distance to an ignition source. Vapors are more dense than air and may settle in low lying areas.

Methylpentadiene (CAS # 54363-49-4, DOT # 2461, RTECS # RZ2473000) has similar chemical, physical and toxicological properties.

METHYLPENTANE - CH3CH2CH(CH3)CH2CH3

Synonyms: 3-methylpentane

CAS Number: 96-14-0   **Description:** Colorless Liquid
DOT Number: 1208, 2462   **DOT Classification:** Flammable Liquid

Molecular Weight: 86.2
Melting Point: -83°C (-118°F)   **Vapor Density:** 3.0   **Vapor Pressure:** 7 (180 mm Hg)
Boiling Point: 63°C (146°F)   **Specific Gravity:** 0.7   **Water Solubility:** Insoluble

Chemical Incompatibilities or Instabilities: Oxidizers.

FLAMMABILITY

NFPA Hazard Code: 1-3-0   **NFPA Classification:** Class IB Flammable
Flash Point: -7°C (20°F)   **LEL:** 1.2%
Autoignition Temp: 278°C (532°F) **UEL:** 7.0%

Fire Extinguishing Methods: Alcohol Resistant Foam, Carbon Dioxide, Dry Chemical, Water Spray or Fog.

Special Fire Fighting Considerations: Isolate for 1/2 mile if rail or tank truck is involved in a fire. Water spray should be used to keep closed containers cool. Continue to cool container after fire is extinguished. For large fires, if possible, withdraw and allow to burn. Immediately withdraw if rising sound from venting device is heard or if fire is causing discoloration to the tank.

TOXICOLOGY

Odor: **Odor Threshold:** N.A.
Physical Contact: Irritant
TLV = N.D.   **STEL** = N.D.   **IDLH** = N.D.

Routes of Entry and Relative LD$_{50}$ (or LC$_{50}$)   **RTECS #** SA2995500
  Inhalation   N.D.
  Ingestion   N.D.
  Skin Absorption   N.D.

Symptoms of Exposure: Possible irritation to skin, eyes and upper respiratory tract; headache; nausea.

Emergency/First Aid Treatment: Remove to ventilated area; immediately remove any contaminated clothing and wash contaminated areas for 15 minutes using water. Treat supportively and observe for possible shock. If ingested, seek immediate medical aid.

Recommended Clean-Up Procedures

Personal Protection: Level B Ensemble   **Recommended Material** N.D.

RCRA Waste # None   **Reportable Quantities:** None

Spills: Remove all potential ignition sources. Dike to contain spill, absorb with non-combustible absorbent and take up using non-sparking tools. Decontaminate spill area using soapy water. Keep area isolated until vapors have dissipated.

Special Emergency Information: May be harmful if inhaled, swallowed or absorbed through the skin. Vapors are heavier than air and may travel some distance to an ignition source. Vapors are more dense than air and may settle in low lying areas.

2-Methylpentane (CAS # 107-83-5, RTECS # SA2995000) has similar chemical, physical and toxicological properties.

METHYLPHOSPHONIC DICHLORIDE - CH3P(O)Cl2

Synonyms:

CAS Number: 676-97-1	**Description:** White powder or colorless liquid
DOT Number: 9206	**DOT Classification:** Corrosive Material

Molecular Weight: 132.9

Melting Point:	35°C (95°F)	**Vapor Density:** N.D.	**Vapor Pressure:**	N.D.
Boiling Point:	163°C (325°F)	**Specific Gravity:** N.A.	**Water Solubility:**	Decomposes

Chemical Incompatibilities or Instabilities: Oxidizers.

FLAMMABILITY

NFPA Hazard Code:	**NFPA Classification:**
Flash Point: 110°C (230°F)	**LEL:** N.D.
Autoignition Temp: N.D.	**UEL:** N.D.

Fire Extinguishing Methods: Carbon Dioxide, Dry Chemical or Flooding Quantities of Water

Special Fire Fighting Considerations: Structural fire fighter protective clothing will not provide adequate protection. Water spray should be used to keep closed containers cool. Continue to cool container after fire is extinguished. Do not direct water water directly at spilled material. May release toxic fumes upon heating (phosphorous oxides, phosphine).

DOT Recommended Isolation Zones:	Small Spill: 150 ft	Large Spill: 150 ft
DOT Recommended Down Wind Take Cover Distance	Small Spill: 0.2 miles	Large Spill: 0.2 miles

TOXICOLOGY

Odor: Sharp	**Odor Threshold:** N.A.

Physical Contact: Material is extremely destructive to human tissues.

TLV = N.D.	**STEL** = N.D.	**IDLH** = N.D.

Routes of Entry and Relative LD$_{50}$ (or LC$_{50}$) **RTECS # TA1840000**

Inhalation	10	(565 mg/m^3/H)
Ingestion	N.D.	
Skin Absorption	N.D.	

Symptoms of Exposure: Possible burns or irritation to skin, eyes and upper respiratory tract; headache; nausea. Inhalation of vapors may be fatal by causing glottis to spasm and suffocation. Exposure may result in chemical pneumonitis or pulmonary edema.

Emergency/First Aid Treatment: Remove to ventilated area; immediately remove any contaminated clothing and wash contaminated areas for 15 minutes using water. Treat supportively and observe for possible shock. If ingested, seek immediate medical aid.

Recommended Clean-Up Procedures

Personal Protection: Level A Ensemble	**Recommended Material** N.D.
RCRA Waste # None	**Reportable Quantities:** None

Spills: Dike to contain spill and collect liquid for later disposal. Decontaminate spill area using flooding quantities of water.

Special Emergency Information: May be fatal in inhaled, ingested or absorbed through the skin.

Methylphosphonic difluoride (CAS # 676-99-3, DOT # 9266, RTECS # TA1840700) has similar chemical, physical and toxicological properties.

N-METHYLPIPERIDINE

Synonyms:

CAS Number: 626-67-5   **Description:** Colorless Liquid
DOT Number: 2399   **DOT Classification:** Flammable Liquid

Molecular Weight: 99.2
Melting Point: °C   (°F)   **Vapor Density:** N.D.   **Vapor Pressure:** N.D.
Boiling Point: 107°C (225°F)   **Specific Gravity:** 0.8   **Water Solubility:** Soluble

Chemical Incompatibilities or Instabilities: Strong oxidizers.

FLAMMABILITY

NFPA Hazard Code: **NFPA Classification:** Class 1B Flammable
Flash Point: 3°C   (38°F)   **LEL:** N.D.
Autoignition Temp: N.D.   **UEL:** N.D.

Fire Extinguishing Methods: Alcohol Resistant Foam, Carbon Dioxide, Dry Chemical, Water Spray or Fog.

Special Fire Fighting Considerations: Isolate for 1/2 mile if rail or tank truck is involved in a fire. Water spray should be used to keep closed containers cool. Continue to cool container after fire is extinguished. For large fires, if possible, withdraw and allow to burn. Immediately withdraw if rising sound from venting device is heard or if fire is causing discoloration to the tank.

TOXICOLOGY

Odor: **Odor Threshold:** N.D.
Physical Contact: Irritant
TLV = N.D.   **STEL** = N.D.   **IDLH** = N.D.

Routes of Entry and Relative LD$_{50}$ (or LC$_{50}$)   **RTECS #** TN1225000
    Inhalation   N.D.
    Ingestion   N.D.
    Skin Absorption   N.D.

Symptoms of Exposure: Possible irritation to skin, eyes and upper respiratory tract; headache; nausea.

Emergency/First Aid Treatment: Remove to ventilated area; immediately remove any contaminated clothing and wash contaminated areas for 15 minutes using water. Treat supportively and observe for possible shock. If ingested, seek immediate medical aid.

Recommended Clean-Up Procedures

Personal Protection: Level B Ensemble   **Recommended Material**   N.D.

RCRA Waste #   None   **Reportable Quantities:**   None

Spills: Remove all potential ignition sources. Decontaminate spill area using water. Dike to contain spill, absorb with non-combustible absorbent and take up using non-sparking tools.

Special Emergency Information: May be harmful if inhaled, swallowed or absorbed through the skin. Vapors are heavier than air and may travel some distance to an ignition source. Vapors are more dense than air and may settle in low lying areas.

METHYL PROPIONATE - CH₃CH₂C(O)OCH₃

Synonyms: propanioc acid, methyl ester

CAS Number: 554-12-1
DOT Number: 1248

Description: Colorless Liquid
DOT Classification: Flammable Liquid

Molecular Weight: 88.1
Melting Point: -88°C (-126°F)
Boiling Point: 85°C (185°F)

Vapor Density: 2.8
Specific Gravity: 0.9

Vapor Pressure: 5 (40 mm Hg)
Water Solubility: Insoluble

Chemical Incompatibilities or Instabilities: Strong oxidizers.

FLAMMABILITY

NFPA Hazard Code: ?-3-?
Flash Point: -2°C (28°F)
Autoignition Temp: 469°C (876°F)

NFPA Classification: Class IB Flammable
LEL: 2.5%
UEL: 13.0%

Fire Extinguishing Methods: Alcohol Resistant Foam, Carbon Dioxide, Dry Chemical, Water Spray or Fog.

Special Fire Fighting Considerations: Isolate for 1/2 mile if rail or tank truck is involved in a fire. Water spray should be used to keep closed containers cool. Continue to cool container after fire is extinguished. For large fires, if possible, withdraw and allow to burn. Immediately withdraw if rising sound from venting device is heard or if fire is causing discoloration to the tank.

TOXICOLOGY

Odor: Sweet
Physical Contact: Irritant
TLV = N.D.   **STEL** = N.D.

Odor Threshold: N.D.

IDLH = N.D.

Routes of Entry and Relative LD₅₀ (or LC₅₀)
Inhalation   N.D.
Ingestion   N.D.
Skin Absorption   N.D.

RTECS # UF5970000

Symptoms of Exposure: Possible irritation to skin, eyes and upper respiratory tract; headache; nausea.

Emergency/First Aid Treatment: Remove to ventilated area; immediately remove any contaminated clothing and wash contaminated areas for 15 minutes using water. Treat supportively and observe for possible shock. If ingested, seek immediate medical aid.

Recommended Clean-Up Procedures

Personal Protection: Level B Ensemble   **Recommended Material** N.D.

RCRA Waste # None   **Reportable Quantities:** None

Spills: Remove all potential ignition sources. Dike to contain spill, absorb with non-combustible absorbent and take up using non-sparking tools. Decontaminate spill area using soapy water.

Special Emergency Information: May be harmful if inhaled, swallowed or absorbed through the skin. Vapors are heavier than air and may travel some distance to an ignition source. Vapors are more dense than air and may settle in low lying areas.

2-METHYLTETRAHYDROFURAN

Synonyms:

CAS Number: 96-47-9
DOT Number: 2536

Description: Colorless Liquid
DOT Classification: Flammable Liquid

Molecular Weight: 86.2
Melting Point: < -40°C(< -40°F)
Boiling Point: 80°C (176°F)

Vapor Density: 3.0
Specific Gravity: 0.9

Vapor Pressure: 6 (>50 mm Hg)
Water Solubility: < 1%

Chemical Incompatibilities or Instabilities: Strong oxidizers.

FLAMMABILITY

NFPA Hazard Code: 2-3-0
Flash Point: -11°C (2°F)
Autoignition Temp: N.D.

NFPA Classification: Class IB Flammable
LEL: N.D.
UEL: N.D.

Fire Extinguishing Methods: Alcohol Resistant Foam, Carbon Dioxide, Dry Chemical, Water Spray or Fog.

Special Fire Fighting Considerations: Isolate for 1/2 mile if rail or tank truck is involved in a fire. Water spray should be used to keep closed containers cool. Continue to cool container after fire is extinguished. For large fires, if possible, withdraw and allow to burn. Immediately withdraw if rising sound from venting device is heard or if fire is causing discoloration to the tank.

TOXICOLOGY

Odor: Ethereal
Physical Contact: Irritant
TLV = N.D.   **STEL** = N.D.

Odor Threshold: N.D.

IDLH = N.D.

Routes of Entry and Relative LD$_{50}$ (or LC$_{50}$)
  Inhalation   3   (85,000 mg/m^3/H)
  Ingestion   N.D.
  Skin Absorption   N.D.

RTECS # LU2800000

Symptoms of Exposure: Possible irritation to skin, eyes and upper respiratory tract; headache; nausea.

Emergency/First Aid Treatment: Remove to ventilated area; immediately remove any contaminated clothing and wash contaminated areas for 15 minutes using water. Treat supportively and observe for possible shock. If ingested, seek immediate medical aid.

Recommended Clean-Up Procedures

Personal Protection: Level B Ensemble
RCRA Waste # None

Recommended Material N.D.
Reportable Quantities: None

Spills: Remove all potential ignition sources. Dike to contain spill, absorb with non-combustible absorbent and take up using non-sparking tools. Decontaminate spill area using soapy water.

Special Emergency Information: May be harmful if inhaled, swallowed or absorbed through the skin. Vapors are heavier than air and may travel some distance to an ignition source. Vapors are more dense than air and may settle in low lying areas. May form unstable peroxides upon exposure to air.

3-Methyltetrahydrofuran (CAS # 13423-15-9) and
tetrahydrofuran (CAS # 109-99-9, DOT # 2056, RTECS # LU5950000) have similar chemical, physical and toxicological properties.

METHYLTRICHLOROSILANE - CH3SCl3

Synonyms: trichloromethylsilane

CAS Number: 75-79-6
DOT Number: 1250

Description: Colorless Liquid
DOT Classification: Flammable Liquid

Molecular Weight: 149.5
Melting Point: °C (°F)
Boiling Point: 66°C (151°F)

Vapor Density: 5.2
Specific Gravity: 1.3

Vapor Pressure: 7 (150 mm Hg)
Water Solubility: Decomposes

Chemical Incompatibilities or Instabilities: Alcohols, water.

FLAMMABILITY

NFPA Hazard Code: 3-3-2-~~W~~
Flash Point: 9°C (15°F)
Autoignition Temp: 404°C (760°F)

NFPA Classification: Class IB Flammable
LEL: 7.6%
UEL: 20.0%

Fire Extinguishing Methods: Alcohol Resistant Foam, Carbon Dioxide, Dry Chemical, Water Spray or Fog.

Special Fire Fighting Considerations: Isolate for 1/2 mile if rail or tank truck is involved in a fire. Water spray should be used to keep closed containers cool. Continue to cool container after fire is extinguished. Do not allow water to enter container. Immediately withdraw if rising sound from venting device is heard or if fire is causing discoloration to the tank.

TOXICOLOGY

Odor: Sharp, penetrating
Physical Contact: Material is extremely destructive to human tissues.
TLV = N.D.   **STEL** = N.D.

Odor Threshold: N.A.

IDLH = N.D.

RTECS # VV4550000

Routes of Entry and Relative LD$_{50}$ (or LC$_{50}$)
Inhalation   7   (11,000 mg/m^3/H)
Ingestion   N.D.
Skin Absorption   N.D.

Symptoms of Exposure: Possible burns or irritation to skin, eyes and upper respiratory tract; headache; nausea; laryngitis. Inhalation of vapors may be fatal by causing glottis to spasm and suffocation. Exposure may result in chemical pneumonitis or pulmonary edema.

Emergency/First Aid Treatment: Remove to ventilated area; immediately remove any contaminated clothing and wash contaminated areas for 15 minutes using water. Treat supportively and observe for possible shock. If ingested, seek immediate medical aid.

Recommended Clean-Up Procedures

Personal Protection: Level A Ensemble
RCRA Waste # None

Recommended Material N.D.
Reportable Quantities: None

Spills: Remove all potential ignition sources. Dike to contain spill, absorb with non-combustible absorbent and take up using non-sparking tools. Decontaminate spill area using dilute alkaline solution.

Special Emergency Information: May be fatal in inhaled, ingested or absorbed through the skin. Vapors are heavier than air and may travel some distance to an ignition source. Vapors are more dense than air and may settle in low lying areas. Material will react violently with water to release hydrogen chloride vapors.

METHYL VINYL KETONE - $CH_3C(O)CH=CH_2$

Synonyms: 1-buten-3-one

CAS Number: 78-94-4
DOT Number: 1251

Description: Colorless Liquid
DOT Classification: Flammable Liquid

Molecular Weight:	70.1				
Melting Point:	°C	(°F)	**Vapor Density:** 2.4	**Vapor Pressure:**	6 (71 mm Hg)
Boiling Point:	81°C	(177°F)	**Specific Gravity:** 0.8	**Water Solubility:**	Insoluble

Chemical Incompatibilities or Instabilities: Strong alkali, light, heat, oxidizers, reducing agents.

FLAMMABILITY

NFPA Hazard Code:	3-3-2	**NFPA Classification:**	Class IB Flammable
Flash Point:	-7°C (20°F)	**LEL:**	2.1%
Autoignition Temp:	491°C (915°F)	**UEL:**	15.6%

Fire Extinguishing Methods: Alcohol Resistant Foam, Carbon Dioxide, Dry Chemical, Water Spray or Fog.

Special Fire Fighting Considerations: Structural fire fighter protective clothing will not provide adequate protection. Isolate for 1/2 mile if rail or tank truck is involved in a fire. Water spray should be used to keep closed containers cool. Continue to cool container after fire is extinguished. Immediately withdraw if rising sound from venting device is heard or if fire is causing discoloration to the tank. Treat all materials used or generated and equipment used involved as contaminated by hazardous waste.

TOXICOLOGY

Odor:
Physical Contact: Material is extremely destructive to human tissues.
TLV = N.D.   **STEL** = N.D.

Odor Threshold: N.A.

IDLH = N.D.

Routes of Entry and Relative LD_{50} (or LC_{50})

Inhalation	10	(28 mg/m^3/H)
Ingestion	9	(30 mg/kg)
Skin Absorption	N.D.	

RTECS # EM9800000

Symptoms of Exposure: Possible burns or irritation to skin, eyes and upper respiratory tract; headache; nausea; laryngitis, lachrymation. Inhalation of vapors may be fatal by causing glottis to spasm and suffocation. Exposure may result in chemical pneumonitis or pulmonary edema.

Emergency/First Aid Treatment: Remove to ventilated area; immediately remove any contaminated clothing and wash contaminated areas for 15 minutes using water. Treat supportively and observe for possible shock. If ingested, seek immediate medical aid.

Recommended Clean-Up Procedures

Personal Protection:	Level A Ensemble	**Recommended Material**	N.D.
RCRA Waste #	None	**Reportable Quantities:**	None

Spills: Remove all potential ignition sources. Dike to contain spill, absorb with non-combustible absorbent and take up using non-sparking tools. Decontaminate spill area using soapy water. Treat all materials used or generated and equipment involved as contaminated by hazardous waste.

Special Emergency Information: May be fatal in inhaled, ingested or absorbed through the skin. Vapors are heavier than air and may travel some distance to an ignition source. Vapors are more dense than air and may settle in low lying areas. May undergo hazardous polymerization.

MEVINPHOS

Synonyms: apavinphos; phosdrin; 3-hydroxycrotonic acid, methyl ester dimethyl phosphate; fosdrin; duraphos

CAS Number: 7786-34-7   **Description:** Yellow liquid
DOT Number: 2783   **DOT Classification:** Poison B

Molecular Weight: 224.2
Melting Point: °C   (°F)   **Vapor Density:** N.D.   **Vapor Pressure:** N.D.
Boiling Point: Decomposes   **Specific Gravity:** 1.3   **Water Solubility:** Soluble

Chemical Incompatibilities or Instabilities: Oxidizers.

FLAMMABILITY

NFPA Hazard Code: **NFPA Classification:** N.A.
Flash Point: N.A.   **LEL:** N.A.
Autoignition Temp: N.A.   **UEL:** N.A.

Fire Extinguishing Methods: Carbon Dioxide, Dry Chemical, Foam, Water Spray or Fog.

Special Fire Fighting Considerations: Structural fire fighter protective clothing will not provide adequate protection. Fight fire from a distance or protected location, if possible. Treat all materials used or generated and equipment used involved as contaminated by hazardous waste.

TOXICOLOGY

Odor: **Odor Threshold:** N.A.
Physical Contact: Irritant
TLV = 0.09 mg/m^3 $STEL$ = 0.3 mg/m^3 $IDLH$ = N.D.

Routes of Entry and Relative LD$_{50}$ (or LC$_{50}$)   **RTECS # GQ5250000**
   Inhalation   10   (140 mg/m^3)
   Ingestion   10   (3 mg/kg)
   Skin Absorption   10   (5 mg/kg)

Symptoms of Exposure: Possible irritation to skin, eyes and upper respiratory tract; headache; nausea; muscle tremors and fatigue, spasms; coma.

Emergency/First Aid Treatment: Remove to ventilated area; immediately remove any contaminated clothing and wash contaminated areas for 15 minutes using water. Treat supportively and observe for possible shock. If ingested, seek immediate medical aid. Treat for anticholinesterase poisoning.

Recommended Clean-Up Procedures

Personal Protection: Level A Ensemble   **Recommended Material**   N.D.

RCRA Waste #   None   **Reportable Quantities:**   10 lb (1 lb)

Spills: Dike to contain spill and collect liquid for later disposal. Decontaminate spill area using soapy water. Treat all materials used or generated and equipment used involved as contaminated by hazardous waste.

Special Emergency Information: May be fatal in inhaled, ingested or absorbed through the skin.

MORPHOLINE

Synonyms: diethyleneimide oxide; tetrahydro-1,4-isoxazine

CAS Number: 110-91-8	**Description:** Colorless Liquid
DOT Number: 1760, 2054	**DOT Classification:** Flammable Liquid

Molecular Weight: 87.1

Melting Point: -5°C (23°F)	**Vapor Density:** 3.0	**Vapor Pressure:** **3** (8 mm Hg)
Boiling Point: 129°C (264°F)	**Specific Gravity:** 1.0	**Water Solubility:** Soluble

Chemical Incompatibilities or Instabilities: Acids, oxidizers.

FLAMMABILITY

NFPA Hazard Code: 2-3-0	**NFPA Classification:** Class IC Flammable
Flash Point: 35°C (95°F)	**LEL:** 1.8%
Autoignition Temp: 310°C (590°F)	**UEL:** 11.0%

Fire Extinguishing Methods: Carbon Dioxide, Dry Chemical, Foam, Water Spray or Fog.

Special Fire Fighting Considerations: Isolate for 1/2 mile if rail or tank truck is involved in a fire. Water spray should be used to keep closed containers cool. Continue to cool container after fire is extinguished. Immediately withdraw if rising sound from venting device is heard or if fire is causing discoloration to the tank.

TOXICOLOGY

Odor: Fish   **Odor Threshold:** 0.04 mg/m^3

Physical Contact: Material is extremely destructive to human tissues.

TLV = 70 mg/m^3 **STEL** = 105 mg/m^3 **IDLH** = 29,000 mg/m^3

Routes of Entry and Relative LD$_{50}$ (or LC$_{50}$)   **RTECS # QD6475000**

Inhalation	2	(230,000 mg/m^3/H)
Ingestion	3	(1050 mg/kg)
Skin Absorption	7	(500 mg/kg)

Symptoms of Exposure: Possible burns or irritation to skin, eyes and upper respiratory tract; headache; nausea; laryngitis. Inhalation of vapors may be fatal by causing glottis to spasm and suffocation. Exposure may result in chemical pneumonitis or pulmonary edema.

Emergency/First Aid Treatment: Remove to ventilated area; immediately remove any contaminated clothing and wash contaminated areas for 15 minutes using water. Treat supportively and observe for possible shock. If ingested, seek immediate medical aid.

Recommended Clean-Up Procedures

Personal Protection: Level B Ensemble	**Recommended Material** Butyl Rubber (+), Polyvinyl Alcohol (0)
RCRA Waste # None	**Reportable Quantities:** None

Spills: Remove all potential ignition sources. Dike to contain spill, absorb with non-combustible absorbent and take up using non-sparking tools. Decontaminate spill area using water.

Special Emergency Information: May be harmful if inhaled, swallowed or absorbed through the skin. Vapors are heavier than air and may travel some distance to an ignition source. Vapors are more dense than air and may settle in low lying areas.

NAPHTHA

Synonyms: coal tar naphtha

CAS Number: 8030-30-6	**Description:** Colorless Liquid
DOT Number: 1255, 1256, 2553	**DOT Classification:** Flammable Liquid

Molecular Weight: varies

Melting Point: °C (°F)	**Vapor Density:** 2.8	**Vapor Pressure:** **8** (400 mm Hg)
Boiling Point: 149°C (300°F)	**Specific Gravity:** 0.9	**Water Solubility:** Insoluble

Chemical Incompatibilities or Instabilities: Oxidixers.

FLAMMABILITY

NFPA Hazard Code: 1-3-0	**NFPA Classification:** Class II Combustible
Flash Point: 41°C (107°F)	**LEL:** 1.0%
Autoignition Temp: 227°C (531°F)	**UEL:** 6.0%

Fire Extinguishing Methods: Carbon Dioxide, Dry Chemical, Foam, Water Spray or Fog.

Special Fire Fighting Considerations: Isolate for 1/2 mile if rail or tank truck is involved in a fire. Water spray should be used to keep closed containers cool. Continue to cool container after fire is extinguished. For large fires, if possible, withdraw and allow to burn. Immediately withdraw if rising sound from venting device is heard or if fire is causing discoloration to the tank.

TOXICOLOGY

Odor: Sweet	**Odor Threshold:** 360 mg/m^3
Physical Contact: Irritant	
TLV = 400 mg/m^3 **STEL** = N.D.	**IDLH** = 45,000 mg/m^3

Routes of Entry and Relative LD$_{50}$ (or LC$_{50}$)   **RTECS # DE3030000**

Inhalation	N.D.
Ingestion	N.D.
Skin Absorption	N.D.

Symptoms of Exposure: Possible irritation to skin, eyes and upper respiratory tract; headache; nausea.

Emergency/First Aid Treatment: Remove to ventilated area; immediately remove any contaminated clothing and wash contaminated areas for 15 minutes using water. Treat supportively and observe for possible shock. If ingested, seek immediate medical aid.

Recommended Clean-Up Procedures

Personal Protection: Level B Ensemble	**Recommended Material** Nitrile Rubber (+), Polyvinyl Alcohol (0), Viton (0)
RCRA Waste # None	**Reportable Quantities:** None

Spills: Remove all potential ignition sources. Dike to contain spill, absorb with non-combustible absorbent and take up using non-sparking tools. Decontaminate spill area using soapy water.

Special Emergency Information: May be harmful if inhaled, swallowed or absorbed through the skin. Vapors are heavier than air and may travel some distance to an ignition source. Vapors are more dense than air and may settle in low lying areas.

NAPHTHALENE

Synonyms: camphor tar

CAS Number: 91-20-3   **Description:** White powder
DOT Number: 1334 (solid), 2304 (molten) **DOT Classification:** ORM-A

Molecular Weight: 128.2
Melting Point: 80°C (176°F)   **Vapor Density:** 4.4   **Vapor Pressure:**   1 (<1 mm Hg)
Boiling Point: 218°C (424°F)   **Specific Gravity:** 1.2   **Water Solubility:**   Insoluble

Chemical Incompatibilities or Instabilities: Oxidizers.

FLAMMABILITY

NFPA Hazard Code: 2-2-0   **NFPA Classification:**
Flash Point: 79°C (174°F)   **LEL:**   0.9%
Autoignition Temp: 526°C (979°F)   **UEL:**   5.9%

Fire Extinguishing Methods: Carbon Dioxide, Dry Chemical, Foam, Water Spray or Fog.

Special Fire Fighting Considerations: Water spray should be used to keep closed containers cool. Continue to cool container after fire is extinguished. For large fires, if possible, withdraw and allow to burn.

TOXICOLOGY

Odor: Mothballs   **Odor Threshold:** 0.005 mg/m^3
Physical Contact: Irritant or burns (DOT 2304)
TLV = 50 mg/m^3 **STEL** = 75 mg/m^3 **IDLH** = 2,500 mg/m^3

Routes of Entry and Relative LD$_{50}$ (or LC$_{50}$)   **RTECS # QJ0525000**
    Inhalation   N.D.
    Ingestion   **5**   (490 mg/kg)
    Skin Absorption   N.D.

Symptoms of Exposure: Possible irritation to skin, eyes and upper respiratory tract; headache; nausea. Absorption may lead to the formation of methemoglobin resulting in cyanosis several hours after exposure.

Emergency/First Aid Treatment: Remove to ventilated area; immediately remove any contaminated clothing and wash contaminated areas for 15 minutes using water. Treat supportively and observe for possible shock. If ingested, seek immediate medical aid.

Recommended Clean-Up Procedures

Personal Protection: Level B Ensemble   **Recommended Material**   Teflon (+)

RCRA Waste #   U165   **Reportable Quantities:**   100 lbs (5000 lbs)

Spills: Take up in non-combustible absorbent and dispose. Decontaminate spill area using soapy water.

Special Emergency Information: May be harmful if inhaled, swallowed or absorbed through the skin. Burns may occur from contact with molten naphthalene.

1-NAPHTHYLAMINE

Synonyms: 1-aminonaphthalene

CAS Number: 134-32-7

DOT Number: 2077

Description: Light brown powder

DOT Classification: Poison B

Molecular Weight: 143.2

Melting Point:	50°C (122°F)	**Vapor Density:** 4.9	**Vapor Pressure:**	1 (< 1 mm Hg)
Boiling Point:	301°C (574°F)	**Specific Gravity:** 1.1	**Water Solubility:**	< 1%

Chemical Incompatibilities or Instabilities: Oxidizers.

FLAMMABILITY

NFPA Hazard Code: 2-1-0

Flash Point: 157°C (315°F)

Autoignition Temp: N.D.

NFPA Classification:

LEL: N.D.

UEL: N.D.

Fire Extinguishing Methods: Carbon Dioxide, Dry Chemical, Foam, Water Spray or Fog.

Special Fire Fighting Considerations: Structural fire fighter protective clothing will not provide adequate protection. Fight fire from a distance or protected location, if possible. Treat all materials used or generated and equipment used involved as contaminated by hazardous waste.

TOXICOLOGY
CARCINOGEN

Odor: Unpleasant, ammonia-like

Physical Contact: Irritant

TLV = See 29 CFR 1910.1004   **STEL** = N.D.

Odor Threshold: N.D..

IDLH = N.D.

RTECS # QM1400000

Routes of Entry and Relative LD$_{50}$ (or LC$_{50}$)

Inhalation	N.D.	
Ingestion	4	(779 mg/kg)
Skin Absorption	N.D.	

Symptoms of Exposure: Possible irritation to skin, eyes and upper respiratory tract; headache; nausea. Absorption may lead to the formation of methemoglobin resulting in cyanosis several hours after exposure.

Emergency/First Aid Treatment: Remove to ventilated area; immediately remove any contaminated clothing and wash contaminated areas for 15 minutes using water. Treat supportively and observe for possible shock. If ingested, seek immediate medical aid.

Recommended Clean-Up Procedures

Personal Protection: Level B Ensemble   **Recommended Material** N.D.

RCRA Waste # U167   **Reportable Quantities:** 100 lbs (1 lb)

Spills: Take up in non-combustible absorbent and dispose. Decontaminate spill area using soapy water.

Special Emergency Information: May be fatal in inhaled, ingested or absorbed through the skin.

2-naphthylamine (CAS # 91-59-8, DOT # 1650, RTECS # QM2100000, RCRA Waste # U168, Reportable Quantities: 10 lbs [1 lb]) has similar chemical, physical and toxicological properties.

1-NAPHTHYLTHIOUREA

Synonyms: ANTU; alpha-naphthylthiourea; alrato; dirax; naphtox

CAS Number: 86-88-4   **Description:** Grey Powder
DOT Number: 1651   **DOT Classification:** Poison B

Molecular Weight: 202.3
Melting Point: 198°C (388°F)   **Vapor Density:** N.D.   **Vapor Pressure:**   1 (<1 mm Hg)
Boiling Point: °C   (°F)   **Specific Gravity:** N.D.   **Water Solubility:** Insoluble

Chemical Incompatibilities or Instabilities: Oxidizers.

FLAMMABILITY

NFPA Hazard Code: **NFPA Classification:**
Flash Point: N.A.   **LEL:** N.A.
Autoignition Temp: N.A.   **UEL:** N.A.

Fire Extinguishing Methods: Carbon Dioxide, Dry Chemical, Foam, Water Spray or Fog.

Special Fire Fighting Considerations: Structural fire fighter protective clothing will not provide adequate protection. Fight fire from a distance or protected location, if possible. Treat all materials used or generated and equipment involved as contaminated by hazardous waste.

TOXICOLOGY
QUESTIONABLE CARCINOGEN

Odor: **Odor Threshold:** N.D.
Physical Contact: Irritant
TLV = 0.3 mg/m^3 **STEL** = N.D.   **IDLH** = N.D.

Routes of Entry and Relative LD$_{50}$ (or LC$_{50}$)   **RTECS #** YT9625000
   Inhalation   N.D.
   Ingestion   9   (6 mg/kg)
   Skin Absorption   N.D.

Symptoms of Exposure: Possible irritation to skin, eyes and upper respiratory tract; headache; nausea.

Emergency/First Aid Treatment: Remove to ventilated area; immediately remove any contaminated clothing and wash contaminated areas for 15 minutes using water. Treat supportively and observe for possible shock. If ingested, seek immediate medical aid.

Recommended Clean-Up Procedures

Personal Protection: Level A Ensemble   **Recommended Material**   N.D.

RCRA Waste #   P072   **Reportable Quantities:**   100 lbs (1 lb)

Spills: Take up in non-combustible absorbent and dispose. Decontaminate spill area using soapy water. Treat all materials used or generated and equipment used involved as contaminated by hazardous waste.

Special Emergency Information: May be fatal in inhaled, ingested or absorbed through the skin.

NICKEL CARBONYL - Ni(CO)₄

Synonyms: nickel tetracarbonyl

CAS Number: 13463-39-3
DOT Number: 1259

Description: White powder
DOT Classification: Poison B

Molecular Weight: 170.8
Melting Point: -19°C (-2°F)
Boiling Point: 43°C (110°F)

Vapor Density: 5.9
Specific Gravity: 1.3

Vapor Pressure: 9 (400 mm Hg)
Water Solubility: Insoluble

Chemical Incompatibilities or Instabilities: Halogens, oxidizers, air, hydrocarbons.

FLAMMABILITY

NFPA Hazard Code: 4-3-3
Flash Point: -24°C (-4°F)
Autoignition Temp: N.D.

NFPA Classification: Class IB Flammable
LEL: 2.0%
UEL: N.D.

Fire Extinguishing Methods: Carbon Dioxide, Dry Chemical, Foam, Water Spray or Fog.

Special Fire Fighting Considerations: Structural fire fighter protective clothing will not provide adequate protection. Water spray from an unmanned device should be used to keep closed containers cool. Continue to cool container after fire is extinguished. Fight fire from a distance or protected location, if possible. Treat all materials used or generated and equipment used involved as contaminated by hazardous waste.

DOT Recommended Isolation Zones: Small Spill: 1500 ft   Large Spill: 1500 ft
DOT Recommended Down Wind Take Cover Distance   Small Spill: 5 miles   Large Spill: 5 miles

TOXICOLOGY
CARCINOGEN

Odor: Sooty
Physical Contact: Irritant
TLV = 0.007 mg/m³   **STEL** = N.D.

Odor Threshold: 3.5 mg/m³

IDLH = 49 mg/m³

RTECS # QR6300000

Routes of Entry and Relative LD₅₀ (or LC₅₀)
Inhalation   10   (120 mg/m³/H)
Ingestion   N.D.
Skin Absorption   N.D.

Symptoms of Exposure: Possible irritation to skin, eyes and upper respiratory tract; headache; nausea. Exposure may result in chemical pneumonitis or pulmonary edema. Inhalation may lead to the formation of methemoglobin resulting in cyanosis several hours after exposure.

Emergency/First Aid Treatment: Remove to ventilated area; immediately remove any contaminated clothing and wash contaminated areas for 15 minutes using water. Treat supportively and observe for possible shock. If ingested, seek immediate medical aid.

Recommended Clean-Up Procedures

Personal Protection: Level A Ensemble   **Recommended Material**   N.D.

RCRA Waste #   P073   **Reportable Quantities:**   10 lbs (1 lb)

Spills: Remove all potential ignition sources. Absorb with non-combustible absorbent and take up using non-sparking tools. Decontaminate spill area using soapy water. Treat all materials used or generated and equipment used involved as contaminated by hazardous waste.

Special Emergency Information: May be fatal in inhaled, ingested or absorbed through the skin. Vapors are heavier than air and may travel some distance to an ignition source. Vapors are more dense than air and may settle in low lying areas.

NICKEL NITRATE - Ni(NO₃)₂

Synonyms:

CAS Number: 13138-45-9   **Description:** Green powder
DOT Number: 2725   **DOT Classification:** Oxidizer

Molecular Weight: 182.7
Melting Point: 57°C (135°F)   **Vapor Density:**   N.D.   **Vapor Pressure:**   N.D.
Boiling Point: 137°C (279°F)   **Specific Gravity:**   2.1   **Water Solubility:**   Soluble

Chemical Incompatibilities or Instabilities: Reducing agents, cyanides, thiocyanates, isothiocyanates, powdered metals.

FLAMMABILITY

NFPA Hazard Code: **NFPA Classification:**
Flash Point: N.A.   **LEL:** N.A.
Autoignition Temp: N.A.   **UEL:** N.A.

Fire Extinguishing Methods: WATER ONLY

Special Fire Fighting Considerations: Water spray should be used to keep closed containers cool. Continue to cool container after fire is extinguished. For large fires, if possible, withdraw and allow to burn. For large fires, use flooding quantities of water. Treat all materials used or generated and equipment involved as contaminated by hazardous waste.

TOXICOLOGY
SUSPECTED CARCINOGEN

Odor: **Odor Threshold:** N.D.
Physical Contact: Irritant
TLV = 0.1 mg (Ni)/m³   **STEL** = N.D.   **IDLH** = N.D.

Routes of Entry and Relative LD₅₀ (or LC₅₀)   **RTECS #** QR7200000
    Inhalation   N.D.
    Ingestion   N.D.
    Skin Absorption   N.D.

Symptoms of Exposure: Possible irritation to skin, eyes and upper respiratory tract; headache; nausea.

Emergency/First Aid Treatment: Remove to ventilated area; immediately remove any contaminated clothing and wash contaminated areas for 15 minutes using water. Treat supportively and observe for possible shock. If ingested, seek immediate medical aid.

Recommended Clean-Up Procedures

Personal Protection: Level B Ensemble   **Recommended Material**   N.D.

RCRA Waste #   None   **Reportable Quantities:**   None

Spills: Take up in non-combustible absorbent and dispose. Decontaminate spill area using water. Treat all materials used or generated and equipment involved as contaminated by hazardous waste.

Special Emergency Information: May be harmful if inhaled, swallowed or absorbed through the skin.

NICOTINE

Synonyms:

CAS Number: 54-11-5   **Description:** Colorless Liquid
DOT Number: 1654   **DOT Classification:** Poison B

Molecular Weight: 162.3
Melting Point: -80°C (-112°F)   **Vapor Density:** 5.6   **Vapor Pressure:** 1 (< 1%)
Boiling Point: 247°C (477°F)   **Specific Gravity:** 1.0   **Water Solubility:** Soluble

Chemical Incompatibilities or Instabilities: Oxidizers.

FLAMMABILITY

NFPA Hazard Code: 4-1-0   **NFPA Classification:** Class IIIB Combustible
Flash Point: 101°C (215°F)   **LEL:**   0.7%
Autoignition Temp: 244°C (471°F)   **UEL:**   4.0%

Fire Extinguishing Methods: Carbon Dioxide, Dry Chemical, Foam, Water Spray or Fog.

Special Fire Fighting Considerations: Structural fire fighter protective clothing will not provide adequate protection. Fight fire from a distance or protected location, if possible. Treat all materials used or generated and equipment used involved as contaminated by hazardous waste.

TOXICOLOGY

Odor: Fishy or insecticide-like   **Odor Threshold:** N.D.
Physical Contact: Irritant
TLV = 0.5 mg/m^3 (skin)   **STEL** = N.D.   **IDLH** = 35 mg/m^3

Routes of Entry and Relative LD$_{50}$ (or LC$_{50}$)   **RTECS #** QS5250000
   Inhalation   N.D.
   Ingestion   9   (50 mg/kg)
   Skin Absorption   9   (50 mg/kg)

Symptoms of Exposure: Possible irritation to skin, eyes and upper respiratory tract; headache; nausea; vertigo; convulsions. Absorption may lead to the formation of methemoglobin resulting in cyanosis several hours after exposure.

Emergency/First Aid Treatment: Remove to ventilated area; immediately remove any contaminated clothing and wash contaminated areas for 15 minutes using water. Treat supportively and observe for possible shock. If ingested, seek immediate medical aid.

Recommended Clean-Up Procedures

Personal Protection: Level A Ensemble   **Recommended Material**   N.D.

RCRA Waste #   P075   **Reportable Quantities:**   100 lbs (1 lb)

Spills: Take up in non-combustible absorbent and dispose. Decontaminate spill area using water. Treat all materials used or generated and equipment used involved as contaminated by hazardous waste.

Special Emergency Information: May be fatal in inhaled, ingested or absorbed through the skin.

Nicotine hydrochloride (CAS # 2820-51-1, DOT # 1656, RTECS # QS8575000),
nicotine monosalicylate (CAS # 29790-52-1, DOT # 1657, RTECS # QS9600000),
nicotine sulfate (CAS # 65-30-5, DOT # 1658, RTECS # QS9625000), and
nicotine tatrate (CAS # 65-31-6, DOT # 1659, RTECS # QT0350000) have similar chemical, physical and toxicological properties.

NITRIC ACID - HNO₃

Synonyms:

CAS Number: 7697-37-2   **Description:** Colorless Liquid
DOT Number: 1760, 2031   **DOT Classification:** Oxidizer

Molecular Weight: 63.0
Melting Point: -42°C (-44°F)   **Vapor Density:** 1.0   **Vapor Pressure:** 4 (11 mm Hg)
Boiling Point: 83°C (181°F)   **Specific Gravity:** 1.5   **Water Solubility:** Soluble

Chemical Incompatibilities or Instabilities: Oxidizable materials, finely divided metals, alkalis, sulfides, carbides, ordinary combustibles.

FLAMMABILITY

NFPA Hazard Code: 3-0-0-OX   **NFPA Classification:**
Flash Point: N.A.   **LEL:** N.A.
Autoignition Temp: N.A.   **UEL:** N.A.

Fire Extinguishing Methods: Carbon Dioxide, Dry Chemical, Flooding Quantities of Water or Water Fog.

Special Fire Fighting Considerations: Structural fire fighter protective clothing will not provide adequate protection. Water spray should be used to keep closed containers cool. Continue to cool container after fire is extinguished. For large fires, if possible, withdraw and allow to burn.

TOXICOLOGY

Odor: Sharp, penetrating   **Odor Threshold:** 0.8 mg/m³
Physical Contact: Material is extremely destructive to human tissues.
TLV = 5 mg/m³   **STEL** = 10 mg/m³   **IDLH** = 260 mg/m³

Routes of Entry and Relative LD₅₀ (or LC₅₀)   **RTECS #** QU5775000
   Inhalation   N.D.
   Ingestion   N.D.
   Skin Absorption   N.D.

Symptoms of Exposure: Possible burns or irritation to skin, eyes and upper respiratory tract; headache; nausea; laryngitis. Inhalation of vapors may be fatal by causing glottis to spasm and suffocation. Exposure may result in chemical pneumonitis or pulmonary edema.

Emergency/First Aid Treatment: Remove to ventilated area; immediately remove any contaminated clothing and wash contaminated areas for 15 minutes using water. Treat supportively and observe for possible shock. If ingested, seek immediate medical aid.

Recommended Clean-Up Procedures

Personal Protection: Level B Ensemble   **Recommended Material**   Butyl Rubber (+), Saranex (+)

RCRA Waste #   None   **Reportable Quantities:**   1000 lb (1000 lb)

Spills: Dike to contain spill and collect liquid for later disposal. Decontaminate spill area using dilute alkaline solution.

Special Emergency Information: May be fatal in inhaled, ingested or absorbed through the skin.

NITRIC ACID, FUMING - HNO3

Synonyms: red nitric acid

CAS Number: 7697-37-2   **Description:** Colorless to red fuming liquid
DOT Number: 2032   **DOT Classification:** Oxidizer

Molecular Weight: 63.0
Melting Point: -42°C (-44°F)   **Vapor Density:** 1.1   **Vapor Pressure:** 6 (51 mm Hg)
Boiling Point: 83°C (181°F)   **Specific Gravity:** 1.5 **Water Solubility:** Soluble

Chemical Incompatibilities or Instabilities: Finely divided metals, carbides, sulfides, solvents, cyanides, ordinary combustibles, oxidizable materials, alkalis.

FLAMMABILITY

NFPA Hazard Code: 3-0-1-OX   **NFPA Classification:**
Flash Point: N.A.   **LEL:** N.A.
Autoignition Temp: N.A.   **UEL:** N.A.

Fire Extinguishing Methods: Dry chemical, Water

Special Fire Fighting Considerations: Structural fire fighter protective clothing will not provide adequate protection. Water spray should be used to keep closed containers cool. Continue to cool container after fire is extinguished. For large fires, use flooding quantities of water. For large fires, if possible, withdraw and allow to burn.

DOT Recommended Isolation Zones: Small Spill: 150 ft   Large Spill: 150 ft
DOT Recommended Down Wind Take Cover Distance   Small Spill: 0.8 miles   Large Spill: 0.8 miles

TOXICOLOGY

Odor: Sharp, suffocating   **Odor Threshold:** 0.8 mg/m^3
Physical Contact: Material is extremely destructive to human tissues.
TLV = 5 mg/m^3 **STEL** = 10 mg/m^3 **IDLH** = 260 mg/m^3

Routes of Entry and Relative LD$_{50}$ (or LC$_{50}$)   **RTECS #** QU6000000
    Inhalation   8   (700 mg/m^3/H)
    Ingestion   N.D.
    Skin Absorption   N.D.

Symptoms of Exposure: Possible burns or irritation to skin, eyes and upper respiratory tract; headache; nausea. Inhalation of vapors may be fatal by causing glottis to spasm and suffocation. Exposure may result in chemical pneumonitis or pulmonary edema.

Emergency/First Aid Treatment: Remove to ventilated area; immediately remove any contaminated clothing and wash contaminated areas for 15 minutes using water. Treat supportively and observe for possible shock. If ingested, seek immediate medical aid.

Recommended Clean-Up Procedures

Personal Protection: Level B Ensemble   **Recommended Material** Butyl Rubber (+), Saranex (+)

RCRA Waste #   None   **Reportable Quantities:** 1000 lb (1000 lb)

Spills: Dike to contain spill and collect liquid for later disposal. Decontaminate spill area using dilute alkaline solution. Keep area isolated until vapors have dissipated.

Special Emergency Information: May be fatal in inhaled, ingested or absorbed through the skin. Vapors are more dense than air and may settle in low lying areas.

NITRIC OXIDE - NO

Synonyms: nitrogen monoxide

CAS Number: 10102-43-9	**Description:** Colorless Gas		
DOT Number: 1660	**DOT Classification:** Poison A		

Molecular Weight: 30.0

Melting Point:	-161°C (-258°F)	**Vapor Density:** 1.1	**Vapor Pressure:**	Gas
Boiling Point:	-151°C (-240°F)	**Specific Gravity:** N.A.	**Water Solubility:**	5% (Decomposes)

Chemical Incompatibilities or Instabilities: Combustibles, ozone, ammonia, halogens, metals, chlorinated hydrocarbons.

FLAMMABILITY

NFPA Hazard Code:	**NFPA Classification:**	
Flash Point: N.A.	**LEL:** N.A.	
Autoignition Temp: N.A.	**UEL:** N.A.	

Fire Extinguishing Methods: WATER ONLY

Special Fire Fighting Considerations: Structural fire fighter protective clothing will not provide adequate protection. Water spray from unmanned devices should be used to keep closed containers cool. Continue to cool container after fire is extinguished. For large fires, if possible, withdraw and allow to burn.

DOT Recommended Isolation Zones:	Small Spill: 600 ft	Large Spill: 900 ft
DOT Recommended Down Wind Take Cover Distance	Small Spill: 2 miles	Large Spill: 3 miles

TOXICOLOGY

Odor:	**Odor Threshold:** N.D.

Physical Contact: Material is extremely destructive to human tissues.

TLV = 30 mg/m^3	**STEL** = N.D.	**IDLH** = 125 mg/m^3

Routes of Entry and Relative LD$_{50}$ (or LC$_{50}$)   **RTECS # QX0525000**

Inhalation	**8**	(1068 mg/m^3/H)
Ingestion	N.D.	
Skin Absorption	N.D.	

Symptoms of Exposure: Possible burns or irritation to skin, eyes and upper respiratory tract; headache; nausea. Inhalation of vapors may be fatal by causing glottis to spasm and suffocation. Exposure may result in chemical pneumonitis or pulmonary edema.

Emergency/First Aid Treatment: Remove to ventilated area; immediately remove any contaminated clothing and wash contaminated areas for 15 minutes using water. Treat supportively and observe for possible shock. If ingested, seek immediate medical aid.

Recommended Clean-Up Procedures

Personal Protection: Level B Ensemble	**Recommended Material**	N.D.
RCRA Waste # P076	**Reportable Quantities:**	10 lbs (1lb)

Spills: Stop leak if it can safely be done. Keep area isolated until vapors dissipate.

Special Emergency Information: May be fatal in inhaled, ingested or absorbed through the skin. Vapors are more dense than air and may settle in low lying areas.

p-NITROANILINE

Synonyms: 1-amino-4-nitrobenzene; 4-nitrobenzamine; 4-nitroaniline

CAS Number: 100-01-6   **Description:** Yellow powder
DOT Number: 1661   **DOT Classification:** Poison B

Molecular Weight: 138.1
Melting Point: 148°C (298°F)   **Vapor Density:** 4.8   **Vapor Pressure:** 1 (<1 mm Hg)
Boiling Point: 332°C (637°F)   **Specific Gravity:** 1.42   **Water Solubility:** Insoluble

Chemical Incompatibilities or Instabilities: Oxidizers, strong reducing agents, strong acids.

FLAMMABILITY

NFPA Hazard Code: 3-1-2   **NFPA Classification:**
Flash Point: 154°C (309°F)   **LEL:** N.D.
Autoignition Temp: N.D.   **UEL:** N.D.

Fire Extinguishing Methods: Carbon Dioxide, Dry Chemical, Foam, Water Spray or Fog.

Special Fire Fighting Considerations: Structural fire fighter protective clothing will not provide adequate protection. Fight fire from a distance or protected location, if possible. Treat all materials used or generated and equipment involved as contaminated by hazardous waste.

TOXICOLOGY

Odor: Ammonia-like   **Odor Threshold:** N.D.
Physical Contact: Irritant
TLV = 3 mg/m^3 (skin)   **STEL** = N.D.   **IDLH** = 300 mg/m^3

Routes of Entry and Relative LD$_{50}$ (or LC$_{50}$)   **RTECS # BY7000000**
    Inhalation   N.D.
    Ingestion   4   (750 mg/kg)
    Skin Absorption   N.D.

Symptoms of Exposure: Possible irritation to skin, eyes and upper respiratory tract; headache; nausea. Absorption may lead to the formation of methemoglobin resulting in cyanosis several hours after exposure.

Emergency/First Aid Treatment: Remove to ventilated area; immediately remove any contaminated clothing and wash contaminated areas for 15 minutes using water. Treat supportively and observe for possible shock. If ingested, seek immediate medical aid.

Recommended Clean-Up Procedures

Personal Protection: Level B Ensemble   **Recommended Material** N.D.

RCRA Waste # P077   **Reportable Quantities:** 5000 lbs (1 lb)

Spills: Take up in non-combustible absorbent and dispose. Decontaminate spill area using soapy water. Treat all materials used or generated and equipment involved as contaminated by hazardous waste.

Special Emergency Information: May be fatal in inhaled, ingested or absorbed through the skin.

3-nitroaniline (CAS # 99-09-2, RTECS # BY6825000) and
2-nitroaniline (CAS # 88-74-4, RTECS # BY6650000) have similar chemical, physical and toxicological properties.

NITROBENZENE

Synonyms: oil of mirbane

CAS Number: 98-95-3

DOT Number: 1662

Molecular Weight: 123.1

Melting Point: 6°C (42°F)

Boiling Point: 211°C (411°F)

Description: Yellow liquid

DOT Classification: Poison B

Vapor Density: 4.2

Specific Gravity: 1.2

Vapor Pressure: 1 (<1 mm Hg)

Water Solubility: Insoluble

Chemical Incompatibilities or Instabilities: Aluminum chloride, aniline and glycerine mixtures, strong acids, perchlorates, nitrogen tetroxide, oxidizers.

FLAMMABILITY

NFPA Hazard Code: 3-2-1

Flash Point: 88°C (190°F)

Autoignition Temp: 496°C (954°F)

NFPA Classification: Class IIIA Combustible

LEL: 1.8%

UEL: N.D.

Fire Extinguishing Methods: Carbon Dioxide, Dry Chemical, Foam, Water Spray or Fog.

Special Fire Fighting Considerations: Structural fire fighter protective clothing will not provide adequate protection. Fight fire from a distance or protected location, if possible. Treat all materials used or generated and equipment used involved as contaminated by hazardous waste.

TOXICOLOGY

Odor: Almonds, shoe polish

Physical Contact: Irritant

$TLV = 5$ mg/m^3 (skin)   **STEL** = N.D.

Odor Threshold: 0.002 mg/m^3

$IDLH = 1000$ mg/m^3

Routes of Entry and Relative LD$_{50}$ (or LC$_{50}$)

Inhalation	N.D.	
Ingestion	5	(489 mg/kg)
Skin Absorption	N.D.	

RTECS # DA6475000

Symptoms of Exposure: Possible irritation to skin, eyes and upper respiratory tract; headache; nausea. Absorption may lead to the formation of methemoglobin resulting in cyanosis several hours after exposure.

Emergency/First Aid Treatment: Remove to ventilated area; immediately remove any contaminated clothing and wash contaminated areas for 15 minutes using water. Treat supportively and observe for possible shock. If ingested, seek immediate medical aid.

Recommended Clean-Up Procedures

Personal Protection: Level B Ensemble

RCRA Waste # U169

Recommended Material Butyl Rubber (+), Viton (+), Teflon (0)

Reportable Quantities: 1000 lbs (1000 lbs)

Spills: Dike to contain spill and collect liquid for later disposal. Decontaminate spill area using soapy water. Treat all materials used or generated and equipment used involved as contaminated by hazardous waste.

Special Emergency Information: May be fatal in inhaled, ingested or absorbed through the skin.

Nitrochlorobenzene (CAS # 100-00-5, DOT # 1578, RTECS # CZ1050000) has similar chemical, physical and toxicological properties.

NITROBENZOTRIFLUORIDE

Synonyms: m-nitrobenzotrifluoride; m-nitrotrifluorotoluene; 3-trifluoromethylnitrobenzene

CAS Number: 98-46-4
DOT Number: 2306

Description: Pale yellow powder
DOT Classification: Poison B

Molecular Weight: 191.1
Melting Point: -5°C (23°F)
Boiling Point: 203°C (397°F)

Vapor Density: 6.6
Specific Gravity: 1.44

Vapor Pressure: 1 (<1 mm Hg)
Water Solubility: Insoluble

Chemical Incompatibilities or Instabilities: Strong oxidizers, strong reducing agents, strong bases.

FLAMMABILITY

NFPA Hazard Code: ?-1-?
Flash Point: 103°C (217°F)
Autoignition Temp: °C (°F)

NFPA Classification:
LEL: N.D.
UEL: N.D.

Fire Extinguishing Methods: Carbon Dioxide, Dry Chemical, Foam, Water Spray or Fog.

Special Fire Fighting Considerations: Water spray should be used to keep closed containers cool. Continue to cool container after fire is extinguished. Treat all materials used or generated and equipment involved as contaminated by hazardous waste.

TOXICOLOGY

Odor: Aromatic
Physical Contact: Irritant
TLV = N.D.   **STEL** = N.D.

Odor Threshold: N.D.

IDLH = N.D.

Routes of Entry and Relative LD$_{50}$ (or LC$_{50}$)
  Inhalation   N.D.
  Ingestion   4   610 mg/kg)
  Skin Absorption N.D.

RTECS # XT3500000

Symptoms of Exposure: Possible irritation to skin, eyes and upper respiratory tract; headache; nausea.

Emergency/First Aid Treatment: Remove to ventilated area; immediately remove any contaminated clothing and wash contaminated areas for 15 minutes using water. Treat supportively and observe for possible shock. If ingested, seek immediate medical aid.

Recommended Clean-Up Procedures

Personal Protection: Level B Ensemble

RCRA Waste # None

Recommended Material N.D.

Reportable Quantities: None

Spills: Dike to contain spill and collect liquid for later disposal. Decontaminate spill area using soapy water. Treat all materials used or generated and equipment involved as contaminated by hazardous waste.

Special Emergency Information: May be harmful if inhaled, swallowed or absorbed through the skin.

NITROETHANE - CH3CH2NO2

Synonyms:

CAS Number: 79-24-3   **Description:** Colorless Liquid
DOT Number: 2842   **DOT Classification:** Flammable Liquid

Molecular Weight: 75.1
Melting Point: -90°C (-130°F)   **Vapor Density:** 2.6   **Vapor Pressure:** 4 (16 mm Hg)
Boiling Point: 114°C (237°F)   **Specific Gravity:** 1.1   **Water Solubility:** Insoluble

Chemical Incompatibilities or Instabilities: Heat, oxidizers, amines, metal oxides, strong acids, strong bases.

FLAMMABILITY

NFPA Hazard Code: 1-3-3   **NFPA Classification:** Class IC Flammable
Flash Point: 28°C (82°F)   **LEL:** 3.4%
Autoignition Temp: 414°C (778°F) **UEL:** N.D.

Fire Extinguishing Methods: Alcohol Resistant Foam, Carbon Dioxide, Dry Chemical, Water Spray or Fog.

Special Fire Fighting Considerations: Isolate for 1/2 mile if rail or tank truck is involved in a fire. Water spray from unmanned devices should be used to keep closed containers cool. Continue to cool container after fire is extinguished. For large fires, if possible, withdraw and allow to burn. Immediately withdraw if rising sound from venting device is heard or if fire is causing discoloration to the tank.

TOXICOLOGY

Odor: Pleasant   **Odor Threshold:** N.D.
Physical Contact: Irritant
TLV = 310 mg/m^3 **STEL** = N.D.   **IDLH** = 3,100 mg/m^3

Routes of Entry and Relative LD$_{50}$ (or LC$_{50}$)   **RTECS # KI5600000**
    Inhalation   N.D.
    Ingestion   3   (1100 mg/kg)
    Skin Absorption N.D.

Symptoms of Exposure: Possible irritation to skin, eyes and upper respiratory tract; headache; nausea. Absorption may lead to the formation of methemoglobin resulting in cyanosis several hours after exposure.

Emergency/First Aid Treatment: Remove to ventilated area; immediately remove any contaminated clothing and wash contaminated areas for 15 minutes using water. Treat supportively and observe for possible shock. If ingested, seek immediate medical aid.

Recommended Clean-Up Procedures

Personal Protection: Level B Ensemble   **Recommended Material** Butyl Rubber (+)

RCRA Waste # None   **Reportable Quantities:** None

Spills: Remove all potential ignition sources. Dike to contain spill, absorb with non-combustible absorbent and take up using non-sparking tools. Vent to dissipate vapors. Decontaminate spill area using soapy water.

Special Emergency Information: May be fatal in inhaled, ingested or absorbed through the skin. Use extreme caution; may be a shock or temperature sensitive explosive.

Nitromethane (CAS # 75-52-5, DOT # 1261, RTECS # PA9800000) has similar chemical, physical and toxicological properties.

NITROGEN DIOXIDE - NO$_2$

Synonyms: nitrogen peroxide; nitrogen tetroxide

CAS Number: 10102-44-0

DOT Number: 1067

Description: Red-Brown Gas

DOT Classification: Poison A

Molecular Weight: 46.0

Melting Point: -9°C (15°F)

Boiling Point: 15°C (70°F)

Vapor Density: 1.7

Specific Gravity: 2.6

Vapor Pressure: Gas

Water Solubility:

Chemical Incompatibilities or Instabilities: Unsaturated hydrocarbons, halogens, cyclohexane, alcohols, aldehydes, amines, halocarbons, carbon disulfide, nitroaromatics.

FLAMMABILITY

NFPA Hazard Code: 3-3-0-OX

Flash Point: N.D.

Autoignition Temp: N.D.

NFPA Classification:

LEL: N.D.

UEL: N.D.

Fire Extinguishing Methods: WATER ONLY

Special Fire Fighting Considerations: Structural fire fighter protective clothing will not provide adequate protection. Water spray from unmanned devices should be used to keep closed containers cool. Continue to cool container after fire is extinguished. For large fires, if possible, withdraw and allow to burn.

DOT Recommended Isolation Zones: Small Spill: 150 ft Large Spill: 600 ft

DOT Recommended Down Wind Take Cover Distance Small Spill: 0.8 miles Large Spill: 2 miles

TOXICOLOGY

Odor: Bleach, irritating

Physical Contact: Material is extremely destructive to human tissues.

TLV = 1.8 mg/m^3

STEL = N.D.

Odor Threshold: 0.11 mg/m^3

IDLH = 90 mg/m^3

RTECS # QW9800000

Routes of Entry and Relative LD$_{50}$ (or LC$_{50}$)

Inhalation 9 (670 mg/m^3/H)

Ingestion N.D.

Skin Absorption N.D.

Symptoms of Exposure: Possible burns or irritation to skin, eyes and upper respiratory tract; headache; nausea; cyanosis. Inhalation of vapors may be fatal by causing glottis to spasm and suffocation. Exposure may result in chemical pneumonitis or pulmonary edema.

Emergency/First Aid Treatment: Remove to ventilated area; immediately remove any contaminated clothing and wash contaminated areas for 15 minutes using water. Treat supportively and observe for possible shock. If ingested, seek immediate medical aid.

Recommended Clean-Up Procedures

Personal Protection: Level A Ensemble **Recommended Material** Saranex (+)

RCRA Waste #P078 **Reportable Quantities:** 10 lbs (1000 lbs)

Spills: Remove all potential ignition sources. Stop leak if it can safely be done. Keep area isolated until vapors are dispersed.

Special Emergency Information: May be fatal in inhaled, ingested or absorbed through the skin. Vapors are heavier than air and may travel some distance to an ignition source. Vapors are more dense than air and may settle in low lying areas.

NITROGEN TRIFLUORIDE - NF$_3$

Synonyms: nitrogen fluoride

CAS Number: 7783-54-2   **Description:** Colorless Gas
DOT Number: 2451   **DOT Classification:** Poison A

Molecular Weight: 71.0

Melting Point: -206°C (-340°F)	**Vapor Density:** N.D.	**Vapor Pressure:** GAS	
Boiling Point: -134°C (-209°F)	**Specific Gravity:** N.A.	**Water Solubility:** Decomposes	

Chemical Incompatibilities or Instabilities: Reducing agents, hydrocarbons, water, hydrogen, hydrogen sulfide, ammonia, hydrazines, carbon monoxide, diborane.

FLAMMABILITY

NFPA Hazard Code: **NFPA Classification:**
Flash Point: N.D.   **LEL:** N.D.
Autoignition Temp: N.D.   **UEL:** N.D.

Fire Extinguishing Methods: Carbon Dioxide, Dry Chemical, Foam, Water Spray or Fog.

Special Fire Fighting Considerations: Water spray from unmanned devices should be used to keep closed containers cool. Continue to cool container after fire is extinguished. For large fires, if possible, withdraw and allow to burn.

DOT Recommended Isolation Zones: Small Spill: 150 ft   Large Spill: 150 ft
DOT Recommended Down Wind Take Cover Distance   Small Spill: 0.4 miles   Large Spill: 0.8 miles

TOXICOLOGY

Odor: Moldy   **Odor Threshold:** N.A.
Physical Contact: Material is extremely destructive to human tissues.
TLV = 29 mg/m^3 **STEL** = N.D.   **IDLH** = 6,000 mg/m^3

Routes of Entry and Relative LD$_{50}$ (or LC$_{50}$)   **RTECS # QX1925000**
     Inhalation   **5**   (20,000 mg/m^3/H)
     Ingestion   N.D.
     Skin Absorption   N.D.

Symptoms of Exposure: Possible burns or irritation to skin, eyes and upper respiratory tract; headache; nausea. Inhalation of vapors may cause cyanosis and the formation of methemoglobin.

Emergency/First Aid Treatment: Remove to ventilated area; immediately remove any contaminated clothing and wash contaminated areas for 15 minutes using water. Treat supportively and observe for possible shock. If ingested, seek immediate medical aid.

Recommended Clean-Up Procedures

Personal Protection: Level A Ensemble   **Recommended Material** N.D.

RCRA Waste # None   **Reportable Quantities:** None

Spills: Remove all potential ignition sources. Stop leak if it can safely be done. Keep area isolated until vapors dissipate.

Special Emergency Information: May be fatal in inhaled, ingested or absorbed through the skin. Vapors are heavier than air and may travel some distance to an ignition source. Vapors are more dense than air and may settle in low lying areas.

NITROGUANIDINE - NH$_2$C(NH)NHNO$_2$

Synonyms: picrite

CAS Number: 556-88-7

DOT Number: 0282, 1336

Description: White to yellow powder

DOT Classification: Class A Explosive

Molecular Weight: 104.1

Melting Point: 246°C (D)(475°F)

Boiling Point: °C (°F)

Vapor Density: N.A.

Specific Gravity: >1.1

Vapor Pressure: 1 (<1 mm Hg)

Water Solubility: 4.4%

Chemical Incompatibilities or Instabilities: Oxidizers, heat, shock.

FLAMMABILITY

NFPA Hazard Code:

Flash Point: N. A

Autoignition Temp: N.A.

NFPA Classification:

LEL: N.A.

UEL: N.A.

Fire Extinguishing Methods: Flooding quantities of water.

Special Fire Fighting Considerations: Water spray from an unmanned device should be used to keep closed containers cool. Continue to cool container after fire is extinguished. For large fires, if possible, withdraw and allow to burn.

TOXICOLOGY

Odor:

Physical Contact: Irritant

TLV = N.D.

STEL = N.D.

Odor Threshold: N.D.

IDLH = N.D.

Routes of Entry and Relative LD$_{50}$ (or LC$_{50}$)

    Inhalation   N.D.

    Ingestion   N.D.

    Skin Absorption   N.D.

RTECS # MF4600000

Symptoms of Exposure: Possible irritation to skin, eyes and upper respiratory tract; headache; nausea.

Emergency/First Aid Treatment: Remove to ventilated area; immediately remove any contaminated clothing and wash contaminated areas for 15 minutes using water. Treat supportively and observe for possible shock. If ingested, seek immediate medical aid.

Recommended Clean-Up Procedures

Personal Protection: Level B Ensemble

RCRA Waste # None

Recommended Material N.D.

Reportable Quantities: None

Spills: Remove all potential ignition sources. Dike to contain spill, absorb with non-combustible absorbent and take up using non-sparking tools. Decontaminate spill area using soapy water.

Special Emergency Information: May be harmful if inhaled, swallowed or absorbed through the skin. May explode upon exposure to shock or heat.

p-NITROPHENOL

Synonyms: 4-nitrophenol

CAS Number: 100-02-7
DOT Number: 1663

Description: Yellow powder
DOT Classification: ORM-E

Molecular Weight: 139.1
Melting Point: 115°C (239°F)
Boiling Point: 279°C (534°F)

Vapor Density: >2.0
Specific Gravity: 1.5

Vapor Pressure: 1 (< 1 mm Hg)
Water Solubility: Insoluble

Chemical Incompatibilities or Instabilities: Diethylphosphite, strong oxidizers.

FLAMMABILITY

NFPA Hazard Code:
Flash Point: N.A.
Autoignition Temp: 283°C (541°F)

NFPA Classification:
LEL: N.A.
UEL: N.A.

Fire Extinguishing Methods: .Carbon Dioxide, Dry Chemical, Foam, Water Spray or Fog.

Special Fire Fighting Considerations: Structural fire fighter protective clothing will not provide adequate protection. distance. Treat all materials used or generated and equipment used involved as contaminated by hazardous waste.

TOXICOLOGY

Odor: None
Physical Contact: Irritant
TLV = N.D.   **STEL** = N.D.

Odor Threshold: N.A.

IDLH = N.D.

Routes of Entry and Relative LD$_{50}$ (or LC$_{50}$)
    Inhalation   N.D.
    Ingestion   6   (250 mg/kg)
    Skin Absorption   N.D.

RTECS # SM2275000

Symptoms of Exposure: Possible irritation to skin, eyes and upper respiratory tract; headache; nausea. Continued contact may cause chemical burns to the eyes and mucous membranes. Absorption may lead to the formation of methemoglobin resulting in cyanosis several hours after exposure.

Emergency/First Aid Treatment: Remove to ventilated area; immediately remove any contaminated clothing and wash contaminated areas for 15 minutes using water. Treat supportively and observe for possible shock. If ingested, seek immediate medical aid.

Recommended Clean-Up Procedures

Personal Protection: Level B Ensemble   **Recommended Material**   N.D.

RCRA Waste #U170   **Reportable Quantities:**   100 lbs (1000 lbs)

Spills: Take up in non-combustible absorbent and dispose. Decontaminate spill area using soapy water. Treat all materials used or generated and equipment used involved as contaminated by hazardous waste.

Special Emergency Information: May be harmful if inhaled, swallowed or absorbed through the skin.

2-Nitrophenol (CAS # 88-75-5, DOT # 1663, RTECS # SM2100000, no RCRA Waste #) and
3-nitrophenol (CAS # 554-84-7, DOT # 1663, RTECS # SM1925000, no RCRA Waste #) have similar chemical, physical and toxicological properties. They also have similar reportable quantities.

NITROPROPANE - $CH_3CH_2CH_2NO_2$

Synonyms: 1-nitropropane

CAS Number: 108-03-2
DOT Number: 2608

Description: Colorless Liquid
DOT Classification: Flammable Liquid

Molecular Weight: 89.1
Melting Point: -93°C (-135°F)
Boiling Point: 132°C (269°F)

Vapor Density: 3.1
Specific Gravity: 1.0

Vapor Pressure: 4 (13 mm Hg)
Water Solubility: Insoluble

Chemical Incompatibilities or Instabilities: Strong oxidizers, metal oxides, hydrocarbons, bases.

FLAMMABILITY

NFPA Hazard Code: 1-3-2
Flash Point: 34°C (93°F)
Autoignition Temp: 802°C (428°F)

NFPA Classification: Class IC Flammable
LEL: 2.2%
UEL: N.D.

Fire Extinguishing Methods: Alcohol Resistant Foam, Carbon Dioxide, Dry Chemical, Water Spray or Fog.

Special Fire Fighting Considerations: Isolate for 1/2 mile if rail or tank truck is involved in a fire. Water spray from an unmanned device should be used to keep closed containers cool. Continue to cool container after fire is extinguished. For large fires, if possible, withdraw and allow to burn. Immediately withdraw if rising sound from venting device is heard or if fire is causing discoloration to the tank.

TOXICOLOGY

Odor: Disagreeable
Physical Contact: Irritant
TLV = 90 mg/m^3

STEL = N.D.

Odor Threshold: 30 mg/m^3

IDLH = 8,500 mg/m^3

RTECS # TZ5075000

Routes of Entry and Relative LD$_{50}$ (or LC$_{50}$)
Inhalation	N.D.	
Ingestion	5	(455 mg/kg)
Skin Absorption	N.D.	

Symptoms of Exposure: Possible irritation to skin, eyes and upper respiratory tract; headache; nausea.

Emergency/First Aid Treatment: Remove to ventilated area; immediately remove any contaminated clothing and wash contaminated areas for 15 minutes using water. Treat supportively and observe for possible shock. If ingested, seek immediate medical aid.

Recommended Clean-Up Procedures

Personal Protection: Level B Ensemble

Recommended Material Butyl Rubber (+), Polyvinyl Alcohol (+), Teflon (0)

RCRA Waste # U171

Reportable Quantities: 10 lbs (1 lbs)

Spills: Remove all potential ignition sources. Dike to contain spill, absorb with non-combustible absorbent and take up using non-sparking tools. Treat all materials used or generated and equipment used involved as contaminated by hazardous waste.

Special Emergency Information: May be harmful if inhaled, swallowed or absorbed through the skin. Vapors are heavier than air and may travel some distance to an ignition source. Vapors are more dense than air and may settle in low lying areas. May explode upon heatng.

2-Nitropropane (CAS # 79-46-9, RTECS # TZ5250000), a suspected carcinogen, has similar chemical, physical and toxicological properties.

NITROSYL CHLORIDE - NOCl

Synonyms: nitrogen oxychloride

CAS Number: 2696-92-6   **Description:** Yellow Gas
DOT Number: 1069   **DOT Classification:** Nonflammable gas

Molecular Weight: 65.5
Melting Point: -65°C (-85°F)   **Vapor Density:** 2.3   **Vapor Pressure:** Gas
Boiling Point: -6°C (21°F)   **Specific Gravity:** 1.3   **Water Solubility:** Decomposes

Chemical Incompatibilities or Instabilities: Acetone and platinum, hydrogen and oxygen mixtures.

FLAMMABILITY

NFPA Hazard Code: **NFPA Classification:**
Flash Point: N.A.   **LEL:** N.A.
Autoignition Temp: N.A.   **UEL:** N.A.

Fire Extinguishing Methods: Carbon Dioxide, Dry Chemical, Foam, Water Spray or Fog.

Special Fire Fighting Considerations: Water spray should be used to keep closed containers cool. Continue to cool container after fire is extinguished. Do not put water directly on spilled material.

DOT Recommended Isolation Zones:	Small Spill: 150 ft	Large Spill: 600
DOT Recommended Down Wind Take Cover Distance	Small Spill: 0.4 miles	Large Spill: 2 miles

TOXICOLOGY

Odor: Irritating   **Odor Threshold:** N.D.
Physical Contact: Material is extremely destructive to human tissues.
TLV = N.D.   **STEL** = N.D.   **IDLH** = N.D.

Routes of Entry and Relative LD$_{50}$ (or LC$_{50}$)   **RTECS #** QZ7883000
    Inhalation   N.D.
    Ingestion   N.D.
    Skin Absorption   N.D.

Symptoms of Exposure: Possible burns or irritation to skin, eyes and upper respiratory tract; headache; nausea. Inhalation of vapors may be fatal by causing glottis to spasm and suffocation. Exposure may result in chemical pneumonitis or pulmonary edema.

Emergency/First Aid Treatment: Remove to ventilated area; immediately remove any contaminated clothing and wash contaminated areas for 15 minutes using water. Treat supportively and observe for possible shock. If ingested, seek immediate medical aid.

Recommended Clean-Up Procedures

Personal Protection: Level B Ensemble   **Recommended Material** N.D.

RCRA Waste # None   **Reportable Quantities:** None

Spills: Stop leak if it can safely be done. Keep area isolated until vapors dissipate.

Special Emergency Information: May be fatal in inhaled, ingested or absorbed through the skin.

o-NITROTOLUENE

Synonyms: 2-methylnitrobenzene; 2-nitrotoluene

CAS Number: 88-72-2   **Description:** Yellow liquid

DOT Number: 1664   **DOT Classification:** Poison B

Molecular Weight: 137.2

Melting Point:	-9°C (15°F)	**Vapor Density:** 4.7		**Vapor Pressure:**	1 (<1 mm Hg)
Boiling Point:	222°C (432°F)	**Specific Gravity:** 1.2		**Water Solubility:**	Insoluble

Chemical Incompatibilities or Instabilities: Alkali, strong oxidizers.

FLAMMABILITY

NFPA Hazard Code: 3-1-1   **NFPA Classification:** Class IIIB Combustible

Flash Point: 106°C (223°F)   **LEL:** 2.2%

Autoignition Temp: 305°C (581°F)   **UEL:** N.D.

Fire Extinguishing Methods: Carbon Dioxide, Dry Chemical, Foam, Water Spray or Fog.

Special Fire Fighting Considerations: Structural fire fighter protective clothing will not provide adequate protection. Fight fire from a distance or protected location, if possible. Treat all materials used or generated and equipment used involved as contaminated by hazardous waste.

TOXICOLOGY

Odor: **Odor Threshold:** N.D.

Physical Contact: Irritant

TLV = 11 mg/m^3 (skin)   **STEL** = N.D.   **IDLH** = 1,100 mg/m^3

Routes of Entry and Relative LD$_{50}$ (or LC$_{50}$)   **RTECS #** XT3150000

Inhalation	N.D.	
Ingestion	4	(891 mg/kg)
Skin Absorption	N.D.	

Symptoms of Exposure: Possible irritation to skin, eyes and upper respiratory tract; headache; nausea. Absorption may lead to the formation of methemoglobin resulting in cyanosis several hours after exposure.

Emergency/First Aid Treatment: Remove to ventilated area; immediately remove any contaminated clothing and wash contaminated areas for 15 minutes using water. Treat supportively and observe for possible shock. If ingested, seek immediate medical aid.

Recommended Clean-Up Procedures

Personal Protection: Level B Ensemble   **Recommended Material** N.D.

RCRA Waste # None   **Reportable Quantities:** None

Spills: Take up in non-combustible absorbent and dispose. Decontaminate spill area using soapy water. Treat all materials used or generated and equipment used involved as contaminated by hazardous waste.

Special Emergency Information: May be fatal in inhaled, ingested or absorbed through the skin. Use care. May be shock or heat sensitive.

3-Nitrotoluene (CAS # 99-08-1, DOT # 1664, RTECS # XT2975000) and
4-nitrotoluene (CAS # 99-99-0, DOT # 1664, RTECS # XT3325000) have similar chemical, physical and toxicological properties.

NITROUS OXIDE - N$_2$O

Synonyms: nitrogen oxide; laughing gas

CAS Number: 10024-97-2	**Description:** Colorless Gas
DOT Number: 1070, 2201	**DOT Classification:** Nonflammable gas

Molecular Weight: 44.0

Melting Point: -98°C (-144°F)	**Vapor Density:** 1.5	**Vapor Pressure:** Gas
Boiling Point: -88°C (-126°F)	**Specific Gravity:** N.A.	**Water Solubility:** 3%

Chemical Incompatibilities or Instabilities: Aluminum, boron, hydrazine, phosphine, ammonia, metal carbides, carbon monoxide, hydrogen sulfide, hydrogen, acetylene, lithium hydride, heat.

FLAMMABILITY

NFPA Hazard Code: **NFPA Classification:**

Flash Point:	N.A.	**LEL:**	N.A.
Autoignition Temp:	N.A.	**UEL:**	N.A.

Fire Extinguishing Methods: Carbon Dioxide, Dry Chemical, Foam, Water Spray or Fog.

Special Fire Fighting Considerations: Isolate for 1/2 mile if rail or tank truck is involved in a fire. Water spray from unmanned devices should be used to keep closed containers cool. Continue to cool container after fire is extinguished. large.

TOXICOLOGY

Odor: **Odor Threshold:** N.A.

Physical Contact: A simple asphyxiant.

TLV = N.D.   **STEL** = N.D.   **IDLH** = N.D.

Routes of Entry and Relative LD$_{50}$ (or LC$_{50}$)   **RTECS # QX1350000**

Inhalation	6	(4275 mg/m^3/H)
Ingestion	N.D.	
Skin Absorption	N.D.	

Symptoms of Exposure: Intoxication.

Emergency/First Aid Treatment: Remove to ventilated area; immediately remove any contaminated clothing and wash contaminated areas for 15 minutes using water. Treat supportively and observe for possible shock. If ingested, seek immediate medical aid.

Recommended Clean-Up Procedures

Personal Protection: Level B Ensemble	**Recommended Material** N.D.
RCRA Waste # None	**Reportable Quantities:** None

Spills: Stop leak if it can safely be done. Keep area isolated until vapors dissipate.

Special Emergency Information: May be harmful if inhaled, swallowed or absorbed through the skin.

OCTANE - CH₃(CH₂)₆CH₃

Synonyms:

CAS Number: 111-65-9   **Description:** Colorless Liquid
DOT Number: 1262   **DOT Classification:** Flammable Liquid

Molecular Weight: 114.3
Melting Point: -56°C (-70°F)   **Vapor Density:** 3.9   **Vapor Pressure:** 4 (11 mm Hg)
Boiling Point: 126°C (258°F)   **Specific Gravity:** 0.7   **Water Solubility:** Insoluble

Chemical Incompatibilities or Instabilities: Strong oxidizers.

FLAMMABILITY

NFPA Hazard Code: 0-3-0   **NFPA Classification:** Class IB Flammable
Flash Point: 13°C   (56°F)   **LEL:**   1.0%
Autoignition Temp: 206°C (403°F)   **UEL:**   6.5%

Fire Extinguishing Methods: Carbon Dioxide, Dry Chemical, Foam, Water Spray or Fog.

Special Fire Fighting Considerations: Isolate for 1/2 mile if rail or tank truck is involved in a fire. Water spray from unmanned devices should be used to keep closed containers cool. Continue to cool container after fire is extinguished. For large fires, if possible, withdraw and allow to burn. Immediately withdraw if rising sound from venting device is heard or if fire is causing discoloration to the tank.

TOXICOLOGY

Odor: Gasoline   **Odor Threshold:** 75 mg/m³
Physical Contact: Irritant
TLV = 1450 mg/m³   **STEL** = 1800 mg/m³   **IDLH** = 23,750 mg/m³

Routes of Entry and Relative LD₅₀ (or LC₅₀)   **RTECS # RG8400000**
  Inhalation   N.D.
  Ingestion   N.D.
  Skin Absorption   N.D.

Symptoms of Exposure: A simple asphyxiant. May act as a narcotic in high concentrations.

Emergency/First Aid Treatment: Remove to ventilated area; immediately remove any contaminated clothing and wash contaminated areas for 15 minutes using water. Treat supportively and observe for possible shock. If ingested, seek immediate medical aid.

Recommended Clean-Up Procedures

Personal Protection: Level B Ensemble   **Recommended Material** Nitrile Rubber (0), Viton (0)

RCRA Waste #   None   **Reportable Quantities:** None

Spills: Remove all potential ignition sources. Dike to contain spill, absorb with noncombustible absorbent and take up using nonsparking tools. Decontaminate spill area using soapy water. Keep area isolated until vapors dissipate.

Special Emergency Information: May be harmful if inhaled, swallowed or absorbed through the skin. Vapors are heavier than air and may travel some distance to an ignition source. Vapors are more dense than air and may settle in low lying areas.

Nonane (CAS # 111-84-2, DOT # 1920, RTECS # RA61115000) and
decane (CAS # 124-18-5, DOT # 2247, RTECS # HD6550000) have similar chemical, physical and toxicological properties.

tert-OCTYL MERCAPTAN

Synonyms: methyl heptanethiol

CAS Number: 63834-87-7   **Description:** Colorless Liquid
DOT Number: 3023   **DOT Classification:**

Molecular Weight: 146.3
Melting Point: °C   (°F)   **Vapor Density:** 5.0   **Vapor Pressure:** N.D.
Boiling Point: 165°C (329°F)   **Specific Gravity:** 0.8   **Water Solubility:** Insoluble

Chemical Incompatibilities or Instabilities: Strong oxidizers.

FLAMMABILITY

NFPA Hazard Code: 2-2-0   **NFPA Classification:** Class II Combustible
Flash Point: 46°C (115°F)   **LEL:** N.D.
Autoignition Temp: N.D.   **UEL:** N.D.

Fire Extinguishing Methods: Carbon Dioxide, Dry Chemical, Foam, Water Spray or Fog.

Special Fire Fighting Considerations: Structural fire fighter protective clothing will not provide adequate protection. Water spray from an unmanned device should be used to keep closed containers cool. Continue to cool container after fire is extinguished. Fight fire from a distance or protected location, if possible. Treat all materials used or generated and equipment used involved as contaminated by hazardous waste.

DOT Recommended Isolation Zones: Small Spill: 150 ft   Large Spill: 150 ft
DOT Recommended Down Wind Take Cover Distance   Small Spill: 0.2 miles   Large Spill: 0.2 miles

TOXICOLOGY

Odor: **Odor Threshold:** N.A.
Physical Contact: Irritant
TLV = N.D.   **STEL** = N.D.   **IDLH** = N.D.

Routes of Entry and Relative LD$_{50}$ (or LC$_{50}$)   **RTECS # MJ1500000**
   Inhalation   9   (1250 mg/m^3/H)
   Ingestion   8   (85 mg/kg)
   Skin Absorption   3   (1954 mg/kg)

Symptoms of Exposure: Possible irritation to skin, eyes and upper respiratory tract; headache; nausea.

Emergency/First Aid Treatment: Remove to ventilated area; immediately remove any contaminated clothing and wash contaminated areas for 15 minutes using water. Treat supportively and observe for possible shock. If ingested, seek immediate medical aid.

Recommended Clean-Up Procedures

Personal Protection: Level B Ensemble   **Recommended Material** N.D.

RCRA Waste # None   **Reportable Quantities:** None

Spills: Remove all potential ignition sources. Dike to contain spill, absorb with noncombustible absorbent and take up using nonsparking tools. Decontaminate spill area using soapy water.

Special Emergency Information: May be harmful if inhaled, swallowed or absorbed through the skin.

OSMIUM TETROXIDE - OsO₄

Synonyms: osmic acid

CAS Number: 20816-12-0
DOT Number: 2471

Description: Pale yellow -green powder
DOT Classification: Poison B

Molecular Weight: 254.2
Melting Point: 41°C (106°F)
Boiling Point: (s) 130°C (266°F)

Vapor Density: N.D.
Specific Gravity: 4.9

Vapor Pressure: 3 (10 mm Hg)
Water Solubility: 6%

Chemical Incompatibilities or Instabilities: 1-Methylimidazole, strong reducing agents.

FLAMMABILITY

NFPA Hazard Code:
Flash Point: N.A.
Autoignition Temp: N.A.

NFPA Classification:
LEL: N.A.
UEL: N.A.

Fire Extinguishing Methods: Carbon Dioxide, Dry Chemical, Foam, Water Spray or Fog.

Special Fire Fighting Considerations: Structural fire fighter protective clothing will not provide adequate protection. Fight fire from a distance or protected location, if possible. Treat all materials used or generated and equipment used involved as contaminated by hazardous waste.

TOXICOLOGY

Odor:
Physical Contact: Material is extremely destructive to human tissues.
TLV = 0.02 mg/m³   **STEL** = 0.6 mg/m³

Odor Threshold: 20 mg/m³

IDLH = 1 mg/m³

Routes of Entry and Relative LD₅₀ (or LC₅₀)
   Inhalation   N.D.
   Ingestion   N.D.
   Skin Absorption   N.D.

RTECS # RN1140000

Symptoms of Exposure: Possible irritation to skin, eyes and upper respiratory tract; headache; nausea; lachrymation; laryngitis. Some effects may be delayed. Vapors may cause structural and functional changes to the eyes, trachea and lungs.

Emergency/First Aid Treatment: Remove to ventilated area; immediately remove any contaminated clothing and wash contaminated areas for 15 minutes using water. Treat supportively and observe for possible shock. If ingested, seek immediate medical aid.

Recommended Clean-Up Procedures

Personal Protection: Level B Ensemble

RCRA Waste # P087

Recommended Material N.D.

Reportable Quantities: 1000 lbs (1 lbs)

Spills: Take up in non-combustible absorbent and dispose. Decontaminate spill area using water. Treat all materials used or generated and equipment involved as contaminated by hazardous waste.

Special Emergency Information: May be fatal in inhaled, ingested or absorbed through the skin.

OXALIC ACID - (COOH)$_2$

Synonyms: ethaniedioic acid

CAS Number: 144-62-7	**Description:** White powder
DOT Number: 2449	**DOT Classification:** N.D.

Molecular Weight: 90.0

Melting Point: 189°C (372°F)	**Vapor Density:** N.D.	**Vapor Pressure:** 1 (< 1mm Hg)
Boiling Point: (s) 157°C (315°F)	**Specific Gravity:** 1.65	**Water Solubility:** 14%

Chemical Incompatibilities or Instabilities: Furfuryl alcohol, silver, chlorates, chlorites.

FLAMMABILITY

NFPA Hazard Code: 2-1-0	**NFPA Classification:**
Flash Point: N.D.	**LEL:** N.D.
Autoignition Temp: N.D.	**UEL:** N.D.

Fire Extinguishing Methods: Carbon Dioxide, Dry Chemical, Foam, Water Spray or Fog.

Special Fire Fighting Considerations: Structural fire fighter protective clothing will not provide adequate protection.

TOXICOLOGY

Odor: None   **Odor Threshold:** N.D.

Physical Contact: Material is extremely destructive to human tissues.

$TLV = 1$ mg/m^3 $STEL = 2$ mg/m^3 $IDLH = 500$ mg/m^3

Routes of Entry and Relative LD$_{50}$ (or LC$_{50}$)   **RTECS # RO2450000**

Inhalation	N.D.	
Ingestion	2	(7500 mg/kg)
Skin Absorption	N.D.	

Symptoms of Exposure: Possible burns or irritation to skin, eyes and upper respiratory tract; headache; nausea. Exposure may result in chemical pneumonitis or pulmonary edema.

Emergency/First Aid Treatment: Remove to ventilated area; immediately remove any contaminated clothing and wash contaminated areas for 15 minutes using water. Treat supportively and observe for possible shock. If ingested, seek immediate medical aid.

Recommended Clean-Up Procedures

Personal Protection: Level B Ensemble	**Recommended Material** Butyl Rubber (+), Viton (+), Neoprene (+)
RCRA Waste # None	**Reportable Quantities:** None

Spills: Take up in non-combustible absorbent and dispose. Decontaminate spill area using water.

Special Emergency Information: May be harmful if inhaled, swallowed or absorbed through the skin.

OXYGEN - O_2

Synonyms:

CAS Number: 7782-44-7   **Description:** Colorless gas or cryogenic liquid
DOT Number: 1072 (gas), 1073 (liquid)   **DOT Classification:** Nonflammable gas

Molecular Weight: 32.0
Melting Point: -218°C (-360°F)   **Vapor Density:** 1.4   **Vapor Pressure:** gas
Boiling Point: -182°C (-296°F)   **Specific Gravity:** 1.4   **Water Solubility:** 14%

Chemical Incompatibilities or Instabilities: A broad range of oxidizable materials

FLAMMABILITY

NFPA Hazard Code: **NFPA Classification:**
Flash Point: N.A.   **LEL:** N.A.
Autoignition Temp: N.A.   **UEL:** N.A.

Fire Extinguishing Methods: Use agent suitable for surrounding fire.

Special Fire Fighting Considerations: Isolate for 1/2 mile if rail or tank truck is involved in a fire. Water spray from an unmanned device should be used to keep closed containers cool. Continue to cool container after fire is extinguished. For large fires, if possible, withdraw and allow to burn.

TOXICOLOGY

Odor: **Odor Threshold:** N.D.
Physical Contact: Frostbite from cyrogenic liquid, irritant
TLV = N.D.   **STEL** = N.D.   **IDLH** = N.D.

Routes of Entry and Relative LD$_{50}$ (or LC$_{50}$)   **RTECS # RS2060000**
    Inhalation   N.D.
    Ingestion   N.D.
    Skin Absorption   N.D.

Symptoms of Exposure: Hyperoxia, vertigo, respiratory difficulties, convulsions.

Emergency/First Aid Treatment: Remove to ventilated area; immediately remove any contaminated clothing and wash contaminated areas for 15 minutes using water. Treat supportively and observe for possible shock. If ingested, seek immediate medical aid.

Recommended Clean-Up Procedures

Personal Protection: Level B Ensemble   **Recommended Material** N.D.

RCRA Waste # None   **Reportable Quantities:** None

Spills: Remove all potential ignition sources. Stop leak if it can safely be done. Keep area isolated until vapors dissiapte.

Special Emergency Information: May be harmful if inhaled.

PARAFORMALDEHYDE - $(CH_2O)_x$

Synonyms: formagene; triformol; trioxyethylene

CAS Number: 30525-89-4   **Description:** White powder
DOT Number: 2213   **DOT Classification:** ORM-A

Molecular Weight: N.A.
Melting Point: 156°C (313°F)   **Vapor Density:** 1.0   **Vapor Pressure:**   2 (2 mm Hg)
Boiling Point: °C   (°F)   **Specific Gravity:** 1.5   **Water Solubility:**   Decomposes

Chemical Incompatibilities or Instabilities: Isocyanates, copper and alloys, oxides, liquid oxygen, heat.

FLAMMABILITY

NFPA Hazard Code: 2-1-0   **NFPA Classification:** Class IIIA Combustible
Flash Point: 71°C (160°F)   **LEL:**   7.0%
Autoignition Temp: 300°C (572°F)   **UEL:**   73.0%

Fire Extinguishing Methods: Alcohol Resistant Foam, Carbon Dioxide, Dry Chemical, Water Spray or Fog.

Special Fire Fighting Considerations: Water spray from an unmanned device should be used to keep closed containers cool. Continue to cool container after fire is extinguished. For large fires, if possible, withdraw and allow to burn.

TOXICOLOGY

Odor: Formaldehyde   **Odor Threshold:** N.D.
Physical Contact: Material is extremely destructive to human tissues.
TLV = N.D.   **STEL** = N.D.   **IDLH** = N.D.

Routes of Entry and Relative LD$_{50}$ (or LC$_{50}$)   **RTECS # RV0540000**
  Inhalation   N.D.
  Ingestion   4   (800 mg/kg)
  Skin Absorption   N.D.

Symptoms of Exposure: Possible burns or irritation to skin, eyes and upper respiratory tract; headache; nausea. Inhalation of vapors may be fatal by causing glottis to spasm and suffocation. Exposure may result in chemical pneumonitis or pulmonary edema. May be an allergen.

Emergency/First Aid Treatment: Remove to ventilated area; immediately remove any contaminated clothing and wash contaminated areas for 15 minutes using water. Treat supportively and observe for possible shock. If ingested, seek immediate medical aid.

Recommended Clean-Up Procedures

Personal Protection: Level B Ensemble   **Recommended Material** Butyl Rubber (-)

RCRA Waste #   None   **Reportable Quantities:**   None

Spills: Take up spill. Decontaminate spill area using water. Keep area isolated until vapors dissipate.

Special Emergency Information: May be harmful if inhaled, swallowed or absorbed through the skin.

PARALDEHYDE

Synonyms: acetaldehyde trimer; paracetaldehyde

CAS Number: 123-63-7

DOT Number: 1264

Description: Colorless Liquid

DOT Classification: Flammable Liquid

Molecular Weight: 132.2

Melting Point:	13°C (55°F)	**Vapor Density:** 4.6	**Vapor Pressure:** N.A.
Boiling Point:	124°C (255°F)	**Specific Gravity:** 1.0	**Water Solubility:** Soluble

Chemical Incompatibilities or Instabilities: Nitric acid, alkalis, hydrocyanic acid, iodides, oxidizers.

FLAMMABILITY

NFPA Hazard Code: 2-3-1

Flash Point: 36°C (96°F)

Autoignition Temp: 238°C (460°F)

NFPA Classification: Class IC Flammable

LEL: 1.3%

UEL: N.D.

Fire Extinguishing Methods: Alcohol Resistant Foam, Carbon Dioxide, Dry Chemical, Water Spray or Fog.

Special Fire Fighting Considerations: Isolate for 1/2 mile if rail or tank truck is involved in a fire. Water spray from an unmanned device should be used to keep closed containers cool. Continue to cool container after fire is extinguished. For large fires, if possible, withdraw and allow to burn. Immediately withdraw if rising sound from venting device is heard or if fire is causing discoloration to the tank.

TOXICOLOGY

Odor: Aromatic

Physical Contact: Irritant

TLV = N.D.

Odor Threshold: N.D.

STEL = N.D.

IDLH = N.D.

RTECS # YK0525000

Routes of Entry and Relative LD$_{50}$ (or LC$_{50}$)

Inhalation	N.D.	
Ingestion	3	(1530 mg/kg)
Skin Absorption	1	(14,000 mg/kg)

Symptoms of Exposure: Possible irritation to skin, eyes and upper respiratory tract; headache; nausea. Other symptoms may include hypnotic and analgesic effects.

Emergency/First Aid Treatment: Remove to ventilated area; immediately remove any contaminated clothing and wash contaminated areas for 15 minutes using water. Treat supportively and observe for possible shock. If ingested, seek immediate medical aid.

Recommended Clean-Up Procedures

Personal Protection: Level B Ensemble

RCRA Waste # U182

Recommended Material N.D.

Reportable Quantities: 1000 lbs (1 lb)

Spills: Remove all potential ignition sources. Dike to contain spill, absorb with noncombustible absorbent and take up using nonsparking tools. Decontaminate spill area using water. Treat all materials used or generated and equipment used involved as contaminated by hazardous waste.

Special Emergency Information: May be harmful if inhaled, swallowed or absorbed through the skin. Vapors are heavier than air and may travel some distance to an ignition source. Vapors are more dense than air and may settle in low lying areas.

PARAQUAT

Synonyms: dimethylviologen; methyl viologen

CAS Number: 1910-42-5	**Description:** Yellow powder	
DOT Number: 2588	**DOT Classification:** N.A.	

Molecular Weight: 186.3

Melting Point: 125°C (257°F)	**Vapor Density:** N.A.	**Vapor Pressure:** 1 (< 1 mm Hg)
Boiling Point: (d) 180°C (356°F)	**Specific Gravity:** >1	**Water Solubility:** Soluble

Chemical Incompatibilities or Instabilities: Strong oxidizers.

FLAMMABILITY

NFPA Hazard Code:	**NFPA Classification:**	
Flash Point: N.A.	**LEL:** N.A.	
Autoignition Temp: N.A.	**UEL:** N.A.	

Fire Extinguishing Methods: Carbon Dioxide, Dry Chemical, Foam, Water Spray or Fog.

Special Fire Fighting Considerations: Structural fire fighter protective clothing will not provide adequate protection. Fight fire from a distance or protected location, if possible. Treat all materials used or generated and equipment used involved as contaminated by hazardous waste.

TOXICOLOGY

Odor: Ammonia-like   **Odor Threshold:** N.D.
Physical Contact: Irritant

$TLV = 0.1$ mg/m^3 (skin) $STEL = 0.5$ mg/m^3 (skin) $IDLH = 1.5$ mg/m^3

Routes of Entry and Relative LD$_{50}$ (or LC$_{50}$)   **RTECS # DW2275000**

Inhalation	N.D.
Ingestion	7 (150 mg/kg)
Skin Absorption	N.D.

Symptoms of Exposure: Possible irritation to skin, eyes and upper respiratory tract; headache; nausea. Exposure may result in chemical pneumonitis or pulmonary edema.

Emergency/First Aid Treatment: Remove to ventilated area; immediately remove any contaminated clothing and wash contaminated areas for 15 minutes using water. Treat supportively and observe for possible shock. If ingested, seek immediate medical aid.

Recommended Clean-Up Procedures

Personal Protection: Level A Ensemble	**Recommended Material** N.D.	
RCRA Waste # None	**Reportable Quantities:** None	

Spills: Take up in non-combustible absorbent and dispose. Decontaminate spill area using water. Treat all materials used or generated and equipment involved as contaminated by hazardous waste.

Special Emergency Information: May be fatal in inhaled, ingested or absorbed through the skin.

PARATHION

Synonyms:

CAS Number: 56-38-2

DOT Number: 2783

Description: Pale yellow liquid

DOT Classification: Poison B

Molecular Weight: 291.3

Melting Point:	144°C (291°F)	**Vapor Density:** N.D.	**Vapor Pressure:** 1 (<1 mm Hg)
Boiling Point:	375°C (707°F)	**Specific Gravity:** 1.3	**Water Solubility:** Insoluble

Chemical Incompatibilities or Instabilities: Strong oxidizers, endrin.

FLAMMABILITY

NFPA Hazard Code:

Flash Point: N.A.

Autoignition Temp: N.A.

NFPA Classification:

LEL: N.A.

UEL: N.A.

Fire Extinguishing Methods: Carbon Dioxide, Dry Chemical, Foam, Water Spray or Fog.

Special Fire Fighting Considerations: Structural fire fighter protective clothing will not provide adequate protection. Fight fire from a distance or protected location, if possible. Treat all materials used or generated and equipment involved as contaminated by hazardous waste.

TOXICOLOGY
QUESTIONABLE CARCINOGEN

Odor:

Physical Contact: Irritant

TLV = 0.1 mg/m^3 (skin)   **STEL** = N.D.

Odor Threshold: N.D.

IDLH = 20 mg/m^3

RTECS # TF4550000

Routes of Entry and Relative LD$_{50}$ (or LC$_{50}$)

Inhalation	10	(336 mg/m^3/H)
Ingestion	10	(2 mg/kg)
Skin Absorption	N.D.	

Symptoms of Exposure: Possible irritation to skin, eyes and upper respiratory tract; headache; nausea; salivation; loss of muscle control; convulsions; respiratory failure.

Emergency/First Aid Treatment: Remove to ventilated area; immediately remove any contaminated clothing and wash contaminated areas for 15 minutes using water. Treat supportively and observe for possible shock. If ingested, seek immediate medical aid. Treat for anticholinesterase poisoning.

Recommended Clean-Up Procedures

Personal Protection: Level A Ensemble   **Recommended Material** Teflon (0)

RCRA Waste # P089   **Reportable Qunatities:** 10 lbs (1 lb)

Spills: Take up in non-combustible absorbent and dispose. Decontaminate spill area using soapy water. Treat all materials used or generated and equipment involved as contaminated by hazardous waste.

Special Emergency Information: May be fatal in inhaled, ingested or absorbed through the skin.

PENTABORANE -B₅H₉

Synonyms:

CAS Number: 19624-22-7   **Description:** Colorless Liquid
DOT Number: 1380   **DOT Classification:** Flammable Liquid

Molecular Weight: 63.1
Melting Point: -47°C (-53°F)   **Vapor Density:** 2.2   **Vapor Pressure:** 7 (171 mm Hg)
Boiling Point: 63°C (145°F)   **Specific Gravity:** 0.6   **Water Solubility:** Decomposes

Chemical Incompatibilities or Instabilities: Carbonyls, ethers, esters, halogens, oxidizers, halogenated hydrocarbons, water, oxygen.

FLAMMABILITY

NFPA Hazard Code: 4-4-2   **NFPA Classification:** Class IC Flammable
Flash Point: 30°C (86°F)   **LEL:** 0.4%
Autoignition Temp: 35°C (95°F)   **UEL:** 98.0%

Fire Extinguishing Methods: None

Special Fire Fighting Considerations: Let fire burn unless leak can be safely stopped. Structural fire fighter protective clothing will not provide adequate protection. Isolate for 1/2 mile if rail or tank truck is involved in a fire. Water spray from an unmanned device should be used to keep closed containers cool. Continue to cool container after fire is extinguished.

DOT Recommended Isolation Zones: Small Spill: 1500 ft Large Spill: 1500 ft
DOT Recommended Down Wind Take Cover Distance   Small Spill: 5 miles Large Spill: 5 miles

TOXICOLOGY

Odor: Sour milk   **Odor Threshold:** 2.5 mg/m³
Physical Contact: Material is extremely destructive to human tissues.
TLV = 0.01 mg/m³   **STEL** = 0.03 mg/m³   **IDLH** = 8 mg/m³

Routes of Entry and Relative LD₅₀ (or LC₅₀)   **RTECS #** RY8925000
   Inhalation   10   (62 mg/m³/H)
   Ingestion   N.D.
   Skin Absorption N.D.

Symptoms of Exposure: Possible burns or irritation to skin, eyes and upper respiratory tract; headache; nausea; muscle tremors. Inhalation of vapors may be fatal by causing glottis to spasm and suffocation. Exposure may result in chemical pneumonitis or pulmonary edema.

Emergency/First Aid Treatment: Remove to ventilated area; immediately remove any contaminated clothing and wash contaminated areas for 15 minutes using water. Treat supportively and observe for possible shock. If ingested, seek immediate medical aid.

Recommended Clean-Up Procedures

Personal Protection: Level A Ensemble   **Recommended Material** N.D.

RCRA Waste # None   **Reportable Quantities:** None

Spills: Clean up only under expert supervision.

Special Emergency Information: May be fatal in inhaled, ingested or absorbed through the skin. Vapors are heavier than air and may travel some distance to an ignition source. Vapors are more dense than air and may settle in low lying areas. Vapors may spontaneously ingnite in air.

PENTACHLOROETHANE - CCl_3CHCl_2

Synonyms: ethane pentachloride; pentalin

CAS Number: 76-01-7	**Description:** Colorless Liquid	
DOT Number: 1669	**DOT Classification:** Poison B	

Molecular Weight: 202.3

Melting Point: -29°C (-20°F)	**Vapor Density:** N.D.	**Vapor Pressure:** 3 (6 mm Hg)	
Boiling Point: 162°C (324°F)	**Specific Gravity:** 1.7	**Water Solubility:** Insoluble	

Chemical Incompatibilities or Instabilities: Strong oxidizers, alkali metals.

FLAMMABILITY

NFPA Hazard Code:	**NFPA Classification:**
Flash Point: N.D.	**LEL:** N.D.
Autoignition Temp: N.D.	**UEL:** N.D..

Fire Extinguishing Methods: Carbon Dioxide, Dry Chemical, Foam, Water Spray or Fog.

Special Fire Fighting Considerations: Structural fire fighter protective clothing will not provide adequate protection. Fight fire from a distance or protected location, if possible. Treat all materials used or generated and equipment involved as contaminated by hazardous waste.

TOXICOLOGY
QUESTIONABLE CARCINOGEN

Odor: Chloroform-like	**Odor Threshold:** N.D.
Physical Contact: Irritant	
TLV = N.D. **STEL** = N.D.	**IDLH** = N.D.

Routes of Entry and Relative LD_{50} (or LC_{50})   **RTECS # KI6300000**

Inhalation	5 (700,000 mg/m^3/H)
Ingestion	N.D.
Skin Absorption	N.D.

Symptoms of Exposure: Possible irritation to skin, eyes and upper respiratory tract; headache; nausea.

Emergency/First Aid Treatment: Remove to ventilated area; immediately remove any contaminated clothing and wash contaminated areas for 15 minutes using water. Treat supportively and observe for possible shock. If ingested, seek immediate medical aid.

Recommended Clean-Up Procedures

Personal Protection: Level A Ensemble	**Recommended Material** N.D.
RCRA Waste # U184	**Reportable Quantities:** 10 lbs (1 lb)

Spills: Remove all potential ignition sources. Dike to contain spill, absorb with non-combustible absorbent and take up using non-sparking tools. Decontaminate spill area using soapy water. Treat all materials used or generated and equipment involved as contaminated by hazardous waste.

Special Emergency Information: May be harmful if inhaled, swallowed or absorbed through the skin.

PERCHLORIC ACID - HClO$_4$

Synonyms:

CAS Number: 7601-90-3
DOT Number: 1802, 1873

Description: Colorless gas or colorless liquid when dissolved in water
DOT Classification: Oxidizer

Molecular Weight: 100.5

Melting Point: -112°C (-170°F)
Boiling Point: 19°C (66°F)

Vapor Density: 3.5
Specific Gravity: 1.8

Vapor Pressure: Gas
Water Solubility: Soluble

Chemical Incompatibilities or Instabilities: Amines alcohols, anhydrides, halides, strong acids, unsaturated hydrocarbons, ketones, dimethyl sulfoxide, organophosphorous compounds.

FLAMMABILITY

NFPA Hazard Code:

Flash Point: N.A.
Autoignition Temp: N.A.

NFPA Classification:

LEL: N.A.
UEL: N.A.

Fire Extinguishing Methods: WATER ONLY

Special Fire Fighting Considerations: Isolate for 1/2 mile if rail or tank truck is involved in a fire. Water spray from an unmanned device should be used to keep closed containers cool. Continue to cool container after fire is extinguished. For large fires, use flooding quantities of water.

TOXICOLOGY

Odor: Pungent
Physical Contact: Material is extremely destructive to human tissues.
TLV = N.D.   **STEL** = N.D.

Odor Threshold: N.D.

IDLH = N.D.

Routes of Entry and Relative LD$_{50}$ (or LC$_{50}$)

Inhalation	N.D.
Ingestion	3 (1100 mg/kg)
Skin Absorption	N.D.

RTECS # SC7500000

Symptoms of Exposure: Possible burns or irritation to skin, eyes and upper respiratory tract; headache; nausea. Exposure may result in chemical pneumonitis or pulmonary edema.

Emergency/First Aid Treatment: Remove to ventilated area; immediately remove any contaminated clothing and wash contaminated areas for 15 minutes using water. Treat supportively and observe for possible shock. If ingested, seek immediate medical aid.

Recommended Clean-Up Procedures

Personal Protection: Level B Ensemble

Recommended Material Natural Rubber (+), Neoprene (+), Nitrile Rubber (+), Polyvinyl Chloride (+)

RCRA Waste # None

Reportable Quantities: None

Spills: Dike to contain spill and collect liquid for later disposal. Decontaminate spill area using dilute alkaline solution.

Special Emergency Information: May be fatal in inhaled, ingested or absorbed through the skin.

PERCHLOROETHYLENE - $CCl_2=CCl_2$

Synonyms: tetrachloroethylene

CAS Number: 127-18-4
DOT Number: 1897

Description: Colorless Liquid
DOT Classification: Poison B

Molecular Weight: 165.8
Melting Point: -24°C (-11°F)
Boiling Point: 121°C (250°F)

Vapor Density: 5.8
Specific Gravity: 1.6

Vapor Pressure: 4 (15 mm Hg)
Water Solubility: Insoluble

Chemical Incompatibilities or Instabilities: Strong alkali, alkali metals, barium, berrylium.

FLAMMABILITY

NFPA Hazard Code: 2-0-0
Flash Point: N.A.
Autoignition Temp: N.A.

NFPA Classification:
LEL: N.A.
UEL: N.A.

Fire Extinguishing Methods: Use agent suitable for surrounding fire.

Special Fire Fighting Considerations: Isolate for 1/2 mile if rail or tank truck is involved in a fire. Water spray should be used to keep closed containers cool. Continue to cool container after fire is extinguished. Treat all materials used or generated and equipment involved as contaminated by hazardous waste.

TOXICOLOGY
CARCINOGEN

Odor: Chloroform-like
Physical Contact: Irritant
TLV = 170 mg/m^3

STEL = 675 mg/m^3

Odor Threshold: 13.5 mg/m^3

IDLH = N.D.

RTECS # KX3850000

Routes of Entry and Relative LD$_{50}$ (or LC$_{50}$)
Inhalation	2	(275,000 mg/m^3/H)
Ingestion	3	(2629 mg/kg)
Skin Absorption	N.D.	

Symptoms of Exposure: Possible irritation to skin, eyes and upper respiratory tract; headache; nausea. May act as a narcotic in high concentrations.

Emergency/First Aid Treatment: Remove to ventilated area; immediately remove any contaminated clothing and wash contaminated areas for 15 minutes using water. Treat supportively and observe for possible shock. If ingested, seek immediate medical aid.

Recommended Clean-Up Procedures

Personal Protection: Level B Ensemble

RCRA Waste # U210

Recommended Material Polyvinyl Alcohol (+), Teflon (+), Viton (+)

Reportable Quantities: 100 lbs (1 lb)

Spills: Dike to contain spill and collect liquid for later disposal. Decontaminate spill area using soapy water. Treat all materials used or generated and equipment involved as contaminated by hazardous waste.

Special Emergency Information: May be harmful if inhaled, swallowed or absorbed through the skin. Vapors are more dense than air and may settle in low lying areas.

PERCHLOROMETHYL MERCAPTAN - Cl₃CSCl

Synonyms: clairsit; trichloromethane sulfenyl chloride

CAS Number: 594-42-3   **Description:** Yellow liquid
DOT Number: 1670   **DOT Classification:** Poison B

Molecular Weight: 185.9
Melting Point: °C   (°F)   **Vapor Density:** 6.4   **Vapor Pressure:** **2** (2 mm Hg)
Boiling Point: 146°C (295°F)   **Specific Gravity:** 1.7   **Water Solubility:** Decomposes

Chemical Incompatibilities or Instabilities: Alkali, amines, strong oxidizers, water.

FLAMMABILITY

NFPA Hazard Code: **NFPA Classification:**
Flash Point: N.A.   **LEL:** N.A.
Autoignition Temp: N.A.   **UEL:** N.A.

Fire Extinguishing Methods: Use agent suitable for surrounding fire.

Special Fire Fighting Considerations: Structural fire fighter protective clothing will not provide adequate protection. Fight fire from a distance or protected location, if possible. Treat all materials used or generated and equipment involved as contaminated by hazardous waste.

DOT Recommended Isolation Zones: Small Spill: 150 ft   Large Spill: 150 ft
DOT Recommended Down Wind Take Cover Distance   Small Spill: 0.8 miles   Large Spill: 0.8 miles

TOXICOLOGY

Odor: Unbearable   **Odor Threshold:** N.D.
Physical Contact: Material is extremely destructive to human tissues.
TLV = 0.8 mg/m³   **STEL** = N.D.   **IDLH** = 77 mg/m³

Routes of Entry and Relative LD₅₀ (or LC₅₀)   **RTECS # PB0370000**
    Inhalation   N.D.
    Ingestion   **8**   (86 mg/kg)
    Skin Absorption   N.D.

Symptoms of Exposure: Possible burns or irritation to skin, eyes and upper respiratory tract; headache; nausea. Inhalation of vapors may be fatal by causing glottis to spasm and suffocation. Exposure may result in chemical pneumonitis or pulmonary edema.

Emergency/First Aid Treatment: Remove to ventilated area; immediately remove any contaminated clothing and wash contaminated areas for 15 minutes using water. Treat supportively and observe for possible shock. If ingested, seek immediate medical aid.

Recommended Clean-Up Procedures

Personal Protection: Level A Ensemble   **Recommended Material**   N.D.

RCRA Waste #   P118   **Reportable Quantities:**   100 lbs (1 lb)

Spills: Dike to contain spill and collect liquid for later disposal. Treat all materials used or generated and equipment involved as contaminated by hazardous waste.

Special Emergency Information: May be fatal in inhaled, ingested or absorbed through the skin. Vapors are more dense than air and may settle in low lying areas.

PERCHLORYL FLUORIDE - ClO3F

Synonyms: chlorine fluorine oxide; chlorine oxyfluoride

CAS Number: 7616-94-6	**Description:** Colorless Gas	
DOT Number: 3083	**DOT Classification:** N.A.	

Molecular Weight: 102.5

Melting Point: -234°C (-234°F)	**Vapor Density:** >1	**Vapor Pressure:** Gas
Boiling Point: -62°C (-52°F)	**Specific Gravity:** N.A.	**Water Solubility:** <1 %

Chemical Incompatibilities or Instabilities: Combustibles, unsaturated hydrocarbons, amines, finely divided metals, oxidizers, strong bases, alcohols, reducing agents.

FLAMMABILITY

NFPA Hazard Code:	**NFPA Classification:**	
Flash Point: N.A.	**LEL:** N.A.	
Autoignition Temp: N.A.	**UEL:** N.A.	

Fire Extinguishing Methods: WATER ONLY

Special Fire Fighting Considerations: Structural fire fighter protective clothing will not provide adequate protection. If possible, contain fire and allow to burn. Water spray should be used to keep closed containers cool. Continue to cool container after fire is extinguished. Treat all materials used or generated and equipment involved as contaminated by hazardous waste.

DOT Recommended Isolation Zones:	Small Spill: 900 ft	Large Spill: 900 ft
DOT Recommended Down Wind Take Cover Distance	Small Spill: 3 miles	Large Spill: 3 miles

TOXICOLOGY

Odor: Sweet	**Odor Threshold:** 42 mg/m^3

Physical Contact: Material is extremely destructive to human tissues.

TLV = 14 mg/m^3	**STEL** = 28 mg/m^3	**IDLH** = 1600 mg/m^3

Routes of Entry and Relative LD$_{50}$ (or LC$_{50}$)   **RTECS # SD1925000**

Inhalation	N.D.
Ingestion	N.D.
Skin Absorption	N.D.

Symptoms of Exposure: Possible burns or irritation to skin, eyes and upper respiratory tract; headache; nausea. Exposure may result in chemical pneumonitis or pulmonary edema. Absorption may lead to the formation of methemoglobin resulting in cyanosis several hours after exposure.

Emergency/First Aid Treatment: Remove to ventilated area; immediately remove any contaminated clothing and wash contaminated areas for 15 minutes using water. Treat supportively and observe for possible shock. If ingested, seek immediate medical aid.

Recommended Clean-Up Procedures

Personal Protection: Level A Ensemble	**Recommended Material** N.D.	
RCRA Waste # None	**Reportable Quantities:** None	

Spills: Stop leak if it can safely be done. Keep area isolated until vapors dissipate. Treat all materials used or generated and equipment involved as contaminated by hazardous waste.

Special Emergency Information: May be fatal in inhaled, ingested or absorbed through the skin. Vapors are more dense than air and may settle in low lying areas. Water spray or fog may be used to keep vapors down while stopping leak.

PETROLEUM

Synonyms: crude oil; petroluem crude

CAS Number: 8002-05-9	**Description:** Dark, viscous liquid	
DOT Number: 1267, 1270	**DOT Classification:** Combustible Liquid	

Molecular Weight: Varies

Melting Point:	N.A.	**Vapor Density:** N.A.	**Vapor Pressure:**	N.A.
Boiling Point:	N.A.	**Specific Gravity:** <1.0	**Water Solubility:**	Insoluble

Chemical Incompatibilities or Instabilities: Strong oxidizers.

FLAMMABILITY

NFPA Hazard Code:	1-3-0	**NFPA Classification:** Class II Combustible
Flash Point:	-6°C - 32°C (20°F-90°F)	**LEL:** N.D.
Autoignition Temp:	N.D.	**UEL:** N.D.

Fire Extinguishing Methods: Carbon Dioxide, Dry Chemical, Foam, Water Spray or Fog.

Special Fire Fighting Considerations: Isolate for 1/2 mile if rail or tank truck is involved in a fire. Water spray should be used to keep closed containers cool. Continue to cool container after fire is extinguished. For large fires, if possible, withdraw and allow to burn. Immediately withdraw if rising sound from venting device is heard or if fire is causing discoloration to the tank. Treat all materials used or generated and equipment involved as contaminated by hazardous waste.

TOXICOLOGY
QUESTIONABLE CARCINOGEN

Odor:	**Odor Threshold:** N.D.
Physical Contact: Irritant	
TLV = N.D. **STEL** = N.D.	**IDLH** = N.D.

Routes of Entry and Relative LD$_{50}$ (or LC$_{50}$)		**RTECS # SE7175000**
Inhalation	N.D.	
Ingestion	N.D.	
Skin Absorption	N.D.	

Symptoms of Exposure: Possible irritation to skin, eyes and upper respiratory tract; headache; nausea.

Emergency/First Aid Treatment: Remove to ventilated area; immediately remove any contaminated clothing and wash contaminated areas for 15 minutes using water. Treat supportively and observe for possible shock. If ingested, seek immediate medical aid.

Recommended Clean-Up Procedures

Personal Protection:	Level B Ensemble	**Recommended Material** N.D.
RCRA Waste #	None	**Reportable Quantities:** None

Spills: Remove all potential ignition sources. Dike to contain spill, absorb with noncombustible absorbent and take up using nonsparking tools. Decontaminate spill area using soapy water. Treat all materials used or generated and equipment involved as contaminated by hazardous waste.

Special Emergency Information: May be harmful if inhaled, swallowed or absorbed through the skin. Vapors are heavier than air and may travel some distance to an ignition source. Vapors are more dense than air and may settle in low lying areas.

PETROLEUM ETHER

Synonyms: mineral spirits; benzine; ligroin; mineral turpentine

CAS Number: 64475-85-0   **Description:** Colorless Liquid
DOT Number: 1271   **DOT Classification:**

Molecular Weight: Varies

| **Melting Point:** | °C | (°F) | **Vapor Density:** 2.5 | **Vapor Pressure:** 8 (400 mm Hg) |
| **Boiling Point:** | 35°C-60°C | (95°F-140°F) | **Specific Gravity:** 0.6 | **Water Solubility:** Insoluble |

Chemical Incompatibilities or Instabilities: Strong oxidizers.

FLAMMABILITY

NFPA Hazard Code: 1-4-0   **NFPA Classification:** Class IB Flammable
Flash Point: -49°C (-57°F)   **LEL:** 1.1%
Autoignition Temp: 228°C (550°F)   **UEL:** 5.9%

Fire Extinguishing Methods: Alcohol Resistant Foam, Carbon Dioxide, Dry Chemical, Water Spray or Fog.

Special Fire Fighting Considerations: Isolate for 1/2 mile if rail or tank truck is involved in a fire. Water spray should be used to keep closed containers cool. Continue to cool container after fire is extinguished. For large fires, if possible, withdraw and allow to burn. Immediately withdraw if rising sound from venting device is heard or if fire is causing discoloration to the tank.

TOXICOLOGY

Odor: Turpentine   **Odor Threshold:** N.D.
Physical Contact: Irritant
TLV = 1200 mg/m^3 **STEL** = 1600 mg/m^3 **IDLH** = 40,000 mg/m^3

Routes of Entry and Relative LD$_{50}$ (or LC$_{50}$)   **RTECS #** DE3030000
     Inhalation   4   (55,750 mg/m^3/H)
     Ingestion   N.D.
     Skin Absorption   N.D.

Symptoms of Exposure: Possible irritation to skin, eyes and upper respiratory tract; headache; nausea. Exposure may result in chemical pneumonitis or pulmonary edema.

Emergency/First Aid Treatment: Remove to ventilated area; immediately remove any contaminated clothing and wash contaminated areas for 15 minutes using water. Treat supportively and observe for possible shock. If ingested, seek immediate medical aid.

Recommended Clean-Up Procedures

Personal Protection: Level B Ensemble   **Recommended Material** Nitrile Rubber (0), Teflon (0)

RCRA Waste # None   **Reportable Quantities:** None

Spills: Remove all potential ignition sources. Dike to contain spill, absorb with noncombustible absorbent and take up using nonsparking tools. Decontaminate spill area using soapy water. Keep area isolated until vapors dissipate.

Special Emergency Information: May be harmful if inhaled, swallowed or absorbed through the skin. Vapors are heavier than air and may travel some distance to an ignition source. Vapors are more dense than air and may settle in low lying areas.

PHENOL

Synonyms: carbolic acid; hydroxybenzene; phenyl hydroxide

CAS Number: 108-95-2	**Description:** White solid	
DOT Number: 1671, 2312 (molten), 2821 (solution)	**DOT Classification:** Poison B	

Molecular Weight: 94.1

Melting Point: 41°C (106°F)	**Vapor Density:** 3.2	**Vapor Pressure:** 1 (<1 mm Hg)
Boiling Point: 182°C (358°F)	**Specific Gravity:** 1.1	**Water Solubility:** Soluble

Chemical Incompatibilities or Instabilities: Strong oxidizers, butadiene, formaldehyde, aluminum chloride.

FLAMMABILITY

NFPA Hazard Code: 3-2-0	**NFPA Classification:**
Flash Point: 79°C (175°F)	**LEL:** 1.7%
Autoignition Temp: 715°C (1319°F)	**UEL:** 8.6%

Fire Extinguishing Methods: Carbon Dioxide, Dry Chemical, Foam, Water Spray or Fog.

Special Fire Fighting Considerations: Structural fire fighter protective clothing will not provide adequate protection. Fight fire from a distance or protected location, if possible. Water spray or fog can be used to keep vapors down. Treat all materials used or generated and equipment involved as contaminated by hazardous waste.

TOXICOLOGY

Odor: Medicinal	**Odor Threshold:** 0.02 mg/m^3

Physical Contact: Material is extremely destructive to human tissues.

TLV = 19 mg/m^3 (skin)	**STEL** = N.D.	**IDLH** = 950 mg/m^3

Routes of Entry and Relative LD$_{50}$ (or LC$_{50}$)   **RTECS #** SJ3325000

Inhalation	9	(316 mg/m^3/H)
Ingestion	5	(317 mg/kg)
Skin Absorption	8	(850 mg/kg)

Symptoms of Exposure: Possible irritation to skin, eyes and upper respiratory tract; headache; nausea. Inhalation of vapors may be fatal by causing glottis to spasm and suffocation. Exposure may result in chemical pneumonitis or pulmonary edema.

Emergency/First Aid Treatment: Remove to ventilated area; immediately remove any contaminated clothing and wash contaminated areas for 15 minutes using water. Treat supportively and observe for possible shock. If ingested, seek immediate medical aid.

Recommended Clean-Up Procedures

Personal Protection: Level A Ensemble	**Recommended Material** Butyl Rubber (+), Viton (+), Neoprene (0)
RCRA Waste # U188	**Reportable Quantities:** 1000 lbs (1000 lbs)

Spills: Take up in non-combustible absorbent and dispose. Decontaminate spill area using water. Treat all materials used or generated and equipment involved as contaminated by hazardous waste.

Special Emergency Information: May be fatal in inhaled, ingested or absorbed through the skin. Vapors are more dense than air and may settle in low lying areas. Molten phenol may cause thermal burns.

PHENYLACETONITRILE

Synonyms: benzyl cyanide; benzyl nitrile

CAS Number: 140-29-4   **Description:** Colorless Liquid
DOT Number: 2470   **DOT Classification:** Poison B

Molecular Weight: 117.2
Melting Point: -23°C   (-9°F)   **Vapor Density:**   **Vapor Pressure:** 1 (<1 mm Hg)
Boiling Point: 234°C (452°F)   **Specific Gravity:** 1.0   **Water Solubility:** Insoluble

Chemical Incompatibilities or Instabilities: Strong oxidizers, sodium hypochlorite.

FLAMMABILITY

NFPA Hazard Code: 2-1-0   **NFPA Classification:**   Class IIIB Combustible
Flash Point: 113°C (235°F)   **LEL:** N.D.
Autoignition Temp: N.D.   **UEL:** N.D.

Fire Extinguishing Methods: Carbon Dioxide, Dry Chemical, Foam, Water Spray or Fog.

Special Fire Fighting Considerations: Structural fire fighter protective clothing will not provide adequate protection. Fight fire from a distance or protected location, if possible. Treat all materials used or generated and equipment involved as contaminated by hazardous waste.

TOXICOLOGY

Odor: Aromatic   **Odor Threshold:** N.D.
Physical Contact: Irritant
TLV = N.D.   **STEL** = N.D.   **IDLH** = N.D.

Routes of Entry and Relative LD$_{50}$ (or LC$_{50}$)   **RTECS # AM1400000**
   Inhalation   9   (860 mg/m^3/H)
   Ingestion   6   (270 mg/kg)
   Skin Absorption   8   (270 mg/kg)

Symptoms of Exposure: Possible irritation to skin, eyes and upper respiratory tract; headache; nausea.

Emergency/First Aid Treatment: Remove to ventilated area; immediately remove any contaminated clothing and wash contaminated areas for 15 minutes using water. Treat supportively and observe for possible shock. If ingested, seek immediate medical aid.

Recommended Clean-Up Procedures

Personal Protection: Level A Ensemble   **Recommended Material**   N.D.

RCRA Waste #   None   **Reportable Quantities:**   None

Spills: Dike to contain spill and collect liquid for later disposal. Treat all materials used or generated and equipment involved as contaminated by hazardous waste.

Special Emergency Information: May be fatal in inhaled, ingested or absorbed through the skin.

1,4-PHENYLENEDIAMINE

Synonyms: p-aminoaniline; 1,4-benzenediamine; p-diaminobenzene

CAS Number: 106-50-3   **Description:** White powder
DOT Number: 1673   **DOT Classification:** ORM-A

Molecular Weight: 108.2
Melting Point: 146°C (°F)   **Vapor Density:** 3.7   **Vapor Pressure:** **1** (<1 mm Hg)
Boiling Point: 267°C (°F)   **Specific Gravity:** >1.0   **Water Solubility:** 5%

Chemical Incompatibilities or Instabilities: Strong oxidizers.

FLAMMABILITY

NFPA Hazard Code: **NFPA Classification:**
Flash Point: 155°C (312°F)   **LEL:** N.D.
Autoignition Temp: N.D.   **UEL:** N.D.

Fire Extinguishing Methods: Carbon Dioxide, Dry Chemical, Foam, Water Spray or Fog.

Special Fire Fighting Considerations: Water spray should be used to keep closed containers cool. Continue to cool container after fire is extinguished.

TOXICOLOGY
QUESTIONABLE CARCINOGEN

Odor: **Odor Threshold:** N.D.
Physical Contact: Irritant
TLV = 0.1 mg/m^3 (skin)   **STEL** = N.D.   **IDLH** = N.D.

Routes of Entry and Relative LD$_{50}$ (or LC$_{50}$)   **RTECS # SS8050000**
      Inhalation   N.D.
      Ingestion   **8**   (80 mg/kg)
      Skin Absorption   N.D.

Symptoms of Exposure: Possible irritation to skin, eyes and upper respiratory tract; headache; nausea. Absorption may lead to the formation of methemoglobin resulting in cyanosis several hours after exposure.

Emergency/First Aid Treatment: Remove to ventilated area; immediately remove any contaminated clothing and wash contaminated areas for 15 minutes using water. Treat supportively and observe for possible shock. If ingested, seek immediate medical aid.

Recommended Clean-Up Procedures

Personal Protection: Level B Ensemble   **Recommended Material**   N.D.

RCRA Waste #   None   **Reportable Quantities:**   None

Spills: Take up in non-combustible absorbent and dispose.

Special Emergency Information: May be fatal in inhaled, ingested or absorbed through the skin.

1,2-phenylaminediamine (CAS # 95-54-5, RTECS # SS7875000) and
1,3-phenylenediamine (CAS # 108-45-2, RTECS # SS7700000) have similar chemical, physical and toxicological properties.

PHENYLHYDRAZINE

Synonyms:

CAS Number: 100-63-0

DOT Number: 2572

Description: Amber Liquid

DOT Classification: Poison B

Molecular Weight: 108.2

Melting Point: 19°C   (66°F)

Boiling Point: 238°C (460°F)

Vapor Density: 4.3

Specific Gravity: 1.1

Vapor Pressure: 1 (<1 mm Hg)

Water Solubility: <1 %

Chemical Incompatibilities or Instabilities: Oxidizers, lead (IV) oxide.

FLAMMABILITY

NFPA Hazard Code: 3-2-0

Flash Point: 88°C (190°F)

Autoignition Temp: 174°C (345°F)

NFPA Classification: Class IIIA Combustible

LEL: N.D.

UEL: N.D.

Fire Extinguishing Methods: Carbon Dioxide, Dry Chemical, Foam, Water Spray or Fog.

Special Fire Fighting Considerations: Water spray should be used to keep closed containers cool. Continue to cool container after fire is extinguished. Treat all materials used or generated and equipment involved as contaminated by hazardous waste.

TOXICOLOGY
CARCINOGEN

Odor: Aromatic

Physical Contact: Material is extremely destructive to human tissues.

TLV = 20 mg/m^3 **STEL** = 45 mg/m^3 (skin)

Odor Threshold: N.D.

IDLH = 1325 mg/m^3

RTECS # MW8925000

Routes of Entry and Relative LD$_{50}$ (or LC$_{50}$)

Inhalation	N.D.	
Ingestion	7	(188 mg/kg)
Skin Absorption	N.D.	

Symptoms of Exposure: Possible burns or irritation to skin, eyes and upper respiratory tract; headache; nausea. Inhalation of vapors may be fatal by causing glottis to spasm and suffocation. Exposure may result in chemical pneumonitis or pulmonary edema.

Emergency/First Aid Treatment: Remove to ventilated area; immediately remove any contaminated clothing and wash contaminated areas for 15 minutes using water. Treat supportively and observe for possible shock. If ingested, seek immediate medical aid.

Recommended Clean-Up Procedures

Personal Protection: Level A Ensemble

RCRA Waste #   None

Recommended Material   N.D.

Reportable Quantities: None

Spills: Dike to contain spill and collect liquid for later disposal. Take up in non-combustible absorbent and dispose. Treat all materials used or generated and equipment involved as contaminated by hazardous waste.

Special Emergency Information: May be fatal in inhaled, ingested or absorbed through the skin.

PHENYL ISOCYANATE

Synonyms:

CAS Number: 103-71-9	**Description:** Colorless Liquid
DOT Number: 2487	**DOT Classification:** Poison B

Molecular Weight: 119.1

Melting Point:	-30°C (-22°F)	**Vapor Density:** N.A.	**Vapor Pressure:** 2 (2 mm Hg)
Boiling Point:	162°C (324°F)	**Specific Gravity:** 1.1	**Water Solubility:** Decomposes

Chemical Incompatibilities or Instabilities: Alcohols, water, amines, strong oxidizers, heat.

FLAMMABILITY

NFPA Hazard Code:	**NFPA Classification:**
Flash Point: 55°C (132°F)	**LEL:** N.D.
Autoignition Temp: N.D.	**UEL:** N.D.

Fire Extinguishing Methods: Carbon Dioxide, Dry Chemical, Foam, Water Spray or Fog.

Special Fire Fighting Considerations: Structural fire fighter protective clothing will not provide adequate protection. Fight fire from a distance or protected location, if possible. Treat all materials used or generated and equipment involved as contaminated by hazardous waste.

DOT Recommended Isolation Zones:	Small Spill: 150 ft	Large Spill: 150 ft
DOT Recommended Down Wind Take Cover Distance	Small Spill: 0.2 miles	Large Spill: 0.2 miles

TOXICOLOGY

Odor:

Physical Contact: Material is extremely destructive to human tissues.

TLV = N.D.   **STEL** = N.D.

Odor Threshold: N.D.

IDLH = N.D.

Routes of Entry and Relative LD$_{50}$ (or LC$_{50}$)

Inhalation	N.D.	
Ingestion	4	(940 mg/kg)
Skin Absorption	1	(7130 mg/kg)

RTECS # DA3675000

Symptoms of Exposure: Possible burns or irritation to skin, eyes and upper respiratory tract; headache; nausea. May cause an allergic reaction or asthma. Exposure may result in chemical pneumonitis or pulmonary edema.

Emergency/First Aid Treatment: Remove to ventilated area; immediately remove any contaminated clothing and wash contaminated areas for 15 minutes using water. Treat supportively and observe for possible shock. If ingested, seek immediate medical aid.

Recommended Clean-Up Procedures

Personal Protection: Level B Ensemble	**Recommended Material** N.D.
RCRA Waste # None	**Reportable Quantities:** None

Spills: Dike to contain spill and collect liquid for later disposal. Decontaminate spill area using water. Treat all materials used or generated and equipment involved as contaminated by hazardous waste.

Special Emergency Information: May be fatal in inhaled, ingested or absorbed through the skin. Vapors are heavier than air and may travel some distance to an ignition source. Vapors are more dense than air and may settle in low lying areas.

PHENYL MERCAPTAN

Synonyms: thiophenol

CAS Number: 108-98-5
DOT Number: 2337

Description: Colorless Liquid
DOT Classification: Poison B

Molecular Weight: 110.2
Melting Point: -15°C (5°F)
Boiling Point: 168°C (334°F)

Vapor Density: 3.8
Specific Gravity: 1.1

Vapor Pressure: 2 (2 mm Hg)
Water Solubility: Soluble

Chemical Incompatibilities or Instabilities: Strong oxidizers.

FLAMMABILITY

NFPA Hazard Code:
Flash Point: 50°C (123°F)
Autoignition Temp: N.D.

NFPA Classification:
LEL: N.D.
UEL: N.D.

Fire Extinguishing Methods: Carbon Dioxide, Dry Chemical, Foam, Water Spray or Fog.

Special Fire Fighting Considerations: Structural fire fighter protective clothing will not provide adequate protection. Water spray from an unmanned device should be used to keep closed containers cool. Continue to cool container after fire is extinguished. Fight fire from a distance or protected location, if possible. Treat all materials used or generated and equipment involved as contaminated by hazardous waste.

DOT Recommended Isolation Zones:	Small Spill: 150 ft	Large Spill: 150 ft
DOT Recommended Down Wind Take Cover Distance	Small Spill: 0.2 miles	Large Spill: 0.2 miles

TOXICOLOGY

Odor: Repulsive
Physical Contact: Material is extremely destructive to human tissues.
TLV = 2.3 mg/m^3 **STEL** = N.D.

Odor Threshold: 0.00015 mg/m^3

IDLH = N.D.

RTECS # DC0525000

Routes of Entry and Relative LD$_{50}$ (or LC$_{50}$)
Inhalation   N.D.
Ingestion   9   (46 mg/kg)
Skin Absorption   N.D.

Symptoms of Exposure: Possible burns or irritation to skin, eyes and upper respiratory tract; headache; nausea. Inhalation of vapors may be fatal by causing glottis to spasm and suffocation. Exposure may result in chemical pneumonitis or pulmonary edema.

Emergency/First Aid Treatment: Remove to ventilated area; immediately remove any contaminated clothing and wash contaminated areas for 15 minutes using water. Treat supportively and observe for possible shock. If ingested, seek immediate medical aid.

Recommended Clean-Up Procedures

Personal Protection: Level B Ensemble   **Recommended Material** N.D.

RCRA Waste # P014   **Reportable Quantities:** 100 lbs (1 lb)

Spills: Remove all potential ignition sources. Dike to contain spill, absorb with noncombustible absorbent and take up using nonsparking tools. Treat all materials used or generated and equipment involved as contaminated by hazardous waste.

Special Emergency Information: May be fatal in inhaled, ingested or absorbed through the skin. Vapors are heavier than air and may travel some distance to an ignition source. Vapors are more dense than air and may settle in low lying areas.

PHOSGENE - COCl$_2$

Synonyms: carbon oxychloride; carbonyl chloride

CAS Number: 75-44-5   **Description:** Colorless Gas
DOT Number: 1076   **DOT Classification:** Poison A

Molecular Weight: 98.9
Melting Point: -118°C (-180°F)   **Vapor Density:**   3.4   **Vapor Pressure:**   Gas
Boiling Point: 8°C   (46°F)   **Specific Gravity:** 1.4   **Water Solubility:**   <1%

Chemical Incompatibilities or Instabilities: Aluminum, alcohols, alkali metals, ammonia, alkali, azides.

FLAMMABILITY

NFPA Hazard Code: 4-0-1   **NFPA Classification:**
Flash Point: N.A.   **LEL:** N.A.
Autoignition Temp: N.A.   **UEL:** N.A.

Fire Extinguishing Methods: Carbon Dioxide, Dry Chemical, Foam, Water Spray or Fog.

Special Fire Fighting Considerations: Structural fire fighter protective clothing will not provide adequate protection. Water spray should be used to keep closed containers cool. Continue to cool container after fire is extinguished.

DOT Recommended Isolation Zones: Small Spill: 1500 ft   Large Spill: 1500 ft
DOT Recommended Down Wind Take Cover Distance   Small Spill: 5 miles   Large Spill: 5 miles

TOXICOLOGY

Odor: New mown hay   **Odor Threshold:** 1 mg/m^3
Physical Contact: Material is extremely destructive to human tissues.
TLV = 0.4 mg/m^3 **STEL** = 0.8 mg/m^3 **IDLH** = 4 mg/m^3

Routes of Entry and Relative LD$_{50}$ (or LC$_{50}$)   **RTECS # SY5600000**
    Inhalation   N.D.
    Ingestion   N.D.
    Skin Absorption   N.D.

Symptoms of Exposure: Possible burns or irritation to skin, eyes and upper respiratory tract; headache; nausea. Symptoms may be delayed by several hours. Exposure may result in chemical pneumonitis or pulmonary edema.

Emergency/First Aid Treatment: Remove to ventilated area; immediately remove any contaminated clothing and wash contaminated areas for 15 minutes using water. Treat supportively and observe for possible shock. If ingested, seek immediate medical aid.

Recommended Clean-Up Procedures

Personal Protection: Level A Ensemble   **Recommended Material**   N.D.

RCRA Waste #   P095   **Reportable Quantities:**   10 lbs (5000 lbs)

Spills: Stop leak if it can safely be done. Keep area isolated until vapors dissipate.

Special Emergency Information: May be fatal in inhaled, ingested or absorbed through the skin. Vapors are more dense than air and may settle in low lying areas.

PHOSPHINE - PH₃

Synonyms: Hydrogen Phosphide

CAS Number: 7803-51-2
DOT Number: 2199

Description: Colorless Gas
DOT Classification: Poison A

Molecular Weight: 34.0
Melting Point: -133°C (-207°F)
Boiling Point: -88°C (-126°F)

Vapor Density: 1.2
Specific Gravity: 0.76

Vapor Pressure: Gas
Water Solubility: Insoluble

Chemical Incompatibilities or Instabilities: Oxidizers, air, halogens, oxides of nitrogen, metal nitrates.

FLAMMABILITY

NFPA Hazard Code: 3-4-2
Flash Point: N.A.
Autoignition Temp: N.D.

NFPA Classification:
LEL: 1.6%
UEL: 98.0%

Fire Extinguishing Methods: Foam, Water Spray or Fog.

Special Fire Fighting Considerations: Structural fire fighter protective clothing will not provide adequate protection. Isolate for 1/2 mile if rail or tank truck is involved in a fire. Water spray should be used to keep closed containers cool. Continue to cool container after fire is extinguished. For large fires, if possible, withdraw and allow to burn. Immediately withdraw if rising sound from venting device is heard or if fire is causing discoloration to the tank. let fire burn unless leak can safely be stopped.

DOT Recommended Isolation Zones:	Small Spill: 1500 ft	Large Spill: 1500 ft
DOT Recommended Down Wind Take Cover Distance	Small Spill: 5 miles	Large Spill: 5 miles

TOXICOLOGY

Odor: Garlic
Physical Contact: Material is extremely destructive to human tissues.
TLV = 0.4 mg/m³   **STEL** = 1 mg/m³

Odor Threshold: 0.7 mg/m³

IDLH = 280 mg/m³

Routes of Entry and Relative LD₅₀ (or LC₅₀)

Inhalation	10	(60 mg/m³/H)
Ingestion	N.D.	
Skin Absorption	N.D.	

RTECS # SY7525000

Symptoms of Exposure: Possible burns or irritation to skin, eyes and upper respiratory tract; headache; nausea. Exposure may result in chemical pneumonitis or pulmonary edema.

Emergency/First Aid Treatment: Remove to ventilated area; immediately remove any contaminated clothing and wash contaminated areas for 15 minutes using water. Treat supportively and observe for possible shock. If ingested, seek immediate medical aid.

Recommended Clean-Up Procedures

Personal Protection: Level A Ensemble
RCRA Waste # P096

Recommended Material N.D.
Reportable Quantities: 100 lbs (1 lb)

Spills: Stop leak if it can safely be done. Keep area isolated until vapors dissipate.

Special Emergency Information: May be fatal in inhaled, ingested or absorbed through the skin. Vapors are heavier than air and may travel some distance to an ignition source. Vapors are more dense than air and may settle in low lying areas.

PHOSPHORIC ACID -H_3PO_4

Synonyms: orthophosphoric acid

CAS Number: 7664-38-2   **Description:** Colorless Liquid
DOT Number: 1805   **DOT Classification:** Corrosive Material

Molecular Weight: 98.0
Melting Point: 42°C (108°F)   **Vapor Density:** N.D.   **Vapor Pressure:**   1 (<1 mm Hg)
Boiling Point: 213°C (415°F)   **Specific Gravity:** 1.89   **Water Solubility:**   Soluble

Chemical Incompatibilities or Instabilities: Strong alkali, finely divided metals, sodium tetraborate, chlorides and stainless steel (may generate hydrogen gas).

FLAMMABILITY

NFPA Hazard Code: 3-0-0   **NFPA Classification:**
Flash Point: N.A.   **LEL:** N.A.
Autoignition Temp: N.A.   **UEL:** N.A.

Fire Extinguishing Methods: Use agent suitable for surrounding fire.

Special Fire Fighting Considerations: Water spray should be used to keep closed containers cool. Continue to cool container after fire is extinguished.

TOXICOLOGY

Odor: None   **Odor Threshold:** N.D..
Physical Contact: Material is extremely destructive to human tissues.
TLV = 1 mg/m^3 **STEL** = 3 mg/m^3 **IDLH** = 10,000 mg/m^3

Routes of Entry and Relative LD$_{50}$ (or LC$_{50}$)   **RTECS # TB6300000**
    Inhalation   N.D.
    Ingestion   3   (1530 mg/kg)
    Skin Absorption   2   (2740 mg/kg)

Symptoms of Exposure: Possible burns or irritation to skin, eyes and upper respiratory tract; headache; nausea. Inhalation of vapors may be fatal by causing glottis to spasm and suffocation. Exposure may result in chemical pneumonitis or pulmonary edema.

Emergency/First Aid Treatment: Remove to ventilated area; immediately remove any contaminated clothing and wash contaminated areas for 15 minutes using water. Treat supportively and observe for possible shock. If ingested, seek immediate medical aid.

Recommended Clean-Up Procedures

Personal Protection: Level B Ensemble   **Recommended Material**   Natural Rubber (+), Neoprene (+), Nitrile Rubber (+), Polyvinyl chloride (+)

RCRA Waste #   None   **Reportable Quantities:**   None

Spills: Dike to contain spill and collect liquid for later disposal. Decontaminate spill area using dilute alkaline solution.

Special Emergency Information: May be harmful if inhaled, swallowed or absorbed through the skin.

PHOSPHORUS (red)

Synonyms: phosphorus, amorphous

CAS Number: 7723-14-0
DOT Number: 1338

Description: Red-brown powder
DOT Classification: Flammable Solid

Molecular Weight: 31.0
Melting Point: °C (°F)
Boiling Point: 416°C (781°F)

Vapor Density: 4.8
Specific Gravity: 2.3

Vapor Pressure: 1 (<1 mm Hg)
Water Solubility: Insoluble

Chemical Incompatibilities or Instabilities: Halogens, halides, copper and its alloys, sulfur, reducing agents, oxidizers, strong acids, metal oxides, non-metal oxides, non-metal hallides, metal sulfates.

FLAMMABILITY

NFPA Hazard Code: 1-1-1
Flash Point: N.D.
Autoignition Temp: 260°C (500°F)

NFPA Classification:
LEL: N.D.
UEL: N.D.

Fire Extinguishing Methods: Carbon Dioxide, Dry Chemical, Foam, Water Spray or Fog.

Special Fire Fighting Considerations: Water spray should be used to keep closed containers cool. Continue to cool container after fire is extinguished. For large fires, if possible, withdraw and allow to burn.

TOXICOLOGY

Odor: None
Physical Contact: Material is extremely destructive to human tissues.
TLV = N.D.   **STEL** = N.D.

Odor Threshold: N.D.

IDLH = N.D.

Routes of Entry and Relative LD$_{50}$ (or LC$_{50}$)
  Inhalation   N.D.
  Ingestion   N.D.
  Skin Absorption   N.D.

RTECS # TH3495000

Symptoms of Exposure: Possible burns or irritation to skin, eyes and upper respiratory tract; headache; nausea.

Emergency/First Aid Treatment: Remove to ventilated area; immediately remove any contaminated clothing and wash contaminated areas for 15 minutes using water. Treat supportively and observe for possible shock. If ingested, seek immediate medical aid.

Recommended Clean-Up Procedures

Personal Protection: Level B Ensemble

RCRA Waste # None

Recommended Material N.D.

Reportable Quantities: 1 lb (1 lb)

Spills: Take up in non-combustible absorbent and dispose.

Special Emergency Information: May be fatal in inhaled, ingested or absorbed through the skin.

PHOSPHORUS (yellow or white) - P$_4$

Synonyms:

CAS Number: 7723-14-0   **Description:** White to yellow solid
DOT Number: 1381, 2447 (molten)   **DOT Classification:** Flammable solid

Molecular Weight: 123.9
Melting Point: 44°C (111°F)   **Vapor Density:** 4.4   **Vapor Pressure:** 1 (<1 mm Hg)
Boiling Point: 280°C (536°F)   **Specific Gravity:** 1.8 **Water Solubility:** Insoluble

Chemical Incompatibilities or Instabilities: Air, halogens, halides, sulfur, oxidizers, metals, alkaline hydroxides (yields phosphine gas).

FLAMMABILITY

NFPA Hazard Code: 3-4-2   **NFPA Classification:**
Flash Point: N.D.   **LEL:** N.D.
Autoignition Temp: 30°C (86°F)   **UEL:** N.D.

Fire Extinguishing Methods: Dry chemical, Sand, Water spray or fog.

Special Fire Fighting Considerations: Water spray should be used to keep closed containers cool. Continue to cool container after fire is extinguished. For large fires, if possible, withdraw and allow to burn. Do not spread material.

TOXICOLOGY

Odor: None   **Odor Threshold:** N.D.
Physical Contact: Material is extremely destructive to human tissues.
TLV = 0.1 mg/m^3 **STEL** = N.D.   **IDLH** = N.D.

Routes of Entry and Relative LD$_{50}$ (or LC$_{50}$)   **RTECS # TH3500000**
   Inhalation   N.D.
   Ingestion   **10**   (3 mg/kg)
   Skin Absorption   N.D.

Symptoms of Exposure: Possible burns or irritation to skin, eyes and upper respiratory tract; headache; nausea.

Emergency/First Aid Treatment: Remove to ventilated area; immediately remove any contaminated clothing and wash contaminated areas for 15 minutes using water. Treat supportively and observe for possible shock. If ingested, seek immediate medical aid.

Recommended Clean-Up Procedures

Personal Protection: Level A Ensemble   **Recommended Material**   N.D.

RCRA Waste #   None   **Reportable Quantities:**   1 lb (1 lb)

Spills: Take up in non-combustible absorbent and dispose.

Special Emergency Information: May be fatal in inhaled, ingested or absorbed through the skin. Material is pyrophoric and may spontaneously ignite in air.

PHOSPHORUS OXYCHLORIDE - POCl₃

Synonyms: phosphorus oxytrichloride; phosphoryl chloride

CAS Number: 10025-87-3

DOT Number: 1810

Description: Colorless Liquid

DOT Classification: Corrosive Material

Molecular Weight: 153.3

Melting Point:	2°C (36°F)	**Vapor Density:**	5.3	**Vapor Pressure:**	5 (28 mm Hg)
Boiling Point:	105°C (223°F)	**Specific Gravity:**	1.7	**Water Solubility:**	Decomposes

Chemical Incompatibilities or Instabilities: Acids, alkalis, alcohols, alkali metals. May react with water to yield hydrochloric acid and phosgene which will ignite.

FLAMMABILITY

NFPA Hazard Code: 3-0-2-~~W~~

Flash Point: N.A.

Autoignition Temp: N.A.

NFPA Classification:

LEL: N.A.

UEL: N.A.

Fire Extinguishing Methods: Carbon Dioxide, Dry Chemical, Water Spray or Fog.

Special Fire Fighting Considerations: Structural fire fighter protective clothing will not provide adequate protection. Water spray should be used to keep closed containers cool. Continue to cool container after fire is extinguished. Use flooding quantities of water for large fires. May release toxic fumes upon heating (phosphine).

DOT Recommended Isolation Zones:		Small Spill: 600 ft		Large Spill: 900 ft
DOT Recommended Down Wind Take Cover Distance		Small Spill: 2 miles		Large Spill: 3 miles

TOXICOLOGY

Odor: Pungent

Odor Threshold: N.D.

Physical Contact: Material is extremely destructive to human tissues.

TLV = 0.6 mg/m³   **STEL** = N.D.

IDLH = N.D.

RTECS # TH4897000

Routes of Entry and Relative LD₅₀ (or LC₅₀)

Inhalation	10	(1200 mg/m³/H)
Ingestion	5	(380 mg/kg)
Skin Absorption	N.D.	

Symptoms of Exposure: Possible burns or irritation to skin, eyes and upper respiratory tract; headache; nausea. Inhalation of vapors may be fatal by causing glottis to spasm and suffocation. Exposure may result in chemical pneumonitis or pulmonary edema.

Emergency/First Aid Treatment: Remove to ventilated area; immediately remove any contaminated clothing and wash contaminated areas for 15 minutes using water. Treat supportively and observe for possible shock. If ingested, seek immediate medical aid.

Recommended Clean-Up Procedures

Personal Protection: Level A Ensemble

Recommended Material Teflon (0)

RCRA Waste # None

Reportable Quantities: 1000 lb (5000 lb)

Spills: Dike to contain spill and collect liquid for later disposal. Decontaminate spill area using flooding quantities of water. Keep area isolated until vapors have dissipated.

Special Emergency Information: May be fatal in inhaled, ingested or absorbed through the skin. Vapors are more dense than air and may settle in low lying areas.

PHOSPHORUS PENTACHLORIDE - PCl5

Synonyms: phosphorus perchloride

CAS Number: 10026-13-8	**Description:** Amber powder	
DOT Number: 1806	**DOT Classification:** Corrosive Material	

Molecular Weight: 208.2

Melting Point: °C (°F)	**Vapor Density:** N.D.	**Vapor Pressure:** 1 (< 1 mm Hg)	
Boiling Point: S 100°C (212°F)	**Specific Gravity:** 1.6	**Water Solubility:** Decomposes	

Chemical Incompatibilities or Instabilities: Halogens, oxidizers, hydrocarbons, reactive metals, water, carbamates.

FLAMMABILITY

NFPA Hazard Code: 3-0-2-~~W~~	**NFPA Classification:**	
Flash Point: N.A.	**LEL:** N.A.	
Autoignition Temp: N.A.	**UEL:** N.A.	

Fire Extinguishing Methods: Carbon Dioxide, Dry Chemical, Water Spray or Fog.

Special Fire Fighting Considerations: Structural fire fighter protective clothing will not provide adequate protection. Water spray should be used to keep closed containers cool. Continue to cool container after fire is extinguished. Use flooding quantities of water for large fires.

TOXICOLOGY

Odor: Pungent

Odor Threshold: N.D.

Physical Contact: Material is extremely destructive to human tissues.

$TLV = 1$ mg/m^3 $STEL = $ N.D. $IDLH = 200$ mg/m^3

Routes of Entry and Relative LD$_{50}$ (or LC$_{50}$)

Inhalation	9	(1750 mg/m^3/H)
Ingestion	4	(660 mg/kg)
Skin Absorption	N.D.	

RTECS # TB6125000

Symptoms of Exposure: Possible burns or irritation to skin, eyes and upper respiratory tract; headache; nausea. Inhalation of vapors may be fatal by causing glottis to spasm and suffocation. Exposure may result in chemical pneumonitis or pulmonary edema.

Emergency/First Aid Treatment: Remove to ventilated area; immediately remove any contaminated clothing and wash contaminated areas for 15 minutes using water. Treat supportively and observe for possible shock. If ingested, seek immediate medical aid.

Recommended Clean-Up Procedures

Personal Protection: Level A Ensemble	**Recommended Material** N.D.
RCRA Waste # None	**Reportable Quantities:** None

Spills: Dike to contain spill and collect liquid for later disposal. Decontaminate spill area using flooding quantities of water.

Special Emergency Information: May be fatal in inhaled, ingested or absorbed through the skin. Vapors are more dense than air and may settle in low lying areas.

PHOSPHORUS PENTASULFIDE - P$_4$S$_{10}$

Synonyms: sulfur phosphide; thiophosphoric andydride

CAS Number: 1314-80-3   **Description:** Yellow-green powder
DOT Number: 1340   **DOT Classification:** Flammable Solid

Molecular Weight: 222.2
Melting Point: 287°C (527°F)   **Vapor Density:** N.D.   **Vapor Pressure:** 1 (< 1 mm Hg)
Boiling Point: 514°C (995°F)   **Specific Gravity:** 2.1   **Water Solubility:** Decomposes

Chemical Incompatibilities or Instabilities: Alkali metals, alcohols, amines, aluminum, water, oxidizers.

FLAMMABILITY

NFPA Hazard Code: **NFPA Classification:**
Flash Point: N.A.   **LEL:** N.A.
Autoignition Temp: 142°C (287°F)   **UEL:** N.A.

Fire Extinguishing Methods: Dry chemical, Soda ash, Sand, Lime.

Special Fire Fighting Considerations: For large fires, withdraw and allow to burn. **DO NOT USE WATER.**

TOXICOLOGY

Odor: Rotten eggs   **Odor Threshold:** N.D.
Physical Contact: Material is extremely destructive to human tissues.
TLV = 1 mg/m^3 **STEL** = 3 mg/m^3 **IDLH** = 750 mg/m^3

Routes of Entry and Relative LD$_{50}$ (or LC$_{50}$)   **RTECS #**
    Inhalation   N.D.
    Ingestion   **5**   (389 mg/kg)
    Skin Absorption   N.D.

Symptoms of Exposure: Possible burns or irritation to skin, eyes and upper respiratory tract; headache; nausea. Inhalation of vapors may be fatal by causing glottis to spasm and suffocation. Exposure may result in chemical pneumonitis or pulmonary edema.

Emergency/First Aid Treatment: Remove to ventilated area; immediately remove any contaminated clothing and wash contaminated areas for 15 minutes using water. Treat supportively and observe for possible shock. If ingested, seek immediate medical aid.

Recommended Clean-Up Procedures

Personal Protection: Level B Ensemble   **Recommended Material**   N.D.

RCRA Waste #   U189   **Reportable Quantities:**   100 lb (100 lb)

Spills: Take up in non-combustible absorbent and dispose.

Special Emergency Information: May be fatal in inhaled, ingested or absorbed through the skin.

PHOSPHORUS PENTOXIDE - P_2O_5

Synonyms: phosphoric anhydride

CAS Number: 1314-56-3
DOT Number: 1807

Description: White powder
DOT Classification: Corrosive Material

Molecular Weight: 141.9
Melting Point: 340°C (644°F)
Boiling Point: S 360°C (680°F)

Vapor Density: 4.9
Specific Gravity: 2.3

Vapor Pressure: 1 (< 1 mm Hg)
Water Solubility: Decomposes

Chemical Incompatibilities or Instabilities: Alkali metals, aluminum, calcium, hydrogen fluoride, oxidizers, iodides, sulfides, formic acid, sodium carbonate, unsaturated hydrocarbons, calcium oxide.

FLAMMABILITY

NFPA Hazard Code:
Flash Point: N.A.
Autoignition Temp: N.A.

NFPA Classification:
LEL: N.A.
UEL: N.A.

Fire Extinguishing Methods: Dry chemical, Carbon dioxide.

Special Fire Fighting Considerations: Use flooding quantities of water for large fires. Structural fire fighter protective clothing will not provide adequate protection. Water spray should be used to keep closed containers cool. Continue to cool container after fire is extinguished.

TOXICOLOGY

Odor:
Physical Contact: Material is extremely destructive to human tissues.
TLV = N.D.   **STEL** = N.D.

Odor Threshold: N.D.

IDLH = N.D.

Routes of Entry and Relative LD$_{50}$ (or LC$_{50}$)
Inhalation   9   (1217 mg/m^3/H)
Ingestion   N.D.
Skin Absorption   N.D.

RTECS # TH3945000

Symptoms of Exposure: Possible burns or irritation to skin, eyes and upper respiratory tract; headache; nausea. Inhalation of vapors may be fatal by causing glottis to spasm and suffocation. Exposure may result in chemical pneumonitis or pulmonary edema.

Emergency/First Aid Treatment: Remove to ventilated area; immediately remove any contaminated clothing and wash contaminated areas for 15 minutes using water. Treat supportively and observe for possible shock. If ingested, seek immediate medical aid.

Recommended Clean-Up Procedures

Personal Protection: Level A Ensemble   **Recommended Material**   N.D.

RCRA Waste #   None   **Reportable Quantities:**   None

Spills: Take up in non-combustible absorbent and dispose. Decontaminate spill area using flooding quantities of water.

Special Emergency Information: May be fatal in inhaled, ingested or absorbed through the skin.

PHOSPHORUS TRICHLORIDE - PCl₃

Synonyms: phosphorus chloride

CAS Number: 7719-12-2   **Description:** Colorless, fuming liquid
DOT Number: 1809   **DOT Classification:** Corrosive Material

Molecular Weight: 137.3
Melting Point: -112°C (-169°F)   **Vapor Density:** 4.8   **Vapor Pressure:** 6 (100 mm Hg)
Boiling Point: 76°C (168°F)   **Specific Gravity:** 1.6   **Water Solubility:** Decomposes

Chemical Incompatibilities or Instabilities: Alkali metals, amines, ammonia, aluminum, oxidizers, hydroxylamine, unsaturated hydrocarbons, sulfur acids, lead (IV) oxide, halogens, water. Reacts violently with water to yield flammable gases that will ignite.

FLAMMABILITY

NFPA Hazard Code: 3-0-2-~~W~~   **NFPA Classification:**
Flash Point: N.A.   **LEL:** N.A.
Autoignition Temp: N.A.   **UEL:** N.A.

Fire Extinguishing Methods: Dry chemical, Carbon dioxide.

Special Fire Fighting Considerations: For large fires use flooding quantities of water. Structural fire fighter protective clothing will not provide adequate protection. Water spray should be used to keep closed containers cool. Continue to cool container after fire is extinguished. May release toxic fumes upon heating (phosgene).

DOT Recommended Isolation Zones: Small Spill: 900 ft   Large Spill: 1500 ft
DOT Recommended Down Wind Take Cover Distance   Small Spill: 3 miles   Large Spill: 5 miles

TOXICOLOGY

Odor: Sharp, pungent   **Odor Threshold:** N.D.
Physical Contact: Material is extremely destructive to human tissues.
TLV = 1.5 mg/m³   **STEL** = 3 mg/m³   **IDLH** = 285 mg/m³

Routes of Entry and Relative LD₅₀ (or LC₅₀)   **RTECS # TH3675000**
   Inhalation   9   (2400 mg/m³/H)
   Ingestion   4   (550 mg/kg)
   Skin Absorption   N.D.

Symptoms of Exposure: Possible burns or irritation to skin, eyes and upper respiratory tract; headache; nausea; lachrymation. Inhalation of vapors may be fatal by causing glottis to spasm and suffocation. Exposure may result in chemical pneumonitis or pulmonary edema.

Emergency/First Aid Treatment: Remove to ventilated area; immediately remove any contaminated clothing and wash contaminated areas for 15 minutes using water. Treat supportively and observe for possible shock. If ingested, seek immediate medical aid.

Recommended Clean-Up Procedures

Personal Protection: Level B Ensemble   **Recommended Material** Teflon (-)

RCRA Waste # None   **Reportable Quantities:** 1000 lb (5000 lb)

Spills: Clean up only under expert supervision.

Special Emergency Information: May be fatal in inhaled, ingested or absorbed through the skin. Vapors are more dense than air and may settle in low lying areas.

Phosphorus tribromide (CAS # 7789-60-8, DOT # 1808, RTECS # TH4460000) has similar chemical, physical and toxicological properties.

PHTHALIC ANHYDRIDE

Synonyms: 1,2-benzenedicarboxylic acid anhydride; 1,3-isobenzofurandione

CAS Number: 85-44-9   **Description:** White powder
DOT Number: 2214   **DOT Classification:** Corrosive Material

Molecular Weight: 148.1
Melting Point: 131°C (268°F)   **Vapor Density:**   5.1   **Vapor Pressure:**   1 (<1 mm Hg)
Boiling Point: 285°C (544°F)   **Specific Gravity:** 1.5   **Water Solubility:**   Decomposes

Chemical Incompatibilities or Instabilities: Strong oxidizers, nitric acids, sodium nitrate, copper oxide, sulfuric acid.

FLAMMABILITY

NFPA Hazard Code: 2-1-0   **NFPA Classification:**
Flash Point: 152°C (305°F)   **LEL:**   1.7%
Autoignition Temp: 570°C (1058°F)   **UEL:**   10.4%

Fire Extinguishing Methods: Carbon Dioxide, Dry Chemical, Foam, Water Spray or Fog.

Special Fire Fighting Considerations: Water spray should be used to keep closed containers cool. Continue to cool container after fire is extinguished. Treat all materials used or generated and equipment involved as contaminated by hazardous waste.

TOXICOLOGY

Odor: Choking   **Odor Threshold:** 0.3 mg/m^3
Physical Contact: Irritant
TLV = 6 mg/m^3 **STEL** = N.D.   **IDLH** = 10,000 mg/m^3

Routes of Entry and Relative LD$_{50}$ (or LC$_{50}$)   **RTECS #** TI3150000
    Inhalation   N.D.
    Ingestion   **3**   (4020 mg/kg)
    Skin Absorption   N.D.

Symptoms of Exposure: Possible irritation to skin, eyes and upper respiratory tract; headache; nausea. May cause an allergic reaction or asthma.

Emergency/First Aid Treatment: Remove to ventilated area; immediately remove any contaminated clothing and wash contaminated areas for 15 minutes using water. Treat supportively and observe for possible shock. If ingested, seek immediate medical aid.

Recommended Clean-Up Procedures

Personal Protection: Level B Ensemble   **Recommended Material**   N.D.

RCRA Waste #   U190   **Reportable Quantities:**   5000 lbs (1 lb)

Spills: Take up in non-combustible absorbent and dispose. Decontaminate spill area using dilute alkaline solution. Treat all materials used or generated and equipment involved as contaminated by hazardous waste.

Special Emergency Information: May be harmful if inhaled, swallowed or absorbed through the skin.

3-PICOLINE

Synonyms: 3-methylpyridine; m-picoline

CAS Number: 108-99-6	**Description:** Colorless Liquid	
DOT Number: 2313	**DOT Classification:** Flammable Liquid	

Molecular Weight: 93.1

Melting Point: °C (°F)	**Vapor Density:** 3.2	**Vapor Pressure:** 2 (4 mm Hg)
Boiling Point: 144°C (291°F)	**Specific Gravity:** 0.9	**Water Solubility:** Soluble

Chemical Incompatibilities or Instabilities: Strong oxidizers.

FLAMMABILITY

NFPA Hazard Code: 2-2-0	**NFPA Classification:**	
Flash Point: 36°C (97°F)	**LEL:** N.D.	
Autoignition Temp: N.D.	**UEL:** N.D.	

Fire Extinguishing Methods: Carbon Dioxide, Dry Chemical, Foam, Water Spray or Fog.

Special Fire Fighting Considerations: Isolate for 1/2 mile if rail or tank truck is involved in a fire. Water spray should be used to keep closed containers cool. Continue to cool container after fire is extinguished. For large fires, if possible, withdraw and allow to burn. Immediately withdraw if rising sound from venting device is heard or if fire is causing discoloration to the tank.

TOXICOLOGY

Odor: Sweet　　　　　　　　　　　　　　　　　　　　　　**Odor Threshold:** N.D.

Physical Contact: Material is extremely destructive to human tissues.

TLV = N.D.	**STEL** = N.D.	**IDLH** = N.D.

Routes of Entry and Relative LD$_{50}$ (or LC$_{50}$)　　　　**RTECS # TJ5000000**

Inhalation	N.D.
Ingestion	N.D.
Skin Absorption	N.D.

Symptoms of Exposure: Possible irritation to skin, eyes and upper respiratory tract; headache; nausea. Inhalation of vapors may be fatal by causing glottis to spasm and suffocation. Exposure may result in chemical pneumonitis or pulmonary edema.

Emergency/First Aid Treatment: Remove to ventilated area; immediately remove any contaminated clothing and wash contaminated areas for 15 minutes using water. Treat supportively and observe for possible shock. If ingested, seek immediate medical aid.

Recommended Clean-Up Procedures

Personal Protection: Level B Ensemble	**Recommended Material** N.D.	
RCRA Waste # None	**Reportable Quantities:** None	

Spills: Remove all potential ignition sources. Dike to contain spill, absorb with noncombustible absorbent and take up using nonsparking tools. Decontaminate spill area using soapy water.

Special Emergency Information: May be harmful if inhaled, swallowed or absorbed through the skin. Vapors are heavier than air and may travel some distance to an ignition source. Vapors are more dense than air and may settle in low lying areas.

2-picoline (CAS # 109-06-8, RTECS # TJ4900000, RCRA Waste # U191, RQ: 5000 lb, [1 lb]) and
4-picoline (CAS # 108-89-4, RTECS # UT5425000) have similar chemical, physical and toxicological properties.

PICRIC ACID

Synonyms: 1,3,5-trinitrophenol

CAS Number: 88-89-1   **Description:** Yellow, moist powder
DOT Number: 0154, 1344, 1336   **DOT Classification:** Class A Explosive (0154) or Flammable Solid

Molecular Weight: 229.1
Melting Point: 122°C (252°F)   **Vapor Density:** 7.9   **Vapor Pressure:**   1 (<1 mm Hg)
Boiling Point: Explodes   **Specific Gravity:** 1.8   **Water Solubility:**   Insoluble

Chemical Incompatibilities or Instabilities: Finely divided metals, calcium, aluminum, ammonia, amines, heat, reducing agents, concrete.

FLAMMABILITY

NFPA Hazard Code: 3-4-4   **NFPA Classification:**
Flash Point: 150°C (302°F)   **LEL:** N.D.
Autoignition Temp: 300°C (572°F)   **UEL:** N.D.

Fire Extinguishing Methods: Flooding quantities of water.

Special Fire Fighting Considerations: For large fires, if possible, withdraw and allow to burn. Water spray from unmanned devices should be used to keep closed containers cool. Continue to cool container after fire is extinguished.

TOXICOLOGY

Odor: None   **Odor Threshold:** N.D.
Physical Contact: Irritant
TLV = 0.1 mg/m^3 (skin)   **STEL** = N.D.   **IDLH** = 100 mg/m^3

Routes of Entry and Relative LD$_{50}$ (or LC$_{50}$)   **RTECS # TJ8750000**
   Inhalation   N.D.
   Ingestion   N.D.
   Skin Absorption   N.D.

Symptoms of Exposure: Possible irritation to skin, eyes and upper respiratory tract; headache; nausea.

Emergency/First Aid Treatment: Remove to ventilated area; immediately remove any contaminated clothing and wash contaminated areas for 15 minutes using water. Treat supportively and observe for possible shock. If ingested, seek immediate medical aid.

Recommended Clean-Up Procedures

Personal Protection: Level B Ensemble   **Recommended Material**   N.D.

RCRA Waste #   None   **Reportable Quantities:**   None

Spills: Clean up only under expert supervision.

Special Emergency Information: May be harmful if inhaled, swallowed or absorbed through the skin. Treat as an explosion hazard.

PINDONE

Synonyms: tert-butyl valone; 1,3-dioxo-2-pivaloy-lindane; pivalyl valone

CAS Number: 83-26-1	**Description:** Yellow powder	
DOT Number: 2472	**DOT Classification:** Poison B	

Molecular Weight: 230.3

Melting Point: 110°C (230°F)	**Vapor Density:** N.A.	**Vapor Pressure:** 1 (<1 mm Hg)
Boiling Point: °C (°F)	**Specific Gravity:** 1.1	**Water Solubility:** 0.002%

Chemical Incompatibilities or Instabilities: Strong oxidizers.

FLAMMABILITY

NFPA Hazard Code:	**NFPA Classification:**
Flash Point: N.A.	**LEL:** N.A.
Autoignition Temp: N.A.	**UEL:** N.A.

Fire Extinguishing Methods: Use agent suitable for surrounding fire.

Special Fire Fighting Considerations: Structural fire fighter protective clothing will not provide adequate protection. Fight fire from a distance or protected location, if possible. Water spray from an unmanned device should be used to keep closed containers cool. Continue to cool container after fire is extinguished. Treat all materials used or generated and equipment involved as contaminated by hazardous waste.

TOXICOLOGY

Odor: None	**Odor Threshold:** N.D.

Physical Contact: Irritant

TLV = 0.1 mg/m^3	**STEL** = N.D.	**IDLH** = 200 mg/m^3

Routes of Entry and Relative LD$_{50}$ (or LC$_{50}$)   **RTECS # NK6300000**

Inhalation	N.D.	
Ingestion	6	(260 mg/kg)
Skin Absorption	N.D.	

Symptoms of Exposure: Possible irritation to skin, eyes and upper respiratory tract; headache; nausea; bruising; excessive bleeding from wounds.

Emergency/First Aid Treatment: Remove to ventilated area; immediately remove any contaminated clothing and wash contaminated areas for 15 minutes using water. Treat supportively and observe for possible shock. If ingested, seek immediate medical aid.

Recommended Clean-Up Procedures

Personal Protection: Level B Ensemble	**Recommended Material** N.D.	
RCRA Waste # None	**Reportable Quantities:** None	

Spills: Take up in non-combustible absorbent and dispose. Decontaminate spill area using soapy water. Treat all materials used or generated and equipment involved as contaminated by hazardous waste.

Special Emergency Information: May be harmful if inhaled, swallowed or absorbed through the skin. Pindone is extremely toxic via a mechanism similar to wargarin: reducing the blood's ability to clot.

2-PINENE

Synonyms: alpha-pinene

CAS Number: 80-56-8

DOT Number: 2368

Description: Colorless Liquid

DOT Classification: Flammable Liquid

Molecular Weight: 136.3

Melting Point:	-55°C (-67°F)	**Vapor Density:**	4.7	**Vapor Pressure:**	**3** (10 mm Hg)
Boiling Point:	156°C (312°F)	**Specific Gravity:**	0.9	**Water Solubility:**	Insoluble

Chemical Incompatibilities or Instabilities: Strong oxidizers.

FLAMMABILITY

NFPA Hazard Code: 1-3-0

Flash Point: 33°C (91°F)

Autoignition Temp: 255°C (491°F)

NFPA Classification: Class IC Flammable

LEL: N.D.

UEL: N.D.

Fire Extinguishing Methods: Alcohol Resistant Foam, Carbon Dioxide, Dry Chemical, Water Spray or Fog.

Special Fire Fighting Considerations: Isolate for 1/2 mile if rail or tank truck is involved in a fire. Water spray should be used to keep closed containers cool. Continue to cool container after fire is extinguished. For large fires, if possible, withdraw and allow to burn. Immediately withdraw if rising sound from venting device is heard or if fire is causing discoloration to the tank.

TOXICOLOGY

Odor: Turpentine

Physical Contact: Material is destructive to human tissues.

TLV = N.D.   **STEL** = N.D.

Odor Threshold: N.D.

IDLH = N.D.

Routes of Entry and Relative LD$_{50}$ (or LC$_{50}$)

Inhalation	N.D.	
Ingestion	**3**	(3700 mg/kg)
Skin Absorption	N.D.	

RTECS # DT7000000

Symptoms of Exposure: Possible irritation to skin, eyes and upper respiratory tract; headache; nausea.

Emergency/First Aid Treatment: Remove to ventilated area; immediately remove any contaminated clothing and wash contaminated areas for 15 minutes using water. Treat supportively and observe for possible shock. If ingested, seek immediate medical aid.

Recommended Clean-Up Procedures

Personal Protection: Level B Ensemble

RCRA Waste # None

Recommended Material N.D.

Reportable Quantities: None

Spills: Remove all potential ignition sources. Dike to contain spill, absorb with noncombustible absorbent and take up using nonsparking tools. Keep area isolated until vapors have dissipated.

Special Emergency Information: May be harmful if swallowed or absorbed through the skin and extremely toxic via inhalation. Vapors are heavier than air and may travel some distance to an ignition source. Vapors are more dense than air and may settle in low lying areas.

PIPERAZINE

Synonyms: 1,4-diethylenediamine; N,N-diethylene diamine

CAS Number: 110-85-0	**Description:** Moist, white crystals
DOT Number: 2579	**DOT Classification:** Corrosive Material

Molecular Weight: 86.2

Melting Point: 110°C (230°F)	**Vapor Density:** 3.0	**Vapor Pressure:** N.D.
Boiling Point: 146°C (294°F)	**Specific Gravity:** 1.1	**Water Solubility:** Soluble

Chemical Incompatibilities or Instabilities: Strong oxidizers, dicyanofurazan.

FLAMMABILITY

NFPA Hazard Code: 2-2-0	**NFPA Classification:** Class IIIA Combustible
Flash Point: 81°C (178°F)	**LEL:** N.D.
Autoignition Temp: °C (°F)	**UEL:** N.D.

Fire Extinguishing Methods: Alcohol Resistant Foam, Carbon Dioxide, Dry Chemical, Water Spray or Fog.

Special Fire Fighting Considerations: Water spray should be used to keep closed containers cool. Continue to cool container after fire is extinguished.

TOXICOLOGY

Odor: Pungent	**Odor Threshold:** N.D.

Physical Contact: Material is extremely destructive to human tissues.

TLV = N.D.	**STEL** = N.D.	**IDLH** = N.D.

Routes of Entry and Relative LD_{50} (or LC_{50})   **RTECS # TK7800000**

Inhalation	N.D.	
Ingestion	3	(1900 mg/kg)
Skin Absorption	2	(4000 mg/kg)

Symptoms of Exposure: Possible burns or irritation to skin, eyes and upper respiratory tract; headache; nausea. Inhalation of vapors may be fatal by causing glottis to spasm and suffocation. Exposure may result in chemical pneumonitis or pulmonary edema.

Emergency/First Aid Treatment: Remove to ventilated area; immediately remove any contaminated clothing and wash contaminated areas for 15 minutes using water. Treat supportively and observe for possible shock. If ingested, seek immediate medical aid.

Recommended Clean-Up Procedures

Personal Protection: Level B Ensemble	**Recommended Material** N.D.
RCRA Waste # None	**Reportable Quantities:** None

Spills: Remove all potential ignition sources. Dike to contain spill, absorb with non-combustible absorbent and take up using non-sparking tools. Decontaminate spill area using soapy water.

Special Emergency Information: May be harmful if inhaled, swallowed or absorbed through the skin.

PIPERIDINE

Synonyms: cyclopentimine; hexazane; pentamethyleneimine

CAS Number: 110-89-4   **Description:** Colorless Liquid
DOT Number: 2401   **DOT Classification:** Flammable Liquid

Molecular Weight: 85.2
Melting Point: -7°C (°F)   **Vapor Density:**   3.0   **Vapor Pressure:**   **5** (40 mm Hg)
Boiling Point: 106°C (223°F)   **Specific Gravity:** 0.9   **Water Solubility:**   Soluble

Chemical Incompatibilities or Instabilities: Oxidizers, 1-perchloryl-piperidine, N-nitrosoacetanilide, dicyanofurazan.

FLAMMABILITY

NFPA Hazard Code: 2-3-3   **NFPA Classification:**   Class IB Flammable
Flash Point: 16°C (61°F)   **LEL:** N.D.
Autoignition Temp: °C   (°F)   **UEL:** N.D.

Fire Extinguishing Methods: Carbon Dioxide, Dry Chemical, Foam, Water Spray or Fog.

Special Fire Fighting Considerations: Isolate for 1/2 mile if rail or tank truck is involved in a fire. Water spray should be used to keep closed containers cool. Continue to cool container after fire is extinguished. Immediately withdraw if rising sound from venting device is heard or if fire is causing discoloration to the tank. Do not allow water to enter container.

TOXICOLOGY

Odor: Pungent, amine-like   **Odor Threshold:** N.D.
Physical Contact: Material is extremely destructive to human tissues.
TLV = N.D.   **STEL** = N.D.   **IDLH** = N.D.

Routes of Entry and Relative LD$_{50}$ (or LC$_{50}$)   **RTECS #** TM3500000
    Inhalation   N.D.
    Ingestion   **5**   (400 mg/kg)
    Skin Absorption   **8**   (320 mg/kg)

Symptoms of Exposure: Possible burns or irritation to skin, eyes and upper respiratory tract; headache; nausea. Inhalation of vapors may be fatal by causing glottis to spasm and suffocation. Exposure may result in chemical pneumonitis or pulmonary edema.

Emergency/First Aid Treatment: Remove to ventilated area; immediately remove any contaminated clothing and wash contaminated areas for 15 minutes using water. Treat supportively and observe for possible shock. If ingested, seek immediate medical aid.

Recommended Clean-Up Procedures

Personal Protection: Level B Ensemble   **Recommended Material**   N.D.

RCRA Waste #   None   **Reportable Quantities:**   None

Spills: Remove all potential ignition sources. Dike to contain spill, absorb with noncombustible absorbent and take up using nonsparking tools. Decontaminate spill area using water.

Special Emergency Information: May be fatal in inhaled, ingested or absorbed through the skin. Vapors are heavier than air and may travel some distance to an ignition source. Vapors are more dense than air and may settle in low lying areas.

POTASSIUM - K

Synonyms:

CAS Number: 7440-09-7   **Description:** Silver metal
DOT Number: 1420, 2257   **DOT Classification:** Flammable Solid

Molecular Weight: 39.1
Melting Point: 63°C (146°F)   **Vapor Density:** N.A.   **Vapor Pressure:** 1 (< 1 mm Hg)
Boiling Point: 766°C (1410°F)   **Specific Gravity:** 0.9   **Water Solubility:** Decomposes

Chemical Incompatibilities or Instabilities: Air, water, halocarbons, oxidizers, carbon monoxide, carbon dioxide.

FLAMMABILITY

NFPA Hazard Code: 3-3-2-W   **NFPA Classification:**
Flash Point: N.A.   **LEL:** N.A.
Autoignition Temp: N.A.   **UEL:** N.A.

Fire Extinguishing Methods: Dry chemical, Soda ash, Sand, **DO NOT USE WATER OR FOAM.**

Special Fire Fighting Considerations: Will react violently with water to yield hydrogen gas.

TOXICOLOGY

Odor: None   **Odor Threshold:** N.D.
Physical Contact: Material is extremely destructive to human tissues.
TLV = N.D.   **STEL** = N.D.   **IDLH** = N.D.

Routes of Entry and Relative LD_{50} (or LC_{50})   **RTECS # TS6460000**
    Inhalation   N.D.
    Ingestion   N.D.
    Skin Absorption   N.D.

Symptoms of Exposure: Possible burns or irritation to skin, eyes and upper respiratory tract; headache; nausea.

Emergency/First Aid Treatment: Remove to ventilated area; immediately remove any contaminated clothing and wash contaminated areas for 15 minutes using water. Treat supportively and observe for possible shock. If ingested, seek immediate medical aid.

Recommended Clean-Up Procedures

Personal Protection: Level B Ensemble   **Recommended Material**   N.D.

RCRA Waste #   None   **Reportable Quantities:**   None

Spills: Clean up only under expert supervision.

Special Emergency Information: May be harmful if inhaled, swallowed or absorbed through the skin. Potentially explosive peroxides may form upon the exposure of potassium to air. Care should be used if spilled material is not black or silver in appearance.

POTASSIUM BROMATE - KBrO₃

Synonyms:

CAS Number: 7758-01-2	**Description:** White powder
DOT Number: 1484	**DOT Classification:** Oxidizer

Molecular Weight: 167.0

Melting Point: 350°C (662°F)	**Vapor Density:** N.A.	**Vapor Pressure:** 1 (<1 mm Hg)
Boiling Point: D 370°C (698°F)	**Specific Gravity:** 3.3	**Water Solubility:** Soluble

Chemical Incompatibilities or Instabilities: Sulfur, sulfur compounds, aluminum, selenium, lead acetate, hydrocarbons, carbon, reducing agents, arsenic, phosphorus, metal sulfides, ordinary combustibles.

FLAMMABILITY

NFPA Hazard Code:		**NFPA Classification:**
Flash Point:	N.A.	**LEL:** N.A.
Autoignition Temp:	N.A.	**UEL:** N.A.

Fire Extinguishing Methods: WATER ONLY.

Special Fire Fighting Considerations: For large fires, use flooding amounts of water. Water spray from unmanned devices should be used to keep closed containers cool. Continue to cool container after fire is extinguished. For large fires, if possible, withdraw and allow to burn.

TOXICOLOGY

Odor: None	**Odor Threshold:** N.A.
Physical Contact: Irritant	

TLV = N.D.   **STEL** = N.D.   **IDLH** = N.D.

Routes of Entry and Relative LD₅₀ (or LC₅₀)   **RTECS # EF8725000**

Inhalation	N.D.	
Ingestion	5	(325 mg/kg)
Skin Absorption	N.D.	

Symptoms of Exposure: Possible burns or irritation to skin, eyes and upper respiratory tract; headache; nausea.

Emergency/First Aid Treatment: Remove to ventilated area; immediately remove any contaminated clothing and wash contaminated areas for 15 minutes using water. Treat supportively and observe for possible shock. If ingested, seek immediate medical aid.

Recommended Clean-Up Procedures

Personal Protection: Level B Ensemble	**Recommended Material** N.D.
RCRA Waste # None	**Reportable Quantities:** None

Spills: Take up in non-combustible absorbent and dispose. Decontaminate spill area using water.

Special Emergency Information: May be harmful if inhaled, swallowed or absorbed through the skin.

Sodium bromate (CAS # 7789-38-0, DOT # 1494, RTECS # EF8750000) has similar chemical, physical and toxicological properties.

POTASSIUM CHLORATE - KClO3

Synonyms: chlorate of potash; potash chlorate; salt of tarter; potcrate; potassium oxymuriate

CAS Number: 3811-04-9

DOT Number: 1485, 2427

Description: White powder

DOT Classification: Oxidizer

Molecular Weight: 122.6

Melting Point: 368°C (694°F)

Boiling Point: D 400°C (752°F)

Vapor Density: N.A.

Specific Gravity: 2.3

Vapor Pressure: 1 (<1 mm Hg)

Water Solubility: Insoluble

Chemical Incompatibilities or Instabilities: halogens, hydrocarbons, cyanides, chromium, germanium, zirconium, titanium, sulfuric acid, ammonia, ammonium salts, nitric acid, strong bases, finely divided metals, phosphorous compounds, metal sulfides, reducing agents, ordinary combustibles.

FLAMMABILITY

NFPA Hazard Code: 1-0-1-OX

Flash Point: N.A.

Autoignition Temp: N.A.

NFPA Classification:

LEL: N.A.

UEL: N.A.

Fire Extinguishing Methods: WATER ONLY

Special Fire Fighting Considerations: For large fires, use flooding quantities of water. Water spray from an unanned device should be used to keep closed containers cool. Continue to cool container after fire is extinguished. For large fires, if possible, withdraw and allow to burn.

TOXICOLOGY

Odor: None

Physical Contact: Irritant

TLV = N.D.

Odor Threshold: N.A.

STEL = N.D.

IDLH = N.D.

Routes of Entry and Relative LD$_{50}$ (or LC$_{50}$)

Inhalation	N.D.	
Ingestion	N.D.	
Skin Absorption	N.D.	

RTECS # FO0350000

Symptoms of Exposure: Possible burns or irritation to skin, eyes and upper respiratory tract; headache; nausea.

Emergency/First Aid Treatment: Remove to ventilated area; immediately remove any contaminated clothing and wash contaminated areas for 15 minutes using water. Treat supportively and observe for possible shock. If ingested, seek immediate medical aid.

Recommended Clean-Up Procedures

Personal Protection: Level B Ensemble

RCRA Waste # None

Recommended Material N.D.

Reportable Quantities: None

Spills: Take up in non-combustible absorbent and dispose. Decontaminate spill area using water.

Special Emergency Information: May be harmful if inhaled, swallowed or absorbed through the skin.

Sodium chlorate (CAS # 7775-09-9, DOT # 1495, RTECS # FO0525000) has similar chemical, physical and toxicological properties.

POTASSIUM CYANIDE - KCN

Synonyms:

CAS Number: 151-50-8   **Description:** White powder
DOT Number: 1680   **DOT Classification:** Poison B

Molecular Weight: 65.1
Melting Point: 634°C (1174°F)   **Vapor Density:**   N.A.   **Vapor Pressure:**   1 (< 1 mm Hg)
Boiling Point: 1625°C (2957°F)   **Specific Gravity:** 1.5   **Water Solubility:**   Soluble

Chemical Incompatibilities or Instabilities: Chlorates, nitrates, will react with acids to yield highly flammable and toxic hydrogen cyanide gas.

FLAMMABILITY

NFPA Hazard Code: 3-0-0   **NFPA Classification:**
Flash Point: N.A.   **LEL:** N.A.
Autoignition Temp: N.A.   **UEL:** N.A.

Fire Extinguishing Methods: Carbon Dioxide, Dry Chemical, Foam, Water Spray or Fog.

Special Fire Fighting Considerations: Structural fire fighter protective clothing will not provide adequate protection. Fight fire from a distance or protected location, if possible. Treat all materials used or generated and equipment involved as contaminated by hazardous waste. May release toxic fumes upon heating (HCN).

TOXICOLOGY

Odor: Almonds   **Odor Threshold:** N.D.
Physical Contact: Irritant
TLV = 5 mg/m^3 **STEL** = N.D.   **IDLH** = 50 mg/m^3

Routes of Entry and Relative LD$_{50}$ (or LC$_{50}$)   **RTECS # TS8750000**
   Inhalation   N.D.
   Ingestion   **10**   (5 mg/kg)
   Skin Absorption   N.D.

Symptoms of Exposure: Possible irritation to skin, eyes and upper respiratory tract; headache; nausea; cyanosis.

Emergency/First Aid Treatment: Remove to ventilated area; immediately remove any contaminated clothing and wash contaminated areas for 15 minutes using water. Treat supportively and observe for possible shock. If ingested, seek immediate medical aid. Immediately obtain mediacl assistance and treat for cyanosis.

Recommended Clean-Up Procedures

Personal Protection: Level A Ensemble   **Recommended Material**   Natural Rubber (-), Neoprene (-),
                                                                           Nitrile Rubber (-)

RCRA Waste #   P098   **Reportable Quantities:**   10 lbs (10 lbs)

Spills: Take up in non-combustible absorbent and dispose. Treat all materials used or generated and equipment involved as contaminated by hazardous waste.

Special Emergency Information: May be fatal in inhaled, ingested or absorbed through the skin.

Sodium cyanide (CAS # 143-33-9, DOT # 1689, RTECS # VZ7525000, RCRA Waste # P106) has similar chemical, physical and toxicological properties and reportable quantities.

Silver cyanide (CAS # 506-64-9, DOT # 1684, RTECS # VW3850000, RCRA Waste # P104) has similar chemical, physical and toxicological properties, but lower reportable quantities (1 lb[1 lb]).

POTASSIUM DICHLOROISOCYANURATE

Synonyms: potassium dichloro-s-triazinetrione; troclosene

CAS Number: 2244-21-5	**Description:** White powder	
DOT Number: 2465	**DOT Classification:** Oxidizer	

Molecular Weight: 236.1

Melting Point: D 250°C (482°F)	**Vapor Density:** N.D.	**Vapor Pressure:** 1 (< 1 mm Hg)
Boiling Point: N.A.	**Specific Gravity:** 1.0	**Water Solubility:** Decomposes

Chemical Incompatibilities or Instabilities: Ammonia, ammonium salts, water, reducing agents.

FLAMMABILITY

NFPA Hazard Code: 3-0-2-OX	**NFPA Classification:**	
Flash Point: N.A.	**LEL:** N.A.	
Autoignition Temp: N.A.	**UEL:** N.A.	

Fire Extinguishing Methods: WATER ONLY.

Special Fire Fighting Considerations: Use flooding quantities of water. Water spray from unmanned devices should be used to keep closed containers cool. Continue to cool container after fire is extinguished. Fight fire from a distance or protected location, if possible. For large fires, if possible, withdraw and allow to burn.

TOXICOLOGY

Odor: Chlorine	**Odor Threshold:** N.D.

Physical Contact: Material is extremely destructive to human tissues.

TLV = N.D.	**STEL** = N.D.	**IDLH** = N.D.

Routes of Entry and Relative LD_{50} (or LC_{50}) **RTECS # XZ1850000**

Inhalation	N.D.	
Ingestion	3	(1215 mg/kg)
Skin Absorption	N.D.	

Symptoms: Possible burns or irritation to skin, eyes and upper respiratory tract; headache; nausea. Exposure may result in chemical pneumonitis or pulmonary edema.

Emergency/First Aid Treatment: Remove to ventilated area; immediately remove any contaminated clothing and wash contaminated areas for 15 minutes using water. Treat supportively and observe for possible shock. If ingested, seek immediate medical aid.

Recommended Clean-Up Procedures

Personal Protection: Level B Ensemble	**Recommended Material** N.D.
RCRA Waste # None	**Reportable Quantities:** None

Spills: Take up in non-combustible absorbent and dispose. Decontaminate spill area using flooding quantities of water. Keep area isolated until vapors have dissipated.

Special Emergency Information: May be harmful if inhaled, swallowed or absorbed through the skin.

POTASSIUM DICHROMATE - K₂Cr₂O₇

Synonyms: potassium bichromate

CAS Number: 7778-50-9   **Description:** Orange powder
DOT Number: 1479   **DOT Classification:** ORM-A

Molecular Weight: 294.2
Melting Point: 398°C (748°F)   **Vapor Density:**   N.A.   **Vapor Pressure:**   **1** (<1 mm Hg)
Boiling Point: D   500°C (932°F)   **Specific Gravity:** 2.7   **Water Solubility:**   Soluble

Chemical Incompatibilities or Instabilities: Reducing agents, organics, ordinary combustibles.

FLAMMABILITY

NFPA Hazard Code: **NFPA Classification:**
Flash Point: N.A.   **LEL:** N.A.
Autoignition Temp: N.A.   **UEL:** N.A.

Fire Extinguishing Methods: WATER ONLY.

Special Fire Fighting Considerations: Use flooding quantities of water. Water spray from unmanned devices should be used to keep closed containers cool. Continue to cool container after fire is extinguished. For large fires, if possible, withdraw and allow to burn. Treat all materials used or generated and equipment involved as contaminated by hazardous waste.

TOXICOLOGY
CARCINOGEN

Odor: None   **Odor Threshold:** N.A.
Physical Contact: Material is extremely destructive to human tissues.
TLV = 0.1 mg ($CrO_3$0/m^3) (Ceiling) **STEL** = N.D.   **IDLH** = 30 mg/m^3

Routes of Entry and Relative LD$_{50}$ (or LC$_{50}$)   **RTECS # HX7680000**
   Inhalation   N.D.
   Ingestion   N.D.
   Skin Absorption   N.D.

Symptoms of Exposure: Possible burns or irritation to skin, eyes and upper respiratory tract; headache; nausea. Inhalation of vapors may be fatal by causing glottis to spasm and suffocation. Exposure may result in chemical pneumonitis or pulmonary edema.

Emergency/First Aid Treatment: Remove to ventilated area; immediately remove any contaminated clothing and wash contaminated areas for 15 minutes using water. Treat supportively and observe for possible shock. If ingested, seek immediate medical aid.

Recommended Clean-Up Procedures

Personal Protection: Level B Ensemble   **Recommended Material**   N.D.

RCRA Waste #   None   **Reportable Quantities:**   None

Spills: Take up in non-combustible absorbent and dispose. Decontaminate spill area using water. Treat all materials used or generated and equipment involved as contaminated by hazardous waste.

Special Emergency Information: May be fatal in inhaled, ingested or absorbed through the skin.

Sodium dichromate (CAS # 7789-12-0, DOT # 1479, RTECS # HX7750000) has similar chemical, physical and toxicological properties.

POTASSIUM HYDROGEN FLUORIDE - KHF$_2$

Synonyms: potassium acid fluoride

CAS Number: 7789-29-9	**Description:** White powder	
DOT Number: 1811	**DOT Classification:** N.D.	

Molecular Weight: 78.1

Melting Point: D 238°C (460°F)	**Vapor Density:** N.A.	**Vapor Pressure:** N.A.
Boiling Point: N.A.	**Specific Gravity:** 2.3	**Water Solubility:** Soluble

Chemical Incompatibilities or Instabilities: Strong acids.

FLAMMABILITY

NFPA Hazard Code:	**NFPA Classification:**	
Flash Point: N.A.	**LEL:** N.A.	
Autoignition Temp: N.A.	**UEL:** N.A.	

Fire Extinguishing Methods: Carbon Dioxide, Dry Chemical, Foam, Water Spray or Fog.

Special Fire Fighting Considerations: Structural fire fighter protective clothing will not provide adequate protection. Water spray should be used to keep closed containers cool. Continue to cool container after fire is extinguished.

TOXICOLOGY

Odor:	**Odor Threshold:** N.D.

Physical Contact: Material is extremely destructive to human tissues.

TLV = 2.5 mg/m^3	**STEL** = N.D.	**IDLH** = 500 mg/m^3

Routes of Entry and Relative LD$_{50}$ (or LC$_{50}$)   **RTECS # TS6650000**

Inhalation	N.D.
Ingestion	N.D.
Skin Absorption	N.D.

Symptoms of Exposure: Possible burns or irritation to skin, eyes and upper respiratory tract; headache; nausea. Inhalation of vapors may be fatal by causing glottis to spasm and suffocation. Exposure may result in chemical pneumonitis or pulmonary edema.

Emergency/First Aid Treatment: Remove to ventilated area; immediately remove any contaminated clothing and wash contaminated areas for 15 minutes using water. Treat supportively and observe for possible shock. If ingested, seek immediate medical aid.

Recommended Clean-Up Procedures

Personal Protection: Level A Ensemble	**Recommended Material** N.D.	
RCRA Waste # None	**Reportable Quantities:** None	

Spills: Take up in non-combustible absorbent and dispose. Decontaminate spill area using water. Treat all materials used or generated and equipment involved as contaminated by hazardous waste.

Special Emergency Information: May be fatal in inhaled, ingested or absorbed through the skin.

Sodium bifluoride (CAS # 1333-83-1, DOT # 1439, RTECS # WB4180000) has similar chemical, physical and toxicological properties.

POTASSIUM PERMANGANATE - KMnO$_4$

Synonyms:

CAS Number: 7722-64-7   **Description:** Purple crystals
DOT Number: 1490   **DOT Classification:** Oxidizer

Molecular Weight: 158.0
Melting Point: 240°C (464°F)   **Vapor Density:**   N.A.   **Vapor Pressure:**   1 (< 1 mm Hg)
Boiling Point: N.A.   **Specific Gravity:** 2.7   **Water Solubility:**   Soluble

Chemical Incompatibilities or Instabilities: Strong acids, hydrocarbons, iron salts (II), mercury salts (I), arsenites, hypophosphites, sulfur, phosphorous, finely divided metals, arsenic, halogens, ammonia, amines.

FLAMMABILITY

NFPA Hazard Code: **NFPA Classification:**
Flash Point: N.A.   **LEL:**   N.A.
Autoignition Temp: N.A.   **UEL:**   N.A.

Fire Extinguishing Methods: WATER ONLY.

Special Fire Fighting Considerations: Use flooding quantities of water. Water spray should be used to keep closed containers cool. Continue to cool container after fire is extinguished. For large fires, if possible, withdraw and allow to burn.

TOXICOLOGY

Odor: None   **Odor Threshold:** N.A.
Physical Contact: Material is extremely destructive to human tissues.
TLV = 5 mg (Mn)/m^3 **STEL** = N.D.   **IDLH** = N.D.

Routes of Entry and Relative LD$_{50}$ (or LC$_{50}$)   **RTECS # SD6475000**
   Inhalation   N.D.
   Ingestion   **3**   (1090 mg/kg)
   Skin Absorption   N.D.

Symptoms of Exposure: Possible burns or irritation to skin, eyes and upper respiratory tract; headache; nausea; laryngitis. LaryngitisInhalation of vapors may be fatal by causing glottis to spasm and suffocation. Exposure may result in chemical pneumonitis or pulmonary edema.

Emergency/First Aid Treatment: Remove to ventilated area; immediately remove any contaminated clothing and wash contaminated areas for 15 minutes using water. Treat supportively and observe for possible shock. If ingested, seek immediate medical aid.

Recommended Clean-Up Procedures

Personal Protection: Level B Ensemble   **Recommended Material**   N.D.

RCRA Waste #   None   **Reportable Quantities:**   100 lb (100 lb)

Spills: Take up in non-combustible absorbent and dispose.

Special Emergency Information: May be harmful if inhaled, swallowed or absorbed through the skin. Treat as an explosion hazard.

Sodium permanganate (CAS # 10101-50-5, DOT # 1503, RTECS # SD6650000) has similar chemical, physical and toxicological properties.

POTASSIUM PERSULFATE - K$_2$S$_2$O$_8$

Synonyms: dipotassium persulfate; potassium peroxydisulfate

CAS Number: 7727-21-1	**Description:** White powder
DOT Number: 1492	**DOT Classification:** Oxidizer

Molecular Weight: 272.3

Melting Point: D 100°C (212°F)	**Vapor Density:** N.D.	**Vapor Pressure:** 1 (<1 mm Hg)
Boiling Point: N.D.	**Specific Gravity:** 2.5	**Water Solubility:** Soluble

Chemical Incompatibilities or Instabilities: Finely divided metals, hydrazine, strong bases, reducing agents.

FLAMMABILITY

NFPA Hazard Code:		**NFPA Classification:**
Flash Point:	N.A.	**LEL:** N.A.
Autoignition Temp:	N.A.	**UEL:** N.A.

Fire Extinguishing Methods: WATER ONLY.

Special Fire Fighting Considerations: Use flooding quantities of water. Water spray should be used to keep closed containers cool. Continue to cool container after fire is extinguished. For large fires, if possible, withdraw and allow to burn.

TOXICOLOGY

Odor: None	**Odor Threshold:** N.A.

Physical Contact: Irritant

TLV = 5 mg (S$_2$O$_8$)/m^3 **STEL** = N.D.   **IDLH** = N.D.

Routes of Entry and Relative LD$_{50}$ (or LC$_{50}$)   **RTECS # SE0400000**

Inhalation	N.D.
Ingestion	N.D.
Skin Absorption	N.D.

Symptoms of Exposure: Possible irritation to skin, eyes and upper respiratory tract; headache; nausea. May cause an allergic reaction.

Emergency/First Aid Treatment: Remove to ventilated area; immediately remove any contaminated clothing and wash contaminated areas for 15 minutes using water. Treat supportively and observe for possible shock. If ingested, seek immediate medical aid.

Recommended Clean-Up Procedures

Personal Protection:	Level B Ensemble	**Recommended Material** N.D.
RCRA Waste #	None	**Reportable Quantities:** None

Spills: Take up in non-combustible absorbent and dispose.

Special Emergency Information: May be harmful if inhaled, swallowed or absorbed through the skin.

Sodium persulfate (CAS # 7775-27-1, DOT # 1505, RTECS # SE0525000) has similar chemical, physical and toxicological properties.

POTASSIUM SULFIDE - K₂S

Synonyms:

CAS Number: 1312-73-8   **Description:** White to red powder
DOT Number: 1382 (dry), 1847 (wet)   **DOT Classification:** Flammable solid (1382), Corrosive Material (1847)

Molecular Weight: 110.3
Melting Point: 912°C (1674°F)   **Vapor Density:** N.A.   **Vapor Pressure:** 1 (< 1 mm Hg)
Boiling Point: N.A.   **Specific Gravity:** 1.8   **Water Solubility:** Soluble

Chemical Incompatibilities or Instabilities: Nitrogen oxide; reacts with acids to yield highly toxic and flammable hydrogen sulfide gas.

FLAMMABILITY \

NFPA Hazard Code: 3-1-0   **NFPA Classification:**
Flash Point: N.A.   **LEL:** N.A.
Autoignition Temp: N.A.   **UEL:** N.A.

Fire Extinguishing Methods: Carbon Dioxide, Dry Chemical, Foam, Water Spray or Fog.

Special Fire Fighting Considerations: Water spray should be used to keep closed containers cool. Continue to cool container after fire is extinguished. For large fires, if possible, withdraw and allow to burn. Contact with steam or acids may liberate hydrogen sulfide gas. May explode from rapid heating or percussion.

TOXICOLOGY

Odor: None   **Odor Threshold:** N.A.
Physical Contact: Material is extremely destructive to human tissues.
TLV = N.D.   **STEL** = N.D.   **IDLH** = N.D.

Routes of Entry and Relative LD₅₀ (or LC₅₀)   **RTECS # TT6000000**
   Inhalation   N.D.
   Ingestion   N.D.
   Skin Absorption   N.D.

Symptoms of Exposure: Possible irritation to skin, eyes and upper respiratory tract; headache; nausea. Inhalation of vapors may be fatal by causing glottis to spasm and suffocation. Exposure may result in chemical pneumonitis or pulmonary edema.

Emergency/First Aid Treatment: Remove to ventilated area; immediately remove any contaminated clothing and wash contaminated areas for 15 minutes using water. Treat supportively and observe for possible shock. If ingested, seek immediate medical aid.

Recommended Clean-Up Procedures

Personal Protection: Level B Ensemble   **Recommended Material**   N.D.

RCRA Waste #   None   **Reportable Quantities:**   None

Spills: Remove all potential ignition sources. Take up in non-combustible absorbent and dispose. Decontaminate spill area using dilute alkaline solution.

Special Emergency Information: May be harmful if inhaled, swallowed or absorbed through the skin. This material reaacts slowly with water and oxygen to form hydrogen sulfide and potassium hydroxide.

Sodium sulfide (CAS # 1313-84-4, DOT # 1385 (dry) 1849 (wet), RTECS # WE1925000) has similar chemical, physical and toxicological properties.

PROPANE - CH₃CH₂CH₃

Synonyms:

CAS Number: 74-98-6
DOT Number: 1075 (liquid), 1978 (gas)

Description: Colorless Gas
DOT Classification: Flammable Gas

Molecular Weight: 44.1
Melting Point: -188°C (-306°F)
Boiling Point: -42°C (-44°F)

Vapor Density: 1.6
Specific Gravity: N.A.

Vapor Pressure: Gas
Water Solubility: Insoluble

Chemical Incompatibilities or Instabilities: Oxidizers.

FLAMMABILITY

NFPA Hazard Code: 1-4-0
Flash Point: Gas
Autoignition Temp: 450°C (842°F)

NFPA Classification:
LEL: 2.1%
UEL: 9.5%

Fire Extinguishing Methods: Carbon Dioxide, Dry Chemical, Water Spray or Fog.

Special Fire Fighting Considerations: Isolate for 1/2 mile if rail or tank truck is involved in a fire. Water spray from an unmanned device should be used to keep closed containers cool. Continue to cool container after fire is extinguished. For large fires, if possible, withdraw and allow to burn. Immediately withdraw if rising sound from venting device is heard or if fire is causing discoloration to the tank.

TOXICOLOGY

Odor: Natural Gas
Physical Contact: Frostbite from liquefied gas.
TLV = 1800 mg/m³   **STEL** = N.D.

Odor Threshold: 22,000 mg/m³

IDLH = 20,000 mg/m³

Routes of Entry and Relative LD₅₀ (or LC₅₀)
Inhalation   N.D.
Ingestion   N.D.
Skin Absorption N.D.

RTECS # TX2275000

Symptoms of Exposure: A simple asphyxiant.

Emergency/First Aid Treatment: Remove to ventilated area; immediately remove any contaminated clothing and wash contaminated areas for 15 minutes using water. Treat supportively and observe for possible shock. If ingested, seek immediate medical aid.

Recommended Clean-Up Procedures

Personal Protection: Level B Ensemble

Recommended Material N.D.

RCRA Waste # None

Reportable Quantities: None

Spills: Remove all potential ignition sources. Stop leak if it can safely be done. Vent to dissipate vapors. Keep area isolated until vapors have dissipated.

Special Emergency Information: May be harmful if inhaled, swallowed or absorbed through the skin. Vapors are heavier than air and may travel some distance to an ignition source. Vapors are more dense than air and may settle in low lying areas.

Methane (CAS # 74-82-8; DOT # 1971 (gas), 1972 (cryogenic liquid); RTECS # PA1490000) and
ethane (CAS # 74-84-0, DOT # 1035 (gas), 1961 (cryogenic liquid); RTECS # KH3800000) have similar chemical, physical and toxicological properties.

1-PROPANETHIOL -CH$_3$CH$_2$CH$_2$SH

Synonyms: n-propyl mercaptan; propane-1-thiol

CAS Number: 107-03-0   **Description:** Colorless Liquid
DOT Number: 2402   **DOT Classification:** Flammable Liquid

Molecular Weight: 76.2
Melting Point: -113°C (-171°F)   **Vapor Density:** 2.5   **Vapor Pressure:** 7 (122 mm Hg)
Boiling Point: 68°C (154°F)   **Specific Gravity:** 0.8   **Water Solubility:** Soluble

Chemical Incompatibilities or Instabilities: Oxidizers, alkali metals, calcium hypochlorite.

FLAMMABILITY

NFPA Hazard Code: **NFPA Classification:** Class IB Flammable
Flash Point: -20°C   (-5°F)   **LEL:** N.D.
Autoignition Temp: N.D.   **UEL:** N.D.

Fire Extinguishing Methods: Carbon Dioxide, Dry Chemical, Foam, Water Spray or Fog.

Special Fire Fighting Considerations: Isolate for 1/2 mile if rail or tank truck is involved in a fire. Water spray should be used to keep closed containers cool. Continue to cool container after fire is extinguished. For large fires, if possible, withdraw and allow to burn. Immediately withdraw if rising sound from venting device is heard or if fire is causing discoloration to the tank.

TOXICOLOGY

Odor: **Odor Threshold:** N.D..
Physical Contact: Irritant
TLV = N.D.   **STEL** = N.D.   **IDLH** = N.D.

Routes of Entry and Relative LD$_{50}$ (or LC$_{50}$)   **RTECS # TZ7300000**
     Inhalation   3   (88,000 mg/m^3/H)
     Ingestion   3   (1790 mg/kg)
     Skin Absorption   N.D.

Symptoms of Exposure: Possible irritation to skin, eyes and upper respiratory tract; headache; nausea.

Emergency/First Aid Treatment: Remove to ventilated area; immediately remove any contaminated clothing and wash contaminated areas for 15 minutes using water. Treat supportively and observe for possible shock. If ingested, seek immediate medical aid.

Recommended Clean-Up Procedures

Personal Protection: Level B Ensemble   **Recommended Material**   N.D.

RCRA Waste #   None   **Reportable Quantities:**   None

Spills: Remove all potential ignition sources. Dike to contain spill, absorb with non-combustible absorbent and take up using non-sparking tools. Decontaminate spill area using water. Keep area isolated until vapors have dissipated.

Special Emergency Information: May be harmful if inhaled, swallowed or absorbed through the skin. Vapors are heavier than air and may travel some distance to an ignition source. Vapors are more dense than air and may settle in low lying areas.

2-propanethiol (CAS # 75-33-2, RTECS # TZ7302000) has similar chemical, physical and toxicological properties.

PROPARGYL ALCOHOL - $HC\equiv CCH_2OH$

Synonyms: 1-propyn-3-ol

CAS Number: 107-19-7   **Description:** Colorless Liquid
DOT Number: 1986   **DOT Classification:** Flammable Liquid

Molecular Weight: 56.1
Melting Point: -53°C (-63°F)   **Vapor Density:** 1.9   **Vapor Pressure:** 4 (12 mm Hg)
Boiling Point: 115°C (239°F)   **Specific Gravity:** 1.0   **Water Solubility:** Soluble

Chemical Incompatibilities or Instabilities: Alkalis, oxidizers, sulfuric acid, phosphorus pentoxide, heat.

FLAMMABILITY

NFPA Hazard Code: 3-3-3   **NFPA Classification:** Class II Combustible
Flash Point: 36°C (97°F)   **LEL:** N.D.
Autoignition Temp: °C (°F)   **UEL:** N.D.

Fire Extinguishing Methods: Alcohol Resistant Foam, Carbon Dioxide, Dry Chemical, Water Spray or Fog.

Special Fire Fighting Considerations: Isolate for 1/2 mile if rail or tank truck is involved in a fire. Water spray should be used to keep closed containers cool. Continue to cool container after fire is extinguished. Immediately withdraw if rising sound from venting device is heard or if fire is causing discoloration to the tank. Treat all materials used or generated and equipment involved as contaminated by hazardous waste.

TOXICOLOGY

Odor: Geraniums   **Odor Threshold:** N.D.
Physical Contact: Material is extremely destructive to human tissues.
TLV = 2.3 mg/m^3 (skin)   **STEL** = N.D.   **IDLH** = N.D.

RTECS # UK5075000

Routes of Entry and Relative LD$_{50}$ (or LC$_{50}$)
    Inhalation   7   (4,000 mg/m^3/H)
    Ingestion   8   (55 mg/kg)
    Skin Absorption   9   (88 mg/kg)

Symptoms of Exposure: Possible irritation to skin, eyes and upper respiratory tract; headache; nausea; laryngitis.

Emergency/First Aid Treatment: Remove to ventilated area; immediately remove any contaminated clothing and wash contaminated areas for 15 minutes using water. Treat supportively and observe for possible shock. If ingested, seek immediate medical aid.

Recommended Clean-Up Procedures

Personal Protection: Level A Ensemble   **Recommended Material** N.D.

RCRA Waste # P102   **Reportable Quantities:** 1000 lbs (1 lb)

Spills: Remove all potential ignition sources. Dike to contain spill, absorb with non-combustible absorbent and take up using non-sparking tools. Decontaminate spill area using water. Keep area isolated until vapors have dissipated. Treat all materials used or generated and equipment involved as contaminated by hazardous waste.

Special Emergency Information: May be fatal in inhaled, ingested or absorbed through the skin. Vapors are heavier than air and may travel some distance to an ignition source. Vapors are more dense than air and may settle in low lying areas. May undergo hazardous polymerization.

PROPIONALDEHYDE - CH3CH2CHO

Synonyms: propanal; propyl aldehyde; propylic aldehyde

CAS Number: 123-38-6	**Description:** Colorless Liquid	
DOT Number: 1275	**DOT Classification:** Flammable Liquid	

Molecular Weight: 58.1

Melting Point: -81°C (-114°F)	**Vapor Density:** 2.0	**Vapor Pressure:** 8 (300 mm Hg)	
Boiling Point: 49°C (120°F)	**Specific Gravity:** 0.8	**Water Solubility:** 20%	

Chemical Incompatibilities or Instabilities: Oxidizers, methyl methacrylate.

FLAMMABILITY

NFPA Hazard Code: 2-3-2	**NFPA Classification:** Class IB Flammable
Flash Point: -9°C (16°F)	**LEL:** 2.9%
Autoignition Temp: 207°C (405°F)	**UEL:** 17.0%

Fire Extinguishing Methods: Alcohol Resistant Foam, Carbon Dioxide, Dry Chemical, Water Spray or Fog.

Special Fire Fighting Considerations: Keep area isolated until vapors have dissipated. Water spray should be used to keep closed containers cool. Continue to cool container after fire is extinguished. For large fires, if possible, withdraw and allow to burn. Immediately withdraw if rising sound from venting device is heard or if fire is causing discoloration to the tank. Treat all materials used or generated and equipment involved as contaminated by hazardous waste.

TOXICOLOGY

Odor: Suffocating   **Odor Threshold:** N.D.

Physical Contact: Material is extremely destructive to human tissues.

TLV = N.D.   **STEL** = N.D.   **IDLH** = N.D.

Routes of Entry and Relative LD$_{50}$ (or LC$_{50}$)   **RTECS # UE0350000**

Inhalation	N.D.	
Ingestion	4	(1410 mg/kg)
Skin Absorption	1	(5040 mg/kg)

Symptoms of Exposure: Possible irritation to skin, eyes and upper respiratory tract; headache; nausea. Inhalation of vapors may be fatal by causing glottis to spasm and suffocation. Exposure may result in chemical pneumonitis or pulmonary edema.

Emergency/First Aid Treatment: Remove to ventilated area; immediately remove any contaminated clothing and wash contaminated areas for 15 minutes using water. Treat supportively and observe for possible shock. If ingested, seek immediate medical aid.

Recommended Clean-Up Procedures

Personal Protection: Level B Ensemble	**Recommended Material** Butyl Rubber (+)
RCRA Waste # None	**Reportable Quantities:** None

Spills: Remove all potential ignition sources. Dike to contain spill, absorb with non-combustible absorbent and take up using non-sparking tools. Decontaminate spill area using water. Keep area isolated until vapors have dissipated.

Special Emergency Information: May be fatal in inhaled, ingested or absorbed through the skin. Vapors are heavier than air and may travel some distance to an ignition source. Vapors are more dense than air and may settle in low lying areas.

PROPIONIC ACID - CH₃CH₂COOH

Synonyms:

CAS Number: 79-09-4
DOT Number: 1848

Description: Colorless Liquid
DOT Classification: Corrosive Material

Molecular Weight: 74.1
Melting Point: -21°C (-6°F)
Boiling Point: 141°C (286°F)

Vapor Density: 2.6
Specific Gravity: 1.0

Vapor Pressure: 2 (2 mm Hg)
Water Solubility: Soluble

Chemical Incompatibilities or Instabilities: Oxidizers.

FLAMMABILITY

NFPA Hazard Code: 2-2-0
Flash Point: 52°C (126°F)
Autoignition Temp: 513°C (955°F)

NFPA Classification: Class IIIB Combustible
LEL: 2.9%
UEL: 12.1%

Fire Extinguishing Methods: Alcohol Resistant Foam, Carbon Dioxide, Dry Chemical, Water Spray or Fog.

Special Fire Fighting Considerations: Water spray should be used to keep closed containers cool. Continue to cool container after fire is extinguished.

TOXICOLOGY

Odor: Rancid
Physical Contact: Material is extremely destructive to human tissues.
TLV = 30 mg/m³   **STEL** = N.D.

Odor Threshold: 0.003 mg/m³

IDLH = N.D.

RTECS # UE5950000

Routes of Entry and Relative LD₅₀ (or LC₅₀)

Route		
Inhalation	N.D.	
Ingestion	3	(3500 mg/kg)
Skin Absorption	7	(500 mg/kg)

Symptoms of Exposure: Possible irritation to skin, eyes and upper respiratory tract; headache; nausea. Inhalation of vapors may be fatal by causing glottis to spasm and suffocation. Exposure may result in chemical pneumonitis or pulmonary edema.

Emergency/First Aid Treatment: Remove to ventilated area; immediately remove any contaminated clothing and wash contaminated areas for 15 minutes using water. Treat supportively and observe for possible shock. If ingested, seek immediate medical aid.

Recommended Clean-Up Procedures

Personal Protection: Level B Ensemble
RCRA Waste # None

Recommended Material Teflon (0)
Reportable Quantities: 5000 lb (5000 lb)

Spills: Remove all potential ignition sources. Dike to contain spill, absorb with non-combustible absorbent and take up using non-sparking tools. Decontaminate spill area using water.

Special Emergency Information: May be harmful if inhaled, swallowed or absorbed through the skin. Vapors are heavier than air and may travel some distance to an ignition source. Vapors are more dense than air and may settle in low lying areas.

PROPIONIC ANHYDRIDE - CH3CH2C(O)O(O)CCH2CH3

Synonyms: propionic acid anhydride

CAS Number: 123-62-6
DOT Number: 2496

Description: Colorless Liquid
DOT Classification: Corrosive Material

Molecular Weight: 130.2
Melting Point: -45°C (-49°F)
Boiling Point: 167°C (333°F)

Vapor Density: 4.5
Specific Gravity: 1.0

Vapor Pressure: **3** (6 mm Hg)
Water Solubility: Decomposes

Chemical Incompatibilities or Instabilities: Oxidizers

FLAMMABILITY

NFPA Hazard Code: **2-2-1**
Flash Point: 63°C (145°F)
Autoignition Temp: 285°C (540°F)

NFPA Classification: Class IIIA Combustible
LEL: N.D.
UEL: N.D.

Fire Extinguishing Methods: Carbon Dioxide, Dry Chemical, Foam, Water Spray or Fog.

Special Fire Fighting Considerations: Isolate for 1/2 mile if rail or tank truck is involved in a fire. Water spray should be used to keep closed containers cool. Continue to cool container after fire is extinguished. Immediately withdraw if rising sound from venting device is heard or if fire is causing discoloration to the tank. Do not allow water to enter to enter the container.

TOXICOLOGY

Odor: Rancid
Physical Contact: Material is extremely destructive to human tissues.
TLV = N.D.   **STEL** = N.D.

Odor Threshold: N.D.

IDLH = N.D.

Routes of Entry and Relative LD$_{50}$ (or LC$_{50}$)

Inhalation	N.D.	
Ingestion	3	(2360 mg/kg)
Skin Absorption	1	(10,000 mg/kg)

RTECS # UF9100000

Symptoms of Exposure: Possible irritation to skin, eyes and upper respiratory tract; headache; nausea. Inhalation of vapors may be fatal by causing glottis to spasm and suffocation. Exposure may result in chemical pneumonitis or pulmonary edema.

Emergency/First Aid Treatment: Remove to ventilated area; immediately remove any contaminated clothing and wash contaminated areas for 15 minutes using water. Treat supportively and observe for possible shock. If ingested, seek immediate medical aid.

Recommended Clean-Up Procedures

Personal Protection: Level B Ensemble

RCRA Waste #   None

Recommended Material   N.D.

Reportable Quantities: 5000 lb (5000 lb)

Spills: Remove all potential ignition sources. Dike to contain spill, absorb with non-combustible absorbent and take up using non-sparking tools. Decontaminate spill area using dilute alkaline solution.

Special Emergency Information: May be harmful if inhaled, swallowed or absorbed through the skin. Vapors are heavier than air and may travel some distance to an ignition source. Vapors are more dense than air and may settle in low lying areas.

PROPIONITRILE - CH3CH2CN

Synonyms: cyanoethane; ethyl cyanide; propanenitrile

CAS Number: 107-12-0   **Description:** Colorless Liquid
DOT Number: 2404   **DOT Classification:** Flammable Liquid

Molecular Weight: 55.1
Melting Point: °C   (°F)   **Vapor Density:** 1.9   **Vapor Pressure:** 5 (40 mm Hg)
Boiling Point: 97°C (207°F)   **Specific Gravity:** 0.8   **Water Solubility:** 10%

Chemical Incompatibilities or Instabilities: Oxidizers, acids.

FLAMMABILITY

NFPA Hazard Code: 4-3-1   **NFPA Classification:** Class IB Flammable
Flash Point: 2°C (36°F)   **LEL:** 3.1%
Autoignition Temp: N.D.   **UEL:** N.D.

Fire Extinguishing Methods: Alcohol Resistant Foam, Carbon Dioxide, Dry Chemical, Water Spray or Fog.

Special Fire Fighting Considerations: Structural fire fighter protective clothing will not provide adequate protection. Isolate for 1/2 mile if rail or tank truck is involved in a fire. Water spray should be used to keep closed containers cool. Continue to cool container after fire is extinguished. Immediately withdraw if rising sound from venting device is heard or if fire is causing discoloration to the tank. Treat all materials used or generated and equipment involved as contaminated by hazardous waste.

TOXICOLOGY

Odor: Ethereal   **Odor Threshold:** N.D.
Physical Contact: Irritant
TLV = N.D.   **STEL** = N.D.   **IDLH** = N.D.

Routes of Entry and Relative LD$_{50}$ (or LC$_{50}$)   **RTECS # UF9625000**
  Inhalation   N.D.
  Ingestion   9   (39 mg/kg)
  Skin Absorption   8   (210 mg/kg)

Symptoms of Exposure: Possible irritation to skin, eyes and upper respiratory tract; headache; nausea; cyanosis.

Emergency/First Aid Treatment: Remove to ventilated area; immediately remove any contaminated clothing and wash contaminated areas for 15 minutes using water. Treat supportively and observe for possible shock. If ingested, seek immediate medical aid. If necessary, treat for cyanosis.

Recommended Clean-Up Procedures

Personal Protection: Level A Ensemble   **Recommended Material** Polyvinyl Alcohol (+)

RCRA Waste # P101   **Reportable Quantities:** 10 lbs (1 lb)

Spills: Absorb with non-combustible absorbent and take up using non-sparking tools. Dike to contain spill, absorb with non-combustible absorbent and take up using non-sparking tools. Decontaminate spill area using water. Keep area isolated until vapors have dissipated. Treat all materials used or generated and equipment involved as contaminated by hazardous waste.

Special Emergency Information: May be fatal in inhaled, ingested or absorbed through the skin. Vapors are heavier than air and may travel some distance to an ignition source. Vapors are more dense than air and may settle in low lying areas.

PROPIONYL CHLORIDE - $CH_3CH_2C(O)Cl$

Synonyms: propionic acid chloride

CAS Number: 79-03-8   **Description:** Colorless Liquid
DOT Number: 1815   **DOT Classification:** Flammable Liquid

Molecular Weight: 92.5
Melting Point: °C   (°F)   **Vapor Density:** 3.2   **Vapor Pressure:** N.D.
Boiling Point: 80°C (176°F)   **Specific Gravity:** 1.1 **Water Solubility:** Decomposes

Chemical Incompatibilities or Instabilities: Water, alcohols, diisopropyl ether.

FLAMMABILITY

NFPA Hazard Code: 3-3-1   **NFPA Classification:** Class IB Flammable
Flash Point: 12°C   (54°F)   **LEL:** N.D.
Autoignition Temp: °C   (°F)   **UEL:** N.D.

Fire Extinguishing Methods: Alcohol Resistant Foam, Carbon Dioxide, Dry Chemical, Water Spray or Fog.

Special Fire Fighting Considerations: Isolate for 1/2 mile if rail or tank truck is involved in a fire. Water spray should be used to keep closed containers cool. Continue to cool container after fire is extinguished. For large fires, if possible, withdraw and allow to burn. Immediately withdraw if rising sound from venting device is heard or if fire is causing discoloration to the tank. Do not allow water to enter container.

TOXICOLOGY

Odor: Sharp, pungent   **Odor Threshold:** N.D.
Physical Contact: Material is extremely destructive to human tissues.
TLV = N.D.   **STEL** = N.D.   **IDLH** = N.D.

Routes of Entry and Relative LD$_{50}$ (or LC$_{50}$)   **RTECS # UG6657000**
    Inhalation   N.D.
    Ingestion   N.D.
    Skin Absorption   N.D.

Symptoms of Exposure: Possible irritation to skin, eyes and upper respiratory tract; headache; nausea. Inhalation of vapors may be fatal by causing glottis to spasm and suffocation. Exposure may result in chemical pneumonitis or pulmonary edema.

Emergency/First Aid Treatment: Remove to ventilated area; immediately remove any contaminated clothing and wash contaminated areas for 15 minutes using water. Treat supportively and observe for possible shock. If ingested, seek immediate medical aid.

Recommended Clean-Up Procedures

Personal Protection: Level B Ensemble   **Recommended Material**   N.D.

RCRA Waste #   None   **Reportable Quantities:** None

Spills: Remove all potential ignition sources. Dike to contain spill, absorb with non-combustible absorbent and take up using non-sparking tools. Decontaminate spill area using dilute alkaline solution.

Special Emergency Information: May be fatal in inhaled, ingested or absorbed through the skin. Vapors are heavier than air and may travel some distance to an ignition source. Vapors are more dense than air and may settle in low lying areas.

n-PROPYL CHLOROFORMATE - CH₃CH₂CH₂OC(O)Cl

Synonyms: propyl chlorocarbonate; chloroformic acid, propyl ester

CAS Number: 109-61-5
DOT Number: 2740

Description: Colorless Liquid
DOT Classification: Flammable Liquid

Molecular Weight: 122.6
Melting Point: °C (°F)
Boiling Point: 105°C (221°F)

Vapor Density: N.D.
Specific Gravity: 1.1

Vapor Pressure: 5 (26 mm Hg)
Water Solubility: Decomposes

Chemical Incompatibilities or Instabilities: Oxidizers, water, acids.

FLAMMABILITY

NFPA Hazard Code:
Flash Point: 28°C (84°F)
Autoignition Temp: N.D.

NFPA Classification: Class IC Flammable
LEL: N.D.
UEL: N.D.

Fire Extinguishing Methods: Carbon Dioxide, Dry Chemical, Foam, Water Spray or Fog.

Special Fire Fighting Considerations: Structural fire fighter protective clothing will not provide adequate protection. Water spray from an unmanned device should be used to keep closed containers cool. Continue to cool container after fire is extinguished. Fight fire from a distance or protected location, if possible. Treat all materials used or generated and equipment involved as contaminated by hazardous waste.

DOT Recommended Isolation Zones:
DOT Recommended Down Wind Take CoverDistance

Small Spill: 150 ft   Large Spill: 150 ft
Small Spill: 0.2 miles   Large Spill: 0.2 miles

TOXICOLOGY

Odor: Sharp
Physical Contact: Material is extremely destructive to human tissues.
TLV = N.D.   **STEL** = N.D.

Odor Threshold: N.D.

IDLH = N.D.

RTECS # LQ6830000

Routes of Entry and Relative LD₅₀ (or LC₅₀)
Inhalation   N.D.
Ingestion   N.D.
Skin Absorption   N.D.

Symptoms of Exposure: Possible irritation to skin, eyes and upper respiratory tract; headache; nausea; lachrymation. Inhalation of vapors may be fatal by causing glottis to spasm and suffocation. Exposure may result in chemical pneumonitis or pulmonary edema.

Emergency/First Aid Treatment: Remove to ventilated area; immediately remove any contaminated clothing and wash contaminated areas for 15 minutes using water. Treat supportively and observe for possible shock. If ingested, seek immediate medical aid.

Recommended Clean-Up Procedures

Personal Protection: Level A Ensemble   **Recommended Material**   N.D.

RCRA Waste #   None   **Reportable Quantities:**   None

Spills: Remove all potential ignition sources. Dike to contain spill, absorb with non-combustible absorbent and take up using non-sparking tools. Decontaminate spill area using dilute alkaline solution. Keep area isolated until vapors have dissipated. Treat all materials used or generated and equipment involved as contaminated by hazardous waste.

Special Emergency Information: May be harmful if inhaled, swallowed or absorbed through the skin. Vapors are heavier than air and may travel some distance to an ignition source. Vapors are more dense than air and may settle in low lying areas.

PROPYLENE - CH_2=$CHCH_3$

Synonyms: propene

CAS Number: 115-07-1
DOT Number: 1077

Description: Colorless Gas
DOT Classification: Flammable Gas

Molecular Weight: 42.1

Melting Point: -185°C (-301°F)	**Vapor Density:** 1.45	**Vapor Pressure:** Gas
Boiling Point: -48°C (-54°F)	**Specific Gravity:** N.A.	**Water Solubility:** Insoluble

Chemical Incompatibilities or Instabilities: Oxidizers, nitrogen oxides, sulfur dioxide.

FLAMMABILITY

NFPA Hazard Code: 1-4-1
Flash Point: -108°C (-162°F)
Autoignition Temp: 460°C (860°F)

NFPA Classification:
LEL: 2.4%
UEL: 10.3%

Fire Extinguishing Methods: Carbon Dioxide, Dry Chemical, Water Spray or Fog.

Special Fire Fighting Considerations: Isolate for 1/2 mile if rail or tank truck is involved in a fire. Water spray from an unmanned device should be used to keep closed containers cool. Continue to cool container after fire is extinguished. For large fires, if possible, withdraw and allow to burn. Immediately withdraw if rising sound from venting device is heard or if fire is causing discoloration to the tank.

TOXICOLOGY
QUESTIONABLE CARCINOGEN

Odor: Aromatic
Physical Contact: Possible frostbite from rapidly expanding gas.
TLV = N.D.   **STEL** = N.D.

Odor Threshold: 17 mg/m^3

IDLH = N.D.

Routes of Entry and Relative LD$_{50}$ (or LC$_{50}$)
  Inhalation   N.D.
  Ingestion   N.D.
  Skin Absorption   N.D.

RTECS # UC6740000

Symptoms of Exposure: A simple asphyxiant.

Emergency/First Aid Treatment: Remove to ventilated area; immediately remove any contaminated clothing and wash contaminated areas for 15 minutes using water. Treat supportively and observe for possible shock. If ingested, seek immediate medical aid.

Recommended Clean-Up Procedures

Personal Protection: Level C Ensemble   **Recommended Material**   N.D.

RCRA Waste #   None   **Reportable Quantities:**   None

Spills: Remove all potential ignition sources. Stop leak if it can safely be done. Keep area isolated until vapors have dissipated.

Special Emergency Information: May be harmful if inhaled, swallowed or absorbed through the skin. Vapors are heavier than air and may travel some distance to an ignition source. Vapors are more dense than air and may settle in low lying areas.

PROPYLENE DICHLORIDE - ClCH$_2$CHClCH$_3$

Synonyms: 1,2-dichloropropane

CAS Number: 78-87-5	**Description:** Colorless Liquid	
DOT Number: 1279	**DOT Classification:** Flammable Liquid	

Molecular Weight: 113.0

Melting Point: °C (°F)	**Vapor Density:** 3.9	**Vapor Pressure:** **5** (40 mm Hg)
Boiling Point: 96°C (205°F)	**Specific Gravity:** 1.2	**Water Solubility:** Insoluble

Chemical Incompatibilities or Instabilities: Oxidizers, aluminum and its alloys.

FLAMMABILITY

NFPA Hazard Code: 2-3-0	**NFPA Classification:** Class IB Flammable	
Flash Point: 16°C (60°F)	**LEL:** 3.4%	
Autoignition Temp: 557°C (1035°F)	**UEL:** 14.5%	

Fire Extinguishing Methods: Carbon Dioxide, Dry Chemical, Foam, Water Spray or Fog.

Special Fire Fighting Considerations: Isolate for 1/2 mile if rail or tank truck is involved in a fire. Water spray from an unmanned device should be used to keep closed containers cool. Continue to cool container after fire is extinguished. For large fires, if possible, withdraw and allow to burn. Immediately withdraw if rising sound from venting device is heard or if fire is causing discoloration to the tank. Treat all materials used or generated and equipment involved as contaminated by hazardous waste.

TOXICOLOGY
QUESTIONABLE CARCINOGEN

Odor: Sweet	**Odor Threshold:** 1.2 mg/m^3
Physical Contact: Irritant	
TLV = 347 mg/m^3 **STEL** = 553 mg/m^3	**IDLH** = 9,400 mg/m^3

Routes of Entry and Relative LD$_{50}$ (or LC$_{50}$)   **RTECS # TX9625000**

Inhalation	1	(112,000 mg/m^3/H)
Ingestion	3	(1910 mg/kg)
Skin Absorption	1	(8750 mg/kg)

Symptoms of Exposure: Possible irritation to skin, eyes and upper respiratory tract; headache; nausea; depression of the central nervous system.

Emergency/First Aid Treatment: Remove to ventilated area; immediately remove any contaminated clothing and wash contaminated areas for 15 minutes using water. Treat supportively and observe for possible shock. If ingested, seek immediate medical aid.

Recommended Clean-Up Procedures

Personal Protection: Level A Ensemble	**Recommended Material**	N.D.
RCRA Waste # U083	**Reportable Quantities:**	1000 lbs (5000 lbs)

Spills: Remove all potential ignition sources. Dike to contain spill, absorb with non-combustible absorbent and take up using non-sparking tools. Decontaminate spill area using soapy water. Keep area isolated until vapors have dissipated.

Special Emergency Information: May be harmful if inhaled, swallowed or absorbed through the skin. Vapors are heavier than air and may travel some distance to an ignition source. Vapors are more dense than air and may settle in low lying areas.

1,3-dichloropropylene (CAS # 142-28-9, RTECS # TX9660000) and
2,2-dichloropropylene (CAS # 594-20-7) have similar chemical, physical and toxicological properties.

PROPYLENE IMINE

Synonyms: 2-methylazacyclopropane; 2-methylaziradine

CAS Number: 75-55-8   **Description:** Colorless Liquid
DOT Number: 1921   **DOT Classification:** Flammable Liquid

Molecular Weight: 57.1
Melting Point: -65°C (-85°F)   **Vapor Density:** N.D.   **Vapor Pressure:** 7 (112 mm Hg)
Boiling Point: 67°C (152°F)   **Specific Gravity:** 0.82 **Water Solubility:** Decomposes

Chemical Incompatibilities or Instabilities: Acids, oxidizers, water.

FLAMMABILITY

NFPA Hazard Code: **NFPA Classification:** Class IB Flammable
Flash Point: -15°C   (5°F)   **LEL:** N.D.
Autoignition Temp: °C   (°F)   **UEL:** N.D.

Fire Extinguishing Methods: Carbon Dioxide, Dry Chemical, Foam, Water Spray or Fog.

Special Fire Fighting Considerations: Structural fire fighter protective clothing will not provide adequate protection. Isolate for 1/2 mile if rail or tank truck is involved in a fire. Water spray should be used to keep closed containers cool. Continue to cool container after fire is extinguished. For large fires, if possible, withdraw and allow to burn. Immediately withdraw if rising sound from venting device is heard or if fire is causing discoloration to the tank. Do not allow to enter container.

TOXICOLOGY
CARCINOGEN

Odor: Ammonia   **Odor Threshold:** N.D.
Physical Contact: Material is extremely destructive to human tissues.
TLV = 5 mg/m^3 (skin)   **STEL** = N.D.   **IDLH** = 1250 mg/m^3

Routes of Entry and Relative LD$_{50}$ (or LC$_{50}$)   **RTECS # CM8050000**
   Inhalation   N.D.
   Ingestion   9   (19 mg/kg)
   Skin Absorption   N.D.

Symptoms of Exposure: Possible burns or irritation to skin, eyes and upper respiratory tract; headache; nausea. Inhalation of vapors may be fatal by causing glottis to spasm and suffocation. Exposure may result in chemical pneumonitis or pulmonary edema.

Emergency/First Aid Treatment: Remove to ventilated area; immediately remove any contaminated clothing and wash contaminated areas for 15 minutes using water. Treat supportively and observe for possible shock. If ingested, seek immediate medical aid.

Recommended Clean-Up Procedures

Personal Protection: Level A Ensemble   **Recommended Material**   N.D.

RCRA Waste #   P067   **Reportable Quantities:**   1 lb (1 lb)

Spills: Remove all potential ignition sources. Dike to contain spill, absorb with non-combustible absorbent and take up using non-sparking tools. Decontaminate spill area using dilute alkaline solution. Treat all materials used or generated and equipment involved as contaminated by hazardous waste.

Special Emergency Information: May be fatal in inhaled, ingested or absorbed through the skin. May undergo hazardous polymerization if exposed to acids.

PROPYLENE OXIDE

Synonyms: 1,2-epoxypropane; propene oxide; propylene epoxide

CAS Number: 75-56-9

DOT Number: 1280

Description: Colorless Liquid

DOT Classification: Flammable Liquid

Molecular Weight: 58.1

Melting Point: -112°C (-170°F)

Boiling Point: 34°C (94°F)

Vapor Density: 2.0

Specific Gravity: 0.9

Vapor Pressure: 9 (440 mm Hg)

Water Solubility: Soluble

Chemical Incompatibilities or Instabilities: Acids, oxidizers, strong acids, ammonium hydroxide.

FLAMMABILITY

NFPA Hazard Code: 4-2-2

Flash Point: -37°C (-35°F)

Autoignition Temp: 465°C (869°F)

NFPA Classification: Class IA Flammable

LEL: 2.1%

UEL: 37.0%

Fire Extinguishing Methods: Alcohol Resistant Foam, Carbon Dioxide, Dry Chemical, Water Spray or Fog.

Special Fire Fighting Considerations: Isolate for 1/2 mile if rail or tank truck is involved in a fire. Water spray should be used to keep closed containers cool. Continue to cool container after fire is extinguished. For large fires, if possible, withdraw and allow to burn. Immediately withdraw if rising sound from venting device is heard or if fire is causing discoloration to the tank.

TOXICOLOGY
CARCINOGEN

Odor: Aromatic

Physical Contact: Material is extremely destructive to human tissues.

TLV = 50 mg/m^3　　　　**STEL** = N.D.

Odor Threshold: 24 mg/m^3

IDLH = 4800 mg/m^3

RTECS # TZ2975000

Routes of Entry and Relative LD$_{50}$ (or LC$_{50}$)

Inhalation	N.D.	
Ingestion	5	(380 mg/kg)
Skin Absorption	4	(1240 mg/kg)

Symptoms of Exposure: Possible irritation to skin, eyes and upper respiratory tract; headache; nausea; laryngitis.

Emergency/First Aid Treatment: Remove to ventilated area; immediately remove any contaminated clothing and wash contaminated areas for 15 minutes using water. Treat supportively and observe for possible shock. If ingested, seek immediate medical aid.

Recommended Clean-Up Procedures

Personal Protection: Level A Ensemble

RCRA Waste # None

Recommended Material N.D.

Reportable Quantities: 100 lb (5000 lb)

Spills: Remove all potential ignition sources. Dike to contain spill, absorb with non-combustible absorbent and take up using non-sparking tools. Decontaminate spill area using water. Keep area isolated until vapors have dissipated.

Special Emergency Information: May be harmful if inhaled, swallowed or absorbed through the skin. Vapors are heavier than air and may travel some distance to an ignition source. Vapors are more dense than air and may settle in low lying areas.

PROPYLTRICHLOROSILANE - CH3CH2CH2SiCl3

Synonyms: trichloropropylsilane

CAS Number: 141-57-1   **Description:** Colorless Liquid
DOT Number: 1816   **DOT Classification:** Corrosive Material

Molecular Weight: 177.5
Melting Point: °C (°F)   **Vapor Density:** 6.1   **Vapor Pressure:** N.D.
Boiling Point: 122°C (252°F)   **Specific Gravity:** 1.2   **Water Solubility:** Decomposes

Chemical Incompatibilities or Instabilities: Oxidizers, water.

FLAMMABILITY

NFPA Hazard Code: 3-3-1   **NFPA Classification:** Class IC Flammable
Flash Point: 37°C (98°F)   **LEL:** N.D.
Autoignition Temp: N.D.   **UEL:** N.D.

Fire Extinguishing Methods: Carbon Dioxide, Dry Chemical, Foam, Water Spray or Fog.

Special Fire Fighting Considerations: Isolate for 1/2 mile if rail or tank truck is involved in a fire. Water spray should be used to keep closed containers cool. Continue to cool container after fire is extinguished. Immediately withdraw if rising sound from venting device is heard or if fire is causing discoloration to the tank. Do not get water inside container.

TOXICOLOGY

Odor: Acrid   **Odor Threshold:** N.D.
Physical Contact: Material is extremely destructive to human tissues.
TLV = N.D.   **STEL** = N.D.   **IDLH** = N.D.

Routes of Entry and Relative LD$_{50}$ (or LC$_{50}$)   **RTECS #** VV5300000
    Inhalation   N.D.
    Ingestion   N.D.
    Skin Absorption   N.D.

Symptoms of Exposure: Possible burns or irritation to skin, eyes and upper respiratory tract; headache; nausea. Inhalation of vapors may be fatal by causing glottis to spasm and suffocation. Exposure may result in chemical pneumonitis or pulmonary edema.

Emergency/First Aid Treatment: Remove to ventilated area; immediately remove any contaminated clothing and wash contaminated areas for 15 minutes using water. Treat supportively and observe for possible shock. If ingested, seek immediate medical aid.

Recommended Clean-Up Procedures

Personal Protection: Level B Ensemble   **Recommended Material** N.D.

RCRA Waste # None   **Reportable Quantities:** None

Spills: Remove all potential ignition sources. Dike to contain spill, absorb with non-combustible absorbent and take up using non-sparking tools. Decontaminate spill area using dilute alkaline solution. Keep area isolated until vapors have dissipated.

Special Emergency Information: May be harmful if inhaled, swallowed or absorbed through the skin. Vapors are heavier than air and may travel some distance to an ignition source. Vapors are more dense than air and may settle in low lying areas.

PYRIDINE

Synonyms: azine

CAS Number: 110-86-1   **Description:** Colorless Liquid
DOT Number: 1282   **DOT Classification:** Flammable Liquid

Molecular Weight: 79.1
Melting Point: -42°C (-44°F)   **Vapor Density:** 2.7   **Vapor Pressure:** 4 (18 mm Hg)
Boiling Point: 115°C (240°F)   **Specific Gravity:** 1.0   **Water Solubility:** Soluble

Chemical Incompatibilities or Instabilities: Strong acids, oxidizers, chlorosulfonic acid, chloroformates.

FLAMMABILITY

NFPA Hazard Code: 2-3-0   **NFPA Classification:** Class IB Flammable
Flash Point: 20°C (68°F)   **LEL:** 1.8%
Autoignition Temp: 482°C (900°F)   **UEL:** 12.4%

Fire Extinguishing Methods: Alcohol Resistant Foam, Carbon Dioxide, Dry Chemical, Water Spray or Fog.

Special Fire Fighting Considerations: Isolate for 1/2 mile if rail or tank truck is involved in a fire. Water spray should be used to keep closed containers cool. Continue to cool container after fire is extinguished. For large fires, if possible, withdraw and allow to burn. Immediately withdraw if rising sound from venting device is heard or if fire is causing discoloration to the tank. Treat all materials used or generated and equipment involved as contaminated by hazardous waste.

TOXICOLOGY

Odor: Penetrating, nauseating   **Odor Threshold:** 0.1 mg/m^3
Physical Contact: Irritant
TLV = 15 mg/m^3 **STEL** = N.D.   **IDLH** = 10,000 mg/m^3

Routes of Entry and Relative LD$_{50}$ (or LC$_{50}$)   **RTECS # UR8400000**
   Inhalation   4   (52,000 mg/m^3/H)
   Ingestion   4   (891 mg/kg)
   Skin Absorption 4   (1121 mg/kg)

Symptoms of Exposure: Possible irritation to skin, eyes and upper respiratory tract; headache; nausea.

Emergency/First Aid Treatment: Remove to ventilated area; immediately remove any contaminated clothing and wash contaminated areas for 15 minutes using water. Treat supportively and observe for possible shock. If ingested, seek immediate medical aid.

Recommended Clean-Up Procedures

Personal Protection: Level B Ensemble   **Recommended Material** Butyl Rubber (+)

RCRA Waste # U196   **Reportable Quantities:** 1000 lbs (1 lb)

Spills: Remove all potential ignition sources. Dike to contain spill, absorb with non-combustible absorbent and take up using non-sparking tools. Decontaminate spill area using water. Keep area isolated until vapors have dissipated. Treat all materials used or generated and equipment involved as contaminated by hazardous waste.

Special Emergency Information: May be harmful if inhaled, swallowed or absorbed through the skin. Vapors are heavier than air and may travel some distance to an ignition source. Vapors are more dense than air and may settle in low lying areas.

QUINOLINE

Synonyms: 1-azanaphthalene

CAS Number: 91-22-5	**Description:** Colorless Liquid
DOT Number: 2656	**DOT Classification:** Poison B

Molecular Weight: 129.2

Melting Point:	-15°C (5°F)	**Vapor Density:** 4.5	**Vapor Pressure:**	1 (<1 mm Hg)
Boiling Point:	238°C (460°F)	**Specific Gravity:** 1.1	**Water Solubility:**	Insoluble

Chemical Incompatibilities or Instabilities: Oxidizer, perchromates, nitrogen tetroxide, hydrogen peroxide, maleic anhydride.

FLAMMABILITY

NFPA Hazard Code: 2-1-0	**NFPA Classification:** Class IIIB Combustible
Flash Point: 101°C (214°F)	**LEL:** N.D.
Autoignition Temp: 480°C (896°F)	**UEL:** N.D.

Fire Extinguishing Methods: Alcohol Resistant Foam, Carbon Dioxide, Dry Chemical, Water Spray or Fog.

Special Fire Fighting Considerations: Structural fire fighter protective clothing will not provide adequate protection. Isolate for 1/2 mile if rail or tank truck is involved in a fire. Water spray should be used to keep closed containers cool. Continue to cool container after fire is extinguished. Immediately withdraw if rising sound from venting device is heard or if fire is causing discoloration to the tank. Treat all materials used or generated and equipment involved as contaminated by hazardous waste.

TOXICOLOGY
QUESTIONABLE CARCINOGEN

Odor: Peculiar	**Odor Threshold:** N.D.
Physical Contact: Irritant	
TLV = N.D. **STEL** = N.D.	**IDLH** = N.D.

Routes of Entry and Relative LD$_{50}$ (or LC$_{50}$)   **RTECS # VA9275000**

Inhalation	N.D.	
Ingestion	5	(331 mg/kg)
Skin Absorption	7	(540 mg/kg)

Symptoms of Exposure: Possible irritation to skin, eyes and upper respiratory tract; headache; nausea.

Emergency/First Aid Treatment: Remove to ventilated area; immediately remove any contaminated clothing and wash contaminated areas for 15 minutes using water. Treat supportively and observe for possible shock. If ingested, seek immediate medical aid.

Recommended Clean-Up Procedures

Personal Protection: Level A Ensemble	**Recommended Material** N.D.
RCRA Waste # None	**Reportable Quantities:** 5000 lbs (1000 lbs)

Spills: Dike to contain spill and collect liquid for later disposal. Decontaminate spill area using soapy water.

Special Emergency Information: May be harmful if inhaled, swallowed or absorbed through the skin.

RESORCINOL

Synonyms: 1,3-benzenediol; 1,3-dihydroxybenzene

CAS Number: 108-46-3	**Description:** White powder	
DOT Number: 2876	**DOT Classification:** ORM-E	

Molecular Weight: 110.1

Melting Point: 111°C (200°F)	**Vapor Density:** 3.8	**Vapor Pressure:** 1 (1 mm Hg)
Boiling Point: 277°C (531°F)	**Specific Gravity:** 1.3	**Water Solubility:** Soluble

Chemical Incompatibilities or Instabilities: Nitric acid, acetanilide, alkali.

FLAMMABILITY

NFPA Hazard Code: ?-1-0	**NFPA Classification:**	
Flash Point: 127°C (261°F)	**LEL:** N.D.	
Autoignition Temp: 608°C (1126°F)	**UEL:** N.D.	

Fire Extinguishing Methods: Carbon Dioxide, Dry Chemical, Foam, Water Spray or Fog.

Special Fire Fighting Considerations: Structural fire fighter protective clothing will not provide adequate protection. Fight fire from a distance or protected location, if possible. Treat all materials used or generated and equipment involved as contaminated by hazardous waste.

TOXICOLOGY

Odor: Musty	**Odor Threshold:** N.D.
Physical Contact: Irritant	
TLV = 45 mg/m^3 **STEL** = 90 mg/m^3	**IDLH** = N.D.

Routes of Entry and Relative LD$_{50}$ (or LC$_{50}$)   **RTECS # VG9625000**

Inhalation	N.D.	
Ingestion	5	(331 mg/kg)
Skin Absorption	2	(3360 mg/kg)

Symptoms of Exposure: Possible irritation to skin, eyes and upper respiratory tract; headache; nausea. Absorption may lead to the formation of methemoglobin resulting in cyanosis several hours after exposure.

Emergency/First Aid Treatment: Remove to ventilated area; immediately remove any contaminated clothing and wash contaminated areas for 15 minutes using water. Treat supportively and observe for possible shock. If ingested, seek immediate medical aid.

Recommended Clean-Up Procedures

Personal Protection:	Level B Ensemble	**Recommended Material**	N.D.
RCRA Waste #	U201	**Reportable Quantities:**	5000 lbs (1000 lbs)

Spills: Take up in non-combustible absorbent and dispose. Decontaminate spill area using water. Treat all materials used or generated and equipment involved as contaminated by hazardous waste.

Special Emergency Information: May be harmful if inhaled, swallowed or absorbed through the skin.

RUBIDIUM - Rb

Synonyms:

CAS Number: 7440-17-7   **Description:** Silvery metal
DOT Number: 1423   **DOT Classification:** Flammable Solid

Molecular Weight: 85.5
Melting Point: 39°C (102°F)   **Vapor Density:** N.A.   **Vapor Pressure:** **1** (<1 mm Hg)
Boiling Point: 688°C (1270°F)   **Specific Gravity:** 1.5   **Water Solubility:** Decomposes

Chemical Incompatibilities or Instabilities: Acids, water, halogens, halogenated hydrocarbons, air, oxidizers, mercury.

FLAMMABILITY

NFPA Hazard Code: **NFPA Classification:**
Flash Point: N.A.   **LEL:** N.A.
Autoignition Temp: N.A.   **UEL:** N.A.

Fire Extinguishing Methods: Dry sand, soda ash, lime, dry chemical.

Special Fire Fighting Considerations: DO NOT USE WATER.

TOXICOLOGY

Odor: None   **Odor Threshold:** N.A.
Physical Contact: Material is extremely destructive to human tissues.
TLV = N.D.   **STEL** = N.D.   **IDLH** = N.D.

Routes of Entry and Relative LD$_{50}$ (or LC$_{50}$)   **RTECS # VL8500000**
   Inhalation   N.D.
   Ingestion   N.D.
   Skin Absorption   N.D.

Symptoms of Exposure: Possible burns or irritation to skin, eyes and upper respiratory tract; headache; nausea; laryngitis.

Emergency/First Aid Treatment: Remove to ventilated area; immediately remove any contaminated clothing and wash contaminated areas for 15 minutes using water. Treat supportively and observe for possible shock. If ingested, seek immediate medical aid.

Recommended Clean-Up Procedures

Personal Protection: Level B Ensemble   **Recommended Material**   N.D.

RCRA Waste #   None   **Reportable Quantities:**   None

Spills: Take up in non-combustible absorbent and and place under an inert atmosphere. Decontaminate spill area using a dilute bicarbonate solution. Vent to dissipate vapors.

Special Emergency Information: May be harmful if inhaled, swallowed or absorbed through the skin. This material may spontaneously ignite in air.

RUBIDIUM HYDROXIDE - RbOH

Synonyms:

CAS Number: 1310-82-3   **Description:** White powder
DOT Number: 2677 (solution), 2678 (solid) **DOT Classification:** Corrosive Material

Molecular Weight: 102.5
Melting Point: 300°C (572°F)   **Vapor Density:**   N.D.   **Vapor Pressure:**   1 (<1 mm Hg)
Boiling Point: N.D.   **Specific Gravity:** 3.2   **Water Solubility:**   Soluble

Chemical Incompatibilities or Instabilities: Strong acids.

FLAMMABILITY

NFPA Hazard Code: **NFPA Classification:**
Flash Point: N.A.   **LEL:**   N.A.
Autoignition Temp: N.A.   **UEL:**   N.A.

Fire Extinguishing Methods: Use agent suitable for surrounding fire.

Special Fire Fighting Considerations: Water spray should be used to keep closed containers cool. Continue to cool container after fire is extinguished.

TOXICOLOGY

Odor: None   **Odor Threshold:** N.A.
Physical Contact: Material is extremely destructive to human tissues.
TLV = N.D.   **STEL** = N.D.   **IDLH** = N.D.

Routes of Entry and Relative LD$_{50}$ (or LC$_{50}$)   **RTECS # VL8750000**
    Inhalation   N.D.
    Ingestion   4   (586 mg/kg)
    Skin Absorption N.D.

Symptoms of Exposure: Possible burns or irritation to skin, eyes and upper respiratory tract; headache; nausea. Inhalation of vapors may be fatal by causing glottis to spasm and suffocation. Exposure may result in chemical pneumonitis or pulmonary edema.

Emergency/First Aid Treatment: Remove to ventilated area; immediately remove any contaminated clothing and wash contaminated areas for 15 minutes using water. Treat supportively and observe for possible shock. If ingested, seek immediate medical aid.

Recommended Clean-Up Procedures

Personal Protection: Level B Ensemble   **Recommended Material**   Butyl Rubber (-), Neoprene (-)

RCRA Waste #   None   **Reportable Quantities:**   None

Large Spills: Take up in non-combustible absorbent and dispose.

Special Emergency Information: May be harmful if inhaled, swallowed or absorbed through the skin.

SELENIC ACID - H$_2$SeO$_4$

Synonyms:

CAS Number: 7783-08-6

DOT Number: 1905

Description: White powder

DOT Classification: Corrosive Material

Molecular Weight: 145.0

Melting Point: 58°C (136°F)

Boiling Point: D 260°C (500°F)

Vapor Density: N.D.

Specific Gravity: 3.0

Vapor Pressure: 1 (<1 mm Hg)

Water Solubility: Soluble

Chemical Incompatibilities or Instabilities: Strong bases.

FLAMMABILITY

NFPA Hazard Code:

Flash Point: N.A.

Autoignition Temp: N.A.

NFPA Classification:

LEL: N.A.

UEL: N.A.

Fire Extinguishing Methods: Use agent suitable for surrounding fire.

Special Fire Fighting Considerations: Water spray should be used to keep closed containers cool. Continue to cool container after fire is extinguished. Treat all materials used or generated and equipment involved as contaminated by hazardous waste.

TOXICOLOGY
QUESTIONABLE CARCINOGEN

Odor:

Physical Contact: Material is extremely destructive to human tissues.

TLV = 0.2 mg (Se)/m^3 **STEL** = N.D.

Odor Threshold: N.D.

IDLH = N.D.

Routes of Entry and Relative LD$_{50}$ (or LC$_{50}$)

Inhalation   N.D.

Ingestion   N.D.

Skin Absorption   N.D.

RTECS # VS6575000

Symptoms of Exposure: Possible burns or irritation to skin, eyes and upper respiratory tract; headache; nausea.

Emergency/First Aid Treatment: Remove to ventilated area; immediately remove any contaminated clothing and wash contaminated areas for 15 minutes using water. Treat supportively and observe for possible shock. If ingested, seek immediate medical aid.

Recommended Clean-Up Procedures

Personal Protection: Level B Ensemble

RCRA Waste # None

Recommended Material N.D.

Reportable Quantities: None

Spills: Take up in non-combustible absorbent and dispose. Decontaminate spill area using water. Treat all materials used or generated and equipment involved as contaminated by hazardous waste.

Special Emergency Information: May be harmful if inhaled, swallowed or absorbed through the skin.

SELENIUM - Se

Synonyms:

CAS Number: 7782-49-2   **Description:** Grey to black metal
DOT Number: 2658   **DOT Classification:** Poison B

Molecular Weight: 79.0
Melting Point: 217°C (423°F)   **Vapor Density:** N.A.   **Vapor Pressure:** 1 (<1 mm Hg)
Boiling Point: 690°C (914°F)   **Specific Gravity:** 4.8   **Water Solubility:** Insoluble

Chemical Incompatibilities or Instabilities: Acids, carbides, halogens, oxidizers, uranium, nickel, zinc.

FLAMMABILITY

NFPA Hazard Code: **NFPA Classification:**
Flash Point: N.A.   **LEL:** N.A.
Autoignition Temp: N.A.   **UEL:** N.A.

Fire Extinguishing Methods: Carbon Dioxide, Dry Chemical, Foam, Water Spray or Fog.

Special Fire Fighting Considerations: Treat all materials used or generated and equipment involved as contaminated by hazardous waste.

TOXICOLOGY

Odor: None   **Odor Threshold:** N.A.
Physical Contact: Irritant
TLV = 0.2 mg (Se)/m^3 **STEL** = N.D.   **IDLH** = N.D.

Routes of Entry and Relative LD$_{50}$ (or LC$_{50}$)   **RTECS #** VS7700000
    Inhalation   N.D.
    Ingestion   **2**   (6700 mg/kg)
    Skin Absorption   N.D.

Symptoms of Exposure: Stomach pain, nausea, diarrhea, breathing difficulties, dermatitis.

Emergency/First Aid Treatment: Remove to ventilated area; immediately remove any contaminated clothing and wash contaminated areas for 15 minutes using water. Treat supportively and observe for possible shock. If ingested, seek immediate medical aid.

Recommended Clean-Up Procedures

Personal Protection: Level B Ensemble   **Recommended Material** N.D.

RCRA Waste #   None   **Reportable Quantities:** 100 lb (1 lb)

Spills: Take up in non-combustible absorbent and dispose. Decontaminate spill area using soapy water. Treat all materials used or generated and equipment involved as contaminated by hazardous waste.

Special Emergency Information: May be harmful if inhaled, swallowed or absorbed through the skin.

SELENIUM HEXAFLUORIDE - SeF$_6$

Synonyms: selenium fluoride

CAS Number: 7483-79-1   **Description:** Colorless Gas
DOT Number: 2194   **DOT Classification:** Poison A

Molecular Weight: 193.0
Melting Point: -39°C (-38°F)   **Vapor Density:** N.D.   **Vapor Pressure:**   Gas
Boiling Point: -35°C (-31°F)   **Specific Gravity:** 3.3   **Water Solubility:**   Decomposes

Chemical Incompatibilities or Instabilities: Oxidizers, silver (I) oxide.

FLAMMABILITY

NFPA Hazard Code: **NFPA Classification:**
Flash Point: N.D.   **LEL:** N.D.
Autoignition Temp: N.D.   **UEL:** N.D.

Fire Extinguishing Methods: Carbon Dioxide, Dry Chemical, Foam, Water Spray or Fog.

Special Fire Fighting Considerations: Structural fire fighter protective clothing will not provide adequate protection. Stop leak if it can safely be done. Water spray should be used to keep closed containers cool. Continue to cool container after fire is extinguished. Keep area isolated until vapors have dissipated.

DOT Recommended Isolation Zones: Small Spill: 1500 ft   Large Spill: 1500 ft
DOT Recommended Down Wind Take CoverDistance   Small Spill: 5 miles   Large Spill: 5 miles

TOXICOLOGY

Odor: **Odor Threshold:** N.D.
Physical Contact: Irritant
TLV = 0.4 mg/m^3 **STEL** = N.D.   **IDLH** = 40 mg/m^3

Routes of Entry and Relative LD$_{50}$ (or LC$_{50}$)   **RTECS # VS9450000**
    Inhalation   N.D.
    Ingestion   N.D.
    Skin Absorption   N.D.

Symptoms of Exposure: Possible irritation to skin, eyes and upper respiratory tract; headache; nausea. Exposure may result in chemical pneumonitis or pulmonary edema.

Emergency/First Aid Treatment: Remove to ventilated area; immediately remove any contaminated clothing and wash contaminated areas for 15 minutes using water. Treat supportively and observe for possible shock. If ingested, seek immediate medical aid.

Recommended Clean-Up Procedures

Personal Protection: Level A Ensemble   **Recommended Material**   N.D.

RCRA Waste #   None   **Reportable Quantities:**   None

Spills: Remove all potential ignition sources. Stop leak if it can safely be done. Keep area isolated until vapors have dissipated.

Special Emergency Information: May be fatal in inhaled, ingested or absorbed through the skin.

SELENIUM OXYCHLORIDE - SeOCl₂

Synonyms: seleninyl chloride; selenium chloride oxide

CAS Number: 7791-23-3   **Description:** Colorless to yellow liquid
DOT Number: 2879   **DOT Classification:** Corrosive Material

Molecular Weight: 165.9
Melting Point: 9°C   (48°F)   **Vapor Density:**   N.D.   **Vapor Pressure:**   1 (<1 mm Hg)
Boiling Point: 176°C (349°F)   **Specific Gravity:** 2.4   **Water Solubility:**   Decomposes

Chemical Incompatibilities or Instabilities: metal oxides, antimony, potassium, phosphorous.

FLAMMABILITY

NFPA Hazard Code: **NFPA Classification:**
Flash Point: N.A.   **LEL:**   N.A.
Autoignition Temp: N.A.   **UEL:**   N.A.

Fire Extinguishing Methods: Carbon Dioxide, Dry Chemical, Foam, Water Spray or Fog.

Special Fire Fighting Considerations: Water spray should be used to keep closed containers cool. Continue to cool container after fire is extinguished.

TOXICOLOGY

Odor: **Odor Threshold:** N.D.
Physical Contact: Material is extremely destructive to human tissues.
TLV = 0.2 mg (Se)/m³   **STEL** = N.D.   **IDLH** = N.D.

Routes of Entry and Relative LD₅₀ (or LC₅₀)   **RTECS # VS7000000**
    Inhalation   N.D.
    Ingestion   N.D.
    Skin Absorption   N.D.

Symptoms of Exposure: Possible burns or irritation to skin, eyes and upper respiratory tract; headache; nausea; stomach pains; diarrhea; breathing difficulties. Exposure may result in chemical pneumonitis or pulmonary edema.

Emergency/First Aid Treatment: Remove to ventilated area; immediately remove any contaminated clothing and wash contaminated areas for 15 minutes using water. Treat supportively and observe for possible shock. If ingested, seek immediate medical aid.

Recommended Clean-Up Procedures

Personal Protection: Level A Ensemble   **Recommended Material**   N.D.

RCRA Waste #   None   **Reportable Quantities:**   None

Spills: Dike to contain spill and collect liquid for later disposal.

Special Emergency Information: May be fatal in inhaled, ingested or absorbed through the skin.

SILANE - SiH$_4$

Synonyms: silicon tetrahydride

CAS Number: 7803-62-5	**Description:** Colorless Gas
DOT Number: 2203	**DOT Classification:** Flammable Gas

Molecular Weight: 32.1

Melting Point:	-185°C (-301°F)	**Vapor Density:** 1.1	**Vapor Pressure:**	Gas
Boiling Point:	-112°C (-170°F)	**Specific Gravity:** 0.7	**Water Solubility:**	Decomposes

Chemical Incompatibilities or Instabilities: Halogens, chlorides, oxygen, air.

FLAMMABILITY

NFPA Hazard Code:	2-4-3	**NFPA Classification:**	
Flash Point:	N.D.	**LEL:**	1.4%
Autoignition Temp:	N.D.	**UEL:**	96%

Fire Extinguishing Methods: Carbon Dioxide, Dry Chemical, Foam, Water Spray or Fog.

Special Fire Fighting Considerations: Isolate for 1/2 mile if rail or tank truck is involved in a fire. Water spray from an unmanned device should be used to keep closed containers cool. Continue to cool container after fire is extinguished. Stop leak if it can safely be done. For large fires, if possible, withdraw and allow to burn. Immediately withdraw if rising sound from venting device is heard or if fire is causing discoloration to the tank.

TOXICOLOGY

Odor: Repulsive	**Odor Threshold:** N.D.
Physical Contact: Irritant	
TLV = 6.6 mg/m^3 **STEL** = N.D.	**IDLH** = N.D.

Routes of Entry and Relative LD$_{50}$ (or LC$_{50}$)   **RTECS # VV1400000**

Inhalation	2	(50,000 mg/m^3/H)
Ingestion	N.D.	
Skin Absorption	N.D.	

Symptoms of Exposure: Dizziness, headache, nausea.

Emergency/First Aid Treatment: Remove to ventilated area; immediately remove any contaminated clothing and wash contaminated areas for 15 minutes using water. Treat supportively and observe for possible shock. If ingested, seek immediate medical aid.

Recommended Clean-Up Procedures

Personal Protection:	Level B Ensemble	**Recommended Material** N.D.
RCRA Waste #	None	**Reportable Quantities:** None

Spills: Stop leak if it can safely be done. Keep area isolated until vapors have dissipated.

Special Emergency Information: May be harmful if inhaled, swallowed or absorbed through the skin. Material may spontaneously ignite upon exposure to air.

SILICON CHLORIDE - SiCl₄

Synonyms: silicon tetrachloride; tetrachlorosilane

CAS Number: 10026-04-7		**Description:** Colorless Liquid	
DOT Number: 1818		**DOT Classification:** Corrosive Material	

Molecular Weight: 169.9

Melting Point:	-57°C (-70°F)	**Vapor Density:** 5.9	**Vapor Pressure:** 7 (194 mm Hg)
Boiling Point:	59°C (138°F)	**Specific Gravity:** 1.5	**Water Solubility:** Decomposes

Chemical Incompatibilities or Instabilities: Alkali metals, water, dimethyl sulfoxide, dimethyl formamide.

FLAMMABILITY

NFPA Hazard Code:	3-0-2-W̶	**NFPA Classification:**	
Flash Point:	N.A.	**LEL:**	N.A.
Autoignition Temp:	N.A.	**UEL:**	N.A.

Fire Extinguishing Methods: Carbon Dioxide or Dry Chemical.

Special Fire Fighting Considerations: Structural fire fighter protective clothing will not provide adequate protection. Water spray should be used to keep closed containers cool. Continue to cool container after fire is extinguished. Do not get water inside container.

TOXICOLOGY

Odor: Suffocating   **Odor Threshold:** N.D.

Physical Contact: Material is extremely destructive to human tissues.

TLV = N.D.   **STEL** = N.D.   **IDLH** = N.D.

Routes of Entry and Relative LD₅₀ (or LC₅₀)   **RTECS # VW0525000**

Inhalation	2 (250,000 mg/m³/H)
Ingestion	N.D.
Skin Absorption	N.D.

Symptoms of Exposure: Possible burns or irritation to skin, eyes and upper respiratory tract; headache; nausea. Inhalation of vapors may be fatal by causing glottis to spasm and suffocation. Exposure may result in chemical pneumonitis or pulmonary edema.

Emergency/First Aid Treatment: Remove to ventilated area; immediately remove any contaminated clothing and wash contaminated areas for 15 minutes using water. Treat supportively and observe for possible shock. If ingested, seek immediate medical aid.

Recommended Clean-Up Procedures

Personal Protection:	Level B Ensemble	**Recommended Material**	N.D.
RCRA Waste #	None	**Reportable Quantities:**	None

Spills: Dike to contain spill and collect liquid for later disposal. Keep area isolated until vapors have dissipated.

Special Emergency Information: May be harmful if inhaled, swallowed or absorbed through the skin.

SILICON TETRAFLUORIDE - SiF$_4$

Synonyms: silicon fluoride; tetrafluorosilane

CAS Number: 7783-61-1   **Description:** Colorless Gas
DOT Number: 1859   **DOT Classification:** Nonflammable Gas

Molecular Weight: 104.1
Melting Point: -90°C (-130°F)   **Vapor Density:**   4.7   **Vapor Pressure:**   Gas
Boiling Point: 65°C (123°F)   **Specific Gravity:** N.A.   **Water Solubility:**   Decomposes

Chemical Incompatibilities or Instabilities: Water.

FLAMMABILITY

NFPA Hazard Code: 3-0-2-W   **NFPA Classification:**
Flash Point: N.A.   **LEL:**   N.A.
Autoignition Temp: N.A.   **UEL:**   N.A.

Fire Extinguishing Methods: Carbon Dioxide, Dry Chemical, Foam, Water Spray or Fog.

Special Fire Fighting Considerations: Water spray should be used to keep closed containers cool. Continue to cool container after fire is extinguished. Stop leak if it can safely be done. Keep area isolated until vapors have dissipated.

DOT Recommended Isolation Zones: Small Spill: 1200 ft   Large Spill: 1500 ft
DOT Recommended Down Wind Take CoverDistance   Small Spill: 4 miles   Large Spill: 5 miles

TOXICOLOGY

Odor: Repulsive   **Odor Threshold:** N.D.
Physical Contact: Material is extremely destructive to human tissues.
TLV = 2.5 mg (F)/m^3 **STEL** = N.D.   **IDLH** = 500 mg/m^3

Routes of Entry and Relative LD$_{50}$ (or LC$_{50}$)   **RTECS # VW2327000**
   Inhalation   N.D.
   Ingestion   N.D.
   Skin Absorption   N.D.

Symptoms of Exposure: Possible burns or irritation to skin, eyes and upper respiratory tract; headache; nausea. Inhalation of vapors may be fatal by causing glottis to spasm and suffocation. Exposure may result in chemical pneumonitis or pulmonary edema.

Emergency/First Aid Treatment: Remove to ventilated area; immediately remove any contaminated clothing and wash contaminated areas for 15 minutes using water. Treat supportively and observe for possible shock. If ingested, seek immediate medical aid.

Recommended Clean-Up Procedures

Personal Protection: Level A Ensemble   **Recommended Material**   N.D.

RCRA Waste #   None   **Reportable Quantities:**   None

Spills: Stop leak if it can safely be done. Keep area isolated until vapors have dissipated.

Special Emergency Information: May be fatal in inhaled, ingested or absorbed through the skin. Reacts with water to yield hydrogen fluoride.

SILVER NITRATE -AgNO₃

Synonyms:

CAS Number: 7761-88-8   **Description:** White powder
DOT Number: 1493   **DOT Classification:** Oxidizer

Molecular Weight: 169.9
Melting Point: 212°C (414°F)   **Vapor Density:** N.A.   **Vapor Pressure:** 1 (< 1 mm Hg)
Boiling Point: D 444°C (831°F)   **Specific Gravity:** 4.4   **Water Solubility:** Soluble

Chemical Incompatibilities or Instabilities: Reducing agents, alkenes, alkynes, phosphine, phosphorous, sulfur, ammonia, alkali

FLAMMABILITY

NFPA Hazard Code: **NFPA Classification:**
Flash Point: N.A.   **LEL:** N.A.
Autoignition Temp: N.A.   **UEL:** N.A.

Fire Extinguishing Methods: WATER ONLY

Special Fire Fighting Considerations: Water spray should be used to keep closed containers cool. Continue to cool container after fire is extinguished. For large fires, if possible, withdraw and allow to burn. Treat all materials used or generated and equipment involved as contaminated by hazardous waste.

TOXICOLOGY

Odor: None   **Odor Threshold:** N.A.
Physical Contact: Material is extremely destructive to human tissues.
TLV = 0.01 mg (Ag)/m³   **STEL** = N.D.   **IDLH** = N.D.

Routes of Entry and Relative LD₅₀ (or LC₅₀)   **RTECS # VW4725000**
   Inhalation   N.D.
   Ingestion   N.D.
   Skin Absorption   N.D.

Symptoms of Exposure: Possible burns or irritation to skin, eyes and upper respiratory tract; headache; nausea. Inhalation of vapors may be fatal by causing glottis to spasm and suffocation. Exposure may result in chemical pneumonitis or pulmonary edema.

Emergency/First Aid Treatment: Remove to ventilated area; immediately remove any contaminated clothing and wash contaminated areas for 15 minutes using water. Treat supportively and observe for possible shock. If ingested, seek immediate medical aid.

Recommended Clean-Up Procedures

Personal Protection: Level B Ensemble   **Recommended Material** N.D.

RCRA Waste # None   **Reportable Quantities:** 1 lb (1 lb)

Spills: Take up in non-combustible absorbent and dispose. Decontaminate spill area using water. Treat all materials used or generated and equipment involved as contaminated by hazardous waste.

Special Emergency Information: May be fatal in inhaled, ingested or absorbed through the skin.

SODIUM - Na

Synonyms: natrium

CAS Number: 7440-23-5   **Description:** Grey solid
DOT Number: 1428, 1429   **DOT Classification:** Flammable Solid

Molecular Weight: 22.9
Melting Point: 98°C (208°F)   **Vapor Density:** N.D.   **Vapor Pressure:** 1 (<1 mm Hg)
Boiling Point: 881°C (1619°F)   **Specific Gravity:** 1.0   **Water Solubility:** Decomposes

Chemical Incompatibilities or Instabilities: Water, metal chlorides, metal bromides, ammonia, ammonium nitrate, halogens, halogenated hydrocarbons, metal oxides, oxidizers.

FLAMMABILITY

NFPA Hazard Code: 3-3-2-W   **NFPA Classification:**
Flash Point: N.D.   **LEL:** N.D.
Autoignition Temp: N.D.   **UEL:** N.D.

Fire Extinguishing Methods: Dry Sand, Class D Extinguishers, Dry Clay, Dry Limestone.

Special Fire Fighting Considerations: For large fires, if possible, withdraw and allow to burn.

TOXICOLOGY

Odor: None   **Odor Threshold:** N.A.
Physical Contact: Material is extremely destructive to human tissues.
TLV = N.D.   **STEL** = N.D.   **IDLH** = N.D.

Routes of Entry and Relative LD$_{50}$ (or LC$_{50}$)   **RTECS # VY0686000**
   Inhalation   N.D.
   Ingestion   N.D.
   Skin Absorption   N.D.

Symptoms of Exposure: Possible burns or irritation to skin, eyes and upper respiratory tract; headache; nausea. Inhalation of vapors may be fatal by causing glottis to spasm and suffocation. Exposure may result in chemical pneumonitis or pulmonary edema.

Emergency/First Aid Treatment: Remove to ventilated area; immediately remove any contaminated clothing and wash contaminated areas for 15 minutes using water. Treat supportively and observe for possible shock. If ingested, seek immediate medical aid.

Recommended Clean-Up Procedures

Personal Protection: Level B Ensemble   **Recommended Material** N.D.

RCRA Waste #   None   **Reportable Quantities:** 10 lb (1000 lb)

Spills: Take up in non-combustible absorbent and dispose. Keep material under a dry inert atmosphere.

Special Emergency Information: May be fatal in inhaled, ingested or absorbed through the skin. Sodium will react with water to liberate extremely flammable hydrogen gas.

SODIUM AMIDE - NaNH$_2$

Synonyms: sodamide

CAS Number: 7782-92-5

DOT Number: 1425

Description: White powder (typically in an organic suspension)

DOT Classification: Flammable Solid

Molecular Weight: 39.0

Melting Point: 210°C (410°F)	**Vapor Density:** N.D.	**Vapor Pressure:** N.D.
Boiling Point: 400°C (752°F)	**Specific Gravity:** N.A.	**Water Solubility:** Decomposes

Chemical Incompatibilities or Instabilities: Water, oxidizers, halogens, halocarbons, heat.

FLAMMABILITY

NFPA Hazard Code:

Flash Point: 29°C (85°F)

Autoignition Temp: N.D.

NFPA Classification:

LEL: N.D.

UEL: N.D.

Fire Extinguishing Methods: Dry Sand, Soda Ash, Dry Chemical, Limestone.

Special Fire Fighting Considerations: **DO NOT USE WATER.** Do not allow water to enter container. For large fires, if possible, withdraw and allow to burn.

TOXICOLOGY

Odor: Ammonia

Odor Threshold: N.D.

Physical Contact: Material is extremely destructive to human tissues.

TLV = N.D.   **STEL** = N.D.   **IDLH** = N.D.

Routes of Entry and Relative LD$_{50}$ (or LC$_{50}$)

Inhalation	N.D.	
Ingestion	N.D.	
Skin Absorption	N.D.	

RTECS # VY2775000

Symptoms of Exposure: Possible burns or irritation to skin, eyes and upper respiratory tract; headache; nausea.

Emergency/First Aid Treatment: Remove to ventilated area; immediately remove any contaminated clothing and wash contaminated areas for 15 minutes using water. Treat supportively and observe for possible shock. If ingested, seek immediate medical aid.

Recommended Clean-Up Procedures

Personal Protection: Level B Ensemble

Recommended Material N.D.

RCRA Waste # None

Reportable Quantities: None

Spills: Remove all potential ignition sources. Absorb with non-combustible absorbent and take up using non-sparking tools. Vent to dissipate vapors.

Special Emergency Information: May be harmful if inhaled, swallowed or absorbed through the skin. Sodium amide typically comes as a suspension in a flammable solvent such as toluene or xylene. May ignite or explode upon heating or grinding. May become an explosion hazard upon prolonged storage.

SODIUM AZIDE - NaN$_3$

Synonyms:

CAS Number: 26628-22-8   **Description:** White powder
DOT Number: 1687   **DOT Classification:** Poison B

Molecular Weight: 65.0
Melting Point: Decomposes   **Vapor Density:**   N.A.   **Vapor Pressure:**   **1** (< 1 mm Hg)
Boiling Point: N.A.   **Specific Gravity:** 1.9   **Water Solubility:**   Soluble

Chemical Incompatibilities or Instabilities: Acids, metals, metal halides, halogens, hydrazine, dimethyl sulfate, carbon disulfide, barium carbonate, strong alkali.

FLAMMABILITY

NFPA Hazard Code: **NFPA Classification:**
Flash Point: N.A.   **LEL:**   N.A.
Autoignition Temp: N.A.   **UEL:**   N.A.

Fire Extinguishing Methods: Carbon Dioxide, Dry Chemical, Foam, Water Spray or Fog.

Special Fire Fighting Considerations: Structural fire fighter protective clothing will not provide adequate protection. Water spray should be used to keep closed containers cool. Continue to cool container after fire is extinguished. For large fires, if possible, withdraw and allow to burn.

TOXICOLOGY
QUESTIONABLE CARCINOGEN

Odor: None   **Odor Threshold:** N.A.
Physical Contact: Irritant
TLV = 0.3 mg/m^3 (ceiling)   **STEL** = N.D.   **IDLH** = N.D.

Routes of Entry and Relative LD$_{50}$ (or LC$_{50}$)   **RTECS # VY8050000**
    Inhalation   N.D.
    Ingestion   **9**   (27 mg/kg)
    Skin Absorption   **10**   (20 mg/kg)

Symptoms of Exposure: Irritation.

Emergency/First Aid Treatment: Remove to ventilated area; immediately remove any contaminated clothing and wash contaminated areas for 15 minutes using water. Treat supportively and observe for possible shock. If ingested, seek immediate medical aid.

Recommended Clean-Up Procedures

Personal Protection: Level A Ensemble   **Recommended Material**   N.D.

RCRA Waste #   P105   **Reportable Quantities:**   1000 lb (1 lb)

Spills: Absorb with non-combustible absorbent and take up using non-sparking tools. Decontaminate spill area using water. Treat all materials used or generated and equipment involved as contaminated by hazardous waste.

Special Emergency Information: May be fatal in inhaled, ingested or absorbed through the skin.

SODIUM BISULFATE - NaHSO₃

Synonyms:

CAS Number: 7631-90-5	**Description:** White powder
DOT Number: 2693, 2837, 1821	**DOT Classification:** ORM-B

Molecular Weight: 104.1

Melting Point: Decomposes	**Vapor Density:** N.A.	**Vapor Pressure:**	1 (< 1 mm Hg)
Boiling Point: N.A.	**Specific Gravity:** 1.5	**Water Solubility:**	Soluble

Chemical Incompatibilities or Instabilities: Strong acids, oxidizers.

FLAMMABILITY

NFPA Hazard Code:	**NFPA Classification:**
Flash Point: N.A.	**LEL:** N.A.
Autoignition Temp: N.A.	**UEL:** N.A.

Fire Extinguishing Methods: Carbon Dioxide, Dry Chemical, Foam, Water Spray or Fog.

Special Fire Fighting Considerations: Water spray should be used to keep closed containers cool. Continue to cool container after fire is extinguished.

TOXICOLOGY

Odor: None **Odor Threshold:** N.A.

Physical Contact: Material is extremely destructive to human tissues.

TLV = 5 mg/m³ **STEL** = N.D. **IDLH** = N.D.

Routes of Entry and Relative LD₅₀ (or LC₅₀) **RTECS # VZ2000000**

Inhalation	N.D.
Ingestion	**3** (2000 mg/kg)
Skin Absorption	N.D.

Symptoms of Exposure: Possible burns or irritation to skin, eyes and upper respiratory tract; headache; nausea, Inhalation of vapors may be fatal by causing glottis to spasm and suffocation. Exposure may result in chemical pneumonitis or pulmonary edema.

Emergency/First Aid Treatment: Remove to ventilated area; immediately remove any contaminated clothing and wash contaminated areas for 15 minutes using water. Treat supportively and observe for possible shock. If ingested, seek immediate medical aid.

Recommended Clean-Up Procedures

Personal Protection:	Level B Ensemble	**Recommended Material** N.D.
RCRA Waste #	None	**Reportable Quantities:** None

Spills: Take up in non-combustible absorbent and dispose. Decontaminate spill area using water.

Special Emergency Information: May be harmful if inhaled, swallowed or absorbed through the skin.

Potassium bisulfate (CAS # 7646-93-7, DOT # 2509, RTECS # TS7200000) has similar chemical, physical and toxicological properties.

SODIUM BOROHYDRIDE - NaBH$_4$

Synonyms:

CAS Number: 16940-66-2   **Description:** White powder
DOT Number: 1426   **DOT Classification:** Flammable Solid

Molecular Weight: 37.8
Melting Point: >300°C(°F) D   **Vapor Density:** N.D.   **Vapor Pressure:** 1 (<1 mm Hg)
Boiling Point: N.A.   **Specific Gravity:** 1.1   **Water Solubility:** Decomposes

Chemical Incompatibilities or Instabilities: Alkali, water, alcohols, finely divided metals, strong acids, oxidizers, ruthenium salts, dimethylformamide, metal salts.

FLAMMABILITY

NFPA Hazard Code: **NFPA Classification:**
Flash Point: N.A.   **LEL:** N.A.
Autoignition Temp: N.A.   **UEL:** N.A.

Fire Extinguishing Methods: Carbon Dioxide, Dry Chemical, Foam, Water Spray or Fog.

Special Fire Fighting Considerations: Water spray should be used to keep closed containers cool. Continue to cool container after fire is extinguished. For large fires, if possible, withdraw and allow to burn.

TOXICOLOGY

Odor: None   **Odor Threshold:** N.A.
Physical Contact: Material is extremely destructive to human tissues.
TLV = N.D.   **STEL** = N.D.   **IDLH** = N.D.

Routes of Entry and Relative LD$_{50}$ (or LC$_{50}$)   **RTECS # ED3325000**
   Inhalation   N.D.
   Ingestion   N.D.
   Skin Absorption   N.D.

Symptoms of Exposure: Possible burns or irritation to skin, eyes and upper respiratory tract; headache; nausea. Inhalation of vapors may be fatal by causing glottis to spasm and suffocation. Exposure may result in chemical pneumonitis or pulmonary edema.

Emergency/First Aid Treatment: Remove to ventilated area; immediately remove any contaminated clothing and wash contaminated areas for 15 minutes using water. Treat supportively and observe for possible shock. If ingested, seek immediate medical aid.

Recommended Clean-Up Procedures

Personal Protection: Level B Ensemble   **Recommended Material** N.D.

RCRA Waste #   None   **Reportable Quantities:**   None

Spills: Remove all potential ignition sources. Absorb with non-combustible absorbent and take up using non-sparking tools. Decontaminate spill area using a dilute acid solution.

Special Emergency Information: May be harmful if inhaled, swallowed or absorbed through the skin. Contact with aqueous alkaline or neutral solutions may liberate flammable hydrogen gas.

Potassium borohydride (CAS # 13762-51-1, DOT # 1870, RTECS # TS7525000) has similar chemical, physical and toxicological properties.

SODIUM CACODYLATE - $(CH_3)_2AsO_2Na$

Synonyms: cacodylic acid, sodium salt; hydroxydimethylarsine oxide

CAS Number: 124-65-2		**Description:** White powder		
DOT Number: 1688		**DOT Classification:** Poison B		

Molecular Weight: 160.0

Melting Point:	N.A.	**Vapor Density:** N.D.	**Vapor Pressure:**	1 (<1 mm Hg)
Boiling Point:	N.A.	**Specific Gravity:** N.A.	**Water Solubility:**	Soluble

Chemical Incompatibilities or Instabilities: Acids, strong oxidizers.

FLAMMABILITY

NFPA Hazard Code:		**NFPA Classification:**
Flash Point:	N.A.	**LEL:** N.A.
Autoignition Temp:	N.A.	**UEL:** N.A.

Fire Extinguishing Methods: Carbon Dioxide, Dry Chemical, Foam, Water Spray or Fog.

Special Fire Fighting Considerations: Structural fire fighter protective clothing will not provide adequate protection. Treat all materials used or generated and equipment involved as contaminated by hazardous waste.

TOXICOLOGY
CARCINOGEN

Odor: None	**Odor Threshold:** N.A.

Physical Contact: Irritant

TLV = 0.5 mg (As)/m^3	**STEL** = N.D.	**IDLH** = N.D.

Routes of Entry and Relative LD$_{50}$ (or LC$_{50}$)   **RTECS # CH7700000**

Inhalation	N.D.	
Ingestion	3	(2600 mg/kg)
Skin Absorption	N.D.	

Symptoms of Exposure: Possible irritation.

Emergency/First Aid Treatment: Remove to ventilated area; immediately remove any contaminated clothing and wash contaminated areas for 15 minutes using water. Treat supportively and observe for possible shock. If ingested, seek immediate medical aid. Symptoms of arsenic poisoning may be delayed for several days. Anyone potentially exposed should be examined by a physician and kept under observation if necessary.

Recommended Clean-Up Procedures

Personal Protection: Level A Ensemble	**Recommended Material** N.D.
RCRA Waste # None	**Reportable Quantities:** None

Spills: Take up in non-combustible absorbent and dispose. Decontaminate spill area using water. Treat all materials used or generated and equipment involved as contaminated by hazardous waste.

Special Emergency Information: May be fatal in inhaled, ingested or absorbed through the skin.

Cacodylic acid (CAS # 75-60-5, DOT # 1572, RTECS # CH7525000) has similar chemical, physical and toxicological properties.

SODIUM CHLORITE

Synonyms:

CAS Number: 7758-19-2	**Description:** White powder		
DOT Number: 1496, 1908	**DOT Classification:** Oxidizer		

Molecular Weight: 90.4

Melting Point:	180°C (356°F) D	**Vapor Density:**	N.D.	**Vapor Pressure:**	1 (<1 mm Hg)
Boiling Point:	°C (°F)	**Specific Gravity:**	N.D.	**Water Solubility:**	Soluble

Chemical Incompatibilities or Instabilities: Acids, phosphorous, sulfur, zinc, ammonia, hydrocarbons, heat, reducing agents.

FLAMMABILITY

NFPA Hazard Code:		**NFPA Classification:**	
Flash Point:	N.A.	**LEL:**	N.A.
Autoignition Temp:	N.A.	**UEL:**	N.A.

Fire Extinguishing Methods: WATER ONLY

Special Fire Fighting Considerations: Use flooding quantities of water from a distance. Structural fire fighter protective clothing will not provide adequate protection. Isolate for 1/2 mile if rail or tank truck is involved in a fire. Fight fire from a distance or protected location, if possible. For large fires, if possible, withdraw and allow to burn. Do not attempt to remove heated containers.

TOXICOLOGY

Odor: Faint chlorine-like	**Odor Threshold:** N.D.
Physical Contact: Irritant	

TLV = N.D.	**STEL** = N.D.	**IDLH** = N.D.

Routes of Entry and Relative LD$_{50}$ (or LC$_{50}$)   **RTECS # VZ4800000**

Inhalation	N.D.
Ingestion	7 (165 mg/kg)
Skin Absorption	N.D.

Symptoms of Exposure: Possible irritation to skin, eyes and upper respiratory tract; headache; nausea. Absorption may lead to the formation of methemoglobin resulting in cyanosis several hours after exposure.

Emergency/First Aid Treatment: Remove to ventilated area; immediately remove any contaminated clothing and wash contaminated areas for 15 minutes using water. Treat supportively and observe for possible shock. If ingested, seek immediate medical aid.

Recommended Clean-Up Procedures

Personal Protection:	Level B Ensemble	**Recommended Material** N.D.
RCRA Waste #	None	**Reportable Quantities:** None

Spills: Take up in non-combustible absorbent and dispose. Decontaminate spill area using water.

Special Emergency Information: May be harmful if inhaled, swallowed or absorbed through the skin. Contact with acids may yield chlorine dioxide gas.

Potassium chlorite (CAS # 14314-27-3) has similar chemical, physical and toxicological properties.

SODIUM CYANIDE - HC≡N

Synonyms:

CAS Number: 143-33-9

DOT Number: 1689

Description: White powder

DOT Classification: Poison B

Molecular Weight: 49.0

Melting Point:	564°C (1047°F)	**Vapor Density:** N.D.	**Vapor Pressure:**	1 (<1 mm Hg)
Boiling Point:	1496°C (2725°F)	**Specific Gravity:** 1.6	**Water Solubility:**	Soluble

Chemical Incompatibilities or Instabilities: Acids, oxidizers.

FLAMMABILITY

NFPA Hazard Code:

Flash Point: N.A.

Autoignition Temp: N.A.

NFPA Classification:

LEL: N.A.

UEL: N.A.

Fire Extinguishing Methods: Use agent suitable for surrounding fire.

Special Fire Fighting Considerations: Structural fire fighter protective clothing will not provide adequate protection. Fight fire from a distance or protected location, if possible. Treat all materials used or generated and equipment involved as contaminated by hazardous waste.

TOXICOLOGY

Odor: None

Physical Contact: Irritant

TLV = 5 mg (CN)/m^3 **STEL** = N.D.

Odor Threshold: N.D.

IDLH = 50 mg (CN)/m^3

Routes of Entry and Relative LD$_{50}$ (or LC$_{50}$)

Inhalation	N.D.	
Ingestion	**9**	(6 mg/kg)
Skin Absorption	N.D.	

RTECS # VZ7525000

Symptoms of Exposure: Possible irritation to skin, eyes and upper respiratory tract; headache; nausea, cyanosis.

Emergency/First Aid Treatment: Remove to ventilated area; immediately remove any contaminated clothing and wash contaminated areas for 15 minutes using water. Treat supportively and observe for possible shock. If ingested, seek immediate medical aid.

Recommended Clean-Up Procedures

Personal Protection: Level A Ensemble   **Recommended Material** N.D.

RCRA Waste # P106   **Reportable Quantities:** 10 lbs (10 lbs)

Spills: Take up in non-combustible absorbent and dispose. Decontaminate spill area using an alkalike hypochlorite solution. Treat all materials used or generated and equipment involved as contaminated by hazardous waste.

Special Emergency Information: May be fatal in inhaled, ingested or absorbed through the skin. Will react with acids to yield highly toxic and flammable hydrogen cyanide vapors.

Potassium cyanide (CAS # 151-50-8, DOT # 1680, RTECS # TS8750000) has similar chemical, physical and toxicological properties.

SODIUM FLUORIDE - NaF

Synonyms:

CAS Number: 7681-49-4   **Description:** White powder
DOT Number: 1690   **DOT Classification:** ORM-B

Molecular Weight: 42.0
Melting Point: 993°C (1819°F)   **Vapor Density:**   N.D.   **Vapor Pressure:**   1 (<1 mm Hg)
Boiling Point: 1700°C (3092°F)   **Specific Gravity:** 2.0   **Water Solubility:**   Soluble

Chemical Incompatibilities or Instabilities: Acid.

FLAMMABILITY

NFPA Hazard Code: 2-0-0   **NFPA Classification:**
Flash Point: N.A.   **LEL:** N.A.
Autoignition Temp: N.A.   **UEL:** N.A.

Fire Extinguishing Methods: Use agent suitable for surrounding fire.

Special Fire Fighting Considerations: Structural fire fighter protective clothing will not provide adequate protection. Treat all materials used or generated and equipment involved as contaminated by hazardous waste.

TOXICOLOGY

Odor: None
Physical Contact: Irritant   **Odor Threshold:** N.D.
TLV = 2.5 mg (F)/m^3 **STEL** = N.D.   **IDLH** = 500 mg (F)/m^3

Routes of Entry and Relative LD$_{50}$ (or LC$_{50}$)   **RTECS # WB0350000**
   Inhalation   N.D.
   Ingestion   **8**   (52 mg/kg)
   Skin Absorption   N.D.

Symptoms of Exposure: Possible irritation to skin, eyes and upper respiratory tract; headache; nausea, salivation, breathing difficulties, abdominal pain.

Emergency/First Aid Treatment: Remove to ventilated area; immediately remove any contaminated clothing and wash contaminated areas for 15 minutes using water. Treat supportively and observe for possible shock. If ingested, seek immediate medical aid.

Recommended Clean-Up Procedures

Personal Protection: Level B Ensemble   **Recommended Material**   Natural Rubber (+), Neoprene (+),
                                                                            Nitrile Rubber (+), Polyvinyl Chloride (+)

RCRA Waste #   None   **Reportable Quantities:**   1000 lbs (1000 lbs)

Spills: Take up in non-combustible absorbent and dispose. Decontaminate spill area using watse. Treat all materials used or generated and equipment involved as contaminated by hazardous waste.

Special Emergency Information: May be fatal in inhaled, ingested or absorbed through the skin. Reactions with acids will yield hydrofluoric acid.

Potassium fluoride (CAS # 7789-23-2, DOT # 1812, RTECS # TT0700000) has similar chemical, physical and toxicological properties.

SODIUM FLUOROACETATE - CH2FCOONa

Synonyms: flouroacetic acid, sodium salt; fratol; compound #1080; ratbane 1080

CAS Number: 62-74-8

DOT Number: 2629

Description: White powder (may be dyed black)

DOT Classification: Poison B

Molecular Weight: 100.0

Melting Point: 35°C (95°F)

Boiling Point: 178°C (352°F)

Vapor Density: N.D.

Specific Gravity: N.D.

Vapor Pressure: 1 (<1 mm Hg)

Water Solubility: Soluble

Chemical Incompatibilities or Instabilities: Acids, oxidizers.

FLAMMABILITY

NFPA Hazard Code:

Flash Point: N.A.

Autoignition Temp: N.A.

NFPA Classification:

LEL: N.A.

UEL: N.A.

Fire Extinguishing Methods: Use agent suitable for surrounding fire.

Special Fire Fighting Considerations: Structural fire fighter protective clothing will not provide adequate protection. Treat all materials used or generated and equipment involved as contaminated by hazardous waste.

TOXICOLOGY

Odor: None

Physical Contact: Irritant

TLV = 0.05 mg/m^3

Odor Threshold: N.D.

STEL = 0.15 mg/m^3

IDLH = 5 mg/m^3

Routes of Entry and Relative LD$_{50}$ (or LC$_{50}$)

RTECS # AH9100000

Inhalation	N.D.	
Ingestion	**10**	(0.1 mg/kg)
Skin Absorption	N.D.	

Symptoms of Exposure: Nausea, muscle spasms, low blood pressure, convulsions, circulatory collapse.

Emergency/First Aid Treatment: Remove to ventilated area; immediately remove any contaminated clothing and wash contaminated areas for 15 minutes using water. Treat supportively and observe for possible shock. If ingested, seek immediate medical aid.

Recommended Clean-Up Procedures

Personal Protection: Level A Ensemble

RCRA Waste # None

Recommended Material N.D.

Reportable Quantities: None

Spills: Take up in non-combustible absorbent and dispose. Decontaminate spill area using water. Treat all materials used or generated and equipment involved as contaminated by hazardous waste.

Special Emergency Information: May be fatal in inhaled, ingested or absorbed through the skin.

Potassium fluoroacetate (CAS # 23745-86-0, DOT # 2628, RTECS # AH8800000) has similar chemical, physical and toxicological properties.

SODIUM HYDRIDE - NaH

Synonyms:

CAS Number: 7646-69-7

DOT Number: 1427

Description: Grey powder

DOT Classification: Flammable Solid

Molecular Weight: 24.0

Melting Point: D 800°C (1472°F)

Boiling Point: N.A.

Vapor Density: N.D.

Specific Gravity: 0.9

Vapor Pressure: 1 (<1 mm Hg)

Water Solubility: Decomposes

Chemical Incompatibilities or Instabilities: Water, dimethylformamide, acetylene, halogens, sulfur, alcohols.

FLAMMABILITY

NFPA Hazard Code: 3-3-2-W

Flash Point: N.A.

Autoignition Temp: N.A.

NFPA Classification:

LEL: N.A.

UEL: N.A.

Fire Extinguishing Methods: Dry Sand, Soda Ash, Crushed Limestone.

Special Fire Fighting Considerations: DO NOT USE WATER. Reacts violently with water to yield hydrogen gas. Do not allow water to enter container.

TOXICOLOGY

Odor: Kerosene

Physical Contact: Material is extremely destructive to human tissues.

TLV = N.D.   **STEL** = N.D.

Odor Threshold: N.D.

IDLH = N.D.

Routes of Entry and Relative LD$_{50}$ (or LC$_{50}$)

Inhalation	N.D.
Ingestion	N.D.
Skin Absorption	N.D.

RTECS # WB3910000

Symptoms of Exposure: Possible burns or irritation to skin, eyes and upper respiratory tract; headache; nausea. Laryngitis.

Emergency/First Aid Treatment: Remove to ventilated area; immediately remove any contaminated clothing and wash contaminated areas for 15 minutes using water. Treat supportively and observe for possible shock. If ingested, seek immediate medical aid.

Recommended Clean-Up Procedures

Personal Protection: Level B Ensemble

RCRA Waste # None

Recommended Material N.D.

Reportable Quantities: None

Spills: Remove all potential ignition sources. Absorb with non-combustible absorbent and take up using non-sparking tools. Decontaminate spill area using water. Vent to dissipate vapors.

Special Emergency Information: May be harmful if inhaled, swallowed or absorbed through the skin. This product is typically found as a mixture containing mineral oil, kerosene or other hydrocarbons.

Potassium hydride (CAS # 7693-26-7) has similar chemical, physical and toxicological properties.

SODIUM HYDROSULFIDE - NaSH

Synonyms: sodium bisulfide; sodium hydrogen sulfide; sodium mercaptan; sodium sulfhydrate

CAS Number: 16721-80-5 **Description:** Yellow powder
DOT Number: 2949, 2922, 2923, 2318 **DOT Classification:** Corrosive Material/Flammable Solid

Molecular Weight: 56.1
Melting Point: 52°C (126°F) **Vapor Density:** N.D. **Vapor Pressure:** 1 (<1 mm Hg)
Boiling Point: °C (°F) **Specific Gravity:** N.D. **Water Solubility:** Soluble

Chemical Incompatibilities or Instabilities: Air, oxidizers, pure metals and alloys containing zinc, aluminum or copper, daizonium salts.

FLAMMABILITY

NFPA Hazard Code: **NFPA Classification:**
Flash Point: N.D. **LEL:** N.D.
Autoignition Temp: N.D. **UEL:** N.D.

Fire Extinguishing Methods: Use agent suitable for surrounding fire.

Special Fire Fighting Considerations: Structural fire fighter protective clothing will not provide adequate protection. Water spray should be used to keep closed containers cool. Continue to cool container after fire is extinguished. For large fires, if possible, withdraw and allow to burn. Treat all materials used or generated and equipment involved as contaminated by hazardous waste. May release toxic fumes upon heating (H_2S).

TOXICOLOGY

Odor: Stench **Odor Threshold:** N.D.
Physical Contact: Material is extremely destructive to human tissues.
TLV = N.D. **STEL** = N.D. **IDLH** = N.D.

Routes of Entry and Relative LD$_{50}$ (or LC$_{50}$) **RTECS # WE1900000**
    Inhalation N.D.
    Ingestion N.D.
    Skin Absorption N.D.

Symptoms of Exposure: Possible burns or irritation to skin, eyes and upper respiratory tract; headache; nausea. Inhalation of vapors may be fatal by causing glottis to spasm and suffocation. Exposure may result in chemical pneumonitis or pulmonary edema. Direct contact with eye tissue may cause permanent blindness.

Emergency/First Aid Treatment: Remove to ventilated area; immediately remove any contaminated clothing and wash contaminated areas for 15 minutes using water. Treat supportively and observe for possible shock. If ingested, seek immediate medical aid.

Recommended Clean-Up Procedures

Personal Protection: Level A Ensemble **Recommended Material** N.D.

RCRA Waste # None **Reportable Quantities:** 5000 lbs (5000 lbs)

Spills: Remove all potential ignition sources. Absorb with non-combustible absorbent and take up using non-sparking tools. Decontaminate spill area using water. Treat all materials used or generated and equipment involved as contaminated by hazardous waste.

Special Emergency Information: May be fatal in inhaled, ingested or absorbed through the skin. May undergo spontaneous combustion.

SODIUM HYDROSULFITE - Na₂S₂O₄

Synonyms: hydrolin; sodium dithionite; sodium hydrosulphite; virtex

CAS Number: 7775-14-6   **Description:** White to yellow powder
DOT Number: 1384   **DOT Classification:** Flammable Solid

Molecular Weight: 174.1
Melting Point: D   70°C (158°F)   **Vapor Density:** N.D.   **Vapor Pressure:** 1 (<1 mm Hg)
Boiling Point: N.A.   **Specific Gravity:** 1.4   **Water Solubility:** Soluble

Chemical Incompatibilities or Instabilities: Oxidizers, moisture.

FLAMMABILITY

NFPA Hazard Code: 3-1-2   **NFPA Classification:**
Flash Point: N.A.   **LEL:** N.A.
Autoignition Temp: N.A.   **UEL:** N.A.

Fire Extinguishing Methods: WATER ONLY

Special Fire Fighting Considerations: Fight fire from a distance or protected location, if possible. Water spray should be used to keep closed containers cool. Continue to cool container after fire is extinguished. For large fires, if possible, withdraw and allow to burn. May release toxic fumes upon heating (SO$_x$).

TOXICOLOGY

Odor: Faint sulfurous   **Odor Threshold:** N.D.
Physical Contact: Irritant
TLV = N.D.   **STEL** = N.D.   **IDLH** = N.D.

Routes of Entry and Relative LD$_{50}$ (or LC$_{50}$)   **RTECS # JP2100000**
     Inhalation   N.D.
     Ingestion   N.D.
     Skin Absorption   N.D.

Symptoms of Exposure: Possible irritation to skin, eyes and upper respiratory tract; headache; nausea. May cause an allergic reaction or sensitization.

Emergency/First Aid Treatment: Remove to ventilated area; immediately remove any contaminated clothing and wash contaminated areas for 15 minutes using water. Treat supportively and observe for possible shock. If ingested, seek immediate medical aid.

Recommended Clean-Up Procedures

Personal Protection: Level B Ensemble   **Recommended Material**   N.D.

RCRA Waste #   None   **Reportable Quantities:**   None

Spills: Remove all potential ignition sources. Absorb with non-combustible absorbent and take up using non-sparking tools. Decontaminate spill area using water.

Special Emergency Information: May be harmful if inhaled, swallowed or absorbed through the skin. Small amounts of moisture may initiate a self-sustaining thermal decomposition.

Potassium hydrosulfite (CAS # 10294-66-3, DOT # 1929) has similar chemical, physical and toxicological properties.

SODIUM HYDROXIDE - NaOH

Synonyms: cuastic soda; lye; sodium hydrate

CAS Number: 1310-73-2
DOT Number: 1823, 1824

Description: White pellet
DOT Classification: Corrosive Material

Molecular Weight: 40.0
Melting Point: 318°C (604°F)
Boiling Point: 1390°C (2534°F)

Vapor Density: 1.0
Specific Gravity: 2.1

Vapor Pressure: 1 (<1 mm Hg)
Water Solubility: Soluble

Chemical Incompatibilities or Instabilities: Aluminum, tin, zinc, nitrated organics, stong acids, oxidizers, halogens, w,w,w-trichloroethanol, sodium tetraborate, 2,2,4,5-tetrachlorobenzene.

FLAMMABILITY

NFPA Hazard Code: 3-0-1
Flash Point: N.A.
Autoignition Temp: N.A.

NFPA Classification:
LEL: N.A.
UEL: N.A.

Fire Extinguishing Methods: Use agent suitable for surrounding fire.

Special Fire Fighting Considerations: Water spray should be used to keep closed containers cool. Continue to cool container after fire is extinguished.

TOXICOLOGY

Odor: None
Physical Contact: Material is extremely destructive to human tissues.
TLV = 2 mg/m^3 (ceiling)　　　**STEL** = N.D.

Odor Threshold: N.D.

IDLH = 250 mg/m^3

RTECS # WB4900000

Routes of Entry and Relative LD$_{50}$ (or LC$_{50}$)
　　Inhalation　　　N.D.
　　Ingestion　　　N.D.
　　Skin Absorption　N.D.

Symptoms of Exposure: Possible burns or irritation to skin, eyes and upper respiratory tract; headache; nausea. Inhalation of vapors may be fatal by causing glottis to spasm and suffocation. Exposure may result in chemical pneumonitis or pulmonary edema.

Emergency/First Aid Treatment: Remove to ventilated area; immediately remove any contaminated clothing and wash contaminated areas for 15 minutes using water. Treat supportively and observe for possible shock. If ingested, seek immediate medical aid.

Recommended Clean-Up Procedures

Personal Protection: Level B Ensemble
RCRA Waste #　　None

Recommended Material Neoprene (+), Polyvinyl Chloride (+)
Reportable Quantities: 1000 lbs (1000 lbs)

Spills: Take up in non-combustible absorbent and dispose. Decontaminate spill area using a dilute acid solution.

Special Emergency Information: May be harmful if inhaled, swallowed or absorbed through the skin. Will react with aluminum, tin and zinc to generate hydrogen gas.

Potassium hydroxide (CAS # 1310-58-3, DOT # 1813, RTECS # TT2100000) has similar chemical, physical and toxicological properties.

SODIUM HYPOCHLORITE - NaOCl

Synonyms: chlorox

CAS Number: 7681-52-9
DOT Number: 1791

Description: Colorless Liquid
DOT Classification: ORM-B

Molecular Weight: 74.4
Melting Point: N.A.
Boiling Point: N.A.

Vapor Density: N.D.
Specific Gravity: 1.1

Vapor Pressure: N.D.
Water Solubility: Soluble

Chemical Incompatibilities or Instabilities: Ammonium salts, aziridine, amines, nitriles, alcohols, phenylacetonitrile, methanol, ethyleneimne, celluose.

FLAMMABILITY

NFPA Hazard Code:
Flash Point: N.A.
Autoignition Temp: N.A.

NFPA Classification:
LEL: N.A.
UEL: N.A.

Fire Extinguishing Methods: Use agent suitable for surrounding fire.

Special Fire Fighting Considerations: Water spray should be used to keep closed containers cool. Continue to cool container after fire is extinguished.

TOXICOLOGY

Odor: Bleach
Physical Contact: Material is extremely destructive to human tissues.
TLV = N.D.   **STEL** = N.D.

Odor Threshold: N.D.

IDLH = N.D.

Routes of Entry and Relative LD$_{50}$ (or LC$_{50}$)
   Inhalation   N.D.
   Ingestion   N.D.
   Skin Absorption   N.D.

RTECS # NH3486300

Symptoms of Exposure: Possible burns or irritation to skin, eyes and upper respiratory tract; headache; nausea. Inhalation of vapors may be fatal by causing glottis to spasm and suffocation. Exposure may result in chemical pneumonitis or pulmonary edema.

Emergency/First Aid Treatment: Remove to ventilated area; immediately remove any contaminated clothing and wash contaminated areas for 15 minutes using water. Treat supportively and observe for possible shock. If ingested, seek immediate medical aid.

Recommended Clean-Up Procedures

Personal Protection: Level B Ensemble

RCRA Waste # None

Recommended Material N.D.

Reportable Quantities: None

Spills: Dike to contain spill and collect liquid for later disposal. Decontaminate spill area using water. Keep area isolated until vapors have dissipated.

Special Emergency Information: May be harmful if inhaled, swallowed or absorbed through the skin. This product is usually present as an aqueous solution. Solutions stored for extended periods may generate oxygen.

SODIUM METHYLATE (dry) - CH3ONa

Synonyms: sodium methoxide

CAS Number: 124-41-4
DOT Number: 1431

Description: White powder
DOT Classification: Flammable Solid

Molecular Weight: 54.0
Melting Point: D 300°C (572°F)
Boiling Point: °C (°F)

Vapor Density: 1.1
Specific Gravity: N.D.

Vapor Pressure: 6 (96 mm Hg)
Water Solubility: Decomposes

Chemical Incompatibilities or Instabilities: Acids, water, chloroform and methanol mixtures.

FLAMMABILITY

NFPA Hazard Code:
Flash Point: N.D.
Autoignition Temp: N.D.

NFPA Classification:
LEL: 7.3%
UEL: 36.0%

Fire Extinguishing Methods: Dry Sand, Soda Ash, Crushed Limestone, Dry Chemical.

Special Fire Fighting Considerations: DO NOT USE WATER. Treat all materials used or generated and equipment involved as contaminated by hazardous waste.

TOXICOLOGY

Odor: N.D.
Physical Contact: Material is extremely destructive to human tissues.
TLV = N.D.   **STEL** = N.D.

Odor Threshold: N.D.

IDLH = N.D.

Routes of Entry and Relative LD$_{50}$ (or LC$_{50}$)
Inhalation   N.D.
Ingestion   N.D.
Skin Absorption   N.D.

RTECS # PC3570000

Symptoms of Exposure: Possible burns or irritation to skin, eyes and upper respiratory tract; headache; nausea. Inhalation of vapors may be fatal by causing glottis to spasm and suffocation. Exposure may result in chemical pneumonitis or pulmonary edema.

Emergency/First Aid Treatment: Remove to ventilated area; immediately remove any contaminated clothing and wash contaminated areas for 15 minutes using water. Treat supportively and observe for possible shock. If ingested, seek immediate medical aid.

Recommended Clean-Up Procedures

Personal Protection: Level A Ensemble

Recommended Material N.D.

RCRA Waste # None

Reportable Quantities: 1000 lbs (1000 lbs)

Spills: Remove all potential ignition sources. Absorb with non-combustible absorbent and take up using non-sparking tools. Treat all materials used or generated and equipment involved as contaminated by hazardous waste.

Special Emergency Information: May be fatal in inhaled, ingested or absorbed through the skin. May ignite upon contact with water.

SODIUM METHYLATE (solution) - CH3ONa

Synonyms: sodium methoxide

CAS Number: 124-41-4	**Description:** Colorless Liquid
DOT Number: 1289	**DOT Classification:** Combustible Liquid

Molecular Weight: 54.0

Melting Point: N.D.	**Vapor Density:** 1.1	**Vapor Pressure:** **6** (96 mm Hg)
Boiling Point: N.D.	**Specific Gravity:** 0.9	**Water Solubility:** Decomposes

Chemical Incompatibilities or Instabilities: Acids, water.

FLAMMABILITY

NFPA Hazard Code:	**NFPA Classification:** Class IB Flammable
Flash Point: 15°C (60°F)	**LEL:** 7.3%
Autoignition Temp: 470°C (878°F)	**UEL:** 36.0%

Fire Extinguishing Methods: Alcohol Resistant Foam, Carbon Dioxide, Dry Chemical, Water Spray or Fog.

Special Fire Fighting Considerations: Isolate for 1/2 mile if rail or tank truck is involved in a fire. For large fires, if possible, withdraw and allow to burn. Immediately withdraw if rising sound from venting device is heard or if fire is causing discoloration to the tank. Do not allow water to enter container.

TOXICOLOGY

Odor: N.D.   **Odor Threshold:** N.D.

Physical Contact: Material is extremely destructive to human tissues.

TLV = N.D.   **STEL** = N.D.   **IDLH** = N.D.

Routes of Entry and Relative LD$_{50}$ (or LC$_{50}$)   **RTECS # WD2455000**

Inhalation	N.D.
Ingestion	N.D.
Skin Absorption	N.D.

Symptoms of Exposure: Possible burns or irritation to skin, eyes and upper respiratory tract; headache; nausea. Inhalation of vapors may be fatal by causing glottis to spasm and suffocation. Exposure may result in chemical pneumonitis or pulmonary edema.

Emergency/First Aid Treatment: Remove to ventilated area; immediately remove any contaminated clothing and wash contaminated areas for 15 minutes using water. Treat supportively and observe for possible shock. If ingested, seek immediate medical aid.

Recommended Clean-Up Procedures

Personal Protection: Level A Ensemble	**Recommended Material** N.D.
RCRA Waste # None	**Reportable Quantities:** 1000 lbs (1000 lbs)

Spills: Remove all potential ignition sources. Dike to contain spill, absorb with non-combustible absorbent and take up using non-sparking tools. Treat all materials used or generated and equipment involved as contaminated by hazardous waste.

Special Emergency Information: May be fatal in inhaled, ingested or absorbed through the skin. Solutions used for this product are usually flammable and their properties should be determined before any action is taken. Product may ignite upon contact with water.

SODIUM MONOXIDE - Na$_2$O

Synonyms: sodium oxide; disodium oxide

CAS Number: 1313-59-3	**Description:** White to grey powder
DOT Number: 1825	**DOT Classification:** Corrosive Material

Molecular Weight: 62.0

Melting Point: °C (°F)	**Vapor Density:** N.D.	**Vapor Pressure:** 1 (<1 mm Hg)
Boiling Point: S 1278°C (2332°F)	**Specific Gravity:** 2.3	**Water Solubility:** Decomposes

Chemical Incompatibilities or Instabilities: Acids, water, nitrated hydrocarbons.

FLAMMABILITY

NFPA Hazard Code: **NFPA Classification:**

Flash Point:	N.A.	**LEL:** N.A.
Autoignition Temp:	N.A.	**UEL:** N.A.

Fire Extinguishing Methods: Dry Sand, Soda Ash, Crushed Limestone, Dry Chemical.

Special Fire Fighting Considerations: Water spray should be used to keep closed containers cool. Continue to cool container after fire is extinguished. Do not get water inside container.

TOXICOLOGY

Odor: N.D.   **Odor Threshold:** N.D.

Physical Contact: Material is extremely destructive to human tissues.

TLV = N.D.   **STEL** = N.D.   **IDLH** = N.D.

Routes of Entry and Relative LD$_{50}$ (or LC$_{50}$)   **RTECS # WC4800000**

Inhalation	N.D.
Ingestion	N.D.
Skin Absorption	N.D.

Symptoms of Exposure: Possible burns or irritation to skin, eyes and upper respiratory tract; headache; nausea. Inhalation of vapors may be fatal by causing glottis to spasm and suffocation. Exposure may result in chemical pneumonitis or pulmonary edema.

Emergency/First Aid Treatment: Remove to ventilated area; immediately remove any contaminated clothing and wash contaminated areas for 15 minutes using water. Treat supportively and observe for possible shock. If ingested, seek immediate medical aid.

Recommended Clean-Up Procedures

Personal Protection: Level A Ensemble	**Recommended Material** N.D.
RCRA Waste # None	**Reportable Quantities:** None

Spills: Take up in non-combustible absorbent and dispose.

Special Emergency Information: May be fatal in inhaled, ingested or absorbed through the skin.

Potassium oxide (DOT # 2033) has similar chemical, physical and toxicological properties.

SODIUM NITRATE - NaNO₃

Synonyms: sodium niter

CAS Number: 7631-99-4
DOT Number: 1498

Description: White powder
DOT Classification: Oxidizer

Molecular Weight: 85.0
Melting Point: 306°C (583°F)
Boiling Point: D 380°C (716°F)

Vapor Density: N.D.
Specific Gravity: 2.3

Vapor Pressure: 1 (<1 mm Hg)
Water Solubility: Soluble

Chemical Incompatibilities or Instabilities: Finely divided metals, reducing agents, cyanides, organic matter, alkali metals, ordinary combustibles.

FLAMMABILITY

NFPA Hazard Code:
Flash Point: N.A.
Autoignition Temp: N.A.

NFPA Classification:
LEL: N.A.
UEL: N.A.

Fire Extinguishing Methods: WATER ONLY. Use flooding quantities for large fires.

Special Fire Fighting Considerations: Water spray from an unmanned device should be used to keep closed containers cool. Continue to cool container after fire is extinguished. For large fires, if possible, withdraw and allow to burn.

TOXICOLOGY

Odor: None
Physical Contact: Irritant
TLV = N.D.

STEL = N.D.

Odor Threshold: N.A.

IDLH = N.D.

Routes of Entry and Relative LD$_{50}$ (or LC$_{50}$)

Inhalation	N.D.
Ingestion	3 (3236 mg/kg)
Skin Absorption	N.D.

RTECS # WC5600000

Symptoms of Exposure: Possible irritation to skin, eyes and upper respiratory tract; headache; nausea.

Emergency/First Aid Treatment: Remove to ventilated area; immediately remove any contaminated clothing and wash contaminated areas for 15 minutes using water. Treat supportively and observe for possible shock. If ingested, seek immediate medical aid.

Recommended Clean-Up Procedures

Personal Protection: Level B Ensemble
RCRA Waste # None

Recommended Material N.D.
Reportable Quantities: None

Spills: Take up in non-combustible absorbent and dispose. Decontaminate spill area using water.

Special Emergency Information: May be harmful if inhaled, swallowed or absorbed through the skin.

Potassium nitrate (CAS # 7757-79-1, DOT # 1486, RTECS # TT3700000) has similar chemical, physical and toxicological properties.

SODIUM NITRITE - $NaNO_2$

Synonyms: diazotizing salts

CAS Number: 7632-00-0	**Description:** White powder
DOT Number: 1500	**DOT Classification:** Oxidizer

Molecular Weight: 69.0

Melting Point:	271°C (520°F)	**Vapor Density:** N.D.	**Vapor Pressure:**	1 (<1 mm Hg)
Boiling Point:	320°C (608°F)	**Specific Gravity:** 2.3	**Water Solubility:**	Soluble

Chemical Incompatibilities or Instabilities: Oxidizers, ammonia, amines, ammonium salts, cyanides, thiocyanates, thiosulfates, iodides, mercury salts.

FLAMMABILITY

NFPA Hazard Code:		**NFPA Classification:**	
Flash Point:	N.A.	**LEL:**	N.A.
Autoignition Temp:	N.A.	**UEL:**	N.A.

Fire Extinguishing Methods: WATER ONLY. Use flooding quantities for large fires.

Special Fire Fighting Considerations: Water spray should be used to keep closed containers cool. Continue to cool container after fire is extinguished. For large fires, if possible, withdraw and allow to burn. Treat all materials used or generated and equipment involved as contaminated by hazardous waste.

TOXICOLOGY

Odor: None		**Odor Threshold:** N.A.
Physical Contact: Irritant		
TLV = N.D.	**STEL** = N.D.	**IDLH** = N.D.

Routes of Entry and Relative LD_{50} (or LC_{50})   **RTECS # RA1225000**

Inhalation	10	(5.5 mg/m^3/H)
Ingestion	8	(85 mg/kg)
Skin Absorption	N.D.	

Symptoms of Exposure: Possible irritation to skin, eyes and upper respiratory tract; headache; nausea. Absorption may lead to the formation of methemoglobin resulting in cyanosis several hours after exposure.

Emergency/First Aid Treatment: Remove to ventilated area; immediately remove any contaminated clothing and wash contaminated areas for 15 minutes using water. Treat supportively and observe for possible shock. If ingested, seek immediate medical aid.

Recommended Clean-Up Procedures

Personal Protection:	Level A Ensemble	**Recommended Material** N.D.
RCRA Waste #	None	**Reportable Quantities:** 100 lbs (100 lbs)

Spills: Take up in non-combustible absorbent and dispose. Decontaminate spill area using water. Treat all materials used or generated and equipment involved as contaminated by hazardous waste.

Special Emergency Information: May be harmful if inhaled, swallowed or absorbed through the skin.

Potassium nitrite (CAS # 7758-09-0, DOT # 1488, RTECS # TT3750000) has similar chemical, physical and toxicological properties.

SODIUM PERCHLORATE - NaClO$_4$

Synonyms: perchloric acid, sodium salt

CAS Number: 7601-89-0   **Description:** White powder
DOT Number: 1502   **DOT Classification:** Oxidizer

Molecular Weight: 122.4
Melting Point: D   482°C (900°F)   **Vapor Density:** N.D.   **Vapor Pressure:** 1 (<1 mm Hg)
Boiling Point: °C   (°F)   **Specific Gravity:** 2.0   **Water Solubility:** Soluble

Chemical Incompatibilities or Instabilities: Finely divided metals, reducing agents, alcohols, butylene glycols, acetone, calcium hydride, amines, magnesium, ammonium nitrate.

FLAMMABILITY

NFPA Hazard Code: 2-0-1-OX   **NFPA Classification:**
Flash Point: N.A.   **LEL:** N.A.
Autoignition Temp: N.A.   **UEL:** N.A.

Fire Extinguishing Methods: WATER ONLY. Use flooding quantities for large fires.

Special Fire Fighting Considerations: Water spray should be used to keep closed containers cool. Continue to cool container after fire is extinguished. For large fires, if possible, withdraw and allow to burn.

TOXICOLOGY

Odor: None   **Odor Threshold:** N.A.
Physical Contact: Irritant
TLV = N.D.   **STEL** = N.D.   **IDLH** = N.D.

Routes of Entry and Relative LD$_{50}$ (or LC$_{50}$)   **RTECS # SC9800000**
   Inhalation   N.D.
   Ingestion   3   (2100 mg/kg)
   Skin Absorption   N.D.

Symptoms of Exposure: Possible irritation to skin, eyes and upper respiratory tract; headache; nausea.

Emergency/First Aid Treatment: Remove to ventilated area; immediately remove any contaminated clothing and wash contaminated areas for 15 minutes using water. Treat supportively and observe for possible shock. If ingested, seek immediate medical aid.

Recommended Clean-Up Procedures

Personal Protection: Level B Ensemble   **Recommended Material**   N.D.

RCRA Waste #   None   **Reportable Quantities:**   None

Spills: Take up in non-combustible absorbent and dispose. Decontaminate spill area using water.

Special Emergency Information: May be harmful if inhaled, swallowed or absorbed through the skin.

Potassium perchlorate (CAS # 7778-74-7, DOT # 1489, RTECS # SC9700000) has similar chemical, physical and toxicological properties.

SODIUM PENTACHLOROPHENATE

Synonyms: pentachlorophenol, sodium salt

CAS Number: 131-52-2   **Description:** Light brown powder
DOT Number: 2567   **DOT Classification:** ORM-A

Molecular Weight: 288.2

Melting Point:	N.D.	**Vapor Density:**	N.D.	**Vapor Pressure:**	1 (<1 mm Hg)
Boiling Point:	N.D.	**Specific Gravity:**	N.D.	**Water Solubility:**	Soluble

Chemical Incompatibilities or Instabilities: Strong oxidizers.

FLAMMABILITY

NFPA Hazard Code: **NFPA Classification:**
Flash Point: N.A.   **LEL:** N.A.
Autoignition Temp: N.A.   **UEL:** N.A

Fire Extinguishing Methods: Carbon Dioxide, Dry Chemical, Foam, Water Spray or Fog.

Special Fire Fighting Considerations: Treat all materials used or generated and equipment involved as contaminated by hazardous waste.

TOXICOLOGY

Odor: N.D.   **Odor Threshold:** N.D.
Physical Contact: Irritant
TLV = N.D.   **STEL** = N.D.   **IDLH** = N.D.

Routes of Entry and Relative LD$_{50}$ (or LC$_{50}$)   **RTECS # SM6490000**
　　Inhalation   10   (12 mg/m^3/H)
　　Ingestion   7   (126 mg/kg)
　　Skin Absorption   N.D.

Symptoms of Exposure: Possible irritation to skin, eyes and upper respiratory tract; headache; nausea.

Emergency/First Aid Treatment: Remove to ventilated area; immediately remove any contaminated clothing and wash contaminated areas for 15 minutes using water. Treat supportively and observe for possible shock. If ingested, seek immediate medical aid.

Recommended Clean-Up Procedures

Personal Protection: Level A Ensemble   **Recommended Material** N.D.

RCRA Waste #   None   **Reportable Quantities:** None

Spills: Take up in non-combustible absorbant and dispose. Decontaminate spill area using water. Treat all materials used or generated and equipment involved as contaminated by hazardous waste.

Special Emergency Information: May be fatal in inhaled, ingested or absorbed through the skin.

SODIUM PEROXIDE - Na$_2$O$_2$

Synonyms: sodium superoxide

CAS Number: 1313-60-6	**Description:** White to yellow powder
DOT Number: 1504	**DOT Classification:** Oxidizer

Molecular Weight: 78.0

Melting Point: D 460°C (860°F)	**Vapor Density:** N.D.	**Vapor Pressure:** 1 (<1 mm Hg)
Boiling Point: N.A.	**Specific Gravity:** 2.8	**Water Solubility:** Soluble

Chemical Incompatibilities or Instabilities: Finely divided metals, acids, organic material, hydrogen sulfide, organics.

FLAMMABILITY

NFPA Hazard Code: 3-0-1-OX	**NFPA Classification:**
Flash Point: N.A.	**LEL:** N.A.
Autoignition Temp: N.A.	**UEL:** N.A.

Fire Extinguishing Methods: WATER ONLY

Special Fire Fighting Considerations: Isolate for 1/2 mile if rail or tank truck is involved in a fire. Water spray should be used to keep closed containers cool. Continue to cool container after fire is extinguished. For large fires, if possible, withdraw and allow to burn.

TOXICOLOGY

Odor: None	**Odor Threshold:** N.A.

Physical Contact: Material is extremely destructive to human tissues.

TLV = N.D.	**STEL** = N.D.	**IDLH** = N.D.

Routes of Entry and Relative LD$_{50}$ (or LC$_{50}$)   **RTECS # WD3450000**

Inhalation	N.D.
Ingestion	N.D.
Skin Absorption	N.D.

Symptoms of Exposure: Possible burns or irritation to skin, eyes and upper respiratory tract; headache; nausea. Inhalation of vapors may be fatal by causing glottis to spasm and suffocation. Exposure may result in chemical pneumonitis or pulmonary edema.

Emergency/First Aid Treatment: Remove to ventilated area; immediately remove any contaminated clothing and wash contaminated areas for 15 minutes using water. Treat supportively and observe for possible shock. If ingested, seek immediate medical aid.

Recommended Clean-Up Procedures

Personal Protection: Level B Ensemble	**Recommended Material** N.D.
RCRA Waste # None	**Reportable Quantities:** None

Spills: Take up in non-combustible absorbant and dispose. Decontaminate spill area using water. Treat all materials used or generated and equipment involved as contaminated by hazardous waste.

Special Emergency Information: May be harmful if inhaled, swallowed or absorbed through the skin.

Potassium peroxide (CAS # 17014-71-0, DOT # 1504, RTECS # TT4450000) has similar chemical, physical and toxicological properties.

SODIUM SELENATE - Na_2SeO_4

Synonyms:

CAS Number: 13410-01-0   **Description:** White powder
DOT Number: 2630   **DOT Classification:** Corrosive Material

Molecular Weight: 188.9
Melting Point: N.D.   **Vapor Density:** N.D.   **Vapor Pressure:** 1 (<1 mm Hg)
Boiling Point: N.D.   **Specific Gravity:** N.D.   **Water Solubility:** Soluble

Chemical Incompatibilities or Instabilities: Acids.

FLAMMABILITY

NFPA Hazard Code: **NFPA Classification:**
Flash Point: N.A.   **LEL:** N.A.
Autoignition Temp: N.A.   **UEL:** N.A.

Fire Extinguishing Methods: Use agent suitable for surrounding fire.

Special Fire Fighting Considerations: Treat all materials used or generated and equipment involved as contaminated by hazardous waste.

TOXICOLOGY
QUESTIONABLE CARCINOGEN

Odor: None   **Odor Threshold:** N.D.
Physical Contact: Irritant
TLV = 0.2 mg (Se)/m^3 **STEL** = N.D.   **IDLH** = N.D.

Routes of Entry and Relative LD$_{50}$ (or LC$_{50}$)   **RTECS # VV6650000**
　　Inhalation　　N.D.
　　Ingestion　　**10**　　(1.6 mg/kg)
　　Skin Absorption　N.D.

Symptoms of Exposure: Possible irritation to skin, eyes and upper respiratory tract; headache; nausea.

Emergency/First Aid Treatment: Remove to ventilated area; immediately remove any contaminated clothing and wash contaminated areas for 15 minutes using water. Treat supportively and observe for possible shock. If ingested, seek immediate medical aid.

Recommended Clean-Up Procedures

Personal Protection: Level A Ensemble   **Recommended Material** N.D.

RCRA Waste # None   **Reportable Quantities:** None

Spills: Take up in non-combustible absorbant and dispose. Treat all materials used or generated and equipment involved as contaminated by hazardous waste. Decontaminate spill area using water.

Special Emergency Information: May be fatal in inhaled, ingested or absorbed through the skin.

Potassium selenate (CAS # 7790-59-2, DOT # 2630, RTECS # VS6600000) has similar chemical, physical and toxicological properties.

SODIUM SELENITE - Na₂SeO₃

Synonyms:

CAS Number: 10102-18-8	**Description:** White powder	
DOT Number: 2630	**DOT Classification:** Poison B	

Molecular Weight: 172.9

Melting Point:	N.D.	**Vapor Density:** N.D.	**Vapor Pressure:**	1 (<1 mm Hg)
Boiling Point:	N.D.	**Specific Gravity:** N.D.	**Water Solubility:**	Soluble

Chemical Incompatibilities or Instabilities: Acids.

FLAMMABILITY

NFPA Hazard Code:		**NFPA Classification:**	
Flash Point:	N.A.	**LEL:**	N.A.
Autoignition Temp:	N.A.	**UEL:**	N.A.

Fire Extinguishing Methods: Use agent suitable for surrounding fire.

Special Fire Fighting Considerations: Treat all materials used or generated and equipment involved as contaminated by hazardous waste.

TOXICOLOGY

Odor: None		**Odor Threshold:** N.A.
Physical Contact: Irritant		
TLV = 0.2 mg (Se)/m³	**STEL** = N.D.	**IDLH** = N.D.

Routes of Entry and Relative LD₅₀ (or LC₅₀)   **RTECS # VS7350000**

Inhalation	N.D.	
Ingestion	9	(7 mg/kg)
Skin Absorption	N.D.	

Symptoms of Exposure: Possible irritation to skin, eyes and upper respiratory tract; headache; nausea.

Emergency/First Aid Treatment: Remove to ventilated area; immediately remove any contaminated clothing and wash contaminated areas for 15 minutes using water. Treat supportively and observe for possible shock. If ingested, seek immediate medical aid.

Recommended Clean-Up Procedures

Personal Protection:	Level A Ensemble	**Recommended Material**	N.D.
RCRA Waste #	None	**Reportable Quantities:**	100 lb (1000 lb)

Spills: Take up in non-combustible absorbant and dispose. Treat all materials used or generated and equipment involved as contaminated by hazardous waste. Decontaminate spill area using water.

Special Emergency Information: May be fatal in inhaled, ingested or absorbed through the skin.

STANNIC CHLORIDE - SnCl$_4$

Synonyms: tin (IV) chloride

CAS Number: 7646-78-8

DOT Number: 1827

Description: Colorless, fuming liquid

DOT Classification: Corrosive Material

Molecular Weight: 260.5

Melting Point: -33°C (-27°F)

Boiling Point: 114°C (237°F)

Vapor Density: N.D.

Specific Gravity: 2.2

Vapor Pressure: 4 (20 mm Hg)

Water Solubility: Decomposes

Chemical Incompatibilities or Instabilities: Alkali metals, organic nitrates, ethylene oxide, turpentine.

FLAMMABILITY

NFPA Hazard Code:

Flash Point: N.A.

Autoignition Temp: N.A.

NFPA Classification:

LEL: N.A.

UEL: N.A.

Fire Extinguishing Methods: Dry Chemical, Carbon Dioxide, Flooding Quantities of Water.

Special Fire Fighting Considerations: Structural fire fighter protective clothing will not provide adequate protection. Isolate for 150 feet if rail or tank truck is involved in a fire. Water spray should be used to keep closed containers cool. Continue to cool container after fire is extinguished. Do not allow water to enter container.

TOXICOLOGY

Odor: N.D.

Odor Threshold: N.D.

Physical Contact: Material is extremely destructive to human tissues.

TLV = 2 mg (Sn)/m^3

STEL = N.D.

IDLH = 400 mg (Sn)/m^3

Routes of Entry and Relative LD$_{50}$ (or LC$_{50}$)

Inhalation	10	(383 mg/m^3/H)
Ingestion	N.D.	
Skin Absorption	N.D.	

RTECS # XP8750000

Symptoms of Exposure: Possible burns or irritation to skin, eyes and upper respiratory tract; headache; nausea. Inhalation of vapors may be fatal by causing glottis to spasm and suffocation. Exposure may result in chemical pneumonitis or pulmonary edema.

Emergency/First Aid Treatment: Remove to ventilated area; immediately remove any contaminated clothing and wash contaminated areas for 15 minutes using water. Treat supportively and observe for possible shock. If ingested, seek immediate medical aid.

Recommended Clean-Up Procedures

Personal Protection: Level A Ensemble

Recommended Material N.D.

RCRA Waste # None

Reportable Quantities: None

Spills: Dike to contain spill and collect liquid for later disposal. Decontaminate spill area using a dilute alkaline solution.

Special Emergency Information: May be fatal in inhaled, ingested or absorbed through the skin. Contact with water will liberate hydrogen chloride vapors.

STANNIC PHOSPHIDE - SnP

Synonyms: tin (IV) phosphide

CAS Number: 25324-56-5
DOT Number: 1433

Description: Silver-white powder
DOT Classification: Flammable Solid

Molecular Weight: 149.7

Melting Point: Decomposes	**Vapor Density:** N.D.	**Vapor Pressure:** N.D.
Boiling Point: N.A.	**Specific Gravity:** 6.6	**Water Solubility:** Decomposes

Chemical Incompatibilities or Instabilities: Water.

FLAMMABILITY

NFPA Hazard Code:
Flash Point: N.A.
Autoignition Temp: N.A.

NFPA Classification:
LEL: N.A.
UEL: N.A.

Fire Extinguishing Methods: Dry Chemical, Clay, Crushed Limestone, Sand.

Special Fire Fighting Considerations: DO NOT USE WATER. Structural fire fighter protective clothing will not provide adequate protection.

TOXICOLOGY

Odor: N.D.
Physical Contact: Material is extremely destructive to human tissues.
$TLV = 2$ mg (Se)/m^3 **STEL** = N.D.

Odor Threshold: N.D.

$IDLH = 400$ mg (Sn)/m

Routes of Entry and Relative LD$_{50}$ (or LC$_{50}$)
Inhalation N.D.
Ingestion N.D.
Skin Absorption N.D.

RTECS # XQ4050000

Symptoms of Exposure: Possible burns or irritation to skin, eyes and upper respiratory tract; headache; nausea. Inhalation of vapors may be fatal by causing glottis to spasm and suffocation. Exposure may result in chemical pneumonitis or pulmonary edema.

Emergency/First Aid Treatment: Remove to ventilated area; immediately remove any contaminated clothing and wash contaminated areas for 15 minutes using water. Treat supportively and observe for possible shock. If ingested, seek immediate medical aid.

Recommended Clean-Up Procedures

Personal Protection: Level A Ensemble
RCRA Waste # None

Recommended Material N.D.
Reportable Quantities: None

Spills: Take up in non-combustible absorbant and dispose under the supervision of an expert. Decontaminate spill area using dry techniques.

Special Emergency Information: May be fatal in inhaled, ingested or absorbed through the skin. Reacts with water to yield highly toxic and flammable phosphine gas.

STANNOUS CHLORIDE - $SnCl_2$

Synonyms: tin (II) chloride

CAS Number: 7772-99-8   **Description:** White powder
DOT Number: 1759   **DOT Classification:** ORM-B

Molecular Weight: 189.6
Melting Point: 246°C (475°F)   **Vapor Density:** N.A.   **Vapor Pressure:** 1 (<1 mm Hg)
Boiling Point: 652°C (1206°F)   **Specific Gravity:** 4.0   **Water Solubility:** Soluble

Chemical Incompatibilities or Instabilities: Alkali metals, hydrogen peroxide, ethylene oxide, hydrazine, nitrates, calcium carbide, interhalogens.

FLAMMABILITY

NFPA Hazard Code: **NFPA Classification:**
Flash Point: N.A.   **LEL:** N.A.
Autoignition Temp: N.A.   **UEL:** N.A.

Fire Extinguishing Methods: Use agent suitable for surrounding fire.

Special Fire Fighting Considerations: Water spray should be used to keep closed containers cool. Continue to cool container after fire is extinguished.

TOXICOLOGY

Odor: None   **Odor Threshold:** N.D.
Physical Contact: Material is extremely destructive to human tissues.
TLV = 2 mg (Sn)/m^3 **STEL** = N.D.   **IDLH** = 400 mg (Sn)/m^3

Routes of Entry and Relative LD$_{50}$ (or LC$_{50}$)   **RTECS #** XP8700000
    Inhalation   N.D.
    Ingestion   4   (700 mg/kg)
    Skin Absorption   N.D.

Symptoms of Exposure: Possible burns or irritation to skin, eyes and upper respiratory tract; headache; nausea. Inhalation of vapors may be fatal by causing glottis to spasm and suffocation. Exposure may result in chemical pneumonitis or pulmonary edema.

Emergency/First Aid Treatment: Remove to ventilated area; immediately remove any contaminated clothing and wash contaminated areas for 15 minutes using water. Treat supportively and observe for possible shock. If ingested, seek immediate medical aid.

Recommended Clean-Up Procedures

Personal Protection: Level B Ensemble   **Recommended Material** N.D.

RCRA Waste #   None   **Reportable Quantities:** None

Spills: Take up in non-combustible absorbant and dispose. Decontaminate spill area using water.

Special Emergency Information: May be harmful if inhaled, swallowed or absorbed through the skin.

STIBINE - SbH₃

Synonyms: antimony hydride

CAS Number: 7803-52-3   **Description:** Colorless Gas
DOT Number: 2676   **DOT Classification:** Poison A

Molecular Weight: 124.8
Melting Point: -88°C (-126°F)   **Vapor Density:** 4.4   **Vapor Pressure:**   Gas
Boiling Point: -17°C   (-1°F)   **Specific Gravity:** N.A.   **Water Solubility:**   Insoluble

Chemical Incompatibilities or Instabilities: Halogens, nitric acid, ozone, oxidizers, ammonia.

FLAMMABILITY

NFPA Hazard Code: 4-4-2   **NFPA Classification:**
Flash Point: N.A.   **LEL:**   N.A.
Autoignition Temp: N.A.   **UEL:**   N.A.

Fire Extinguishing Methods: Water spray or fog.

Special Fire Fighting Considerations: Stop leak if it can safely be done. Structural fire fighter protective clothing will not provide adequate protection. Isolate for 1/2 mile if rail or tank truck is involved in a fire. Water spray should be used to keep closed containers cool. Continue to cool container after fire is extinguished. For large fires, if possible, withdraw and allow to burn. Immediately withdraw if rising sound from venting device is heard or if fire is causing discoloration to the tank.

DOT Recommended Isolation Zones: Small Spill: 1500 ft   Large Spill: 1500 ft
DOT Recommended Down Wind Take Cover Distance   Small Spill: 5 miles   Large Spill: 5 miles

TOXICOLOGY

Odor: N.D.   **Odor Threshold:** N.D.
Physical Contact: Irritant
TLV = 0.5 mg/m³   **STEL** = N.D.   **IDLH** = 208 mg/m³

Routes of Entry and Relative LD₅₀ (or LC₅₀)   **RTECS # WJ0700000**
    Inhalation   N.D.
    Ingestion   N.D.
    Skin Absorption   N.D.

Symptoms of Exposure: Possible irritation to skin, eyes and upper respiratory tract; headache; nausea.

Emergency/First Aid Treatment: Remove to ventilated area; immediately remove any contaminated clothing and wash contaminated areas for 15 minutes using water. Treat supportively and observe for possible shock. If ingested, seek immediate medical aid.

Recommended Clean-Up Procedures

Personal Protection: Level A Ensemble   **Recommended Material**   N.D.

RCRA Waste #   None   **Reportable Quantities:**   None

Spills: Stop leak if it can safely be done. Keep area isolated until vapors have dissipated.

Special Emergency Information: May be fatal in inhaled, ingested or absorbed through the skin.

STRYCHNINE

Synonyms:

CAS Number: 57-24-9	**Description:** White powder	
DOT Number: 1692	**DOT Classification:** Poison B	

Molecular Weight: 370.9

Melting Point:	268°C (514°F)	**Vapor Density:** N.D.	**Vapor Pressure:**	1 (<1 mm Hg)
Boiling Point:	270°C (518°F)	**Specific Gravity:** 1.4	**Water Solubility:**	Insoluble

Chemical Incompatibilities or Instabilities: Oxidizers.

FLAMMABILITY

NFPA Hazard Code:		**NFPA Classification:**	
Flash Point:	N.A.	**LEL:**	N.A.
Autoignition Temp:	N.A.	**UEL:**	N.A.

Fire Extinguishing Methods: Use agent suitable for surrounding fire.

Special Fire Fighting Considerations: Treat all materials used or generated and equipment involved as contaminated by hazardous waste.

TOXICOLOGY

Odor: None	**Odor Threshold:** N.A.

Physical Contact: Irritant

TLV = 0.15 mg/m^3	**STEL** = N.D.	**IDLH** = 3 mg/m^3

Routes of Entry and Relative LD$_{50}$ (or LC$_{50}$)   **RTECS # WL2275000**

Inhalation	N.D.	
Ingestion	**10**	(2.3 mg/kg)
Skin Absorption	N.D.	

Symptoms of Exposure: Irritability, muscle twitching, breathing difficulties, muscle spasms. May cause an allergic reaction.

Emergency/First Aid Treatment: Remove to ventilated area; immediately remove any contaminated clothing and wash contaminated areas for 15 minutes using water. Treat supportively and observe for possible shock. If ingested, seek immediate medical aid.

Recommended Clean-Up Procedures

Personal Protection:	Level A Ensemble	**Recommended Material** N.D.
RCRA Waste #	P108	**Reportable Quantities:** 10 lbs (10 lbs)

Spills: Take up in non-combustible absorbant and dispose. Treat all materials used or generated and equipment involved as contaminated by hazardous waste. Decontaminate spill area using soapy water.

Special Emergency Information: May be fatal in inhaled, ingested or absorbed through the skin. May cause an allergic reaction.

STYRENE

Synonyms: vinyl benzene; phenyl ethylene

CAS Number:	100-42-5	**Description:**	Colorless Liquid
DOT Number:	2055	**DOT Classification:**	Flammable Liquid

Molecular Weight: 104.2

Melting Point:	-31°C (-23°F)	**Vapor Density:**	3.6	**Vapor Pressure:**	3 (6 mm Hg)
Boiling Point:	145°C (293°F)	**Specific Gravity:**	0.9	**Water Solubility:**	Insoluble

Chemical Incompatibilities or Instabilities: Oxidizers, peroxides, strong acids, aluminum chloride, copper and its alloys.

FLAMMABILITY

NFPA Hazard Code:	2-3-2	**NFPA Classification:**	Class IC Flammable
Flash Point:	31°C (88°F)	**LEL:**	1.1%
Autoignition Temp:	496°C (914°F)	**UEL:**	7.0%

Fire Extinguishing Methods: Carbon Dioxide, Dry Chemical, Foam, Water Spray or Fog.

Special Fire Fighting Considerations: Isolate for 1/2 mile if rail or tank truck is involved in a fire. Water spray should be used to keep closed containers cool. Continue to cool container after fire is extinguished. For large fires, if possible, withdraw and allow to burn. Immediately withdraw if rising sound from venting device is heard or if fire is causing discoloration to the tank.

TOXICOLOGY
SUSPECTED CARCINOGEN

Odor: Sharp, sweet   **Odor Threshold:** 0.02 mg/m^3
Physical Contact: Irritant
TLV = 215 mg/m^3 **STEL** = 425 mg/m^3 **IDLH** = 21,500 mg/m^3

Routes of Entry and Relative LD$_{50}$ (or LC$_{50}$)   **RTECS #** WL3675000
   Inhalation   2   (72,000 mg/m^3/H)
   Ingestion   3   (5,000 mg/kg)
   Skin Absorption N.D.

Symptoms of Exposure: Possible irritation to skin, eyes and upper respiratory tract; headache; nausea; CNS depression.

Emergency/First Aid Treatment: Remove to ventilated area; immediately remove any contaminated clothing and wash contaminated areas for 15 minutes using water. Treat supportively and observe for possible shock. If ingested, seek immediate medical aid.

Recommended Clean-Up Procedures

Personal Protection:	Level B Ensemble	**Recommended Material**	Polyvinyl Alcohol (0), Teflon (0)
RCRA Waste #	None	**Reportable Quantities:**	1000 lb (1000 lb)

Spills: Remove all potential ignition sources. Dike to contain spill, absorb with noncombustible absorbant and take up using nonsparking tools. Decontaminate spill area using soapy water. Keep area isolated until vapors have dissipated.

Special Emergency Information: May be harmful if inhaled, swallowed or absorbed through the skin. May undergo hazardous polymerization. Vapors are heavier than air and may travel some distance to an ignition source. Vapors are more dense than air and may settle in low lying areas.

SULFAMIC ACID - NH₂SO₂OH

Synonyms:

CAS Number: 5329-14-6   **Description:** White powder
DOT Number: 2967   **DOT Classification:** Corrosive Material

Molecular Weight: 97.1
Melting Point: D   205°C (401°F)   **Vapor Density:** N.D.   **Vapor Pressure:** 1 (<1 mm Hg)
Boiling Point: N.A.   **Specific Gravity:** 2.2   **Water Solubility:** Soluble

Chemical Incompatibilities or Instabilities: Ammonia, nitrates, nitrites, nitric acid.

FLAMMABILITY

NFPA Hazard Code: **NFPA Classification:**
Flash Point: N.A.   **LEL:** N.A.
Autoignition Temp: N.A.   **UEL:** N.A.

Fire Extinguishing Methods: Use agent suitable for surrounding fire.

Special Fire Fighting Considerations: Water spray should be used to keep closed containers cool. Continue to cool container after fire is extinguished.

TOXICOLOGY

Odor: None   **Odor Threshold:** N.D.
Physical Contact: Irritant
TLV = N.D.   **STEL** = N.D.   **IDLH** = N.D.

Routes of Entry and Relative LD$_{50}$ (or LC$_{50}$)   **RTECS #** WO5950000
    Inhalation   N.D.
    Ingestion   3   (3160 mg/kg)
    Skin Absorption   N.D.

Symptoms of Exposure: Possible irritation to skin, eyes and upper respiratory tract; headache; nausea.

Emergency/First Aid Treatment: Remove to ventilated area; immediately remove any contaminated clothing and wash contaminated areas for 15 minutes using water. Treat supportively and observe for possible shock. If ingested, seek immediate medical aid.

Recommended Clean-Up Procedures

Personal Protection: Level B Ensemble   **Recommended Material**   N.D.

RCRA Waste #   None   **Reportable Quantities:**   None

Spills: Take up in non-combustible absorbant and dispose. Decontaminate spill area using a dilute alkaline solution.

Special Emergency Information: May be harmful if inhaled, swallowed or absorbed through the skin.

SULFUR - S

Synonyms: sulphur; brimstone

CAS Number: 7704-34-9   **Description:** Yellow powder
DOT Number: 1305 (dry), 2448 (molten)   **DOT Classification:** ORM-C

Molecular Weight: 32.1
Melting Point: 115°C (239°F)   **Vapor Density:** N.A.   **Vapor Pressure:** 1 (<1 mm Hg)
Boiling Point: 445°C (833°F)   **Specific Gravity:** 2.1   **Water Solubility:** Insoluble

Chemical Incompatibilities or Instabilities: Strong oxidizers, halogens, ammonia, calcium, aluminum, boron, interhalogens, carbides, alkali metals, nickel, paladium, phosphorous, alkali earth metals.

FLAMMABILITY

NFPA Hazard Code: 1-1-0 (dry), 2-1-0 (molten)   **NFPA Classification:**
Flash Point: 207°C (405°F)   **LEL:** N.D.
Autoignition Temp: 232°C (450°F)   **UEL:** N.D.

Fire Extinguishing Methods: Use agent suitable for surrounding fire.

Special Fire Fighting Considerations: Water spray should be used to keep closed containers cool. Continue to cool container after fire is extinguished.

TOXICOLOGY

Odor: None   **Odor Threshold:** N.D.
Physical Contact: Irritant or thermal burns
TLV = N.D.   **STEL** = N.D.   **IDLH** = N.D.

Routes of Entry and Relative LD$_{50}$ (or LC$_{50}$)   **RTECS # WS4250000**
   Inhalation   N.D.
   Ingestion   N.D.
   Skin Absorption   N.D.

Symptoms of Exposure: Possible irritation to skin, eyes and upper respiratory tract; headache; nausea. May cause an allergic reaction. Molten sulfur will cause thermal burns.

Emergency/First Aid Treatment: Remove to ventilated area; immediately remove any contaminated clothing and wash contaminated areas for 15 minutes using water. Treat supportively and observe for possible shock. If ingested, seek immediate medical aid.

Recommended Clean-Up Procedures

Personal Protection: Level B Ensemble   **Recommended Material**   N.D.

RCRA Waste #   None   **Reportable Quantities:**   None

Spills: Use water to cool molten sulfur. Take up in non-combustible absorbant and dispose. Decontaminate spill area using soapy water. Keep area isolated until vapors have dissipated.

Special Emergency Information: May be harmful if inhaled, swallowed or absorbed through the skin. Molten sulfur will react with air to yield sulfur dioxide and will react with organic materials to yield hydrogen sulfide and carbon disulfide.

SULFUR CHLORIDE - S$_2$Cl$_2$

Synonyms: sulfur monochloride; disulfur dichloride; sulfur subchloride

CAS Number: 10025-67-9   **Description:** Yellow liquid

DOT Number: 1828   **DOT Classification:** Corrosive Material

Molecular Weight: 135.0

Melting Point:	-77°C (-107°F)	**Vapor Density:** 4.7	**Vapor Pressure:**	3 (7 mm Hg)
Boiling Point:	138°C (280°F)	**Specific Gravity:** 1.7	**Water Solubility:**	Decomposes

Chemical Incompatibilities or Instabilities: Hydrocarbons, phosphorous oxides, water, peroxides, metals, oxidizers.

FLAMMABILITY

NFPA Hazard Code: 2-1-1   **NFPA Classification:** Class IIIB Combustible

Flash Point: 118°C (245°F)   **LEL:** N.D.

Autoignition Temp: 234°C (453°F)   **UEL:** N.D.

Fire Extinguishing Methods: Dry Chemical, Carbon Dioxide, Flooding Quantities of Water.

Special Fire Fighting Considerations: Structural fire fighter protective clothing will not provide adequate protection. Isolate for 150 feet if rail or tank truck is involved in a fire. Water spray should be used to keep closed containers cool. Continue to cool container after fire is extinguished. Do not allow water to enter container.

DOT Recommended Isolation Zones: Small Spill: 600 ft   Large Spill: 600 ft

DOT Recommended Down Wind Take Cover Distance   Small Spill: 2 miles   Large Spill: 2 miles

TOXICOLOGY

Odor: Pungent, nauseating   **Odor Threshold:** N.D.

Physical Contact: Material is extremely destructive to human tissues.

TLV = 6 mg/m^3 (ceiling)   **STEL** = N.D.   **IDLH** = 56 mg/m^3

Routes of Entry and Relative LD$_{50}$ (or LC$_{50}$)   **RTECS # WS4300000**

Inhalation	N.D.
Ingestion	N.D.
Skin Absorption	N.D.

Symptoms of Exposure: Possible burns or irritation to skin, eyes and upper respiratory tract; headache; nausea; lachrymation. Inhalation of vapors may be fatal by causing glottis to spasm and suffocation. Exposure may result in chemical pneumonitis or pulmonary edema.

Emergency/First Aid Treatment: Remove to ventilated area; immediately remove any contaminated clothing and wash contaminated areas for 15 minutes using water. Treat supportively and observe for possible shock. If ingested, seek immediate medical aid.

Recommended Clean-Up Procedures

Personal Protection: Level A Ensemble   **Recommended Material** N.D.

RCRA Waste # None   **Reportable Quantities:** 1000 lbs (1000 lbs)

Spills: Dike to contain spill and collect liquid for later disposal. Decontaminate spill area using a dilute alkaline solution.

Special Emergency Information: May be fatal in inhaled, ingested or absorbed through the skin. Reacts with water to yield hydrogen chloride, thiosulfuric acid and elemental sulfur.

SULFUR DICHLORIDE - SCl$_2$

Synonyms:

CAS Number: 10545-99-0	**Description:** Red-brown liquid		
DOT Number: 1828	**DOT Classification:** Corrosive Material		

Molecular Weight: 103.0

Melting Point: -78°C (-108°F)	**Vapor Density:** 3.6	**Vapor Pressure:** 7 (170 mm Hg)
Boiling Point: 59°C (138°F)	**Specific Gravity:** 1.6	**Water Solubility:** Decomposes

Chemical Incompatibilities or Instabilities: Aluminum, ammonia, alkali metals, dimethyl sulfoxide, dimethyl formamide, acetone, nitric acid, metals, oxidizers, toluene.

FLAMMABILITY

NFPA Hazard Code:	**NFPA Classification:**	
Flash Point: N.A.	**LEL:** N.A.	
Autoignition Temp: N.A.	**UEL:** N.A.	

Fire Extinguishing Methods: Dry Chemical, Carbon Dioxide, Flooding Quantities of Water.

Special Fire Fighting Considerations: Isolate for 150 feet if rail or tank truck is involved in a fire. Structural fire fighter protective clothing will not provide adequate protection. Water spray should be used to keep closed containers cool. Continue to cool container after fire is extinguished. Do not allow water to enter container.

DOT Recommended Isolation Zones:	Small Spill: 600 ft	Large Spill: 600 ft
DOT Recommended Down Wind Take Cover Distance	Small Spill: 2 miles	Large Spill: 2 miles

TOXICOLOGY

Odor: N.D.	**Odor Threshold:** N.D.

Physical Contact: Material is extremely destructive to human tissues.

TLV = N.D.	**STEL** = N.D.	**IDLH** = N.D.

Routes of Entry and Relative LD$_{50}$ (or LC$_{50}$)   **RTECS # WS4500000**

Inhalation	N.D.
Ingestion	N.D.
Skin Absorption	N.D.

Symptoms of Exposure: Possible burns or irritation to skin, eyes and upper respiratory tract; headache; nausea; lachrymation. Inhalation of vapors may be fatal by causing glottis to spasm and suffocation. Exposure may result in chemical pneumonitis or pulmonary edema.

Emergency/First Aid Treatment: Remove to ventilated area; immediately remove any contaminated clothing and wash contaminated areas for 15 minutes using water. Treat supportively and observe for possible shock. If ingested, seek immediate medical aid.

Recommended Clean-Up Procedures

Personal Protection: Level A Ensemble	**Recommended Material** N.D.	
RCRA Waste # None	**Reportable Quantities:** None	

Spills: Take up in non-combustible absorbant and dispose. Decontaminate spill area using a dilute alkaline solution.

Special Emergency Information: May be fatal in inhaled, ingested or absorbed through the skin.

SULFUR DIOXIDE - SO$_2$

Synonyms:

CAS Number: 7446-09-5   **Description:** Colorless Gas
DOT Number: 1079   **DOT Classification:** Nonflammable gas

Molecular Weight: 64.1
Melting Point: -76°C (-104°F)   **Vapor Density:** 2.2   **Vapor Pressure:** Gas
Boiling Point: -10°C (14°F)   **Specific Gravity:** N.A.   **Water Solubility:** Decomposes

Chemical Incompatibilities or Instabilities: Halogens, alkali metals, azides, hydrides, finely divided metals, alkenes, oxidizers, interhalogens, manganese, chrome, aluminum.

FLAMMABILITY

NFPA Hazard Code: 3-0-0   **NFPA Classification:**
Flash Point: N.A.   **LEL:** N.A.
Autoignition Temp: N.A.   **UEL:** N.A.

Fire Extinguishing Methods: Use agent suitable for surrounding fire.

Special Fire Fighting Considerations: Isolate for 150 feet if rail or tank truck is involved in a fire.

DOT Recommended Isolation Zones: Small Spill: 600 ft   Large Spill: 1500 ft
DOT Recommended Down Wind Take Cover Distance   Small Spill: 2 miles   Large Spill: 5 miles

TOXICOLOGY
QUESTIONABLE CARCINOGEN

Odor: Sharp   **Odor Threshold:**
Physical Contact: Material is extremely destructive to human tissues.
TLV = 5 mg/m^3 **STEL** = 10 mg/m^3 **IDLH** = 250 mg/m^3

Routes of Entry and Relative LD$_{50}$ (or LC$_{50}$)   **RTECS # WS4550000**
  Inhalation   6   (6,500 mg/m^3/H)
  Ingestion   N.D.
  Skin Absorption   N.D.

Symptoms of Exposure: Possible burns or irritation to skin, eyes and upper respiratory tract; headache; nausea. Inhalation of vapors may be fatal by causing glottis to spasm and suffocation. Exposure may result in chemical pneumonitis or pulmonary edema.

Emergency/First Aid Treatment: Remove to ventilated area; immediately remove any contaminated clothing and wash contaminated areas for 15 minutes using water. Treat supportively and observe for possible shock. If ingested, seek immediate medical aid.

Recommended Clean-Up Procedures

Personal Protection: Level A Ensemble   **Recommended Material** N.D.

RCRA Waste #   None   **Reportable Quantities:** None

Spills: Stop leak if it can safely be done. Keep area isolated until vapors have dissipated.

Special Emergency Information: May be fatal in inhaled, ingested or absorbed through the skin.

SULFURIC ACID - H₂SO₄

Synonyms: battery acid; oil of vitrol

CAS Number: 7664-93-9   **Description:** Colorless Liquid
DOT Number: 1830, 1831 (fuming), 1832 (spent) **DOT Classification:** Corrosive Material

Molecular Weight: 98.1
Melting Point: 10°C   (50°F)   **Vapor Density:**   3.4   **Vapor Pressure:**   **6** (73 mm Hg)
Boiling Point: 330°C (626°F)   **Specific Gravity:** 1.8   **Water Solubility:**   Soluble

Chemical Incompatibilities or Instabilities: Carbides, chlorates, ful,onates, picrates, ocidizers, alkyl halides, phosphorous, organic nitrates, unsaturated hydrocarbons.

FLAMMABILITY

NFPA Hazard Code: 3-0-2-~~W~~   **NFPA Classification:**
Flash Point: N.A.   **LEL:** N.A.
Autoignition Temp: N.A.   **UEL:** N.A.

Fire Extinguishing Methods: Use agent suitable for surrounding fire.

Special Fire Fighting Considerations: Structural fire fighter protective clothing will not provide adequate protection. Isolate for 150 feet if rail or tank truck is involved in a fire. water, do not allow water to enter container.

TOXICOLOGY

Odor: Sharp   **Odor Threshold:** 0.6 mg/m³
Physical Contact: Material is extremely destructive to human tissues.
TLV = 1 mg/m³   **STEL** = 3 mg/m³   **IDLH** = 80 mg/m³

Routes of Entry and Relative LD₅₀ (or LC₅₀)   **RTECS # WS5600000**
    Inhalation   9   (1020 mg/m³/H)
    Ingestion   N.D.
    Skin Absorption   N.D.

Symptoms of Exposure: Possible burns or irritation to skin, eyes and upper respiratory tract; headache; nausea. Inhalation of vapors may be fatal by causing glottis to spasm and suffocation. Exposure may result in chemical pneumonitis or pulmonary edema.

Emergency/First Aid Treatment: Remove to ventilated area; immediately remove any contaminated clothing and wash contaminated areas for 15 minutes using water. Treat supportively and observe for possible shock. If ingested, seek immediate medical aid.

Recommended Clean-Up Procedures

Personal Protection: Level B Ensemble   **Recommended Material**   Butyl Rubber (+), Polyethylene (-),Teflon (+), Saranex (-)

RCRA Waste #   None   **Reportable Quantities:**   1000 lbs (1000 lbs)

Spills: Take up in non-combustible absorbant and dispose. Decontaminate spill area using a dilute alkaline solution.

Special Emergency Information: May be fatal in inhaled, ingested or absorbed through the skin. Reaction with water is extremely exothermic.

SULFUR TRIOXIDE - SO₃

Synonyms: sulfuric anhydride

CAS Number: 7446-11-9   **Description:** Colorless crystals
DOT Number: 1829   **DOT Classification:** Corrosive Material

Molecular Weight: 80.1
Melting Point: 17°C (63°F)   **Vapor Density:** 2.8   **Vapor Pressure:** **8** (280 mm Hg)
Boiling Point: 45°C (113°F)   **Specific Gravity:** 2.0   **Water Solubility:** Decomposes

Chemical Incompatibilities or Instabilities: Cyanides, oxidizers, dimethyl sulfoxide, phosphorous, barium oxide, lead oxide, diphenyl mercury, nitryl chloride.

FLAMMABILITY

NFPA Hazard Code: **NFPA Classification:**
Flash Point: N.A.   **LEL:** N.A.
Autoignition Temp: N.A.   **UEL:** N.A.

Fire Extinguishing Methods: Use agent suitable for surrounding fire.

Special Fire Fighting Considerations: Structural fire fighter protective clothing will not provide adequate protection. Isolate for 150 feet if rail or tank truck is involved in a fire. Do not allow water to enter container. Do not put a direct water stream on spilled material.

DOT Recommended Isolation Zones: Small Spill: 150 ft   Large Spill: 150 ft
DOT Recommended Down Wind Take Cover Distance   Small Spill: 0.4 miles   Large Spill: 0.8 miles

TOXICOLOGY

Odor: Sharp   **Odor Threshold:** N.D.
Physical Contact: Material is extremely destructive to human tissues.
TLV = N.D.   **STEL** = N.D.   **IDLH** = N.D.

Routes of Entry and Relative LD₅₀ (or LC₅₀)   **RTECS # WT4830000**
    Inhalation   N.D.
    Ingestion   N.D.
    Skin Absorption   N.D.

Symptoms of Exposure: Possible burns or irritation to skin, eyes and upper respiratory tract; headache; nausea. Laryngitis. Inhalation of vapors may be fatal by causing glottis to spasm and suffocation. Exposure may result in chemical pneumonitis or pulmonary edema.

Emergency/First Aid Treatment: Remove to ventilated area; immediately remove any contaminated clothing and wash contaminated areas for 15 minutes using water. Treat supportively and observe for possible shock. If ingested, seek immediate medical aid.

Recommended Clean-Up Procedures

Personal Protection: Level A Ensemble   **Recommended Material** N.D.

RCRA Waste #   None   **Reportable Quantities:** None

Spills: Take up in non-combustible absorbent and dispose. Decontaminate spill area using a dilute alkaline solution.

Special Emergency Information: May be fatal in inhaled, ingested or absorbed through the skin.

SULFUROUS ACID - H₂SO₃

Synonyms: sulfur dioxide solution

CAS Number: 7782-99-2   **Description:** Colorless Liquid
DOT Number: 1833   **DOT Classification:** Corrosive Material

Molecular Weight: 82.1
Melting Point: N.D.   **Vapor Density:** N.D.   **Vapor Pressure:**   N.D.
Boiling Point: N.D.   **Specific Gravity:** 1.0   **Water Solubility:** Soluble

Chemical Incompatibilities or Instabilities: Strong alkali.

FLAMMABILITY

NFPA Hazard Code: **NFPA Classification:**
Flash Point: N.A.   **LEL:** N.A.
Autoignition Temp: N.A.   **UEL:** N.A.

Fire Extinguishing Methods: Use agent suitable for surrounding fire.

Special Fire Fighting Considerations: Water spray should be used to keep closed containers cool. Continue to cool container after fire is extinguished.

TOXICOLOGY

Odor: N.D.   **Odor Threshold:** N.D.
Physical Contact: Material is extremely destructive to human tissues.
TLV = N.D.   **STEL** = N.D.   **IDLH** = N.D.

Routes of Entry and Relative LD₅₀ (or LC₅₀)   **RTECS #** WT2775000
   Inhalation   N.D.
   Ingestion   N.D.
   Skin Absorption   N.D.

Symptoms of Exposure: Possible burns or irritation to skin, eyes and upper respiratory tract; headache; nausea. Inhalation of vapors may be fatal by causing glottis to spasm and suffocation. Exposure may result in chemical pneumonitis or pulmonary edema.

Emergency/First Aid Treatment: Remove to ventilated area; immediately remove any contaminated clothing and wash contaminated areas for 15 minutes using water. Treat supportively and observe for possible shock. If ingested, seek immediate medical aid.

Recommended Clean-Up Procedures

Personal Protection: Level B Ensemble   **Recommended Material**   N.D.

RCRA Waste #   None   **Reportable Quantities:**   None

Spills: Take up in non-combustible absorbent and dispose. Decontaminate spill area using a dilute alkaline solution.

Special Emergency Information: May be fatal in inhaled, ingested or absorbed through the skin. This product is a mixture of sulfur dioxide and water. See "sulfur dioxide."

SULFURYL CHLORIDE - SO$_2$Cl$_2$

Synonyms: sulfonyl chloride; sulfuric oxychloride

CAS Number: 7791-25-5

DOT Number: 1834

Description: Colorless Liquid

DOT Classification: Corrosive Material

Molecular Weight: 135.0

Melting Point: -54°C (-65°F)	**Vapor Density:** 4.7	**Vapor Pressure:** 7 (105 mm Hg)
Boiling Point: 69°C (156°F)	**Specific Gravity:** 1.7	**Water Solubility:** Decomposes

Chemical Incompatibilities or Instabilities: Dimethylsulfoxide, dimethylformamide, lead dioxide, peroxides, phosphorous, alkalis, ethers, dinitrogen pentoxide, alkali metals, ammonia, aluminum.

FLAMMABILITY

NFPA Hazard Code: 3-0-1

Flash Point: N.A.

Autoignition Temp: N.A.

NFPA Classification:

LEL: N.A.

UEL: N.A.

Fire Extinguishing Methods: Dry Chemical, Carbon Dioxide, Flooding Quantities of Water.

Special Fire Fighting Considerations: Structural fire fighter protective clothing will not provide adequate protection. Isolate for 150 feet if rail or tank truck is involved in a fire. Water spray should be used to keep closed containers cool. Continue to cool container after fire is extinguished.

TOXICOLOGY
QUESTIONABLE CARCINOGEN

Odor: Pungent

Physical Contact: Material is extremely destructive to human tissues.

TLV = N.D.   **STEL** = N.D.

Odor Threshold: N.D.

IDLH = N.D.

Routes of Entry and Relative LD$_{50}$ (or LC$_{50}$)

Inhalation	N.D.	
Ingestion	N.D.	
Skin Absorption	N.D.	

RTECS # WT4870000

Symptoms of Exposure: Possible burns or irritation to skin, eyes and upper respiratory tract; headache; nausea; lachrymation. Inhalation of vapors may be fatal by causing glottis to spasm and suffocation. Exposure may result in chemical pneumonitis or pulmonary edema.

Emergency/First Aid Treatment: Remove to ventilated area; immediately remove any contaminated clothing and wash contaminated areas for 15 minutes using water. Treat supportively and observe for possible shock. If ingested, seek immediate medical aid.

Recommended Clean-Up Procedures

Personal Protection: Level A Ensemble

RCRA Waste # None

Recommended Material N.D.

Reportable Quantities: None

Spills: Dike to contain spill and collect liquid for later disposal. Decontaminate spill area using a dilute alkaline solution.

Special Emergency Information: May be harmful if inhaled, swallowed or absorbed through the skin.

SULFURYL FLUORIDE - SO$_2$F$_2$

Synonyms: sulfur dioxide difluoride; sulfuric oxyfluoride

CAS Number: 2699-79-8	**Description:** Colorless Gas
DOT Number: 2191	**DOT Classification:** Nonflammable Gas

Molecular Weight: 102.1

Melting Point: -135°C (-212°F)	**Vapor Density:** 3.7	**Vapor Pressure:** Gas
Boiling Point: -90°C (-68°F)	**Specific Gravity:** N.A.	**Water Solubility:** < 1 %

Chemical Incompatibilities or Instabilities: Alkali metals, ammonia, aluminum.

FLAMMABILITY

NFPA Hazard Code:	**NFPA Classification:**
Flash Point: N.A.	**LEL:** N.A.
Autoignition Temp: N.A.	**UEL:** N.A.

Fire Extinguishing Methods: Dry Chemical, Carbon Dioxide, Water Spray or Fog.

Special Fire Fighting Considerations: Structural fire fighter protective clothing will not provide adequate protection. Isolate for 150 feet if rail or tank truck is involved in a fire. Water spray should be used to keep closed containers cool. Continue to cool container after fire is extinguished. Do not allow water to enter container.

DOT Recommended Isolation Zones:	Small Spill: 900 ft	Large Spill: 1500 ft
DOT Recommended Down Wind Take Cover Distance	Small Spill: 3 miles	Large Spill: 5 miles

TOXICOLOGY

Odor: N.D.   **Odor Threshold:** N.D.

Physical Contact: Material is extremely destructive to human tissues.

TLV = 20 mg/m^3 **STEL** = 40 mg/m^3 **IDLH** = 4000 mg/m^3

Routes of Entry and Relative LD$_{50}$ (or LC$_{50}$)   **RTECS #** WT5075000

Inhalation	6	(15,000 mg/m^3/H)
Ingestion	8	(100 mg/kg)
Skin Absorption	N.D.	

Symptoms of Exposure: Possible burns or irritation to skin, eyes and upper respiratory tract; headache; nausea; lachrymation. Inhalation of vapors may be fatal by causing glottis to spasm and suffocation. Exposure may result in chemical pneumonitis or pulmonary edema.

Emergency/First Aid Treatment: Remove to ventilated area; immediately remove any contaminated clothing and wash contaminated areas for 15 minutes using water. Treat supportively and observe for possible shock. If ingested, seek immediate medical aid.

Recommended Clean-Up Procedures

Personal Protection: Level A Ensemble	**Recommended Material** N.D.
RCRA Waste # None	**Reportable Quantities:** None

Spills: Stop leak if it can safely be done. Keep area isolated until vapors have dissipated.

Special Emergency Information: May be fatal in inhaled, ingested or absorbed through the skin.

2,4,5-T

Synonyms: 2,4,5-trichlorophenoxyacetic acid

CAS Number: 93-76-5	**Description:** Light brown solid
DOT Number: 2765	**DOT Classification:** ORM-A

Molecular Weight: 255.5

Melting Point:	153°C (307°F)	**Vapor Density:** N.D.	**Vapor Pressure:**	1 (<1 mm Hg)
Boiling Point:	Decomposes	**Specific Gravity:** N.D.	**Water Solubility:**	< 1%

Chemical Incompatibilities or Instabilities: Oxidizers.

FLAMMABILITY

NFPA Hazard Code: **NFPA Classification:**

Flash Point:	N.A.	**LEL:**	N.A.
Autoignition Temp:	N.A.	**UEL:**	N.A.

Fire Extinguishing Methods: Carbon Dioxide, Dry Chemical, Foam, Water Spray or Fog.

Special Fire Fighting Considerations: Structural fire fighter protective clothing will not provide adequate protection. Fight fire from a distance or protected location, if possible. Treat all materials used or generated and equipment involved as contaminated by hazardous waste.

TOXICOLOGY
SUSPECTED CARCINOGEN

Odor: None   **Odor Threshold:** N.A.

Physical Contact: Irritant

TLV = 10 mg/m^3 **STEL** = N.D.   **IDLH** = N.D.

Routes of Entry and Relative LD$_{50}$ (or LC$_{50}$)   **RTECS # AJ8400000**

Inhalation	N.D.	
Ingestion	6	(300 mg/kg)
Skin Absorption	3	(1535 mg/kg)

Symptoms of Exposure: Stomach pains, lethargy, nausea, diarrhea.

Emergency/First Aid Treatment: Remove to ventilated area; immediately remove any contaminated clothing and wash contaminated areas for 15 minutes using water. Treat supportively and observe for possible shock. If ingested, seek immediate medical aid.

Recommended Clean-Up Procedures

Personal Protection:	Level B Ensemble	**Recommended Material**	N.D.
RCRA Waste #	U232	**Reportable Quantities:**	1000 lb (100 lb)

Spills: Take up in non-combustible absorbent and dispose. Decontaminate spill area using soapy water. Treat all materials used or generated and equipment involved as contaminated by hazardous waste.

Special Emergency Information: May be harmful if inhaled, swallowed or absorbed through the skin.

TELLURIUM HEXAFLUORIDE - TeF$_6$

Synonyms: tellurium fluoride

CAS Number: 7783-80-4
DOT Number: 2195

Description: Colorless Gas
DOT Classification: Poison A

Molecular Weight: 241.6
Melting Point: -36°C (-33°F)
Boiling Point: Sublimes 191°C(376°F)

Vapor Density: N.D.
Specific Gravity: N.A.

Vapor Pressure: Gas
Water Solubility: Decomposes

Chemical Incompatibilities or Instabilities:

FLAMMABILITY

NFPA Hazard Code:
Flash Point: N.A.
Autoignition Temp: N.A.

NFPA Classification:
LEL: N.A.
UEL: N.A.

Fire Extinguishing Methods: Carbon Dioxide, Dry Chemical, Foam, Water Spray or Fog.

Special Fire Fighting Considerations: Structural fire fighter protective clothing will not provide adequate protection. Water spray should be used to keep closed containers cool. Continue to cool container after fire is extinguished. Do not allow water to enter container. Keep area isolated until vapors have dissipated.

TOXICOLOGY

Odor: Repulsive
Physical Contact: Material is extremely destructive to human tissues.
TLV = 0.2 mg (Te)/m^3 **STEL** = N.D.

Odor Threshold: N.D.

IDLH = 20 mg/m^3

Routes of Entry and Relative LD$_{50}$ (or LC$_{50}$)
    Inhalation   N.D.
    Ingestion   N.D.
    Skin Absorption   N.D.

RTECS # WY2800000

Symptoms of Exposure: Possible burns or irritation to skin, eyes and upper respiratory tract; headache; nausea; garlic breath odor. Inhalation of vapors may be fatal by causing glottis to spasm and suffocation. Exposure may result in chemical pneumonitis or pulmonary edema.

Emergency/First Aid Treatment: Remove to ventilated area; immediately remove any contaminated clothing and wash contaminated areas for 15 minutes using water. Treat supportively and observe for possible shock. If ingested, seek immediate medical aid.

Recommended Clean-Up Procedures

Personal Protection: Level A Ensemble
RCRA Waste #   None

Recommended Material   N.D.
Reportable Quantities: None

Spills: Stop leak if it can safely be done. Keep area isolated until vapors have dissipated.

Special Emergency Information: May be fatal in inhaled, ingested or absorbed through the skin.

TETRACHLOROETHANE

Synonyms: acetylene tetrachloride; 1,1,2,2-tetrachloroethane

CAS Number: 79-34-5   **Description:** Colorless Liquid
DOT Number: 1702   **DOT Classification:** IMO: Poison B

Molecular Weight: 167.8
Melting Point: -27°C (-33°F)   **Vapor Density:** N.D.   **Vapor Pressure:** 3 (9 mm Hg)
Boiling Point: 146°C (296°F)   **Specific Gravity:** 1.6   **Water Solubility:** Insoluble

Chemical Incompatibilities or Instabilities: Nitrates, 2,4-dinitrophenyldisulfide, alkali metals, strong alkali, dinitrogen tetroxide.

FLAMMABILITY

NFPA Hazard Code: **NFPA Classification:**
Flash Point: N.A.   **LEL:** N.A.
Autoignition Temp: N.A.   **UEL:** N.A.

Fire Extinguishing Methods: Carbon Dioxide, Dry Chemical, Foam, Water Spray or Fog.

Special Fire Fighting Considerations: Structural fire fighter protective clothing will not provide adequate protection. Fight fire from a distance or protected location, if possible. Treat all materials used or generated and equipment involved as contaminated by hazardous waste.

TOXICOLOGY
CARCINOGEN

Odor: Chloroform-like   **Odor Threshold:** 1.6 mg/m^3
Physical Contact: Irritant
TLV = 7 mg/m^3 **STEL** = N.D.   **IDLH** = 1505 mg/m^3

Routes of Entry and Relative LD$_{50}$ (or LC$_{50}$)   **RTECS # KI8575000**
    Inhalation   N.D.
    Ingestion   4   (800 mg/kg)
    Skin Absorption   N.D.

Symptoms of Exposure: Possible irritation to skin, eyes and upper respiratory tract; headache; nausea; CNS depression. Lachrymation.

Emergency/First Aid Treatment: Remove to ventilated area; immediately remove any contaminated clothing and wash contaminated areas for 15 minutes using water. Treat supportively and observe for possible shock. If ingested, seek immediate medical aid.

Recommended Clean-Up Procedures

Personal Protection: Level A Ensemble   **Recommended Material** Polyvinyl Alcohol (+), Viton (+)

RCRA Waste #   U209   **Reportable Quantities:**   100 lb (1 lb)

Spills: Dike to contain spill and collect liquid for later disposal. Decontaminate spill area using soapy water. Treat all materials used or generated and equipment involved as contaminated by hazardous waste.

Special Emergency Information: May be fatal in inhaled, ingested or absorbed through the skin.

1,1,1,2-tetrachloroethane (CAS # 630-20-6, RTECS # KI8450000, RCRA Waste # U208, RQ: 100 lb [1lb]) has similar chemical, physical and toxicological properties.

TETRAETHYLENEPENTAMINE - $NH(CH_2CH_2NHCH_2CH_2NH_2)_2$

Synonyms:

CAS Number: 112-57-2	**Description:** Colorless Liquid	
DOT Number: 2320	**DOT Classification:** Corrosive Material	

Molecular Weight: 189.3

Melting Point: -40°C (-40°F)   **Vapor Density:** 6.5   **Vapor Pressure:** 1 (<1 mm Hg)

Boiling Point: 340°C (644°F)   **Specific Gravity:** 1.0 **Water**

Solubility: Soluble

Chemical Incompatibilities or Instabilities: Oxidizers.

FLAMMABILITY

NFPA Hazard Code: 2-1-0   **NFPA Classification:** Class IIIB Combustible

Flash Point: 185°C (365°F)   **LEL:** N.D.

Autoignition Temp: 321°C (610°F)   **UEL:** N.D.

Fire Extinguishing Methods: Carbon Dioxide, Dry Chemical, Foam, Water Spray or Fog.

Special Fire Fighting Considerations: Water spray should be used to keep closed containers cool. Continue to cool container after fire is extinguished. Treat all materials used or generated and equipment involved as contaminated by hazardous waste.

TOXICOLOGY

Odor: N.D.   **Odor Threshold:** N.D.

Physical Contact: Material is extremely destructive to human tissues.

TLV = N.D.   **STEL** = N.D.   **IDLH** = N.D.

Routes of Entry and Relative LD$_{50}$ (or LC$_{50}$)   **RTECS # KH8585000**

Inhalation	N.D.	
Ingestion	6	(205 mg/kg)
Skin Absorption	6	(660 mg/kg)

Symptoms of Exposure: Possible burns or irritation to skin, eyes and upper respiratory tract; headache; nausea; laryngitis. Inhalation of vapors may be fatal by causing glottis to spasm and suffocation. Exposure may result in chemical pneumonitis or pulmonary edema.

Emergency/First Aid Treatment: Remove to ventilated area; immediately remove any contaminated clothing and wash contaminated areas for 15 minutes using water. Treat supportively and observe for possible shock. If ingested, seek immediate medical aid.

Recommended Clean-Up Procedures

Personal Protection: Level A Ensemble   **Recommended Material** Butyl Rubber (+), Neoprene (+), Viton (+)

RCRA Waste # None   **Reportable Quantities:** None

Spills: Dike to contain spill and collect liquid for later disposal. Decontaminate spill area using water. Treat all materials used or generated and equipment involved as contaminated by hazardous waste.

Special Emergency Information: May be harmful if inhaled, swallowed or absorbed through the skin. May be a sensitizer.

TETRAETHYL PYROPHOSPHATE

Synonyms: diphosphoric acid, tetraethyl ester; fosvex; grisol; killax; tetrastigmine; tetron; vapotone

CAS Number: 107-49-3   **Description:** White to amber liquid
DOT Number: 2783   **DOT Classification:** Poison B

Molecular Weight: 290.2
Melting Point: °C   (°F)   **Vapor Density:** N.D.   **Vapor Pressure:** 1 (<1 mm Hg)
Boiling Point: D   170°C   (338°F)   **Specific Gravity:** 1.2   **Water Solubility:** Soluble

Chemical Incompatibilities or Instabilities: Water, oxidizers.

FLAMMABILITY

NFPA Hazard Code: **NFPA Classification:**
Flash Point: N.A.   **LEL:** N.A.
Autoignition Temp: N.A.   **UEL:** N.A.

Fire Extinguishing Methods: Use agent suitable for surrounding fire.

Special Fire Fighting Considerations: Structural fire fighter protective clothing will not provide adequate protection. Isolate for 150 feet if rail or tank truck is involved in a fire. Fight fire from a distance or protected location, if possible. Treat all materials used or generated and equipment involved as contaminated by hazardous waste.

TOXICOLOGY

Odor: Agreeable   **Odor Threshold:** N.D.
Physical Contact: Irritant
TLV = 0.05 mg /m^3 **STEL** = N.D.   **IDLH** = N.D.

Routes of Entry and Relative LD$_{50}$ (or LC$_{50}$)   **RTECS # UX6825000**
  Inhalation   N.D.
  Ingestion   **10**   (0.5 mg/kg)
  Skin Absorption   N.D.

Symptoms of Exposure: Excitement, muscle spasms, nausea, stomach disturbances.

Emergency/First Aid Treatment: Remove to ventilated area; immediately remove any contaminated clothing and wash contaminated areas for 15 minutes using water. Treat supportively and observe for possible shock. If ingested, seek immediate medical aid. Treat for acetylcholine esterase inhibitor poisoning.

Recommended Clean-Up Procedures

Personal Protection: Level A Ensemble   **Recommended Material**   N.D.

RCRA Waste #   P111   **Reportable Quantities:**   10 lb (100 lb)

Spills: Take up in non-combustible absorbent and dispose. Decontaminate spill area using water. Treat all materials used or generated and equipment involved as contaminated by hazardous waste.

Special Emergency Information: May be fatal in inhaled, ingested or absorbed through the skin. Material slowly hydrolyzes in water.

TETRAFLUOROETHYLENE - $F_2C=CF_2$

Synonyms: perfluoroethylene; perfluoroethene; tetrafluoroethene

CAS Number: 116-14-3
DOT Number: 1081

Description: Colorless Gas
DOT Classification: Flammable Gas

Molecular Weight: 100.0
Melting Point: -142°C (-224°F)
Boiling Point: -78°C (-78°F)

Vapor Density: 3.5
Specific Gravity: N.A.

Vapor Pressure: Gas
Water Solubility: Insoluble

Chemical Incompatibilities or Instabilities: Oxidizers, sulfur trioxide, difluoromethylene, oxygen, air, interhalogens, dioxygen difluoride.

FLAMMABILITY

NFPA Hazard Code: 2-4-3
Flash Point: 0°C (32°F)
Autoignition Temp: 188°C (370°F)

NFPA Classification:
LEL: 11.0%
UEL: 60.0%

Fire Extinguishing Methods: Carbon Dioxide, Dry Chemical, Foam, Water Spray or Fog.

Special Fire Fighting Considerations: Isolate for 1/2 mile if rail or tank truck is involved in a fire. Water spray from an unmanned device should be used to keep closed containers cool. Continue to cool container after fire is extinguished. For large fires, if possible, withdraw and allow to burn. Immediately withdraw if rising sound from venting device is heard or if fire is causing discoloration to the tank.

TOXICOLOGY
QUESTIONABLE CARCINOGEN

Odor: Sweet
Physical Contact: Irritant
TLV = N.D.   **STEL** = N.D.

Odor Threshold: N.D.

IDLH = N.D.

Routes of Entry and Relative LD$_{50}$ (or LC$_{50}$)
  Inhalation   1   (500,000 mg/m^3/H)
  Ingestion   N.D.
  Skin Absorption   N.D.

RTECS # KX4000000

Symptoms of Exposure: A simple asphyxiant.

Emergency/First Aid Treatment: Remove to ventilated area; immediately remove any contaminated clothing and wash contaminated areas for 15 minutes using water. Treat supportively and observe for possible shock. If ingested, seek immediate medical aid.

Recommended Clean-Up Procedures

Personal Protection: Level B Ensemble

Recommended Material Butyl Rubber (+), Neoprene (+), Polyvinyl Alcohol (+), Viton (+)

RCRA Waste # None

Reportable Quantities: None

Spills: Stop leak if it can safely be done. Keep area isolated until vapors have dissipated.

Special Emergency Information: May be harmful if inhaled, swallowed or absorbed through the skin. May undergo hazardous polymerization. Vapors are heavier than air and may travel some distance to an ignition source. Vapors are more dense than air and may settle in low lying areas.

TETRAHYDROTHIOPHENE

Synonyms:

CAS Number: 110-01-0   **Description:** Colorless Liquid
DOT Number: 2412   **DOT Classification:** Flammable Liquid

Molecular Weight: 88.2
Melting Point: -96°C (-141°F)   **Vapor Density:** N.D.   **Vapor Pressure:** N.D.
Boiling Point: 119°C (246°F)   **Specific Gravity:** 1.0   **Water Solubility:** Insoluble

Chemical Incompatibilities or Instabilities: Hydrogen peroxide, oxidizers.

FLAMMABILITY

NFPA Hazard Code: **NFPA Classification:** Class 1B Flammable
Flash Point: 12°C (55°F)   **LEL:** N.D.
Autoignition Temp: N.D.   **UEL:** N.D.

Fire Extinguishing Methods: Alcohol Resistant Foam, Carbon Dioxide, Dry Chemical, Water Spray or Fog.

Special Fire Fighting Considerations: Isolate for 1/2 mile if rail or tank truck is involved in a fire. Water spray should be used to keep closed containers cool. Continue to cool container after fire is extinguished. For large fires, if possible, withdraw and allow to burn. Immediately withdraw if rising sound from venting device is heard or if fire is causing discoloration to the tank.

TOXICOLOGY

Odor: N.D.   **Odor Threshold:** N.D.
Physical Contact: Irritant
TLV = N.D.   **STEL** = N.D.   **IDLH** = N.D.

Routes of Entry and Relative LD$_{50}$ (or LC$_{50}$)   **RTECS #** N.D.
    Inhalation   N.D.
    Ingestion   N.D.
    Skin Absorption   N.D.

Symptoms of Exposure: Possible irritation to skin, eyes and upper respiratory tract; headache; nausea.

Emergency/First Aid Treatment: Remove to ventilated area; immediately remove any contaminated clothing and wash contaminated areas for 15 minutes using water. Treat supportively and observe for possible shock. If ingested, seek immediate medical aid.

Recommended Clean-Up Procedures

Personal Protection: Level B Ensemble   **Recommended Material** N.D.

RCRA Waste # None   **Reportable Quantities:** None

Spills: Remove all potential ignition sources. Dike to contain spill, absorb with non-combustible absorbent and take up using non-sparking tools. Decontaminate spill area using soapy water.

Special Emergency Information: May be harmful if inhaled, swallowed or absorbed through the skin. May form unstable peroxides upon prolonged exposure to air.

TETRAMETHYLAMMONIUM HYDROXIDE - (CH3)4NOH

Synonyms:

CAS Number: 75-59-2
DOT Number: 1835

Description: Colorless Liquid
DOT Classification: Corrosive Material

Molecular Weight: 91.2
Melting Point: °C (°F)
Boiling Point: °C (°F)

Vapor Density: N.D.
Specific Gravity: 0.9

Vapor Pressure: 1 (<1 mm Hg)
Water Solubility: Soluble

Chemical Incompatibilities or Instabilities: Oxidizers.

FLAMMABILITY

NFPA Hazard Code:
Flash Point: N.D.
Autoignition Temp: N.D.

NFPA Classification:
LEL: N.D.
UEL: N.D.

Fire Extinguishing Methods: Carbon Dioxide, Dry Chemical, Foam, Water Spray or Fog.

Special Fire Fighting Considerations: Water spray should be used to keep closed containers cool. Continue to cool container after fire is extinguished.

TOXICOLOGY

Odor: N.D.
Physical Contact: Material is extremely destructive to human tissues.
TLV = N.D.   **STEL** = N.D.

Odor Threshold: N.D.

IDLH = N.D.

Routes of Entry and Relative LD$_{50}$ (or LC$_{50}$)
  Inhalation   N.D.
  Ingestion   N.D.
  Skin Absorption N.D.

RTECS # PA0875000

Symptoms of Exposure: Possible burns or irritation to skin, eyes and upper respiratory tract; headache; nausea. Laryngitis. Inhalation of vapors may be fatal by causing glottis to spasm and suffocation. Exposure may result in chemical pneumonitis or pulmonary edema.

Emergency/First Aid Treatment: Remove to ventilated area; immediately remove any contaminated clothing and wash contaminated areas for 15 minutes using water. Treat supportively and observe for possible shock. If ingested, seek immediate medical aid.

Recommended Clean-Up Procedures

Personal Protection: Level B Ensemble
RCRA Waste # None

Recommended Material N.D.
Reportable Quantities: None

Spills: Take up in non-combustible absorbent and dispose. Decontaminate spill area using water.

Special Emergency Information: May be harmful if inhaled, swallowed or absorbed through the skin.

TETRAMETHYLMETHANEDIAMINE - $(CH_3)_2NCH_2N(CH_3)_2$

Synonyms: N,N,N',N'-tetramethyldiaminomethane

CAS Number: 51-80-9
DOT Number: 9069

Description: Colorless Liquid
DOT Classification: ORM-A

Molecular Weight: 102.2
Melting Point: °C (°F)
Boiling Point: 85°C (185°F)

Vapor Density: N.D.
Specific Gravity: 0.7

Vapor Pressure: N.D.
Water Solubility: N.D.

Chemical Incompatibilities or Instabilities: Strong Oxidizers.

FLAMMABILITY

NFPA Hazard Code:
Flash Point: 1°C (35°F)
Autoignition Temp: N.D.

NFPA Classification: Class 1B Flammable
LEL: N.D.
UEL: N.D.

Fire Extinguishing Methods: Carbon Dioxide, Dry Chemical, Foam, Water Spray or Fog.

Special Fire Fighting Considerations: Isolate for 1/2 mile if rail or tank truck is involved in a fire. Water spray should be used to keep closed containers cool. Continue to cool container after fire is extinguished. Do not allow water to enter container. Immediately withdraw if rising sound from venting device is heard or if fire is causing discoloration to the tank.

TOXICOLOGY

Odor: N.D.
Physical Contact: Irritant
TLV = N.D.

STEL = N.D.

Odor Threshold: N.D.

IDLH = N.D.

Routes of Entry and Relative LD$_{50}$ (or LC$_{50}$)

Inhalation	N.D.
Ingestion	N.D.
Skin Absorption	N.D.

RTECS # PA6700000

Symptoms of Exposure: Possible irritation to skin, eyes and upper respiratory tract; headache; nausea.

Emergency/First Aid Treatment: Remove to ventilated area; immediately remove any contaminated clothing and wash contaminated areas for 15 minutes using water. Treat supportively and observe for possible shock. If ingested, seek immediate medical aid.

Recommended Clean-Up Procedures

Personal Protection: Level B Ensemble

RCRA Waste # None

Recommended Material N.D.

Reportable Quantities: None

Spills: Remove all potential ignition sources. Dike to contain spill, absorb with non-combustible absorbent and take up using non-sparking tools. Decontaminate spill area using soapy water.

Special Emergency Information: May be harmful if inhaled, swallowed or absorbed through the skin. Vapors are heavier than air and may travel some distance to an ignition source. Vapors are more dense than air and may settle in low lying areas.

TETRAMETHYL ORTHOSILICATE - Si(CH₃O)₄

Synonyms: tetramethoxysilane

CAS Number: 681-84-5   **Description:** Colorless Liquid
DOT Number: 2606   **DOT Classification:** N.D.

Molecular Weight: 152.2
Melting Point: -4°C   (25°F)   **Vapor Density:** 5.3   **Vapor Pressure:** N.D.
Boiling Point: 121°C (250°F)   **Specific Gravity:** 1.0   **Water Solubility:** Decomposes

Chemical Incompatibilities or Instabilities: Water.

FLAMMABILITY

NFPA Hazard Code: 3-3-1   **NFPA Classification:** Class IB Flammable
Flash Point: 20°C   (69°F)   **LEL:** N.D.
Autoignition Temp: N.D.   **UEL:** N.D.

Fire Extinguishing Methods: Carbon Dioxide, Dry Chemical, Foam, Water Spray or Fog.

Special Fire Fighting Considerations: Structural fire fighter protective clothing will not provide adequate protection. Water spray should be used to keep closed containers cool. Continue to cool container after fire is extinguished. For large fires, if possible, withdraw and allow to burn.

DOT Recommended Isolation Zones: Small Spill: 150 ft   Large Spill: 150 ft
DOT Recommended Down Wind Take Cover Distance   Small Spill: 0.4 miles   Large Spill: 0.8 miles

TOXICOLOGY

Odor: Pungent   **Odor Threshold:** N.D.
Physical Contact: Material is extremely destructive to human tissues.
TLV = N.D.   **STEL** = N.D.   **IDLH** = N.D.

Routes of Entry and Relative LD₅₀ (or LC₅₀)   **RTECS # VV9800000**
   Inhalation   N.D.
   Ingestion   N.D.
   Skin Absorption   1   (17,000 mg/kg)

Symptoms of Exposure: Possible burns or irritation to skin, eyes and upper respiratory tract; headache; nausea. Inhalation of vapors may be fatal by causing glottis to spasm and suffocation. Exposure may result in chemical pneumonitis or pulmonary edema.

Emergency/First Aid Treatment: Remove to ventilated area; immediately remove any contaminated clothing and wash contaminated areas for 15 minutes using water. Treat supportively and observe for possible shock. If ingested, seek immediate medical aid.

Recommended Clean-Up Procedures

Personal Protection: Level B Ensemble   **Recommended Material**   N.D.

RCRA Waste #   None   **Reportable Quantities:**   None

Spills: Remove all potential ignition sources. Dike to contain spill, absorb with non-combustible absorbent and take up using non-sparking tools. Decontaminate spill area using water.

Special Emergency Information: May be harmful if inhaled, swallowed or absorbed through the skin. Vapors are heavier than air and may travel some distance to an ignition source. Vapors are more dense than air and may settle in low lying areas. Reacts with water to yield methanol.

TETRAMETHYLSILANE - (CH3)4Si

Synonyms:

CAS Number: 75-76-3

DOT Number: 2749

Molecular Weight: 88.2

Description: Colorless Liquid

DOT Classification: Flammable Liquid

Melting Point:	-99°C (-146°F)	**Vapor Density:** N.D.	**Vapor Pressure:** 9 (560 mm Hg)
Boiling Point:	27°C (81°F)	**Specific Gravity:** 0.6	**Water Solubility:** Decomposes

Chemical Incompatibilities or Instabilities: Water, strong oxidizers, air, halogens.

FLAMMABILITY

NFPA Hazard Code:

Flash Point: -27°C (-17°F)

Autoignition Temp: 450°C (842°F)

NFPA Classification: Class IA Flammable

LEL: N.D.

UEL: N.D.

Fire Extinguishing Methods: Carbon Dioxide, Dry Chemical, Foam, Water Spray or Fog.

Special Fire Fighting Considerations: Isolate for 1/2 mile if rail or tank truck is involved in a fire. Water spray should be used to keep closed containers cool. Continue to cool container after fire is extinguished. Do not allow water to enter container. Immediately withdraw if rising sound from venting device is heard or if fire is causing discoloration to the tank.

TOXICOLOGY

Odor: N.D.

Physical Contact: Irritant

TLV = N.D.   **STEL** = N.D.

Odor Threshold: N.D.

IDLH = N.D.

Routes of Entry and Relative LD$_{50}$ (or LC$_{50}$)

    Inhalation   N.D.

    Ingestion   N.D.

    Skin Absorption N.D.

RTECS # VV5705400

Symptoms of Exposure: Possible irritation to skin, eyes and upper respiratory tract; headache; nausea.

Emergency/First Aid Treatment: Remove to ventilated area; immediately remove any contaminated clothing and wash contaminated areas for 15 minutes using water. Treat supportively and observe for possible shock. If ingested, seek immediate medical aid.

Recommended Clean-Up Procedures

Personal Protection: Level B Ensemble

RCRA Waste # None

Recommended Material N.D.

Reportable Quantities: None

Spills: Remove all potential ignition sources. Dike to contain spill, absorb with non-combustible absorbent and take up using non-sparking tools. Decontaminate spill area using soapy water. Keep area isolated until vapors have dissipated.

Special Emergency Information: May be harmful if inhaled, swallowed or absorbed through the skin. Vapors are heavier than air and may travel some distance to an ignition source. Vapors are more dense than air and may settle in low lying areas. May ignite upon exposure to air.

TETRANITROMETHANE - (NO₂)₄C

Synonyms:

CAS Number: 509-14-8	**Description:** Pale yellow liquid	
DOT Number: 1510	**DOT Classification:** Oxidizer	

Molecular Weight: 196.1

Melting Point:	13°C (55°F)	**Vapor Density:** N.D.	**Vapor Pressure:** 3 (10 mm Hg)
Boiling Point:	126°C (259°F)	**Specific Gravity:** 1.7	**Water Solubility:** Insoluble

Chemical Incompatibilities or Instabilities: Oxidizers, nitrated hydrocarbons, amines, pyridine, sodium ethoxide, aluminum.

FLAMMABILITY

NFPA Hazard Code: **NFPA Classification:** Class IIIB Combustible

Flash Point:	110°C (230°F)	**LEL:** N.D.
Autoignition Temp:	N.D.	**UEL:** N.D.

Fire Extinguishing Methods: WATER ONLY

Special Fire Fighting Considerations: Isolate for 1/2 mile if rail or tank truck is involved in a fire. Water spray should be used to keep closed containers cool. Continue to cool container after fire is extinguished. For large fires, if possible, withdraw and allow to burn.

DOT Recommended Isolation Zones:	Small Spill: 150 ft	Large Spill: 150 ft
DOT Recommended Down Wind Take Cover Distance	Small Spill: 0.4 miles	Large Spill: 0.8 miles

TOXICOLOGY

Odor: Pungent   **Odor Threshold:** N.D.
Physical Contact: Irritant
TLV = 8 mg/m³   **STEL** = N.D.   **IDLH** = 40 mg/m³

Routes of Entry and Relative LD₅₀ (or LC₅₀)   **RTECS # PB4025000**

Inhalation	9	(576 mg/m³/H)
Ingestion	7	(130 mg/kg)
Skin Absorption	N.D.	

Symptoms of Exposure: Possible irritation to skin, eyes and upper respiratory tract; headache; nausea. Absorption may lead to the formation of methemoglobin resulting in cyanosis several hours after exposure.

Emergency/First Aid Treatment: Remove to ventilated area; immediately remove any contaminated clothing and wash contaminated areas for 15 minutes using water. Treat supportively and observe for possible shock. If ingested, seek immediate medical aid.

Recommended Clean-Up Procedures

Personal Protection:	Level A Ensemble	**Recommended Material** N.D.
RCRA Waste #	P112	**Reportable Quantities:** 100 lb (1 lb)

Spills: Take up in non-combustible absorbent and dispose. Decontaminate spill area using soapy water. Treat all materials used or generated and equipment involved as contaminated by hazardous waste.

Special Emergency Information: May be harmful if inhaled, swallowed or absorbed through the skin. May detonate from heat shock or friction.

THALLIUM NITRATE - TlNO$_3$

Synonyms:

CAS Number: 10102-45-1
DOT Number: 2727

Description: White powder
DOT Classification: Poison B

Molecular Weight: 266.4
Melting Point: 206°C (403°F)
Boiling Point: 430°C (806°F)

Vapor Density: N.D.
Specific Gravity: 5.6

Vapor Pressure: 1 (<1 mm Hg)
Water Solubility: Soluble

Chemical Incompatibilities or Instabilities: Finely divided metals, reducing agents.

FLAMMABILITY

NFPA Hazard Code:
Flash Point: N.A.
Autoignition Temp: N.A.

NFPA Classification:
LEL: N.A.
UEL: N.A.

Fire Extinguishing Methods: WATER ONLY

Special Fire Fighting Considerations: Water spray should be used to keep closed containers cool. Continue to cool container after fire is extinguished. Treat all materials used or generated and equipment involved as contaminated by hazardous waste.

TOXICOLOGY

Odor: None
Physical Contact: Irritant
TLV = 0.1 mg (Tl)/m^3 (skin)   **STEL** = N.D.

Odor Threshold: N.D.

IDLH = 20 mg/m^3

RTECS # XG5950000

Routes of Entry and Relative LD$_{50}$ (or LC$_{50}$)
    Inhalation   N.D.
    Ingestion   N.D.
    Skin Absorption   N.D.

Symptoms of Exposure: Possible irritation to skin, eyes and upper respiratory tract; headache; nausea.

Emergency/First Aid Treatment: Remove to ventilated area; immediately remove any contaminated clothing and wash contaminated areas for 15 minutes using water. Treat supportively and observe for possible shock. If ingested, seek immediate medical aid.

Recommended Clean-Up Procedures

Personal Protection: Level B Ensemble

RCRA Waste # U217

Recommended Material N.D.

Reportable Quantities: 100 lb (1 lb)

Spills: Take up in non-combustible absorbent and dispose. Decontaminate spill area using water. Treat all materials used or generated and equipment involved as contaminated by hazardous waste.

Special Emergency Information: May be fatal in inhaled, ingested or absorbed through the skin..

TIN TETRACHLORIDE - SnCl₄

Synonyms: tin chloride (fuming); tin perchloride

CAS Number: 7646-78-8
DOT Number: 1827

Description: Colorless Liquid
DOT Classification: Corrosive Material

Molecular Weight: 260.5
Melting Point: -33°C (-27°F)
Boiling Point: 114°C (237°F)

Vapor Density: N.D.
Specific Gravity: 2.2

Vapor Pressure: 4 (20 mm Hg)
Water Solubility: Decomposes

Chemical Incompatibilities or Instabilities: ethylene oxide, water, turpentine, alkyl nitrates, alkali metals.

FLAMMABILITY

NFPA Hazard Code: 3-0-1
Flash Point: N.A.
Autoignition Temp: N.A.

NFPA Classification:
LEL: N.A.
UEL: N.A.

Fire Extinguishing Methods: Dry Chemical, Carbon Dioxide, Flooding Quantities of Water.

Special Fire Fighting Considerations: Structural fire fighter protective clothing will not provide adequate protection. Isolate for 150 feet if rail or tank truck is involved in a fire. Do not allow water to enter container. Water spray should be used to keep closed containers cool. Continue to cool container after fire is extinguished. Reacts with water to liberate hydrogen chloride gas.

TOXICOLOGY

Odor: Sharp
Physical Contact: Material is extremely destructive to human tissues.
$TLV = 2 \text{ mg/m}^3$ **STEL** = N.D.

Odor Threshold: N.D.

IDLH = N.D.

Routes of Entry and Relative LD₅₀ (or LC₅₀)
Inhalation	10	(38 mg/m³/H)
Ingestion	N.D.	
Skin Absorption	N.D.	

RTECS # XP8750000

Symptoms of Exposure: Possible burns or irritation to skin, eyes and upper respiratory tract; headache; nausea. Laryngitis. Inhalation of vapors may be fatal by causing glottis to spasm and suffocation. Exposure may result in chemical pneumonitis or pulmonary edema.

Emergency/First Aid Treatment: Remove to ventilated area; immediately remove any contaminated clothing and wash contaminated areas for 15 minutes using water. Treat supportively and observe for possible shock. If ingested, seek immediate medical aid.

Recommended Clean-Up Procedures

Personal Protection: Level A Ensemble
RCRA Waste # None

Recommended Material N.D.
Reportable Quantities: None

Spills: Dike to contain spill and collect liquid for later disposal. Decontaminate spill area using a dilute alkaline solution. Keep area isolated until vapors have dissipated.

Special Emergency Information: May be fatal in inhaled, ingested or absorbed through the skin.

TITANIUM - Ti

Synonyms:

CAS Number: 7440-32-6   **Description:** Gray metal or powder
DOT Number: 1352, 2546, 2878   **DOT Classification:** Flammable Solid

Molecular Weight: 47.9
Melting Point: 1677°C (3051°F)   **Vapor Density:** N.D.   **Vapor Pressure:** 1 (< 1 mm Hg)
Boiling Point: 3277°C (5931°F)   **Specific Gravity:** 4.5   **Water Solubility:** Insoluble

Chemical Incompatibilities or Instabilities: Oxygen, halogens, halocarbons, strong oxidizers, boron trifluorid.

FLAMMABILITY

NFPA Hazard Code: **NFPA Classification:**
Flash Point: N.A.   **LEL:** N.A.
Autoignition Temp: 250°C (482°F) (in air)   **UEL:** N.A.

Fire Extinguishing Methods: Dry Chemical, Dry Sand, Clay, Water Spray, Foam.

Special Fire Fighting Considerations: Water spray should be used to keep closed containers cool. Continue to cool container after fire is extinguished. For large fires, if possible, withdraw and allow to burn.

TOXICOLOGY

Odor: None   **Odor Threshold:** N.D.
Physical Contact: Irritant
TLV = N.D.   **STEL** = N.D.   **IDLH** = N.D.

Routes of Entry and Relative LD$_{50}$ (or LC$_{50}$)   **RTECS #** XR1700000
    Inhalation   N.D.
    Ingestion   N.D.
    Skin Absorption   N.D.

Symptoms of Exposure: Possible irritation to skin, eyes and upper respiratory tract.

Emergency/First Aid Treatment: Remove to ventilated area; immediately remove any contaminated clothing and wash contaminated areas for 15 minutes using water. Treat supportively and observe for possible shock. If ingested, seek immediate medical aid.

Recommended Clean-Up Procedures

Personal Protection: Level B Ensemble   **Recommended Material** N.D.

RCRA Waste # None   **Reportable Quantities:** None

Spills: Take up in non-combustible absorbent and dispose.

Special Emergency Information: May be harmful if inhaled, swallowed or absorbed through the skin.

TITANIUM TETRACHLORIDE - TiCl₄

Synonyms: titanium (IV) chloride

CAS Number: 7550-45-0	**Description:** Colorless Liquid
DOT Number: 1838	**DOT Classification:** Corrosive Material

Molecular Weight: 189.7

Melting Point:	-11°C (-24°F)	**Vapor Density:** N.D.	**Vapor Pressure:**	**3** (9 mm Hg)
Boiling Point:	136°C (278°F)	**Specific Gravity:** 2.2	**Water Solubility:**	Decomposes

Chemical Incompatibilities or Instabilities: Water.

FLAMMABILITY

NFPA Hazard Code:	3-0-2-~~W~~	**NFPA Classification:**	
Flash Point:	N.A.	**LEL:**	N.A.
Autoignition Temp:	N.A.	**UEL:**	N.A.

Fire Extinguishing Methods: Dry Chemical, Carbon Dioxide, Flooding Quantities of Water.

Special Fire Fighting Considerations: Structural fire fighter protective clothing will not provide adequate protection. Isolate for 150 feet if rail or tank truck is involved in a fire. Water spray should be used to keep closed containers cool. Continue to cool container after fire is extinguished. Do not allow water to enter container.

DOT Recommended Isolation Zones:	Small Spill: 150 ft	Large Spill: 150 ft
DOT Recommended Down Wind Take Cover Distance	Small Spill: 0.2 miles	Large Spill: 0.2 miles

TOXICOLOGY

Odor: Sharp, penetrating	**Odor Threshold:** N.D.

Physical Contact: Material is extremely destructive to human tissues.

TLV = N.D.	**STEL** = N.D.	**IDLH** = N.D.

Routes of Entry and Relative LD$_{50}$ (or LC$_{50}$)   **RTECS # XR1925000**

Inhalation	9	(400 mg/m^3/H)
Ingestion	N.D.	
Skin Absorption	N.D.	

Symptoms of Exposure: Possible burns or irritation to skin, eyes and upper respiratory tract; headache; nausea; laryngitis; lachrymation. Inhalation of vapors may be fatal by causing glottis to spasm and suffocation. Exposure may result in chemical pneumonitis or pulmonary edema.

Emergency/First Aid Treatment: Remove to ventilated area; immediately remove any contaminated clothing and wash contaminated areas for 15 minutes using water. Treat supportively and observe for possible shock. If ingested, seek immediate medical aid.

Recommended Clean-Up Procedures

Personal Protection:	Level A Ensemble	**Recommended Material** Saranex (+)
RCRA Waste #	None	**Reportable Quantities:** None

Spills: Clean up only under expert supervision. Keep area isolated until vapors have dissipated.

Special Emergency Information: May be fatal in inhaled, ingested or absorbed through the skin.

TITANIUM TRICHLORIDE - TiCl$_3$

Synonyms: titanium chloride; titanium (III) chloride; trichlorotitanium

CAS Number: 7705-07-9   **Description:** Dark red to purple powder
DOT Number: 2441, 2869, 2441   **DOT Classification:** Flammable Solid

Molecular Weight: 154.3
Melting Point: D   440°C (824°F)   **Vapor Density:** N.D.   **Vapor Pressure:** 1 (<1 mm Hg)
Boiling Point: N.A.   **Specific Gravity:** 2.6   **Water Solubility:** Decomposes

Chemical Incompatibilities or Instabilities: Oxidizers, water, air, alkali metals, hydrofluoric acid.

FLAMMABILITY

NFPA Hazard Code: **NFPA Classification:**
Flash Point: N.A.   **LEL:** N.A.
Autoignition Temp: N.A.   **UEL:** N.A.

Fire Extinguishing Methods: Dry Chemical, Limestone, Dry Sand, Flooding Quantities of Water.

Special Fire Fighting Considerations: Do not allow water to enter container. Water spray should be used to keep closed containers cool. Continue to cool container after fire is extinguished. For large fires, if possible, withdraw and allow to burn.

TOXICOLOGY

Odor: N.D.   **Odor Threshold:** N.D.
Physical Contact: Material is extremely destructive to human tissues.
TLV = N.D.   **STEL** = N.D.   **IDLH** = N.D.

Routes of Entry and Relative LD$_{50}$ (or LC$_{50}$)   **RTECS # XR1924000**
    Inhalation   N.D.
    Ingestion   N.D.
    Skin Absorption   N.D.

Symptoms of Exposure: Possible burns or irritation to skin, eyes and upper respiratory tract; headache; nausea. Inhalation of vapors may be fatal by causing glottis to spasm and suffocation. Exposure may result in chemical pneumonitis or pulmonary edema.

Emergency/First Aid Treatment: Remove to ventilated area; immediately remove any contaminated clothing and wash contaminated areas for 15 minutes using water. Treat supportively and observe for possible shock. If ingested, seek immediate medical aid.

Recommended Clean-Up Procedures

Personal Protection: Level B Ensemble   **Recommended Material**   N.D.

RCRA Waste #   None   **Reportable Quantities:**   None

Spills: Take up in non-combustible absorbent and dispose. Decontaminate spill area using dry decontamination procedures.

Special Emergency Information: May be fatal in inhaled, ingested or absorbed through the skin. Reacts with water to yield hydrogen chloride vapors.

TOLUENE

Synonyms: methylbenzene; phenylmethane

CAS Number: 108-88-3	**Description:** Colorless Liquid
DOT Number: 1294	**DOT Classification:** Flammable Liquid

Molecular Weight: 92.2

Melting Point: -95°C (-139°F)	**Vapor Density:** 3.2	**Vapor Pressure:** 5 (22 mm Hg)
Boiling Point: 111°C (232°F)	**Specific Gravity:** 0.9	**Water Solubility:** Insoluble

Chemical Incompatibilities or Instabilities: Oxidizers, dinitromethane, interhalogens, strong acids.

FLAMMABILITY

NFPA Hazard Code: 2-3-0	**NFPA Classification:** Class IB Flammable
Flash Point: 4°C (40°F)	**LEL:** 1.2%
Autoignition Temp: 480°C (896°F)	**UEL:** 7.1%

Fire Extinguishing Methods: Carbon Dioxide, Dry Chemical, Foam, Water Spray or Fog.

Special Fire Fighting Considerations: Isolate for 1/2 mile if rail or tank truck is involved in a fire. Water spray from an unmanned device should be used to keep closed containers cool. Continue to cool container after fire is extinguished. For large fires, if possible, withdraw and allow to burn. Immediately withdraw if rising sound from venting device is heard or if fire is causing discoloration to the tank.

TOXICOLOGY

Odor: Airplane glue

Physical Contact: Irritant

Odor Threshold: 0.08 mg/m^3

$TLV = 375$ mg/m^3 $STEL = 560$ mg/m^3 $IDLH = 7,500$ mg/m^3

Routes of Entry and Relative LD$_{50}$ (or LC$_{50}$)   **RTECS # XS5250000**

Inhalation	N.D.	
Ingestion	3	(5,000 mg/kg)
Skin Absorption	1	(12,124 mg/kg)

Symptoms of Exposure: Possible irritation to skin, eyes and upper respiratory tract; headache; nausea; laryngitis. Exposure may result in chemical pneumonitis or pulmonary edema.

Emergency/First Aid Treatment: Remove to ventilated area; immediately remove any contaminated clothing and wash contaminated areas for 15 minutes using water. Treat supportively and observe for possible shock. If ingested, seek immediate medical aid.

Recommended Clean-Up Procedures

Personal Protection: Level B Ensemble	**Recommended Material**	Polyvinyl Alcohol (+), Teflon (+), Viton (+)
RCRA Waste # U220	**Reportable Quantities:**	1000 lb (1000 lb)

Spills: Remove all potential ignition sources. Dike to contain spill, absorb with non-combustible absorbent and take up using non-sparking tools. Decontaminate spill area using soapy water. Keep area isolated until vapors have dissipated.

Special Emergency Information: May be harmful if inhaled, swallowed or absorbed through the skin. Vapors are heavier than air and may travel some distance to an ignition source. Vapors are more dense than air and may settle in low lying areas.

TOLUENE-2,4-DIAMINE

Synonyms: 2,4-diaminotoluene

CAS Number: 95-80-7   **Description:** Tan powder
DOT Number: 1709   **DOT Classification:** Poison B

Molecular Weight: 122.2

Melting Point:	99 C (210 F)	**Vapor Density:**	N.D.	**Vapor Pressure:**	1 (<1 mm Hg)
Boiling Point:	285 C (545 F)	**Specific Gravity:**	N.D.	**Water Solubility:**	Soluble

Chemical Incompatibilities or Instabilities: Oxidizers.

FLAMMABILITY

NFPA Hazard Code:		**NFPA Classification:**	
Flash Point:	N.D.	**LEL:**	N.D.
Autoignition Temp:	N.D.	**UEL:**	N.D.

Fire Extinguishing Methods: Carbon Dioxide, Dry Chemical, Foam, Water Spray or Fog.

Special Fire Fighting Considerations: Treat all materials used or generated and equipment involved as contaminated by hazardous waste.

TOXICOLOGY
CARCINOGEN

Odor: **Odor Threshold:** N.D.
Physical Contact: Irritant
TLV = N.D.   **STEL** = N.D.   **IDLH** = N.D.

Routes of Entry and Relative LD$_{50}$ (or LC$_{50}$)   **RTECS # XS9625000**
     Inhalation   N.D.
     Ingestion   N.D.
     Skin Absorption   N.D.

Symptoms of Exposure: Possible irritation to skin, eyes and upper respiratory tract; headache; nausea. Absorption may lead to the formation of methemoglobin resulting in cyanosis several hours after exposure.

Emergency/First Aid Treatment: Remove to ventilated area; immediately remove any contaminated clothing and wash contaminated areas for 15 minutes using water. Treat supportively and observe for possible shock. If ingested, seek immediate medical aid.

Recommended Clean-Up Procedures

Personal Protection:	Level B Ensemble	**Recommended Material**	N.D.
RCRA Waste #	U221	**Reportable Quantities:**	10 lb (1 lb)

Spills: Take up in non-combustible absorbent and dispose. Decontaminate spill area using water. Decontaminate spill area using water. Treat all materials used or generated and equipment involved as contaminated by hazardous waste.

Special Emergency Information: May be harmful if inhaled, swallowed or absorbed through the skin.

2,5 diaminotoluene (CAS # 95-70-5, RTECS # XS9700000) and
2,6-diaminotoluene (CAS # 823-40-5, RTECS # XS9750000) have similar chemical, physical and toxicological properties, and the same RCRA Waste # and Reportable Quantities.

TOLUENE-2,4-DIISOCYANATE

Synonyms: di-iso-cyanatoluene; 2,4-toluenediisocyanate; tolylene 2,4-diisocyanate

CAS Number: 584-84-9   **Description:** Colorless Liquid
DOT Number: 2078   **DOT Classification:** Poison B

Molecular Weight: 174.2
Melting Point: 22 C   (71 F)   **Vapor Density:** 6.0   **Vapor Pressure:** **1** (<1 mm Hg)
Boiling Point: 251 C   (484 F)   **Specific Gravity:** 1.22   **Water Solubility:** Decomposes

Chemical Incompatibilities or Instabilities: Alkali, acyl chlorides, water, alcohols, organic acids, organometallics, heat.

FLAMMABILITY

NFPA Hazard Code: **3-1-2**   **NFPA Classification:** Class IIIB Combustible
Flash Point: 132 C   (270 F)   **LEL:** 0.9%
Autoignition Temp: C   (F)   **UEL:** 9.5%

Fire Extinguishing Methods: Carbon Dioxide, Dry Chemical, Foam, Water Spray or Fog.

Special Fire Fighting Considerations: Structural fire fighter protective clothing will not provide adequate protection. Water spray from an unmanned device should be used to keep closed containers cool. Continue to cool container after fire is extinguished. Fight fire from a distance or protected location, if possible. Treat all materials used or generated and equipment involved as contaminated by hazardous waste.

TOXICOLOGY
CARCINOGEN

Odor: Sharp, pungent   **Odor Threshold:** N.D.
Physical Contact: Material is extremely destructive to human tissues.
TLV = 0.04 mg/m^3 **STEL** = 0.15 mg/m^3 **IDLH** = 72.4 mg/m^3

Routes of Entry and Relative LD$_{50}$ (or LC$_{50}$)   **RTECS # CZ6300000**
   Inhalation   10   (405 mg/m^3/H)
   Ingestion   2   (5800 mg/kg)
   Skin Absorption   N.D.

Symptoms of Exposure: Possible burns or irritation to skin, eyes and upper respiratory tract; headache; nausea; lachrymation. May cause an allergic reaction. Exposure may result in chemical pneumonitis or pulmonary edema.

Emergency/First Aid Treatment: Remove to ventilated area; immediately remove any contaminated clothing and wash contaminated areas for 15 minutes using water. Treat supportively and observe for possible shock. If ingested, seek immediate medical aid.

Recommended Clean-Up Procedures

Personal Protection: Level A Ensemble   **Recommended Material**   Butyl Rubber (+), Polyvinyl Alcohol (+), Nitrile Rubber (+), Viton (+), Saranex (+)

RCRA Waste #   U223   **Reportable Quantities:**   100 lb (1 lb)

Spills: Dike to contain spill and collect liquid for later disposal. Decontaminate spill area using water. Keep area isolated until vapors have dissipated. Treat all materials used or generated and equipment involved as contaminated by hazardous waste.

Special Emergency Information: May be fatal in inhaled, ingested or absorbed through the skin. May under hazardous polymerization. Reaction with water yields carbon dioxide.

Toluene-1,3-diisocyanate (CAS # 26471-62-5, RCRA WASTE # U223) and toluene-2,6-diisocyanate (CAS # 91-08-7, RTECS # CZ6310000, RCRA WASTE #U223) have similar chemical, physical and toxicological properties, and the same RCRA Waste # and Reportable Quantities.

4-TOLUENESULFONIC ACID

Synonyms: p-methylbenzenesulfonic acid; 4-methylbenzene sulfonic acid; p-toluenesulfonic acid; tosic acid

CAS Number: 104-15-4
DOT Number: 2583, 2584, 2585, 2586

Description: White powder
DOT Classification: Corrosive Material

Molecular Weight: 172.2
Melting Point: 107°C (225°F)
Boiling Point: °C (°F)

Vapor Density: N.D.
Specific Gravity: N.D.

Vapor Pressure: 1 (<1 mm Hg)
Water Solubility: Soluble

Chemical Incompatibilities or Instabilities: Acetic anhydride.

FLAMMABILITY

NFPA Hazard Code:
Flash Point: N.A.
Autoignition Temp: N.A.

NFPA Classification:
LEL: N.A.
UEL: N.A.

Fire Extinguishing Methods: Carbon Dioxide, Dry Chemical, Foam, Water Spray or Fog.

Special Fire Fighting Considerations: Water spray should be used to keep closed containers cool. Continue to cool container after fire is extinguished.

TOXICOLOGY

Odor:
Physical Contact: Irritant
TLV = N.D.   **STEL** = N.D.

Odor Threshold: N.D.

IDLH = N.D.

Routes of Entry and Relative LD$_{50}$ (or LC$_{50}$)
Inhalation   N.D.
Ingestion   N.D.
Skin Absorption   N.D.

RTECS # XT6300000

Symptoms of Exposure: Possible irritation to skin, eyes and upper respiratory tract; headache; nausea.

Emergency/First Aid Treatment: Remove to ventilated area; immediately remove any contaminated clothing and wash contaminated areas for 15 minutes using water. Treat supportively and observe for possible shock. If ingested, seek immediate medical aid.

Recommended Clean-Up Procedures

Personal Protection: Level B Ensemble
RCRA Waste # None

Recommended Material Neoprene (0), Polyvinyl Chloride (0)
Reportable Quantities: None

Spills: Take up in non-combustible absorbent and dispose. Decontaminate spill area using water.

Special Emergency Information: May be harmful if inhaled, swallowed or absorbed through the skin.

o-TOLUIDINE

Synonyms: 1-amino-2-methylbenzene; 2-amino-1-methylbenzene; 2-aminotoluene; 2-methylaniline; o-tolylamine

CAS Number: 95-53-4	**Description:** Colorless Liquid	
DOT Number: 1708	**DOT Classification:** Poison B	

Molecular Weight: 107.2

Melting Point:	-16 C	(-3 F)	**Vapor Density:** 3.7	**Vapor Pressure:**	1 (<1 mm Hg)
Boiling Point:	200 C	(391 F)	**Specific Gravity:** 1.0	**Water Solubility:**	Insoluble

Chemical Incompatibilities or Instabilities: Oxidizers.

FLAMMABILITY

NFPA Hazard Code:	3-2-0	**NFPA Classification:**	Class IIIA Combustible
Flash Point:	85 C (185 F)	**LEL:**	1.5%
Autoignition Temp:	482 C (900 F)	**UEL:**	N.D.

Fire Extinguishing Methods: Carbon Dioxide, Dry Chemical, Foam, Water Spray or Fog.

Special Fire Fighting Considerations: Structural fire fighter protective clothing will not provide adequate protection. Fight fire from a distance or protected location, if possible. Treat all materials used or generated and equipment involved as contaminated by hazardous waste.

TOXICOLOGY
CARCINOGEN

Odor: Amine		**Odor Threshold:** 0.1 mg/m^3
Physical Contact: Irritant		
TLV = 22 mg/m^3 (skin)	**STEL** = N.D.	**IDLH** = 450 mg/m^3

Routes of Entry and Relative LD$_{50}$ (or LC$_{50}$)   **RTECS #**

Inhalation	N.D.	
Ingestion	4	(670 mg/kg)
Skin Absorption	2	(3250 mg/kg)

Symptoms of Exposure: Possible irritation to skin, eyes and upper respiratory tract; headache; nausea. Absorption may lead to the formation of methemoglobin resulting in cyanosis several hours after exposure.

Emergency/First Aid Treatment: Remove to ventilated area; immediately remove any contaminated clothing and wash contaminated areas for 15 minutes using water. Treat supportively and observe for possible shock. If ingested, seek immediate medical aid.

Recommended Clean-Up Procedures

Personal Protection:	Level A Ensemble	**Recommended Material**	Teflon (0)
RCRA Waste #	U328	**Reportable Quantities:**	100 lb (1 lb)

Spills: Dike to contain spill and collect liquid for later disposal. Decontaminate spill area using soapy water. Treat all materials used or generated and equipment involved as contaminated by hazardous waste.

Special Emergency Information: May be fatal in inhaled, ingested or absorbed through the skin.

m-toluidine (CAS # 108-44-1, DOT # 1708, RTECS # XU2800000) and
p-toluidine (CAS # 106-49-0, DOT # 1708, RTECS # XU3150000, RCRA Waste # U353, Reportable Quantities: 100 lb [1 lb]) have similar chemical, physical and toxicological properties.

TRIBUTYLAMINE - (CH₃CH₂CH₂CH₂)₃N

Synonyms:

CAS Number: 102-82-9 **Description:** Colorless Liquid
DOT Number: 2542 **DOT Classification:** Corrosive Material

Molecular Weight: 185.4
Melting Point: -70°C (-94°F) **Vapor Density:** 6.4 **Vapor Pressure:** 1 (<1 mm Hg)
Boiling Point: 214°C (417°F) **Specific Gravity:** 0.8 **Water Solubility:** Insoluble

Chemical Incompatibilities or Instabilities: Oxidizers.

FLAMMABILITY

NFPA Hazard Code: 3-2-0 **NFPA Classification:** Class IIIA Combustible
Flash Point: 63°C (145°F) **LEL:** 1.4%
Autoignition Temp: 410°C (770°F) **UEL:** 6.0%

Fire Extinguishing Methods: Carbon Dioxide, Dry Chemical, Foam, Water Spray or Fog.

Special Fire Fighting Considerations: Isolate for 1/2 mile if rail or tank truck is involved in a fire. Water spray should be used to keep closed containers cool. Continue to cool container after fire is extinguished. Immediately withdraw if rising sound from venting device is heard or if fire is causing discoloration to the tank.

DOT Recommended Isolation Zones:	Small Spill: 150 ft	Large Spill: 150 ft
DOT Recommended Down Wind Take Cover Distance	Small Spill: 0.2 miles	Large Spill: 0.2 miles

TOXICOLOGY

Odor: Ammonia **Odor Threshold:** N.D.
Physical Contact: Material is extremely destructive to human tissues.
TLV = N.D. **STEL** = N.D. **IDLH** = N.D.

Routes of Entry and Relative LD₅₀ (or LC₅₀) **RTECS # YA0350000**

Inhalation	N.D.	
Ingestion	4	(540 mg/kg)
Skin Absorption	8	(240 mg/kg)

Symptoms of Exposure: Possible burns or irritation to skin, eyes and upper respiratory tract; headache; nausea. Inhalation of vapors may be fatal by causing glottis to spasm and suffocation. Exposure may result in chemical pneumonitis or pulmonary edema.

Emergency/First Aid Treatment: Remove to ventilated area; immediately remove any contaminated clothing and wash contaminated areas for 15 minutes using water. Treat supportively and observe for possible shock. If ingested, seek immediate medical aid.

Recommended Clean-Up Procedures

Personal Protection: Level A Ensemble **Recommended Material** N.D.

RCRA Waste # None **Reportable Quantities:** None

Spills: Dike to contain spill and collect liquid for later disposal. Decontaminate spill area using soapy water.

Special Emergency Information: May be harmful if inhaled, swallowed or absorbed through the skin.

TRICHLOROACETIC ACID - CCl₃COOH

Synonyms: TCA; trichloroethanoic acid

CAS Number: 76-03-9　　　　**Description:** White powder
DOT Number: 1839, 2564　　　**DOT Classification:** Corrosive Material

Molecular Weight: 163.4
Melting Point:　58 C　(136 F)　**Vapor Density:** N.D.　**Vapor Pressure:** 1 (<1 mm Hg)
Boiling Point: 198 C　(388 F)　**Specific Gravity:** 1.6　**Water Solubility:** Soluble

Chemical Incompatibilities or Instabilities: Oxidizers, copper wool in dimethyl sulfoxide.

FLAMMABILITY

NFPA Hazard Code:　　　　　　　　**NFPA Classification:**
Flash Point:　110 C　(230 F)　　**LEL:** N.A.
Autoignition Temp:　N.D.　　　　**UEL:** N.A.

Fire Extinguishing Methods: Carbon Dioxide, Dry Chemical, Foam, Water Spray or Fog.

Special Fire Fighting Considerations: Structural fire fighter protective clothing will not provide adequate protection. Water spray should be used to keep closed containers cool. Continue to cool container after fire is extinguished.

TOXICOLOGY

Odor: None　　　　　　　　　　　**Odor Threshold:** N.A.
Physical Contact: Material is extremely destructive to human tissues.
TLV = 13 mg/m³　　　**STEL** = N.D.　　　**IDLH** = N.D.

Routes of Entry and Relative LD₅₀ (or LC₅₀)　　**RTECS # AJ7875000**
　Inhalation　　N.D.
　Ingestion　　**5**　　(400 mg/kg)
　Skin Absorption　N.D.

Symptoms of Exposure: Possible burns or irritation to skin, eyes and upper respiratory tract; headache; nausea. Inhalation of vapors may be fatal by causing glottis to spasm and suffocation. Exposure may result in chemical pneumonitis or pulmonary edema.

Emergency/First Aid Treatment: Remove to ventilated area; immediately remove any contaminated clothing and wash contaminated areas for 15 minutes using water. Treat supportively and observe for possible shock. If ingested, seek immediate medical aid.

Recommended Clean-Up Procedures

Personal Protection:　Level B Ensemble　**Recommended Material**　Nitrile Rubber (0)

RCRA Waste #　None　　　**Reportable Quantities:**　None

Spills: Take up in non-combustible absorbent and dispose. Decontaminate spill area using water.

Special Emergency Information: May be fatal in inhaled, ingested or absorbed through the skin.

TRICHLOROACETYL CHLORIDE - CCl3COCl

Synonyms:

CAS Number: 76-02-8   **Description:** Colorless Liquid
DOT Number: 2442   **DOT Classification:** Corrosive Material

Molecular Weight: 181.2
Melting Point: -146 C (-231 F)   **Vapor Density:** N.D.   **Vapor Pressure:** 4 (16 mm Hg)
Boiling Point: 116 C (241 F)   **Specific Gravity:** 1.6   **Water Solubility:** Decomposes

Chemical Incompatibilities or Instabilities: Water, alcohols, oxidizers.

FLAMMABILITY

NFPA Hazard Code: **NFPA Classification:**
Flash Point: N.A.   **LEL:** N.A.
Autoignition Temp: N.A.   **UEL:** N.A.

Fire Extinguishing Methods: Carbon Dioxide, Dry Chemical, Foam, Water Spray or Fog.

Special Fire Fighting Considerations: Structural fire fighter protective clothing will not provide adequate protection. Water spray should be used to keep closed containers cool. Continue to cool container after fire is extinguished.

TOXICOLOGY

Odor: Acrid   **Odor Threshold:** N.D.
Physical Contact: Material is extremely destructive to human tissues.
TLV = N.D.   **STEL** = N.D.   **IDLH** = N.D.

Routes of Entry and Relative LD_{50} (or LC_{50})   **RTECS # AO7140000**
     Inhalation   9   ($1900 \ mg/m^3/H$)
     Ingestion   4   (600 mg/kg)
     Skin Absorption   N.D.

Symptoms of Exposure: Possible burns or irritation to skin, eyes and upper respiratory tract; headache; nausea; lachrymation; laryngitis. Inhalation of vapors may be fatal by causing glottis to spasm and suffocation. Exposure may result in chemical pneumonitis or pulmonary edema.

Emergency/First Aid Treatment: Remove to ventilated area; immediately remove any contaminated clothing and wash contaminated areas for 15 minutes using water. Treat supportively and observe for possible shock. If ingested, seek immediate medical aid.

Recommended Clean-Up Procedures

Personal Protection: Level A Ensemble   **Recommended Material**   N.D.

RCRA Waste #   None   **Reportable Quantities:**   None

Spills: Dike to contain spill and collect liquid for later disposal. Decontaminate spill area using water. Keep area isolated until vapors have dissipated.

Special Emergency Information: May be fatal in inhaled, ingested or absorbed through the skin. Vapors are more dense than air and may settle in low lying areas.

1,2,4-TRICHLOROBENZENE

Synonyms:

CAS Number: 120-82-1   **Description:** Colorless Liquid
DOT Number: 2321   **DOT Classification:** Poison B

Molecular Weight: 181.4
Melting Point: 16 C   (F)   **Vapor Density:**   6.3   **Vapor Pressure:**   1 (<1 mm Hg)
Boiling Point: 213 C (415 F)   **Specific Gravity:** 1.45   **Water Solubility:** Insoluble

Chemical Incompatibilities or Instabilities: Oxidizers

FLAMMABILITY

NFPA Hazard Code: 2-1-0   **NFPA Classification:** Class IIIB Combustible
Flash Point: 105 C   (222 F)   **LEL:** N.D.
Autoignition Temp: 571 C (1060 F)   **UEL:** N.D.

Fire Extinguishing Methods: Carbon Dioxide, Dry Chemical, Foam, Water Spray or Fog.

Special Fire Fighting Considerations: Water spray should be used to keep closed containers cool. Continue to cool container after fire is extinguished.

TOXICOLOGY

Odor: Aromatic   **Odor Threshold:** 22 mg/m^3
Physical Contact: Irritant
TLV = 37 mg/m^3 (ceiling)   **STEL** = N.D.   **IDLH** = N.D.

Routes of Entry and Relative LD$_{50}$ (or LC$_{50}$)   **RTECS #**   DC2100000
   Inhalation   N.D.
   Ingestion   **4**   (756 mg/kg)
   Skin Absorption   N.D.

Symptoms of Exposure: Possible irritation to skin, eyes and upper respiratory tract; headache; nausea.

Emergency/First Aid Treatment: Remove to ventilated area; immediately remove any contaminated clothing and wash contaminated areas for 15 minutes using water. Treat supportively and observe for possible shock. If ingested, seek immediate medical aid.

Recommended Clean-Up Procedures

Personal Protection: Level B Ensemble   **Recommended Material**   Teflon (-)

RCRA Waste #   None   **Reportable Quantities:**   100 lb (1 lb)

Spills: Dike to contain spill and collect liquid for later disposal. Decontaminate spill area using soapy water.

Special Emergency Information: May be harmful if inhaled, swallowed or absorbed through the skin.

1,3,5-trichlorobenzene (CAS # 108-70-3, RTECS # DC2100100) and 1,2,3-trichlorobenzene (CAS # 87-61-6, RTECS # DC2095000) have similar chemical, physical and toxicological properties.

TRICHLOROETHYLENE - $Cl_2C{=}CHCl$

Synonyms: acetylene trichloride; 1-chloro-2,2-dichloroethylene; 1,2,2-trichloroethylene

CAS Number: 79-01-6   **Description:** Colorless Liquid
DOT Number: 1710   **DOT Classification:** ORM-A

Molecular Weight: 131.4
Melting Point: -85 C (-121 F)   **Vapor Density:** 4.5   **Vapor Pressure:** **6** (58 mm Hg)
Boiling Point: 87 C (189 F)   **Specific Gravity:** 1.5   **Water Solubility:** Insoluble

Chemical Incompatibilities or Instabilities: Oxidizers, aluminum, magnesium, alkali, epoxides, barium, alkali metals, titanium, dinitrogen tetroxide, epoxides, ozone, dichloroacetylene.

FLAMMABILITY

NFPA Hazard Code: **2-2-0**   **NFPA Classification:** Class IC Flammable
Flash Point: 32 C (90 F)   **LEL:** 8.0%
Autoignition Temp: 420 C (788 F)   **UEL:** 10.5%

Fire Extinguishing Methods: Carbon Dioxide, Dry Chemical, Foam, Water Spray or Fog.

Special Fire Fighting Considerations: Isolate for 1/2 mile if rail or tank truck is involved in a fire. Water spray should be used to keep closed containers cool. Continue to cool container after fire is extinguished. Treat all materials used or generated and equipment involved as contaminated by hazardous waste.

TOXICOLOGY
SUSPECTED CARCINOGEN

Odor: Ether-like   **Odor Threshold:** 3 mg/m^3
Physical Contact: Irritant
TLV = 270 mg/m^3 **STEL** = 1080 mg/m^3 **IDLH** = 5500 mg/m^3

Routes of Entry and Relative LD$_{50}$ (or LC$_{50}$)   **RTECS # KX4550000**
   Inhalation   3   (140,000 mg/m^3/H)
   Ingestion   N.D.
   Skin Absorption   N.D.

Symptoms of Exposure: Possible irritation to skin, eyes and upper respiratory tract; headache; nausea. Consumption of alcohol may increase toxic effects.

Emergency/First Aid Treatment: Remove to ventilated area; immediately remove any contaminated clothing and wash contaminated areas for 15 minutes using water. Treat supportively and observe for possible shock. If ingested, seek immediate medical aid.

Recommended Clean-Up Procedures

Personal Protection: Level B Ensemble   **Recommended Material** Polyvinyl Alcohol (+), Viton (+)

RCRA Waste # U228   **Reportable Quantities:** 100 lb (1000 lb)

Spills: Remove all potential ignition sources. Dike to contain spill, absorb with non-combustible absorbent and take up using non-sparking tools. Decontaminate spill area using soapy water. Keep area isolated until vapors have dissipated. Treat all materials used or generated and equipment involved as contaminated by hazardous waste.

Special Emergency Information: May be harmful if inhaled, swallowed or absorbed through the skin. Vapors are heavier than air and may travel some distance to an ignition source. Vapors are more dense than air and may settle in low lying areas.

TRICHLOROISICYANURIC ACID

Synonyms: trichloro-S-triazinetrione; trichloroiminocyanuric acid; symclosene; 1,3,5-trichloro-2,4,6-trioxohexahydro-S-triazine

CAS Number: 87-90-1   **Description:** White powder
DOT Number: 2468   **DOT Classification:** Oxidizer

Molecular Weight: 232.4
Melting Point: C   (F)   **Vapor Density:** N.D.   **Vapor Pressure:** 1 (<1 mm Hg)
Boiling Point: 230 C (446 F)   **Specific Gravity:** 1.2   **Water Solubility:** Decomposes

Chemical Incompatibilities or Instabilities: Hydrocarbons, reducing agents, amines.

FLAMMABILITY

NFPA Hazard Code: 3-0-2-OX   **NFPA Classification:**
Flash Point: N.A.   **LEL:** N.A.
Autoignition Temp: N.A.   **UEL:** N.A.

Fire Extinguishing Methods: WATER ONLY

Special Fire Fighting Considerations: Water spray should be used to keep closed containers cool. Continue to cool container after fire is extinguished. For large fires, if possible, withdraw and allow to burn. Do not allow water to enter container.

TOXICOLOGY

Odor: Chlorine   **Odor Threshold:** N.D.
Physical Contact: Material is extremely destructive to human tissues.
TLV = N.D.   **STEL** = N.D.   **IDLH** = N.D.

Routes of Entry and Relative LD$_{50}$ (or LC$_{50}$)   **RTECS # XZ1925000**
   Inhalation   N.D.
   Ingestion   5   (406 mg/kg)
   Skin Absorption   1   (20,000 mg/kg)

Symptoms of Exposure: Possible burns or irritation to skin, eyes and upper respiratory tract; headache; nausea; laryngitis. Inhalation of vapors may be fatal by causing glottis to spasm and suffocation. Exposure may result in chemical pneumonitis or pulmonary edema.

Emergency/First Aid Treatment: Remove to ventilated area; immediately remove any contaminated clothing and wash contaminated areas for 15 minutes using water. Treat supportively and observe for possible shock. If ingested, seek immediate medical aid.

Recommended Clean-Up Procedures

Personal Protection: Level B Ensemble   **Recommended Material**   N.D.

RCRA Waste #   None   **Reportable Quantities:**   None

Spills: Take up spill. Wash contaminated areas with water. Decontaminate spill area using water. Keep area isolated until vapors have dissipated.

Special Emergency Information: May be harmful if inhaled, swallowed or absorbed through the skin.

2,4,6-TRICHLOROPHENOL

Synonyms: phenachlor

CAS Number: 88-06-2   **Description:** White to orange powder
DOT Number: 2020   **DOT Classification:** N.D.

Molecular Weight: 197.4
Melting Point: 68 C   (154 F)   **Vapor Density:**   N.D.   **Vapor Pressure:**   1 (<1 mm Hg)
Boiling Point: 245 C   (473 F)   **Specific Gravity:** 1.5   **Water Solubility:** Soluble

Chemical Incompatibilities or Instabilities: Oxidizers.

FLAMMABILITY

NFPA Hazard Code: **NFPA Classification:**
Flash Point: 62 C   (144 F)   **LEL:** N.D.
Autoignition Temp: N.D.   **UEL:** N.D.

Fire Extinguishing Methods: Carbon Dioxide, Dry Chemical, Foam, Water Spray or Fog.

Special Fire Fighting Considerations: Water spray should be used to keep closed containers cool. Continue to cool container after fire is extinguished. Treat all materials used or generated and equipment involved as contaminated by hazardous waste.

TOXICOLOGY
CARCINOGEN

Odor: Phenolic   **Odor Threshold:** N.D.
Physical Contact: Irritant
TLV = N.D.   **STEL** = N.D.   **IDLH** = N.D.

Routes of Entry and Relative LD$_{50}$ (or LC$_{50}$)   **RTECS # SN1575000**
    Inhalation   N.D.
    Ingestion   4   (820 mg/kg)
    Skin Absorption   N.D.

Symptoms of Exposure: Possible burns or irritation to skin, eyes and upper respiratory tract; headache; nausea. Direct contact can cause eye damage.

Emergency/First Aid Treatment: Remove to ventilated area; immediately remove any contaminated clothing and wash contaminated areas for 15 minutes using water. Treat supportively and observe for possible shock. If ingested, seek immediate medical aid.

Recommended Clean-Up Procedures

Personal Protection: Level A Ensemble   **Recommended Material**   N.D.

RCRA Waste #   U231   **Reportable Quantities:**   10 lb (10 lb)

Spills: Take up in non-combustible absorbent and dispose. Decontaminate spill area using soapy water. Treat all materials used or generated and equipment involved as contaminated by hazardous waste.

Special Emergency Information: May be harmful if inhaled, swallowed or absorbed through the skin.

2,3,4-trichlorophenol (CAS # 15950-66-0),
2,3,5-trichlorophenol (CAS # 933-78-8),
2,3,6-trichlorophenol (CAS # 933-75-5, RTECS # SN1300000),
2,4,5-trichlorophenol (CAS # 95-95-4, RTECS # SN1400000, RCRA Waste # U230, Reportable Quantities; 10 lb [10 lb]), and
3,4,5-trichlorophenol (CAS # 609-19-8, RTECS # SN1650000) have similar chemical, physical and toxicological properties.

TRICHLOROSILANE - HSiCl₃

Synonyms: silicochloroform; trichloromonosilane

CAS Number: 10025-78-2   **Description:** Colorless Liquid
DOT Number: 1295   **DOT Classification:** Flammable Liquid

Molecular Weight: 135.5
Melting Point: -127 C (-196 F)   **Vapor Density:**   4.7   **Vapor Pressure:**   9 (500 mm Hg)
Boiling Point: 32 C   (90 F)   **Specific Gravity:** 1.34   **Water Solubility:** Decomposes

Chemical Incompatibilities or Instabilities: Air, water, oxidizers, hydrocarbons.

FLAMMABILITY

NFPA Hazard Code: 3-4-2-W   **NFPA Classification:** Class 1B Flammable
Flash Point: -28 C   (-18 F)   **LEL:**   7.0%
Autoignition Temp: 182 C   (360 F)   **UEL:**   83.0%

Fire Extinguishing Methods: Dry Chemical, Sand, Flooding Quantities of Water.

Special Fire Fighting Considerations: Water spray from an unmanned device should be used to keep closed containers cool. Continue to cool container after fire is extinguished. For large fires, if possible, withdraw and allow to burn. Treat all materials used or generated and equipment involved as contaminated by hazardous waste.

TOXICOLOGY

Odor: Suffocating   **Odor Threshold:** N.D.
Physical Contact: Material is extremely destructive to human tissues.
TLV = N.D.   **STEL** = N.D.   **IDLH** = N.D.

Routes of Entry and Relative LD₅₀ (or LC₅₀)   **RTECS # VV5950000**
  Inhalation   N.D.
  Ingestion   3   (1030 mg/kg)
  Skin Absorption   N.D.

Symptoms of Exposure: Possible burns or irritation to skin, eyes and upper respiratory tract; headache; nausea; laryngitis; lachrymation. Inhalation of vapors may be fatal by causing glottis to spasm and suffocation. Exposure may result in chemical pneumonitis or pulmonary edema.

Emergency/First Aid Treatment: Remove to ventilated area; immediately remove any contaminated clothing and wash contaminated areas for 15 minutes using water. Treat supportively and observe for possible shock. If ingested, seek immediate medical aid.

Recommended Clean-Up Procedures

Personal Protection: Level A Ensemble   **Recommended Material** N.D.

RCRA Waste #   None   **Reportable Quantities:** None

Spills: Dike to contain spill and collect liquid for later disposal. Decontaminate spill area using dry decontamination techniques.

Special Emergency Information: May be fatal in inhaled, ingested or absorbed through the skin. Vapors are heavier than air and may travel some distance to an ignition source. Vapors are more dense than air and may settle in low lying areas. Reacts violently with water to yield hydrogen chloride. May spontaneously ignite in air.

TRIETHYLAMINE - $(CH_3CH_2)_3N$

Synonyms: (diethylamino)ethane; N,N-diethylethylamine; TEA; TEN

CAS Number: 121-44-8
DOT Number: 1296

Description: Colorless Liquid
DOT Classification: Flammable Liquid

Molecular Weight: 101.2
Melting Point: -115 C (-175 F)
Boiling Point: 89 C (193 F)

Vapor Density: 3.5
Specific Gravity: 0.7

Vapor Pressure: 6 (54 mm Hg)
Water Solubility: Soluble

Chemical Incompatibilities or Instabilities: Oxidizing, dinitrogen tetroxide.

FLAMMABILITY

NFPA Hazard Code: 2-3-0
Flash Point: -9 C (16 F)
Autoignition Temp: 249 C (480 F)

NFPA Classification: Class IB Flammable
LEL: 1.2%
UEL: 8.0%

Fire Extinguishing Methods: Carbon Dioxide, Dry Chemical, Foam, Water Spray or Fog.

Special Fire Fighting Considerations: Isolate for 1/2 mile if rail or tank truck is involved in a fire. Water spray should be used to keep closed containers cool. Continue to cool container after fire is extinguished. Immediately withdraw if rising sound from venting device is heard or if fire is causing discoloration to the tank. Treat all materials used or generated and equipment involved as contaminated by hazardous waste.

TOXICOLOGY

Odor: Ammonia
Physical Contact: Material is extremely destructive to human tissues.
TLV = 40 mg/m^3
STEL = 60 mg/m^3

Odor Threshold: 0.5 mg/m^3

IDLH = 4200 mg/m^3

RTECS # YE1750000

Routes of Entry and Relative LD$_{50}$ (or LC$_{50}$)

Inhalation	N.D.	
Ingestion	5	(460 mg/kg)
Skin Absorption	7	(570 mg/kg)

Symptoms of Exposure: Possible irritation to skin, eyes and upper respiratory tract; headache; nausea, laryngitis; lachramation. Inhalation of vapors may be fatal by causing glottis to spasm and suffocation. Exposure may result in chemical pneumonitis or pulmonary edema.

Emergency/First Aid Treatment: Remove to ventilated area; immediately remove any contaminated clothing and wash contaminated areas for 15 minutes using water. Treat supportively and observe for possible shock. If ingested, seek immediate medical aid.

Recommended Clean-Up Procedures

Personal Protection: Level B Ensemble
Recommended Material: Nitrile Rubber (+), Viton (+), Saranex (+)

RCRA Waste # None
Reportable Quantities: 5000 lb (5000 lb)

Spills: Remove all potential ignition sources. Dike to contain spill, absorb with non-combustible absorbent and take up using non-sparking tools. Decontaminate spill area using soapy water. Keep area isolated until vapors have dissipated. Treat all materials used or generated and equipment involved as contaminated by hazardous waste.

Special Emergency Information: May be harmful if inhaled, swallowed or absorbed through the skin. Vapors are heavier than air and may travel some distance to an ignition source. Vapors are more dense than air and may settle in low lying areas.

TRIETHYLAMINETETRAMINE

Synonyms: N,N'-bis(2-aminoethyl)-1,2diaaminoethane; TETA; trien; trientine

CAS Number:	112-24-3	**Description:** Colorless Liquid	
DOT Number:	2259	**DOT Classification:** Corrosive Material	

Molecular Weight: 146.2

Melting Point:	12 C (54 F)	**Vapor Density:** N.D.	**Vapor Pressure:**	1 (< 1 mm Hg)
Boiling Point:	267 C (513 F)	**Specific Gravity:** 1.0	**Water Solubility:**	Soluble

Chemical Incompatibilities or Instabilities: Oxidizers.

FLAMMABILITY

NFPA Hazard Code:		**NFPA Classification:** Class IIIB Combustible	
Flash Point:	N.D.	**LEL:** N.D.	
Autoignition Temp:	338 C (640 F)	**UEL:** N.D.	

Fire Extinguishing Methods: Carbon Dioxide, Dry Chemical, Foam, Water Spray or Fog.

Special Fire Fighting Considerations: Water spray should be used to keep closed containers cool. Continue to cool container after fire is extinguished. Do not allow water to enter container.

TOXICOLOGY

Odor:

Odor Threshold: N.D.

Physical Contact: Material is extremely destructive to human tissues.

TLV = N.D.   **STEL** = N.D.   **IDLH** = N.D.

Routes of Entry and Relative LD$_{50}$ (or LC$_{50}$)   **RTECS # YE6650000**

Inhalation	N.D.	
Ingestion	3	(2500 mg/kg)
Skin Absorption	5	(805 mg/kg)

Symptoms of Exposure: Possible burns or irritation to skin, eyes and upper respiratory tract; headache; nausea; laryngitis. Inhalation of vapors may be fatal by causing glottis to spasm and suffocation. Exposure may result in chemical pneumonitis or pulmonary edema.

Emergency/First Aid Treatment: Remove to ventilated area; immediately remove any contaminated clothing and wash contaminated areas for 15 minutes using water. Treat supportively and observe for possible shock. If ingested, seek immediate medical aid.

Recommended Clean-Up Procedures

Personal Protection:	Level A Ensemble	**Recommended Material**	N.D.
RCRA Waste #	None	**Reportable Quantities:**	None

Spills: Dike to contain spill and collect liquid for later disposal. Decontaminate spill area using water.

Special Emergency Information: May be harmful if inhaled, swallowed or absorbed through the skin. May act as a sensitizer.

TRIISOBUTYL ALUMINUM - ((CH₃)₂CHCH₂)₃Al

Synonyms:

CAS Number: 100-99-2	**Description:** Colorless Liquid		
DOT Number: 1930	**DOT Classification:** Flammable Solid		

Molecular Weight: 198.3

Melting Point: 4°C (39°F)	**Vapor Density:** N.D.	**Vapor Pressure:** 1(<! mm Hg)
Boiling Point: Decomposes	**Specific Gravity:** 0.8	**Water Solubility:** Decomposes

Chemical Incompatibilities or Instabilities: Water, acids, air, alcohols, amines, halogens.

FLAMMABILITY

NFPA Hazard Code: 3-4-3-W	**NFPA Classification:**
Flash Point: 4°C (39°F)	**LEL:** N.D.
Autoignition Temp: N.A.	**UEL:** N.D.

Fire Extinguishing Methods: Dry Chemical, Soda Ash, Limestone, Dry Sand.

Special Fire Fighting Considerations: DO NOT USE WATER OR FOAM. For large fires, if possible, withdraw and allow to burn.

TOXICOLOGY

Odor: **Odor Threshold:** N.D.

Physical Contact: Material is extremely destructive to human tissues.

TLV = N.D.   **STEL** = N.D.   **IDLH** = N.D.

Routes of Entry and Relative LD₅₀ (or LC₅₀)   **RTECS # BD2203500**

    Inhalation   N.D.

    Ingestion   N.D.

    Skin Absorption   N.D.

Symptoms of Exposure: Possible burns or irritation to skin, eyes and upper respiratory tract; headache; nausea; laryngitis. Inhalation of vapors may be fatal by causing glottis to spasm and suffocation. Exposure may result in chemical pneumonitis or pulmonary edema.

Emergency/First Aid Treatment: Remove to ventilated area; immediately remove any contaminated clothing and wash contaminated areas for 15 minutes using water. Treat supportively and observe for possible shock. If ingested, seek immediate medical aid.

Recommended Clean-Up Procedures

Personal Protection: Level A Ensemble	**Recommended Material** N.D.
RCRA Waste # None	**Reportable Quantities:** None

Spills: Take up in non-combustible absorbent and dispose. Decontaminate spill area using dry decontamination techniques.

Special Emergency Information: May be fatal in inhaled, ingested or absorbed through the skin. Reacts violently with water to yield flammable gases. Material may be packaged as a flammable solvent solution. material is pyrophoric. Material may ignite upon exposure to air. Material may react violently to yield flammable gases.

TRIMETHOXYSILANE

Synonyms:

CAS Number: 2487-90-3   **Description:** Colorless Liquid
DOT Number: 9269   **DOT Classification:** N.D.

Molecular Weight: 122.2
Melting Point: -115°C (-175°F)   **Vapor Density:** N.D.   **Vapor Pressure:** **6** (57 mm Hg)
Boiling Point: 81°C (178°F)   **Specific Gravity:** 0.1   **Water Solubility:** Decomposes

Chemical Incompatibilities or Instabilities: Water, alcohols, acids, alkali, heat.

FLAMMABILITY

NFPA Hazard Code: 3-3-2   **NFPA Classification:** Class IB Flammable
Flash Point: 7°C (45°F)   **LEL:** N.D.
Autoignition Temp: °C (°F)   **UEL:** N.D.

Fire Extinguishing Methods: Carbon Dioxide, Dry Chemical, Foam, Water Spray or Fog.

Special Fire Fighting Considerations: Structural fire fighter protective clothing will not provide adequate protection. Water spray should be used to keep closed containers cool. Continue to cool container after fire is extinguished. Do not allow water to enter container. Treat all materials used or generated and equipment involved as contaminated by hazardous waste.

DOT Recommended Isolation Zones: Small Spill: 150 ft   Large Spill: 150 ft
DOT Recommended Down Wind Take Cover Distance   Small Spill: 0.4 miles   Large Spill: 0.8 miles

TOXICOLOGY

Odor: Suffocating   **Odor Threshold:** N.D.
Physical Contact: Irritant
TLV = N.D.   **STEL** = N.D.   **IDLH** = N.D.

Routes of Entry and Relative LD$_{50}$ (or LC$_{50}$)   **RTECS #** VV6750000
   Inhalation   7   (2500 mg/m^3/H)
   Ingestion   2   (9330 mg/kg)
   Skin Absorption   1   (6300 mg/kg)

Symptoms of Exposure: Possible irritation to skin, eyes and upper respiratory tract; headache; nausea.

Emergency/First Aid Treatment: Remove to ventilated area; immediately remove any contaminated clothing and wash contaminated areas for 15 minutes using water. Treat supportively and observe for possible shock. If ingested, seek immediate medical aid.

Recommended Clean-Up Procedures

Personal Protection: Level B Ensemble   **Recommended Material** N.D.

RCRA Waste # None   **Reportable Quantities:** None

Spills: Remove all potential ignition sources. Dike to contain spill, absorb with non-combustible absorbent and take up using non-sparking tools. Decontaminate spill area using a dilute alkaline solution.

Special Emergency Information: May be harmful if inhaled, swallowed or absorbed through the skin. Vapors are heavier than air and may travel some distance to an ignition source. Vapors are more dense than air and may settle in low lying areas.

TRIMETHYLAMINE - (CH₃)₃N

Synonyms:

CAS Number: 75-50-3   **Description:** Colorless Gas
DOT Number: 1083 (gas), 1297 (solution)   **DOT Classification:** Flammable Gas

Molecular Weight: 59.1
Melting Point: -118°C (-179°F)   **Vapor Density:** 2.0   **Vapor Pressure:** Gas
Boiling Point: 3°C (37°F)   **Specific Gravity:** N.A.   **Water Solubility:** Soluble

Chemical Incompatibilities or Instabilities: Oxidizers, halogens, halogenated hydrocarbons, acids, ethylene oxide, mercury, ethynyl aluminum.

FLAMMABILITY

NFPA Hazard Code: 3-4-0   **NFPA Classification:**
Flash Point: N.A.   **LEL:** 2.0%
Autoignition Temp: 190°C (374°F)   **UEL:** 11.6%

Fire Extinguishing Methods: Carbon Dioxide, Dry Chemical, Foam, Water Spray or Fog.

Special Fire Fighting Considerations: Isolate for 1/2 mile if rail or tank truck is involved in a fire. Water spray should be used to keep closed containers cool. Continue to cool container after fire is extinguished. Do not allow water to enter container. Immediately withdraw if rising sound from venting device is heard or if fire is causing discoloration to the tank.

TOXICOLOGY

Odor: Fishy   **Odor Threshold:** 0.0003 mg/m³
Physical Contact: Material is extremely destructive to human tissues.
TLV = 24 mg/m³   **STEL** = 38 mg/m³   **IDLH** = N.D.

Routes of Entry and Relative LD₅₀ (or LC₅₀)   **RTECS #** PA0350000
    Inhalation   N.D.
    Ingestion   N.D.
    Skin Absorption   N.D.

Symptoms of Exposure: Possible burns or irritation to skin, eyes and upper respiratory tract; headache; nausea; laryngitis; lachrymation. Inhalation of vapors may be fatal by causing glottis to spasm and suffocation. Exposure may result in chemical pneumonitis or pulmonary edema.

Emergency/First Aid Treatment: Remove to ventilated area; immediately remove any contaminated clothing and wash contaminated areas for 15 minutes using water. Treat supportively and observe for possible shock. If ingested, seek immediate medical aid.

Recommended Clean-Up Procedures

Personal Protection: Level B Ensemble   **Recommended Material** N.D.

RCRA Waste # None   **Reportable Quantities:** 100 lb (1000 lb)

Spills: Gas: Remove all potential ignition sources. Stop leak if it can safely be done. Solution: Remove all potential ignition sources. Dike to contain spill, absorb with non-combustible absorbent and take up using non-sparking tools. Decontaminate spill area using a dilute acid solution.

Special Emergency Information: May be harmful if inhaled, swallowed or absorbed through the skin. Vapors are heavier than air and may travel some distance to an ignition source. Vapors are more dense than air and may settle in low lying areas.

TRIMETHYL BORATE - (CH3O)3B

Synonyms: methyl borate

CAS Number: 121-43-7
DOT Number: 2416

Description: Colorless Liquid
DOT Classification: Flammable Liquid

Molecular Weight: 103.9
Melting Point: -34°C (-29°F)
Boiling Point: 69°C (156°F)

Vapor Density: 3.6
Specific Gravity: 0.9

Vapor Pressure: N.D.
Water Solubility: Decomposes

Chemical Incompatibilities or Instabilities: Oxidizers.

FLAMMABILITY

NFPA Hazard Code: 2-3-1
Flash Point: -8°C (16°F)
Autoignition Temp: N.D.

NFPA Classification: Class IB Flammable
LEL: N.D.
UEL: N.D.

Fire Extinguishing Methods: Alcohol Resistant Foam, Carbon Dioxide, Dry Chemical, Water Spray or Fog.

Special Fire Fighting Considerations: Isolate for 1/2 mile if rail or tank truck is involved in a fire. Water spray from an unmanned device should be used to keep closed containers cool. Continue to cool container after fire is extinguished. For large fires, if possible, withdraw and allow to burn. Immediately withdraw if rising sound from venting device is heard or if fire is causing discoloration to the tank.

TOXICOLOGY

Odor:
Physical Contact: Irritant
TLV = N.D.   **STEL** = N.D.

Odor Threshold: N.D.

IDLH = N.D.

Routes of Entry and Relative LD$_{50}$ (or LC$_{50}$)
Inhalation	N.D.	
Ingestion	2	(6140 mg/kg)
Skin Absorption	3	(1830 mg/kg)

RTECS # ED560000

Symptoms of Exposure: Possible irritation to skin, eyes and upper respiratory tract; headache; nausea.

Emergency/First Aid Treatment: Remove to ventilated area; immediately remove any contaminated clothing and wash contaminated areas for 15 minutes using water. Treat supportively and observe for possible shock. If ingested, seek immediate medical aid.

Recommended Clean-Up Procedures

Personal Protection: Level B Ensemble
RCRA Waste # None

Recommended Material N.D.
Reportable Quantities: None

Spills: Remove all potential ignition sources. Dike to contain spill, absorb with non-combustible absorbent and take up using non-sparking tools. Decontaminate spill area using water.

Special Emergency Information: May be harmful if inhaled, swallowed or absorbed through the skin. Vapors are heavier than air and may travel some distance to an ignition source. Vapors are more dense than air and may settle in low lying areas.

TRIMETHYLCHLOROSILANE - (CH₃)₃SiCl

Synonyms: chlorotrimethylsilane

CAS Number: 75-77-4　　　　**Description:** Colorless Liquid
DOT Number: 1298　　　　　　**DOT Classification:** Flammable Liquid

Molecular Weight: 108.7
Melting Point: -40°C (-40°F)　**Vapor Density:** 3.8　**Vapor Pressure:** 7 (100 mm Hg)
Boiling Point: 57°C (135°F)　**Specific Gravity:** 0.9　**Water Solubility:** Decomposes

Chemical Incompatibilities or Instabilities: Water, strong oxidizers.

FLAMMABILITY

NFPA Hazard Code: 3-3-2-W̶　　　　**NFPA Classification:** Class IB Flammable
Flash Point: -28°C (-18°F)　　　　**LEL:** N.D.
Autoignition Temp: °C (°F)　　　　**UEL:** N.D.

Fire Extinguishing Methods: Alcohol Resistant Foam, Carbon Dioxide, Dry Chemical, Water Spray or Fog.

Special Fire Fighting Considerations: Water spray should be used to keep closed containers cool. Continue to cool container after fire is extinguished. Do not allow water to enter container. Immediately withdraw if rising sound from venting device is heard or if fire is causing discoloration to the tank. May react violently with water.

TOXICOLOGY
QUESTIONABLE CARCINOGEN

Odor: Suffocating　　　　　　　　　　　**Odor Threshold:** N.D.
Physical Contact: Material is extremely destructive to human tissues.
TLV = N.D.　　　　**STEL** = N.D.　　　　**IDLH** = N.D.

Routes of Entry and Relative LD₅₀ (or LC₅₀)　　**RTECS # VV2710000**
　Inhalation　　N.D.
　Ingestion　　N.D.
　Skin Absorption　N.D.

Symptoms of Exposure: Possible burns or irritation to skin, eyes and upper respiratory tract; headache; nausea; laryngitis. Inhalation of vapors may be fatal by causing glottis to spasm and suffocation. Exposure may result in chemical pneumonitis or pulmonary edema.

Emergency/First Aid Treatment: Remove to ventilated area; immediately remove any contaminated clothing and wash contaminated areas for 15 minutes using water. Treat supportively and observe for possible shock. If ingested, seek immediate medical aid.

Recommended Clean-Up Procedures

Personal Protection: Level A Ensemble　**Recommended Material** N.D.

RCRA Waste # None　　　　　　　**Reportable Quantities:** None

Spills: Remove all potential ignition sources. Dike to contain spill, absorb with non-combustible absorbent and take up using non-sparking tools. Decontaminate spill area using dry decontamination techniques.

Special Emergency Information: May be harmful if inhaled, swallowed or absorbed through the skin. Vapors are heavier than air and may travel some distance to an ignition source. Vapors are more dense than air and may settle in low lying areas.

TRIMETHYL PHOSPHITE - (CH₃O)₃P

Synonyms: methyl phosphite; trimethoxyphosphine

CAS Number: 121-45-9   **Description:** Colorless Liquid
DOT Number: 2329   **DOT Classification:** Flammable Liquid

Molecular Weight: 124.1
Melting Point: -78°C (-108°F)   **Vapor Density:** 4.3   **Vapor Pressure:** **4** (17 mm Hg)
Boiling Point: 234°C (453°F)   **Specific Gravity:** 0.9   **Water Solubility:** Insoluble

Chemical Incompatibilities or Instabilities: Strong oxidizers.

FLAMMABILITY

NFPA Hazard Code: 0-2-0   **NFPA Classification:** Class IIIB Combustible
Flash Point: 171°C (340°F)   **LEL:** N.D.
Autoignition Temp: N.D.   **UEL:** N.D.

Fire Extinguishing Methods: Carbon Dioxide, Dry Chemical, Foam, Water Spray or Fog.

Special Fire Fighting Considerations: Isolate for 1/2 mile if rail or tank truck is involved in a fire. Water spray from an unmanned device should be used to keep closed containers cool. Continue to cool container after fire is extinguished. For large fires, if possible, withdraw and allow to burn. Immediately withdraw if rising sound from venting device is heard or if fire is causing discoloration to the tank.

TOXICOLOGY

Odor: Stench   **Odor Threshold:** 0.005 mg/m³
Physical Contact: Material is extremely destructive to human tissues.
TLV = 10 mg/m³   **STEL** = N.D.   **IDLH** = N.D.

Routes of Entry and Relative LD₅₀ (or LC₅₀)   **RTECS #** TH1400000
    Inhalation   N.D.
    Ingestion   **3**   (1600 mg/kg)
    Skin Absorption   N.D.

Symptoms of Exposure: Possible burns or irritation to skin, eyes and upper respiratory tract; headache; nausea; laryngitis. Inhalation of vapors may be fatal by causing glottis to spasm and suffocation. Exposure may result in chemical pneumonitis or pulmonary edema.

Emergency/First Aid Treatment: Remove to ventilated area; immediately remove any contaminated clothing and wash contaminated areas for 15 minutes using water. Treat supportively and observe for possible shock. If ingested, seek immediate medical aid.

Recommended Clean-Up Procedures

Personal Protection: Level B Ensemble   **Recommended Material** N.D.

RCRA Waste # None   **Reportable Quantities:** None

Spills: Remove all potential ignition sources. Dike to contain spill, absorb with non-combustible absorbent and take up using non-sparking tools. Decontaminate spill area using soapy water.

Special Emergency Information: May be harmful if inhaled, swallowed or absorbed through the skin. Vapors are heavier than air and may travel some distance to an ignition source. Vapors are more dense than air and may settle in low lying areas. May undergo hydrolysis to yield a questionable carcinogen.

TRIPROPYLALUMINUM - (CH3CH2CH2)3Al

Synonyms:

CAS Number: 102-67-0
DOT Number: 2718

Description: Colorless Liquid
DOT Classification: Flammable Solid

Molecular Weight: 156.3
Melting Point: -107°C (-161°F)
Boiling Point: °C   (°F)

Vapor Density: N.D.
Specific Gravity: 0.8

Vapor Pressure: 1 (<1 mm Hg)
Water Solubility: Decomposes

Chemical Incompatibilities or Instabilities: Alcohols, air, oxygen, acids, oxidizers, water, chloroform, carbon tetrachloride.

FLAMMABILITY

NFPA Hazard Code: ?-3-3-~~W~~
Flash Point: N.A.
Autoignition Temp: N.A.

NFPA Classification:
LEL: N.D.
UEL: N.D.

Fire Extinguishing Methods: Dry Chemical, Soda Ash, Sand, Limestone.

Special Fire Fighting Considerations: DO NOT USE FOAM OR WATER. Reacts violently with water.

TOXICOLOGY

Odor: N.D.

Odor Threshold: N.D.

Physical Contact: Material is extremely destructive to human tissues.
TLV = N.D.   **STEL** = N.D.

IDLH = N.D.

Routes of Entry and Relative LD$_{50}$ (or LC$_{50}$)
  Inhalation   N.D.
  Ingestion   N.D.
  Skin Absorption   N.D.

RTECS # BD2208000

Symptoms of Exposure: Possible burns or irritation to skin, eyes and upper respiratory tract; headache; nausea; laryngitis. Inhalation of vapors may be fatal by causing glottis to spasm and suffocation. Exposure may result in chemical pneumonitis or pulmonary edema.

Emergency/First Aid Treatment: Remove to ventilated area; immediately remove any contaminated clothing and wash contaminated areas for 15 minutes using water. Treat supportively and observe for possible shock. If ingested, seek immediate medical aid.

Recommended Clean-Up Procedures

Personal Protection: Level A Ensemble
Recommended Material   N.D.

RCRA Waste #   None
Reportable Quantities: None

Spills: Remove all potential ignition sources. Dike to contain spill, absorb with non-combustible absorbent and take up using non-sparking tools. Place in a metal container under an inert atmosphere. Decontaminate spill area using dry decontamination techniques.

Special Emergency Information: May be fatal in inhaled, ingested or absorbed through the skin. Reacts with water to yield flammable gas. May be pyrophoric.

TRIPROPYLAMINE - (CH₃CH₂CH₂)₃N

Synonyms:

CAS Number: 102-69-2   **Description:** Colorless Liquid
DOT Number: 2260   **DOT Classification:** Flammable Liquid

Molecular Weight: 143.3
Melting Point: -94°C (-137°F)   **Vapor Density:**   4.9   **Vapor Pressure:**   2 (3 mm Hg)
Boiling Point: 156°C (313°F)   **Specific Gravity:** 0.8   **Water Solubility:**   < 1%

Chemical Incompatibilities or Instabilities: Oxidizers.

FLAMMABILITY

NFPA Hazard Code: 2-2-0   **NFPA Classification:**   Class II Combustible
Flash Point: 41°C (105°F)   **LEL:** N.D.
Autoignition Temp: °C   (°F)   **UEL:** N.D.

Fire Extinguishing Methods: Carbon Dioxide, Dry Chemical, Foam, Water Spray or Fog.

Special Fire Fighting Considerations: Water spray should be used to keep closed containers cool. Continue to cool container after fire is extinguished. Immediately withdraw if rising sound from venting device is heard or if fire is causing discoloration to the tank. Treat all materials used or generated and equipment involved as contaminated by hazardous waste.

TOXICOLOGY

Odor: **Odor Threshold:** N.D.
Physical Contact: Irritant
TLV = N.D.   **STEL** = N.D.   **IDLH** = N.D.

Routes of Entry and Relative LD₅₀ (or LC₅₀)   **RTECS # TX1575000**
    Inhalation   N.D.
    Ingestion   8   (72 mg/kg)
    Skin Absorption   7   (429 mg/kg)

Symptoms of Exposure: Possible irritation to skin, eyes and upper respiratory tract; headache; nausea.

Emergency/First Aid Treatment: Remove to ventilated area; immediately remove any contaminated clothing and wash contaminated areas for 15 minutes using water. Treat supportively and observe for possible shock. If ingested, seek immediate medical aid.

Recommended Clean-Up Procedures

Personal Protection: Level B Ensemble   **Recommended Material**   Neoprene (+), Nitrile Rubber (+), Polyvinyl Alcohol (+), Viton (+)

RCRA Waste #   None   **Reportable Quantities:**   None

Spills: Remove all potential ignition sources. Dike to contain spill, absorb with non-combustible absorbent and take up using non-sparking tools. Decontaminate spill area using soapy water. Treat all materials used or generated and equipment involved as contaminated by hazardous waste.

Special Emergency Information: May be harmful if inhaled, swallowed or absorbed through the skin. Vapors are heavier than air and may travel some distance to an ignition source. Vapors are more dense than air and may settle in low lying areas.

TURPENTINE

Synonyms: gum spirits; spirits of turpentine

CAS Number: 8006-64-2

DOT Number: 1299

Description: Colorless Liquid

DOT Classification: Flammable Liquid

Molecular Weight: N.A.

Melting Point: -58°C (-72°F)	**Vapor Density:** 4.8	**Vapor Pressure:** **2** (5 mm Hg)
Boiling Point: 149°C (300°F)	**Specific Gravity:** <1.0	**Water Solubility:** Insoluble

Chemical Incompatibilities or Instabilities: Strong oxidizers.

FLAMMABILITY

NFPA Hazard Code: 1-3-0

Flash Point: 35°C (95°F)

Autoignition Temp: 253°C (488°F)

NFPA Classification: Class IC Flammable

LEL: 0.8

UEL: N.D.

Fire Extinguishing Methods: Carbon Dioxide, Dry Chemical, Foam, Water Spray or Fog.

Special Fire Fighting Considerations: Isolate for 1/2 mile if rail or tank truck is involved in a fire. Water spray from an unmanned device should be used to keep closed containers cool. Continue to cool container after fire is extinguished. For large fires, if possible, withdraw and allow to burn. Immediately withdraw if rising sound from venting device is heard or if fire is causing discoloration to the tank.

TOXICOLOGY

Odor: Characteristic

Physical Contact: Irritant

$TLV = 560$ mg/m^3

Odor Threshold: N.D.

$STEL = $ N.D.

$IDLH = 8,500$ mg/m^3

Routes of Entry and Relative LD$_{50}$ (or LC$_{50}$)

Inhalation	N.D.
Ingestion	**2** (5760 mg/kg)
Skin Absorption	N.D.

RTECS # YO8400000

Symptoms of Exposure: Possible irritation to skin, eyes and upper respiratory tract; headache; nausea.

Emergency/First Aid Treatment: Remove to ventilated area; immediately remove any contaminated clothing and wash contaminated areas for 15 minutes using water. Treat supportively and observe for possible shock. If ingested, seek immediate medical aid.

Recommended Clean-Up Procedures

Personal Protection: Level B Ensemble

Recommended Material Polyvinyl Alcohol (+), Teflon (0)

RCRA Waste # None

Reportable Quantities: None

Spills: Remove all potential ignition sources. Dike to contain spill, absorb with non-combustible absorbent and take up using non-sparking tools. Decontaminate spill area using soapy water. Keep area isolated until vapors have dissipated.

Special Emergency Information: May be harmful if inhaled, swallowed or absorbed through the skin. Vapors are heavier than air and may travel some distance to an ignition source. Vapors are more dense than air and may settle in low lying areas.

VALERALDEHYDE

Synonyms: amyl aldehyde, pentanal, valerylaldehyde

CAS Number: 110-62-3   **Description:** Colorless Liquid
DOT Number: 2058   **DOT Classification:** Flammable Liquid

Molecular Weight: 86.2
Melting Point: °C   (°F)   **Vapor Density:** 3.0   **Vapor Pressure:** N.D.
Boiling Point: 103°C (217°F)   **Specific Gravity:** 0.8   **Water Solubility:** Insoluble

Chemical Incompatibilities or Instabilities: Oxidizers.

FLAMMABILITY

NFPA Hazard Code: 1-3-0   **NFPA Classification:** Class IB Flammable
Flash Point: 12°C   (54°F)   **LEL:** N.D.
Autoignition Temp: 222°C (432°F)   **UEL:** N.D.

Fire Extinguishing Methods: Alcohol Resistant Foam, Carbon Dioxide, Dry Chemical, Water Spray or Fog.

Special Fire Fighting Considerations: Isolate for 1/2 mile if rail or tank truck is involved in a fire. Water spray from an unmanned device should be used to keep closed containers cool. Continue to cool container after fire is extinguished. For large fires, if possible, withdraw and allow to burn. Immediately withdraw if rising sound from venting device is heard or if fire is causing discoloration to the tank.

TOXICOLOGY

Odor: Rancid   **Odor Threshold:** 0.003 mg/m^3
Physical Contact: Irritant
TLV = 200 mg/m^3 **STEL** = N.D.   **IDLH** = N.D.

Routes of Entry and Relative LD$_{50}$ (or LC$_{50}$)   **RTECS # YV3600000**
　　Inhalation   N.D.
　　Ingestion   3   (3200 mg/kg)
　　Skin Absorption   2   (4857 mg/kg)

Symptoms of Exposure: Possible irritation to skin, eyes and upper respiratory tract; headache; nausea.

Emergency/First Aid Treatment: Remove to ventilated area; immediately remove any contaminated clothing and wash contaminated areas for 15 minutes using water. Treat supportively and observe for possible shock. If ingested, seek immediate medical aid.

Recommended Clean-Up Procedures

Personal Protection: Level B Ensemble   **Recommended Material**   N.D.

RCRA Waste #   None   **Reportable Quantities:**   None

Spills: Remove all potential ignition sources. Dike to contain spill, absorb with non-combustible absorbent and take up using non-sparking tools. Decontaminate spill area using soapy water. Keep area isolated until vapors have dissipated.

Special Emergency Information: May be harmful if inhaled, swallowed or absorbed through the skin. Vapors are heavier than air and may travel some distance to an ignition source. Vapors are more dense than air and may settle in low lying areas.

VALERYL CHLORIDE - CH$_3$CH$_2$CH$_2$CH$_2$COCl

Synonyms:

CAS Number: 638-29-9   **Description:** Colorless Liquid
DOT Number: 2502   **DOT Classification:** Corrosive Material

Molecular Weight: 120.6
Melting Point: 32°C (90°F)   **Vapor Density:** N.D.   **Vapor Pressure:** N.D.
Boiling Point: 127°C (261°F)   **Specific Gravity:** 1.0   **Water Solubility:** Decomposes

Chemical Incompatibilities or Instabilities: Oxidizers, alcohols, water.

FLAMMABILITY

NFPA Hazard Code: **NFPA Classification:** Class II Combustible
Flash Point: 31°C (91°F)   **LEL:** N.D.
Autoignition Temp: N.D.   **UEL:** N.D.

Fire Extinguishing Methods: Carbon Dioxide, Dry Chemical, Foam, Water Spray or Fog.

Special Fire Fighting Considerations: Water spray should be used to keep closed containers cool. Continue to cool container after fire is extinguished.

TOXICOLOGY

Odor: **Odor Threshold:** N.D.
Physical Contact: Material is extremely destructive to human tissues.
TLV = N.D.   **STEL** = N.D.   **IDLH** = N.D.

Routes of Entry and Relative LD$_{50}$ (or LC$_{50}$)   **RTECS #** YV91075000
    Inhalation   N.D.
    Ingestion   N.D.
    Skin Absorption   N.D.

Symptoms of Exposure: Possible irritation to skin, eyes and upper respiratory tract; headache; nausea.

Emergency/First Aid Treatment: Remove to ventilated area; immediately remove any contaminated clothing and wash contaminated areas for 15 minutes using water. Treat supportively and observe for possible shock. If ingested, seek immediate medical aid.

Recommended Clean-Up Procedures

Personal Protection: Level B Ensemble   **Recommended Material**   N.D.

RCRA Waste #   None   **Reportable Quantities:**   None

Spills: Remove all potential ignition sources. Dike to contain spill, absorb with non-combustible absorbent and take up using non-sparking tools. Decontaminate spill area using a dilute alkaline solution.

Special Emergency Information: May be harmful if inhaled, swallowed or absorbed through the skin. Vapors are heavier than air and may travel some distance to an ignition source. Vapors are more dense than air and may settle in low lying areas. Reacts exothermically to yield valeric acid.

VANADIUM OXYTRICHLORIDE - VOCl₃

Synonyms: vanadyl chloride

CAS Number: 7727-18-6
DOT Number: 2443

Description: Yellow to orange liquid
DOT Classification: Corrosive Material

Molecular Weight: 173.3
Melting Point: -77°C (-107°F)
Boiling Point: 127°C (261°F)

Vapor Density: N.D.
Specific Gravity: 1.8

Vapor Pressure: 4 (14 mm Hg)
Water Solubility: Decomposes

Chemical Incompatibilities or Instabilities: Acids, alcohols, water, alkali metals.

FLAMMABILITY

NFPA Hazard Code:
Flash Point: N.D.
Autoignition Temp: N.D.

NFPA Classification:
LEL: N.D.
UEL: N.D.

Fire Extinguishing Methods: Dry Chemical, Carbon Dioxide or Flooding Quantities of Water.

Special Fire Fighting Considerations: Structural fire fighter protective clothing will not provide adequate protection. Water spray should be used to keep closed containers cool. Continue to cool container after fire is extinguished. Do not allow water to enter container. Treat all materials used or generated and equipment involved as contaminated by hazardous waste.

TOXICOLOGY

Odor:
Physical Contact: Material is extremely destructive to human tissues.
TLV = N.D.   **STEL** = N.D.

Odor Threshold: N.D.

IDLH = N.D.

Routes of Entry and Relative LD₅₀ (or LC₅₀)
  Inhalation   N.D.
  Ingestion   7   (140 mg/kg)
  Skin Absorption   N.D.

RTECS # YW2975000

Symptoms of Exposure: Possible burns or irritation to skin, eyes and upper respiratory tract; headache; nausea; laryngitis. Inhalation of vapors may be fatal by causing glottis to spasm and suffocation. Exposure may result in chemical pneumonitis or pulmonary edema.

Emergency/First Aid Treatment: Remove to ventilated area; immediately remove any contaminated clothing and wash contaminated areas for 15 minutes using water. Treat supportively and observe for possible shock. If ingested, seek immediate medical aid.

Recommended Clean-Up Procedures

Personal Protection: Level B Ensemble
RCRA Waste #   None

Recommended Material N.D.
Reportable Quantities: None

Spills: Dike to contain spill and collect liquid for later disposal. Decontaminate spill area using dry decontamination techniques. Treat all materials used or generated and equipment involved as contaminated by hazardous waste.

Special Emergency Information: May be harmful if inhaled, swallowed or absorbed through the skin. May react violently with water.

VANADIUM PENTOXIDE - V$_2$O$_5$

Synonyms: vanadic anhydride; vanadium (V) oxide

CAS Number: 1314-62-1   **Description:** Yellow to red powder
DOT Number: 2862   **DOT Classification:** ORM-E

Molecular Weight: 181.9
Melting Point: 690°C (1274°F)   **Vapor Density:** N.A.   **Vapor Pressure:** 1 (0 mm Hg)
Boiling Point: 1750°C (3182°F)   **Specific Gravity:** 3.4   **Water Solubility:** 0.1%

Chemical Incompatibilities or Instabilities: Alkali metals, chlorine trifluoride, peroxyformic acid, interhalogens.

FLAMMABILITY

NFPA Hazard Code: **NFPA Classification:**
Flash Point: N.A.   **LEL:** N.A.
Autoignition Temp: N.A.   **UEL:** N.A.

Fire Extinguishing Methods: Dry Chemical, Carbon Dioxide, Flooding Quantities of Water.

Special Fire Fighting Considerations: Structural fire fighter protective clothing will not provide adequate protection. Isolate for 150 feet if rail or tank truck is involved in a fire. Water spray should be used to keep closed containers cool. Continue to cool container after fire is extinguished. Treat all materials used or generated and equipment involved as contaminated by hazardous waste.

TOXICOLOGY

Odor: None   **Odor Threshold:** N.A.
Physical Contact: Irritant
TLV = 0.05 mg/m^3 **STEL** = N.D.   **IDLH** = 70 mg/m^3

Routes of Entry and Relative LD$_{50}$ (or LC$_{50}$)   **RTECS # YW2460000**
     Inhalation   N.D.
     Ingestion   9   (10 mg/kg)
     Skin Absorption   N.D.

Symptoms of Exposure: Irritation to eyes and upper respiratory tract, green tongue, breathing difficulties, skin rash. Possible teratogen or mutagen. Green color of gums and tongue.

Emergency/First Aid Treatment: Remove to ventilated area; immediately remove any contaminated clothing and wash contaminated areas for 15 minutes using water. Treat supportively and observe for possible shock. If ingested, seek immediate medical aid.

Recommended Clean-Up Procedures

Personal Protection: Level A Ensemble   **Recommended Material** N.D.

RCRA Waste # P120   **Reportable Quantities:** 1000 lb (1000 lb)

Spills: Take up in non-combustible absorbent and dispose. Decontaminate spill area using soapy water. Treat all materials used or generated and equipment involved as contaminated by hazardous waste.

Special Emergency Information: May be fatal in inhaled, ingested or absorbed through the skin.

VANADIUM TETRACHLORIDE - VCl₄

Synonyms: vanadium (IV) chloride

CAS Number: 7632-51-1   **Description:** Dark red liquid
DOT Number: 2444   **DOT Classification:** Corrosive Material

Molecular Weight: 192.7
Melting Point: -28°C (-18°F)   **Vapor Density:** N.D.   **Vapor Pressure:** N.D.
Boiling Point: 148°C (299°F)   **Specific Gravity:** 1.8   **Water Solubility:** Decomposes

Chemical Incompatibilities or Instabilities: Water.

FLAMMABILITY

NFPA Hazard Code: 3-0-2- W   **NFPA Classification:**
Flash Point: N.A.   **LEL:** N.A.
Autoignition Temp: N.A.   **UEL:** N.A.

Fire Extinguishing Methods: Dry Chemical, Carbon Dioxide

Special Fire Fighting Considerations: DO NOT USE WATER. Structural fire fighter protective clothing will not provide adequate protection. Isolate for 150 feet if rail or tank truck is involved in a fire. Water spray should be used to keep closed containers cool. Continue to cool container after fire is extinguished. Do not allow water to enter container.

TOXICOLOGY

Odor: Sharp, irritating   **Odor Threshold:** N.D.
Physical Contact: Material is extremely destructive to human tissues.
TLV = N.D.   **STEL** = N.D.   **IDLH** = N.D.

Routes of Entry and Relative LD$_{50}$ (or LC$_{50}$)   **RTECS # YW2625000**
  Inhalation   N.D.
  Ingestion   7   (160 mg/kg)
  Skin Absorption   N.D.

Symptoms of Exposure: Possible burns or irritation to skin, eyes and upper respiratory tract; headache; nausea. Inhalation of vapors may be fatal by causing glottis to spasm and suffocation. Exposure may result in chemical pneumonitis or pulmonary edema.

Emergency/First Aid Treatment: Remove to ventilated area; immediately remove any contaminated clothing and wash contaminated areas for 15 minutes using water. Treat supportively and observe for possible shock. If ingested, seek immediate medical aid.

Recommended Clean-Up Procedures

Personal Protection: Level A Ensemble   **Recommended Material**   N.D.

RCRA Waste #   None   **Reportable Quantities:**   None

Spills: Dike to contain spill, take up in non-combustible absorbent and dispose. Decontaminate spill area using dry decontamination techniques.

Special Emergency Information: May be fatal in inhaled, ingested or absorbed through the skin. Reacts violently with water to yield hydrogen chloride, vanadium trichloride and vanadium oxychloride.

VINYL ACETATE - CH$_3$C(O)OCH=CH$_3$

Synonyms: 1-acetoxyacetylene; vinyl a monomer

CAS Number: 108-05-4	**Description:** Colorless Liquid	
DOT Number: 1301	**DOT Classification:** Flammable Liquid	

Molecular Weight: 86.1

Melting Point: -93°C (-135°F)	**Vapor Density:** 3.0	**Vapor Pressure:** **6** (88 mm Hg)	
Boiling Point: 73°C (163°F)	**Specific Gravity:** 0.9	**Water Solubility:** Insoluble	

Chemical Incompatibilities or Instabilities: Peroxides, ozone, strong acids, desiccants, ethylene diamine, ethyleneimine, 2-aminoethanol, light.

FLAMMABILITY

NFPA Hazard Code: **2-3-2**	**NFPA Classification:** Class IB Flammable	
Flash Point: -6°C (18°F)	**LEL:** 2.6%	
Autoignition Temp: 402°C (756°F)	**UEL:** 13.4%	

Fire Extinguishing Methods: Carbon Dioxide, Dry Chemical, Foam, Water Spray or Fog.

Special Fire Fighting Considerations: Isolate for 1/2 mile if rail or tank truck is involved in a fire. Water spray from an unmanned device should be used to keep closed containers cool. Continue to cool container after fire is extinguished. For large fires, if possible, withdraw and allow to burn. Immediately withdraw if rising sound from venting device is heard or if fire is causing discoloration to the tank.

TOXICOLOGY

Odor: Sour	**Odor Threshold:** 0.5 mg/m^3
Physical Contact: Irritant	
TLV = 35 mg/m^3 **STEL** = 70 mg/m^3	**IDLH** = N.D.

Routes of Entry and Relative LD$_{50}$ (or LC$_{50}$)   **RTECS # AK0875000**

Inhalation	5	(28,000 mg/m^3/H)
Ingestion	3	(2920 mg/kg)
Skin Absorption	3	(2335 mg/kg)

Symptoms of Exposure: Possible irritation to skin, eyes and upper respiratory tract; headache; nausea. May act as a narcotic in high concentrations or after prolonged exposures.

Emergency/First Aid Treatment: Remove to ventilated area; immediately remove any contaminated clothing and wash contaminated areas for 15 minutes using water. Treat supportively and observe for possible shock. If ingested, seek immediate medical aid.

Recommended Clean-Up Procedures

Personal Protection: Level B Ensemble	**Recommended Material** N.D.
RCRA Waste # None	**Reportable Quantities:** 5000 lb (100 lb)

Spills: Remove all potential ignition sources. Dike to contain spill, absorb with non-combustible absorbent and take up using non-sparking tools. Keep area isolated until vapors have dissipated. Decontaminate spill area using soapy water.

Special Emergency Information: May be harmful if inhaled, swallowed or absorbed through the skin. Vapors are heavier than air and may travel some distance to an ignition source. Vapors are more dense than air and may settle in low lying areas. May undergo hazardous polymerization upon exposure to light. May form unstable peroxides upon prolonged exposure to air.

VINYL BROMIDE - CH$_2$=CHBr

Synonyms: bromoethene; bromoethylene

CAS Number: 593-60-2	**Description:** Colorless gas
DOT Number: 1085	**DOT Classification:** Flammable Liquid

Molecular Weight:

Melting Point: -139 C (-218 F)	**Vapor Density:** 3.7	**Vapor Pressure:** Gas
Boiling Point: 15 C (60 F)	**Specific Gravity:** 1.5 (<60 F)	**Water Solubility:** Insoluble

Chemical Incompatibilities or Instabilities: Sunlight, oxidizers, peroxides, copper and its alloys.

FLAMMABILITY

NFPA Hazard Code: 2-0-1	**NFPA Classification:**
Flash Point: N.A.	**LEL:** 6.0%
Autoignition Temp: 530 C (986 F)	**UEL:** 15.0%

Fire Extinguishing Methods: Carbon Dioxide, Dry Chemical, Foam, Water Spray or Fog.

Special Fire Fighting Considerations: Water spray should be used to keep closed containers cool. Continue to cool container after fire is extinguished. Immediately withdraw if rising sound from venting device is heard or if fire is causing discoloration to the tank.

TOXICOLOGY
SUSPECTED CARCINOGEN

Odor: **Odor Threshold:** N.D.

Physical Contact: Irritant

TLV = 22 mg/m^3 **STEL** = N.D.   **IDLH** = N.D.

Routes of Entry and Relative LD$_{50}$ (or LC$_{50}$)   **RTECS # KU8400000**

Inhalation	N.D.	
Ingestion	5	(500 mg/kg)
Skin Absorption	N.D.	

Symptoms of Exposure: Possible irritation to skin, eyes and upper respiratory tract; headache; nausea.

Emergency/First Aid Treatment: Remove to ventilated area; immediately remove any contaminated clothing and wash contaminated areas for 15 minutes using water. Treat supportively and observe for possible shock. If ingested, seek immediate medical aid.

Recommended Clean-Up Procedures

Personal Protection: Level B Ensemble	**Recommended Material**	N.D.
RCRA Waste # None	**Reportable Quantities:**	None

Spills: Stop leak if it can safely be done. Keep area isolated until vapors have dissipated.

Special Emergency Information: May be harmful if inhaled, swallowed or absorbed through the skin. Vapors are heavier than air and may travel some distance to an ignition source. Vapors are more dense than air and may settle in low lying areas. May undergo hazardous polymerization upon exposure to light. May form unstable peroxides upon prolonged exposure to air.

VINYL CHLORIDE - CH₂=CHCl
$$\text{VINYL CHLORIDE - } CH_2{=}CHCl$$

Synonyms: chloroethene; chloroethylene; monochloroethene; vinyl monomer

CAS Number: 75-01-4   **Description:** Colorless gas
DOT Number: 1086   **DOT Classification:** Flammable Gas

Molecular Weight:
Melting Point: -154 C (-245 F)	**Vapor Density:** 2.2	**Vapor Pressure:** Gas
Boiling Point: -14 C (7 F)	**Specific Gravity:** N.A.	**Water Solubility:** Insoluble

Chemical Incompatibilities or Instabilities: Oxidizers, nitrogen oxides, copper, aluminum, steel.

FLAMMABILITY

NFPA Hazard Code: 2-4-2   **NFPA Classification:**
Flash Point: N.A.   **LEL:** 1.7%
Autoignition Temp: 360 C (680 F)   **UEL:** 27.0%

Fire Extinguishing Methods: Carbon Dioxide, Dry Chemical, Foam, Water Spray or Fog.

Special Fire Fighting Considerations: Isolate for 1/2 mile if rail or tank truck is involved in a fire. Water spray from an unmanned device should be used to keep closed containers cool. Continue to cool container after fire is extinguished. For large fires, if possible, withdraw and allow to burn. Immediately withdraw if rising sound from venting device is heard or if fire is causing discoloration to the tank.

TOXICOLOGY
CARCINOGEN

Odor: Sweet   **Odor Threshold:** 26 mg/m³
Physical Contact: Irritant
TLV = 2.6 mg/m³   **STEL** = 13 mg/m³   **IDLH** = N.D.

Routes of Entry and Relative LD₅₀ (or LC₅₀)   **RTECS # KU9625000**
Inhalation	N.D.	
Ingestion	5	(500 mg/kg)
Skin Absorption	N.D.	

Symptoms of Exposure: Possible irritation to skin, eyes and upper respiratory tract; headache; nausea. Rapidly evaporating or liquefied gas may cause frostbite.

Emergency/First Aid Treatment: Remove to ventilated area; immediately remove any contaminated clothing and wash contaminated areas for 15 minutes using water. Treat supportively and observe for possible shock. If ingested, seek immediate medical aid.

Recommended Clean-Up Procedures

Personal Protection: Level B Ensemble   **Recommended Material** Nitrile Rubber (0), Viton (0)

RCRA Waste # U043   **Reportable Quantities:** 1 lb (1 lb)

Spills: Remove all potential ignition sources. Stop leak if it can safely be done. Keep area isolated until vapors have dissipated.

Special Emergency Information: May be harmful if inhaled, swallowed or absorbed through the skin. Vapors are heavier than air and may travel some distance to an ignition source. Vapors are more dense than air and may settle in low lying areas. May undergo hazardous polymerization. May form unstable peroxides upon prolonged exposure to air.

Vinyl fluoride (DOT # 1860) has similar chemical, physical and toxicological properties, but does not have a RCRA Waste number or reportable quantities.

VINYL ETHER - $(CH_2=CH)_2O$

Synonyms: divinyl ether; divinyl oxide; vinether

CAS Number: 109-93-3	**Description:** Colorless Liquid	
DOT Number: 1167	**DOT Classification:** Flammable Liquid	

Molecular Weight: 70.1

Melting Point:	-101 C (-150 F)	**Vapor Density:** 2.4	**Vapor Pressure:**	9 (500 mm Hg)
Boiling Point:	28 C (83 F)	**Specific Gravity:** 0.8	**Water Solubility:**	Insoluble

Chemical Incompatibilities or Instabilities: Oxidizers, peroxides, nitric acid.

FLAMMABILITY

NFPA Hazard Code:	2-3-2	**NFPA Classification:**	Class IA Flammable
Flash Point:	<-30 C (<-22 F)	**LEL:**	1.0%
Autoignition Temp:	360 C (680 F)	**UEL:**	27.0%

Fire Extinguishing Methods: Carbon Dioxide, Dry Chemical, Foam, Water Spray or Fog.

Special Fire Fighting Considerations: Isolate for 1/2 mile if rail or tank truck is involved in a fire. Water spray from an unmanned device should be used to keep closed containers cool. Continue to cool container after fire is extinguished. For large fires, if possible, withdraw and allow to burn. Immediately withdraw if rising sound from venting device is heard or if fire is causing discoloration to the tank.

TOXICOLOGY

Odor: Characteristic	**Odor Threshold:** N.D.

Physical Contact: Irritant

TLV = N.D.	**STEL** = N.D.	**IDLH** = N.D.

Routes of Entry and Relative LD$_{50}$ (or LC$_{50}$)   **RTECS #** YZ6700000

Inhalation	N.D.
Ingestion	N.D.
Skin Absorption	N.D.

Symptoms of Exposure: Possible irritation to skin, eyes and upper respiratory tract; headache; nausea.

Emergency/First Aid Treatment: Remove to ventilated area; immediately remove any contaminated clothing and wash contaminated areas for 15 minutes using water. Treat supportively and observe for possible shock. If ingested, seek immediate medical aid.

Recommended Clean-Up Procedures

Personal Protection:	Level B Ensemble	**Recommended Material** N.D.
RCRA Waste #	None	**Reportable Quantities:** None

Spills: Remove all potential ignition sources. Dike to contain spill, absorb with non-combustible absorbent and take up using non-sparking tools. Decontaminate spill area using soapy water. Keep area isolated until vapors have dissipated.

Special Emergency Information: May be harmful if inhaled, swallowed or absorbed through the skin. Vapors are heavier than air and may travel some distance to an ignition source. Vapors are more dense than air and may settle in low lying areas. May undergo hazardous polymerization. May form unstable peroxides upon prolonged exposure to air.

VINYLIDINE CHLORIDE - CH$_2$=CCl$_2$

Synonyms: 1,1-dichloroethene; 1,1-dichloroethylene

CAS Number: 75-35-4	**Description:** Colorless Liquid	
DOT Number: 1303	**DOT Classification:** Flammable Liquid	

Molecular Weight: 96.9

Melting Point: -122 C (-189 F)	**Vapor Density:** 3.3	**Vapor Pressure:** **8** (400 mm Hg)
Boiling Point: 32 C (89 F)	**Specific Gravity:** 1.2	**Water Solubility:** Insoluble

Chemical Incompatibilities or Instabilities: Oxidizers, strong acids, copper and its alloys, aluminum and its alloys.

FLAMMABILITY

NFPA Hazard Code: 2-4-2	**NFPA Classification:** Class IB Flammable
Flash Point: -18 C (0 F)	**LEL:** 7.3%
Autoignition Temp: 570 C (1058 F)	**UEL:** 16.0%

Fire Extinguishing Methods: Alcohol Resistant Foam, Carbon Dioxide, Dry Chemical, Water Spray or Fog.

Special Fire Fighting Considerations: Isolate for 1/2 mile if rail or tank truck is involved in a fire. Water spray from an unmanned device should be used to keep closed containers cool. Continue to cool container after fire is extinguished. For large fires, if possible, withdraw and allow to burn. Immediately withdraw if rising sound from venting device is heard or if fire is causing discoloration to the tank. Treat all materials used or generated and equipment involved as contaminated by hazardous waste.

TOXICOLOGY
SUSPECTED CARCINOGEN

Odor: Sweet, chloroform-like	**Odor Threshold:** N.D.
Physical Contact: Irritant	
TLV = 4 mg/m^3 **STEL** = N.D.	**IDLH** = N.D.

Routes of Entry and Relative LD$_{50}$ (or LC$_{50}$)   **RTECS # KV9275100**

Inhalation	2	(100,000 mg/m^3/H)
Ingestion	7	(200 mg/kg)
Skin Absorption	N.D.	

Symptoms of Exposure: Possible irritation to skin, eyes and upper respiratory tract; headache; nausea; laryngitis.

Emergency/First Aid Treatment: Remove to ventilated area; immediately remove any contaminated clothing and wash contaminated areas for 15 minutes using water. Treat supportively and observe for possible shock. If ingested, seek immediate medical aid.

Recommended Clean-Up Procedures

Personal Protection: Level A Ensemble	**Recommended Material** Polyvinyl Alcohol (0). Teflon (0)
RCRA Waste # U078	**Reportable Quantities:** 100 lb (5000 lb)

Spills: Remove all potential ignition sources. Dike to contain spill, absorb with non-combustible absorbent and take up using non-sparking tools. Decontaminate spill area using soapy water. Keep area isolated until vapors have dissipated. Treat all materials used or generated and equipment involved as contaminated by hazardous waste.

Special Emergency Information: May be harmful if inhaled, swallowed or absorbed through the skin. Vapors are heavier than air and may travel some distance to an ignition source. Vapors are more dense than air and may settle in low lying areas. May undergo hazardous polymerization. May form unstable peroxides upon prolonged exposure to air.

VINYL TOLUENE

Synonyms: methyl styrene

CAS Number: 25013-15-4	**Description:** Colorless Liquid
DOT Number: 2618	**DOT Classification:** Flammable Liquid

Molecular Weight: 118.2

Melting Point:	-77 C (-107 F)	**Vapor Density:** 4.1	**Vapor Pressure:** 2 (1.1 mm Hg)
Boiling Point:	168 C (334 F)	**Specific Gravity:** 0.9	**Water Solubility:** Insoluble

Chemical Incompatibilities or Instabilities: Oxidizers, acids, peroxides, metal salts.

FLAMMABILITY

NFPA Hazard Code:	2-2-2	**NFPA Classification:** Class II Combustible
Flash Point:	52 C (125 F)	**LEL:** 0.8%
Autoignition Temp:	494 C (921 F)	**UEL:** 11.0%

Fire Extinguishing Methods: Carbon Dioxide, Dry Chemical, Foam, Water Spray or Fog.

Special Fire Fighting Considerations: Isolate for 1/2 mile if rail or tank truck is involved in a fire. Water spray from an unmanned device should be used to keep closed containers cool. Continue to cool container after fire is extinguished. For large fires, if possible, withdraw and allow to burn. Immediately withdraw if rising sound from venting device is heard or if fire is causing discoloration to the tank.

TOXICOLOGY

Odor: Pungent	**Odor Threshold:** N.D.
Physical Contact: Irritant	
TLV = 480 mg/m^3 **STEL** = N.D.	**IDLH** = N.D.

Routes of Entry and Relative LD$_{50}$ (or LC$_{50}$)   **RTECS # WL5075000**

Inhalation	N.D.	
Ingestion	3	(4000 mg/kg)
Skin Absorption	N.D.	

Symptoms of Exposure: Possible irritation to skin, eyes and upper respiratory tract; headache; nausea.

Emergency/First Aid Treatment: Remove to ventilated area; immediately remove any contaminated clothing and wash contaminated areas for 15 minutes using water. Treat supportively and observe for possible shock. If ingested, seek immediate medical aid.

Recommended Clean-Up Procedures

Personal Protection: Level A Ensemble	**Recommended Material** N.D.
RCRA Waste # None	**Reportable Quantities:** None

Spills: Remove all potential ignition sources. Dike to contain spill, absorb with non-combustible absorbent and take up using non-sparking tools. Decontaminate spill area using soapy water. Keep area isolated until vapors have dissipated.

Special Emergency Information: May be harmful if inhaled, swallowed or absorbed through the skin. Vapors are heavier than air and may travel some distance to an ignition source. Vapors are more dense than air and may settle in low lying areas. May undergo hazardous polymerization. May form unstable peroxides upon prolonged exposure to air.

XYLENE

Synonyms:

CAS Number: 1330-20-7		**Description:** Colorless Liquid	
DOT Number: 1307		**DOT Classification:** Flammable Liquid	

Molecular Weight: 106.2

Melting Point: C (F)	**Vapor Density:** 3.6	**Vapor Pressure:** 3 (7 mm Hg)	
Boiling Point: 139 C (282 F)	**Specific Gravity:** 0.9	**Water Solubility:** Insoluble	

Chemical Incompatibilities or Instabilities: Strong oxidizers.

FLAMMABILITY

NFPA Hazard Code: 2-3-0	**NFPA Classification:** Class IB Flammable	
Flash Point: 29 C (85 F)	**LEL:** 1.1%	
Autoignition Temp: 464 C (867 F)	**UEL:** 7.0%	

Fire Extinguishing Methods: Carbon Dioxide, Dry Chemical, Foam, Water Spray or Fog.

Special Fire Fighting Considerations: Isolate for 1/2 mile if rail or tank truck is involved in a fire. Water spray from an unmanned device should be used to keep closed containers cool. Continue to cool container after fire is extinguished. For large fires, if possible, withdraw and allow to burn. Immediately withdraw if rising sound from venting device is heard or if fire is causing discoloration to the tank.

TOXICOLOGY

Odor: Sweet, aromatic

Odor Threshold: 5 mg/m^3

Physical Contact: Irritant

TLV = 435 mg/m^3 **STEL** = 655 mg/m^3 **IDLH** = 4350 mg/m^3

Routes of Entry and Relative LD$_{50}$ (or LC$_{50}$)   **RTECS # ZE2190000**

Inhalation	4	(88,000 mg/m^3/H)
Ingestion	3	(4300 mg/kg)
Skin Absorption	N.D.	

Symptoms of Exposure: Possible irritation to skin, eyes and upper respiratory tract; headache; nausea. May act as a narcotic in high concentrations. Exposure may result in chemical pneumonitis or pulmonary edema.

Emergency/First Aid Treatment: Remove to ventilated area; immediately remove any contaminated clothing and wash contaminated areas for 15 minutes using water. Treat supportively and observe for possible shock. If ingested, seek immediate medical aid.

Recommended Clean-Up Procedures

Personal Protection: Level B Ensemble	**Recommended Material** Polyvinyl Alcohol (+), Viton (+), Teflon (0)	
RCRA Waste # U239	**Reportable Quantities:** 1000 lb (1000 lb)	

Spills: Remove all potential ignition sources. Dike to contain spill, absorb with non-combustible absorbent and take up using non-sparking tools. Decontaminate spill area using soapy water. Keep area isolated until vapors have dissipated.

Special Emergency Information: May be harmful if inhaled, swallowed or absorbed through the skin. Vapors are heavier than air and may travel some distance to an ignition source. Vapors are more dense than air and may settle in low lying areas.

o-Xylene (CAS # 95-47-6, DOT # 1307, RTECS # ZE2450000),
m-xylene (CAS # 108-38-3, DOT # 1307, RTECS # ZE2275000), and
p-xylene (CAS # 106-42-3, DOT # 1307, RTECS # ZE2625000) have similar chemical, physical and toxicological properties, no RCRA Waste # and the same Reportable Quantities.

o-XYLIDINE

Synonyms: 2,3-dimethylaniline; 2,3-dimethylbenzamine; 2,3-dimethylphenylamine

CAS Number: 87-59-2   **Description:** Yellow to brown liquid
DOT Number: 1711   **DOT Classification:** Poison B

Molecular Weight: 121.2
Melting Point: 2 C   (36 F)   **Vapor Density:**   **Vapor Pressure:**
Boiling Point: 221 C   (430 F)   **Specific Gravity:** 1.0   **Water Solubility:**   Insoluble

Chemical Incompatibilities or Instabilities: Oxidizers, hypochlorites.

FLAMMABILITY

NFPA Hazard Code: 3-1-0   **NFPA Classification:** Class IIIB Combustible
Flash Point: 96 C   (205 F)   **LEL:**   1.0%
Autoignition Temp: N.D.   **UEL:** N.D.

Fire Extinguishing Methods: Carbon Dioxide, Dry Chemical, Foam, Water Spray or Fog.

Special Fire Fighting Considerations: Structural fire fighter protective clothing will not provide adequate protection. Fight fire from a distance or protected location, if possible. Treat all materials used or generated and equipment involved as contaminated by hazardous waste.

TOXICOLOGY

Odor: Amine-like   **Odor Threshold:** N.D.
Physical Contact: Material is extremely destructive to human tissues.
TLV = 10 mg/m^3 (skin)   **STEL** = N.D.   **IDLH** = 750 mg/m^3

Routes of Entry and Relative LD$_{50}$ (or LC$_{50}$)   **RTECS # ZE8575000**
    Inhalation   N.D.
    Ingestion   N.D.
    Skin Absorption   N.D.

Symptoms of Exposure: Possible burns or irritation to skin, eyes and upper respiratory tract; headache; nausea; laryngitis. Inhalation of vapors may be fatal by causing glottis to spasm and suffocation. Exposure may result in chemical pneumonitis or pulmonary edema. Absorption may lead to the formation of methemoglobin resulting in cyanosis several hours after exposure.

Emergency/First Aid Treatment: Remove to ventilated area; immediately remove any contaminated clothing and wash contaminated areas for 15 minutes using water. Treat supportively and observe for possible shock. If ingested, seek immediate medical aid.

Recommended Clean-Up Procedures

Personal Protection: Level A Ensemble   **Recommended Material**   N.D.

RCRA Waste #   None   **Reportable Quantities:**   None

Spills: Dike to contain spill and collect liquid for later disposal. Decontaminate spill area using soapy water. Treat all materials used or generated and equipment involved as contaminated by hazardous waste.

Special Emergency Information: May be fatal in inhaled, ingested or absorbed through the skin.

ZINC CHLORATE - Zn(ClO$_3$)$_2$

Synonyms:

CAS Number: 10361-95-2
DOT Number: 1513

Description: White to yellow powder
DOT Classification: Oxidizer

Molecular Weight: 122.6
Melting Point: D   60 C   (140 F)
Boiling Point: N.D.

Vapor Density: N.D.
Specific Gravity: 2.2

Vapor Pressure: 1 (<1 mm Hg)
Water Solubility: Soluble

Chemical Incompatibilities or Instabilities: Aluminum, organics, copper, metal sulfides, manganese dioxide, acids, sulfur, phosphorous, heat, ordinary combustibles, arsenic, stron acids.

FLAMMABILITY

NFPA Hazard Code: **1-0-1-OX**
Flash Point: N.A.
Autoignition Temp: N.A.

NFPA Classification:
LEL: N.A.
UEL: N.A.

Fire Extinguishing Methods: WATER ONLY

Special Fire Fighting Considerations: Water spray from an unmanned device should be used to keep closed containers cool. Continue to cool container after fire is extinguished. For large fires, if possible, withdraw and allow to burn.

TOXICOLOGY

Odor:
Physical Contact: Material is extremely destructive to human tissues.
TLV = N.D.   **STEL** = N.D.

Odor Threshold: N.D.

IDLH = N.D.

Routes of Entry and Relative LD$_{50}$ (or LC$_{50}$)

Inhalation	N.D.
Ingestion	N.D.
Skin Absorption	N.D.

RTECS # ZH1350000

Symptoms of Exposure: Possible burns or irritation to skin, eyes and upper respiratory tract; headache; nausea. Exposure may result in chemical pneumonitis or pulmonary edema.

Emergency/First Aid Treatment: Remove to ventilated area; immediately remove any contaminated clothing and wash contaminated areas for 15 minutes using water. Treat supportively and observe for possible shock. If ingested, seek immediate medical aid.

Recommended Clean-Up Procedures

Personal Protection: Level B Ensemble
RCRA Waste #   None

Recommended Material   N.D.
Reportable Quantities: None

Spills: Take up in non-combustible absorbent and dispose. Decontaminate spill area using water.

Special Emergency Information: May be fatal in inhaled, ingested or absorbed through the skin.

ZINC CHLORIDE - ZnCl$_2$

Synonyms: butter of zinc; tinning glux; zinc dichloride

CAS Number: 7646-85-7	**Description:** White powder	
DOT Number: 1840, 2331	**DOT Classification:** Corrosive Material	

Molecular Weight: 136.3

Melting Point: 290 C (554 F)	**Vapor Density:** N.A.	**Vapor Pressure:** 1 (0 mm Hg)
Boiling Point: 732 C (1350 F)	**Specific Gravity:** 2.9	**Water Solubility:** Soluble

Chemical Incompatibilities or Instabilities: Alkali metals.

FLAMMABILITY

NFPA Hazard Code:		**NFPA Classification:**
Flash Point:	N.A.	**LEL:** N.A.
Autoignition Temp:	N.A.	**UEL:** N.A.

Fire Extinguishing Methods: Carbon Dioxide, Dry Chemical, Foam, Water Spray or Fog.

Special Fire Fighting Considerations: Water spray should be used to keep closed containers cool. Continue to cool container after fire is extinguished.

TOXICOLOGY
QUESTIONABLE CARCINOGEN

Odor: None   **Odor Threshold:** N.A.

Physical Contact: Material is extremely destructive to human tissues.

TLV = 1 mg/m^3 **STEL** = 2 mg/m^3 **IDLH** = 4800 mg/m^3

Routes of Entry and Relative LD$_{50}$ (or LC$_{50}$)   **RTECS # ZH1400000**

Inhalation	N.D.	
Ingestion	5	(350 mg/kg)
Skin Absorption	N.D.	

Symptoms of Exposure: Possible burns or irritation to skin, eyes and upper respiratory tract; headache; nausea. Inhalation of vapors may be fatal by causing glottis to spasm and suffocation. Exposure may result in chemical pneumonitis or pulmonary edema.

Emergency/First Aid Treatment: Remove to ventilated area; immediately remove any contaminated clothing and wash contaminated areas for 15 minutes using water. Treat supportively and observe for possible shock. If ingested, seek immediate medical aid.

Recommended Clean-Up Procedures

Personal Protection: Level B Ensemble	**Recommended Material** N.D.	
RCRA Waste # None	**Reportable Quantities:** 1000 lb (5000 lb)	

Spills: Take up in non-combustible absorbent and dispose. Decontaminate spill area using soapy water. Treat all materials used or generated and equipment involved as contaminated by hazardous waste.

Special Emergency Information: May be harmful if inhaled, swallowed or absorbed through the skin.

ZINC NITRATE - Zn(NO3)2

Synonyms:

CAS Number: 7779-88-6   **Description:** White powder
DOT Number: 1514   **DOT Classification:** Oxidizer

Molecular Weight: 189.4
Melting Point: 43 C (109 F)   **Vapor Density:** N.A.   **Vapor Pressure:** 1 (0 mm Hg)
Boiling Point: C (F)   **Specific Gravity:** 2.1   **Water Solubility:** Soluble

Chemical Incompatibilities or Instabilities: Organics, metal sulfides, phosphorus, sulfur, aluminum, alkyl esters, tin (II) chloride, phosphinates, cyanides, isocyanates, isothiocyanates, hypophosphites.

FLAMMABILITY

NFPA Hazard Code: **NFPA Classification:**
Flash Point: N.A.   **LEL:** N.A.
Autoignition Temp: N.A.   **UEL:** N.A.

Fire Extinguishing Methods: WATER ONLY

Special Fire Fighting Considerations: Water spray from an unmanned device should be used to keep closed containers cool. Continue to cool container after fire is extinguished. For large fires, if possible, withdraw and allow to burn.

TOXICOLOGY

Odor: None   **Odor Threshold:** N.A.
Physical Contact: Material is extremely destructive to human tissues.
TLV = N.D.   **STEL** = N.D.   **IDLH** = N.D.

Routes of Entry and Relative LD$_{50}$ (or LC$_{50}$)   **RTECS # ZH4775000**
    Inhalation   N.D.
    Ingestion   N.D.
    Skin Absorption   N.D.

Symptoms of Exposure: Possible burns or irritation to skin, eyes and upper respiratory tract; headache; nausea. Inhalation of vapors may be fatal by causing glottis to spasm and suffocation. Exposure may result in chemical pneumonitis or pulmonary edema.

Emergency/First Aid Treatment: Remove to ventilated area; immediately remove any contaminated clothing and wash contaminated areas for 15 minutes using water. Treat supportively and observe for possible shock. If ingested, seek immediate medical aid.

Recommended Clean-Up Procedures

Personal Protection: Level B Ensemble   **Recommended Material** N.D.

RCRA Waste # None   **Reportable Quantities:** 1000 lb (5000 lb)

Spills: Take up in non-combustible absorbent and dispose. Decontaminate spill area using water.

Special Emergency Information: May be harmful if inhaled, swallowed or absorbed through the skin.

ZINC PHOSPHIDE - Zn$_3$P$_2$

Synonyms:

CAS Number:	1314-84-7	**Description:** Dark gray powder
DOT Number:	1714	**DOT Classification:** Flammable Solid

Molecular Weight: 258.1

Melting Point:	420°C (788°F)	**Vapor Density:** N.A.	**Vapor Pressure:** 1 (< 1 mm Hg)
Boiling Point:	1100°C (2012°F)	**Specific Gravity:** 4.55	**Water Solubility:** Insoluble

Chemical Incompatibilities or Instabilities: Acids, oxidizers.

FLAMMABILITY

NFPA Hazard Code:	3-3-1	**NFPA Classification:**	
Flash Point:	N.A.	**LEL:**	N.A.
Autoignition Temp:	N.A.	**UEL:**	N.A.

Fire Extinguishing Methods: Dry Chemical, Sand, Soda Ash, Crushed Limestone.

Special Fire Fighting Considerations: Structural fire fighter protective clothing will not provide adequate protection. For large fires, if possible, withdraw and allow to burn. Treat all materials used or generated and equipment involved as contaminated by hazardous waste. Do not allow water to enter container. May release toxic fumes upon heating (phosphine gas).

TOXICOLOGY

Odor: None	**Odor Threshold:** N.A.
Physical Contact: Irritant	

TLV = N.D.	**STEL** = N.D.	**IDLH** = N.D.

Routes of Entry and Relative LD$_{50}$ (or LC$_{50}$)   **RTECS #** ZH4900000

Inhalation	N.D.	
Ingestion	**9**	(12 mg/kg)
Skin Absorption	N.D.	

Symptoms of Exposure: Possible irritation to skin, eyes and upper respiratory tract; headache; nausea. Exposure may result in chemical pneumonitis or pulmonary edema. When ingested, zinc phosphide reacts with stomach acid to yield extremely toxic phosphine.

Emergency/First Aid Treatment: Remove to ventilated area; immediately remove any contaminated clothing and wash contaminated areas for 15 minutes using water. Treat supportively and observe for possible shock. If ingested, seek immediate medical aid.

Recommended Clean-Up Procedures

Personal Protection:	Level A Ensemble	**Recommended Material** N.D.
RCRA Waste # P122	**Reportable Quantities:** 100 lb (1000 lb)	

Spills: Take up in non-combustible absorbent and dispose. Decontaminate spill area using soapy water. Treat all materials used or generated and equipment involved as contaminated by hazardous waste.

Special Emergency Information: May be fatal in inhaled, ingested or absorbed through the skin.

ZIRCONIUM - Zr

Synonyms:

CAS Number: 7440-67-7   **Description:** Grayishg - white metal or powder

DOT Number: 1308, 1358, 2008, 2009, 2858   **DOT Classification:** Flammable Solid

Molecular Weight: 91.2

Melting Point: 1852°C (3366°F)   **Vapor Density:** N.A.   **Vapor Pressure:** 1 (o mm Hg)

Boiling Point: 3577°C (6471°F)   **Specific Gravity:** 6.5   **Water Solubility:** Insoluble

Chemical Incompatibilities or Instabilities: Air, alkali metals, oxidizers, molybdates, sulfates, tungstates, borax, chloroform, metal hydroxides, carbon tetrachloride, copper oxide, lead oxide, lead, nitrylfluoride, strong acids.

FLAMMABILITY

NFPA Hazard Code: **NFPA Classification:**

Flash Point: N.D.   **LEL:** N.D.

Autoignition Temp: N.D.   **UEL:** N.D.

Fire Extinguishing Methods: Dry Chemical, Sand, Soda Ash, Crushed Limestone, Flooding Quantities of Water.

Special Fire Fighting Considerations: Water spray from an unmanned device should be used to keep closed containers cool. Continue to cool container after fire is extinguished. For large fires, if possible, withdraw and allow to burn.

TOXICOLOGY

Odor: None   **Odor Threshold:** N.A.

Physical Contact: Irritant

TLV = 5 mg (Zr)/m^3 **STEL** = 10 mg (Zr)/m^3 **IDLH** = 500 mg (Zr)/m^3

Routes of Entry and Relative LD$_{50}$ (or LC$_{50}$)   **RTECS #** ZH7070000

     Inhalation   N.D.

     Ingestion   N.D.

     Skin Absorption   N.D.

Symptoms of Exposure: Possible irritation to skin, eyes and upper respiratory tract; headache; nausea.

Emergency/First Aid Treatment: Remove to ventilated area; immediately remove any contaminated clothing and wash contaminated areas for 15 minutes using water. Treat supportively and observe for possible shock. If ingested, seek immediate medical aid.

Recommended Clean-Up Procedures

Personal Protection: Level B Ensemble   **Recommended Material**   N.D.

RCRA Waste # None   **Reportable Quantities:** None

Spills: Take up in non-combustible absorbent and dispose. Decontaminate spill area using water.

Special Emergency Information: May be harmful if inhaled, swallowed or absorbed through the skin. Powdered zirconium (DOT # 2008) may spontaneously ignite in air.

Powdered zirconium (DOT # 2008) is an explosion hazard if the powder is allowed to become suspended in air.

ZIRCONIUM TETRACHLORIDE - ZrCL$_4$

Synonyms: zirconium chloride

CAS Number:	10026-11-6	**Description:** White to pink powder
DOT Number:	2503	**DOT Classification:** Corrosive Material

Molecular Weight: 233.0

Melting Point: S 300°C (572°F)	**Vapor Density:** N.D.	**Vapor Pressure:**	1 (< 1 mm Hg)
Boiling Point: N.A.	**Specific Gravity:** 2.8	**Water Solubility:**	
Decomposes			

Chemical Incompatibilities or Instabilities: Air, acids, amines, alcohols.

FLAMMABILITY

NFPA Hazard Code: 3-0-2-W		**NFPA Classification:**	
Flash Point:	N.A.	**LEL:**	N.A.
Autoignition Temp:	N.A.	**UEL:**	N.A.

Fire Extinguishing Methods: Dry Chemical, Carbon Dioxide, Alcohol Resistant Foam.

Special Fire Fighting Considerations: Structural fire fighter protective clothing will not provide adequate protection. Isolate for 150 feet if rail or tank truck is involved in a fire. Water spray should be used to keep closed containers cool. Continue to cool container after fire is extinguished.

TOXICOLOGY

Odor: **Odor Threshold:** N.D.

Physical Contact: Material is extremely destructive to human tissues.

TLV = 5 mg (Zr)/m^3 **STEL** = 10 mg (Zr)/m^3 **IDLH** = 500 mg (Zr)/m^3

Routes of Entry and Relative LD$_{50}$ (or LC$_{50}$)   **RTECS #**   ZH7175000

Inhalation	N.D.	
Ingestion	3	(1688 mg/kg)
Skin Absorption	N.D.	

Symptoms of Exposure: Possible burns or irritation to skin, eyes and upper respiratory tract; headache; nausea; laryngitis. Inhalation of vapors may be fatal by causing glottis to spasm and suffocation. Exposure may result in chemical pneumonitis or pulmonary edema. May cause an allergic reaction.

Emergency/First Aid Treatment: Remove to ventilated area; immediately remove any contaminated clothing and wash contaminated areas for 15 minutes using water. Treat supportively and observe for possible shock. If ingested, seek immediate medical aid.

Recommended Clean-Up Procedures

Personal Protection:	Level B Ensemble	**Recommended Material** N.D.
RCRA Waste # None	**Reportable Quantities:** None	

Spills: Take up in non-combustible absorbent and dispose. Decontaminate spill area using dry decontamination techniques.

Special Emergency Information: May be harmful if inhaled, swallowed or absorbed through the skin. Reacts exothermically with water to produce hydrogen chloride and zirconium oxide vapors. May be pyrophoric.

ACETALDEHYDE AMMONIA $CH_3CH(OH)NH_2$

acetaldehyde, amine salt; 1-aminoethanol; α-aminoethyl alcohol

| **Molecular Wt:** | 61.1 | **CAS #** | 75-39-8 |
| **DOT #** | 1841 | **DOT Class:** | ORM-A |

A odorless, colorless to yellow-brown solid that is soluble in water. When involved in a fire, use agents suitable for extinguishing surrounding materials. Acetaldehyde ammonia may cause chemical burns after exposures to high concentrations, but generally has a low toxicity. Body contamination should be treated by washing with water for a minimum of 15 minutes. If ingested, prompt medical aid should be obtained. Level B or C ensembles (depending upon the severity of the spill) should be used to clean up spills. Spill areas and equipment can be decontaminated using water.

ADIPIC ACID $HOOCCH_2CH_2CH_2CH_2COOH$

adipinic acid; 1,4-butanedicarboxylic acid; 1,6-hexanedioic acid

| **Molecular Wt:** | 146.1 | **CAS #** | 124-04-9 |
| **DOT #** | 9077 | **DOT Class:** | N.D. |

A white solid that is slightly water soluble. When involved in a fire, use agent suitable for extinguishing surrounding materials. Adipic acid may cause severe eye irritation. Contaminated areas of the body should be rinsed with water for a minimum of 15 minutes. If ingested, prompt medical aid should be obtained. Level B or C ensembles (depending upon the severity of the spill) should be used to clean up spills. Spill areas and equipment can be decontaminated using water.

ALUMINUM CHLORIDE HEXAHYDRATE $AlCl_3 \cdot 6H_2O$

| **Molecular Wt:** | 241.43 | **CAS #** | 7784-13-6 |
| **DOT #** | N.D. | **DOT Class:** | N.D. |

A white, nonflammable, water soluble solid that may cause irritation. When involved in a fire, use agent suitable for extinguishing surrounding materials. Contaminated body areas should be rinsed with water for a minimum of 15 minutes. If ingested, prompt medical aid should be obtained. Level B or C ensembles (depending upon the severity of the spill) should be used to clean up spills. Spill areas and equipment can be decontaminated using water.

ALUMINUM HYDROXIDE HYDRATE $Al(OH)_3 \cdot xH_2O$

alumigel; alumina hydrate; alumina trihydrate; aluminic hydrate

| **Molecular Wt:** | 78.0 | **CAS #** | 21645-51-2 |
| **DOT #** | N.D. | **DOT Class:** | N.D |

An odorless, white solid that is virtually insoluble in water. When involved in a fire, use agent suitable for extinguishing surrounding materials. Contamination may result in irritation and should be treated by rinsing with water for a minimum of 15 minutes. If ingested, prompt medical aid should be obtained. Level B or C ensembles (depending upon the severity of the spill) should be used to clean up spills. Spill areas and equipment can be decontaminated using soapy water.

AMMONIUM CARBONATE $(NH_4)_2CO_2$

carbonic acid, ammonium salt; ammonium carbonate, diamonium salt; diammonium carbonate

Molecular Wt:	96.1	**CAS #**	506-87-6
DOT #	9084	**DOT Class:**	ORM-A

An ordorless, white solid that is water soluble. When involved in a fire, use agent suitable for extinguishing surrounding materials. Contamination may result in irritation and should be treated by rinsing with water for a minimum of 15 minutes. If ingested, prompt medical aid should be obtained. Level B or C ensembles (depending upon the severity of the spill) should be used to clean up spills. Spill areas and equipment can be decontaminated using water.

AMMONIUM CHLORIDE NH_4Cl

ammonium muriate

Molecular Wt.:	53.5	**CAS #**	12125-02-9
DOT #	N.D.	**DOT Class:**	ORM-E

An ordorless, white solid that is water soluble. When involved in a fire, use agent suitable for extinguishing surrounding materials. Contamination may result in irritation and should be treated by rinsing with water for a minimum of 15 minutes. If ingested, prompt medical aid should be obtained. Level B or C ensembles (depending upon the severity of the spill) should be used to clean up spills. Spill areas and equipment can be decontaminated using water.

AMMONIUM SULFATE $(NH_4)_2SO_4$

ammonium sulphate; diammonium sulfate; sulfuric acid, diammonium salt

Molecular Wt.:	132.1	**CAS #**	7783-20-2
DOT #	N.D.	**DOT Class:**	N.D.

An ordorless, white solid that is water soluble. When involved in a fire, use agent suitable for extinguishing surrounding materials. Contamination may result in irritation and should be treated by rinsing with water for a minimum of 15 minutes. If ingested, prompt medical aid should be obtained. Level B or C ensembles (depending upon the severity of the spill) should be used to clean up spills. Spill areas and equipment can be decontaminated using water.

BENZOIN

benzoylphenylcarbinol; α-hydrozybenzyl phenyl ketone; α-hydroxy-α-phenylphenylacetylphenone

Molecular Wt.:	212.3	**CAS #**	119-53-9
DOT #	N.D.	**DOT Class:**	N.D.

A light yellow powder that in almost water insoluble. When involved in a fire, use agent suitable for extinguishing surrounding materials. Keep exposed containers cool. Contamination may result in irritation and should be treated by rinsing with water for a minimum of 15 minutes. If ingested, prompt medical aid should be obtained. Level B or C ensembles (depending upon the severity of the spill) should be used to clean up spills. Spill areas and equipment can be decontaminated using soapy water.

BISMUTH SUBNITRATE

bismuth hydroxide nitrate oxide; bismuth nitrate, basic: bismuth oxynitrate

Molecular Wt.: N.A.
DOT # N.D.

CAS # 1304-85-4
DOT Class: N.D.

An odorless, white powder that is virtually water insoluble. When heated, material will decompose yielding nitrogen oxides. When involved in a fire, use agent suitable for extinguishing surrounding materials. Keep exposed containers cool. Contamination may result in irritation and should be treated by rinsing with water for a minimum of 15 minutes. If ingested, prompt medical aid should be obtained. Level B or C ensembles (depending upon the severity of the spill) should be used to clean up spills. Spill areas and equipment can be decontaminated using soapy water.

BORIC ACID H_3BO_3

borofax; orthoboric acid

Molecular Wt.: 61.8
DOT # N.D.

CAS # 10043-35-3
DOT Class: N.D.

An ordorless, white powder that is water insoluble. When involved in a fire, use agent suitable for extinguishing surrounding materials. Contamination may result in irritation and should be treated by rinsing with water for a minimum of 15 minutes. Boric acid is readily absorbed through the skin and may cause methemoglobin formation. If ingested, prompt medical aid should be obtained. Level B or C ensembles (depending upon the severity of the spill) should be used to clean up spills. Spill areas and equipment can be decontaminated using soapy water.

CALCIUM CARBONATE $CaCO_3$

calcite; chalk, dolomite; limestone; marble; vaterite

Molecular Wt.: 100.1
DOT # N.D.

CAS # 1317-65-3
DOT Class: N.D.

An odorless, white powder that is almost water insoluble. When involved in a fire, use agent suitable for extinguishing surrounding materials. Calcium carbonate ignite upon contact with fluorine and is incompatible acids, alum and ammonium salts. Contamination may result in irritation and should be treated by rinsing with water for a minimum of 15 minutes. If ingested, prompt medical aid should be obtained. The OSHA PEL is 15 mg/m^3 (5 mg/m^3 for respirable dusts). Level B or C ensembles (depending upon the severity of the spill) should be used to clean up spills. Spill areas and equipment can be decontaminated using soapy water.

CALCIUM CHLORIDE $CaCl_2$

Molecular Wt.: 111.0
DOT # N.D.

CAS # 10043-52-4
DOT Class: N.D.

An odorless, colorless powder that is water soluble. When involved in a fire, use agent suitable for extinguishing surrounding materials. Calcium chloride reacts violently with interhalogens and boric anhydride/calcium oxide mixtures. It will react with zinc to release hydrogen gas and will catalyze the violent polymerization of methyl vinyl ether. Contamination may result in irritation and should be treated by rinsing with water for a minimum of 15 minutes. If ingested, prompt medical aid should be obtained. Level B or C ensembles (depending upon the severity of the spill) should be used to clean up spills. Spill areas and equipment can be decontaminated using water.

CALCIUM HYDROXIDE $Ca(OH)_2$

calcium hydrate; hydrated lime; slaked lime

Molecular Wt.: 74.1		**CAS #**	1305-62-0
DOT #	N.D.	**DOT Class:**	N.D.

An odorless, white powder that is water soluble. When involved in a fire, use agent suitable for extinguishing surrounding materials. Calcium hydroxide will react violently with phosphorous, maleic anhydride and nitroparaffins. Reactions of calcium hydroxide with aluminum will yield hydrogen gas and with chlorinated phenyls and inorganic nitrates will yield highly toxic products. Contamination may result in irritation and should be treated by rinsing with water for a minimum of 15 minutes. If ingested, prompt medical aid should be obtained. The OSHA TWA is 5 mg/m^3. Level B or C ensembles (depending upon the severity of the spill) should be used to clean up spills. Spill areas and equipment can be decontaminated using water.

CALCIUM SULFATE DIHYDRATE $CaSO_4 \cdot 2H_2O$

alabaster; gypsum; magnesia white; mineral white

Molecular Wt.: 172.2		**CAS #**	10101-41-4
DOT #	N.D.	**DOT Class:**	N.D.

An odorless, white powder that is water soluble. When involved in a fire, use agent suitable for extinguishing surrounding materials. Calcium sulfate may react violently with phosphorous or diazomethane. Contamination may result in irritation and should be treated by rinsing with water for a minimum of 15 minutes. If ingested, prompt medical aid should be obtained. Calcium sulfate is considered a nuisance dust with an OSHA PEL OF 15 mg/m^3 (5 mg/m^3 respirable dust). Level B or C ensembles (depending upon the severity of the spill) should be used to clean up spills. Spill areas and equipment can be decontaminated using water. The anhydrous form (CAS # 7778-18-9) is less water soluble, but otherwise has very similar chemical, physical properties.

CARBON   C

graphite, charcoal

Molecular Wt.: 12.0		**CAS #**	7440-44-0
DOT #	1361, 1362	**DOT Class:**	Flammable Solid

An odorless, black powder. Dust clouds are an explosion hazard. When involved in a fire, use agent suitable for extinguishing surrounding materials. Carbon dust may react violently with oxidants, interhalogens, oxygen, unsaturated hydrocarbons and metals. Contamination may result in irritation and should be treated by rinsing with water for a minimum of 15 minutes. If ingested, prompt medical aid should be obtained. Carbon powder is considered a nuisance dust with an OSHA PEL of 15 mg/m^3 (5 mg/m^3 respirable fraction). Level B or C ensembles (depending upon the severity of the spill) should be used to clean up spills. Avoid raising any dust clouds dusting fire extinguishing or spill clean up. Spill areas and equipment can be decontaminated using soapy water.

CELLULOSE

Molecular Wt.:	N.A.	**CAS #**	9004-34-6
DOT #	N.D.	**DOT Class:**	N.D.

An ordorless, white powder that is insoluble in water. When involved in a fire, use agent suitable for extinguishing surrounding materials. When powdered, cellulose in incompatible with oxidizers. Contamination may result in irritation and should be treated by rinsing with water for a minimum of 15 minutes. If ingested, prompt medical aid should be obtained. Powdered cellulose is considered a nuisance dust with an OSHA PEL of 15 mg/m^3 (5 mg/m^3 respirable fraction). Level B or C ensembles (depending upon the severity of the spill) should be used to clean up spills. Spill areas and equipment can be decontaminated using water.

CELLULOSE ACETATE

Molecular Wt.:	N.A.	**CAS #**	9004-35-7
DOT #	N.D.	**DOT Class:**	N.D.

An odorless, white powder that is insoluble in water. When involved in a fire, use agent suitable for extinguishing surrounding materials. Contamination may result in irritation and should be treated by rinsing with water for a minimum of 15 minutes. If ingested, prompt medical aid should be obtained. Level B or C ensembles (depending upon the severity of the spill) should be used to clean up spills. Spill areas and equipment can be decontaminated using soapy water.

3-CHLOROANISIDINE

3-chloro-4-methoxy-benzenamine; orthochloroparanisidine

Molecular Wt.:	157.6	**CAS #**	5345-54-0
DOT #	N.D.	**DOT Class:**	N.D.

A dark brown solid that is slightly water soluble. When involved in a fire, use agent suitable for extinguishing surrounding materials. Contamination may result in irritation and should be treated by rinsing with water for a minimum of 15 minutes. If ingested, prompt medical aid should be obtained. Absorption may lead to the formation of methemoglobin up to 4 hours after exposure. Level B or C ensembles (depending upon the severity of the spill) should be used to clean up spills. Spill areas and equipment can be decontaminated using soapy water.

4-CHLORO-m-CRESOL

aptal; baktol; p-chloro-m-cresol; chlorocresol; 2-chlorohydroxytoluene; 3-methyl-4-chlorophenol; parmetol

Molecular Wt.:	142.6	**CAS #**	59-50-7
DOT #	2669	**DOT Class:**	N.D.

An odorless, white solid that is moderately water soluble. When involved in a fire, use agent suitable for extinguishing surrounding materials. Firefighter's turnouts may not provide adequate protection when fighting fires involving this material. Contamination may result in irritation and should be treated by rinsing with water for a minimum of 15 minutes. If ingested, prompt medical aid should be obtained. Level B ensemble should be used to clean up spills. Spill areas and equipment can be decontaminated using water. 4-Chloro-m-cresol (RCRA Waste # U039) has a statutory RQ of 1 lb and a final RQ of 5000 lb.

2-CHLORO-4-NITROTOLUENE

Molecular Wt.:	171.6	**CAS #**	121-86-8
DOT #	2238	**DOT Class:**	N.D.

A pale yellow solid that is water insoluble. When involved in a fire, use agent suitable for extinguishing surrounding materials. Isolate for 1/2 mile in all directions if a tank truck, tank or rail car is involved. Water spray should be used to keep closed containers cool. Continue to cool container after fire is extinguished. Contamination may result in irritation and should be treated by rinsing with water for a minimum of 15 minutes. If ingested, prompt medical aid should be obtained. Level B or C ensembles (depending upon the severity of the spill) should be used to clean up spills. Spill areas and equipment can be decontaminated using soapy water.

CHLOROPLATINIC ACID H_2PtCl_6

dihydrogen hexachloroplatinate; hexachloroplatinic acid; platinic chloride

Molecular Wt.:	409.8	**CAS #**	16941-12-1
DOT #	2507	**DOT Class:**	ORM-B

Yellow to brown water soluble crystals. When involved in a fire, use agent suitable for extinguishing surrounding materials. Contamination may result in irritation and should be treated by rinsing with water for a minimum of 15 minutes. If ingested, prompt medical aid should be obtained. Exposure may cause an allergic reaction. Level B or C ensembles (depending upon the severity of the spill) should be used to clean up spills. Spill areas and equipment can be decontaminated using water.

3-CHLOROPROPIONIC ACID $ClCH_2CH_2COOH$

β-propionic acid; β-monochloropropionic acid

Molecular Wt.:	108.5	**CAS #**	107-94-8
DOT #	2511	**DOT Class:**	N.D.

A white powder that is water soluble. When involved in a fire, use agent suitable for extinguishing surrounding materials. Contamination may result in irritation or burns and should be treated by rinsing with water for a minimum of 15 minutes. If ingested, prompt medical aid should be obtained. Exposure may result in chemical pneumonitis or pulmonary edema. Inhalation of vapors may be fatal by causing the glottis to spasm and suffocation. Level B or C ensembles (depending upon the severity of the spill) should be used to clean up spills. Spill areas and equipment can be decontaminated using water.

2-CHLOROPROPIONIC ACID $CH_3CHClCOOH$

α-chloropropionic acid; α-monochloropropionic acid

Molecular Wt.:	108.5	**CAS #**	598-78-7
DOT #	2511	**DOT Class:**	N.D.

A white powder that is water soluble. When involved in a fire, use agent suitable for extinguishing surrounding materials. Contamination may result in irritation or burns and should be treated by rinsing with water for a minimum of 15 minutes. If ingested, prompt medical aid should be obtained. Exposure may result in chemical pneumonitis or pulmonary edema. Inhalation of vapors may be fatal by causing the glottis to spasm and suffocation. Level B or C ensembles (depending upon the severity of the spill) should be used to clean up spills. Spill areas and equipment can be decontaminated using water.

CHLOROTETRAFLUOROETHANE $ClHFCCF_3$

monochlorotetrafluoroethane

Molecular Wt.: 136.5	**CAS #**	63938-10-3
DOT # 1021	**DOT Class:**	Nonflammable Gas

A colorless, water insoluble gas. When involved in a fire, use agent suitable for extinguishing surrounding materials. This gas acts as a simple asphyxiant. Exposure to leaks from a liquefied or pressurized tanks may result in frostbite. Spills may be controlled by stopping leak and allowing vapors to dissipate. Spill areas and equipment can be decontaminated using soapy water.

CHLOROTRIFLUOROETHYLENE $ClFC=CF_2$

1-chloro-1,2,2-trifluoroethene; trifluorovinyl chloride

Molecular Wt.: 116.5	**CAS #**	79-38-9
DOT # 1082	**DOT Class:**	Nonflammable Gas

A colorless, water insoluble gas. When involved in a fire, use agent suitable for extinguishing surrounding materials. This gas acts as a simple asphyxiant. Exposure to leaks from a liquefied or pressurized tanks may result in frostbite. Spills may be controlled by stopping leak and allowing vapors to dissipate. Spill areas and equipment can be decontaminated using soapy water.

CHLOROTRIFLUOROMETHANE ClF_3C

freon 13; halocarbon 13; trifluorochloromethane; trifluoromethyl chloride

Molecular Wt.: 104.5	**CAS #**	75-72-9
DOT # 1022	**DOT Class:**	Nonflammable Gas

A colorless, water insoluble gas. When involved in a fire, use agent suitable for extinguishing surrounding materials. This gas acts as a simple asphyxiant. Exposure to leaks from a liquefied or pressurized tanks may result in frostbite. Spills may be controlled by stopping leak and allowing vapors to dissipate. Spill areas and equipment can be decontaminated using soapy water.

CINNAMALDEHYDE

benzylideneacetaldehyde; cinnimal; phenylacrolein; 3-phenylpropenal

Molecular Wt.: 132.2	**CAS #**	104-55-2
DOT # N.D.	**DOT Class:**	N.D.

A pale yellow, oily liquid that has a strong odor of cinnamon and is slightly soluble in water. When involved in a fire, use agent suitable for extinguishing surrounding materials. Isolate for 1/2 mile if rail or tank truck is involved in a fire. Contamination may result in irritation and should be treated by rinsing with water for a minimum of 15 minutes. If ingested, prompt medical aid should be obtained. Level B or C ensembles (depending upon the severity of the spill) should be used to clean up spills. Ensure that all ignition sources have been eliminated. Spill areas and equipment can be decontaminated using soapy water.

CINNAMIC ACID

phenylacrylic acid; 3-phenyl-2-propenoic acid

Molecular Wt.:	148.2	**CAS #**	621-82-9
DOT #	N.D.	**DOT Class:**	N.D.

A white powder that is slightly soluble in water. When involved in a fire, use agent suitable for extinguishing surrounding materials. Isolate for 1/2 mile if rail or tank truck is involved in a fire. Contamination may result in irritation and should be treated by rinsing with water for a minimum of 15 minutes. If ingested, prompt medical aid should be obtained. Level B or C ensembles (depending upon the severity of the spill) should be used to clean up spills. Ensure that all ignition sources have been eliminated. Spill areas and equipment can be decontaminated using soapy water.

COBALT (II) CHLORIDE $CoCl_2$

cobalt dichloride; cobaltous chloride

Molecular Wt:	129.9	**CAS #**	7646-79-9
DOT #	N.D.	**DOT Class:**	N.D.

A light blue solid that is water soluble. When involved in a fire, use agent suitable for extinguishing surrounding materials. Contamination may result in irritation and should be treated by rinsing with water for a minimum of 15 minutes. If ingested, prompt medical aid should be obtained. Exposure may result in an allergic reaction. Level B or C ensembles (depending upon the severity of the spill) should be used to clean up spills. Spill areas and equipment can be decontaminated using water. Cobalt (II) chloride hexahydrate (CAS #7791-13-1) has similar physical, chemical and toxicological properties.

COPPER CHLORATE $Cu(ClO_3)_2$

cupric chlorate

Molecular Wt:	230.5	**CAS #**	14721-21-2
DOT #	2721	**DOT Class:**	Oxidizer

A light blue, water soluble solid. When involved in a fire, use water only. Cupric chlorate is an oxidizer and should not be allowed to come into contact with combustibles. Contamination may result in irritation or burns and should be treated by rinsing with water for a minimum of 15 minutes. If ingested, prompt medical aid should be obtained. Level B or C ensembles (depending upon the severity of the spill) should be used to clean up spills. Spill areas and equipment can be decontaminated using water.

COPPER CHLORIDE $CuCl_2$

cupric chloride

Molecular Wt:	134.4	**CAS #**	1344-67-8
DOT #	2802	**DOT Class:**	ORM-B

A yellow to brown, hygroscopic powder that is water soluble. When involved in a fire, use agent suitable for extinguishing surrounding materials. Contamination may result in irritation and should be treated by rinsing with water for a minimum of 15 minutes. If ingested, prompt medical aid should be obtained. Level B or C ensembles (depending upon the severity of the spill) should be used to clean up spills. Spill areas and equipment can be decontaminated using water.

COPPER CYANIDE $Cu(CN)_2$

copper (II) cyanide; cupric cyanide

Molecular Wt:	115.6	**CAS #**	14763-77-0
DOT #	1587	**DOT Class:**	Poison B

A yellow-green, water soluble powder. When involved in a fire, use agent suitable for extinguishing surrounding materials. Firefighter's turnout gear will not provide adequate protection when this material is involved in a fire. Copper cyanide is extremely toxic by all routes of entry. Exposures should be treated immediately for cyanide poisoning. Level A or B ensembles (depending upon the severity of the spill) should be used to clean up spills. Spill areas and equipment can be decontaminated using an alkaline hypochlorite solution.

COPRA (OIL)

coconut butter; coconut oil; coconut palm oil; copra pellets

Molecular Wt:	N.A.	**CAS #**	8001-31-8
DOT #	1363	**DOT Class:**	ORM-C

A fatty solid or oil that is water insoluble. When involved in a fire, use agent suitable for extinguishing surrounding materials. Water spray should be used to keep closed containers cool. Continue to cool container after fire is extinguished. Contamination may result in irritation and should be treated by rinsing with water for a minimum of 15 minutes. If ingested, prompt medical aid should be obtained. Level B or C ensembles (depending upon the severity of the spill) should be used to clean up spills. Spill areas and equipment can be decontaminated using soapy water.

CYMENE

cymol; p-isopropyltoluene; p-methyl cumene

Molecular Wt.:	134.2	**CAS #**	99-87-6
DOT #	2046	**DOT Class:**	Flammable Liquid

Colorless to pale yellow liquid that is water soluble. When involved in a fire, use agent suitable for extinguishing surrounding materials. Water spray should be used to keep closed containers cool. Continue to cool container after fire is extinguished. Isolate for 1/2 mile if rail or tank truck is involved in a fire. Contamination may result in irritation and should be treated by rinsing with water for a minimum of 15 minutes. If ingested, prompt medical aid should be obtained. Level B or C ensembles (depending upon the severity of the spill) should be used to clean up spills. Ensure that all ignition sources have been eliminated. Spill areas and equipment can be decontaminated using soapy water.

DIBROMODIFLUOROMETHANE F_2Br_2C

difluorodibromomethane; freon 12-B2; halon 1202

Molecular Wt: 209.8   **CAS #**   75-61-6
DOT #   1941   **DOT Class:**   ORM-A

A colorless, water insoluble liquid that is water soluble. When involved in a fire, use agent suitable for extinguishing surrounding materials. Contamination may result in irritation and should be treated by rinsing with water for a minimum of 15 minutes. If ingested, prompt medical aid should be obtained. The OSHA PEL is 850 mg/m^3. Level B or C ensembles (depending upon the severity of the spill) should be used to clean up spills. Spill areas and equipment can be decontaminated using soapy water.

DICHLORODIFLUOROETHYLENE $C_2Cl_2F_2$

Molecular Wt.: 132.9   **CAS #**   27156-03-2
DOT #   9018   **DOT Class:**   ORM-A

A colorless, water insoluble liquid. When involved in a fire, use agent suitable for extinguishing surrounding materials. Water spray should be used to keep closed containers cool. Continue to cool container after fire is extinguished. Contamination may result in irritation and should be treated by rinsing with water for a minimum of 15 minutes. If ingested, prompt medical aid should be obtained. Level B or C ensembles (depending upon the severity of the spill) should be used to clean up spills. Spill areas and equipment can be decontaminated using soapy water. 1,1-dichloro-2,2-difluoroethylene (CAS # 79-35-6) has similar chemical, physical and toxicological properties.

DICHLORODIFLUOROMETHANE Cl_2F_sC

difluorodichloromethane; electro-cf 12; fluorocarbon 12; freon f-12

Molecular Wt.: 120.9   **CAS #**   75-71-8
DOT #   1028   **DOT Class:**   Nonflammable Gas

A colorless, odorless gas that is water insoluble. When involved in a fire, use agent suitable for extinguishing surrounding materials. Contamination may result in irritation and should be treated by rinsing with water for a minimum of 15 minutes. Exposure to leaks from compressed or liquefied gas tanks may result in frostbite. Spills should be cleaned up by stopping the leak and allowing the vapors to dissipate. Dichlorodiflouromethane has RCRA waste # U075, a statutory RQ of 1 lb and a final RQ of 5000 lb. Spill areas and equipment can be decontaminated using water.

1,2-DICHLOROTETRAFLUOROETHANE CF_2ClCF_2Cl

Molecular Wt.: 170.9   **CAS #**   76-14-2
DOT #   1958   **DOT Class:**   Nonflammable Gas

A colorless gas that is water insoluble. When involved in a fire, use agent suitable for extinguishing surrounding materials. Isolate for 1/2 mile if rail or tank truck is involved in a fire. Water spray should be used to keep closed containers cool. Continue to cool container after fire is extinguished. Immediately withdraw if rising sound from venting device is heard or if fire is causing discoloration to the tank. May react violently with alcohols. Acts as a simple asphyxiant. Exposure to leaks from compressed or liquefied gas tanks may result in frostbite. Spills should be cleaned up by stopping the leak and allowing the vapors to dissipate. Level B or C ensembles (depending upon the severity of the spill) should be used to clean up spills. Spill areas and equipment can be decontaminated using soapy water.

DICUMYL PEROXIDE

comene peroxide; isopropylbenzene peroxide

Molecular Wt.: 270.4   **CAS #**   80-43-3
DOT #   2121   **DOT Class:**   Organic Peroxide

A white solid that is slightly water soluble. When involved in a fire, use agent suitable for extinguishing surrounding materials. Water spray should be used to keep closed containers cool. Continue to cool container after fire is extinguished. Undergoes a self-accelerating decomposition at temperatures above 91 F. Contamination may result in irritation and should be treated by rinsing with water for a minimum of 15 minutes. If ingested, prompt medical aid should be obtained. Level B or C ensembles (depending upon the severity of the spill) should be used to clean up spills. Spill areas and equipment can be decontaminated using soapy water.

DIPROPYL KETONE $CH_3CH_2CH_2C(O)CH_2CH_2CH_3$

butyrone; 4-heptanone

Molecular Wt.: 114.2   **CAS #**   123-19-3
DOT #   2710   **DOT Class:**   Flammable Liquid

A colorless, water insoluble liquid. When involved in a fire, use agent suitable for extinguishing surrounding materials. Contamination may result in irritation and should be treated by rinsing with water for a minimum of 15 minutes. If ingested, prompt medical aid should be obtained. The OSHA PEL is 230 mg/m^3. Level B or C ensembles (depending upon the severity of the spill) should be used to clean up spills. Ensure that all ignition sources have been eliminated. Vapors are heavier than air and may travel some distance to an ignition source. Spill areas and equipment can be decontaminated using soapy water.

ETHANE CH_3CH_3

Molecular Wt.: 30.1   **CAS #**   74-84-0
DOT #   1035, 1961   **DOT Class:**   Flammable Gas

A colorless, odorless, water insoluble gas that is extremely flammable When involved in a fire, use agent suitable for extinguishing surrounding materials. Isolate for 1/2 mile if rail or tank truck is involved in a fire. Water spray from an unmanned device should be used to keep closed containers cool. Continue to cool container after fire is extinguished. Immediately withdraw if rising sound from venting device is heard or if fire is causing discoloration to the tank. For large fires, if possible, withdraw and allow to burn. Level B ensembles should be used to clean up spills. Ensure that all ignition sources have been eliminated. Leaks should be stopped if it can safely be done. Ethane is a simple asphyxiant. Area should be evacuated until vapors dissipate. Spill areas and equipment can be decontaminated using soapy water.

ETHOXYETHANOL $CH_3CH_2OCH_2CH_2OH$

cellosolve; ethyl cellosolve; ethylene glycol nonoethyl ether; monoethyl ether; hydroxyether

Molecular Wt.: 90.1		**CAS #**	110-80-5
DOT # 1171		**DOT Class:**	Combustible Liquid.

A colorless, almost odorless water soluble liquid. When involved in a fire, use agent suitable for extinguishing surrounding materials. May form unstable peroxides upon prolonged exposure to air. Contamination may result in irritation and should be treated by rinsing with water for a minimum of 15 minutes. If ingested, prompt medical aid should be obtained. The OSHA PEL is 750 mg/m^3 (skin). Level B or C ensembles (depending upon the severity of the spill) should be used to clean up spills. Spill areas and equipment can be decontaminated using water.

ETHYL AMYL KETONE $CH_3CH_2CH(CH_3)CH_2C(O)CH_2CH_3$

amyl ethyl ketone; 3-methyl-5-heptanone; 5-methyl-3-heptanone

Molecular Wt.: 128.2		**CAS #**	541-85-5
DOT # 2271		**DOT Class:**	Flammable Liquid

A colorless, water insoluble liquid with a fruity odor. When involved in a fire, use agent suitable for extinguishing surrounding materials. When involved in a fire, use agent suitable for extinguishing surrounding materials. Contamination may result in irritation and should be treated by rinsing with water for a minimum of 15 minutes. If ingested, prompt medical aid should be obtained. May act as a narcotic in high concentrations. The OSHA PEL is 130 mg/m^3. Level B or C ensembles (depending upon the severity of the spill) should be used to clean up spills. Ensure that all ignition sources have been eliminated. Vapors are heavier than air and may travel some distance to an ignition source. Spill areas and equipment can be decontaminated using soapy water.

2-ETHYLBUTANOL $(CH_3CH_2)_2CCH_2OH$

2-ethyl-1-butanol; 2-ethyl butyl alcohol; sec-hexanol

Molecular Wt.: 102.2		**CAS #**	97-95-0
DOT # 2275, 2282		**DOT Class:**	Flammable Liquid

A colorless liquid that is slightly water soluble. When involved in a fire, use agent suitable for extinguishing surrounding materials. Isolate for 1/2 mile if rail or tank truck is involved in a fire. Water spray from an unmanned device should be used to keep closed containers cool. Continue to cool container after fire is extinguished. Immediately withdraw if rising sound from venting device is heard or if fire is causing discoloration to the tank. Contamination may result in irritation and should be treated by rinsing with water for a minimum of 15 minutes. If ingested, prompt medical aid should be obtained. Level B or C ensembles (depending upon the severity of the spill) should be used to clean up spills. Ensure that all ignition sources have been eliminated. Vapors are heavier than air and may travel some distance to an ignition source. Spill areas and equipment can be decontaminated using soapy water.

ETHYL CAPROATE $CH_3(CH_2)_5CH_2C(O)CH_2CH_3$

ethyl butylacetate; ethyl hexanoate

Molecular Wt.:	144.2	**CAS #**	123-66-0
DOT #	1177	**DOT Class:**	Flammable Liquid

A colorless, water insoluble liquid that has an wine odor. When involved in a fire, use agent suitable for extinguishing surrounding materials. Isolate for 1/2 mile if rail or tank truck is involved in a fire. Water spray from an unmanned device should be used to keep closed containers cool. Continue to cool container after fire is extinguished. For large fires, if possible, withdraw and allow to burn. Immediately withdraw if rising sound from venting device is heard or if fire is causing discoloration to the tank. Contamination may result in irritation and should be treated by rinsing with water for a minimum of 15 minutes. If ingested, prompt medical aid should be obtained. Level B or C ensembles (depending upon the severity of the spill) should be used to clean up spills. Ensure that all ignition sources have been eliminated. Vapors are heavier than air and may travel some distance to an ignition source. Spill areas and equipment can be decontaminated using soapy water.

ETHYL CYANOACETATE $CH_3CH_2OC(O)CH_2C{\equiv}N$

cyanoacetic acid, ethyl ester; cyanoacetic ester

Molecular Wt.:	113.1	**CAS #**	105-56-6
DOT #	2666	**DOT Class:**	Poison B

A colorless to tan colored liquid that is water insoluble. When involved in a fire, use agent suitable for extinguishing surrounding materials. Structural fire fighter protective clothing will not provide adequate protection. Fight fire from a distance or protected location, if possible. Treat all materials used or generated and equipment involved as contaminated by hazardous waste. Contamination may result in irritation and should be treated by rinsing with water for a minimum of 15 minutes. If ingested, prompt medical aid should be obtained. Level B or C ensembles (depending upon the severity of the spill) should be used to clean up spills. Spill areas and equipment can be decontaminated using an alkaline hypochlorate solution.

ETHYL LACTATE $CH_3CH(OH)C(O)OCH_2CH_3$

ethyl 2-hydroxypropionate; lactic acid, ethyl ester

Molecular Wt.:	118.2	**CAS #**	97-64-3
DOT #	1192	**DOT Class:**	Flammable Liquid

A colorless, water soluble liquid. When involved in a fire, use agent suitable for extinguishing surrounding materials. Isolate for 1/2 mile if rail or tank truck is involved in a fire. Water spray from an unmanned device should be used to keep closed containers cool. Continue to cool container after fire is extinguished. For large fires, if possible, withdraw and allow to burn. Immediately withdraw if rising sound from venting device is heard or if fire is causing discoloration to the tank. Contamination may result in irritation and should be treated by rinsing with water for a minimum of 15 minutes. If ingested, prompt medical aid should be obtained. Level B or C ensembles (depending upon the severity of the spill) should be used to clean up spills. Ensure that all ignition sources have been eliminated. Vapors are heavier than air and may travel some distance to an ignition source. Spill areas and equipment can be decontaminated using soapy water.

FERRIC NITRATE NONAHYDRATE $Fe(NO_3)_3 \cdot 9H_2O$

iron (III) nitrate nonahydrate

Molecular Wt.: 404.0   **CAS #**   7782-61-6
DOT #   1466   **DOT Class:**   Oxidizer

White to pink, water soluble solid. When involved in a fire, use agent suitable for extinguishing surrounding materials. Ferric nitrate is not compatible with ordinary combustibles or organics. Contamination may result in irritation and should be treated by rinsing with water for a minimum of 15 minutes. If ingested, prompt medical aid should be obtained. The OSHA PEL is 1 mg $(Fe)/m^3$. Level B or C ensembles (depending upon the severity of the spill) should be used to clean up spills. Spill areas and equipment can be decontaminated using water.

FLUOROETHANE CH_3CH_2F

ethyl fluoride; monofluoroethane

Molecular Wt.: 48.1   **CAS #**   353-36-6
DOT #   2453   **DOT Class:**   Flammable Gas

A colorless, odorless, water insoluble gas that is extremely flammable When involved in a fire, use agent suitable for extinguishing surrounding materials. Isolate for 1/2 mile if rail or tank truck is involved in a fire. Water spray from an unmanned device should be used to keep closed containers cool. Continue to cool container after fire is extinguished. Immediately withdraw if rising sound from venting device is heard or if fire is causing discoloration to the tank. For large fires, if possible, withdraw and allow to burn. Ensure that all ignition sources have been eliminated. Level B ensembles should be used to clean up spills. Leaks should be stopped if it can safely be done. Ethane is a simple asphyxiant. Area should be evacuated until vapors dissipate. Spill areas and equipment can be decontaminated using soapy water.

FLUOROPHOSPHORIC ACID FH_2PO_3

phosphofluoridic acid

Molecular Wt.: 100.0   **CAS #**   13537-32-1
DOT #   1776   **DOT Class:**   Corrosive Material

A colorless, corrosive, water soluble liquid. When involved in a fire, use agent suitable for extinguishing surrounding materials. Structural fire fighter protective clothing will not provide adequate protection. Contamination may result in burns or irritation and should be treated by rinsing with water for a minimum of 15 minutes. If ingested, prompt medical aid should be obtained. Level B or A ensembles (depending upon the severity of the spill) should be used to clean up spills. Spill areas and equipment can be decontaminated using a dilute alkaline solution.

FUEL OIL

diesel fuel

Molecular Wt.:	N.A.	**CAS #**	N.A.
DOT #	1201	**DOT Class:**	Combustible Liquid.

A colorless to brown liquid that is water insoluble. When involved in a fire, use agent suitable for extinguishing surrounding materials. Isolate for 1/2 mile if rail or tank truck is involved in a fire. Water spray from an unmanned device should be used to keep closed containers cool. Continue to cool container after fire is extinguished. For large fires, if possible, withdraw and allow to burn. Immediately withdraw if rising sound from venting device is heard or if fire is causing discoloration to the tank. Contamination may result in irritation and should be treated by rinsing with water for a minimum of 15 minutes. If ingested, prompt medical aid should be obtained. Level B or C ensembles (depending upon the severity of the spill) should be used to clean up spills. Ensure that all ignition sources have been eliminated. Vapors are heavier than air and may travel some distance to an ignition source. Spill areas and equipment can be decontaminated using soapy water.

FURFURYLAMINE

2-furanmethylamine

Molecular Wt.:	97.1	**CAS #**	617-89-0
DOT #	2526	**DOT Class:**	Flammable Liquid

A light tan colored liquid that is water soluble. When involved in a fire, use agent suitable for extinguishing surrounding materials. Structural fire fighter protective clothing will not provide adequate protection. Isolate for 1/2 mile if rail or tank truck is involved in a fire. Water spray from an unmanned device should be used to keep closed containers cool. Continue to cool container after fire is extinguished. Immediately withdraw if rising sound from venting device is heard or if fire is causing discoloration to the tank. Treat all materials used or generated and equipment involved as contaminated by hazardous waste. Contamination may result in burns or irritation and should be treated by rinsing with water for a minimum of 15 minutes. If ingested, prompt medical aid should be obtained. Level B or A ensembles (depending upon the severity of the spill) should be used to clean up spills. Ensure that all ignition sources have been eliminated. Vapors are heavier than air and may travel some distance to an ignition source. Spill areas and equipment can be decontaminated using soapy water.

HELIUM   He

A colorless, odorless that is water insoluble. When involved in a fire, use agent suitable for extinguishing surrounding materials. Helium acts as a simple asphyxiant. Helium leaking from a compressed or cryogenic storage tank may cause frostbite. Level B or C ensembles (depending upon the severity of the spill) should be used to clean up spills.

HEXADECYLTRICHLOROSILANE $CH_3(CH_2)_{14}CH_2SiCl_3$

trichlorohexadecylsilane

Molecular Wt.: 359.9   **CAS #**   5894-60-0
DOT #   1781   **DOT Class:**   Corrosive Material

A colorless to yellow liquid that decomposes in water. When involved in a fire, use agent suitable for extinguishing surrounding materials. Water spray should be used to keep closed containers cool. Continue to cool container after fire is extinguished. Contamination may result in burns or irritation and should be treated by rinsing with water for a minimum of 15 minutes. If ingested, prompt medical aid should be obtained. Level B or C ensembles (depending upon the severity of the spill) should be used to clean up spills. Spill areas and equipment can be decontaminated using an alkaline solution.

LEAD ACETATE $(CH_3C(O)O)_2Pb$

Molecular Wt.: 325.3   **CAS #**   301-04-2
DOT #   1616   **DOT Class:**   N.D.

A white, water soluble solid. When involved in a fire, use agent suitable for extinguishing surrounding materials. Contamination may result in irritation and should be treated by rinsing with water for a minimum of 15 minutes. If ingested, prompt medical aid should be obtained. Lead acetate is a confirmed carcinogen. The OSHA TWA is 0.05 mg (Pb)/m^3. Level B or A ensembles (depending upon the severity of the spill) should be used to clean up spills. Spill areas and equipment can be decontaminated using water. Treat all materials used or generated and equipment involved as contaminated by hazardous waste. Lead acetate trihydrate (CAS #6080-56-4) has similar chemical, physical and toxicological properties.

LEAD FLUOBORATE $Pb(BF_4)_2$

Molecular Wt.: 380.8   **CAS #**   13814-96-5
DOT #   2291   **DOT Class:**   ORM-B

A colorless, water soluble solid. When involved in a fire, use agent suitable for extinguishing surrounding materials. Contamination may result in irritation and should be treated by rinsing with water for a minimum of 15 minutes. If ingested, prompt medical aid should be obtained. The OSHA TWA is 0.05 mg (Pb)/m^3. Level B or C ensembles (depending upon the severity of the spill) should be used to clean up spills. Spill areas and equipment can be decontaminated using water.

LEAD (II) FLUORIDE PbF_2

lead difluoride; lead fluoride

Molecular Wt.: 245.2   **CAS #**   7783-46-2
DOT #   2811   **DOT Class:**   OMB-B

A white solid that is marginally soluble in water. When involved in a fire, use agent suitable for extinguishing surrounding materials. Lead fluoride in not compatible with fluorine or calcium carbide. Contamination may result in irritation and should be treated by rinsing with water for a minimum of 15 minutes. If ingested, prompt medical aid should be obtained. The OSHA TWA is 0.05 mg (Pb)/m^3 and 0.15 mg (F)/m^3. Level B or C ensembles (depending upon the severity of the spill) should be used to clean up spills. Spill areas and equipment can be decontaminated using water.

LITHIUM HYPOCHLORITE   LiClO

Molecular Wt.:	58.4	**CAS #**	13840
DOT #	1471	**DOT Class:**	Oxidizer

A white, water soluble solid. When involved in a fire, use water only. Water spray should be used to keep closed containers cool. Continue to cool container after fire is extinguished. For large fires, if possible, withdraw and allow to burn. Lithium hypochlorite is an oxidizer that is incompatible with ordinary combustibles and organics. Contamination may result in irritation and should be treated by rinsing with water for a minimum of 15 minutes. If ingested, prompt medical aid should be obtained. Level B or C ensembles (depending upon the severity of the spill) should be used to clean up spills. Spill areas and equipment can be decontaminated using water.

LITHIUM NITRATE $LiNO_3$

Molecular Wt.:	68.9	**CAS #**	7790-69-4
DOT #	2722	**DOT Class:**	Oxidizer

A white water soluble solid. When involved in a fire, use water only. Water spray should be used to keep closed containers cool. Continue to cool container after fire is extinguished. For large fires, if possible, withdraw and allow to burn. Lithium nitrate is an oxidizer that is incompatible with ordinary combustibles and organics. Contamination may result in irritation and should be treated by rinsing with water for a minimum of 15 minutes. If ingested, prompt medical aid should be obtained. Level B or C ensembles (depending upon the severity of the spill) should be used to clean up spills. Spill areas and equipment can be decontaminated using water.

LITHIUM PEROXIDE Li_2O_2

Molecular Wt.:	45.9	**CAS #**	12031-80-0
DOT #	1472	**DOT Class:**	Oxidizer

A white to yellow solid that decomposes upon exposure to water. When involved in a fire, use water only. Isolate for 1/2 mile if rail or tank truck is involved in a fire. Lithium hydride will react with ordinary combustibles, organics and reducing agents. Contamination may result in irritation and should be treated by rinsing with water for a minimum of 15 minutes. If ingested, prompt medical aid should be obtained. Level B or C ensembles (depending upon the severity of the spill) should be used to clean up spills. Spill areas and equipment can be decontaminated using water.

MALEIC ACID   HOOCH=HCOOH

cis-butenoic acid; cis-ethylenedicarboxylic acid; toxilic acid

Molecular Wt.:	116.1	**CAS #**	110-16-7
DOT #	2215	**DOT Class:**	ORM-A

A white, water soluble solid. When involved in a fire, use agent suitable for extinguishing surrounding materials. Water spray should be used to keep closed containers cool. Continue to cool container after fire is extinguished. Contamination may result in irritation and should be treated by rinsing with water for a minimum of 15 minutes. If ingested, prompt medical aid should be obtained. Level B or C ensembles (depending upon the severity of the spill) should be used to clean up spills. Spill areas and equipment can be decontaminated using water.

METHANE   CH$_4$

marsh gas

Molecular Wt.: 16.1		**CAS #**	74-82-8
DOT # 1971, 1972		**DOT Class:**	Flammable Gas

A colorless, odorless, water insoluble gas that is extremely flammable When involved in a fire, use agent suitable for extinguishing surrounding materials. Isolate for 1/2 mile if rail or tank truck is involved in a fire. Water spray from an unmanned device should be used to keep closed containers cool. Continue to cool container after fire is extinguished. Immediately withdraw if rising sound from venting device is heard or if fire is causing discoloration to the tank. For large fires, if possible, withdraw and allow to burn. Level B ensembles should be used to clean up spills. Ensure that all ignition sources have been eliminated. Leaks should be stopped if it can safely be done. Ethane is a simple asphyxiant. Area should be evacuated until vapors dissipate. Spill areas and equipment can be decontaminated using soapy water.

4-METHOXY-4-METHYL-2-PENTANONE

Molecular Wt.: 130.2		**CAS #**	107-70-0
DOT # 2293		**DOT Class:**	Flammable Liquid

A colorless, water insoluble liquid. When involved in a fire, use agent suitable for extinguishing surrounding materials. Isolate for 1/2 mile if rail or tank truck is involved in a fire. Water spray from an unmanned device should be used to keep closed containers cool. Continue to cool container after fire is extinguished. For large fires, if possible, withdraw and allow to burn. Immediately withdraw if rising sound from venting device is heard or if fire is causing discoloration to the tank. Contamination may result in irritation and should be treated by rinsing with water for a minimum of 15 minutes. If ingested, prompt medical aid should be obtained. Level B or A ensembles (depending upon the severity of the spill) should be used to clean up spills. Spill areas and equipment can be decontaminated using soapy water.

1-METHOXY-2-PROPANOL   CH$_3$OCH$_3$CH(OH)CH$_3$

propylene glycol, monomethyl ether

Molecular Wt.: 90.1		**CAS #**	107-98-2
DOT # 3092		**DOT Class:**	N.D.

A colorless liquid that is moderately soluble in water. When involved in a fire, use agent suitable for extinguishing surrounding materials. Isolate for 1/2 mile if rail or tank truck is involved in a fire. Water spray from an unmanned device should be used to keep closed containers cool. Continue to cool container after fire is extinguished. For large fires, if possible, withdraw and allow to burn. Immediately withdraw if rising sound from venting device is heard or if fire is causing discoloration to the tank. May form unstable peroxides upon prolonged exposure to air. Contamination may result in irritation and should be treated by rinsing with water for a minimum of 15 minutes. If ingested, prompt medical aid should be obtained. The OSHA TWA is 375 mg/m^3 and STEL is 560 mg/m^3. Level B or A ensembles (depending upon the severity of the spill) should be used to clean up spills. Ensure that all ignition sources have been eliminated. Vapors are heavier than air and may travel some distance to an ignition source. Spill areas and equipment can be decontaminated using soapy water.

METHYL-n-AMYL KETONE $CH_3C(O)CH_2CH_2CH_2CH_2CH_3$

n-amyl methyl ketone; amyl methyl ketone; methyl amyl ketone; methyl pentyl ketone

Molecular Wt.: 114.2	**CAS #**	110-43-0	
DOT # 1110	**DOT Class:**	Flammable Liquid	

A colorless, fruity smelling liquid that is slightly water soluble. When involved in a fire, use agent suitable for extinguishing surrounding materials. Isolate for 1/2 mile if rail or tank truck is involved in a fire. Water spray from an unmanned device should be used to keep closed containers cool. Continue to cool container after fire is extinguished. For large fires, if possible, withdraw and allow to burn. Immediately withdraw if rising sound from venting device is heard or if fire is causing discoloration to the tank. Contamination may result in irritation and should be treated by rinsing with water for a minimum of 15 minutes. If ingested, prompt medical aid should be obtained. Level B or C ensembles (depending upon the severity of the spill) should be used to clean up spills. Spill areas and equipment can be decontaminated using soapy water.

METHYL BENZOATE

methyl benzenecarboxylate; niobe oil

Molecular Wt.: 136.2	**CAS #**	93-53-3	
DOT # 2938	**DOT Class:**	Poison B	

A colorless, water insoluble liquid with a pleasant, fragrant odor. When involved in a fire, use agent suitable for extinguishing surrounding materials. Contamination may result in irritation and should be treated by rinsing with water for a minimum of 15 minutes. If ingested, prompt medical aid should be obtained. Level B or C ensembles (depending upon the severity of the spill) should be used to clean up spills. Spill areas and equipment can be decontaminated using soapy water.

α-ETHYLBENZYL ALCOHOL

1-phenol ethanol; α-phenethyl alcohol; methyl phenyl methanol; stryallyl alcohol

Molecular Wt.: 122.2	**CAS #**	98-85-1	
DOT # 2937	**DOT Class:**	Poison B	

A colorless liquid that is slightly water soluble. When involved in a fire, use agent suitable for extinguishing surrounding materials. Structural fire fighter protective clothing will not provide adequate protection. Fight fire from a distance or protected location, if possible. Treat all materials used or generated and equipment involved as contaminated by hazardous waste. Contamination may result in irritation and should be treated by rinsing with water for a minimum of 15 minutes. If ingested, prompt medical aid should be obtained. Level B or C ensembles (depending upon the severity of the spill) should be used to clean up spills. Spill areas and equipment can be decontaminated using soapy water.

METHYLCYCLOHEXANOL

A colorless, viscous, slightly water soluble liquid that a menthol-like odor. When involved in a fire, use agent suitable for extinguishing surrounding materials. Isolate for 1/2 mile if rail or tank truck is involved in a fire. Water spray should be used to keep closed containers cool. Continue to cool container after fire is extinguished. For large fires, if possible, withdraw and allow to burn. Immediately withdraw if rising sound from venting device is heard or if fire is causing discoloration to the tank. Contamination may result in irritation and should be treated by rinsing with water for a minimum of 15 minutes. If ingested, prompt medical aid should be obtained. The OSHA TWA is 235 mg/m^3. Level B or C ensembles (depending upon the severity of the spill) should be used to clean up spills. Spill areas and equipment can be decontaminated using soapy water.

METHYL FLUORIDE $\qquad$ CH$_3$F

fluoromethane

Molecular Wt.:	34.0	**CAS #**	593-53-3
DOT #	2454	**DOT Class:**	Flammable Gas

A colorless, water insoluble gas that has a ether-like odor. When involved in a fire, use agent suitable for extinguishing surrounding materials. Isolate for 1/2 mile if rail or tank truck is involved in a fire. Water spray from an unmanned device should be used to keep closed containers cool. Continue to cool container after fire is extinguished. For large fires, if possible, withdraw and allow to burn. Immediately withdraw if rising sound from venting device is heard or if fire is causing discoloration to the tank. Contamination may result in irritation and should be treated by rinsing with water for a minimum of 15 minutes. If ingested, prompt medical aid should be obtained. Level B or C ensembles (depending upon the severity of the spill) should be used to clean up spills. Ensure that all ignition sources have been eliminated. leaks should be stopped if it can safely be done. Keep area isolated until vapors dissipate. Spill areas and equipment can be decontaminated using soapy water.

METHYL ISOAMYL KETONE $\qquad$ (CH$_3$)$_2$CH$_2$CH$_2$CH$_2$C(O)CH$_3$

2-methyl-5-hexanone; 5 methyl-2-hexanone

Molecular Wt.:	114.2	**CAS #**	110-12-3
DOT #	2302	**DOT Class:**	Flammable Liquid

A colorless liquid that is slightly water soluble. When involved in a fire, use agent suitable for extinguishing surrounding materials. Isolate for 1/2 mile if rail or tank truck is involved in a fire. Water spray from an unmanned device should be used to keep closed containers cool. Continue to cool container after fire is extinguished. For large fires, if possible, withdraw and allow to burn. Immediately withdraw if rising sound from venting device is heard or if fire is causing discoloration to the tank. Contamination may result in irritation and should be treated by rinsing with water for a minimum of 15 minutes. If ingested, prompt medical aid should be obtained. Level B or C ensembles (depending upon the severity of the spill) should be used to clean up spills. Spill areas and equipment can be decontaminated using soapy water.

NEON $\qquad$ Ne

Molecular Wt.:	20.1	**CAS #**	7440-01-9
DOT #	1065, 1913	**DOT Class:**	Nonflammable Gas

A colorless, odorless that is water insoluble. When involved in a fire, use agent suitable for extinguishing surrounding materials. Neon acts as a simple asphyxiant. Neon leaking from a compressed or cryogenic storage tank may cause frostbite. Level B or C ensembles (depending upon the severity of the spill) should be used to clean up spills.

2-NITROCRESOL

4-methyl-2-nitrophenol

Molecular Wt.:	153.2	**CAS #**	119-33-5
DOT #	2446	**DOT Class:**	N.D.

A pale yellow, slightly water soluble material that is a liquid above 32 C (90 F). When involved in a fire, use agent suitable for extinguishing surrounding materials. Structural fire fighter protective clothing will not provide adequate protection. Fight fire from a distance or protected location, if possible. Treat all materials used or generated and equipment involved as contaminated by hazardous waste. Contamination may result in irritation and should be treated by rinsing with water for a minimum of 15 minutes. If ingested, prompt medical aid should be obtained. Level B or C ensembles (depending upon the severity of the spill) should be used to clean up spills. Spill areas and equipment can be decontaminated using soapy water.

NITROGEN

Molecular Wt.:	28.0	**CAS #**	7727-37-9
DOT #	1066, 1977	**DOT Class:**	Nonflammable Gas

A colorless, odorless that is water insoluble. When involved in a fire, use agent suitable for extinguishing surrounding materials. Nitrogen acts as a simple asphyxiant. Nitrogen leaking from a compressed or cryogenic storage tank may cause frostbite. Level B or C ensembles (depending upon the severity of the spill) should be used to clean up spills.

POTASSIUM BISULFATE $HKSO_4$

Acid potassium sulfate; monopotassium sulfate; potassium acid sulfate; potassium hydrogen sulfate; sulfuric acid, monopotassium salt

Molecular Wt.:	136.2	**CAS #**	7646-93-7
DOT #	2509	**DOT Class:**	ORM-B

A white, odorless water soluble solid. When involved in a fire, use agent suitable for extinguishing surrounding materials. Contamination may result in irritation and should be treated by rinsing with water for a minimum of 15 minutes. If ingested, prompt medical aid should be obtained. Level B or C ensembles (depending upon the severity of the spill) should be used to clean up spills. Spill areas and equipment can be decontaminated using water. Sodium bisulfate (CAS # 7631-90-5, DOT # 2693) has similar chemical, physical and toxicological properties.

SULFUR HEXAFLUORIDE SF_6

Molecular Wt.:	146.1	**CAS #**	2551-62-4
DOT #	1080	**DOT Class:**	Nonflammable Gas

A colorless gas that may contain toxic or reactive impurities. When involved in a fire, use agent suitable for extinguishing surrounding materials. Isolate for 1/2 mile if rail or tank truck is involved in a fire. Water spray should be used to keep closed containers cool. Continue to cool container after fire is extinguished. Immediately withdraw if rising sound from venting device is heard or if fire is causing discoloration to the tank. Incompatible with disilane. Contamination may result in irritation and should be treated by rinsing with water for a minimum of 15 minutes. If ingested, prompt medical aid should be obtained. The OSHA TWA is 6000 mg/m^3. Level B or C ensembles (depending upon the severity of the spill) should be used to clean up spills. Leaks should be stopped only if it can safely be done. Spill areas and equipment can be decontaminated using soapy water.

TETRAHYDROHYDROPHTHALIC ACID ANHYDRIDE

Molecular Wt.:	152.2	**CAS #**	85-43-8
DOT #	2698	**DOT Class:**	Corrosive Material

A white odorless powder that decomposes in water. When involved in a fire, use agent suitable for extinguishing surrounding materials. Water spray should be used to keep closed containers cool. Continue to cool container after fire is extinguished. Contamination may result in irritation and should be treated by rinsing with water for a minimum of 15 minutes. If ingested, prompt medical aid should be obtained. Level B or C ensembles (depending upon the severity of the spill) should be used to clean up spills. Spill areas and equipment can be decontaminated using water.

UNDECANE $CH_3(CH_2)_9CH_3$

Molecular Wt.:	156.4	**CAS #**	1120-21-4
DOT #	2330	**DOT Class:**	Flammable Liquid

A colorless, water insoluble liquid. When involved in a fire, use agent suitable for extinguishing surrounding materials. Isolate for 1/2 mile if rail or tank truck is involved in a fire. Water spray from an unmanned device should be used to keep closed containers cool. Continue to cool container after fire is extinguished. For large fires, if possible, withdraw and allow to burn. Immediately withdraw if rising sound from venting device is heard or if fire is causing discoloration to the tank. Contamination may result in irritation and should be treated by rinsing with water for a minimum of 15 minutes. If ingested, prompt medical aid should be obtained. Level B or C ensembles (depending upon the severity of the spill) should be used to clean up spills. Spill areas and equipment can be decontaminated using soapy water.

XENON   Xe

A colorless, odorless that is water insoluble. When involved in a fire, use agent suitable for extinguishing surrounding materials. Xenon acts as a simple asphyxiant. Xenon leaking from a compressed or cryogenic storage tank may cause frostbite. Level B or C ensembles (depending upon the severity of the spill) should be used to clean up spills.

108-45-2, 500	110-87-2, 273	124-40-3, 260, 279
108-46-3, 547	110-89-4, 520	124-41-4, 573, 574
108-60-1, 245	110-91-8, 458	124-63-0, 413
108-70-3, 624	110-96-3, 260	124-65-2, 563
108-77-0, 212	111-14-8, 353	126-98-7, 411
108-83-8, 274	111-15-9, 304	126-99-8, 190
108-86-1, 113	111-27-3, 354	127-18-4, 493
108-87-2, 433	111-36-4, 131	127-20-8, 250
108-88-3, 616	111-40-0, 268	131-52-2, 579
108-89-4, 515	111-42-2, 258	131-74-8, 52
108-90-7, 113	111-44-4, 241	134-32-7, 461
108-91-8, 216	111-65-9, 481	135-98-8, 128
108-93-0, 214	111-69-3, 22	140-29-4, 499
108-94-1, 215	111-71-7, 350	140-31-8, 38
108-95-2, 498	111-84-2, 481	140-88-5, 305
108-98-5, 503	111-92-2, 236, 260	141-32-3, 125
108-99-6, 515	112-24-3, 630	141-43-5, 302
109-02-4, 447	112-57-2, 602	141-57-1, 544
109-06-8, 515	115-07-1, 540	141-78-6, 304
109-09-1, 193	115-10-6, 269, 286	141-79-7, 408
109-52-4, 353	115-21-9, 327	141-93-5, 265
109-60-4, 304, 377	116-14-3, 604	142-28-9, 541
109-61-5, 539	116-16-5, 343	142-62-1, 353
109-65-9, 114	116-54-1, 436	142-82-5, 352
109-66-0, 352	118-74-1, 344	142-84-7, 260
109-67-1, 64	119-33-5, 679	142-96-1, 130
109-69-3, 115, 129	119-53-9, 660	143-33-9, 524, 565
109-73-9, 126	120-92-3, 221	144-62-7, 484
109-77-3, 402	121-14-2, 296	151-50-8, 524, 565
109-86-4, 317	121-43-7, 634	151-56-4, 318
109-87-5, 417	121-44-8, 629	156-59-2, 244
109-89-7, 260	121-45-9, 636	156-60-5, 244
109-90-0, 322	121-69-7, 281	260-94-6, 17
109-93-3, 298, 648	121-73-3, 186	287-23-0, 122
109-94-4, 320	121-75-5, 400	287-92-3, 220
109-95-5, 324	121-82-4, 217	298-00-0, 448
109-99-9, 454	121-86-8, 664	301-04-2, 674
110-00-9, 336	122-86-4, 124	302-01-2, 356
110-12-3, 678	123-00-2, 40	302-17-0, 169
110-16-7, 675	123-19-3, 669	309-00-2, 24
110-43-0, 677	123-30-8, 39	329-71-5, 295
110-49-6, 317	123-38-6, 534	333-41-5, 231
110-54-3, 352	123-42-2, 227	348-54-9, 329
110-58-7, 62	123-54-6, 14	352-32-9, 332
110-62-3, 56, 640	123-62-6, 536	352-70-5, 332
110-66-7, 59	123-63-7, 487	353-36-6, 672
110-68-9, 425	123-66-0, 671	353-50-4, 162
110-71-4, 278	123-72-8, 127	371-40-4, 329
110-80-5, 670	123-91-1, 297	372-19-0, 329
110-82-7, 213	124-02-7, 229	431-03-8, 228
110-83-8, 213	124-04-9, 659	460-19-5, 209
110-85-0, 519	124-18-5, 481	462-06-6, 330
110-86-1, 545	124-38-4, 157	462-08-8, 41

1616, 674	1733, 69, 71	1828, 591, 592
1617, 73, 386	1736, 95	1829, 595
1618, 386	1737, 97	1830, 594
1622, 73	1738, 98	1831, 594
1625, 406	1741, 106	1832, 594
1627, 406	1744, 108	1833, 596
1629, 403	1745, 109	1834, 597
1636, 405	1746, 110	1835, 606
1648, 9	1748, 143	1838, 614
1649, 389	1749, 176	1839, 622
1650, 461	1750, 179	1840, 654
1651, 462	1751, 179	1841, 659
1654, 465	1752, 182	1845, 157
1656, 465	1754, 194	1846, 161
1657, 465	1755, 196	1847, 530
1658, 465	1759, 328, 585	1848, 535
1659, 465	1760, 37, 40, 250, 328, 353,	1849, 530
1660, 468	458, 466	1854, 78
1661, 469	1764, 237	1859, 556
1662, 470	1765, 238	1860, 647
1663, 476	1766, 248	1865, 384
1664, 479	1767, 266	1868, 225
1669, 491	1773, 328	1869, 396
1670, 494	1776, 672	1870, 562
1671, 498	1777, 331	1873, 492
1673, 500	1779, 334	1884, 83
1680, 524, 565	1780, 335	1885, 88
1684, 524	1781, 674	1888, 185
1687, 560	1782, 349	1889, 210
1688, 563	1787, 366	1891, 311
1689, 524, 565	1788, 357	1897, 493
1690, 566	1789, 357	1905, 550
1692, 587	1790, 362	1908, 564
1695, 180	1791, 572	1910, 145
1697, 181	1792, 369	1911, 232
1702, 601	1802, 492	1913, 678
1706, 353	1805, 506	1915, 215
1708, 620	1806, 510	1916, 241
1709, 617	1807, 512	1917, 305
1710, 625	1808, 513	1918, 207
1711, 652	1809, 513	1919, 416
1714, 656	1810, 509	1920, 481
1715, 5	1811, 527	1921, 542
1716, 12	1812, 566	1927, 419
1717, 13	1813, 571	1929, 570
1722, 29	1815, 538	1930, 631
1724, 32	1816, 544	1938, 111
1725, 34	1818, 555	1941, 668
1726, 34	1821, 561	1942, 48
1727, 44	1823, 571	1958, 668
1729, 68	1824, 571	1961, 531, 669
1730, 70	1825, 575	1962, 313
1731, 70	1827, 583, 612	1966, 358

2802, 666
2805, 394
2806, 395
2809, 404
2811, 674
2815, 38
2817, 44
2820, 132
2821, 498
2822, 193
2823, 205
2831, 431
2833, 25
2837, 561
2839, 23
2841, 229, 322
2842, 472
2845, 272
2858, 657
2859, 53
2862, 643
2869, 615
2872, 233
2874, 338
2876, 547
2878, 613
2879, 553
2880, 143
2922, 569
2923, 569
2937, 677
2938, 677
2941, 329
2944, 329
2945, 425
2949, 569
2950, 396
2967, 589
3023, 482
3056, 350
3078, 164
3079, 411
3083, 495
3092, 676
9011, 149
9018, 668
9037, 347
9069, 607
9077, 659
9084, 660
9190, 50
9191, 175
9192, 329

9202, 159
9264, 252
9265, 413
9266, 451
9269, 632

SYMPTOMS

GLOSSARY

Acids: Materials that will decrease the pH of a solution.

Acid Chlorides: A water reactive organic acid where the hydroxyl group (OH) is replaced by a chloride (Cl). Example: propionyl chloride [$CH_3CH_2C(O)Cl$].

Alcohols: Hydrocarbons containing an -O-H functionality.

Aqueous: A solution in water.

Aromatic: Any six membered ring system that contains three double bounds.

Acute Exposure: A single exposure to large concentrations of a given chemical mixture.

Alkali Earth Metals: Beryllium, magnesium, calcium, strontium, barium, radium.

Alkali Metals: Lithium, sodium, potassium, rubidium, cesium, francium.

Alkali: A material that will increase the pH of a solution. Examples: sodium hydroxide, ammonium hydroxide, potassium hydroxide, etc.

Alkyl Group: A saturated hydrocarbon that is a part of a larger molecule. Examples: methyl, ethyl, propyl, isopropyl, etc.

Amines: Chemicals containing -NH_2 functional groups.

Anhydrous: Without water.

Azide: A molecule that has a N_3 (-N=N≡N) present.

Central Nervous System (CNS): The nerves that control movement and sensation. The brain and spinal cord are apart of this system.

Chronic exposures: Exposures to a chemical or mixtures, typically at low concentrations, over long periods of time.

Cryogenic: Extremely cold, typically less than -100 F. Short exposures may result in frostbite or freeze burns.

Edema: Abnormal collection of fluids in body tissues. Pulmonary edema is the collection of fluids within the lungs causing respiratory distress or death.

Endothermic: Energy consuming. An endothermic reaction is one that becomes cool or cold (energy consuming).

Epoxide: A highly energetic, three sided cyclical functional group in which two adjacent carbons are bound to one oxygen.

Ethers: A molecule containing R-O-R, where R is a hydrocarbon. Ethers are typically very flammable and represent a peroxide threat upon prolonged storage after it has been exposed to air.

Exothermic: Energy releasing. An exothermic reaction is one that becomes hot (energy releasing). The more energy released over a short period of time will determine the extent of the exothermic hazard.

Functional Groups: Any species, excluding hydrogen, that is bound to a hydrocarbon. These include double bonds, triple bonds, halogens alcohols, etc.

Halides: The salt of a halogen.

Hazardous Polymerization: A reaction that convert liquids or gases to a solid while releasing great amounts of energy. This may be a highly energetic, almost explosive reaction.

Halocarbons: Any hydrocarbon that has a halogen bound to it.

Halogens: Fluorine, chlorine, bromine, iodine, astatine.

Hydrous: Having water present.

Hydration State: The number of water molecules associated with another molecule.

Hydrocarbon: A molecule that is primarily made up of carbon and hydrogen.

Hypergolic Reaction: A reaction between two chemicals that results in an immediate fire or explosion.

Ignition Energy: The energy required to ignite a flammable or combustible material. The lower the ignition energy, the easier it is to ignite.

Interhalogens: Extremely reactive chemicals that are composed of halogens. Examples: bromine trifluoride (BrF_3), chlorine trifluoride (ClF_3), etc.

Lachrymation: The uncontrolled secretion of tears.

Olfactory Fatigue: The inability to smell certain odors after an initial exposure.

Ordinary Combustibles: Common combustibles such as paper, cardboard, etc.

Organics: Purified organic chemicals such as organic solvents, waxes, oils, etc.

Oxidizers: A class of chemicals that, when mixed with flammable or combustible materials, will either initiate combustion or will accelerate the rate of combustion.

Peroxides: Materials containing a -O-O-H group. These are usually quite reactive and unstable. When peroxides are present, one should use extreme caution.

pH: This is a logarithmic scale that measures the concentrations of H^+ in aqueous solutions. A pH of 7.0 in considered neutral. The human body is at approximately 7.4. Solutions will become more corrosive as pH increases or decreases.

v: A functional group that is bound to a carbon atom that is bound to one other carbon.

A functional group that is bound to a carbon that is bound to two carbon.

mperature and pressure. This can usually be taken as room temperature and pressure.

Sublimes: The transformation of a solid to a gas without ever being a liquid.

Tertiary: A functional group that is bound to a carbon which is bound to three carbons.

Thio-: Containing sulfur.